Materials and Crystallographic Aspects of HT$_c$-Superconductivity

NATO ASI Series

Advanced Science Institutes Series

A Series presenting the results of activities sponsored by the NATO Science Committee, which aims at the dissemination of advanced scientific and technological knowledge, with a view to strengthening links between scientific communities.

The Series is published by an international board of publishers in conjunction with the NATO Scientific Affairs Division

A Life Sciences **B Physics**	Plenum Publishing Corporation London and New York
C Mathematical **and Physical Sciences** **D Behavioural and Social Sciences** **E Applied Sciences**	Kluwer Academic Publishers Dordrecht, Boston and London
F Computer and Systems Sciences **G Ecological Sciences** **H Cell Biology** **I Global Environmental Change**	Springer-Verlag Berlin, Heidelberg, New York, London, Paris and Tokyo

NATO-PCO-DATA BASE

The electronic index to the NATO ASI Series provides full bibliographical references (with keywords and/or abstracts) to more than 30000 contributions from international scientists published in all sections of the NATO ASI Series.
Access to the NATO-PCO-DATA BASE is possible in two ways:

– via online FILE 128 (NATO-PCO-DATA BASE) hosted by ESRIN,
Via Galileo Galilei, I-00044 Frascati, Italy.

– via CD-ROM "NATO-PCO-DATA BASE" with user-friendly retrieval software in English, French and German (© WTV GmbH and DATAWARE Technologies Inc. 1989).

The CD-ROM can be ordered through any member of the Board of Publishers or through NATO-PCO, Overijse, Belgium.

Series E: Applied Sciences - Vol. 263

Materials and Crystallographic Aspects of HT$_C$-Superconductivity

edited by

E. Kaldis

Laboratorium für Festkörperphysik,
Eidgenössische Technische Hochschule Hönggerberg,
Zürich, Switzerland

Springer Science+Business Media, B.V.

Proceedings of the NATO Advanced Study Institute on
Materials and Crystallographic Aspects of HT_c-Superconductivity
Erice, Sicily, Italy
May 17–30, 1993

A C.I.P. Catalogue record for this book is available from the Library of Congress.

ISBN 978-0-7923-2773-8   ISBN 978-94-011-1064-8 (eBook)
DOI 10.1007/978-94-011-1064-8

Printed on acid-free paper

TABLE OF CONTENTS

PREFACE

The present volume contains most lectures given at the 20th course of the International School of Crystallography at the Ettore Majorana Centre in Erice, Sicily, Italy (May 17th-29th, 1993), with subject "Materials and Crystallographic Aspects of High T_C Superconductivity". More than 100 participants and lecturers from 23 countries followed the strongly interdisciplinary program of this course, which covered some aspects of the Structure, Solid State Chemistry, Solid State Physics and Applications of the Cuprate Superconductors, and some highlights of the properties of the organic superconductors. In such innovation rich and multifaceted scientific field like that of HT_c, it is not possible to analyze the scientific value of the program in the preface. Nevertheless, I want to point out at least two new structural topics, which will shatter the prejudice that we have concerning the idealized average structure. Both refer to the local structure, which is most important also for the theoretical understanding of the HT_c Superconductivity.

Measurements of the atomic Pair Distribution Function of cuprate superconductors, show that the real structure deviates clearly in the short range from the idealized average srtructure. The dynamics and correlation of the corresponding atomic displacements sensitively reflect the onset of superconductivity and indicate the possibility of bipolaron mediated superconductivity.

Also the x-ray structure refinement of the chain superstructures in 123 is now possible and gives support to the electron diffraction investigations existing up to now.

In addition to the lectures several round table discussions were organized, whose contents, unfortunately, could not be published due to lack of space. A general message, however, emanating from lectures and discussions was the extreme complexity of these materials and the difficulty to synthesize high quality samples. Progress in this field will certainly change the measured values of some properties important for the theoretical understanding.

As the Director of the School, I am indebted to the Executive Secretary of the International School of Crystallography, Prof. L. Riva di Sanseverino, who with good organization and a lot of humor made the stay of the participants very pleasant.

I take also this opportunity to thank all lecturers for their contribution to the high scientific level of the school and for finding the time to follow this course and participate actively to the round table discussions and at hoc meetings.

Main sponsor of this school was the NATO Advanced Study Institutes Program. Many thanks are due to the director, Dr. L. Veiga da Cunha. Everyone who has participated in a scientific meeting in Erice feels obliged to thank the Director and Founder of the Centre Prof. A. Zichichi, who has created in the medieval city of Erice an ideal centre for scientific exchange.

Emanuel Kaldis
Zurich, Switzerland

Part I

**Structure and
Structure-Properties Relationship**

A CLASSIFICATION OF THE COPPER OXIDE SUPERCONDUCTORS AND THE RELATIONSHIP BETWEEN THE CU VALENCE AND THE SUPERCONDUCTING PROPERTIES

M. MAREZIO and C. CHAILLOUT
Laboratoire de Cristallographie, CNRS-UJF
BP 166, 38042 Grenoble cedex 09, France

ABSTRACT. A classification based on the feature that all high T_c superconducting compounds have a layered structure, is proposed. These compounds can be viewed as made of alternating layers having either the AO (A is a relatively large cation) or BO_2 stoichiometry (B is a medium size cation). In these layers the cations or the anions can be partially or totally removed. Adjacent layers are shifted of (1/2,1/2,0). In the compounds containing crystallographic shears, the shift may be (1/2,0,0) or (0,1/2,0). This description is useful for understanding the structural properties of the Cu-based high T_c compounds and to compare them to one another.
    The metallic behavior and the superconducting properties of the Cu mixed oxides is controlled by the formal valence of the Cu cations, which is closely related to the oxygen content. In an oxide compound the formal valence can be calculated by the bond length-bond strength method. The relationship between the formal Cu valence, the oxygen content and the superconducting properties is discussed by illustrating the $YBa_2Cu_3O_{6+x}$ and $Pb_2Sr_2Y_{1-x}Ca_xCu_3O_{8+\delta}$ systems.

1. Introduction

In mixed oxides, the Cu cations may have three stable valence states, namely 1+, 2+, and 3+. Moreover, they may be found in several different coordinations: octahedral, pyramidal, tetrahedral, square, and linear-two-fold. A given Cu cation cannot assume any one of the five coordinations, as some specificity exists; for example, the linear two-fold coordination can only accommodate Cu^{1+}. This variety of possible valences and arrangements gives rise to a large number of mixed Cu oxides.
    Investigations of these compounds were mainly carried out for the understanding of the chemical and physical properties of the Cu cations. The mixed copper oxides became extremely interesting when they were found to exhibit metallic behavior, due to the presence of mixed valence states (between 2+ and 3+). Up to 1980 a high pressure of 60 kbar was needed to stabilize the 3+ state [1]. Later this state was stabilized by introducing an alkali-earth element in the structure. A comprehensive review of this subject can be found in the article published in 1984 by Michel and Raveau [2].
    In 1986, the discovery of a superconducting transition at a relatively high temperature (28 K) in Ba-doped La_2CuO_4 by Bednorz and Müller [3], generated a great revival of the mixed Cu oxides. Since then more than thirty five superconducting copper compounds have been synthesized with transition temperatures ranging between 10 and 133 K. It is of interest, therefore, to determine the structural features common among these oxides and more importantly those linked to the superconducting properties.

2. Classification of the Copper Oxides Superconductors.

2.1 THE PEROVSKITE AND NaCl STRUCTURES.

All superconducting copper mixed oxides are related in a more or less evident fashion to the perovskite structure. This is the structure of ABO_3 compounds in which A is a relatively large cation and B a metal capable of forming a three-dimensional array of corner-sharing BO_6 octahedra. The A cations occupy the cuboctahedral sites generated by such an array. The undistorted structure is cubic and has a truly three-dimensional

3

character. Another way of describing the perovskite structure is to take into consideration the fact that the A and the oxygen atoms form a cubic close-packed array. The B cations occupy 1/4 of the possible octahedral sites, namely those formed by oxygen atoms. The rest, that is 3/4, is formed by 4 oxygen and 2 A atoms. It is evident that such an arrangement is undistorted only when the A atoms have an ionic radius very close to that of the oxygen, that is 1.40 Å. There are only a few cations, Ba^{2+}, K^{1+}, Rb^{1+}, for example [4], which have such a large radius and, therefore, only a few compounds are known to have the perovskite cubic structure. When the radius of the A atoms is somewhat smaller than 1.40 Å, the AO_3 network undergoes a distortion and the symmetry is no longer cubic. A variety of different distortions of the perovskite structure is known [5], the most common one is an orthorhombic distortion which was first found in $GdFeO_3$ [6].

In the middle 50's the discovery of the magnetic properties in the rare earth orthoferrites led to systematic studies of $REMO_3$ compounds where RE is a rare-earth element and M a trivalent 3d transition metal or any other trivalent metal whose ionic radius is comparable to that of Fe^{3+}. All these compounds have the orthorhombic distortion of $GdFeO_3$. This structure belongs to space group Pbnm with $a \approx b \approx \sqrt{2}a_p$ and $c \approx 2a_p$ (a_p being the simple cubic perovskite parameter). Even though this structure is severely distorted, the oxygen octahedra around the B atoms are only slightly so. The individual octahedra behave like rigid entities, but the octahedral network is so distorted that the A atoms have only eight first-nearest oxygen atoms and four second-nearest ones. Such a structure has a two-dimensional character.

Along any of the three four-fold axes of the cubic symmetry, the cubic perovskite structure can be considered as built of alternate layers of stoichiometry AO and BO_2. These layers can be stacked if the A-O and B-O distances are related by the relationship: A-O/B-O $\approx \sqrt{2}$. This relationship was first proposed by Goldschmidt [7] who called it the tolerance factor for the existence of the perovskite structure.

2.2 THE AO AND BO_2 BASIC LAYERS

As stated above, all known oxide superconductors belong to a class of inorganic compounds more or less related to the perovskite structure. They lose the three-dimensional character by cation ordering, by oxygen vacancy ordering, or by ordered stacking faults. The feature of the perovskite structure of being built of AO and BO_2 layers can be used for the purpose of classification of the oxide superconductors and of comparison to one another [8].

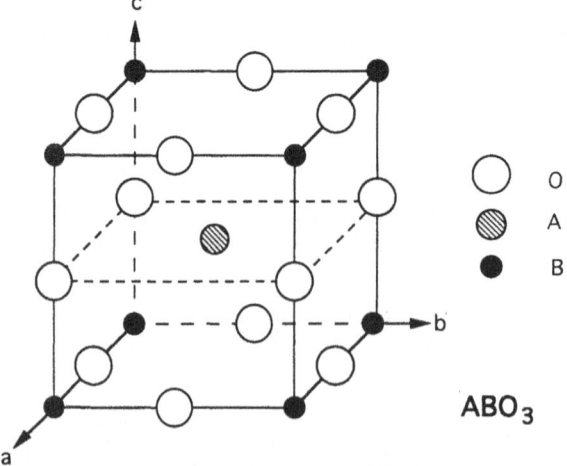

Fig. 1. The perovskite structure

It can be seen from Fig. 1, showing the cubic ABO_3 perovskite structure, that the first layer perpendicular to the c axis has composition BO_2, with the B atoms located at the corners of a square mesh and the O atoms at the edge midpoints. The next layer along the c axis has composition AO, with the A atoms at the center of the square mesh and the oxygen atoms at the corner. The perovskite structure contains then the following sequence:

$$(BO_2)(AO)(BO_2)(AO)(BO_2)$$

In order to specify the structural relationship between adjacent layers along the c axis, one may use the symbol $(BO_2)_o$ or $(BO_2)_c$ which specifies whether the B cation is at the origin or at center of the mesh. If we use the same notation for the AO-type layers, the sequence of the perovskite structure can be written:

$$[(BO_2)_o(AO)_c(BO_2)_o](AO)_c(BO_2)_o.$$

If we take as a reference the cation of each layer, a shift of 1/2,1/2,0 occurs on going from a given layer to the adjacent one.

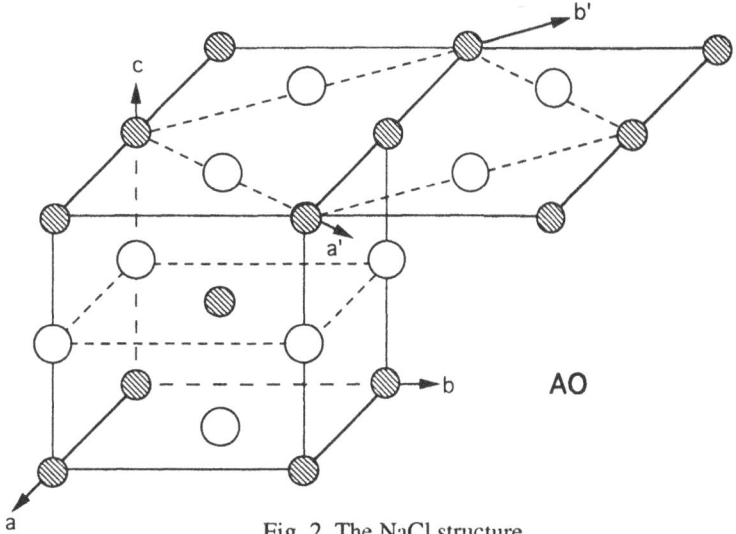

Fig. 2. The NaCl structure

Fig. 2 represents the NaCl-type structure. It comprises the sequence:

$$[(AO)_c(AO)_o](AO)_c(AO)_o.$$

From such a description it is possible to deduce the unit cell of the compound and the coordination of the different atoms. It cannot, though, take into account any distortion. The unit cell of the perovskite and NaCl structures are between brackets. In the case of the perovskite, the atom A of a layer $(AO)_c$ is surrounded by 12 O atoms, 4 located at the corners of the same mesh, and 8 at the edge midpoints of the layers $(BO_2)_o$ above and below.

In order to be able to describe all the different sequences found in the oxide superconductors we have to define other types of planar lattices. The origin of an o or c layers may also be shifted by 1/2 along x or 1/2 along y, which may be indicated by the subscripts ox, oy, cx, or cy. It is easy to verify that:

$$(BO_2)_{ox}=(BO_2)_{cy} \text{ and } (BO_2)_{cx}=(BO_2)_{oy}$$

and

$$(AO)_{cx}=(AO)_{oy} \text{ and } (AO)_{ox}=(AO)_{cy}.$$

In addition, the composition of the layers AO and BO_2 may vary and some or all the A or B or O atoms may be missing. Layers of this type may be called defective. These layers have compositions such as (A), (BO_x), (O_2) etc. All possible meshes of full or defective layers are illustrated in Fig. 3.

6

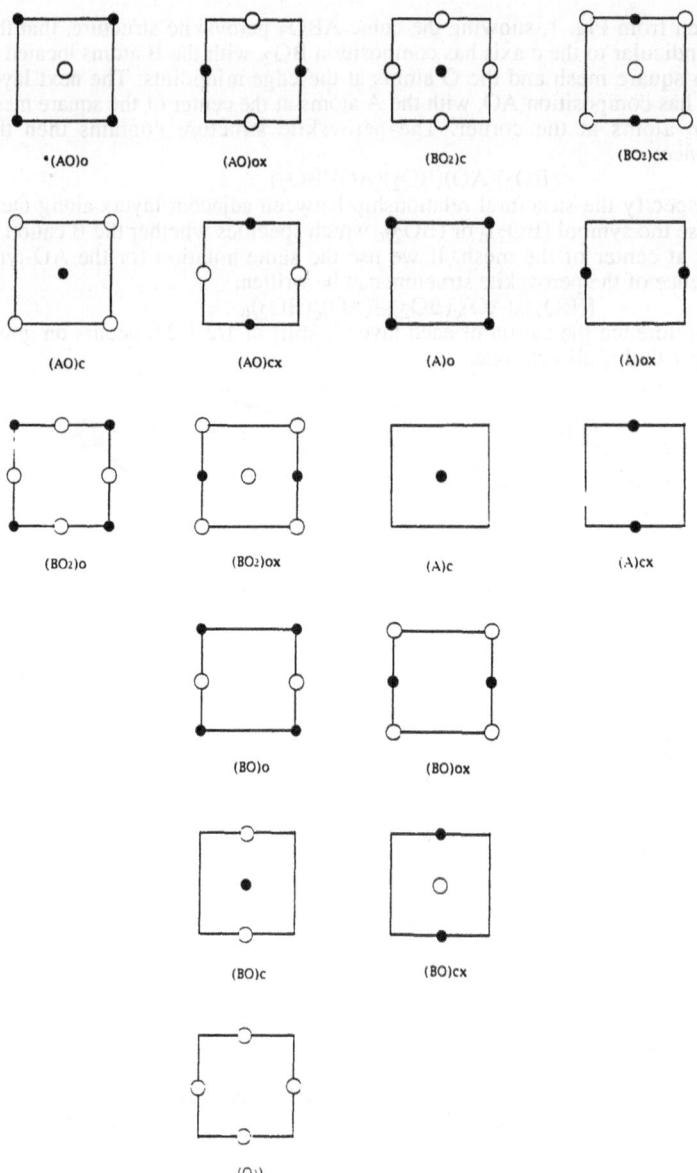

Fig. 3. All possible meshes used to describe the structures
of the Cu-based superconductors.

2.3 La$_2$CuO$_4$

The structure of La$_2$CuO$_4$ (K$_2$NiF$_4$-type) can be described by the sequence:
$$...[(CuO_2)_o(LaO)_c(LaO)_o(CuO_2)_c(LaO)_o(LaO)_c(CuO_2)_o](LaO)_c...$$
It can be easily seen that this structure is related to the perovskite structure. The main structural difference is the double (LaO) layers inserted between the (CuO$_2$) ones. This doubling breaks the three-dimensional character of the perovskite structure, in fact, the

K_2NiF_4 structures are definitely two-dimensional. As for the cation coordinations, one can see that the La cations are 9-coordinated (to 4 oxygen atoms at the same level, to 4 of the CuO_2 layer above or below, and to one of the LaO layer below or above). The Cu cations are located at the center of the octahedra which form corner-sharing layers.

2.4 $YBa_2Cu_3O_7$

Several months after the discovery of the superconducting properties in doped La_2CuO_4, the same property with a much higher T_c (≈ 93 K) was reported in a second system, namely $YBa_2Cu_3O_{6+x}$ [9]. The sequence corresponding to this compound is as follows:

$$...[(Y)_c(CuO_2)_o(BaO)_c(CuO_x)_o(BaO)_c(CuO_2)_o(Y)_c](CuO_2)_o(BaO)_c...$$

It can be seen that this is the sequence of a perovskite structure. The two-dimensionality is due to the ordering between the Y and the Ba cations on the A sites along one of the three pseudo four-fold axes and to the fact that some of the layers are oxygen-deficient. The Y layer, which is of AO type, is totally oxygen-depleted. The (CuO_x) layer is of BO_2 type, however, x can only vary between 0 and 1. There are four types of cation sites: 1 Y, 1 Ba, and 2 Cu sites. The coordination of the Y cations is 8-fold while that of Ba depends upon the value of x. It varies between 8 in $YBa_2Cu_3O_6$ and 9 in $YBa_2Cu_3O_7$. The Cu cations of the (CuO_2) layers are pyramidally coordinated and the pyramids form corner-sharing layers. The Cu cations of the oxygen depleted layers which are sandwiched between two BaO layers, are 2-coordinated (linearly) in $YBa_2Cu_3O_6$ and 4-coordinated (square) in $YBa_2Cu_3O_7$. The squares form corner-sharing chains along one of the basal axes. Schematic representations of the $YBa_2Cu_3O_6$ and $YBa_2Cu_3O_7$ structures are shown in Fig. 4. The oxygen-depleted (Cu) layer sandwiched between two (BaO) layers in $YBa_2Cu_3O_6$, a semiconductor, can incorporate extra oxygen atoms, and at about $x\approx 0.35$ $YBa_2Cu_3O_{6+x}$ becomes a metal and eventually superconducting. The formal valence of the Cu cations of the (CuO_2) layers is 2+ in $YBa_2Cu_3O_6$. As oxygen is incorporated, the valence of all Cu cations increases, and at a certain critical value of the mixed valence state the compound exhibits a metallic behavior. This oxidation process which is closely linked to the appearance of superconductivity will be discussed in detail below.

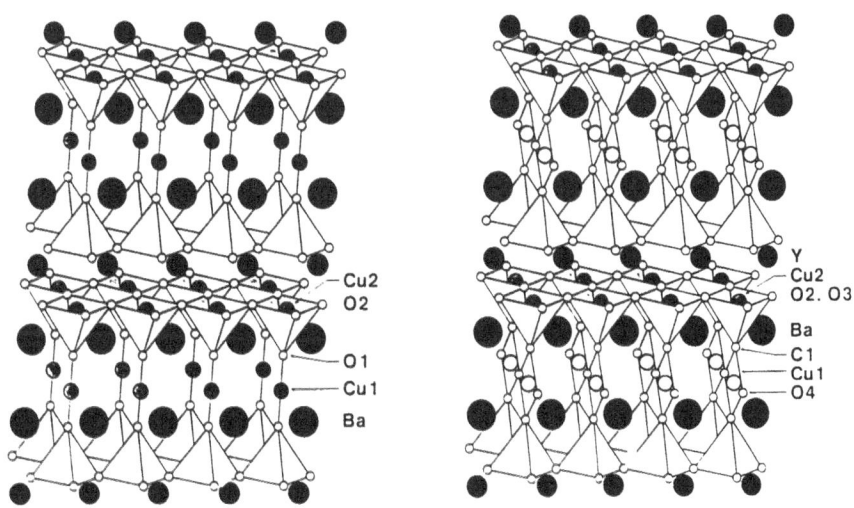

Fig. 4. The structures of $YBa_2Cu_3O_6$ (left) and $YBa_2Cu_3O_7$ (right)

2.5 MATERIALS WITH CRYSTALLOGRAPHIC SHEAR

In 1988, besides $YBa_2Cu_3O_7$, two other superconducting materials were discovered in the Y-Ba-Cu-O quaternary system, namely $YBa_2Cu_4O_8$ [10, 11] and $Y_2Ba_4Cu_7O_{14+x}$ [12, 13]. The structures of these compounds are closely related to that of $YBa_2Cu_3O_7$, and can be described with the same mechanism based on the stacking AO and BO_2 layers. However, to describe these materials we have to make use of the mesh $(BO_2)_{ox}$. In these compounds there exist two consecutive, oxygen deficient layers (CuO) whose stacking is accompanied by the shift of origin of (1/2,0,0). This configuration may be represented by the sequence: $(CuO)_0(CuO)_{ox}$ in which the symbol $(CuO)_{ox}$ means that the layer is the same as $(CuO)_0$ with the shift of origin of (1/2,0,0). The sequence of $YBa_2Cu_4O_8$ can be represented as:
$[(Y)_c(CuO_2)_0(BaO)_c(CuO)_0(CuO)_{ox}(BaO)_{cx}(CuO_2)_{ox}(Y)_{cx}(CuO_2)_{ox}(BaO)_{cx}(CuO)_{ox}$
$(CuO)_0(BaO)_c(CuO_2)_0(Y)_c]$

2.6 $N_mM_2R_{n-1}Cu_nO_{3n+m+1}$

At the beginning of 1988, three important series of high Tc superconductors were discovered within approximately a two month period [14-20]. They were actually homologous series whose general formula can be written: $N_mM_2R_{n-1}Cu_nO_{3n+m+1}$ in which N=Tl or Bi, M=Ba or Sr, R=Ca or one of the small rare-earths. For N=Tl, m can be either 1 or 2, while n varies from 0 to 4 or more. When N=Bi, n can only be 2, while n varies as in the case of thallium compounds. The sequences of these quaternary oxides are:

m=2	n=1	n=2	n=3	n=4
	$(CuO_2)_0$	$(CuO_2)_0$	$(CuO_2)_0$	$(CuO_2)_0$
	$(MO)_c$	$(MO)_c$	$(MO)c$	$(MO)_c$
	$(NO)_0$	$(NO)_0$	$(NO)_0$	$(MO)_0$
	$(NO)_c$	$(NO)c$	$(NO)_c$	$(NO)_c$
	$(MO)_0$	$(MO)_0$	$(MO)_0$	$(MO)_0$
	$(CuO_2)_c$	$(CuO_2)_c$	$(CuO_2)_c$	$(CuO_2)_c$
	$(MO)_0$	$(R)_0$	$(R)_0$	$(R)_0$
	$(NO)_c$	$(CuO_2)_c$	$(CuO_2)_c$	$(CuO_2)_c$
	$(NO)_0$	$(MO)_0$	$(R)_0$	$(R)_0$
	$(MO)_c$	$(NO)_c$	$(CuO_2)_c$	$(CuO_2)_c$
	$(CuO_2)_0$	$(NO)_0$	$MO)_0$	$(R)_0$
		$(MO)_c$	$(NO)_c$	$(CuO_2)_c$
		$(CuO_2)_0$	$(NO)_0$	$(MO)_0$
			$(MO)_c$	$(NO)_c$
			$(CuO_2)_0$	$(NO)_0$
				$(MO)_c$
				$(CuO_2)_0$

Until a few months ago, only the thallium-based compounds were known to exist for m=1. Very recently, it has been shown that the compounds with m=1 and N=Hg can be synthesized as well [21, 22]. The sequences of these compounds are quite similar to those shown above, the difference being that only one NO layer, instead of two, is sandwiched between two MO layers. For example, the sequence of the member with m=1, n=2, M=Ba, and R=Ca is:
$(CuO_2)_0(BaO)_c(TlO)_0(MO)_c(CuO_2)_0(Ca)_c(CuO_2)_0(BaO)_c(TlO)_0(BaO)_c(CuO_2)_0$
From the sequences shown above it is evident that an easy description of these compounds is that they contain blocks of perovskite, $(CuO_2)_0(Ca)_c(CuO_2)_0$, and NaCl ones, $(BaO)_c(TlO)_0(BaO)_c$ or $(BaO)_c(TlO)_0(TlO)_c(BaO)_0$. It can also be easily seen that for n=1 the Cu cations are octahedrally coordinated, for n=2 they are surrounded by pyramids and for n>2 there are two types of coordination: pyramidal and square. For n=3 the ratio of these two coordinations is 2:1, it decreases for increasing n.

Because of the high values of the critical temperatures, these series proved to be extremely important from the superconductivity point of view. The thallium compound with m=2 and n=3 established the record for T_c (125 K) [18], which remained unbeaten for five years. The recently prepared $HgBa_2Ca_2Cu_3O_{10+\delta}$ has $T_c \approx 133$ K [22]. For each series the superconducting transition increases with the number of layers per unit cell, that is the value of n. This proved to be true only up to n=3.

2.7 $Ca_{0.86}Sr_{0.14}CuO_2$

The compound $Ca_{0.86}Sr_{0.14}CuO_2$, containing the sequence:
$$...(Ca,Sr)_o(CuO_2)_c(Ca,Sr)_o(CuO_2)_c(Ca,Sr)_o...$$
can be considered the member of the homologous series containing an infinite number of (CuO_2) layers. The structure is built up of infinite layers of corner-sharing squares centered around the Cu cations, intercalated by (Ca,Sr) layers [23]. These are of AO type, but all the oxygen atoms have been removed. This compound as such has not been found to become superconducting, even though it possesses the structural features essential for high T_c superconductivity. Recently, Takano et al. [24] have shown that superconducting (110 K) samples with the general chemical formula $(Ca_{1-y}Sr_y)_{1-x}CuO_{2-x}$ can be prepared under high pressure.

2.8 $Pb_2Sr_2YCu_3O_8$

An important compound, also discovered in 1988 [25], is $Pb_2Sr_2YCu_3O_8$. The corresponding sequence is:
$$(CuO_2)_o(Y)_c(CuO_2)_o(SrO)_c(PbO)_o(Cu)_c(PbO)_o(SrO)_c(CuO_2)_o(Y)_c(CuO_2)_o$$
Besides the oxygen depleted Cu layer sandwiched between two (PbO) layers, this compound has all the features of the m=2, n=2 member of the thallium or bismuth homologous series. The two blocks $(TlO)_2$ and $(BiO)_2$ possess a 2+ charge, whereas the hypothetical $(PbO)_2$ block is not charged. It becomes so by insertion of a Cu layer which has a charge of 1+. These latter Cu cations have the linear two-fold coordination which indicates that they are Cu^{1+} cations. The other Cu cations, namely those of (CuO_2) layers are in the 2+ state. The necessary features for superconductivity are indeed present in this compound, but the valence of the Cu cations of these layers is not high enough and it has not been found to become superconducting. A schematic representation of the $Pb_2Sr_2YCu_3O_8$ structure is shown in Fig. 5.

3. The Cu Valence

An observation made at the very beginning of high T_c superconductivity which proved to be extremely important, was that the average formal Cu valence, in both the alkali metal-doped La_2CuO_4 and $YBa_2Cu_3O_{6+x}$ superconductors, was greater than 2+. When the first members of the homologous series $N_mM_2R_{n-1}Cu_nO_{3n+m+1}$ were synthesized, it was readily seen that the mixed valence model was not applicable. In all compounds of the homologous series the formal valence of Cu was exactly 2+. For example, if one took into consideration the average chemical formula for the bismuth compound with m=2 and n=2, $Bi_2Sr_2CaCu_2O_9$, one could see that the Cu cations were in the 2+ state. However, electron microscopy and diffraction together with neutron diffraction [26-28] showed that the structures of these componds were much more complicated than the simple models

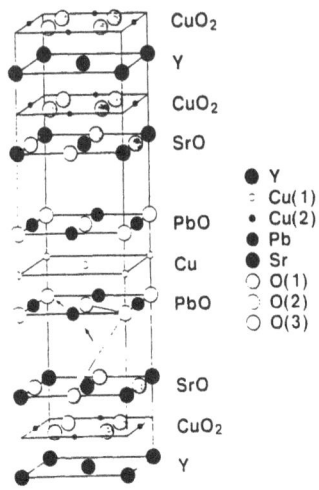

Fig. 5. The structure of $Pb_2Sr_2YCu_3O_8$

which can be derived from the sequences showed above. They all contain structural modulations which are due to the insertion of extra oxygen atoms on the BiO or TlO layers [29]. Furthermore, anomalous dispersion X-ray scattering experiments [30] and microprobe analyses [31] also gave evidence of a substantial cation substitution. It is generally accepted today that the formal Cu valence in these compounds is also greater than 2+, and this larger value is attained through oxygen insertion in the NaCl blocks and cation substitution.

Although chemists refer to the presence of Cu^{3+} cations when discussing the structural features of the high T_c copper oxide superconductors, XANES experiments clearly showed that the $3d^8$ state of the Cu cation does not exist in these compounds [32]. The Cu cations are in the Cu^{2+} state $(3d^9)$ with a positive hole located on the oxygen sublattice forming the (CuO_2) layers. This strongly indicates that the most important structural feature for the conductivity of these compounds is the (CuO_2) layer which is present in every high T_c superconductor. The transition to a metallic state is always the result of an oxidation process and the resulting structural changes take place in the blocks between the (CuO_2) layers. These blocks were believed to act as charge reservoirs.

As shown by Cava et al. [25], $Pb_2Sr_2YCu_3O_8$ becomes superconducting at about 80 K when some of the trivalent Y cations are replaced by divalent Ca. This means that the positive charges which become available on the (Y, Ca) layers are transferred to the (CuO_2) conduction layers. This is an important observation, because for this compound the extra positive charges do not seem to come from the charge reservoir as in the case of $YBa_2Cu_3O_{6+x}$. Thus, the concept of charge reservoir is not an intrinsic feature of the high Tc oxide superconductors.

Because of the oxygen depleted (Cu) layer, $Pb_2Sr_2Y_{1-x}Ca_xCu_3O_8$, can incorporate extra oxygen atoms, however, this incorporation suppresses the superconducting state. This is exactly the opposite of what happens in the $YBa_2Cu_3O_{6+x}$ system. It should also be pointed out that the incorporation of oxygen into $Pb_2Sr_2YCu_3O_8$ does not induce the superconducting state. This strongly indicates that from the structural point of view, each system must have its appropriate charge transfer which induces the structural features necessary for the establishment of the superconducting state.

3.1 THE BOND LENGTH-BOND STRENGTH METHOD

The charge transfer process can be unveiled by the detailed determination of crystal structures. The formal cation and anion valences can be calculated from the precise interatomic distances.

Pauling [33] was the first to assign in an ionic crystal a bond strength to a cation-anion bond. The bond strength was defined as $s=z/N$, in which z is the cation formal valence and N the coordination number, that is the number of first-nearest anions. Pauling also demonstrated that the anion valence could be obtained by summing the bond strengths over the cations bonded to that given oxygen, $\Sigma_c s = v_a$. In the early thirties such principle, known as the second Pauling rule, was mainly used to check if a newly determined structure was correct from the crystallochemical point of view. Later, when more complicated structures, as those containing hydrogen bonds or water molecules, were determined, the second Pauling rule was used to determine which were the oxygen atoms forming the O-H bonds and which were those belonging the H_2O molecules. Thus, one could guess where the hydrogen atoms were located.

Pauling assumed that a given bond length was given by the sum of the ionic radii and did not take into consideration the interdependence between the bond length and the bond strength. A few years after the establishment of the Pauling rules, Zachariasen [34] demonstrated that this interdependence existed and had to be taken into account if one wanted to explain the large variation of the bond lengths observed for the same cation and anion.

The Pauling rule of local charge balance was generalized by Byström and Wilhelmi [35] and by Zachariasen [36, 37] in order to explain the variations in bond lengths observed in vanadates, and chromates [35], uranyl compounds [36], and borates [37]. The principles established by Zachariasen can be summarized as follows:

1) A bond $s_{ij} = s_{ji}$ is assigned to a bond between the ith and jth atoms of a structure so that $\Sigma_j s_{ij} = v_i$ and $\Sigma_i s_{ji} = v_j$, where v_i and v_j are the valences of the two atoms.

2) The length of an A-B bond is a function only of the strength of the bond. Thus, the existence of a universal function $d_{A-B}=f(s)$ valid for all structures containing the A-B bond, was postulated.

A logarithmic variation of the covalent radius of a given cation with the bond strength had already been proposed by Pauling [38]: $r(s) = r(1) - 0.30log_{10}s$, where $r(1)$ is the radius of unity strength. Byström and Wilhelmi found that the vanadates data fitted the equation $d(s) = 1.77 - 0.78log_{10}s$ which corresponds to a coefficient of 0.39 Å rather than the value of 0.30 Å used by Pauling.

In the late seventies Zachariasen [39] found that the logrithmic function

$$d(s) = d(1)(1-Alns) = d(1) - Blns$$

gave good agreement with experiments not only for the vanadates but also for a large number of oxygen and halogen compounds of d and f elements. The $d(1)$ values were obtained from precisely determined structures and corresponded to the interatomic distance of unity strength. These values together with those for A (or B) can be found in references [39, 40]. By using the Zachariasen formula and constants one can calculate the cation and anion valences in most of the mixed oxides or fluorides or chlorides of d and f elements. These calculations are of extreme importance for those compounds containing the same element in more than one valence state.

In 1985 Brown and Altermatt [41] pointed out that a better agreement would be obtained if a different $d(1)$ value was assigned to each valence state of a given cation. With this assumption the bond length-bond strength method loses some of its generalization, but better fits are usually obtained. The new $d(1)$ values are commonly used today to calculate the cation and anion valences in inorganic materials and they have been used to calculate the valences in the mixed copper oxide superconductors.

By using individual bond lengths the distortion of the coordination polyhedra is somewhat taken into account, but not entirely. In some cases other parameters should be taken into account. Brown [42] pointed out, for example, that in layered compounds stresses and strains due to the misfit between adjacent layers must be taken into account when calculating cation valences. This aspect is extremely important in the high T_c superconductors.

3.2 THE OXYGEN CONTENT AND THE Cu VALENCE

All copper oxide superconducting systems contain Cu cations in a mixed valence state. The metallic behavior of these compounds is associated with a critical valence value of the Cu cations of the (CuO_2) layers. This critical valence is related to a given oxygen stoichiometry. Thus, the oxygen stoichiometry is one of the most important parameters governing the physical properties of high-T_c superconducting parameters. The mechanism for attaining the appropriate valence state for the copper cations which is the same as the appropriate oxygen content, is not a general feature for all high-T_c superconductors, but a specific one for almost each system. As an example, the oxidation processes of the systems $YBa_2Cu_3O_{6+x}$ and $Pb_2Sr_2Y_{1-x}Ca_xCu_3O_8$ will be illustrated.

3.2.1 $YBa_2Cu_3O_7$.
The compound $YBa_2Cu_3O_6$ is an antiferromagnetic semiconductor and has a tetragonal structure in which there are 2 Cu sites. The Cu2 sites (herein planar Cu) are surrounded by oxygen pyramids forming infinite layers via corner-sharing, while the Cu1 sites (herein chain Cu) are two-coordinated and they link together the blocks $(BaO)(CuO2)(Y)(CuO2)(BaO)$. Because of this coordination the Cu1 cations are in the 1+ state. It is readily calculated then that the Cu2 cations are instead in the 2+ state. Extra oxygen atoms up to O_7 can be inserted in $YBa_2Cu_3O_6$. The extra anions are located on the Cu1 layers as to form squares around the Cu1 cations. In $YBa_2Cu_3O_7$ the squares form infinite chains via corner-sharing. In this orthorhombic compound there

are also two Cu sites, one pyramidal and one squarely coordinated. These coordinations are not specific of any valence state so we can only say that the average Cu valence in $YBa_2Cu_3O_7$ is 2.33+.

It is of interest to follow the oxidation process of this compound by plotting the Cu valence calculated by the bond length-bond strength method as a function of x. These studies have been carried out by Cava et al. [43] and by Jorgensen et al. [44]. Fig. 6 shows the data of reference [43]. The average valence of the chain Cu cations increases linearly from 1.3 for x=0 to 2.4 for x=1, independently of the symmetry of the phase and whether the material is semiconducting or superconducting. The calculated value of 1.3 v.u. differs substantially from the value of 1+ quoted above. As stated above, deviations of the bond strength sum from the expected valence of a cation, in this case 1+, are to be considered to reflect internal stresses due to constraints imposed by the remainder of strucure. Thus, the absolute calculated values of the valences may not be precise, but the changes are meaningful.

The valence of the planar Cu cations displays a remarkable non-linear variation with x. It can be seen that, although positive charges are created in the structure, the planar Cu valence does not vary between x=0 and x≈0.35. Between x=0.35 and x=0.45, that is at the appearance of superconductivity, the planar Cu valence exhibits an abrupt increase. There is then a plateau near x=0.5, followed by a gradual increase, and between x=0.8 and x=1 the valence of the planar Cu is constant.

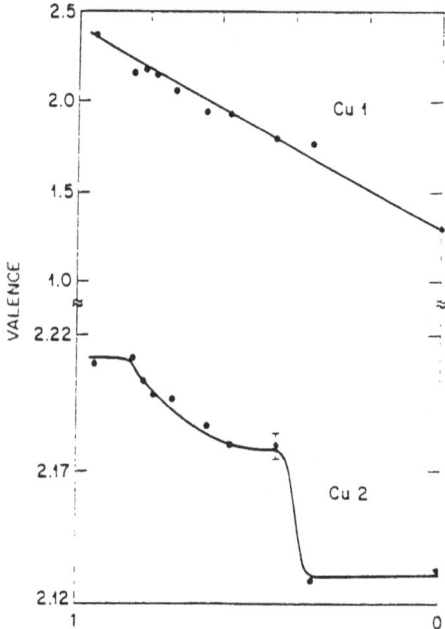

Fig. 6. Calculated valences for the chain Cu1 and planar Cu2 cations as a function of x for $YBa_2Cu_3O_{6+x}$

Fig. 7 shows a comparison between the variation of the valence of the planar Cu cations as a function of x and that of the superconducting transition temperatures for $YBa_2Cu_3O_{6+x}$. The behavior of the planar Cu valence as a function of x can be interpreted then as follows: as oxygen is incorporated between x=0 and x=0.35, all the generated positive charges remain on the chain Cu cations. When about one third of these cations have become squarely coordinated, a charge transfer occurs from the chain to the planar Cu cations and superconductivity appears. At this point, the charge transfer seems to stop and the valence of the planar Cu cations remains approximately constant between x=0.45 and x=0.64. When two thirds of the chain Cu cations have become squarely coordinated, a more gradual increase of the planar Cu valence occurs up to x=0.84. Then the transfer charge stops again and a second plateau is established. Recent results have shown that above x≈0.9 the relationship between T_c and x is not as simple as shown here [45]. Cava et al. [43] studied only one sample with x>0.9 and they could not properly determine the variation of T_c with x in this region. On the other, hand the correlation between the two curves, $T_c=f(x)$ and $v_{Cu2}=f(x)$, up to x=0.9 is striking.

This clearly indicate that the charge transfer model from a charge reservoir block, $(BaO)(CuO_x)(BaO)$, to the superconducting layers is valid for the $YBa_2Cu_3O_{6+x}$ system.

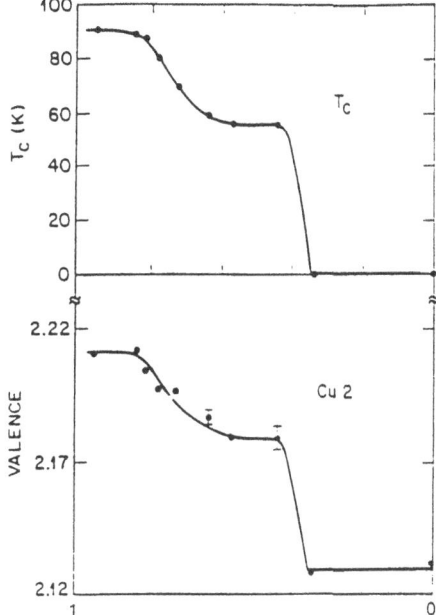

Fig. 7. Comparison of T_c and calculated valence of the planar Cu2 cations as a function x for $YBa_2Cu_3O_{6+x}$.

3.2.2. $Pb_2Sr_2Y_{1-x}Ca_xCu_3O_{8+\delta}$. As stated above this compound becomes superconducting when for $\delta=0$ some of the trivalent Y cations are replaced by divalent Ca cations. For $\delta=0$ the extra positive charges are created in the (Y,Ca) layers and these are transferred to the superconducting (CuO_2) and the compound becomes superconducting. The oxidation/reduction potentials of the Pb^{4+}/Pb^{2+}, Cu^{2+}/Cu^{1+}, and Cu^{3+}/Cu^{2+} systems indicate that before the oxidation of Cu^{2+} to Cu^{3+} (or Cu^{2+} and an electron hole), those of Pb^{2+} to Pb^{4+} and Cu^{1+} to Cu^{2+} should occur. However, the latter oxidations do not take place because they can occur only if extra oxygen is incorporated. For example, Cu^{2+} cations are not stable in a two-fold coordination. The charge transfer model is not valid for the $Pb_2Sr_2Y_{1-x}Ca_xCu_3O_8$ system.

This has been confirmed by detailed structural data [46, 47]. For the $YBa_2Cu_3O_{6+x}$ system the charge transfer occurs between the chain Cu cations to the planar Cu ones, but what actually changes in the structure is the position of O1, that is the oxygen linking the squares to the pyramids. The charge transfer occurs because this oxygen moves away from Cu1 (the chain Cu) toward Cu2 (the planar Cu). Consequently, the distances Cu1-O1 and Cu2-O1 increase and decrease, respectively. Structural refinements of $Pb_2Sr_2YCu_3O_8$ (non supercondcuting) [48] and of $Pb_2Sr_2Y_{0.73}Ca_{0.27}Cu_3O_8$ (T_c=67 K) [47] showed that the apical distance, that is the distance between the pyramidal coordinated Cu to the apical oxygen, is larger in the doped superconducting compound than in the undoped non superconducting one. This is a clear evidence that the apical oxygen is not involved in the charge transfer.

Oxidation of $Pb_2Sr_2YCu_3O_8$ and $Pb_2Sr_2Y_{1-x}Ca_xCu_3O_8$ does not induce superconductivity in the former and suppresses the superconducting state in the latter [25]. The oxygen addition to $Pb_2Sr_2YCu_3O_8$ induces the oxidation of Cu^{1+} and Pb^{2+} cations to Cu^{2+} and Pb^{4+}, respectively [38]. The extra positive charges never reach the (CuO_2) conducting layers, but are localized in the $(PbO)(CuO_\delta)(PbO)$ blocks. For the superconducting $Pb_2Sr_2Y_{1-x}Ca_xCu_3O_8$ the extra positive charges are also tranferred to the $(PbO)(Cu)(PbO)$ blocks inducing the oxidation of the same cations as in $Pb_2Sr_2YCu_3O_8$. The superconductivity is suppressed because, in order to generate the appropriate coordination for Pb^{4+}, the apical oxygen, belonging to the (SrO) layer, is displaced away from the (CuO_2) one, and a charge transfer occurs from the $(CuO2)(Y,Ca)(CuO2)$ blocks to the $(PbO)(CuO_\delta)(PbO)$ ones. These latter seem to act as a reservoir of negative charges.

Concluding Remarks

I have tried to illustrate a simple classification which should enable you to understand and deal with, the important structural features of the Cu mixed oxide superconductors. Since all these compounds have a layered structure the classification is based on the stacking of two basic layers of AO and BO_2 types. In order to avoid short cation-cation (or anion-anion) distances adjacent layers are shifted with respect to each other.

The formal Cu valence in these compounds is one of the most important crystallochemical parameters. It can be calculated from precise interatomic distances by the bond length-bond strength method. These calculations show that it is closely related to the oxygen content and to the superconducting critical temperature. Two examples, $YBa_2Cu_3O_{6+x}$ and $Pb_2Sr_2Y_{1-x}Ca_xO_{8+\delta}$, clearly show these relationships. It was also shown that the mixed Cu oxide superconducting compounds become insulating by means of a charge transfer mechanism. It seems that a specific charge transfer mechanism exists for each system.

References

[1]   G. Demazeau, C. Parent, M. Pouchard, & P. Hagenmuller,
        Mat. Res. Bull., **7**, 913, (1973)
[2]   C. Michel, & B. Raveau, Rev. Chimie Miner., **21**, 407, (1984)
[3]   J.G. Bednorz, & K.A. Müller, Z. Phys. B, **69**, 189, (1986)
[4]   R.D. Shannon, Acta Cryst., **A32**, 751, (1976)
[5]   J.B. Goodenough, & J.L. Longo, Crystallographic and magnetic properties
        of perovskite and perovskite-related compounds. Landolt-Börnstein III/4a
        nouvelle série, p. 312
[6]   S. Geller, Acta Cryst., **10**, 243, (1957)
[7]   V.M. Goldschmidt, Mat. Naturv. Kl. **8**, 1926.
[8]   A. Santoro, F. Beech, M. Marezio, & R.J. Cava, Physica C, **156**, 693, (1988)
[9]   M.K. Wu, R.J. Ashburn, C.J. Torng, P.H. Hor, R.L. Meng, L. Gao, Z.J. Huang,
        Y.Z. Wang, & C.W. Chu, Phys. Rev. Lett., **58**, 908, (1987)
[10]   K. Char, M Lee, R.W. Barton, A.F. Marshall, R.H. Bozovic, R.H. Hammond,
        M.R. Beasley, T.H. Geballe, A. Kapitulnik, & S.S. Laderman, Phys. Rev. B
        38, 834, 1988
[11]   P. Marsh, R.M. Fleming, M.L. Mandich, A.M. DeSantolo, J. Kwo, M. Hong,
        & L.J. Martinez-Miranda, Narure, **334**, 141, 1988.

15

[12]   J. Karpinski, C. Beeli, E. Kaldis, A. Wisard, & E. Jilek, Physica C, **153-155**, L 830, 1988
[13]   P. Bordet, C. Chaillout, J. Chenavas, J.L. Hodeau, M. Marezio, J. Karpinski, & E. Kaldis, Nature, **334**, 596, 1988.
[14]   H. Maeda, Y. Tanaka, M. Fukutomi, & T. Asano, Jpn J. Appl. Phys., **27**, L209, (1988)
[15]   R.M. Hazen, C.T. Prewitt, R.J. Angel, N.L. Ross, L.W. Finger, C.G. Hadidiacos, D.R. Vleben, P.J. Heaney, P.H. Hor, R.L. Meng, Y.Y. Sun, Y.Q. Hwang, Y.Y. Xue, Z.J. Huang, L. Gao, J. Bechtold, & C.W. Chu, Phys. Rev. Lett., **60**, 1174, (19, 332, 55 (1988)
[16]   Z.Z.Sheng, & A.M. Hermann. Nature, **332**, 55, 1988
[17]   Z.Z.Sheng, & A.M. Hermann. Nature, **332**, 138, 1988
[18]   S.S.P. Parkin, V.Y. Lee, A.I. Nazzal, R. Savoy, & R. Beyers, Phys. Rev. Lett., 61, 750, 1988.
[19]   S.S.P. Parkin, V.Y. Lee, A.I. Nazzal, R. Savoy, T.C. Huang, G. Gorman, & R. Beyers, Phys. Rev. B, 38, 6531, 1988
[20]   C. Martin, C. Michel, A. Maignan, M. Hervieu, & B. Raveau. C.R. Acad. Sc. Paris, 307 II, 27, 1988.
[21]   S.N. Putilin, E.V. Antipov, O. Chmaissem, & M. Marezio, Nature **362**, 226, 1993
[22]   A. Schilling, M. Cantoni, J.D. Guo, & H.R. Ott, Nature, submitted (preprint)
[23]   T. Siegrist, S.M. Zahurac, D.W. Murphy, & R.S. Roth, Nature, **334**, 231, 1988
[24]   M. Takano, Y. Takeda, H. Okada, M. Miyamoto, & K. Kusaka, Physica C, **159**, 375, (1989)
[25]   R.J. Cava, B. Batlogg, J.J. Krajewski, L.W. Rupp, L.F. Schneemeyer, T. Siegrist, R.B. von Dover, P. Marsh, W.F. Peck Jr, P.K. Gallagher, S.H. Glarum, J.H. Marshall, R.C. Farrow, J.V. Waszczak, R. Hull, & P. Trevor, Nature, **336**, 211, (1988)
[26]   P. Bordet, J.J. Capponi, C. Chaillout, J. Chenavas, A.W. Hewat, E.A. Hewat, J.L. Hodeau, M. Marezio, J.L. Tholence, & D. Tranqui, Physica C, **153-155**, 623, (1988)
[27]   E.A. Hewat, M. Dupuy, P. Bordet, J.J. Capponi, C. Chaillout, J.L. Hodeau, & M. Marezio, Physica C, **153-155**, 619, (1988)
[28]   E.A. Hewat, P. Bordet, J.J. Capponi, C. Chaillout, J. Chenavas, M. Godinho, A.W. Hewat, J.L. Hodeau, & M. Marezio, Physica C, **156**, 375, (1988)
[29]   Y. LePage, W.R. McKinnon, J.M. Tarascon, & P. Barboux, Phys. Rev. B, **40**, 6810, 1989
[30]   Y. Gao, S. Sheu, V. Petricek, R. Restori, P. Coppens, A. Darovskikh, J.C. Phillips, A.W. Sleight, & M.A. Subramanian, Science, 244, 62, 1989
[31]   A.M. Chippendale, S.J. Hibble, J.A. Hriljac, L. Cowey, D.M.S. Bagguley, P. Day, & A.K. Cheetham, Physica C, **152**, 154, 1988
[32]   A. Bianconi, J. Budnick, A.M. Flank, A. Fontaine, P. Lagarde, A. Marcelli, H. Tolentino, B. Chamberland, C. Michel, B. Raveau, & G. Demazeau, Phys. Lett. A, **127**, 285, 1988.
[33]   L. Pauling, J. Am. Chem. Soc., **51**, 1010, 1929
[34]   W.H. Zachariasen, Z. Kristall., **80**, 137, 1931
[35]   A. Byström, & K.-A. Wilhelmi, Acta Chem. Scand., **5**, 1003, 1951
[36]   W.H. Zachariasen, Acta Crystallogr., **7**, 795, 1954
[37]   W.H. Zachariasen, Acta Crystallogr., **16**, 385, 1963
[38]   L. Pauling, J. Am. Chem. Soc., **69**, 542, 1947
[39]   W.H. Zachariasen, J. Less common Met., **62**, 1, 1978
[40]   W.H. Zachariasen, J. Less common Met., **69**, 369, 1980
[41]   I.D. Brown, & D. Altermatt, Acta Crystallogr. B, **41**, 244, 1985
[42]   I.D. Brown, Proc. of the Int. Conf. on the Chemistry of Electronic Ceramic Materials, Jackson Hole, WY, USA, Aug. 1990, pp. 471-481
[43]   R.J. Cava, A.W. Hewat, E.A. Hewat, B.Batlogg, M. Marezio, K.M. Rabe, J.J. Krajewski, W.F. Peck Jr., & L.W. Rupp Jr., Physica C, **165**, 419, 1990

16

[44]   J.D. Jorgensen, B.W. Veal, A.P. Paulikas, L.J. Nowicki, G.W. Crabtree,
         H. Claus, W.K. Kwok, Phys. Rev. B, **41**, 1863, 1990.
[45]   H. Claus, U. Gebhard, G. Linker, K. Röhberg, S. Riedling, J. Franz, T. Ishida,
         A. Erb, G. Müller-Vogt, & H. Wühl, Physica C, **200**, 271, 1992.
[46]   M. Marezio, A. Santoro, J.J. Capponi, R.J. Cava, O. Chmaissem, & Q. Huang,
         Physica C, **199**, 365, 1992.
[47]   C. Chaillout, O. Chmaissem, J.J. Capponi, T. Fournier, G.J. Mc Intyre, &
         M. Marezio, Physica C, **175**, 293, 1991.
[48]   R.J. Cava, M. Marezio, J.J. Krajewski, W.F. Peck Jr., A. Santoro, & F. Beech,
         Physica C, 157, 272, 1989.

NEUTRON POWDER DIFFRACTION ON THE ILL HIGH FLUX REACTOR AND HIGH TC SUPERCONDUCTORS

A.W. HEWAT
Institut Max von Laue - Paul Langevin,
156X Grenoble Cedex,
38042 France.

ABSTRACT: Neutron powder diffraction, with both reactor and pulsed neutron sources, has been important for our understanding of the structures of the oxide superconductors. These structures appeared very different when seen with neutrons, which emphasized the oxygen lattice, compared with X-rays, which emphasized the heavy metal atoms. Neutron diffraction first demonstrated the two dimensional nature of the early materials, and this lead to the successful search for new 2D superconductors based on the well known 'Aurivillius' structure. Neutron powder diffraction also played an important rôle in the developement of 'charge transfer' ideas of the electronic doping mechanism in oxide superconductors. Although it is not known if these ideas are relevant to the superconducting mechanism itself, they have guided chemists in the discovery of new materials. They have also helped us understand why oxidation sometimes increases Tc, and sometimes kills superconductivity, and how the distribution of electron holes between the different layers of a superconductor can be changed by effects such as pressure, which often produce unexpectedly large changes in Tc.

1. Introduction

1.1. NEUTRON POWDER DIFFRACTION AND HIGH TEMPERATURE SUPERCONDUCTORS

For a few days in March , all of the world's major laboratories struggled to understand the structure of the 90K superconductor, using every possible technique. The results at the end of the month can be summarised by figure 1. It shows a typical X-ray picture of the structure of the superconductor [1], beside a typical neutron picture of the 'same' structure [2].

The X-ray and neutron pictures in figure 1 are both technically correct. They both show defect perovskite structures, with some oxygen missing, and they agree about where this oxygen is missing, and about the positions of all of the cations. In fact, the X-ray structure was used as a starting point for the neutron refinement. Yet the neutron work gives an entirely different picture of the structure. The X-ray drawing shows CuO_6 octahedrae, as believed necessary for high Tc according to Bednorz and Müller's ideas [3]. The neutron picture shows no such octahedrae, but instead 2-dimensional planes of copper oxide pyramids, connected by 1-dimension chains of copper oxide. The absence of octahedrae was at first strongly contested [4], and this controversy helped to make the neutron paper [1] the most cited experimental work in the field during the following year [5]. The one- and two-dimensional nature of the 90K superconductor excited the interest of theoreticians, and stimulated chemists to look for superconductivity in other 1- and 2-D perovskite materials.

17

E. Kaldis (ed.), Materials and Crystallographic Aspects of HTc-Superconductivity, 17–44.
© 1994 *Kluwer Academic Publishers.*

18

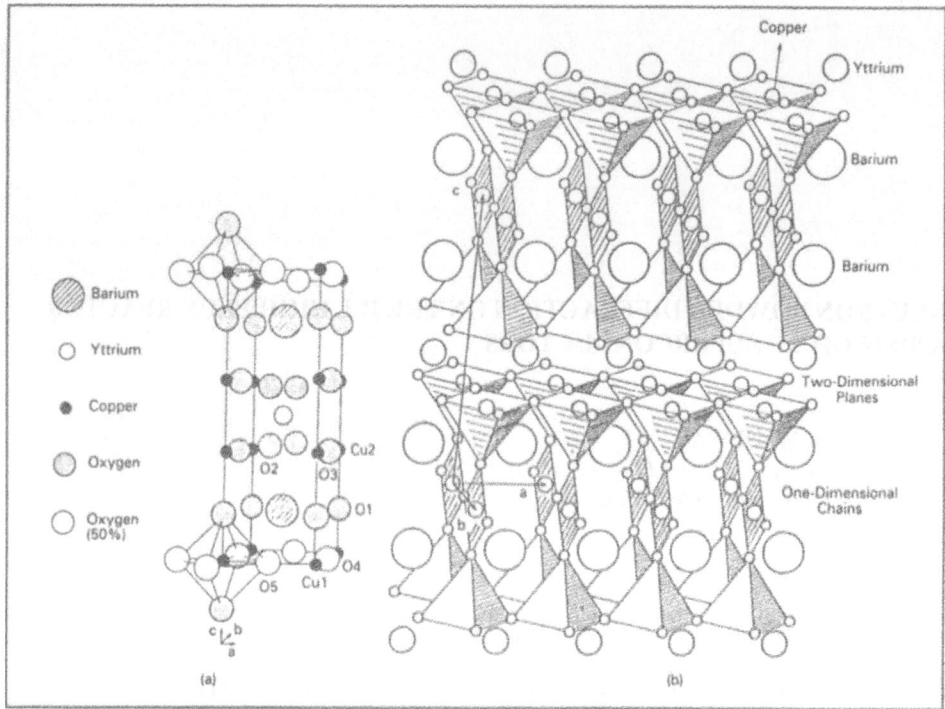

Fig. 1. Structures of the 90K superconductor
(a) 'Ba$_2$YCu$_3$O$_{9-x}$' obtained by X-rays [2] and of
(b) 'Ba$_2$YCu$_3$O$_7$' obtained by neutrons [1].

The differences between the two drawings were due simply to the fact that X-rays could not locate all of the light oxygen atoms, especially in a polycrystalline material. This story would later be repeated, when another famous laboratory, again using X-ray measurements, at first reported the bismuth-copper superconductors to be Aurivillius structures [6], while neutron measurements showed that the true structure of the bismuth oxide layers was quite different [7].

Even the chemical formula for the X-ray structure of the 90K superconductor, given in the paper's title as Ba$_2$YCu$_3$O$_{9-y}$, was uncertain as to the precise number of oxygens, though the oxidation state of the copper ions was the question of crucial interest according to Bednorz and Müller [3]. The formula in the title of the neutron paper was Ba$_2$YCu$_3$O$_7$, now written YBa$_2$Cu$_3$O$_7$. Neutrons further showed how oxygen could be extracted [8, 9], finally destroying superconductivity in YBa$_2$Cu$_3$O$_6$.

1.2. ADVANTAGES OF NEUTRON DIFFRACTION

This is only the most recent example where the extra 'details' provided by neutrons proved essential, leading directly to a fuller understanding of the chemistry, and even pointing the way to more interesting materials. The problem with X-rays is that scattering depends on the number of electrons, so light elements are difficult to see in the presence of heavy atoms. Neutrons are scattered by reaction with the nucleus, which can be just as strong for light elements as for heavy elements. Furthermore, X-ray scattering falls off strongly with angle, because of the large size of the atom (the 'form factor' effect), and high angle scattering is needed for high spatial resolution of the structure. Neutrons, scattered by the much smaller nucleus, have a form factor almost constant with scattering angle.

The advantages of neutrons, strong scattering from light atoms and no 'form factor' fall off with scattering angle, is best illustrated by the relative scattering factors for X-rays, electrons and neutrons for the atoms of high temperature superconductors (fig. 2).

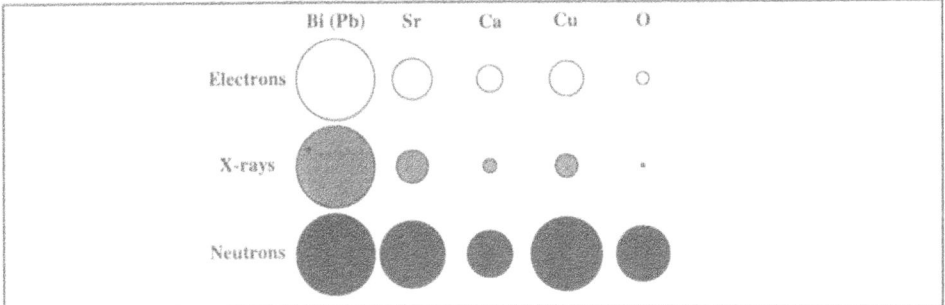

Fig. 2. The scattering power for X-rays, electrons and neutrons of the different atoms in oxide supercon-ductors. The circles are proportional to the scattering amplitude, and the areas to the intensity, for scatter-ing at angles 2θ such that $\sin\theta/\lambda = 1$.

Finally, chemists preparing new compounds often have only polycrystalline material, and X-ray powder diffraction suffers from texture problems that makes the location of light atoms even more difficult.

2. Neutron Powder Diffraction

2.1 THE SIMPLEST NEUTRON DIFFRACTION MACHINES - POWDER DIFFRACTOMETERS

Neutron diffraction conjures up ideas of the complicated and expensive equipement associated with a nuclear reactor. It is true that the neutron source itself is expensive, but if we already have such a source, the additional equipment for the simplest neutron diffraction measurements is actually no more complicated than that needed for X-rays.

Figure 3 shows the essentials of a simple neutron powder diffractometer, the D1A machine at ILL Grenoble on which the data for figure 1 was collected [1]. The white beam of neutrons, extracted through a hole in the reactor, is collimated by α_1 to limit the spread of incident directions. The monochromator selects a band of wavelengths $\Delta\lambda/\lambda$ and directs these neutrons to the powder sample, sometimes through a second collimator α_2. The detector collects the neutrons, scattered in many directions by the sample, and records the scattered intensity as well as the angle 2θ of scattering. This angular detection is achieved using either a position sensitive detector (PSD) or a third collimator α_3. In the latter case, the detector is scanned to cover the whole scattering range. The scale of the machine, related to the size of the sample and neutron beam, is larger than the usual laboratory X-ray machine, but may be smaller than a synchrotron powder diffractometer.

The intensity from the reactor is always smaller than desirable, and much smaller than from any X-ray source. The two design criteria for a neutron powder diffractometer are then to maximise the angular resolution but also to maximise the recorded intensity. These objectives are of course contradictory. The resolution can easily be increased by narrowing the collimators α, but this naturally reduces the intensity. In finding a compromise we must of course make sure that we do not introduce errors in the data.

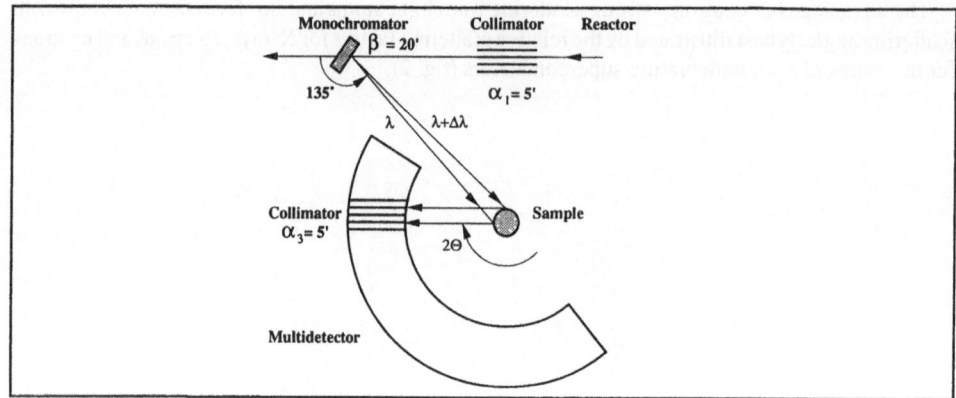

Fig. 3. The principal components of a high resolution powder diffractometer [10]. The white thermal neutron beam from the reactor is tightly collimated by a_1, and a small band of wavelengths $\Delta\lambda/\lambda$ is selected by a mosaic monochromating crystal. These different wavelengths are scattered back into a *parallel* direction ($\alpha_3 \approx \alpha_1$) by the polycrystalline sample when the scattering angle is equal to the (large) monochromator angle (focussing condition).

2.2 THE POWDER SAMPLE AND ENVIRONMENT

In order to gain intensity the samples are larger for neutrons, so positioning and alignment can be less precise. The larger sample size is also a fundamental advantage for powder diffraction. The sample must represent all possible orientations of the constituent microscopic crystallites, and this is often not true for the very small samples used with X-rays, where the high absorption limits sample size. Of course, the larger neutron samples are a disadvantage if only microscopic quantities of material can be prepared, but this is not usually a problem for modern neutron diffractometers on high flux reactors.

It is true that the much higher fluxes available with synchrotron sources permit higher line resolution than is possible for neutrons, even with a high flux neutron source. However, in practice resolution is limited by the particle size of the material. Only ideal materials can be made with large enough particle size to match synchrotron resolution, and in those cases single crystals can almost always be obtained. It is not too much of an exaggeration to say that the most interesting materials do not easily form single crystals ! This is certainly the case for the high Tc superconductors.

The resolution advantage of synchrotron powder diffraction is not so great as first appears from the narrow line width. This is because we are usually interested in the resolution in the spacing of atomic planes — the d-spacing — rather than the diffraction line width. The $\Delta d/d$ resolution depends on the line width $\Delta 2\theta$, but also on the scattering angle 2θ, as shown by differentiating Bragg's equation $2d\sin\theta = \lambda$ for scattering from the sample (fig. 4):.

$$\Delta d/d = \Delta\theta . \cot\theta \qquad (2.2.1)$$

Clearly, $\Delta d/d$ is small (high resolution) for high scattering angles 2θ. But neutrons have a fundamental advantage over X-rays at high scattering angles. Because scattering is from the almost point nucleus, and not from the much larger shells of atomic electrons as with X-rays, neutron scattering does not decrease with scattering angle. This means that usually lines can be measured at very high scattering angles with neutrons (fig. 5), so that even though the lines are broader, the $\Delta d/d$ resolution with backscattered neutrons can rival that of the best X-ray machines [11, 12].

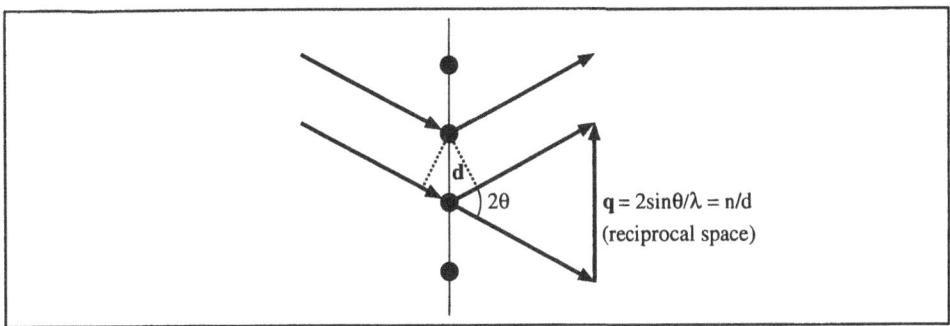

Fig. 4. Bragg's law is simply the condition for constructive interference for scattering by angle 2θ with wavelength λ from a set of planes of d-spacing d.

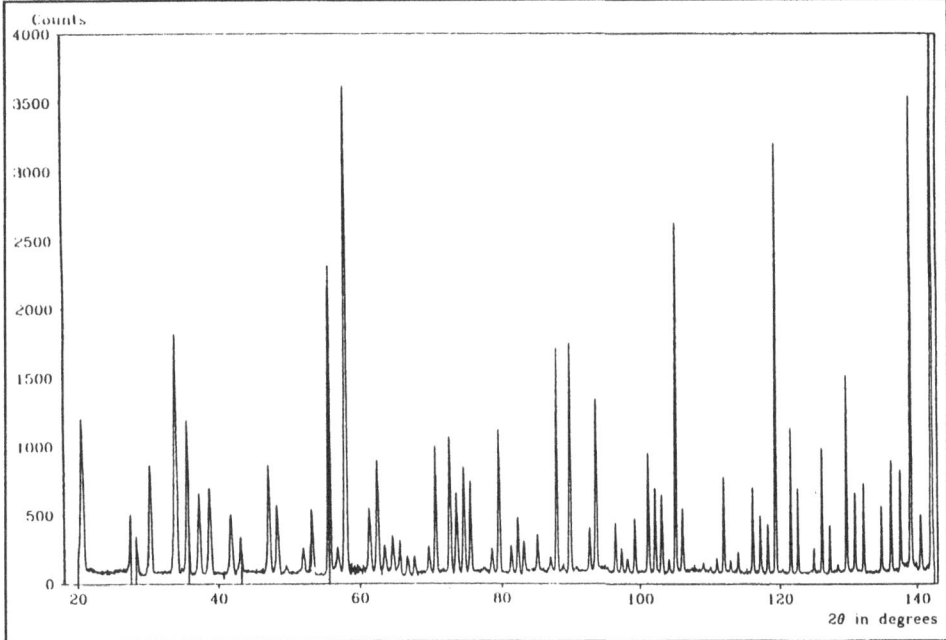

Fig. 5. The results of a high resolution neutron powder scan for the 12Å unit cell of yttrium iron garnet (YIG) on the D2B machine [11]. Notice how sharp the Bragg reflexions are for the entire pattern, and how strong the peaks are at high scattering angle, where X-ray patterns fall off because of the 'form factor'.

Strictly, the line intensity falls with angle because of the smearing of the nucleus by atomic vibration. This Debye-Waller factor can however be greatly reduced by cooling the sample, which for technical reasons to do with the sample size, is easier with neutrons. Neutron samples are usually measured at low temperature, while X-ray samples are usually measured at room temperature.

The ease with which neutron powder measurements can be performed at low temperature is also an advantage for studying the true crystal structure. A surprisingly large number of materials transform to a lower symmetry structure when cooled from room temperature. The most well known examples are the large family of perovskite structures, to which high Tc superconductors belong !

Finally, the larger sample size makes it easier to construct detectors for neutron powder diffraction. Powder samples naturally scatter in many directions simultaneously, so it is a big advantage to have a position sensitive detector (PSD), or a multidetector. The simplest form of PSD is of course the photographic plate, used from the beginning of X-ray diffraction. However, it is not ideally suited to precise measurements of intensity, which are necessary for structure determination. Electronic PSD's have only recently become practical for X-ray powder diffraction, while they have been used for more than 20 years with neutrons. And it is still difficult to make X-ray linear detectors with the necessary precision over a wide angular range.

2.3 THE MONOCHROMATOR

The monochromator, with the detector, is the most important component determining the neutron intensity apart from the size of the sample. We want to use as wide a band of wavelengths $\Delta\lambda/\lambda$ as possible, since this is directly proportional to the intensity. A wide wavelength band is simply achieved by having a large mosaic spread $\Delta\theta_M$ for the monochromator, since again differentiating Bragg's equation $2d\sin\theta_M=\lambda$ for scattering from the monochromator gives:

$$\Delta\lambda/\lambda = \Delta\theta_M.\cot\theta_M \qquad (2.3.1)$$

Since this equation for $\Delta\lambda/\lambda$ is similar to the previous equation for $\Delta d/d$, it would appear that large $\Delta\lambda/\lambda$ and high intensity implies large peak widths $\Delta d/d$. Fortunately this is not so. For the geometry shown in figure 3, the different wavelengths, leaving the monochromator in different directions, are brought back into the same direction when scattered a second time by the sample. This is strictly true only when the angle for scattering from the sample 2θ is equal to the monochromator 'take-off' angle $2\theta_M$, and is called 'reciprocal space' or wavelength focussing.

A small take-off angle θ_M to increase $\cot\theta_M$ would also increase $\Delta\lambda/\lambda$ and intensity, but if we do that, we only have wavelength focussing at small scattering angles. In practice we want to focuss at large angles, so that we need an even larger monochromator mosaic to compensate. A monochromator mosaic of 15 minutes is not excessive for even the highest resolution neutron powder machine. In that case we should use a 30 minute wide collimator to allow all the wavelength band to reach the sample. In practice, we need no collimator at all, so that α_3 is determined by the size of the sample divided by the distance to the collimator.

A second kind of focussing, in real space, is also possible, since the instrument resolution does not depend much on the *vertical* divergence of the incident beam. We may use a vertical divergence of as much as 5 degrees on a machine which still has high resolution in the equatorial plane. Vertical divergence is obtained by placing the machine as close as possible to the source, and cutting a large hole in the reactor. But if this is not sufficient, we must use a vertically focussing monochromator.

Vertical focussing is achieved by cutting the monochromator into horizontal strips, and aligning the strips on a vertically curved backing plate to focus onto the sample. On D2B at ILL Grenoble, 30 x 1cm strips are used to focus a 30cm high neutron beam onto a 4cm high sample [11]. The focal length is constant, even though different wavelengths can be selected. This is made possible by the large fixed monochromator take-off angle.

Germanium in reflexion geometry is a good choice for the monochromator for several reasons. If the (h,h,l) (odd,odd,odd) reflexions are used, then both 2λ and $1/2\lambda$ harmonic contamination is avoided. Futhermore, with the high take-off angle, high order reflexions are used, and there is a useful choice of wavelengths for the same scattering angle. For example, (711), (533), (511), etc. all give different wavelengths. Since they all lie in the same crystallographic plane, changing the neutron wavelength is simply a matter of rotating the monochromator crystal by the small angle seperating these high angle reflexions.

2.4 SÖLLER COLLIMATORS

Söller collimators α_1 serve to define the direction of the beam incident on the monochromator. They consist of a stack of flat absorbing plates. The divergence is determined by the ratio of the distance between the plates and their length, and is therefore decoupled from the beam cross section, which is simply the cross section of the stack. The ratio of the plate thickness to plate seperation determines the neutron loss compared to a perfect collimator.

A big advance in neutron collimators was made when very thin stretched plastic foils were used as the plates [13]. For example, 25μ thick foils, coated with another 25μ of neutron absorber such as gadolinium oxide paint, give 90% efficiency even if the plate seperation is only 0.5mm, as required for a 5 minute collimator 34cm long [14]. Such a collimator may have a beam cross section of 25x100 mm high or more. The beam width reflects the maximum width of the sample, while the beam height is important for obtaining high vertical divergence to increase intensity. Mylar collimators have an almost perfect triangular transmission function, and produce an almost perfect gaussian diffractometer line shape when rocked against each other (fig. 6).

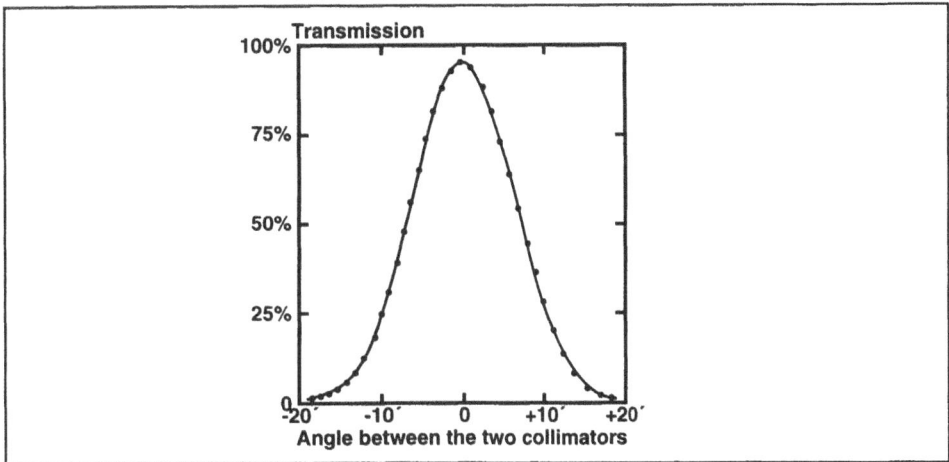

Fig. 6. The rocking curve for two 10' Söller collimators, representing the convolution of triangular functions, is equal to the line width at focussing, and is almost a perfect Gaussian.

2.5 THE PSD DETECTOR

The neutron PSD is simpler to make, mainly because of the size of the sample and the scale of the diffractometer. If the sample is a cylinder of 5mm diameter, which is already small for neutrons, the position resolution of the detector may also be several mm. Neutron PSD's can be made simply from a stack of individual detectors, each 5mm wide [15]. The angular resolution of such a detector depends on the ratio of the sample/detector diameter to the distance between the two, so a 5mm diameter sample and detector seperated by 1500mm means an angular resolution of 0.2 degrees. The line shape will then be at least $0.2\sqrt{2}$ degrees 2θ, since the detector resolution is folded with the resolution of the remainder of the diffractometer, as we shall see.

This would be a medium resolution powder diffractometer, provided that we could work at high scattering angles. Higher resolution with a PSD requires samples smaller than 5mm, and of course the intensity falls as the square of the sample diameter, usually making smaller samples impractical.

The D1B diffractometer at ILL Grenoble [16] was designed with these characteristics. Instead of individual 5mm detectors, a single BF_3 gas container was used, with individual wires spaced at 5mm intervals. This was the first, highly successful, PSD for neutron powder diffraction (Fig.7). 400 wires at 0.2 degree intervals covered an angular range of 80 degrees. Later the poisonous BF_3 gas was replaced by He^3 gas at higher pressure.

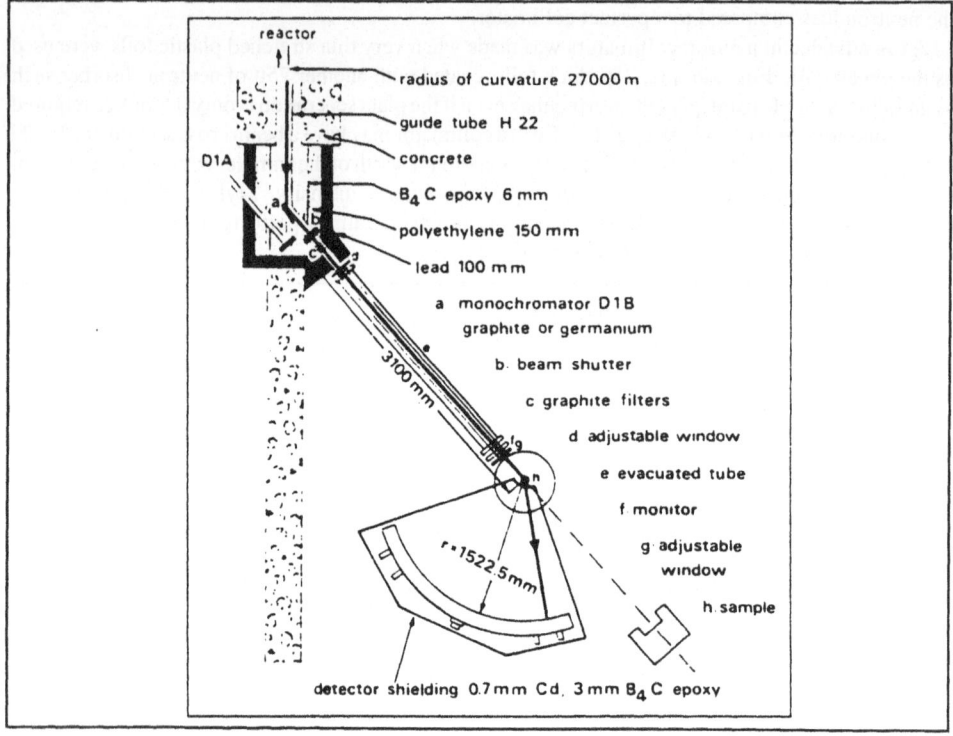

Fig. 7. The 'banana' on D1B at ILL was the first example of the use of a multiwire PSD for neutron powder diffraction.

Each wire of such a PSD produces one point on the diffraction pattern, so the number of points, or *definition* of the pattern, is low. This is a problem for the usual Rietveld refinement of the diffraction pattern profile. The definition of the profile can be increased, without changing the resolution, by displacing the whole detector by 0.1 degrees to collect another 400 points at intermediate angles. Indeed the PSD can be step-scanned to produce a profile with any desired definition.

The sample height can be much greater than the sample diameter, since vertical divergence does not much affect horizontal resolution. Vertical angles of 5 degrees or more can be tolerated, corresponding to samples as high as 125mm. Of course, the sample height is often limited by the available sample volume. The detector height could similarly be increased to 125mm or more, but is limited by the high gas pressures (5-10 bar) that must be supported over a correspondingly large detector window.

Larger samples could be used with larger PSD's to obtain more intensity if these mechanical problems can be solved, but such detectors would be expensive.

Inexpensive PSD's can consist of linear wire detectors, where the single wire is horizontal, and the position of the detected neutron is determined by timing the arrival of the resulting electrical pulse at both ends of the wire. This is the same technique used for X-ray powder PSD's. The X-ray detectors can be curved horizontally, but so far the neutron detectors use straight wires, whis means that parallax errors must be corrected. A number of short straight sections can extend the angular range of the detector, and the vertical divergence can be increased by stacking several detector tubes [17].

Yet another kind of PSD can be constructed using neutron scintillators or other solid state techniques [18]. This has the advantage that the mechanical problems of containing high pressure gas are eliminated and large detectors can be made quite cheaply. The detection efficiency is not as high as for the best gas detectors, but may often be sufficient.

The proceedings of the ILL meeting on position sensitive detectors [19] describes all of these the different kinds of PSD's. The gas PSD developed at ILL has since become the most widely used, but this is a relatively expensive solution.

2.6 THE MULTICOLLIMATOR DETECTOR

An alternative to the PSD becomes attractive if large samples are available and the highest resolution is required. Instead of allowing the sample diameter to determine the resolution with a PSD, a söller collimator α_3 can be placed between the sample and detector to limit the horizontal divergence. Since söller collimators may have a divergence of less than 0.1 degrees, compared to the usual 0.2 degrees imposed by the sample size with a PSD, very high resolution is possible. Since this resolution is independent of sample size, samples up to 15mm diameter or more may be used, increasing the sample volume, and therefore intensity, by an order of magnitude. A bank of söller collimators and individual detectors can cover a wide angular range, of 160 degrees or more (fig. 8) [11, 20].

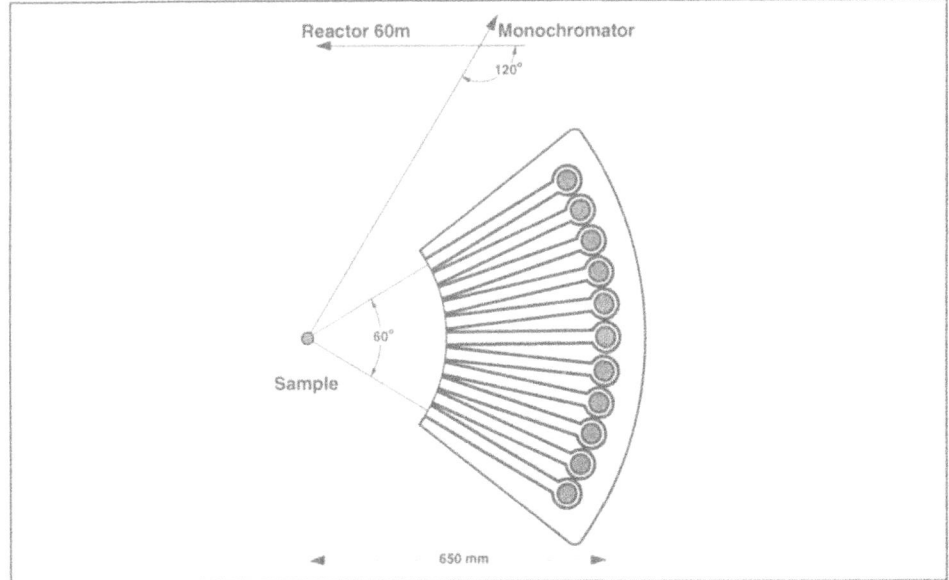

Fig. 8. The multicollimator detector on D1A at ILL (on which the data for fig.1b was collected) permits very high resolution, independent of the sample diameter, and excludes scattering from the sample environment.

A PSD simultaneously collects data over the whole of its range, while each collimator only collects data within the angle α_3. If $\alpha_3=0.1^\circ$ for example, then 64 collimators at 2.5° intervals covering 160° would be equivalent to a PSD covering an angle of only 6.4°. Alternatively, an α_3 of 0.1° could be achieved with a 2.5 mm diameter sample 1500 mm from a PSD covering 160° rather than 6.4°. The x25 larger PSD detector solid angle would be offset by the x36 larger volume of the 15mm diameter sample that could be used on the multicollimator machine.

The high resolution multicollimator is relatively inexpensive and easy to build, while a 160° PSD with a resolution of 0.1° would be very expensive (~\$1M). ILL has however been developing such a detector for a number of years.

The multicollimator machine has other practical advantages apart from cost. The α_3 collimators accept scattering from only a very small volume around the sample, while the PSD accepts scattering from all directions. This makes it much easier to use cryostats and furnaces with the multicollimator machine, and most experiments now require such sample environments. A radial collimator [21] can be used with a PSD detector to exclude scattering from much of the sample environment, but not to improve the resolution. It must be spun or rocked to average out the attenuation of the scattered radiation over the entire detector.

2.7 TIME-OF-FLIGHT POWDER DIFFRACTION

The PSD and multi-collimator machines described above are designed to simultaneously collect monochromatic neutrons scattered at all Bragg angles 2θ, corresponding to all d-spacings of the crystal. Bragg's equation $2d\sin\theta=\lambda$ suggests that all d-spacings might instead be obtained by holding 2θ fixed and scanning through neutron wavelength λ. This is the basis of time-of-flight (TOF) powder diffraction, where the wavelength is measured by timing a pulse of neutrons over a long flight path. This technique can be used by 'chopping' a reactor source of neutrons, but is probably best suited to pulsed neutron sources. Again, high resolution is obtained by using a high scattering angle. Figure 9 shows such a TOF powder diffractometer, HRPD at the ISIS (UK) pulsed neutron source.

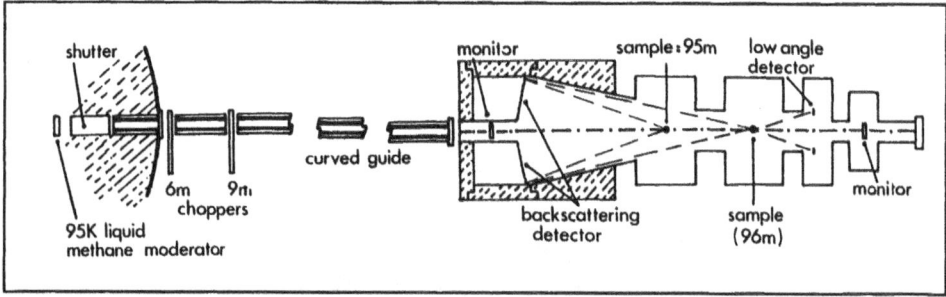

Fig. 9. The high resolution TOF powder diffractometer HRPD at the ISIS (UK) pulsed neutron source, showing how high resolution is again obtained by using high scattering angles (backscattering), while high intensity is gained by using a large detector solid angle.

The resolution of the best TOF and multi-collimator neutron powder machines is often limited only by the powder particle size rather than the diffractometer design, while the rate of data collection depends on the time averaged neutron flux on the sample and the detector solid angle. The TOF machine, and the PSD diffractometer, have an advantage of an order of magnitude larger detector solid angle over the multi-collimator diffractometer, but the time averaged flux on the sample is an order of magnitude smaller for TOF machines with present pulsed sources.

3. Rietveld refinement of powder diffraction data

3.1 PRINCIPLES OF RIETVELD REFINEMENT

However good the resolution of the diffractometer, many Bragg peaks will not be fully resolved. Initially this problem was solved by fitting a number of Gaussians to a group of overlapping peaks, to extract the individual Bragg intensities. The peak shape, being the convolution of a number of instrument parameters is closely approximated by a Gaussian, but more complex functions can be used to account for peak asymmetry, sample strain etc.

The difficulty is that a large number of parameters are required, four for each peak even for a simple Gaussian (position, height, width and background). In cases of severe peak overlap, these parameters are highly correlated, and the errors in the resulting Bragg intensities become important.

Rietveld [22] recognised that it was not necessary to extract the individual Bragg intensities, which was previously assumed by all crystallographers, but that the actual structural parameters could be refined directly to fit the entire diffraction pattern. The structural parameters, having physical reality, are much less correlated among themselves, and there are far fewer of them.

The atom co-ordinates, which are the quantities we want anyway, determine the peak intensities, so they may as well be refined directly. Furthermore, all of the peak positions are determined by at most six lattice constants corresponding to the cell dimensions and angles $(a,b,c,\alpha,\beta,\gamma)$. The peak widths can be obtained from just three parameters U,V,W in Caglioti's formula, and the background can be measured in regions where there are no peaks, and be interpolated since it varies smoothly.

Then instead of four parameters for each of several hundred peaks, we typically have only a few tens of physically real parameters.

For each 2θ point in the diffraction pattern, the Rietveld refinement program [23] calculates the contributions from all of the Bragg peaks in the vicinity, and compares the total $Y_i(calc)$ with the observed count $Y_i(obs)$. The structure parameters, which determine $Y_i(calc)$, are then adjusted to minimize the 'chi-squared' quantity:

$$\aleph^2 = \sum_i W_i.|Y_i(obs)-{}^1/_c.Y_i(calc)|^2 \qquad 3.1.1$$

The summation is over all the 2θ points i and $W_i \propto 1/\sigma_i^2 \approx 1/Y_i(obs)$ is the weight allotted to the count $Y_i(obs)$ while c is a scale factor. Rietveld's weighting scheme can be justified in the following way [23].

Suppose that the counts $Y_i(obs)$ are samples of some population function $Y_i(calc)$, which is completely defined by the crystal structure parameters. The probability p_i that a given sample count $Y_i(obs)$ will differ from $^1/_c.Y_i(calc)$ is given by:

$$p_i \propto (1/\sigma_i).exp\{-(1/2\sigma_i^2).|Y_i(obs)-{}^1/_c.Y_i(calc)|^2\} \qquad 3.1.2$$

since each sample count comes from a normal distribution (a Gaussian peak) centred on $^1/_c.Y_i(calc)$, with standard deviation $\sigma_i^2 = {}^1/_c.Y_i(calc)$. The probability P that all of the counts are samples of the population $^1/_c.Y_i(calc)$ is the product $\prod_i p_i$ of the individual probabilities:

$$P = \prod_i p_i \propto (\prod_i 1/\sigma_i).exp\{-\sum_i (1/2\sigma_i^2).|Y_i(obs)-{}^1/_c.Y_i(calc)|^2\} \propto exp\{-\aleph^2\} \qquad 3.1.3$$

This probability is maximized if the exponent factor $\aleph^2$ is minimized using the weighting scheme $W_i \propto 1/\sigma_i^2 = 1/Y_i(calc) \approx 1/Y_i(obs)$. It should be noted that this statistical weight W_i is calculated from the raw data Y_i *including* the background contribution.

Since the background is a smooth function obtained from the average of many points, the statistical error introduced by background subtraction is usually neglected, as in single crystal measurements. Of course, *systematic* errors may occur; for example, an underestimation of the contribution of thermal diffuse scattering will lead to an underestimation of the overall temperature factor.

4. Applications of Neutron Powder Diffraction to the crystal chemistry of superconductors

4.1 UNDERSTANDING THE EFFECTS OF OXIDATION-REDUCTION

Because neutrons are so sensitive to oxygen, neutron powder diffraction was quickly used to investigate the reduction of superconducting $YBa_2Cu_3O_7$ to non-superconducting $YBa_2Cu_3O_6$ [8, 9]. It was found that the oxygens in the b-axis chains were removed, leaving instead 2D plains of Cu^+. The b-axis then became equivalent to the a-axis, and the structure changed from orthorhombic to tetragonal symmetry.

This effect was seen most strikingly on the PSD machine D1B [24]. The speed of data collection on this machine permitted a complete diffraction pattern to be obtained every minute while the $YBa_2Cu_3O_7$ was being heated, and while it was being cooled again in an oxygen or other atmosphere. In figure 10, these diffraction patterns are plotted one after the other to reveal a 3D plot of intensity, scattering angle and temperature.

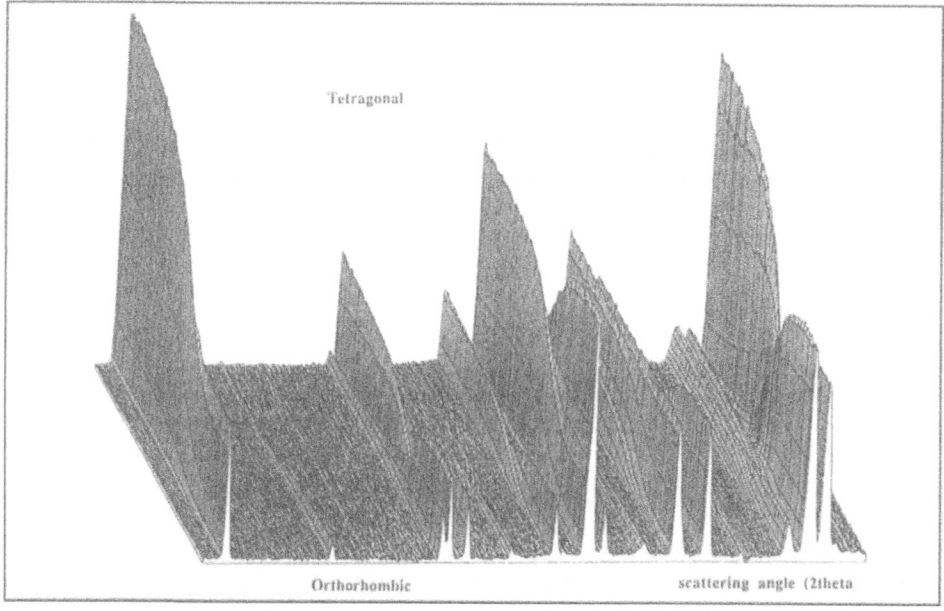

Fig. 10. Changes in the neutron diffraction pattern with reduction of $YBa_2Cu_3O_{7-x}$ to $YBa_2Cu_3O_6$. The complete diffraction pattern was recorded on D1B at ILL at 1 minute intervals while heating.

Each minute, the oxygen occupancy for each site can be obtained by Rietveld refinement (fig. 11). This speed of data collection and refinement means that structural parameters such as the oxygen content of a particular site, can be followed in real time as the external conditions (temperature, atmosphere) are changed.

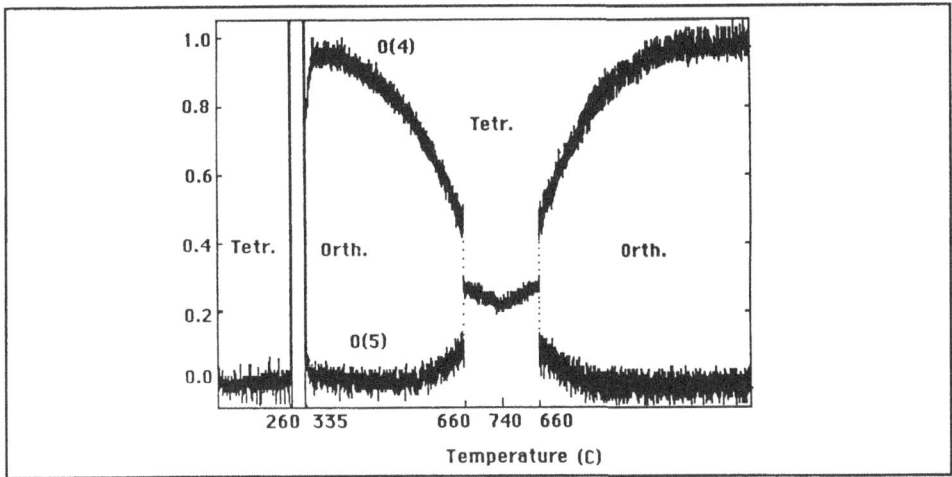

Fig. 11. The occupancy of the oxygen chain sites O(4) and O(5) as oxygen is lost from $YBa_2Cu_3O_{7-x}$ plotted in real time as the temperature is first increased and then reduced.

4.2 THE REFINEMENT OF STRUCTURES PARTLY DETERMINED BY X-RAY METHODS

The most common use of Rietveld refinement, as the name implies, is for the *refinement* of structures already 'determined' by X-ray methods. This was the case for the 90K superconductor. Because X-rays are strongly scattered by the heavy atoms, it is relatively easy to obtain the positions of these heavy atoms — the contribution of the light atoms can be neglected to a good approximation. When the heavy atoms have been located, the remaining atoms can be found approximately by 'difference Fourier' techniques, and their positions refined from the more sensitive neutron data. Solving the structure from the neutron data alone is more difficult, since all of the atoms contribute almost equally, and so (almost) all must be found together in the first step.

In the case of the 90K superconductor, the positions of the heavy atoms determined by X-ray diffraction were assumed to be correct, but the X-ray positions of the oxygen atoms were assumed to be unreliable. Instead, the oxygen atoms were simply placed, as a first approximation, according to the arrangement of oxygen in a perovskite structure, and then their positions and site occupancy refined by the Rietveld method. This immediately lead to the structure shown in fig. 1b.

4.3 PRECISE MEASUREMENT OF INTER-ATOMIC DISTANCES

Not only can the arrangement of the oxygen atoms completely change our conception of the structure, but the precision with which we can measure the atom co-ordinates, and hence the interatomic distances, can help our understanding of the oxidation state of the different cations. As Bednorz and Müller [3] pointed out, the copper-oxygen bond distances will be different for Cu^+, Cu^{++} and Cu^{+++}.

A simple way of understanding this is to note that the copper cation will become smaller when it loses electrons in its higher oxidation states, so the copper-oxygen distances will decrease with oxidation. More precisely, Bednorz and Müller argued that the apical Cu-O distances (along the z-axis) will be large for Cu^{2+} due to the 'Jahn-Teller' distortion, but that these oxygens will move toward copper when it is oxidised to Cu^{3+} (fig. 12). They believed that fluctuations in the electronic structure of copper might therefore be coupled to fluctuations in the positions of the apical oxygen atoms, and that this might provide the electron-phonon coupling thought to be necessary for superconductuctivity.

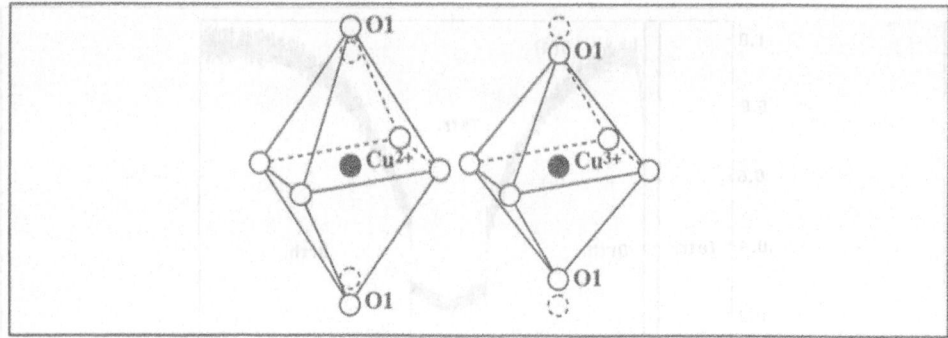

Fig. 12. Expected effect of Cu^{2+}/Cu^{3+} fluctuations on the disorder of apical oxygen O1 in the superconductor $(La,Sr)_2CuO_4$.

Rietveld refinement had already been used in 1984 to look at the apparent fluctuations in the Jahn-Teller distortion of Cu^{++} in copper Tutton's salt [25]. In that case, the fluctuations were not due to fluctuations in the electronic structure, but simply to fluctuations in the *direction* of the Cu-O Jahn-Teller elongated axis. The *average* lengths of the different Cu-O bonds therefore changed with increasing temperature.

In the case of $(La,Sr)_2CuO_4$, neutron powder diffraction [26] showed no evidence for Cu-O 'breathing' displacements, but there was a structural transition involving displacement of the apical oxygen associated with tilting of the CuO_6 octahedra (Fig.13). Such transverse displacements would not be strongly coupled to the electronic structure of copper, since the Cu-O distance changes only to second order.

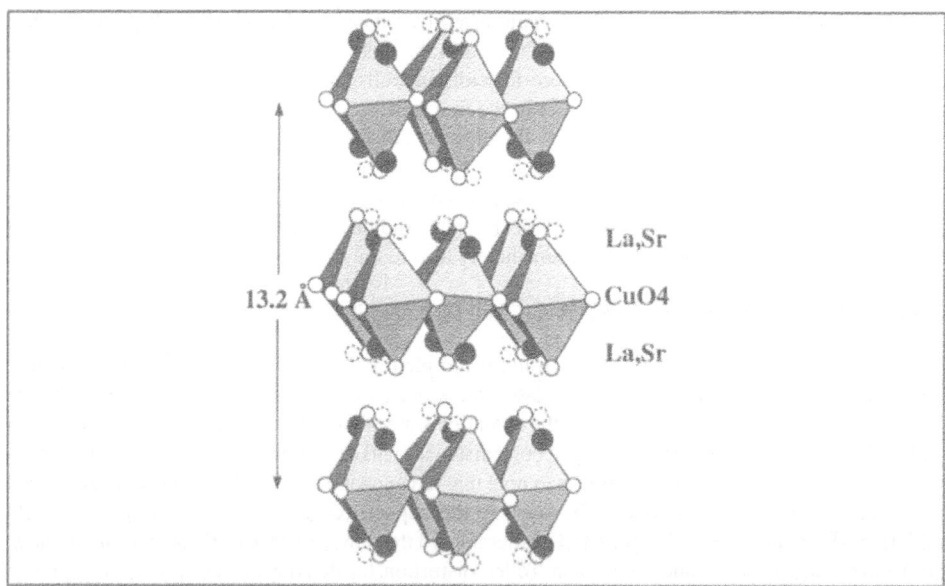

Fig. 13. Observed disorder of apical oxygen in $(La,Sr)_2CuO_4$ due to tilting of the CuO_6 octahedra resulting in low temperature structural transitions.

4.4 EARLIER OXIDE SUPERCONDUCTORS

In 1976 Cox & Sleight [27] had already used Rietveld refinement of neutron powder data to show that in $BaBiO_3$ the bismuth atoms were distributed over two sites of different size: apparently the bismuth valence was dissociated into Bi^{3+} and Bi^{5+} (Fig.14). This required the precise location of oxygen atoms, and the determination of Bi-O bond lengths. In $BaPbO_3$, there is only one site for Pb^{4+}, and when $BaBiO_3$ is doped with Pb^{4+} the two (Bi,Pb)-sites become equivalent, with the Bi^{3+} and Bi^{5+} disordered over this single site. The resulting material is a superconductor with Tc up to 13K.

More recently, Cava et al.[28] have shown that Tc can be increased to 30K if instead Ba is replaced by K. Again the symmetry of the $BaBiO_3$ structure is increased, and Bi^{3+} and Bi^{5+} are distributed over a single site.

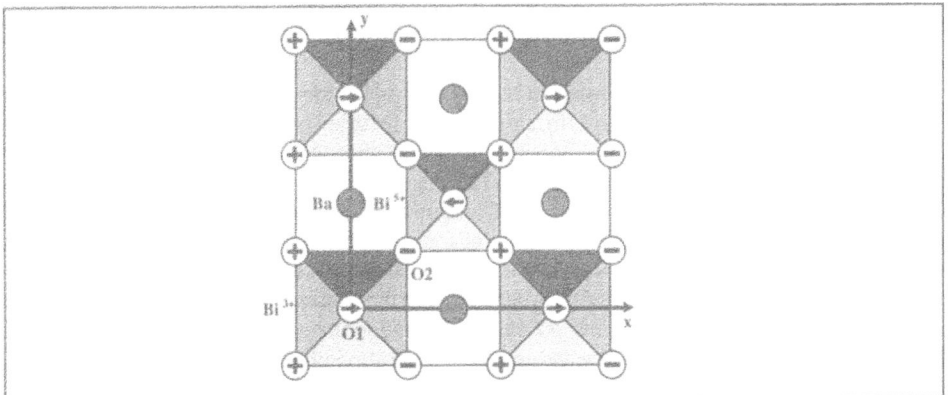

Fig. 14. Ordering of Bi^{3+} and Bi^{5+} on sites of different size in $BaBiO_3$. With Pb/Bi (or K/Ba) doping Bi^{3+}/Bi^{5+} is distributed over a single site, and the material becomes superconducting. (The arrows and +,- signs indicate the directions of tilting of the BiO_6 octahedra.)

4.5 THE PAULING CONCEPT OF BOND STRENGTH AND ZACHARIASEN'S VALENCE CHARGE

In both $(La,Sr)_4CuO_4$ and $Ba(Bi,Pb)O_4$ it was necessary to measure the precise cation-oxygen bond lengths, using Rietveld refinement of neutron powder data, to understand the effective oxidation states of the cations, apparently so important for superconductivity. A quantitative relation between cation oxidation state and bond lengths can be developed from ideas originally due to Pauling and Zachariasen.

Pauling [29] introduced the concept of the bond strength s=z/N, where z is the cation charge and N is its co-ordination number (the number of bonded anions, assuming all to be equivalent for the moment). Pauling's 'local charge balance' concept required that the sum of all the bond strength contributions s_{ij} from the different anions j had to equal the 'valence charge' v_i for cation i, with a similar sum rule for v_j of the anions i.e.

$$\Sigma_j s_{ij} = v_i \quad \text{and} \quad \Sigma_i s_{ij} = v_j. \qquad (4.5.1)$$

Zachariasen [30, 31] found an empirical relation between the bond distances D_{ij} and bond strengths s_{ij}

$$s_{ij} = \exp\{(D1-D_{ij})/(A.D1)\} \qquad (4.5.2)$$

where A and D1 (the distance corresponding to unit bond strength) are empirical constants that can be determined for a given cation-anion pair by precise measurements of bond distances in many compounds containing these cation-anions. In practice, according to Zachariasen, A=0.178 for all X-O pairs for 3d-elements such as Cu, and D1=1.74 for Cu increases to only 1.89 for Sc.

Brown and Altermatt [32] have used the Inorganic Crystal Structure Database to show that $A.D1 = B \approx 0.37$ in all cases to a good approximation, so that a single bond strength parameter $D1 = r_0$ can be tabulated for each of 750 different atom pairs.

Capponi et al. [1] used these relatively constant values of the empirical parameters A and D1 to show that the excess electron holes in the $YBa_2Cu_3O_7$ superconductor were distributed between the chain and plane copper sites, and further [9] that when oxygen was removed to produce non-superconducting $YBa_2Cu_3O_6$, the chain copper was reduced to Cu^+ and the plane copper to Cu^{++}.

4.6 LIMITATIONS TO THE BOND STRENGTH-VALENCE CHARGE CONCEPT

The bond strength-valence charge concept indicates the bond lengths that atoms with a given valence charge would have in an ideal structure. In real structures, there are other constraints that can modify these values: for example, it is not possible to satisfy the ideal bond lengths for all atoms simultaneously in a complex structure. Brown [33] has illustrated this for $YBa_2Cu_3O_7$, and concluded that the Ba-site is smaller than that ideally required for Ba^{++} in $YBa_2Cu_3O_7$, and larger than required in $YBa_2Cu_3O_6$. The calculated valence charge cannot then be expected to be precisely 2.0 for barium. If these internal stresses were larger, a structural phase transition might be expected, such as is common in perovskite structures ABO_3, depending on the relative sizes of the cations and anions.

Jansen and Block [34] have concluded from this limitation that the concept is not applicable to high-Tc superconductors. They derive the empirical bond strength relations from the Born-Mayer model of ionic solids, and deduce that the concept is only valid for regular ionic solids, without disorder or vacancies, with cations and anions of comparable size, and at zero pressure. This appears to rule out most inorganic chemistry !

In fact, the connexion with the Born-Mayer model had already been made by Zachariasen [30] in 1931, who did not claim that bond strength was any less empirical for that, and who did not hesitate to apply the concept to materials that were clearly not ideal Born-Mayer ionic solids.

5. Applications to high temperaturesuperconductors

5.1 THE 90K 123-SUPERCONDUCTOR $YBa_2Cu_3O_7$

Unlike $(La,Sr)_2CuO_4$, there are no obvious low temperature structural transitions in the 90K superconductor $YBa_2Cu_3O_7$, but Capponi et al. [1] showed that the oxygens along the CuO-chain (Fig.1) appear to be disordered perpendicular to the chain axis even at low temperature. François et al. [35] proposed that these CuO-chains may zig-zag for short distances, with tilting of the CuO_4 squares similar to tilting of CuO_6 octahedra in normal perovskites. Again, as in $(La,Sr)_2CuO_4$, such transverse O4 displacements would not couple strongly to the electronic state of copper.

As we have seen, neutron powder diffraction [8, 9] showed a structural transition at *high* temperature to a tetragonal non-superconducting phase as oxygen is removed from the CuO-chains (Fig.15). Cava et al. [36] found that as oxygen is removed, Tc decreases, but with a plateau corresponding to $YBa_2Cu_3O_{6.5}$ (Fig.16). Using the neutron powder measurements of Santoro et al. [37], they demonstrated that a sharp *drop* in Tc was associated with a sharp *increase* of the Cu2-O1 apical oxygen distance.

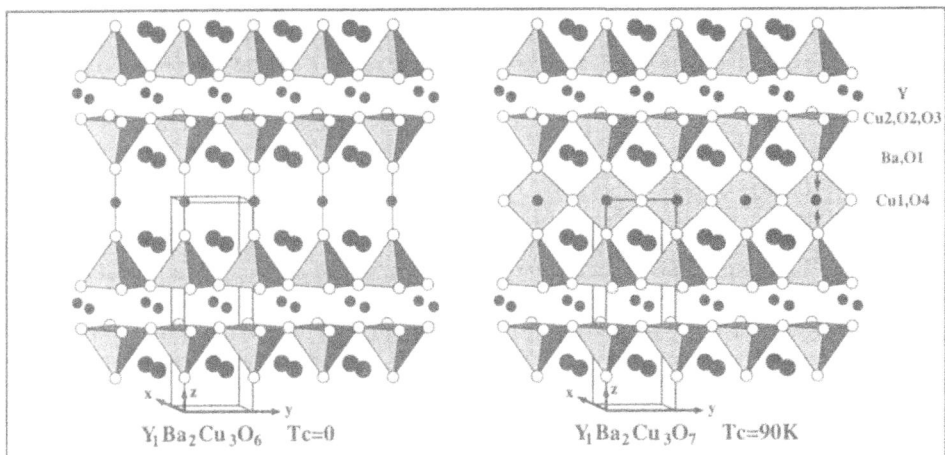

Fig. 15. Structures of YBa$_2$Cu$_3$O$_6$ and YBa$_2$Cu$_3$O$_{7-x}$ showing the displacement of the bridging oxygen O1 associated with a decrease in Tc on removal of the chain oxygen O4.

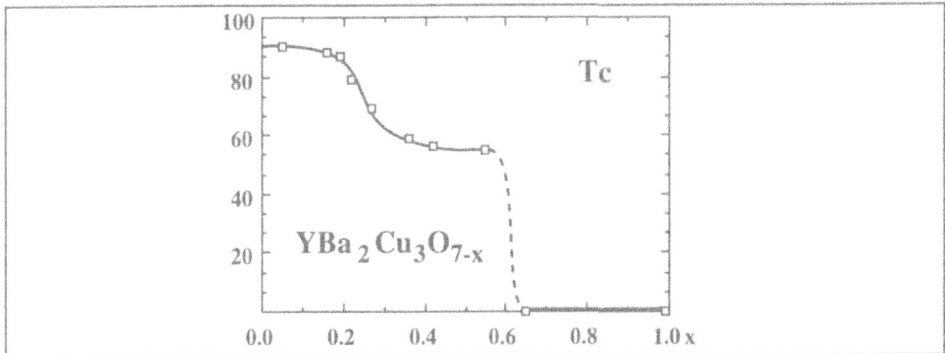

Fig. 16. Variation of Tc with oxygen loss in YBa$_2$Cu$_3$O$_{7-x}$ showing two plateau regions followed by a sharp drop to the non-superconducting phase.

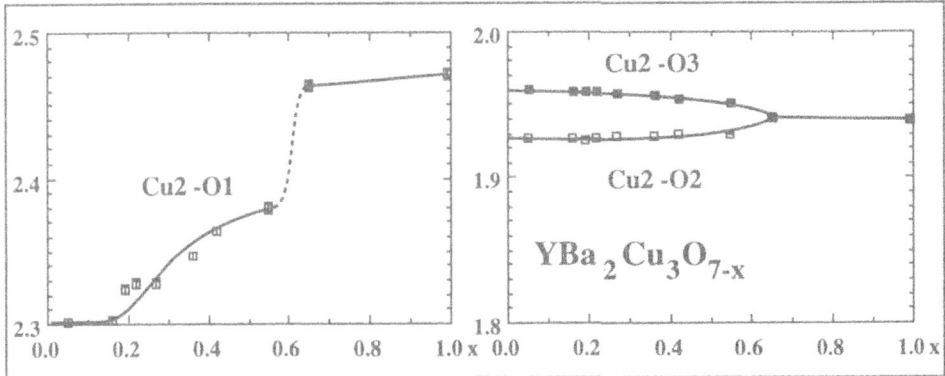

Fig. 17. Cu2-O1 distances in YBa$_2$Cu$_3$O$_{7-x}$, mirroring the changes in Tc, with an initial plateau and a sharp increase with the loss of superconductivity. The other Cu-O distances hardly change.

Cava et al. [38] used neutron powder diffraction to measure all Cu-O distances in $YBa_2Cu_3O_{7-x}$ as a function of oxygen loss x from the Cu-O chains (Fig. 17). Their samples were very well ordered, since they were prepared by low temperature 'gettering' to remove precise amounts of oxygen.

The Cu-O distances *within* the planes and chains varied little with oxygen loss, in striking contrast to the the apical oxygen O1 bridging planes and chains. As oxygen was removed from the CuO-chains, reducing Tc, this bridging oxygen moved away from Cu2 in the planes toward Cu1 in the chains.

The Cu2-O1 distance (Fig.17) seems to remain constant, as does Tc (Fig.16), for small losses of oxygen (x<0.2). It then increases, but starts to level out again between 0.4<x<0.6, corresponding to the plateau in Tc, before increasing abruptly with the loss of superconductivity. Since this distance is only the average of Cu2-O1 distances for more complex intermediate superstructures, an exact correspondence cannot be expected.

These increases can however be associated with loss of Cu^{3+} (or more precisely, electron holes) from the CuO-planes, implying a direct relation between the number of Cu^{3+} in these planes and Tc. The ideas of Bednorz and Müller appear then to be supported by the results of these neutron powder diffraction measurements.

5.2 THE CHARGE TRANSFER PICTURE OF OXIDE SUPERCONDUCTORS

Superconductivity in these materials depends then on the oxidation state of the copper oxide layers, and changes in this oxidation state can be indicated by changes in Cu-O bond lengths that can be measured by neutron diffraction. We will see that these simple chemical ideas can help us understand why superconductivity in some materials requires the addition of oxygen, while in others superconductivity is killed when oxygen is added.

We know that the oxygen content of the conducting layers themselves is not usually changed. We can then describe the different materials as containing essentially identical copper oxide layers, responsible for the superconductivity, with very diverse layers of other materials - oxides of copper, lead, bismuth, thalium etc - acting as 'charge reservoirs' (fig.18). Changes in the oxidation state of the charge reservoir can be transferred to the conducting copper oxide layers, usually through the bridging oxygen.

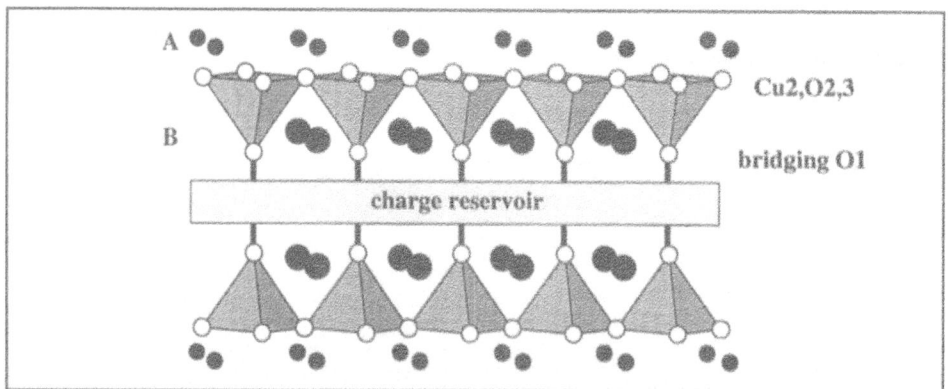

Fig. 18. The structure of copper oxide superconductors can be described as common copper oxide conducting layers whose oxidation state can be controlled by charge transfer from a variety of intermediate charge reservoirs of other oxides.

In $YBa_2Cu_3O_{7-x}$ the charge reservoir is simply the copper oxide chains. Adding oxygen to this charge reservoir also oxidizes the copper oxide planes, which then become superconducting.

5.3 $Pb_2Sr_2Y_{1-x}Ca_xCu_3O_{8+y}$, WHERE SUPERCONDUCTIVITY IS KILLED BY OXIDATION

Now consider the superconductor $Pb_2Sr_2Y_{1-x}Ca_xCu_3O_{8+y}$ [39] whose structure is shown in figure 19. Clearly it contains the usual copper oxide planes, while the charge reservoir layer consists of copper oxide chains as in $YBa_2Cu_3O_{7-x}$, except that now these chains are enclosed by a pair of lead oxide layers. When this material is oxidized, superconductivity is killed, which is the opposite to what happens in $YBa_2Cu_3O_{7-x}$.

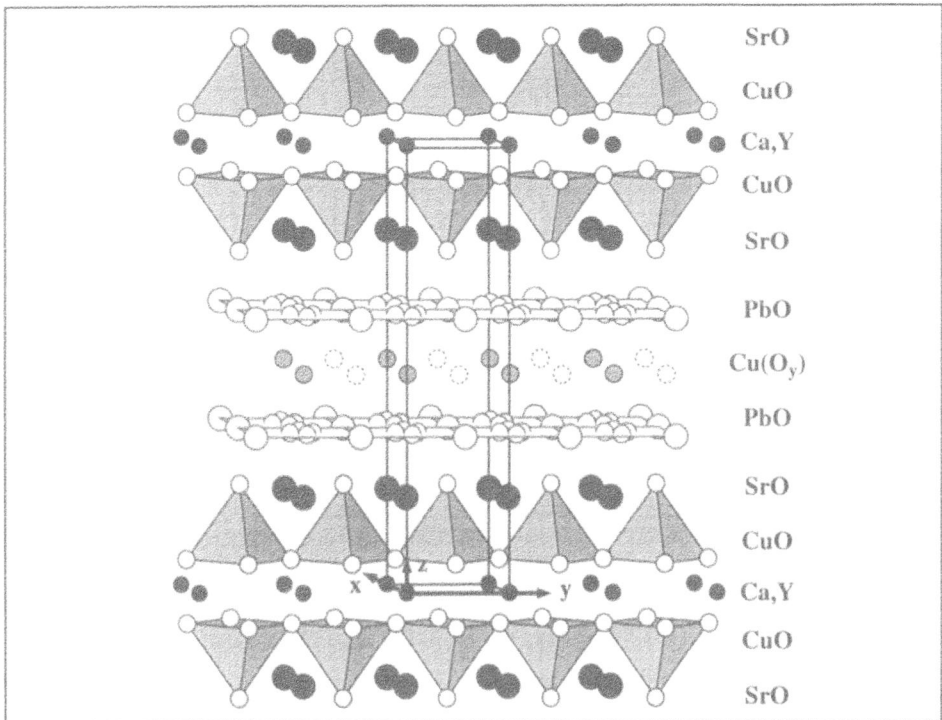

Fig. 19. The superconductor $Pb_2Sr_2Y_{1-x}Ca_xCu_3O_{8+y}$, where the charge reservoir consists of layers of copper oxide and lead oxide, and where superconductivity is killed by oxidation, the opposite to what is normally expected.

Adding oxygen to $Pb_2Sr_2Y_{1-x}Ca_xCu_3O_{8+y}$ oxidizes the copper in the charge reservoir layer, but it also tends to oxidize the lead, from Pb^{2+} to Pb^{4+}. However, not enough oxygen can be added to satisfy the higher oxidation state of Pb^{4+}, which then tends to reduce the superconducting layers, killing superconductivity. Neutron diffraction experiments should show the bridging oxygen moving toward the lead, and away from the conducting copper oxide planes.

5.4 ANOMALOUS PRESSURE EFFECTS IN THE 124-SUPERCONDUCTOR $YBa_2Cu_4O_8$.

$YBa_2Cu_3O_7$ has several practical disadvantages, including the ease with which it can loose oxygen and the twinned nature of the crystals. A more stable phase, $YBa_2Cu_4O_8$, with double CuO chains can be made under oxygen pressure (Fig.20). This material is not normally twinned (though like $YBa_2Cu_3O_7$ it is orthorhombic) and the chain oxygen cannot easily be lost because this would leave planes of copper in contact.

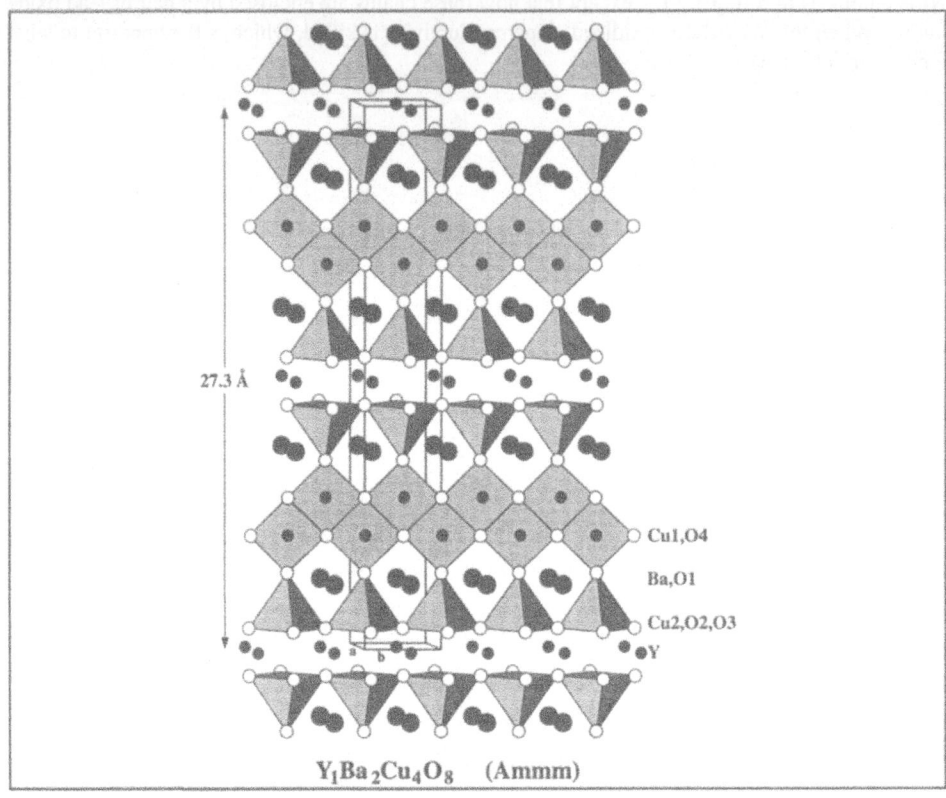

$Y_1Ba_2Cu_4O_8$ (Ammm)

Fig. 20. Structure of $YBa_2Cu_4O_8$ showing the double CuO chains.

Van Eenige et al.[40] have shown that Tc increases strongly with applied pressure in $YBa_2Cu_4O_8$, going from 80K to 108K with very moderate pressures of 10 GPa. Even with the relatively low pressures possible with neutron powder diffraction, Tc can be increased by 5.5K on applying only 10kbar. Normally, changes in Tc due to pressure are very much smaller, and are caused by the compressibility of the electronic structure.

Kaldis et al.[41] proposed instead that in $YBa_2Cu_4O_8$ the double Cu-O chains might be less compressible than the CuO layers, so that with pressure electron-holes would be transferred from the chains to the conducting CuO-layers. The evidence was that the Cu2-O1 bond, bridging the CuO layers to the chains, shortens anomalously with pressure as measured by neutron powder diffraction. More recent measurements, [42] especially on single crystals [43] have supported this idea, obtaining more precise values for the structural changes and extending the measurements to higher pressure.

They confirm that the Cu2-O1 bond is indeed more compressible than the other CuO distances. Figure 21 shows the shortening in this Cu2-O1 bond length as a function of the change in Tc; the points for $YBa_2Cu_4O_8$ under pressure agree well with the average line obtained by Cava et al. for the oxidation of $YBa_2Cu_3O_{7-x}$. Clearly the shortening of the Cu2-O1 distance for a given change in Tc is similar in the two materials, even though in one case Tc is changed by oxidation, and in the other by applying pressure.

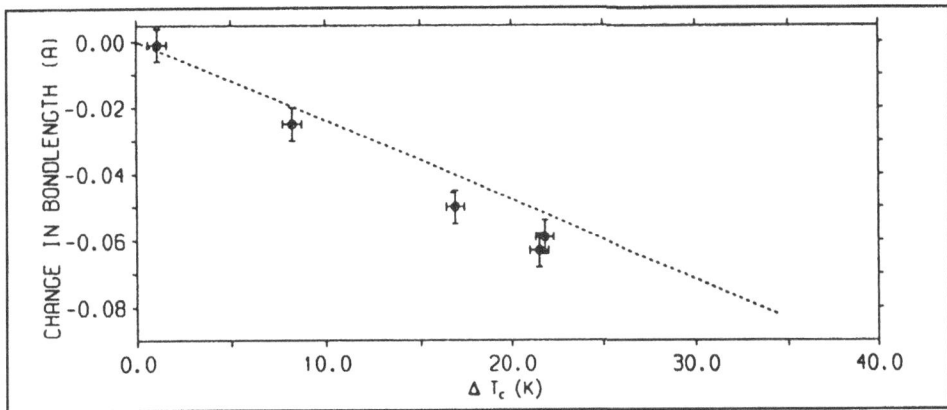

Fig. 21. The changes in the Cu2-O1 distance with pressure in $YBa_2Cu_4O_8$ [43]. The dashed lines are deduced from similar Tc changes on oxidation of $YBa_2Cu_3O_{7-x}$ according to Cava et al [38].

Jansen and Block [34] are particularly critical of the concept of bond strength as applied to materials under pressure, because 'this is not the minimal internal energy condition required by the Born-Mayer model'. They argue that according to the valence charge concept, ionic charges must appear to increase under pressure because the bond lengths decrease. This would be correct if A and D1 in equation 4.5.2 were universal constants independant of pressure, which of course is never assumed.

Jansen and Block apparently do not accept that under pressure, charge may be transferred from one part of a structure to another if one part is more compressible than the other. They prefer instead their own explanation that the increase in Tc with pressure is due to "an increase in the density of oxygen anions, and therewith, an increasing pairing interaction". This implies that the effect of pressure should depend simply on the compressibility of the material.

5.5 $Tl_1Ca_{n-1}Ba_2Cu_nO_{2n+3}$ SUPERCONDUCTORS

It is possible to replace the CuO-chains in the 123- and 124-superconductors by other 'charge reservoir' layers containing BiO, TlO or PbO. The simplest example is the single layer TlO-series (fig. 24) discovered by Parkin.[44] Figure 22 shows how the original Bednorz and Müller 214-superconductor $(La,Sr)_2CuO_4$ becomes the first member of this series, $(Sr,La)_2CuO_4.TlO$.

In fig.23, the second member of this series is seen to be very similar to $YBa_2Cu_3O_7$ (fig.8), with the CuO-chains replaced by a TlO-layer. However, for a stoichiometric layer of $Tl^{3+}O^{2-}$ the formula $Ca_1Ba_2Cu_2O_6.TlO$ would not permit the excess electron holes (Cu^{3+}) necessary for superconductivity. Either some Tl is missing, or present as Tl^{1+}, or there is excess oxygen, or some other such modification of the stoichiometry is necessary. One objective of neutron structure refinement is to resolve this question.

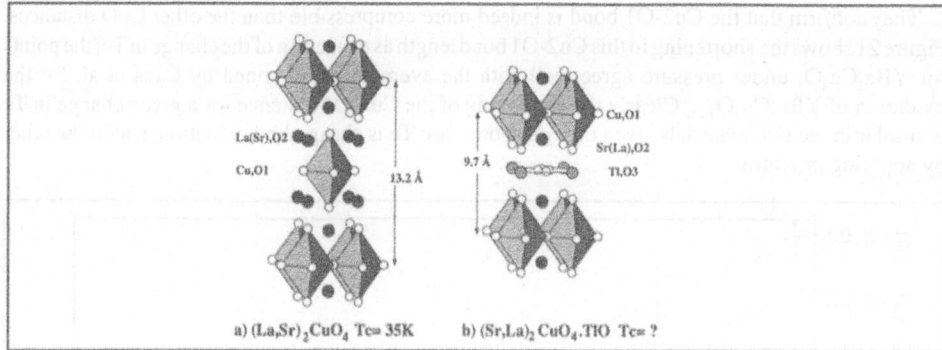

Fig. 22. The first member of the single TlO-layer "Parkin" phases compared to the original Bednorz and Müller 214-superconductor.

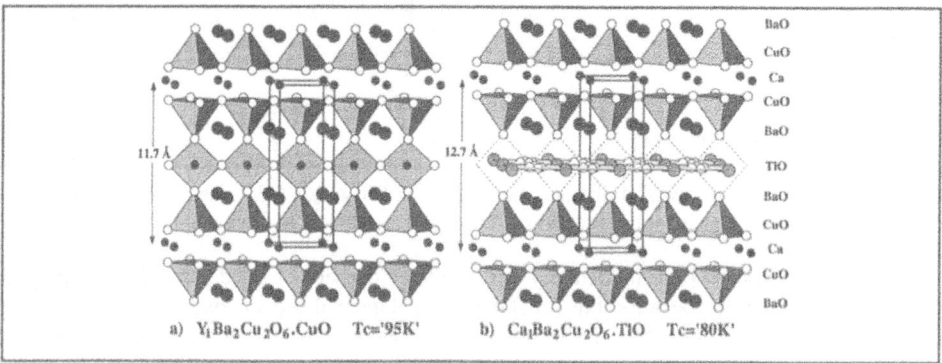

Fig. 23. The second member of the "Parkin"-TlO-phases compared to the 90K 123superconductor.

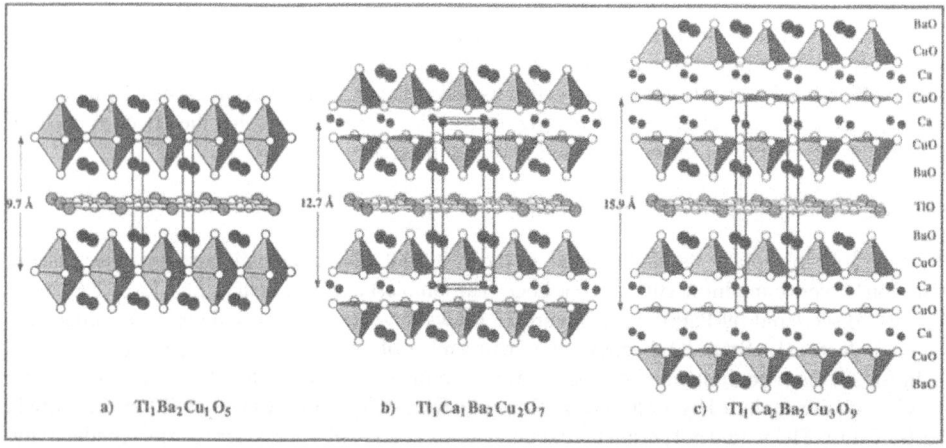

Fig. 24. The single TlO-layer "Parkin" series. Tc increases with the number of CuO-layers up to n=3.

5.6 $Bi_2Ca_{n-1}Sr_2Cu_nO_{2n+4}$ AND $Tl_2Ca_{n-1}Ba_2Cu_nO_{2n+4}$

In fact, more work has been done on the similar double TlO- and BiO-layer materials shown in figure 25. Again the second member of this series may be compared (fig. 26) to $YBa_2Cu_4O_8$, where the double CuO-chains are replaced by double layers of TlO or BiO.

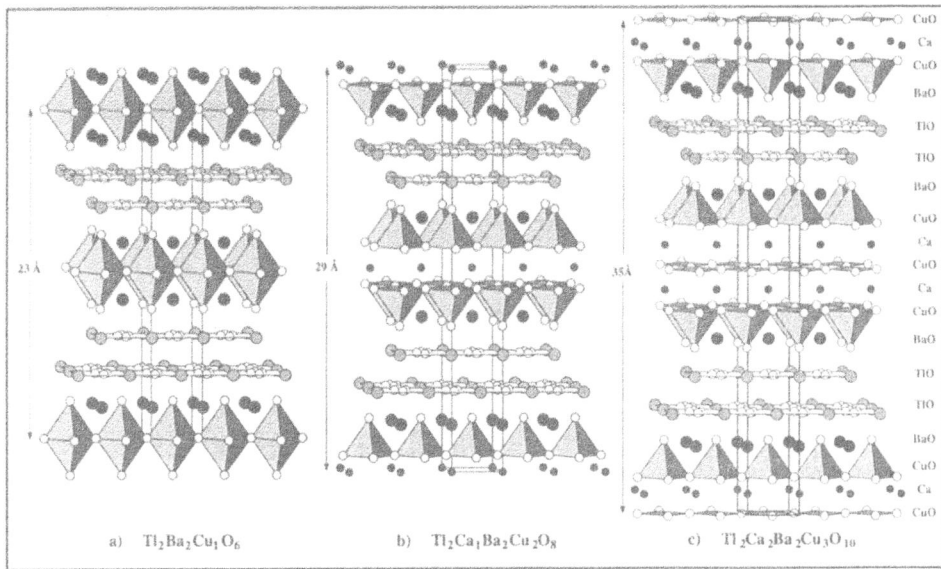

Fig. 25. The double layer TlO- (and BiO-) phases.

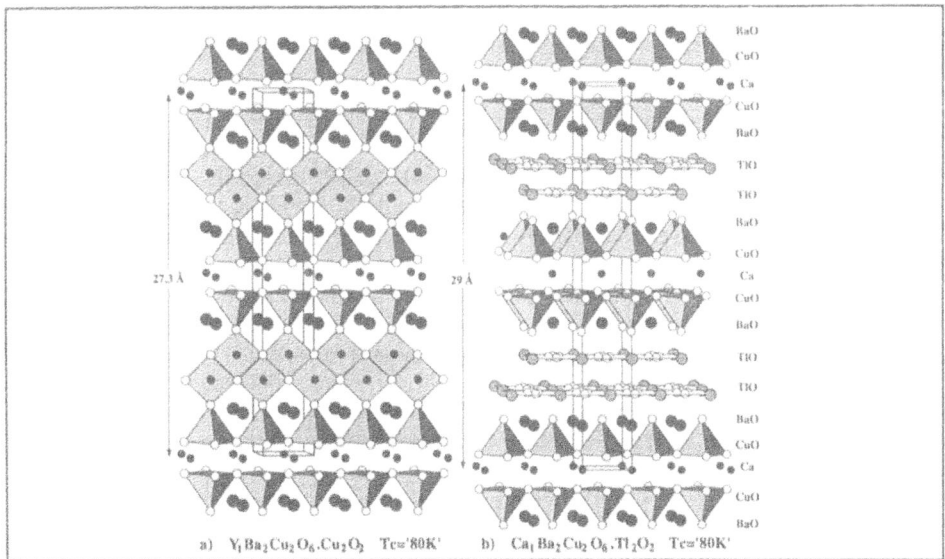

Fig. 26. The second member of the double layer TlO/BiO phases compared to the double chain 124-superconductor.

The main difference between the BiO- and TlO- phases is due to the lone pair electrons on Bi^{3+}, but not on Tl^{3+}. These lone pair electrons lie between pairs of BiO-layers, and force them apart; in fact the material cleaves between these layers. The structures are not then of the classical 'Aurivillius' type (fig. 27) [6] proposed after the first X-ray work. This particular lone-pair arrangement also explains why single layer BiO-phases are not found.

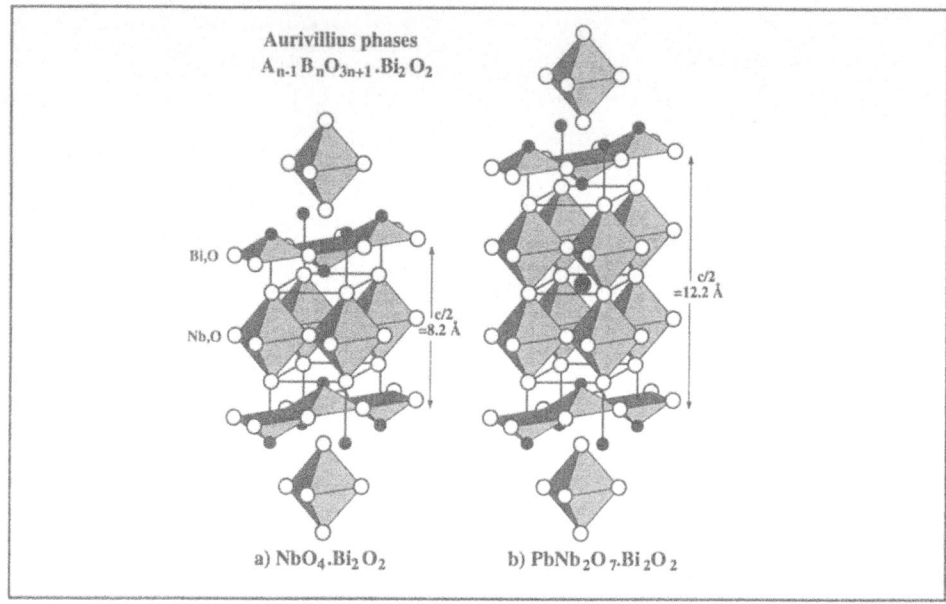

Aurivillius phases
$A_{n-1} B_n O_{3n+1} \cdot Bi_2 O_2$

Bi,O

Nb,O

$c/2 \approx 8.2$ Å

$c/2 = 12.2$ Å

a) $NbO_4 \cdot Bi_2 O_2$

b) $PbNb_2 O_7 \cdot Bi_2 O_2$

Fig. 27. The 'Aurivillius' type structure [6] that was proposed after the first X-ray work shows a structure for the BiO-Layers quite different to that obtained with neutrons. It was difficult to locate the light oxygen atoms in the presence of the heavy Bi-atoms.

5.7 INCOMMENSURABLE STRUCTURES OF $Tl_1 Ca_{n-1} Ba_2 Cu_n O_{2n+3}$ SUPERCONDUCTORS

There remain important questions. Bismuth can be either Bi^{3+} or Bi^{5+}; what is its valence state in these materials, since any oxidation of Bi^{3+} to Bi^{5+} will result in reduction of copper, and loss of superconductivity ? If Bi^{3+} and Bi^{5+} co-exist as in $Ba(Bi,Pb)O_3$, perhaps the BiO-planes rather than the CuO-planes are responsible for the superconductivity ?

There is too much space for the oxygen atom within the BiO-layer, and it appears disordered in fig. 25. This 'disorder' indicates that the true structure is more complex, and that the model structure is only an average structure. As well, the incommensurable superlattice reflections cannot be explained with the average structures of fig. 25.

The possibility of Bi^{5+} was first proposed by von Schnering et al.[45] but Bordet et al.[7] were able to find an average structure that removed much of the apparent oxygen disorder and showed that the oxygen co-ordination of bismuth was typical of Bi^{3+}.

The new structure was based on the determination of the space group by Withers et al.[46], using convergent beam electron diffraction, as Amaa (or the non-centric A2aa). Bordet et al.[47] found that only with A2aa could an average structure be found with oxygen order. In this structure (fig. 28), only one of the four possible positions for BiO-oxygen shown in fig. 25 is occupied.

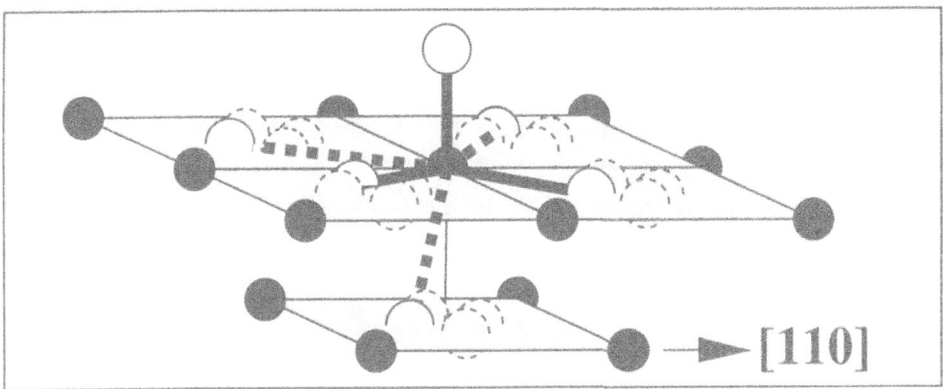

Fig. 28. Average oxygen environment of Bi according to the neutron powder data of Bordet et al.[47], showing co-ordination typical of Bi^{3+}.

Oxygen then moves off the centre of the Bi-square toward a pair of bismuth atoms, so that each bismuth has a total of three close oxygens (including the oxygen bridging to the CuO-layer) plus three more distant oxygens. This co-ordination is typical of Bi^{3+}, with the lone-pair electrons presumably opposite the three close oxygens i.e. between the BiO-layers.

If the Bordet et al.[47] structure is drawn for the BiO-plane (fig. 29), it is seen to consist of zig-zag Bi-O-Bi-O... chains. One of the important questions about the charge content of the reservoir layer has therefore been answered, but in this average structure there is no additional oxygen, and still no explanation of the incommensurable spots.

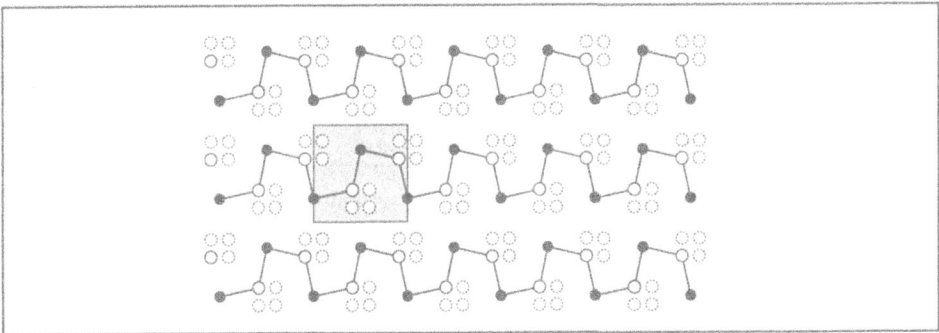

Fig. 29. The Bordet et al. average structure in the BiO-planes. Each Bi has two short O-bonds within the plane. The alternative O-positions are dotted. This model removes most of the apparent oxygen 'disorder' but does not account for the incommensurable structure.

The answer to this second question was first suggested by Le Page et al.[48] from X-ray single crystal measurements on an isomorphous Fe- instead of Cu-compound for which the superstructure was commensurable.

Their model (fig. 30) also shows oxygen moving toward a pair of Bi-atoms as in the Bordet et al.[47] average structure (fig. 29). However, the displacement now increases along the x-axis until oxygen is directly between a pair of Bi-atoms. In this way it is possible to fit one extra oxygen for every 4 or 5 unit cells along the x-axis.

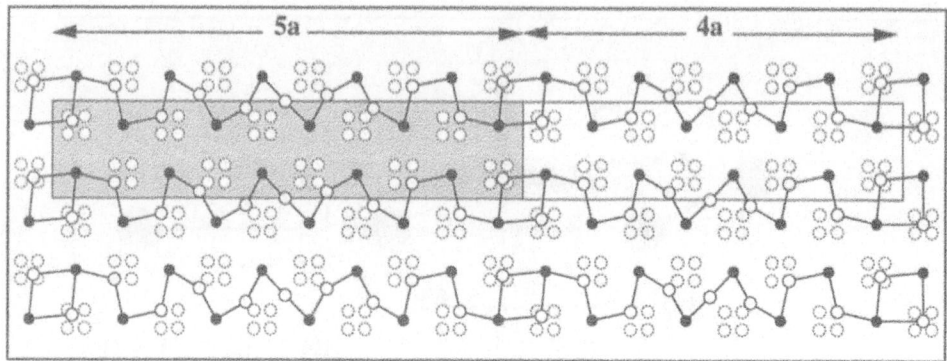

Fig. 30. The superstructure according to the model of LePage et al.[48] One extra oxygen is accommodated for every 4 or 5 cells. The superstructure also consists of expansion around the short Bi-O-Bi bonds (not shown).

The Le Page model shows how additional oxygen can be inserted while retaining the bismuth-oxygen co-ordination typical of Bi^{3+} found by Bordet et al. In particular, there is no obvious 'oxygen interstitial', since all of the oxygen, and all of the bismuth, have similar co-ordination.

Of course the Bi-lattice buckles with this extra oxygen, especially when it is directly between adjacent Bi-atoms, and it is this heavy atom modulation (not shown in fig. 30), together with the associated buckling of the CuO-layers, that produces the superlattice spots.

Figure 23 also shows why there is no obvious reason that this superstructure should be commensurable with an integer number of average unit cells. In the Fe-compound of Le Page et al.[48] the superstructure 'locks-on' to repeat after 5 unit cells, and usually we would expect such a structure to lock-on to the underlying perovskite metric.

Incommensurable superlattice spots would still be obtained with lock-on if there were a pseudo-periodic sequence of 5,5,5,4 etc blocks as suggested by the high resolution electron images,[49] and as illustrated in figure 23. In fact there is no easy way to determine if the superstructure is a single incommensurable wave or a series of discrete blocks.

Yamamoto et al.[50] and Petricek et al.[51] have adapted Rietveld techniques to refine such 'incommensurable' structures, while Dmowski et al.[52] have shown how pair correlation functions can be extracted from neutron powder data and used to obtain models for the short range order in TlO-chains.

In the Tl-compounds the average structure of the TlO-layers is similar.[53] The superstructure though is different, and may be due to Tl-vacancies ,[54] which would be another way of creating electron holes.

However, samples with very similar stoichiometries may have very different Tc's, and the reason for this is not yet clear. It may be related to the details of the order of the structure, which may produce small shifts in the positions of bridging oxygen atoms, apparently so important for the superconductivity of $YBa_2Cu_3O_7$ and $YBa_2Cu_4O_8$.

Conclusion

The discovery of high Tc superconductors can be seen as a kind of test of the relevance of different techniques to a new problem in solid state physics and chemistry. This examination was all the more interesting because it arrived without warning, and because almost nothing was known about the new materials. Everyone could therefore participate in the test, but of course some techniques proved more relevant than others.

Surprisingly perhaps, the details of the crystal structure proved to be very important. They were indeed the basis of Bednorz and Müller's ideas that lead to the discovery of the new materials. Physicists often assume that such details should be irrelevant, or that crystal structures can be 'determined' once and for all by a variety of standard techniques. On the contrary, we have seen that for oxide superconductors, crystal chemistry ideas can help reconcile the apparently contradictory observations of the effects of oxidation/reduction and the anomalous pressure dependence of Tc.

They can help us understand why incommensurable structures may be important, not because they necessarily indicate complicated electronic interactions, but because they provide a means for changing the precise stoichiometry, and hence the electron hole doping and Tc. And most importantly, crystal chemistry ideas can lead to the discovery of other new superconductors.

Neutron powder diffraction proved to be essential, not only for understanding the basic structures, but for understanding how these structures can be changed, and how this affects superconductivity. In this paper we have not attempted to review all of the many contributions of neutron diffraction. Instead we have tried to show some of the unique advantages of the technique, and to explain why, for this particular problem, it was more successful than competing methods, such as for example, diffraction using synchrotron radiation.

All theories or ideas must ultimately be judged by their ability to predict new things, and in this the ideas of Bednorz and Müller, and what we have called 'crystal chemistry', have been singularly successful. The more sophisticated physical theories, however intellectually satisfying, have so far failed to show us how to prepare any new materials, or even how to increase the Tc of the materials that have already been discovered.

References

1.	J. J. Capponi, C. Chaillout, A. W. Hewat, P. Lejay et al.: Europhys.Lett. **3** 1301 (1987)
2.	T. Siegrist, S. Sunshine, D. W. Murphy, R. J. Cava et al.: Phys.Rev.B **35** 7137 (1987)
3.	K. A. Müller, J. G. Bednorz: Science **237** 1133 (1987)
4.	A. Reller, J. G. Bednorz, K. A. Müller: Zeit.Phys.B **67** 285 (1987)
5.	Editorial, The Scientist **May 30** 17 (1988)
6.	B. Aurivillius: Arki. Kemi. **2** 519 (1949)
7.	P. Bordet, J. J. Capponi, C. Chaillout, J. Chenavas et al.: Phys.C **153** 623 (1988)
8.	J. D. Jorgensen, M. A. Beno, D. G. Hinks, L. Soderholm et al.: Phys. Rev. B **36** 3608 (1987)
9.	A. W. Hewat, J. J. Capponi, C. Chaillout, M. Marezio et al.: Solid State Comm. **64** 301 (1987)
10.	A. W. Hewat: Nucl. Inst. Methods **127** 361 (1975)
11.	A. W. Hewat: Mat. Sci. Forum **9** 69 (1986)
12.	W. I. F. David Harrison W T A, Johnson M W: Mat. Sci. Forum **9** 89 (1986)
13.	H. Meister, B. Weckermann: Nucl. Inst. Meth. **108** 107 (1973)
14.	C. J. Carlile, P. D. Hey, B. Mack: J. Phys. E **10** 543 (1977)
15.	A. W. Hewat: pp 316 in *Position Sensitive Detection of Thermal Neutrons* (Academic Press) Ed: P. Convert, J. B. Forsyth (1983)

44

16.   P. Convert, D. Fruchart, E. Roudaut, P. Wolfers: pp 302 in *Position Sensitive Detection of Thermal Neutrons* (Academic Press) Ed: P. Convert, J. B. Forsyth (1983)

17.   C. W. Tompson, D. F. R. Mildner, M. Mehregany, J. Sudol et al.: J. Appl. Cryst. **17** 385 (1984)

18.   W. Schäfer, I. Naday, G. Will: pp 207 in *Position Sensitive Detection of Thermal Neutrons* (Academic Press) Ed: P. Convert, J. B. Forsyth (1983)

19.   P. Convert, J. B. Forsyth: *Position Sensitive Detection of Thermal Neutrons* (Academic Press) (1983)

20.   A. W. Hewat, I. Bailey: Nucl. Instrum. Methods **137** 463 (1976)

21.   A. F. Wright, M. Berneron, S. Heathman: Nucl. Inst. Meth. **180** 655 (1981)

22.   H. M. Rietveld: J. Appl. Cryst. **2** 65 (1969)

23.   A. W. Hewat: *The Rietveld Computer Program for the Profile Analysis of Neutron Diffraction Powder Patterns, Modified for Anisotropic Thermal Vibration.*: UKAEA Research Group Report R7350 (unpublished) (1973)

24.   J. Rodriguez, J. Bassas, X. Obradors, M. Vallet et al.: Physica C **153-155** 1671 (1988)

25.   B. J. Hathaway, A. W. Hewat: in *J. Solid State Chem. 51,364-75.* (1984)

26.   J. D. Jorgensen, H. Schuttler, D. G. Hinks, D. W. Capone et al.: Phys. Rev. Lett. **58** 1024 (1987)

27.   D. E. Cox, A. W. Sleight: Solid State Comm. **19** 969 (1976)

28.   R. J. Cava, B. Batlogg, J. J. Krajewski, R. Farrow et al.: Nature **332** 814 (1988)

29.   L. Pauling: J. Am. Chem. Soc. **51** 1010 (1929)

30.   W. H. Zachariasen: Z. Kristall. **80** 137 (1931)

31.   W. H. Zachariasen: J. Less Common Metals **62** 1 (1978)

32.   I. D. Brown, D. Altermatt: **B41** 244 (1985)

33.   I. D. Brown: J.Solid State Chem. **82** 122 (1989)

34.   L. Jansen, R. Block: Physica C **181** 149 (1991)

35.   M. Francois, A. Junod, K. Yvon, A. W. Hewat et al.: Solid State Comm. **66** 1117 (1988)

36.   R. J. Cava, B. Batlogg, S. A. Sunshine, T. Siegrist et al.: Physica C **153-155** 560 (1988)

37.   A. Santoro, F. Beech, S. A. Sunshine, D. W. Murphy et al.: Mat. Res. Bull. **22** 1007 (1987)

38.   R. J. Cava, A. W. Hewat, E. A. Hewat, B. Batlogg et al.: Physica C **165** 419 (1990)

39.   C. Chaillout, P. Bordet, J. J. Capponi, R. J. Cava et al.: J. Less-Comm. Metals **164** 816 (1990)

40.   E. N. Van Eenige, R. Griessen, R. J. Wijngaarden, J. Karpinski et al.: Physica C **168** 482 (1990)

41.   E. Kaldis, P. Fischer, A. W. Hewat, E. A. Hewat et al.: Physica C **159** 668 (1989)

42.   Y. Yamada, J. D. Jorgensen, S. Pei, P. Lightfoot et al.: Physica C **173** 185 (1991)

43.   R. J. Nelmes, E. Loveday, E. Kaldis, J. Karpinski: Physica C **172** 311 (1990)

44.   S. S. P. Parkin, V. Y. Lee, A. I. Nazzal, R. Savoy et al.: Phys. Rev. B **38** 6531 (1988)

45.   H. G. von Schnering, M. Walz, M. Schwarz, W. Becker et al.: Angew. Chem. Int. Ed. Engl. **27** 574 (1988)

46.   R. L. Withers, J. G. Thompson, L. R. Wallenberg, J. D. FitzGerald et al.: J.Phys.C **21** 6067 (1988)

47.   P. Bordet, J. J. Capponi, C. Chaillout, J. Chenavas et al.: Physica C **156** 189 (1988)

48.   Y. Le Page, W. R. McKinnon, J. M. Tarascon, P. Barboux: Phys. Rev. B **40** 6810 (1989)

49.   E. A. Hewat, M. Dupuy, P. Bordet, J. J. Capponi et al.: Nature **333** 53 (1988)

50.   A. Yamamoto, M. Onoda, E. Takayamamuromachi, F. Izumi et al.: Phys. Rev. B **42** 4228 (1990)

51.   V. Petricek, Y. Gao, P. Lee, P. Coppens: Phys. Rev. B **42** 387 (1990)

52.   W. Dmowski, B. H. Toby, T. Egami, M. A. Subramanian et al.: Phys. Rev. Lett. **61** 2608 (1988)

53.   A. W. Hewat, E. A. Hewat, J. Brynestad, H. A. Mook et al.: Physica C **152** 438 (1988)

54.   A. W. Hewat, P. Bordet, J. J. Capponi, C. Chaillout et al.: Physica C **156** 369 (1988)

LOCAL STRUCTURAL DISTORTION : IMPLICATION TO THE MECHANISM OF HIGH TEMPERATURE SUPERCONDUCTIVITY

T. EGAMI

Institute for Materials Research, Tohoku University,
Sendai 980, Japan, and
Department of Materials Science and Engineering and
Laboratory for Research on the Structure of Matter,
University of Pennsylvania,
Philadelphia, PA 19104-6272, USA

ABSTRACT.   Various local probes indicate that the nano–scale atomic structure of superconducting oxides is not uniform, but is characterized by local dynamic distortions. Furthermore the dynamics of these local distortions appear to change appreciably at the onset of superconductivity, indicating that these distortions are resulting from strong electron–lattice interaction. In this article we describe some details of the interpretation of the pulsed neutron scattering results which led to the identification of local structural distortions, and discuss implications of these observations to the mechanism of high temperature superconductivity.

1. Introduction

The atomic structure of superconducting oxides has been studied in detail by various methods, and their crystallographic structure has been well established [1]. On the other hand, there are a number of experimental results which indicate that the crystallographic structure represents merely an averaged atomic structure, and at a microscopic scale the actual position of an atom is slightly deviated from the crystallographic site, resulting in local lattice distortion.

Of course, structural distortions can occur due to various extrinsic defects. Many of the superconducting oxides are mixed ion oxides in which some of the crystallographic atom site are occupied by different species of atoms. Around such a site the size difference among these substitutional elements can cause local distortion. Also the lattice defects can result in distortion. However, the size and symmetry of the distortions actually observed in superconducting oxides suggest that in many cases they are not due to such extrinsic reasons. Some of them are entirely unexpected from the usual crystal chemistry argument.

Furthermore, surprisingly, much of these distortions appear to be dynamic rather than static, and their dynamics change appreciably at the onset of superconductivity. They therefore apparently belong to a group of symptoms observed recently, the so-called lattice anomalies, a body of which suggests the importance of lattice involvement in superconductivity.   In other words it is possible that these local distortions either are actively involved in high temperature superconductivity, or at least sensitively mirror some action of pair–forming interaction.

In order to determine the details of these local structural distortions conventional crystallographic techniques are of little use. While they do indicate in average how far each atom is deviated from the crystallographic site in terms of thermal parameters as discussed below, they cannot provide an adequate description of the local distortion, because they presume periodicity and therefore entirely neglect short range atomic correlations. Local structural techniques such as the EXAFS method and atomic pair distribution analysis have to be employed to determine local structural distortions. In

45

E. Kaldis (ed.), Materials and Crystallographic Aspects of HTc-Superconductivity, 45–64.
© 1994 *Kluwer Academic Publishers.*

this article we describe some of the results of pulsed neutron atomic pair distribution analysis indicating local structural distortions, supplemented by the results by other techniques.

2. Lattice Anomalies

There have been a number of reports recently on the observation of anomalous effects near the onset of superconductivity which are associated with the lattice. They include the changes near T_C in the critical angle for ion–channelling [2,3], changes in the atomic pair distribution function (PDF) determined by pulsed neutron scattering [4–7] and by EXAFS [8–12], changes in the average phonon energy as determined by neutron resonance [13], and the local symmetry observed by Mössbauer spectroscopy [14]. If one tries to interpret these changes in terms of the changes in the phonon frequencies or elastic constant, these changes correspond to a 10 – 20 % change.

However, the sense of change is confusing. The ion channelling results indicate lattice *hardening* below T_C, while the result of neutron resonance suggests lattice *softening* below T_C. Furthermore, the actual changes in the phonon frequencies directly observed by Raman spectroscopy and neutron inelastic scattering amount to only up to 2 %, which is indeed expected by the strong–coupling theory [15]. Thus these changes mentioned above should not be interpreted in terms of changes in the bulk elastic properties, and are truly *anomalous*. What they suggest is unusual electron–lattice coupling operating in these solids.

3. Experimental Evidence of Local Distortion

3.1. CRYSTALLOGRAPHIC EVIDENCE

In the crystallographic description of the structure, any deviation from the ideal crystal structure, such as thermal vibrations, is described in terms of the thermal, or Debye–Waller, factor, $\exp(-B(Q/4\pi)^2)$, where Q $(= 4\pi\sin\theta/\lambda)$ is the diffraction vector, θ is the diffraction angle and λ is the neutron wavelength. The B factor is related to the displacement away from the crystallographic site, u, by

$$B = \frac{8\pi}{3}<u^2> \qquad (1)$$

The value of B for superconducting oxides determined by the single crystal diffraction or powder diffraction with the Rietveld analysis is usually anomalously large, particularly for oxygen [1]. Table 1 shows some of the thermal factors in typical crystallographic

Table 1. Debye–Waller factor of some HTSC's

Composition	Atom	B(Å²)	<u²>^{1/2} (Å)
$La_{1.85}Sr_{.15}CuO_4$ [16]	La	0.48	0.14
	Sr	0.48	0.14
(T = 10 K)	Cu	0.42	0.13
	O1	1.33	0.22
	O2	0.78	0.17
$Tl_2Ba_2CaCu_2O_8$[17]	Tl	1.82	0.26
	Ba	0.96	0.19
	Ca	1.22	0.22
(T = 300 K)	Cu	0.44	0.13
	O1	0.64	0.16
	O2	1.22	0.22
	O3	3.84	0.38

structures, in terms of the Debye–Waller factor, B, and the corresponding phonon amplitude, $<u^2>^{1/2}$. The normal amplitude, expected from the standard Debye theory, is less than 0.1 Å. Thus these values clearly indicate that either the local elastic constant is very small or there are local atomic displacements. The absence of strong temperature dependence shows that the former is not the case, and favors the latter interpretation.

3.2. LOCAL STRUCTURAL PROBES

There are several non–crystallographic structural methods which provide a *local* view of the structure, i.e., the structure seen from the atoms in the solid. Among them, the EXAFS and PDF analysis offer the most direct evidence of local lattice distortion. Direct lattice imaging by electron diffraction is another local method, but in this case the averaging over the thickness of the sample, usually about 200 Å or more, obscures the local variation.

The EXAFS method is a well established local structural tool widely used in many areas. It provides information regarding the interatomic distance and coordination of near neighbor atoms for a specific atomic species. While X–rays are used in collecting the EXAFS data, diffraction occurs for photoelectrons excited in the sample by X–rays. Since electrons are strongly scattered by atoms, EXAFS is a very sensitive probe of the local structure. Unfortunately, precisely because electrons are strongly scattered, the phase shift for the scattering is large, and the multiple scattering intensity is strong. These effects render the data analysis complex. However, carefully executed EXAFS experiments provide valuable information.

The PDF analysis of diffraction data by X–rays and neutrons has been used almost exclusively for the study of liquids and amorphous solids for which conventional crystallographic methods are powerless [18,19]. But recently we have been applying this method to the study of various disordered crystalline solids including the superconducting oxides [4–6,20]. Unlike the EXAFS method, the data analysis is straightforward and unambiguous. For a long time, the PDF analysis has been suffering from the effect of termination errors which arise from the limited range of Q, which produces spurious peaks in the PDF. However, the advent of synchrotron base sources, such as the pulsed neutron source and synchrotron radiation source, drastically reduced this type of errors. These local probes have established the presence of unharmonic displacements of some of the atoms, notably oxygen atoms, as we describe below.

3.3. OTHER EVIDENCES

In many superconducting oxides the local selection rules for Raman and infra–red (IR) active phonons are violated, so that the IR active phonons can be observed by Raman scattering, and vice versa [21]. This clearly shows that the local inversion symmetry in the average symmetry is removed, due to local distortions. Mössbauer spectroscopy as well as NQR and positron angular correlation (PAC) measurements show local distortions at low temperatures [22,23].

4. Pulsed Neutron PDF Analysis

4.1. EXPERIMENTAL SETUP

Pulsed neutrons are produced by a spallation source such as the Intense Pulsed

Neutron Source (IPNS) of Argonne National Laboratory. Pulses of protons are accelerated to a high speed (750 MV at the IPNS) and hit a heavy metal target, usually U or W, producing bursts of fast neutrons. These fast neutrons are moderated by a solid or liquid moderator, but only about a half become fully thermalized, with the rest remaining epithermal neutrons with energies well above the temperature of the moderator. Thus the spectrum of the pulsed neutrons covers up to several eV. These neutrons are scattered by the sample and detected by a large array of detectors. A computer records the time of flight since the production of a pulse to the detection of a neutron in a histogram for each group of detectors. We use 5 – 8 groups of detectors, about 150 in total, placed at different angles. The dependence of the diffraction intensity on the diffraction angle provides important dynamical information as discussed below.

Now, a fully thermalized neutron with the energy of 25 meV has the wavelength of 2.5 Å, thus the wavevector (momentum), k, of 2.4 Å^{-1}. If one uses this neutron for diffraction, the momentum transfer by diffraction, $Q = 2k \sin \theta$, has to be less than 4.8 Å^{-1}. As shown in Fig. 1, the diffraction intensity from a crystalline solid persists up to much higher values of Q, typically 30 Å^{-1} or so. Therefore much of the scattering intensity remains unprobed if one uses thermal neutrons. On the other hand by using epithermal neutrons from a pulsed source the intensity can be fully determined up to 40 Å^{-1} or more.

4.2. DATA ANALYSIS

The data collected for each detector bank are corrected for the spectrum of the incident neutrons, background due to the detector housing and the sample container, absorption, and multiple scattering, and are normalized to the unit of scattering by an average atom. The normalized intensity thus obtained, I(Q), contains both elastic and inelastic scattering intensities. The elastic scattering intensity consists of the Bragg intensity and diffuse scattering intensity. The conventional crystallographic methods, such as the Rietveld method, retain only the Bragg diffraction intensity, and discard the rest as *background*. Note that this *background* is different from our instrumental background, since the crystallographic *background* to be thrown out during the data processing includes both diffuse and inelastic intensities, while ours does not.

The intensity of neutrons scattered by a sample is given by

$$\frac{\partial^2 I}{\partial \omega \partial \Omega} = \frac{k'}{k} N <b>^2 S(Q,\omega) \tag{2}$$

where N is the number of atoms, <b> is the compositionally averaged neutron scattering length, k is the momentum of the incident neutron, k' is the momentum of the scattered neutron, $Q = k' - k$, ω is the energy transfer, and $S(Q,\omega)$ is the dynamic structure factor, given by,

$$S(Q,\omega) = \frac{1}{N<b>^2} \sum_{i,j} b_i b_j \int \exp(-i\omega t) <\exp[i\vec{Q}\cdot(\vec{r}_i(0) - \vec{r}_j(t))]> dt \tag{3}$$

where <<....>> denotes an ensemble (quantum and thermal) average, b_i is the neutron scattering length of an atom i, and $r_i(t)$ is the position of the i–th atom at time t [24]. Since the time of flight detectors have no energy discrimination, the measured intensity is an integral over energy,

$$S_m(Q_0) = \int S(Q(\omega),\omega)d\omega \tag{4}$$

Here Q weakly depends upon ω, and $Q(0) = Q_0 = 2k \sin\theta$. The ratio of the path lengths from the sample to the detector and from the pulse source to the sample determines the dependence of Q on ω. If Q does not depend upon ω, from eqs. (3) and (4) we see that the measured intensity is determined by the instantaneous (t = 0) correlation. Thus, Fourier transforming $S_m(Q)$ by

$$\rho(r) = \rho_0 + \frac{1}{2\pi^2 r}\int [S_m(Q) - 1]sin(Qr)QdQ \tag{5}$$

where ρ_0 is the average density, yields the atomic pair density function (PDF). The PDF describes the density of an atom separated from an average atom by r. An example for the case of f.c.c. aluminum at T = 10 K is shown in Fig. 2 [25,26]. Now theoretically the integration should be done from 0 to infinity, but experimentally accessible range of Q is limited to be less than $2k_i$, where k_i is the momentum of the incident particle. Thus the integration in eq. (5) has to be terminated at a finite value of Q, and this produces the termination error. However, this problem has largely been eliminated by the use of a pulsed neutron source which provides high intensity of epithermal energetic neutrons as discussed above.

4.3. NEUTRON DYNAMICS

In reality, however, Q does depend upon ω. Thus the deviation of the real Q from Q_0 produces inaccuracy in the Fourier transform. These questions were first addressed by Placzek [27], so we call the difference between Q and Q_0 the Placzek shift. Fortunately the effect of this inaccuracy is not serious. If the Plazcek shift is too large the Fourier transform randomizes the information, so that there is little effect on $\rho(r)$. Thus effectively S(Q) can be replaced by

$$S_m(Q) = \int_{-\omega_\theta}^{\omega_\theta} S(Q,\omega)d\omega \tag{6}$$

where $2\omega_\theta$ is the effective window width which depends upon the diffraction angle θ. In other words vibrational modes with the frequency below ω_θ appear frozen after the Fourier transform, while modes above this frequency are averaged out. Thus when the data from a detector bank located at a high scattering angle are used only low frequency fluctuations are frozen, while in the data from the detector bank at a low angle even high frequency modes appear frozen.

However, this detector angle dependence gives only an indirect evidence of dynamics. A better method is to determine $S(Q,\omega)$ by resolving the energy of the scattered neutrons. This can be done by using a triple–axis–spectrometer with a reactor (cw) source, or by using a pulse chopper with a pulsed source. A chopper, properly timed with the pulse generation, allows only the neutrons within an energy

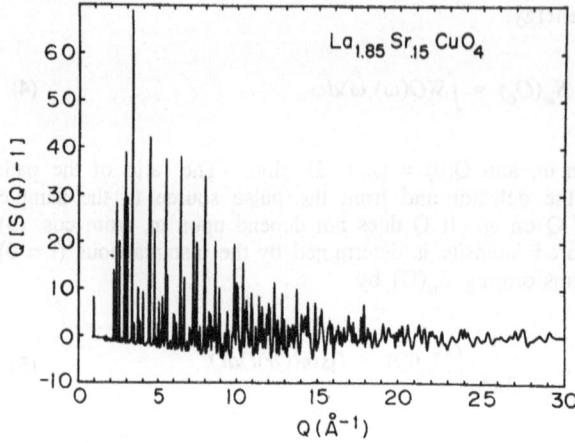

Fig. 1. Structure function S(Q) of $La_{1.85}Sr_{.15}CuO_4$ determined by pulsed neutron scattering [32], shown in the form of Q[S(Q) − 1] as in eq. (5), rather than S(Q) itself. Note that the portions at higher Q are more emphasized by multiplying through Q.

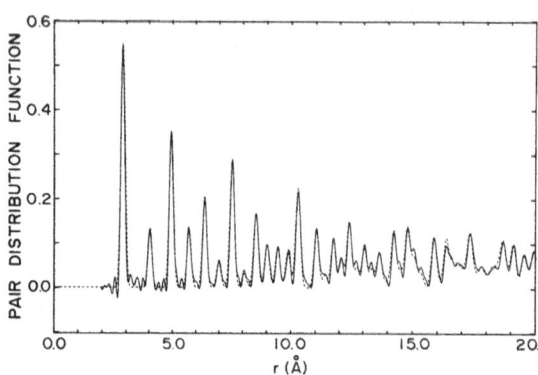

Fig. 2. Pair density function (PDF) of f.c.c. aluminum determined by pulsed neutron scattering [25]. Since the data were taken at T = 10 K, the peak width is due entirely to the zero point lattice vibration. Dashed line is the calculated PDF. For detailed discussion of the accuracy of the data, see Ref. 26.

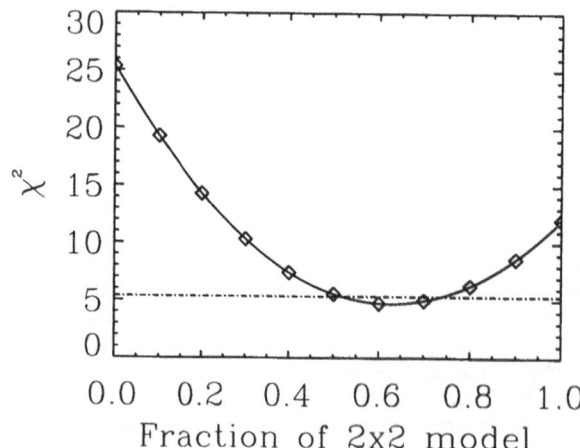

Fig. 3. Goodness of fit, χ^2, for the two phase model for $Nd_{1.835}Ce_{.165}CuO_4$ [34]. The variable is the concentration of the distorted phase in the undistorted matrix. Also the value of χ^2 for the best single phase model is shown by a chained line.

window to go through, and acts as a monochromator, or more accurately a narrow band filter with a moderate to poor resolution. Of course the price to pay for resolving the energy is a reduced scattered intensity. With these methods we can determine $S(Q,\omega)$ integrated over the energy resolution function $W(\omega)$,

$$S_a(Q,\omega_0) = \int W(\omega-\omega_0)S(Q,\omega)d\omega \tag{7}$$

If we set $\omega_0 = 0$, only elastically scattered neutrons are counted, and we measure $S(Q,0)$. Then upon the Fourier transformation by eq. (5) we obtain correlations between time averaged single atom density function, $\rho_s(r)$,

$$\rho_{ave}(\vec{r}) = \frac{1}{N}\int \rho_s(\vec{r'})\rho_s(\vec{r'}+\vec{r})d\vec{r'} \tag{8}$$

Note that $\rho_s(r)$ has a Gaussian spread due to thermal or quantum vibrations, so that even the time averaged correlation function is not composed of delta-functions. It is not the correlation function of the time averaged positions [28].

Now we may write

$$\vec{r}_j(t) = <\vec{r}_j> + \vec{u}_j(t)$$
$$\vec{u}_j(t) = \int \vec{v}_j(\omega)\exp(i\omega t)d\omega \tag{9}$$

where $u_i(t)$ is the time dependent displacement of the i-th atom from the average position, then,

$$<\exp[i\vec{Q}\cdot(\vec{r}_i(0)-\vec{r}_j(t)]> = \exp[i\vec{Q}\cdot(<\vec{r}_i>-<\vec{r}_j>)]<\exp[i\vec{Q}\cdot(\vec{u}_i(0)-\vec{u}_j(t)]>$$
$$= \exp[i\vec{Q}\cdot(<\vec{r}_i>-<\vec{r}_j>)][1 - \frac{1}{2}<(\vec{Q}\cdot\vec{u}_i(0))(\vec{Q}\cdot\vec{u}_j(t))>....] \tag{10}$$

Therefore, $S(Q,\omega)$ is given by

$$S(Q,\omega) = \frac{1}{N<b>^2}\sum_{ij} b_ib_j\exp[i\vec{Q}\cdot(<\vec{r}_i>-<\vec{r}_j>)][\delta(\omega)$$
$$- \frac{1}{2}<(\vec{Q}\cdot\vec{v}_i(\omega))(\vec{Q}\cdot\vec{v}_j(\omega))> +....] \tag{11}$$

Thus the Fourier transform of $S(Q,\omega)/Q^2$ by eq. 5 results in the PDF weighted by the ω component of the displacements [7].

4.4. STRUCTURE MODELING

Since the relation between the PDF and the real three-dimensional structure is equivalent to that of the powder diffraction pattern and the three-dimensional reciprocal lattice structure, in principle it is possible to determine the crystal structure from the PDF alone, just as is possible to determine the structure from the powder pattern alone. For instance the PDF of an f.c.c. crystal shown in Fig. 2 is practically identical to the powder diffraction pattern (corrected for the atomic scattering factor) from a b.c.c. crystal, since the reciprocal lattice of the b.c.c. structure is f.c.c. The only difference is that thermal vibration reduces the resolution in the real space, but reduces the intensity in the reciprocal space via the Debye–Waller factor. Then by indexing the PDF peaks and evaluating the area under the peak the structure can be confirmed.

However, since the spatial resolution of the PDF is relatively poor due to thermal vibrations, peaks start to overlap each other soon beyond the first few peaks. Therefore, just as in the Rietveld method of the powder pattern analysis, it is better to compare directly the experimental PDF with a PDF calculated for the model, and to improve the model by minimizing the agreement factor,

$$A^2 = \int_{r_1}^{r_2} [\rho_{mod}(r) - \rho_{exp}(r)]^2 dr \ / \ \int_{r_1}^{r_2} \rho_0^2 dr \qquad (12)$$

where $\rho_{mod}(r)$ and $\rho_{exp}(r)$ are the PDF calculated for the model structure and the PDF experimentally determined, respectively. The model PDF is convoluted with a Gaussian function representing thermal vibration. The integration range, from r_1 to r_2, is chosen according to the purpose of modeling. Note that the PDF extends up to large distances as determined by the Q resolution of S(Q), dQ, or up to $2\pi/dQ$. Consequently the real space structural analysis of a perfect crystal requires a fitting over a large range of distances.

Thus the structural analysis of a perfect crystal can be done equally well either in real space or in reciprocal space, since they are merely a Fourier transfer of each other. However, **for disordered solids with short range order the real–space method has a clear and unique advantage**. Namely by carrying out the model fitting over different ranges of distances it is possible to separate short, medium and long range order. An example in the case of $Nd_{1.835}Ce_{.165}CuO_4$ will be discussed later.

Modeling can be done either by specifying a parameter and minimizing the A factor with respect to this parameter, or by the Monte–Carlo method. When appropriate parameters cannot be identified it is most convenient and informative to carry out the Monte–Carlo simulation. In this method a starting structure, for instance the ideal periodic structure, is presumed, and each atom is displaced in a random direction. The A factor is evaluated before and after the displacement, and the difference, ΔA, is used to calculate

$$P = \frac{1}{e^{\Delta A/T} + 1} \qquad (13)$$

where T is a fictitious temperature used in order to prevent the system from falling into a local minima of A. Then a random number $0 < R < 1$ is generated, and,
    1) if $R < P$, the displacement is taken, and the model is modified,
    2) if $R > P$, the displacement is rejected.
By repeating this procedure the model can be improved gradually.

One of the disadvantages of the Monte–Carlo method is its slowness to improve the model. The system can often be trapped in a local minimum, and may not reach a better model even with the aid of the fictitious temperature T and a large number of repeats. One way to improve the speed of convergence is the use of the genetic algorithm. In this method two trains of structure are mixed in a manner similar to the exchange of genes, and the one which gives a lower A factor is employed. By this method more collective structural change which would take a very large number of steps by the single atom Monte-Carlo method can be achieved [29].

Thus the best tactics seems to be the combination of these. Initially we perform the Monte-Carlo refinement with a large system to identify the most likely magnitude and direction of displacement and to find out the general features of the model. Several variables are chosen through this process, and the A factor is minimized with respect to these variables using a smaller system. At this stage the possibility of multi-phase mixture can be assessed. An example of this process is shown in Fig. 3. Here the composition of two phases described in detail later is varied as a parameter, and the diamonds show the goodness of fitting for the two-phase model in terms of $\chi^2 = A^2/A_{min}^2$, where A_{min} is defined below. Compared is the χ^2 value for the best single-phase model, shown by a chained line. Clearly the two-phase model wins.

4.5. ACCURACY OF MEASUREMENT AND MODELING

The PDF thus obtained includes various experimental inaccuracies. Before carrying out extensive modeling effort it is imperative to evaluate the accuracy of the result. The experimental inaccuracies include statistical errors and systematic errors. The statistical errors can be readily evaluated by

$$[\Delta\rho(r)]^2 = \frac{1}{(2\pi^2 r)^2}\sum_k [\Delta S(Q_k)\sin(Q_k r)Q_k \Delta Q_k]^2 \tag{14}$$

where k refers to each data point, Q_k is the value of Q for the k-th data point, $\Delta S(Q_k)$ is the statistical error of $S(Q_k)$ due to $N^{-1/2}$ error, and ΔQ_k is the interval of the Q values between each data point [26]. This can be approximately given by

$$\Delta\rho(r) \approx \frac{R}{r} \tag{15}$$

where

$$R^2 = \frac{1}{8\pi^4}\int \frac{S(Q)Q^2}{R(Q)<b>^2}dQ$$

$$R(Q_k)<b>^2 = \frac{N_{obs}(Q_k)/\Delta Q_k}{S(Q_k)} \tag{16}$$

and $N_{obs}(Q_k)$ is the neutron (X-ray photon) count at the k-th data point. Thus the noise **decreases with r** as $1/r$. This is because the statistical error is a white noise in the Fourier integral in eq. (5), thus in $r\rho(r)$. Various factors contribute to the systematic errors in the data processing, such as the inaccuracies of the energy spectrum of incoming neutrons, absorption correction, multiple scattering, and termination of eq. (5). The termination errors can be evaluated theoretically, but other systematic errors can be best evaluated by comparing the PDF of a standard sample, such as Al or Ni powder [26]. Our experience at the IPNS suggests that the systematic errors still dominate the statistical error. However, for data comparison with the identical condition, such as the study of temperature dependence, only the statistical error contributes [5].

In the modeling process the effect of these errors has to be evaluated carefully. Because of these errors the A factor cannot be zero even when the model correctly describe the real structure of the sample. The minimum of the A factor due to statistical error is given by

$$A_{min}^2 = [\int \Delta \rho^2(r)dr] / [\int \rho_0^2 dr] \qquad (17)$$

Since this does not include the effect of systematic errors the actual minimum of the A factor is larger than this value.

The A factor itself suffers from statistical error. The statistical error of the A factor ΔA can be calculated by further propagating the error to the A factor. This defines the minimum recognizable difference in the A factor. In other words if two models yield similar A factors, for one of them to be chosen the difference in the A factor has to be larger than ΔA. Otherwise these two models cannot be differentiated even when there is a small difference in the A factor. For instance in the case shown in Fig. 3, the difference in the A factor between the two–phase and the single–phase models is larger than ΔA by more than a factor of two [30]. Thus the two–phase model is demonstrably better than the single–phase model.

Another factor to be considered in evaluating the accuracy of modeling is the number of variables. It is easy to reduce the A factor, if a large enough number of variables are used. For instance in principle it is possible to reduce the A factor to practically zero. However, in this case one is merely modeling even the errors. The maximum number of meaningful variables can be given by

$$N_{eff} = A_{min}^2 / \Delta A^2 \qquad (18)$$

which is close to

$$N_{max} = [r_2 - r_1] / \Delta r$$
$$\Delta r = \pi / Q_{max} \qquad (19)$$

where Q_{max} is where the integration in eq. (5) is terminated. Because of the termination the minimum resolution of the real space, Δr, is given by π/Q_{max}. Thus the number of information retained in the region from r_1 to r_2 is given by N_{max}. The number of variables should be much smaller than this value. Otherwise one will be modeling the errors.

This results calls for a great care in using the reverse Monte–Carlo method with a large model system. Not only a model with a large number of atoms suffers from the lack of uniqueness, but also potentially the model is erromous because of the error in the data the model was fitted to.

5. Local Distortion in Various Superconducting Oxides

We will now review the details of the local distortions in various superconducting oxides as observed by pulsed neutron PDF analysis.

5.1. $Tl_2Ba_2CaCu_2O_8$

This compound was most extensively studied by our pulsed neutron PDF analysis, even though its crystal structure is rather complex. This is because an anomaly in the PDF near T_C was observed for the first time for this solid, and we felt it was important to establish this effect with high statistical accuracy. As shown in Fig. 4 the peak height of the PDF at 3.4 Å shows anomalous temperature dependence near T_C [5]. In an ordinary solid the PDF peak height is reciprocally proportional to the atomic vibrational amplitude, and thus decreases with increasing temperature. In Fig. 4 the solid line represents the normal temperature dependence calculated from the phonon density of states which was determined by neutron inelastic scattering measurement [28]. Near T_C (= 110 K) the measured peak height deviates significantly from this line.

The peak at 3.4 Å describes the correlation between the apical oxygen (O2) and the in-plane oxygen (O1). This correlation is complex, since both O1 and O2 atoms are deviated from the average site. At low temperatures there are two kinds of Cu environments. In one environment, the oxygen atoms are occupying the normal, undistorted position, i.e. Cu–O2 distance is long (2.7 Å), and O1 is in the Cu–O plane. In the other environment Cu–O2 distance is short (2.4 Å), and O1 is vertically deviated from the Cu–O plane (Fig. 5) [28].

Judged from the detector angle dependence as we discussed above, and from the direct elastic scattering measurement using a triple–axis spectrometer, these deviations are dynamic, and if averaged over time all the Cu atoms are equivalent (Fig. 5) [28]. The time constant associated with the motion of these oxygen atoms is about 10 – 30 meV in energy. Close to T_C the motion of the oxygen atoms becomes faster, giving rise to the increase in the peak height in Fig. 4. Further details of these points will be described elsewhere.

In an earlier report we pointed out that the structure in the Tl–O plane is significantly deviated from the crystallographic structure, and shows short range displacive order [31]. The origin of this effect is the mismatch between the crystallographic structure and the ionic radius. The distance between the Tl site and O3 site in the crystallographic structure is 2.73 Å, while the sum of the ionic radii for Tl and O is only 2.24 Å. Thus atoms rearrange themselves to reduce the Tl–O distance. In the rearranged structure Tl has two O neighbors at 2.4 Å and other two at 3.1 Å, instead of having four neighbors at 2.73 Å as in the crystallographic structure. This reorganization of the Tl–O plane apparently has no effect on superconductivity, and shows no change through T_C. However, it may have some effect on the capability of the Tl–O plane to transfer charge carriers to the Cu–O plane.

5.2. $La_{2-x}Sr_xCuO_4$

Our first PDF analysis of superconducting oxides was made on this system [32]. Extra–peaks not appearing in the model PDF of a perfect crystallographic structure

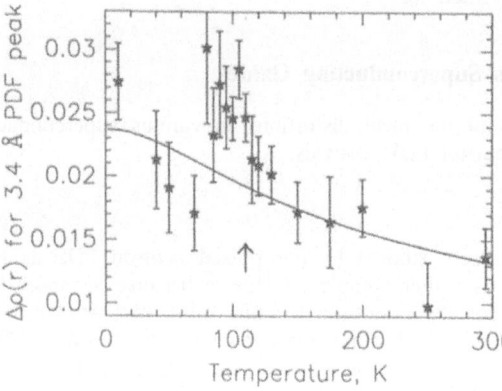

Fig. 4. Anomalous temperature dependence of the PDF peak height at $r = 3.4$ Å, for $Tl_2Ba_2CaCu_2O_8$ [5]. Vertical bars indicate statistical error. The arrow shows T_C. The solid line is the normal dependence calculated from the phonon density of states measured by inelastic neutron scattering.

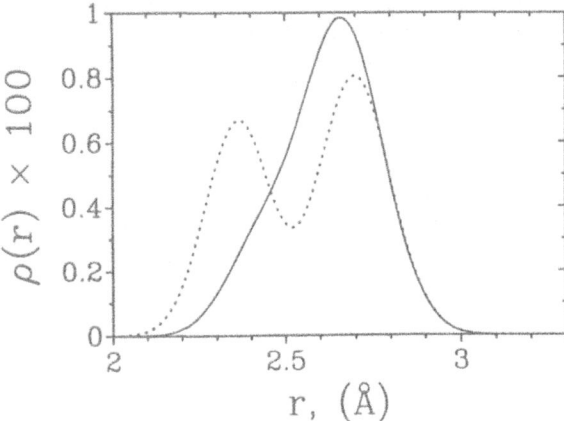

Fig. 5. Distribution of the distances between Cu and apical oxygen; instantaneous correlation (dashed line, determined by modeling the result of pulsed neutron scattering) and correlation between the time averaged densities (solid line, determined by modeling the result of elastic neutron scattering) [28].

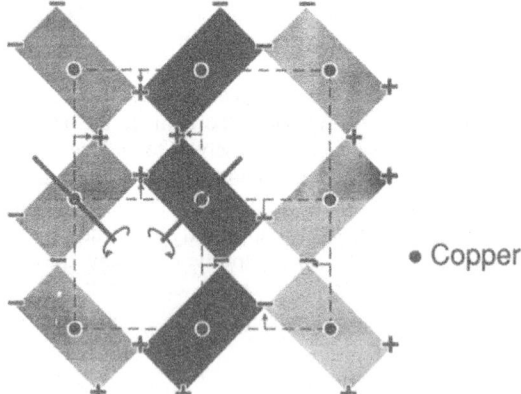

Fig. 6. Pattern of oxygen displacements away from the crystallographic sites in the CuO^2 plane of superconducting $Nd_{1.835}Ce_{.165}CuO_4$ [34]. The signs, + and −, refer to the displacements perpendicular to the Cu−O plane.

were found at 2.2 Å and 2.95 Å. The former may be associated with the reduction in the Cu–O2 distance, where O2 is the apical oxygen. The latter is most likely the O1–O2 distance, where O1 is the in-plane oxygen. At that time it was not clear whether these extra-peaks truly represent the real structure, or at least partly due to noise. Repeated measurements later on and the evaluation of the errors now made it clear that these peaks are a genuine part of the structure. Thus the pattern of oxygen displacement is the same as in $Tl_2Ba_2CaCu_2O_8$, with the apical oxygen coming closer to Cu, and the in-plane oxygen being displaced along the c–axis, to reduce the distance between the apical oxygen and the in-plane oxygen. More detailed analysis is under way using new data.

This system undergoes a tetragonal to orthorhombic phase transition as a function of temperature and composition, x. The crystallographic structure in the orthorhombic phase is characterized by collectively tilted CuO_6 octahedra, while the tilting is absent in the tetragonal phase. However, we found that locally the CuO_6 octahedra are always tilted. In the tetragonal phase, both at high temperatures and just above the critical Sr concentration for O/T transition (x = 0.21), the direction of tilt is random from site to site, resulting in no long range order [32,33].

5.3. $Nd_{2-x}Ce_xCuO_4$

In our most recent and most extensive modeling effort the structure of this compound was studied for x = 0.165 (superconducting), and 0.2 (non–superconducting) [6,34]. Complex displacements of oxygen atoms in the Cu–O plane were found as shown in Fig. 6. As shown in Fig. 3, the A factor in the range of 3.2 – 6 Å is significantly reduced by mixing two phases, one with well ordered large atomic displacements and the other with no displacements. This model accounts for the peaks in the PDF around 4 – 4.5 Å much better than the single–phase model does, as shown in Fig. 7. However, in the range of 6 – 12 Å, the single–phase model with uncorrelated displacements (but in the directions shown in Fig. 6) shows better agreement than the two–phase model, and in the range 12 – 20 Å, the crystallographic, undisplaced, model with a large thermal factor wins.

All these indicate that the superconducting phase with x = 0.165 contains many small domains of about 6 Å in size, within which oxygen atoms are displaced in a highly correlated, well ordered pattern. **The solid is made of microdomains of two well ordered structures.** The domains with distortions occupy about 60 % in volume. This two–phase nature is seen also in $Tl_2Ba_2CaCu_2O_8$ [28].

5.4. $YBa_2Cu_3O_{7-\delta}$

Displacements are rather small in the fully oxygenated state ($\delta = 0$), and our modeling effort has not been successful in determining the details of the pattern of displacements. The Cu–O peak at 1.95 Å and the a, b repeat distance at 3.85 Å show some small anomaly at T_C. From the measurement of $S(Q,\omega)$ [7], however, rather significant displacements were observed at higher frequencies. Thus it is possible that in the O_7 solid oxygen atoms are displaced with high frequencies, putting them outside of our energy window as discussed earlier. We are now performing measurements with a larger energy window. On the other hand, our recent measurements with $\delta = 0.3$ and 0.5 suggest that oxygen depleted samples have much larger displacements. Thus, either the reduced T_C has reduced the frequency, or alternatively the disruption of the Cu–O chain by vacancies resulted in the localization of the displacive modes.

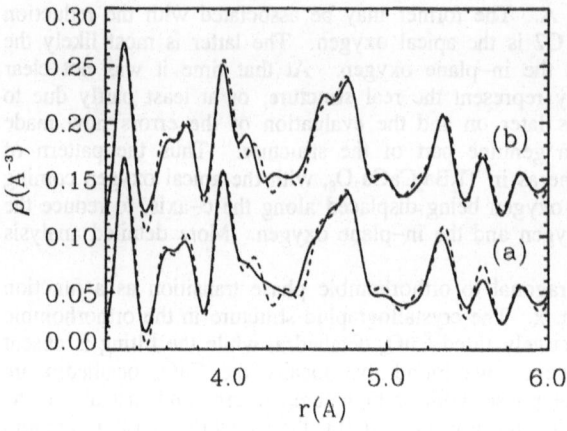

Fig. 7. Experimental PDF of $Nd_{1.835}Ce_{.165}CuO_4$ (dashed line, both top and bottom), and the PDF of the best two–phase model (above, solid line), and that of the best single–phase model (below, solid line) [34].

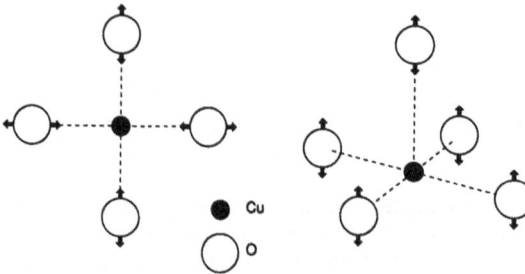

Fig. 8. Patterns of oxygen displacements around Cu: Longitudinal displacements in the Cu–O plane (left), and displacements along the c–axis (right).

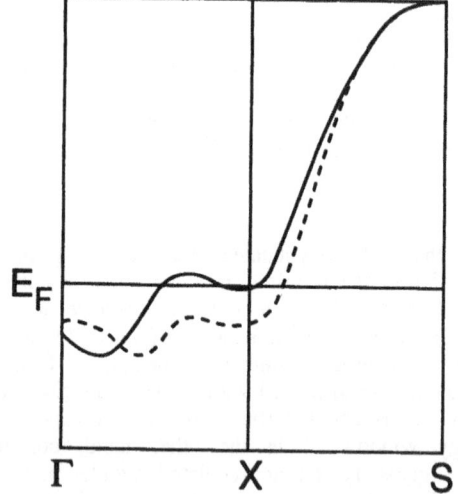

Fig. 9. A portion of the electronic band structure of $YBa_2Cu_3O_7$, without oxygen displacement (dashed line) with oxygen displaced along the c–axis (solid line) [37].

5.5. $YBa_2Cu_4O_8$

The PDF of this solid obtained recently shows several new peaks not seen in the PDF of the crystallographic structure. These peaks are very similar in position and strength to the peaks seen in the reduced $YBa_2Cu_3O_{7-\delta}$. We have not carried out an extensive modeling of this system.

5.6. $(Ca_{.85}Sr_{.15})CuO_2$

This system is the simplest of the layered cuprates, in which CuO_2 layers are repeated infinitely sandwitching Ca, Sr. Thus this is called the "infinite layer" cuprate. While this solid is insulating, doped compounds of this solid are superconducting. Locally the Cu–O planes in this system were found to be buckled, with oxygen atoms displaced along the c-axis without a long range order [35]. While the magnitude of displacement in this solid is smaller than in other superconducting solids, this buckling appears to be an intrinsic nature of the Cu–O plane.

6. Mode of Atomic Displacements

The salient features of the atomic displacements observed in various superconducting oxides could be summarized as follows:

1. Large displacements of oxygen atoms are observed. The magnitudes of displacements of other atoms are at least a factor of two smaller.

2. Displacements are **not random**, but locally well defined, often producing new, sharp peaks in the PDF.

3. Not all the atoms are displaced: Nearly half of the atoms are located at the average crystallographic sites.

4. At low temperatures displacements are **highly correlated**, and occur inside small domains of which size is about 6 – 10 Å. Thus the solid is made of microdomains of two well ordered structures.

5. The dynamics of the domain changes near T_C.

6. So far all the displacements observed by pulsed neutron analysis are along the c–axis (Fig. 8). On the other hand recent EXAFS measurements report some in–plane breathing or quadrupolar modes in some systems. This may be because the in–plane modes have higher frequencies and are outside the energy window of our standard measurement. We are carrying out new measurements using a wider energy window.

7. In all the solids we studied so far an important common feature is that the oxygen–oxygen distance of about 3 – 3.7 Å is reduced by about 0.2 Å inside the microdomain.

7. Origin of Local Distortion and Microdomains

Certain local lattice distortions can be expected purely for steric reasons based upon the mismatch of atomic size and crystal lattice structure. However, the observed distortions are far larger than expected for these reasons. In $Nd_{1.835}Ce_{.165}CuO_4$, for instance, the observed displacements are by a factor of two or more larger than calculated for the effect of Ce substitution for Nd using the shell model [34]. The same applies to $(Ca_{.85}Sr_{.15})CuO_2$ [35]. Moreover the direction and symmetry of the displacements cannot be understood by the size argument.

Other defects such as vacancies could cause atomic displacements. However, the distortion is quite ubiquitous and does not appear to depend critically on the processing conditions. Furthermore the density of the distorted domains is too high to be due to the lattice defects. Also the displacements are *highly correlated*, as discussed before, and the system is actually made of microdomains of two phases. Such a state would not be produced by extrinsic defects. In fact, if the distortions are caused by defects, thus are incoherent, they would result in large electrical resistivity, but the measured resistivity is low, indicating that the distortions are indeed locally highly correlated and nearly coherent.

Finally the changes observed near T_C as shown in Fig. 4 clearly establish a direct link with the superconductivity phenomena. Thus it is most likely that these displacements are not defect related, but intrinsic, and electronic in origin. However, atomic displacements due to electron–lattice interaction usually leads to homogeneous phase transition, usually through some soft phonon modes, rather than to localized distortions of the lattice. It is quite strange that the distortions remain local.

It is well known that O^{2-} ion is unstable by itself due to the large correlation energy in the 2p state, and is stabilized in the solid purely due to the ligand field of the surrounding cations [36]. For this reason electronic bands mainly consisting of oxygen atomic orbitals are always found right beneath the Fermi level. Now, if the distance between two oxygen atoms is reduced as in Fig. 8, these oxygen related states can cross the Fermi level, and become empty. In other words closer oxygen atoms can trap a hole in–between. For instance according to Andersen et al. [37], when O4 (apical oxygen) of $YBa_2Cu_3O_7$ is displaced along the c–axis toward Cu, the saddle point of the σ^* band moves up crossing the Fermi level, as shown in Fig. 9. Similar changes occur when the in–plane oxygen atoms (O2 and O3) are displaced along the c–axis toward O4. The orbital character of this pocket of holes is $p_{O4}(x)$ + $[p_{O2}(z)-p_{O3}(z)]$.

A possible explanation of the localization of strain (displacement) is the non–linear electron–lattice interaction [38,39]. Then the formation of a localized wave–packet due to the states in the hole pocket as in Fig. 9 will stabilize the local distortion, thus resulting in the formation of a *polaron*. In $Nd_{1.835}Ce_{.165}CuO_4$ too, a close O1–O2 distance would produce hole pockets in the oxygen p–band. Thus this solid should have p–type as well as n–type carriers, explaining the reversal of the Hall coefficient as a function of temperature [40].

It is now well established that most of the conduction electrons (holes) in the cuprates are in the σ^*–band, the anti–bonding states made of Cu $d(x^2-y^2)$ and O $p(x)$ orbitals [41]. Most of the simple band models such as the t–J model deal exclusively with this band. However, the mode of displacement most frequently seen so far involves displacements along the c–axis of both the apical oxygen and the in–plane oxygen, as shown in Fig. 8. These displacements would not affect the σ^* states directly, but most strongly affect oxygen p(z) states and Cu $d(z^2)$ states. Indeed holes are found in both of these states, albeit in small proportions, by careful soft X–ray absorption studies [42]. Thus our observations suggest the involvement of minority carriers in these states which may play some role in producing superconductivity.

The states in the σ^* band will primarily be modified by the breathing Cu–O mode, or the a–b longitudinal mode (Fig. 8), and indeed in the early BCS theories were based upon these breathing modes [43]. Our PDF studies indicate that this type of oxygen displacement is either very small or of too high frequencies for us to see. Only in the case of $La_{1.875}Sr_{.125}CuO_4$ an anomaly in the PDF was seen for the Cu–O bond in the CuO_2 plane, suggesting the involvement of the breathing mode at low frequencies [28]. A sample with a similar composition, $La_{1.85}Sr_{.15}CuO_4$, does not show the same anomaly. It is interesting to note that only very close to this Sr composition ($x = 0.125$) T_C is slightly suppressed and a strong isotope effect on T_C is observed [44].

8. Implications to the Mechanism of Superconductivity

As we have discussed so far various experimental results including our pulsed neutron PDF analysis have established the presence of significant local lattice distortions in superconducting oxides. This alone has interesting implications. For instance, sharp van Hove singularities in the electronic states will be smeared by these local displacements, so that a mechanism strongly dependent on the sharpness of the singularity [45] is not likely to be successful. In fact in some cases the distortions may be the structural response to the van Hove singularity [46].

Among the properties of these displacements perhaps the most important of all is that these displacements are not random, but occur locally in a highly correlated manner in the superconducting phase. Thus the solid is made of two microphases of 6 – 10 Å in size:

Phase A: Oxygen atoms are displaced in a regular pattern. This phase may be associated with a higher concentration of the minority carriers, with the majority being in the σ^* band.

Phase B: Oxygen atoms are not displaced, just as in the insulating phase. Magnetic correlations are most likely alive in this phase.

A clear implication is that both the superconducting order parameter and magnetic correlation will be inhomogeneous in space, coexisting but probably each avoiding the other in space. The inhomogeneity of the magnetic correlation is well demonstrated by neutron inelastic scattering measurements [47,48]. For instance the magnetic correlation length for $La_{1.85}Sr_{.15}CuO_4$ is about 10 Å [47], comparable to the size of the domain in question. Also the dependence of the width and the integrated intensity of the magnetic peak on the energy transfer [48] can be explained well in terms of the dynamics of the microdomains, as will be described in detail elsewhere.

Such a phase separation was predicted even with the t–J model [49]. In this case, however, one expects the breathing mode lattice inhomogeneity. The observed oxygen displacements along the c–axis imply the involvement of $p(z)$ and $d(z^2)$ orbitals. While the density of holes in these orbitals cannot be high [42], they could be playing important roles in superconductivity.

One of the key issues which needs to be resolved is whether the lattice distortions are *dynamically* involved in superconductivity or not. Our present tentative answer is negative, for the following reasons:

1) The frequency of anharmonic lattice distortion, ω_0, phenomenologically between the two minima of the double–well potential, is low, 10 – 20 meV, judged from neutron dynamics.
2) The states of the lattice just above and just below T_C are similar.

3) The frequency ω_0 *increases* in the vicinity of T_C, judged from the results of neutron scattering and EXAFS [9]. Thus ω_0 is *within* the superconducting gap, Δ, at low temperatures, and as Δ is decreased near T_C the lattice motion resonates with superconducting carriers and ω_0 increases.

Thus it is most likely that the lattice distortions are *not* dynamically involved, but they setup new electronic states and excitations. In a way this is similar to what has been advocated by J. C. Phillips in a slightly different context [50]. Furthermore if the new states are the bipolarons with the spin–singlet state, or the negative–U centers [51], then either the charge or spin excitations within the negative–U center can pair-up free electrons in the σ^* band. Such a mechanism has been discussed extensively [52–55].

The frequency of the atomic motion, ω_0, is consistent with the size of the domain. If a train of microdomains which are apart by 10 Å are collectively moving with the sound velocity of 5×10^3 m/sec., the frequency of the domain passage which should be equal to ω_0 is 5×10^{12} 1/sec., or about 20 meV. Since the motion of the domains are not totally coherent, while they are correlated (probably due to Coulomb repulsion), the frequency would be widely distributed around 20 meV, as observed.

9. Conclusions

Various experimental observations of the lattice structure and lattice related properties of superconducting oxides show that the following two are the common features of superconducting oxides which may be of importance in identifying the mechanism of superconductivity:

1) Atomic positions are locally deviated from the average crystallographic sites.
2) Certain aspects of the lattice structure change very sensitively at the onset of superconductivity. Observed changes are too large to be attributed to changes in the phonon frequency and lifetime.

Most of the experimental methods provide evidence for either one of these two, while the local structural probes such as the EXAFS and pulsed neutron PDF analysis can detect both of these features, and convincingly demonstrate that the so–called lattice anomalies are due to changes in local atomic displacements.

Moreover through the recent effort of real–space modeling of the PDF, it became clear that these displacements are not random, but are highly correlated in the superconducting phase. Thus the low temperature phase is made of two microphases with the size of 6 – 10 Å. In one phase (A) oxygen atoms are displaced in a well ordered fashion, while in the other phase (B), atoms are at the crystallographic sites. These two phases are not static, but are migrating with the velocity close to the sound velocity. We conjecture that the phase A contains the charge carriers in oxygen p(z) or copper $d(z^2)$ orbital, and the phase B is anti–ferromagnetic.

These results invoke strong skepticism toward the conventional view of electron dynamics in the cuprates based upon the strongly correlated d–electrons on a perfect square lattice, such as the t–J model. It appears that at least some modifications are necessary along the following lines:

1) Inclusion of the lattice modifications by the microdomain model.
2) Inclusion of the z–component states, such as p(z) and $d(z^2)$.

It is unclear if the modification of the conventional model improves its chance for superconductivity or not. An alternative is to develop an argument based upon the microdomains, such as in the negative–U center model. At present direct evidence is unavailable as to the superiority between these two choices, but the presence of the local lattice distortion is more naturally understood by the latter approach.

Acknowledgments

Parts of this work were supported by the National Science Foundation through DMR90–01704 and DMR91–20668. The author is deeply grateful to Prof. M. Tachiki and his group at Tohoku University where he is staying on his sabbatical leave, for hospitality and very helpful discussions. He is thankful to his numerous collaborators, in particular S. J. L. Billinge, W. Dmowski, B. H. Toby, J. D. Jorgensen, D. G. Hinks, M. A. Subramanian, A. W. Sleight and M. K. Crawford. He is also indebted to Y. Bar–Yam, J. Mustre–de Leon, A. R. Bishop, S. D. Conradson, A. Bianconi, H. Kamimura, D. Emin, J. C. Phillips, J. Goodenough, O. K. Andersen, W. E. Pickett, R. E. Cohen, A. Bussmann–Holder, M. Arai and H. Takagi for useful and stimulating discussions.

References:
1. K. Yvon and M. Francois, Z. Physik, **B76**, 413 (1989).
2. R.P. Sharma, L.E. Rehn, P.M. Baldo and J.Z. Liu, Phys. Rev., **B38**, 9287 (1988); Phys. Rev. Lett., **62**, 2869 (1989).
3. T. Haga, K. Yamaya, Y. Abe, Y. Tajima and Y. Hidaka, Phys. Rev., **B41**, 826 (1990).
4. T. Egami, B.H. Toby, W. Dmowski, S. Billinge, P.K. Davies, J.D. Jorgensen, M.A. Subramanian, J. Gopalakrishnan and A.W. Sleight, Physica C, **162–164**, 103 (1989).
5. B.H. Toby, T. Egami, J.D. Jorgensen and M.A. Subramanian, Phys. Rev. Lett., **64**, 2414 (1990).
6. S.J.L. Billinge, T. Egami, D.R. Richards, D.G. Hinks, B. Dabrowski, J.D. Jorgensen and K. Volin, Physica C, **179**, 279 (1991).
7. M. Arai, K. Yamada, Y. Hidaka, S. Itoh, Z.A. Bowden, A.D. Taylor and Y. Endoh, Phys. Rev. lett., **69**, 359 (1992).
8. S.D. Conradson and I.D. Raistrick, Science, **243**, 1340 (1989).
9. J. Mustre–de Leon, S.D. Conradson, I. Batistic and A.R. Bishop, Phys. Rev. Lett., **65**, 1675 (1990).
10. E.A. Stern, M. Qian, Y. Yacoby, S.M. Heald and H. Maeda, in *Lattice Effects in High T_C Superconductors*, eds. Y. Bar–Yam, T. Egami, J. Mustre–de Leon and A.R. Bishop (World Scientific, 1992) p. 51.
11. A. Bianconi, S. Della Longa, M. Missori, I. Pettiti and M. Pompa, ibid, p. 65.
12. J. Röhler, ibid, p. 77.
13. H.A. Mook, M. Nostoller, J.A. Harvey, N.W. Hill, B.C. Chakoumakos, and B.C. Sales, Phys. Rev. Lett., **65**, 2712 (1990).
14. Y. Wu, S. Pradhan and P. Boolchand, Phys. Rev. Lett., **67**, 3184 (1991).
15. R. Zeyher and G. Zwicknagl, Solid St. Commun., **66**, 617 (1988).
16. R.J. Cava, A. Santoro, D.W. Johnson, Jr. and W.W. Rhodes, Phys. Rev., **B35**, 6716 (1987).
17. P. Bordet, J.J. Capponi, C. Chaillout, J. Chenavas, A.W. Hewat, E.A. Hewat, J.L. Hodeau, M. Marezio, J.L. Tholence and D. Tranqui, Physica C, **156**, 189 (1988).
18. B.E. Warren, *X–Ray Diffraction* (Addison–Wesley, 1969).
19. H.P. Klug and L.E. Alexander, *X–Ray Diffraction Procedures for Polycrystalline and Amorphous Materials,* 2nd ed. (Wiley, 1968).
20. T. Egami, Mater. Trans. **31**, 163 (1990).

64

21. S. Sugai, S. Shamoto, M. Sato, T. Ido, H. Takagi and S. Uchida, Solid St. Commun., **76**, 371 (1990).
22. P.C. Hammel, private communication.
23. J. Saylor and C. Hohenemser, Phys. Rev. Lett., **65**, 1824 (1990).
24. S.W. Lovesay, *Theory of Neutron Scattering from Condensed Matter,* (Clarendon Press, 1984).
25. S. Nanao, W. Dmowski, T. Egami, J.W. Richardson, Jr., and J. D. Jorgensen, Phys. Rev., **B35**, 435 (1987).
26. B.H. Toby and T. Egami, Acta Cryst., **A48**, 336 (1992).
27. G. Placzek, Phys. Rev., **86**, 377 (1952).
28. T. Egami, B.H. Toby, S.J.L. Billinge, Chr. Janot, J.D. Jorgensen, D.G. Hinks, M.A. Subramanian, M.K. Crawford, W.E. Farneth and E.M. McCarron, in *High Temperature Superconductivity: Physical Properties, Microscopic Theory and Mechanisms*, eds. J. Ashkenazi et al. (Plenum Press, 1992) p. 389.
29. H.D. Rosenfeld and T. Egami, Ferroelectrics, in press.
30. S.J.L. Billinge, Ph.D. Thesis, University of Pennsylvania (1992).
31. W. Dmowski, B.H. Toby, T. Egami, M.A. Subramanian, J. Gopalakrishnan, and A.W. Sleight, Phys. Rev. Lett., **61**, 2608 (1988).
32. T. Egami, W. Dmowski, J.D. Jorgensen, D.G. Hinks, D.W. Capone, II, C.U. Segre, and K. Zhang, Rev. Solid St. Sci., **1**, 101 (1987).
33. T. Sendyka, T. Egami, in *Lattice Effects in High T_C Superconductors,* eds. Y. Bar-Yam, T. Egami, J. Mustre-de Leon and A.R. Bishop (World Scientific, 1992) p. 111.
34. S.J.L. Billinge and T. Egami, to be published.
35. S.J.L. Billinge, P.K. Davies, T. Egami and C.R.A. Catlow, Phys. Rev., **B40**, 10340 (1991).
36. R.E. Watson, Phys. Rev., **111**, 1108 (1958).
37. O.K. Andersen, A.I. Liechtenstein, O. Rodriguez, I.I. Mazin, O. Jepsen, V.P. Antropov, O. Gunnarsson, and S. Gopalan, Physica C, **185–189**, 147 (1991).
38. A. Bussmann-Holder, A. Simon and H. Büttner, Phys. Rev., **B39**, 207 (1989); A. Bussmann-Holder and A.R. Bishop, Phys. Rev., **B44**, 2853 (1992).
39. J. Ranninger, Solid St. Commun., **85**, 929 (1993).
40. T.W. Jing, Z.Z. Wang, T.R. Chien, N.P. Ong, J.M. Tarascon and E. Wang, in *Advances in Superconductivity*, eds. T. Ishiguro and K. Kajimura (Springer, 1990) p. 499.
41. W.E. Pickett., Rev. Mod. Phys., **61**, 433 (1989).
42. C.T. Chen, L.H. Tjeng, J. Kwo, H.L. Kao, P. Rudolf, F. Sette, and R.M. Flemming, Phys. Rev. Lett., **68**, 2543 (1992).
43. W. Weber, Phys. Rev. Lett., **58**, 1371 (1987).
44. M.K. Crawford, W.E. Farneth, E.M. McCarron, R.L. Harlow and A.H. Moudden, Science, **250**, 1390 (1990).
45. D.M. Newns, H.R. Krishnamurthy, P.C. Pattnaik, C.C. Tsuei and C.L. Kane, Phys. Rev. Lett., **69**, 1264 (1992).
46. R.E. Cohen, W.E. Pickett, D. Papaconstantopoulos and H. Krakauer, in *Lattice Effects in High T_C Superconductors,* eds. Y. Bar-Yam, T. Egami, J. Mustre-de Leon and A. Bishop (World Scientific, 1992) p. 223.
47. T.R. Thurston, R.J. Birgeneau, M.A. Kastner, N.W. Preyer, G. Shirane, Y. Fujii, K. Yamada, Y. Endoh, K. Kakurai, M. Matsuda, Y. Hidaka and T. Murakami, Phys. Rev., **B40**, 4585 (1989).
48. J. Rossat-Mignod, L.P. Regnault, C. Vettier, P. Bourges, P. Burlet, J. Bossy, J.Y. Henry and G. Lapertot, Physica C, **185–189**, 86 (1991).
49. V. Emery, S.A. Kivelson and H.Q. Lin, Phys. Rev. Lett., **64**, 475 (1990).
50. e.g., J.C. Phillips, Phys. Rev., **B45**, 12647 (1992).
51. P. W. Anderson, Phys. Rev. Lett., **34**, 953 (1975).
52. C.S. Ting, D.N. Talwar and K.L. Ngai, Phys. Rev. Lett., **45**, 1213 (1980).
53. R. Micnas, J. Ranninger, and S. Robaszkiewicz, Rev. Mod. Phys., **62**, 113 (1990).
54. Y. Bar-Yam, Phys. Rev., **B43**, 359; 2601 (1991).
55. T. Egami, Ferroelectrics, **130**, 15 (1992).

SUPERSTRUCTURES IN 123 COMPOUNDS
X-RAY AND NEUTRON DIFFRACTION

D. HOHLWEIN
Institut f. Kristallographie der Universität Tübingen
c/o Hahn-Meitner-Institut
Glienicker Str. 100
1000 Berlin 39
Germany

ABSTRACT. Superstructure investigations by X-ray and neutron diffraction of the system Y-Ba-Cu-Oxide are reviewed. More or less ordered superstructures with tetragonal (orthorhombic) $2\sqrt{2}$ a $\times(2)\sqrt{2}$ a cells and 2a×b ortho-II cells are characterized by anisotropic domain sizes. The degree of order has influence on the superconducting transition temperature. In the ortho-II structure displacements of the atoms Ba ,Y ,Cu(2) and the apex oxygen have been determined. The displacements are a consequence of the charge ordering in the basal plane.

1. INTRODUCTION

The system Y-Ba-Cu oxide is very well suited for the study of the relationship between structure and superconducting properties. The superconducting transition temperature T_C can be varied between 0 and 90 K by controlling the oxygen content. The oxygen incorporation takes place as an intercalation process into a well defined layer, the chain layer, and to well-known places, keeping intact the backbone of the structure. The highly mobile oxygen atoms in the layer can build up several superstructures with different degrees of order (long and short range order) which strongly influence T_C. Characteristic displacements of the backbone atoms, e.g. of the apical oxygen atom near the superconducting layer as a function of the oxygen content, have led to the idea of charge transfer from the chains to the superconducting layer.

Detailed structural informations can only be gained by X-ray or neutron diffraction. Up to quite recently only the structures of YBCO 6.0 (no oxygen in the chain) and YBCO 7.0 (all chains filled) as well as the average structures of the compounds with intermediate oxygen concentrations could be determined with these methods /e.g.1,2,3/. Average structure means that ordering phenomena of the oxygen atoms in the chain layer are not determined explicitly (no superstructures) but only the average occupations and positions of the atoms in a small unit cell. Superstructure reflections have intensities too small to be seen in a powder diagram. It is still difficult to produce homogeneous single crystals with a definite type and degree of order.

Superstructures have been observed by electron diffraction in numerous works /e.g.4,5,6/. They were attributed to chain ordering with different periodicities of complete

E. Kaldis (ed.), Materials and Crystallographic Aspects of HTc-Superconductivity, 65–81.
© *1994 Kluwer Academic Publishers.*

Cu-O-Cu and empty Cu-Cu chains. In the ideal ortho-II phase, YBCO 6.5, complete and empty chains alternate, giving a doubling of the unit cell in a-direction 2a×b×c. This structure is connected with the 60 K plateau in the T_C versus x curve of Fig.1. For concentrations above YBCO 6.5 also periodicities of 3a (ortho-III) and 5a (ortho-V) have been detected. At low oxygen concentration another unit cell, 2 √2 a × 2 √2 a × c, has been observed which cannot be explained with chain ordering, see chapter 2.

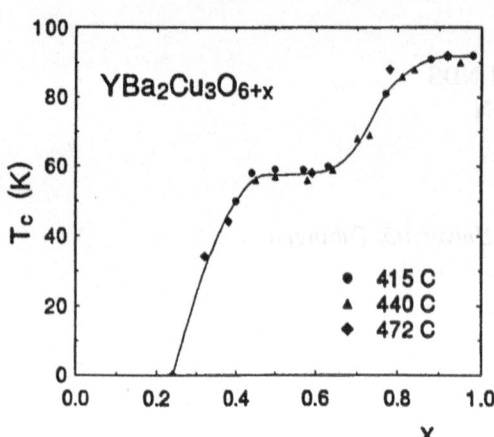

Fig.1. T_C as a function of oxygen concentration, experimental values and a theoretical curve, /8/.

The superconducting and structural properties depend on the way the samples are prepared. In particular, for a given oxygen concentration, the properties are found to depend on the high-temperature annealing and cooling procedure by which the oxygen content is set to the desired value. For example, when oxygen is removed from YBCO 7.0 by zirconium gettering at low temperature /1/, a step-like discontinuity in the c lattice constant is observed at the composition where superconductivity suddenly disappears, while no such discontinuity is seen for oxygen-deficient samples quenched from slightly higher temperatures. Additionally, a more pronounced plateau in T_C versus oxygen concentration, can be seen for samples quenched from lower temperatures.

Also ordering kinetics can play a significant role, explaining the differences between quenched samples and slowly cooled samples initially equilibrated at the same point in the phase diagram. The dramatic effect on superconducting properties that can result fom subtle changes in the state of the sample at fixed composition is evidenced by experiments in which the superconducting and structural properties were shown to evolve while annealing at room temperature after beeing quenched from high temperature /e.g.7/. It was found that T_C increases systematically with the time spent annealing at room temperature immediately following the quench, Fig. 2. During the annealing process changes in the lattice constant occur with no change in the net oxygen content. The changes are attributed to short range ordering of oxygen atoms in the chain layer and the question arises whether short range chain ordering is a requirement for superconductivity.

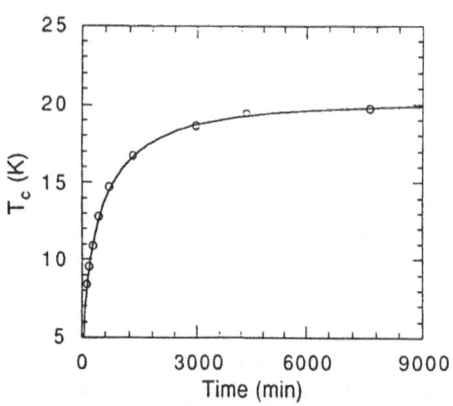

Fig. 2. T_C as a function of the annealing time at room temperature, /7/.

The oxygen ordering in the chain plane has been studied theoretically with the ASYNNNI model, a two-dimensional Ising model with anisotropic next-nearest neighbor (NNN) interactions between the oxygen atoms. In a computer simulation with three interaction parameters /8/ the order, which was only short-ranged, has been determined as a function of the oxygen content. The influence on T_c has been considered by the amount of charge transfer to the superconducting layer. Assuming that only the ordered ortho-I and ortho-II phases contribute to superconductivity, and that the ordered domains have to be at least of a certain size (for ortho-I clusters 4a x 4b and for ortho-II 8a x 8b), Fig. 3, a theoretical prediction of $T_c(x)$ in close quantitative agreement with the two plateaux behavior found experimentally, was achieved, Fig. 1.

A phase diagram with ortho-I, ortho-II and ortho-III regions has been calculated with four interaction parameters, Fig. 4, /9/. Experimental studies with a special tempering procedure and electron microscopic investigations /10/ roughly confirm the phase diagram.

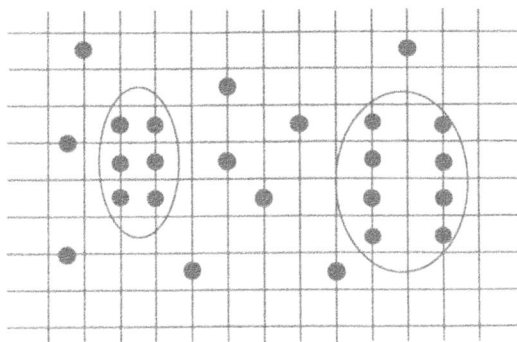

Fig. 3. Small clusters of ortho-I and ortho-II phases

The stability of superstructures was also calculated with a model Hamiltonian which assumes that any two oxygen atoms repel each other with a screened Coulomb interaction, but the repulsion between second-neighbor O ions with a Cu ion in between is reduced by a factor f, /11/.

The screened Coulomb interaction decreases in the metallic phase. Therefore Cu-O chains are only stable in the metallic phase. In the semiconducting region other structures, as described in section 2, can exist.

At the present status it is not clear in which way ordered superstructures can be produced and how the ordering phenomena are correlated with the superconductivity. In the theories displacements of atoms in the ordered structures have not been considered.

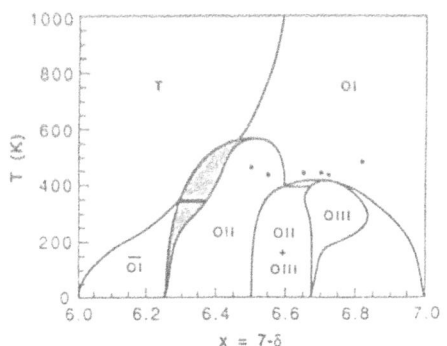

Fig. 4. Calculated phase diagramm with the ASYNNNI model /9/ and experimental values /10/.

In the following we describe in detail the results of neutron and X-ray diffraction investigations of more or less ordered superstructures. The descriptons start with compounds of low oxygen concentration and are arranged in the order of increasing oxygen content. At the end a summary and conclusions are given.

68

2. YBCO 6.35

2.1. PREPARATION

The single crystal was prepared by the flux method using the eutectic composition 7BaO-18CuO with a ratio to YBa$_2$Cu$_3$O$_7$ flux of 0.25/0.75. The mixture was prereacted at 880 °C, melted at 990 °C, and then slowly cooled down to 900 °C in one week and finally to room temperature in two days /12/.

2.2. NEUTRON DIFFRACTION

The investigated crystal had dimensions of 5x5x4 mm^3, an oxygen composition of 6.35(0.05), and lattice constants a = b =3.84 Å and c = 11.75 Å.

Neutron Diffraction measurements were performed on the four-circle diffractometer D10 at the Institut Laue-Langevin in Grenoble which is situated at a thermal neutron guide /12/. The wavelength was 2.4 Å and special care was taken to determine the second, 8x10^{-5}, and fourth order, 2x10^{-5}, harmonic contaminations.

The authors found five superstructure reflections in the (hk0)-layer, Table 1. Scans around the superstructure reflections did not show any broadening in the three directions of reciprocal space. The intensities are about 10^4 - 10^5 times smaller than the main Bragg reflections.

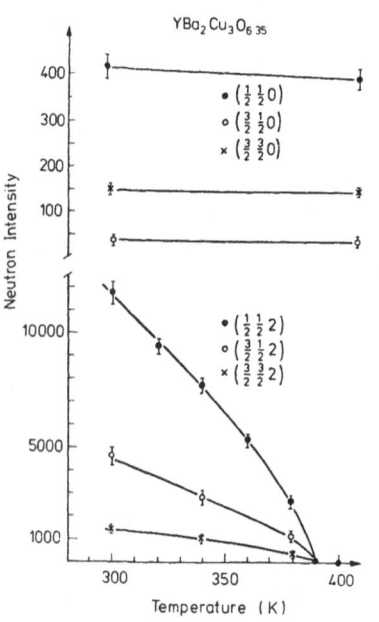

Fig. 5. Temperature behavior of magnetic (lower) and oxygen (upper) superstructure reflections /12/.

Table 1. Observed and calculated integrated intensities of superstructure reflections. Additional 27 reflections with I_{obs} < 10 are omitted.

h,k,l	I_{obs}	I_{calc}
1/2,0,0	40	60
1/2,1/2,0	420	400
1/4,3/4,0	50	40
1/2,3/2,0	30	155
3/2,3/2,0	150	115

To check that the reflections are not of magnetic origin the temperature dependence of the intensities has been measured. Fig.5 shows the different behavior in comparison with magnetic reflections. The intensities of the magnetic reflections at room temperature agree well with the known 3D antiferromagnetic structure of the Cu^{2+} ions in the CuO$_2$ layers with the spin directions along the crystallographic a or b axis /13/. The Neel temperature is 390(5) K and the magnetic moment 0.4(0.05) μ_B per Cu^{2+} ion.

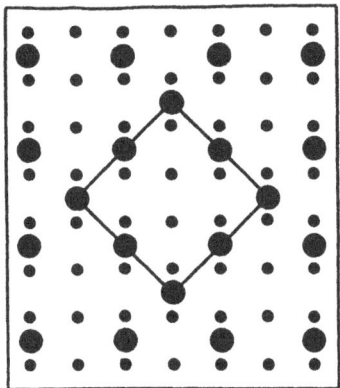

Fig. 6. 2 √2 a×2 √2 a oxygen super-structure in YBCO 6.35 (small circles denote Cu,big circles O) /12/

The observed superstructure reflections in the (hk0) layer can be indexed with a 2 √2 a x 2 √2 a (diagonal) superlattice cell which is 8 times larger than the original one. The best agreement with the observed intensities is reached with the oxygen ordering model given in Fig. 6.

The ordered structure corresponds to an oxygen composition of YBCO 6.375. The model was also proposed on the basis of superstructure reflections observed by electron microscopy /14/. In comparison with the fully occupied CuO chains along the b-direction in YBCO 7.0, there are only half-filled CuO chains alternating with quarter-filled ones. Pure oxygen chains are formed in the 45 degree direction. In this structure each oxygen atom has oxygen vacancies as first and second neighbors.

By comparison with the intensities of the main Bragg reflections one can conclude, that only 15% of all oxygen atoms in the basal plane contribute to the superstructure. These 15% can be situated in fully (super)ordered regions of the crystal while other regions are completly disordered, or, more likely, there is one phase with a fractional long range order or a long-range order parameter less than 1.

2.3. X-RAY DIFFRACTION

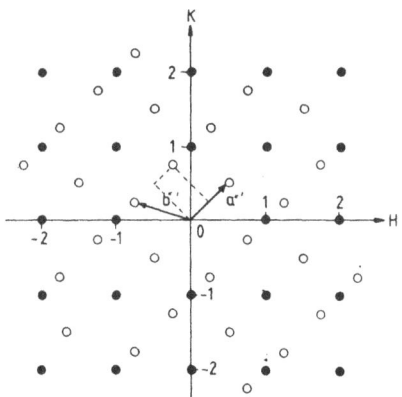

Fig. 7. Observed superstructure reflections (open circles) and fundamental reflections (full circles) in the hk0-plane /15/.

From the same crystal just described in the neutron work a tiny crystal was cleaved for the X-ray work described here /15/. By chance the crystal showed orthorhombic superstructure domains of only one size and orientation. Therefore a more accurate structure determination could be obtained than in the neutron work, where a tetragonal (average) structure was determined.

X-ray diffraction measurements were performed on a four-circle diffractometer with a conventional X-ray tube, graphite monochromator and a wavelength of 0.71 Å. The crystal dimensions were 0.5×0.3×0.06 mm^3 and the lattice constants (main reflections, tetragonal, the superstructure is orthorhombic) a = b = 3.8652(3)Å, c =

11.815(2) Å.

In the (hk0) layer 28 superstructure reflections, 8 symmetry independent, could be observed. Q-scans across the reflections in the directions of the reciprocal axes did not show any broadening compared to the Bragg reflections. Therefore the superstructure is of three dimensional long range order, but probably with diplacement disorder along c, because no higher layer reflections could be observed. Figure 7 illustrates the position of the observed superstructure reflections. A small symmetry adapted cell can be chosen describing a C-centered orthorhombic direct cell of $2 \sqrt{2}$ a $\times \sqrt{2}$ a $\times$ c. Figure 8 shows the two possible orientations of the supercell. The single cystal consisted only of domains in one orientation.

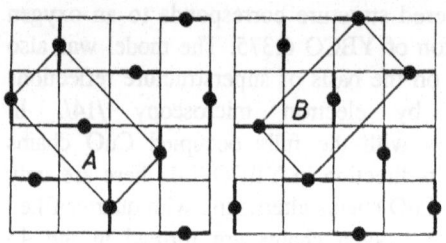

Fig. 8. Two possible orientations of the $2 \sqrt{2}$ a $\times \sqrt{2}$ a superstructure cell relative to the original unit cell of the Cu-squares /15/.

The best agreement between the measured and calculated intensities is achieved by the ordering scheme shown in Fig. 8. It consists of linear Cu-O-Cu dimers on every possible fourth site on both basal axes of the original unit cell. The occupation factor is 0.7 for the composition YBCO 6.35. The electrostatic repulsion between two oxygen ions is in this structure minimized because near neighbor and next near neighbor oxygen sites are vacant.

In Table 2, the calculated intensities for the oxygen ordering scheme, $I_c(O)$, are compared with the observed ones. The agreement is quite good, but there is clear evidence that the calculated intensities with large Q-vectors are systematically too weak.

Therefore in addition displacements of atoms have been considered. A significally better agreement with experiment has been reached by assuming Cu-displacements in the Cu(1) (chains) and Cu(2) layers, shown in Fig. 9 and calculated as $I_c(O+Cu)$ in Table 2.

Table 2. Observed and calculated integrated intensities of superstructure reflections /15/.

h,k,l	I_{obs}	$I_c(O)$	$I_c(O+Cu)$
1/2,1/2,0	390(20)	398	385
1/4,-3/4,0	295(20)	340	305
1/4,5/4,0	165(20)	142	155
1/2,-3/2,0	110(20)	84	121
3/4,7/4,0	125(15)	49	114
3/2,3/2,0	86(15)	33	75
5/4,-7/4,0	74(15)	32	92
3/4,-9/4,0	33(15)	22	35

Due to local strain linked with the oxygen incorporation in the squared Cu-lattice the Cu(1) atoms shift about 0.13 Å perpendicular to their dimer axis and the Cu(2) atoms by 0.06 Å nearly in the opposite direction.. In YBCO 7.0 the local strain is responsible for the difference of 0.07 Å between the a and b lattice constant.

By scaling the intensities of the superstructure reflections with those of the main Bragg reflections one finds that only 12 % of the oxygen atoms in the basal plane contribute to the superstructure. Nearly the same value, 15%, has been found in the neutron work

Also a $2 \sqrt{2}$ a $\times 2 \sqrt{2}$ a cell has been observed by electron diffraction /16/ at samples of YBCO 6.0 and YBCO 7.0 heated in situ in the vacuum. In this case the cell is probably formed by ordered Cu and Ba vacancies. This can not be the case for the described YBCO 6.35 crystal. First, comparing the neutron and X-ray results , the percentage of ordered cation vacancies would be more than a factor 10 different for the two methods. Only the assumption of ordered oxygen atoms give nearly the same result as mentioned above (the contrast between the atoms is quite different for neutrons and X-rays). Second, the observed orthorhombic cell is face-centered and cannot be formed with vacancies in the Cu-plane.

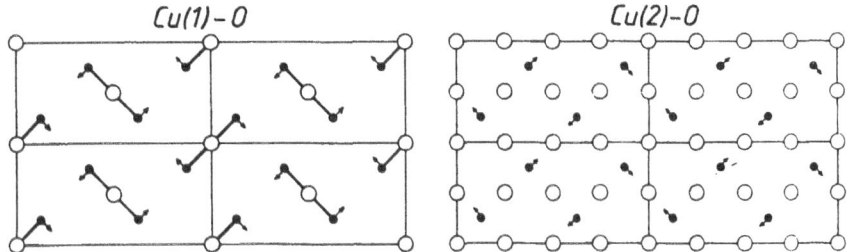

Fig. 9. Oxygen ordering and copper displacement in the basal (chain) layer, Cu(1), and in the superconducting layer, Cu(2). Oxygen atoms are the open circles /15/.

2.4. COMPARISON AND CONCLUSION

Assuming that in the neutron experiment at the large single crystal, equal amounts of the two orthorhombic domains or twins were present, then one can not explain the observed intensities equally well as with the proposed tetragonal structure, Fig. 10a. Therefore also coherent superpositions of amplitudes of different orthorhombic domains (also possibly of antiphase domains) have to occur in the neutron crystal. In a theoretical work /11/, a slightly different structure , Fig. 10c, has been proposed which explains the observed data also quite well. All structures have in common that first and second neighbored oxygen sites are not occupied. So the oxygen repulsion plays a decisive role at this concentration. In the theoretical work just mentioned, this is taken into account by the absence of the screening effect by conduction electrons in the semiconducting region of the phase diagram. The theoretical ASYNNNI model can not explain this structure. All theories have as yet not considered the possibility of atom displacements, in this case of copper atoms.

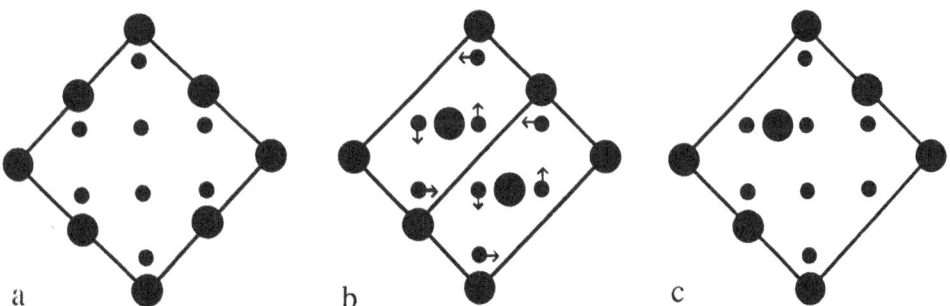

Fig 10. Superstructures in YBCO 6.35 derived from (a) neutron diffraction /12/, (b) X-ray diffraction /15/ and (c) theory /11/.

3. YBCO 6.40

3.1. PREPARATION

A YBCO 7.0 single crystal was grown from CuO-BaO flux in an alumina crucible by a slow cooling method. After annealing in oxygen, the crystal showed a T_C of 87 K. To develop the ortho-II phase, the crystal was reduced to an oxygen concentration of 0.4 by stepwise cooling from 650 °C to 150 °C in 200 h under controlled oxygen equilibrium pressures. At 150 °C the sample was subjected to a long-time (300 h) ageing process to establish oxygen ordering. After this treatment, T_C determined by a.c. susceptibility measurements was 38K with a transition width of 5K. The crystal size was 6x3x2 mm^3 /17/.

3.2. NEUTRON DIFFRACTION

Neutron measurements were performed on the triple axis spectrometer TAS 1 at Risø National Laboratory with wavelengths of 4 and 2.4 Å /17/. To improve the signal-to-noise ratio and to separate inelastic scattering events, the spectrometer was operated in the elastic mode.

The crystal was twinned with equal amounts of different domains with interchanged a and b axes. The orthorhombicity (b-a)/b was 0.0055(12) and the length of the c-axis 11.73(2).

By comparing the measured values of T_C, orthorhombicity, and lattice constant c with literature values for the oxygen concentration dependence one derives the formula YBCO 6.40(5) as a likely oxygen stoichiometry of the sample.

Five superstructure reflections consistent with the ortho-II phase lattice, (1/2,0,0), (1/2,1,0), (1/2,2,0), (1/2,0,1), (3/2,0,2) have been detected. The profile of the (1/2,0,0) reflection along the k direction in reciprocal space is shown in Fig.11. In all directions the width of the superstructure reflection is much broader than the instrumental resolution. From the width finite domain sizes are derived by assuming Gaussian distributions The results show large anisotropy of the ordered regions. The

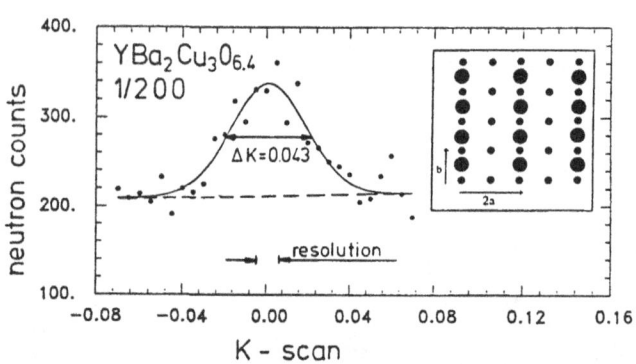

Fig. 11. Scan along the k direction of the (1/2,0,0) reflection /18/.

tendency of the system to build Cu-O chains is reflected in the domain length along b, the chain direction, of 24b (90 Å) which is more than twice as large as along a with 10 a

(40Å). In the c-direction there is only a correlation between nearest and next nearest basal planes, 2c (22 Å).

By scaling the integrated intensity of superstructure reflections with that from a Bragg reflection one can estimate that about 60% of the crystal volume contribute to the superstructure. A detailed structure determination was not presented.

4. YBCO 6.41

4.1. PREPARATION

A single crystal of 34 mm^3 was obtained by a grain growth technique from a ceramic prepared from the oxides Y_2O_3 (99.99%) and hydroxide $Ba(OH)_2 \cdot xH_2O$ (99%). To ensure the oxygen content the following procedure was used. The single crystal and appropiate amounts of YBCO 6.15 and of YBCO 6.99 ceramics were put into a quartz tube which was evacuated for 1 h at 300 °C before it was closed and warmed up to 670 ° C. It was kept at this temperature for 8 h and cooled at 20 °C/h to 620 °C, at 10 °C/h to 520 °C and to 20 °C at 5 °C/h, where it was kept for 10 h before removal of the crystal. The weight variation of the crystal and of the ceramics gave a composition of YBCO 6.48(4). The superconducting transition temperature was 39 K with a transition width of 7 K /19/.

4.2. NEUTRON DIFFRACTION

The neutron measurements were performed at the SILOE reactor in Grenoble. A wavelength of 2.35 Å was used. Only reflections in the (h0l) layer could be measured /19/. From 12 main Bragg reflections an oxygen composition of YBCO 6.41(2) was determined. Superstructure reflections could be detected in l-scans (1/2,0,l) and (3/2,0,l), presented in Fig. 12. The reflections are so braod that they cannot be integrated seperately. Therefore a special linear correlation function, which is a infinite sum of Lorentzians, has been fitted to the data in l direction. The authors estimate domain sizes (here we use the full-width-half-maximum of the correlation function) of 15×(>55)×7Å^3 or 8a×(>30b)×1c, these are the values corrected in a later paper /20 / .

The l-profile in Fig. 12 was refined using the special correlation function with one parameter and developping the structure factor into a series to first order in displacement of

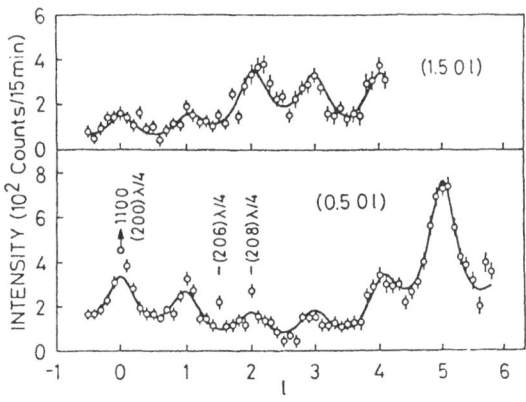

Fig. 12. Scans (1/2,0l) and (3/2,0,l) /19/.

atoms. The best agreement was reached with antiferro-distortive shifts in directions as shown in Fig.15 and values for barium atoms $\Delta x(Ba)=-0.038(2)Å$, for yttrium atoms $\Delta x(Y)=0.010(3)Å$, copper atoms in the superconducting layer $\Delta z(Cu2)=0.012(2)$ Å and the apex oxygen $\Delta z(O)=0.038(2)$ Å.

5. YBCO 6.51

5.1 CRYSTAL PREPARATION

The cystal was grown from CuO-BaO flux in a ZrO_2 crucible by the slow cooling method. A mixture of high-purity (4N) Y_2O_3, $BaCO_3$ and CuO was stirred with a ZrO_2 rod at 1010 °C. Then the crucible was cooled down at a rate of 0.6 °C/h to start the growth. At a temperature of 965 °C the crystal growth was stopped by sucking up the remaining melt with a porous ZrO_2 ceramic which was dipped into the melt. After cooling down to room temperature the crystal was extracted and annealed in flowing oxygen between 590 and 390 °C for 400 h. The T_c of this crystal was 90.5 K with a $\Delta T_c = 1.0$ K. To develop the ortho-II phase the crystal was reduced to an oxygen content near 6.5 by stepwise cooling down from 650 to 150 °C under controlled oxygen equilibrium pressures. At 150 °C the crystal was subjected to a long (400 h) ageing to establish oxygen ordering. The T_c was 55.8 K with a transition width of 2.7 K /21/.

5.2. X-RAY DIFFRACTION

X-ray measurements were performed at a four-circle diffractometer with $\lambda=0.71Å$ and a graphite monochromator /21/. The crystal size was $0.5\times0.3\times0.06$ mm^3.

Because of the orthorhombic twinning the h00-reflections are splitted quadruply. A 2-dimensional plot of such a splitting is shown in Fig.13.(There is further splitting by the Kα doublet). The splitting is due to the interchange of the a and b axis on a macroscopic scale and to microscopic twinning about (110) planes. The crystal consists of about 75% of domains with only two different orientations. The orthorhombicity (b-a)/b is 0.011(1), a=3.830Å, b=3.872Å and the c lattice constant 11.73(1) Å. From these values one can estimate the oxygen stochiometry as YBCO 6.51(5).

The superstrucure reflections are considerably broader than the main Bragg reflections corresponding to domain sizes of $18a \times 135b \times 6c$

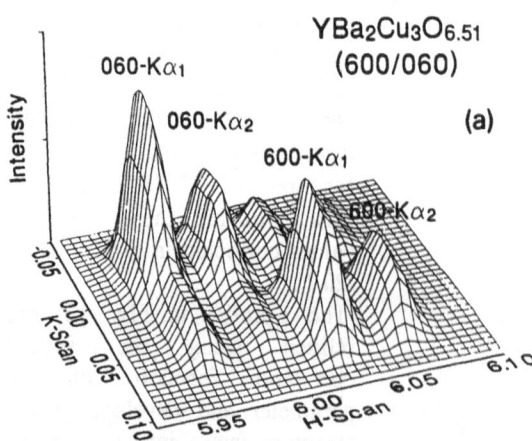

Fig. 13. The domain structure of the YBCO 6.51 crystal /21/.

or 68×520×65 Å³. The correlation length in chain direction is about seven times larger than in a direction.

The ratio between the peak intensities of the superstructure reflections and of the main reflections is about 5×10^{-5} and can be seen in Fig.14.

The intensities of the superstructure reflections show orthorhombic symmetry, space group Pmmm, with a 2a×b×c direct cell. 28 symmetry independent reflections could be observed clearly. Because of their broad intensity distribution in reciprocal space the reflections are not fully integrated by ω-2θ scans. Simulating the movements of the four-circle diffractometer the partialities of the measured integrated intensities have been determined numerically and corrected for.

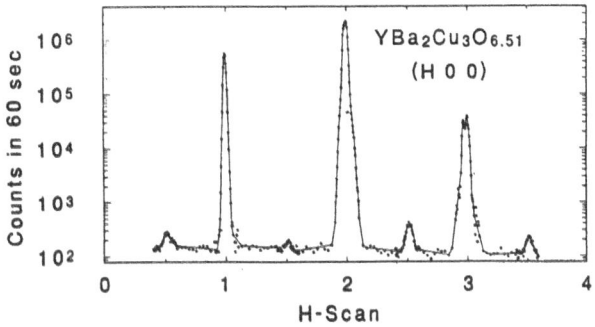

Fig. 14. Superstructure reflections in comparison with main reflections /21/.

The structure factors do not fall off monotonically with increasing scattering vector, as can be seen in Fig. 14, therefore pure oxygen ordering of ortho-II type can not explain the intensity distribution. The intensity is increasing in (100) direction therefore atom displacements in (100) direction have to occur. The best agreement with the observed structure factors is achieved for displacements of the barium atoms. The Ba atoms shift 0.034(3) Å towards the Cu-O chains in an antiferro-distortive way as illustrated in Fig. 15. The refined parameters in the spacegroup Pmmm were the position parameters of the barium atom x and z and the (isotropic) temperature factors of barium and oxygen.

By scaling the integrated intensities of superstructure reflections with those of the main reflections, one finds that about 55% of the crystal volume contributes to the superstructure.

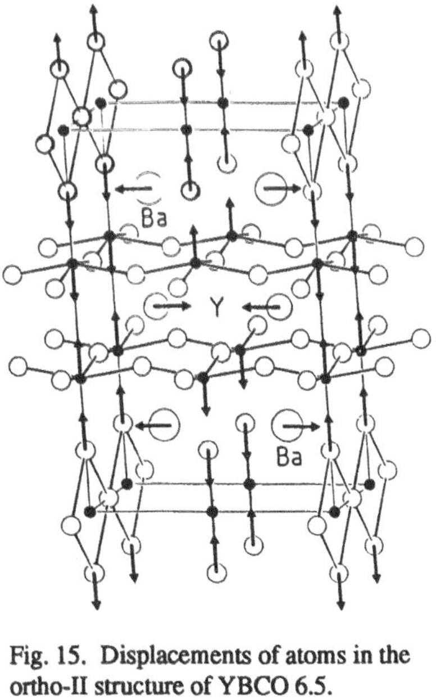

5.3. ORIGIN OF THE BARIUM DISPLACEMENT

In YBCO 6.5 with ortho-II oxygen order the symmetry of nearest and next nearest neighbor coordination is, in comparison with YBCO 7.0, conserved for all cations but the barium ions. The (100) mirror plane through the barium sites is

Fig. 15. Displacements of atoms in the ortho-II structure of YBCO 6.5.

removed. The different charges of alternating Cu-O-Cu and Cu-Cu chains in the basal plane lead to an asymmetry of the electrostatic potential at the Ba places. The minimum of the potential has been calculated in a pure ionic model with effective charges and an empirical Born-Mayer repulsion term /22/. In Fig.16 the lattice energy is shown as a function of the x-coordinate (along the a axis). The minimum of the lattice energy is achieved for x=0.2450 which corresponds to a shift of the barium atoms by 0.039 Å in excellent quantitative agreement with the experimental value of 0.034(3) Å.

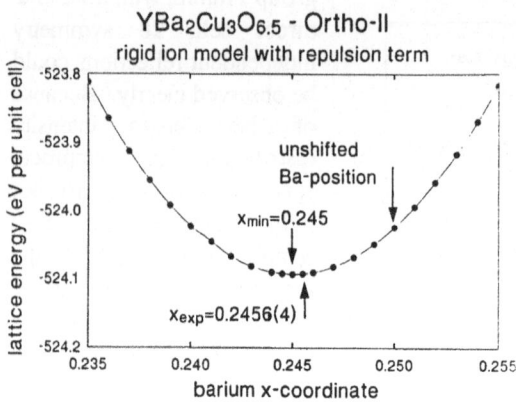

Fig.16. Lattice energy (Coulomb + repulsion) as a function of the Ba coordinate /22/.

5.4 X-RAY ANOMALOUS SCATTERING

A verification of the barium displacements has been given by X-ray anomalous scattering experiments /22/. The experiments were performed at the synchrotron four-circle diffractometer D3 at DESY-HASYLAB in Hamburg.

At two wavelengths, $\lambda_1 = 0.3350$ Å and $\lambda_2 = 0.3325$ Å, near the barium K-absorption edge, Fig. 17, the ortho-II type superstructure reflections (h/2,00), h=5,7,9,11,13,were measured.

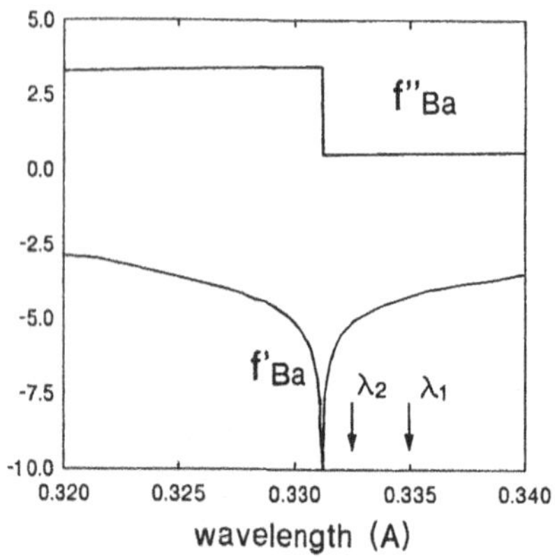

The anomalous scattering is desribed by an energy dependent complex dispersion term $(f'+if'')$ which must be added to the normal scattering factor. From the above described oxygen-order-barium-displacement model one can calculate the contribution of the oxygen and the displaced barium atoms to the reflections (h/2,0,0). Undisplaced barium atoms would not contribute to these reflections. For h>1 the superstructure reflection are dominated by the barium contribution and should therefore be very sensitive to anomalous dispersion at the Ba-K-absorption edge.

The relative changes of the structure factor for the two wavelengths are shown in Fig.18.

Fig. 17. Scattering factors of barium at the absorption edge /22/.

The results are consistent with the determined barium displacements. From the data follows that the intensity contribution from other displaced atoms cannot be more than 5(13)% of the barium fraction.

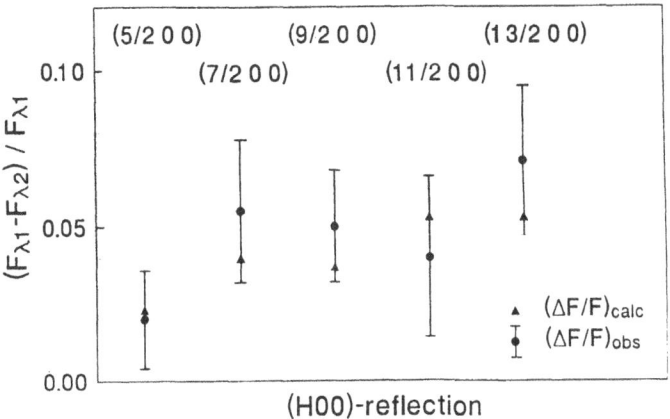

Fig. 18. Relative changes of the structure factors due to anomalous dispersion

5.5. SYNCHROTRON X-RAY STRUCTURE DETERMINATION

At the same crystal already described in section 5.2, Synchrotron X-ray measurements were performed on the D3 four-circle diffractometer at the HASYLAB in Hamburg /23/. Because of the high intensity of the monochromatic beam a much larger number of superstructure reflections could be collected than on a conventional X-ray tube. The wavelength was 0.7100 Å and the intensities of 346 reflections, 227 symmetry independent, were determined. With this data set also small displacements of other atoms , not only barium, could be refined. The directions of the displacements are shown in Fig. 15. In x-directions the barium atom is displaced by -0.041(2) Å and the yttrium atom in antiphase by 0.011(1) Å. Displacements in z-directions have been found for the copper atom in the superconducting layer of 0.018(1) Å and for the apex oxygen atom of 0.026(6) Å.

6. YBCO 6.58

6.1. PREPARATION

The crystal was grown from BaO+CuO flux in ZrO_2 crucibles. An initial negative vertical gradient 5K/cm was created at 1265 K. During the cooling process it was changed to a positive one at 1180 K reaching 3 K/cm at 1150 K. The oxygen content of the crystals quenched from 1125 K (150 K/min) was YBCO 6.15. They were up to 7×7×0.1 mm³ in size and had a very low concentration of growth defects. To obtain the required oxygen content, as grown crystals were annealed in air in a surrounding of the YBCO powder at initial temperature of 975 K. The temperature was decreased to the final value defined by the oxygen content, kept stable during 20 hours, and then the crystals were quenched to room temperature /20/.

6.2. X-RAY DIFFRACTION

X-ray measurements were performed on a conventional X-ray tube with graphite monochromator and AgKα radiation /20/. The transmission geometry was used with the crystal projection on the beam wider than the beam cross-section.

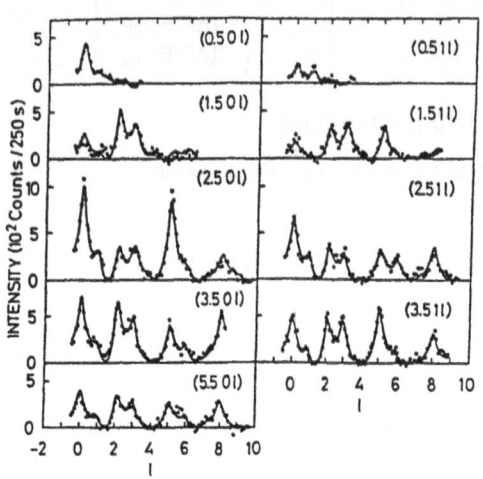

Fig. 19 Scans of superstructure reflections along l for YBCO 6.58 /20/.

The oxygen content was determined by the lattice constant c=11.729 Å and the z-parameter of the barium atom, 0.1883, as YBCO 6.58(3). The transition temperature T_c was 54K with a transition width of 8K.

Domain sizes of the ortho-II phase were determined by Lorentzian fits (along c with a special correlation function) to the broad superstructure reflections giving volumes of $18 \times 90 \times 16$ Å^3 or $5a \times 23b \times 1.3c$.

Superstructure reflections were measured along the l-axis as shown in Fig.19. The intensity distribution was fitted to the structure factor using a linear correlation function as described in section 4.2.

Displacements of atoms in directions as shown in Fig.15 were determined with values of $\Delta x(Ba) = -0.039(1)$Å, $\Delta x(Y) = 0.011(1)$Å , $\Delta z(Cu2) = 0.018(2)$Å and of the apical oxygen atom $\Delta z(O2) = 0.049(6)$Å.

7. YBCO 6.7

7.1. PREPARATION

As grown crystals with an oxygen content smaller than 6.50 were sealed with YBCO 7.0 powder in a quartz tube and heated to 470 °C and tempered for two weeks. The oxygen concentration was then about 6.8 which then was reduced by quenching from 667K in O_2 to a nominal composition of 6.7. The superconducting transition temperature was 57 K with a transition width of 10 K /24/.

7.2. X-RAY DIFFRACTION

X-ray measurements were performed using Cu Kα radiation from a rotating anode source, a singly bent graphite monochromator that focussed radiation in the vertical plane, and a

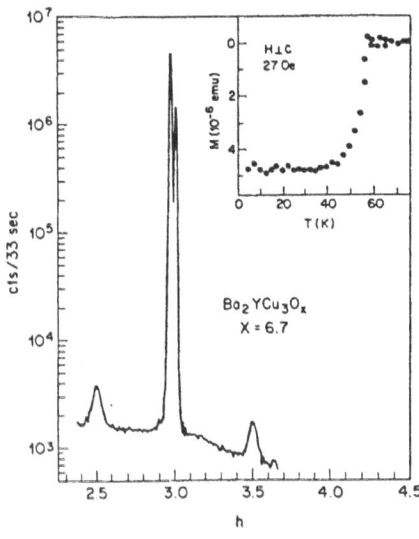

flat graphite analyzer /24/. The sample was mounted with his b-axis vertical on a four-circle diffractometer. The sample size was about $0.7 \times 1.5 \times 0.012$ mm^3.

A scan along (h00) shows Fig.20. Two diffuse superstructure reflections at positions of the ortho-II phase can clearly be seen. Another scan, (2.5,1,1) with overlapping reflections, has been presented in the paper. From the half-width of the reflections the coherence length (fwhm) has been determined by Lorentzian fits. Corresponding domain sizes are $21 \times 16 \times 9$ Å^3 or $5a \times 4b \times 1c$ along the principal axes. No further structural results have been given.

Fig. 20. Scan along h00 in YBCO 6.7 /24/.

8. Summary and Conclusion

The main results of all investigations described above are put together in Table3.

Just below the superconducting region of the phase diagram, a tetragonal $2\sqrt{2}\,a \times 2\sqrt{2}\,a$ superstructure has been found by neutron diffraction and an orthorhombic $2\sqrt{2}\,a \times \sqrt{2}\,a$ structure with copper displacements by X-ray diffraction. The neutron structure is a tetragonal average of the orthorhombic structure. In both structures, there is no tendency to the formation of chains.. The structure is dominated by the Coulomb repulsion between the oxygen ions, i.e., near neighbor and next near neighbor oxygen sites are vacant. From the theoretical point of view, this can be due to the absence of the screening power of conduction electrons which are missing in the semiconducting phase. The ordered fraction of this phase is only about 15 % of the volume but the order is long ranged. Relative large displacements of the copper atoms occur in the Cu-O-Cu dimers of the basal plane.

Table 3. Summary of the results

YBCO	T_c	super-cell	domain size	fract.	atom displacements	method
6.35	0	$2\sqrt{2}\,a \times 2\sqrt{2}a$	>100(a×b×c)	0.15	0	Neutron
6.35	0	$2\sqrt{2}\,a \times \sqrt{2}\,a$	>500(a×b×c)	0.12	Cu1 0.13, Cu2 -0.06	X-ray
6.40	38(5)	2a×b	10a×24b×2c	0.6		Neutron
6.43	39(7)	2a×b	8a×(>30b)×1c		Ba -0.038(2), Y 0.010(3) Cu2 0.012(2), O 0.038(2)	Neutron
6.51	56(3)	2a×b	18a×135b×6c	0.55	Ba -0.034(3)	X-ray
6.51	56(3)	2a×b	18a×135b×6c	0.55	Ba -0.041(2), Y 0.011(1), Cu2 0.018(1), O 0.026(4)	Synchr. X-rays
6.58	54(8)	2a×b	5a×23b×1c		Ba -0.039(1), Y 0.011(1), Cu2 0.018(2), O 0.049(6)	X-ray
6.7	57(10)	2a×b	5a×4b×1c			X-ray

In the superconducting compounds more or less ordered ortho-II phases exist. These phases constitute in general only a fraction , about 0.6, of the crystal volume. One could not decide, whether the phases are homogeneously (short range order) distributed or isolated domains are present. The remaining fraction of 0.4 can be in a more disordered state or a phase seperation between well-ordered ortho-I clusters and strongly disordered ortho-I phases takes place.

The best ordered ortho-II phase has been found in the YBCO 6.51 crystal, with domains strongly elongated along the chain direction , 18a $\times$ 135b $\times$6c. For this crystal also the largest number of superstructure reflections (227) has been measured. Comparing the YBCO 6.51 with the YBCO 6.58 crystal, one can conclude that the T_c value increases with increasing order of the ortho-II phase. There are not sufficient examples studied to give a quantitative relationship.

In four cases, atom displacements in the ortho-II phase have been determined with consistent values. The barium atoms shift about 0.035 Å in the direction to the oxygen filled chains along the a-axis and the yttrium atom by 0.010 Å in the opposite directions. These displacements have been explained with the Coulomb forces which are produced by the charge ordering connected with the filled and empty Cu-O-Cu chains. The displacements in c-direction of the copper atoms in the superconducting layer by 0.020 Å and of the apex oxygen atom by 0.050 Å are in the same directions as the known variations of these distances with oxygen concentration. The difference between the YBCO 6.0 and the YBCO 7.0 compound is for the copper atom 0.120 Å and for the apex oxygen atom 0.050 Å. So, the CuO_2 plane is more rigid and acts more as a whole to the inhomogeneity in the chain layer.

The charge ordering in the ortho-II structure shows through the displacements of atoms the flexibility or the coupling of the backbone atoms to the presence of electrical charges. A positive charge (hole) on an oxygen atom in the superconducting CuO_2-layer will produce similar displacements of atoms on a short length scale. In contrast to the static displacements due to oxygen order in the basal plane, these shiftings should be dynamical. Therefore the observed displacements could be important experimental data for the test of theories describing the interaction of charge carriers with the lattice (structure).

REFERENCES

1. R.J.Cava,A.W.Hewat,E.A.Hewat,B.Batlogg,M.Marezio,K.M.Rabe,
   J.J.Krajewski,W.F.Peck Jr.,L.W.Rupp Jr..Physica C **165**, 419 (1990).
2. J.D.Jorgensen,B.W.Veal,A.P.Paulikas,L.J.Nowicki,G.W.Crabtree,H.Claus,W.K.Kwok.
   Phys.Rev. B **41**,1863 (1990).
3. J.D.Jorgensen,D.G.Hinks,P.G.Radaelli,Shiyou Pei,P.Lightfood,B.Dabrowski,C.U.Segre
   B.A.Hunter. Physica C **185-189**, 184 (1991).
4. M.A.Alario-Franco,C.Chaillout,J.J.Capponi,J.Chenavas,M.Marezio.
   Physica C **156**, 455 (1988).
5. J.Reyes-Gasga,T.Krekels,G.Van Tendeloo,J.Van Landuyt,W.H.M.Bruggink,
   M.Verweij,S.Amelincks. Sol.State Comm. **70**, 269 (1989).

6. R.Beyers,B.T.Ahn,G.Gorman,V.Y.Lee,S.S.P.Parkin,M.L.Ramirez,K.P.Roche, J.E.Vasquez,T.M.Gür,R.A.Huggins. Nature **340**, 619 (1989).
7. J.D.Jorgensen,S.Pei,P.Lightfood,H.Shi,A.P.Paulikas,B.W.Veal. Physica C **167**, 571 (1990).
8. H.F.Poulsen,N.H.Andersen,J.V.Andersen,H.Bohr,O.G.Mouritsen. Nature **349**, 594 (1991).
9. G.Ceder,M.Asta,D.deFontaine. Physica C **177**, 106 (1991).
10. S.Yang,H.Claus,B.W.Veal,R.Wheeler,A.P.Paulikas,J.W.Downey. Physica C **193**, 243 (1992).
11. A.A.Aligia,J.Garces,H.Bonadeo. Physica C **190**, 234 (1992).
12. R.Sonntag,D.Hohlwein,T.Brückel,G.Collin. Phys.Rev.Lett. **66**, 1497 (1991).
13. J.M.Tranquada,D.E.Cox,W.Kunnmann,H.Moudden,G.Shirane,M.Suenaga,P.Zolliker, D.Vaknin,S.K.Sinha,M.S.Alvarez,A.J.Jacobsen. Phys.Rev.Lett. **60**,156 (1988).
14. J.Reyes-Gasga,T.Krekels,G.Van Tendeloo,J. Van Landuyt,S.Amelinckx W.H.M.Bruggink,H.Verweij. Physica C **159**, 831 (1989).
15. Th.Zeiske,D.Hohlwein,R.Sonntag,F.Kubanek,G.Collin. Z.Physik B **86**, 11 (1992).
16. D.J.Werder,C.H.Chen,G.P.Espinosa. Physica C **173**, 285 (1991).
17. Th.Zeiske,R.Sonntag,D.Hohlwein,N.H.Andersen,Th.Wolf. Nature **353**, 542 (1991).
18. R.Sonntag,Th.Zeiske,D.Hohlwein. Physica B **180&181**, 374 (1992).
19. P.Burlet,V.P.Plakhty,C.Marin,J.V.Henry. Physics Lett. A **167**, 401, (1992).
20. V.Plakhty,A.Stratilatov,Yu.Chernenkov,V.Fedorov,S.K.Sinha,ChunK.Loong, B.Gaulin,M.Vlasow,S.Moshkin. Solid State Comm. **84**, 639 (1992).
21 Th.Zeiske,D.Hohlwein,R.Sonntag,F.Kubanek,Th.Wolf. Physica C **194**, 1 (1992).
22. Th.Zeiske,D.Hohlwein,R.Sonntag,J.Grybos,K.Eichhorn,Th.Wolf. Physica C **207**, 333 (1993).
23. J.Grybos,D.Hohlwein,Th.Zeiske,F.Eichhorn. In preparation.
24. R.M.Fleming,L.F.Schneemeyer,P.K.Gallagher,B.Batlogg,L.W.Rupp, J.V.Waszczak. Phys.Rev. B **37**, 7920 (1988).

ACCURATE X-RAY STRUCTURAL INVESTIGATIONS OF SINGLE CRYSTALS OF HIGH-Tc MATERIALS

V.I.SIMONOV
Institute of Crysrallography
Academy of Sciences of Russia
Moscow 117333, Russia

ABSTRACT. The atomic structure of HT_c-superconducting single crystals was refined using X-ray diffraction methods: 1.Different degree of ordering in Sr atom distribution over La sites and T_c dependence on it were revealed in La-phases. 2.Oxygen atom ordered arrangement was determined in Y-phases with various oxygen content. Orthorombic local symmetry of single crystals was detected in the so-called tetragonal superconducting Y-phases. 3.The changes of structural parameters during the superconducting phase transition were established in $Tl_2Ba_2CaCu_2O_8$. 4.Distribution of valence electrons in space was obtained for Nd_2CuO_4.

1. Introduction

In 1911 the Dutch physicist G.Kamerling-Onnes discovered in the course of a study of the behaviour of electrical resistance of mercury upon temperature lowering that for $T < T_c = 4.15$ K the resistance dropped down to zero. It was shown later that at the critical temperature T_c a second-order phase transition occurs. This new state was called the superconducting state. Then Kamerling-Onnes observed that the application of a strong magnetic field eliminated the superconductivity. A fundamental property of superconductors - the Meissner effect - was discovered only 22 years later. The experiments conducted by W.Meissner and R.Oxenfeld , in which an appropriate sample was pushed from the magnetic field, proved that the external field does not penetrate the bulk of the superconducting material. The physical reason for the Meissner effect is as follows. Nonattenuating currents arise in the surface layer of a superconductor hundreds of angstroms thick under the effect

E. Kaldis (ed.), Materials and Crystallographic Aspects of HTc-Superconductivity, 83–128.
© 1994 *Kluwer Academic Publishers*.

of an external magnetic field. These currents compensate for the external field inside the sample. The theoretical understanding of superconductivity was being developed rather slowly. It was only in 1934 that the brothers F. and G.Londons suggested the first version of the phenomenological theory of electrodynamic properties in superconductors. In 1937 L.D.Landau predicted the structure of an intermediate state in superconductors which was later determined experimentally by A.I.Shal'nikov and co-workers. The generalized phenomenological theory of superconductivity was founded by V.L.Ginzburg and L.D.Landau in 1950. Based in their behaviour in a magnetic field superconductors can be divided into two groups: first- and second-order superconductors. In the former, superconductivity in the entire bulk is destroyed by a magnetic field. The latter superconductors were described in 1952 by A.A.Abrikosov. These superconductors are characterized by two different values of the critical magnetic field. At intermediate values of an external magnetic field a second-order superconductor is pierced by the Abrikosov swirls whose density increases with the field and the superconductivity is retained only beyond these swirls. The superconducting state is a quantum state of a macroscopic object. According to the theory, the magnetic flux piercing a superconducting ring with nonattenuating current is quantized. This effect was observed experimentally for the first time in 1961 and enabled the determination of the charge of the particles that are current carriers in superconductors. This charge was twice as large as the charge of an electron. This observation is in conformity with the effect of formation of stable pairs by electrons with opposite spins in a crystal lattice, which was predicted by L.Cooper in 1956. It is these Cooper pairs that are carriers of nonattenuating current in superconductors.

A rigorous microscopic theory of superconductivity was developed in 1957-1958 in the works of J.Bardeen, L.Cooper, G.Schrieffer and, at the same time, by N.N.Bogolyubov. It is based on the following. Electrons of a Cooper pair, which exchange phonons with the lattice, are attracted and form a particle with a zero spin. These particles obey Bose-Einstein statistics. Bose condensation of such particles and their superfluidity takes place in these superconductors. Since each particle carries a 2e charge, superfluidity of a quantum electronic liquid leads to superconductivity.

In the period 1911-1985 superconductivity was found experimentally in dozens of pure metals and in hundreds of various alloys and intermetallic compounds. An unrivalled superconductor was $Nb_3(Ge_{0.8}Nb_{0.2})$ with $T_c = 23.2$ K. The structure of these crystals is shown in Fig.1. Superconductivity was discovered in some strongly doped semiconductors and even in polymers. Organic superconductors hold a special place among other superconductors. A typical member of this family is the β-phase $(BEDT-TTF)_2I_3$ (bisethylenedithio-tetrathiafulvalene two iodine three). The compoundis a two-dimensional organic metal with $T_c \sim 1.5$ K. By varying preparation methods for these compounds, as well as their thermal treatment and the effect of high pressures on them, one can raise T_c in these organic superconductors up to 6-7K.

A new era in superconductivity was opened in 1986 when J.G.Bednorz and K.A.Muller discovered superconductivity in ceramic $(La,Ba)_2CuO_4$ with $T_c = 36$ K [Bednorz,Muller, 1986]. They were awarded the Nobel prize, and this work was an impetus for wide scale investigations in a new branch of

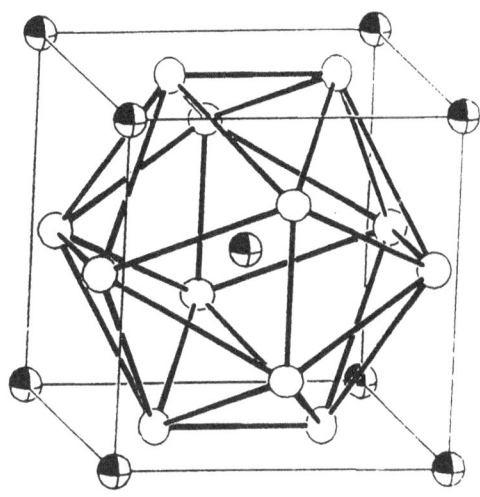

Fig.1. Crystal structure of a classic superconductor $Nb_3(Ge_{0.8}Nb_{0.2})$, $T_c = 23.2K$

solid state physics - high temperature superconductivity. Shortly after this discovery, in 1987, Wu with co-workers published a communication reporting synthesis and study of $YBa_2Cu_3O_{7-\delta}$ with $T_c = 93$ K [Wu et al.,1987]. In the following five years about twenty various crystalline superconductors were prepared at different laboratories all over the world. The total number of superconducting compounds, including isostructural ones which differ greatly in their chemical composition, now exceeds 500-600, and their number is rapidly increasing [Aksenov et al.,1990; Shelton et al.,1989; Narlikar,1991; Putilin,1993].For quite a long time the record superconducting transition temperature was 125 K and belonged to the Tl-phase of $Tl_2Ca_2Ba_2Cu_3O_{10}$. A paper on a new high-temperature superconductor $HgBa_2CuO_{4+\delta}$ with $T_c = 94$ K has just been published [Putilin,1993]. Then appeared communications about synthesis of other members of the Hg-family of compounds which exhibit the record $T_c = 134$ K and even 140 K.

In the first two years after the discovery of superconductivity in ceramics the list of superconductors was supplemented only by copper-containing compounds. In 1988 a compound with the perovskite structure $(Ba,K)BiO_3$ and $T_c = 30$ K was synthesized. This compound did not contain any copper atoms. In 1975, a study reporting a $T_c = 12$ K of $Ba(Rb,Bi)O_3$ was published, but at that time it did not seem interesting. The discovery of superconductivity in $(Nd,Ce)_2CuO_4$ crystals in 1989 [Tokura et al.,1989] was of principal importance . This compound was the first superconducting material that exhibited electronic conductivity. All the previously obtained materials had hole conductivity. A comparison of the hole $(La,Sr)_2CuO_4$ and electronic $(Nd,Ce)_2CuO_4$ superconductors indicates remarkable differences in their crystal structure. The fact that in the first case some trivalent lanthanum cations are replaced by divalent strontium ions, while in the second one trivalent neodymium is partially replaced by tetravalent cerium is of major significance. Of major importance in the investigations of specific features of the structure and composition of new materials that are responsible for superconductivity are experimental studies of La-phases. The La_2CuO_4 compound of the stoichiometric composition does not possess superconducting properties. These properties do not result only from the replacement of some La atoms by Ba or Sr atoms, as was found by the pioneers. It is sufficient to obtain samples with some lanthanum deficiency, $La_{2-\delta}CuO_4$ or, with oxygen redundancy, $La_2CuO_{4+\delta}$. In other words, there should be some deviation from the stoichiometric composition responsible for the presence of current carriers in the material with certain atomic and electronic structure.

The crystal structures of the high-temperature superconductors $(La,Sr)_2CuO_4$, $(Nd,Ce)_2CuO_4$ and their derivatives $(Nd,Ce)(Nd,Sr)CuO_4$ are shown in Fig.2. In the structure of the first compound with hole conductivity, the (La,Sr)-cations are located in nine- cornered polyhedra, whereas Cu atoms are located in greatly elongated octahedra (the Jahn-Teller effect). In the second electronic conductor the geometry of the cationic arrangement is the same as in the first structure . As for oxygen atoms, half of them occupy totally different sites as compared to the first structure, which leads to the arrangement of (Nd,Ce) cations in cubes, while copper atoms are confined to plane square coordination. The third structure is a combination of the first two. Copper atoms in this structure are located in pyramids, (Nd,Ce) cations are in cubes, while (Nd,Se) cations are in nine-cornered polyhedra.

A new page in the exciting history of superconductivity was connected with the discovery of superconductivity in K doped fullerane K_3C_{60}, T_c = 18 K. Later superconductivity in Rb_2CsC_{60} was reported with T_c = 31.3 K. Such crystals are closely packed spherical molecules of C_{60} and K, Rb or Cs

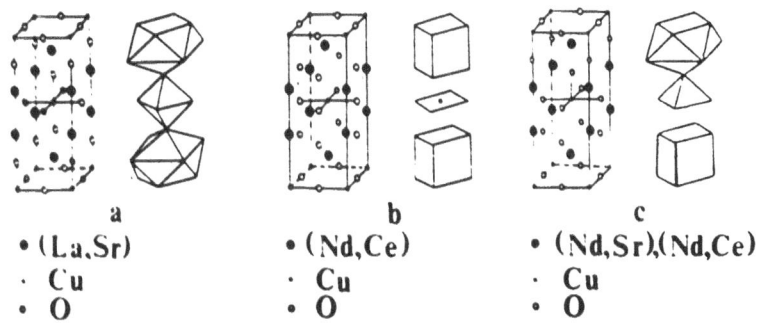

a	b	c
• (La,Sr)	• (Nd,Ce)	• (Nd,Sr),(Nd,Ce)
· Cu	· Cu	· Cu
• O	• O	• O

Fig.2. Crystal structures of models of high-Tc superconductors in spheres and polyhedra:
   a) $(La,Sr)_2CuO_4$ - hole conductivity,
   b) $(Nd,Cl)_2CuO_4$ - electronic conductivity,
   c) $(Nd,Sr)(Nd,Cl)CuO_4$ - derivative structure.

atoms are distributed over cavities typical of close packing.

2. Crystal Structure and the Superconductivity Transition Temperature of $(La,Sr)_2CuO_4$.

A dependence of T_c on the Sr content in $(La,Sr)_2CuO_4$ was established for ceramic materials (Fig.3). As for single crystals, such an unambiguous dependence was not found. Samples with identical Sr contents were found to have essentially different superconducting transition temperatures. The structural studies of single crystals with Sr content ranging from $(La_{0.97}Sr_{0.03})_2CuO_{4-\delta}$ to $(La_{0.88}Sr_{0.12})_2CuO_{4-\delta}$ permitted us to study why there is no regular T_c dependence on Sr content in single crystals.

Samples selected for X-ray diffraction study were machined into spheres, provided their habit permitted it, or they were left as plates. Integrated intensities of the diffraction reflections were measured using AgK_α-radiation, $\lambda=0.5609$Å or MoK_α-radiation, $\lambda=0.71069$Å. The analysis of ϑ-profiles of diffraction reflections did not always allow an unambiguous solution of the problem of twinning in the studied samples. The ϑ-profile of overlapping (10 2 0) and (2 10 0) reflections from domains of a twin crystal of $(La_{0.97}Sr_{0.03})_2CuO_{4-\delta}$ is presented in Fig.4a as an example. This profile can be interpreted as the scattering from a

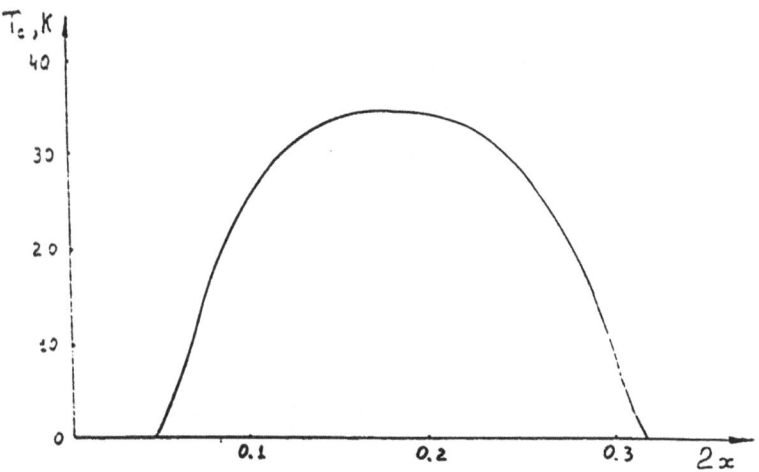

Fig.3. Dependence of Tc on Sr content in $(La_{1-x}Sr_x)_2CuO_{4-\delta}$.

single-domained sample. It should be noted that it is just the ϑ-profiles of reflections that are observed on powder diffraction patterns. The twinning and its character could be reliably determined only from maps of two-dimensional $\omega/2\vartheta$ scanning of reflections. Four maxima due to a doublet of X-ray lines α_1 and α_2 and twin domains are evident on such map for high-angle reflections. In Fig.4b such images

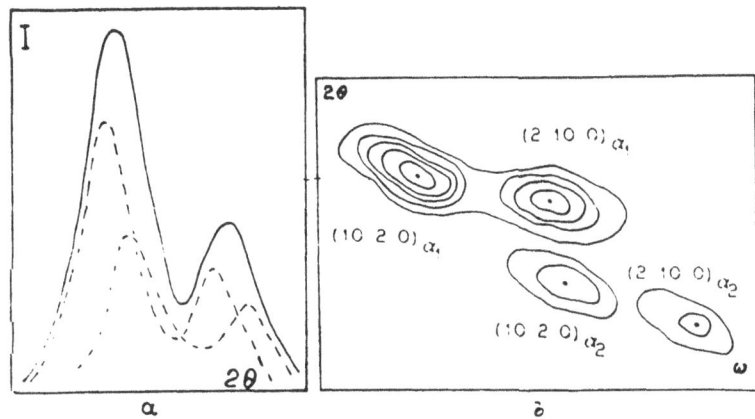

Fig.4. Intensity distribution in diffraction reflection
(2 10 0) and (10 2 0), twin crystal
$(La_{0.97}Sr_{0.03})_2CuO_{4-\delta}$:
   a) one-dimensional observed (solid line) and split
      (dashed line) ϑ-profiles;
   b) two-dimensional $\omega/2\vartheta$-scanning scan of the same
      reflections .

of reflections (10 2 0) and (2 10 0) ($\sin\vartheta/\lambda$ =0.95$\overset{o}{A}^{-1}$) from a twin crystal of $(La_{0.97}Sr_{0.03})_2CuO_{4-\delta}$ are presented. Similar results were obtained for other investigated samples. In all cases there was twinning according to pseudomerohedry law with twinning planes (1 1 0). Experimental ϑ-profiles of diffraction reflections in this case were well approximated by superposition of two Gaussian curves with appropriate displacement and allowance for twin volume ratios (Fig.4a.). Unit cell parameters of all the studied samples were refined from high-angle reflections on an automated RED-4 diffractometer from the data arrays for 25 reflections. The corrections for twinning were introduced into the a and b parameters. These corrections were determined from two-dimensional $\omega/2\vartheta$ scanning of reflections (10 2 0), (2 10 0) and (0 0 12).Twin volume ratios $V_1:V_2$ for each studied sample were determined from the intensities of the split twin diffraction reflections.

Crystals of $(La_{0.94}Sr_{0.06})_2CuO_{4-\delta}$ were an exception, because the analysis of two-dimensional $\omega/2\vartheta$ scanning maps in them did not reveal twinning, while unit cell parameters were, in fact, close to tetragonal. It is these crystals that have the maximum superconducting transition temperature of all of the studied samples. The unit cell parameters of single domain components of all the investigated crystals are listed in Table 1. According to the data presented in Table

Table 1

Unit cell parameters of studied samples of $(La_{1-x}Sr_x)_2Cu\,O_{4-\delta}$.

N	XSA	XMA	a (Å)	b (Å)	c (Å)	V (Å^3)	$\frac{a-b}{a+b} \times 10^4$	$T(K)$
1	–	0.02	5.401	5.359	13.169	381.2	39	<4.2
2	–	0.035	5.377	5.356	13.197	390.1	20	13.5
3	0.03	0.04	5.361	5.348	13.200	378.5	12	10.0
4	0.03	0.04	5.362	5.349	13.200	378.6	12	15.0
5	–	0.05	5.365	5.353	13.211	379.4	11	17.0
6	–	0.065	5.358	5.357	13.229	379.7	1	21.0
7	–	0.065	5.371	5.353	13.210	379.8	17	–
8	0.03	–	5.366	5.353	13.223	379.8	12	15.0
9	0.10	–	5.363	5.348	13.190	378.7	14	15.0
10	0.12	–	5.363	5.338	13.167	376.9	23	5.0

XSA-results of X-ray structural analysis,

XMA-results of X-ray spectral microanalysis.

an increase in Sr content up to x = 0.065 indeed reduced the extent of orthorhombic distortions. However, a further increase in concentration leads to more pronounced orthorhombic distortions. Moreover, during the studies of two samples obtained under the same growth conditions with practically similar compositions $(La_{0.935}Sr_{0.065})_2CuO_{4-\delta}$, it turned out that one of them is tetragonal(within the accuracy of our measurements), while the second one has noticeable orthorhombic distortions. Unfortunately, the quality of these samples did not allow us to carry out a complete X-ray diffraction study. Below we shall discuss these results using complete structural data obtained from other samples. For all the crystals whose quality was high enough for a complete X -ray structural investigation, the integrated intensities of reflections in the reciprocal space hemisphere were measured. Parameters of diffraction experiments and crystal data are presented in Table 2. The

Table 2

Main crystal data on $(La_{1-x}Sr_x)_2CuO_{4-\delta}$ single crystals.

Sample N	(10)	(9)	(8)	(3)	(4)
Sr content, x[a]	0.12	0.10	0.04	0.04	0.04
Size (mm)	r=0.225		0.2 x 0.2 x 0.02 mm^3		
Radiation	Ag K$_\alpha$	Mo K$_\alpha$	Ag K$_\alpha$	Mo K$_\alpha$	Mo K$_\alpha$
Space group	Pbma	Pbma	Abma	Abma	Abma
No. of reflections	1349	1625	1265	1653	1647
No. of independent reflections	340	382	262	300	287
R$_{av}$	0.037	0.051	0.022	0.024	0.047
Reflections violating F centering only					
I(3 0 2)	1853(27)	928(12)	878(13)	856(20)	880(24)
I(5 0 2)	1022(19)	570(11)	527(11)	490(15)	510(19)
I(1 0 4)	587(24)	272(11)	293(11)	268(11)	282(12)
I(3 0 8)	290(3)	183(7)	175(9)	198(12)	173(10)
Reflections violating A and F centering					
I(0 0 15)	420(14)	196(5)	-	-	-
I(0 2 15)	320(36)	142(8)	-	-	-
I(1 1 16)	151(9)	60(4)	-	-	-
I(1 3 16)	140(9)	75(3)	-	-	-
I(1 1 18)	204(6)	142(8)	-	-	-

[a] Sr content determined from X-ray structural analysis.

correction for absorption in a spherical sample was made using the standard programs. The shape of plate crystals was approximated through faceting by crystallographic planes. The absorption in them was taken into account by numerical integrating following the Gaussian technique, covering a sample by a net of partitions of 16x16x16 points. Independent control for absorption was performed from transmission curves for several selected reflections. With allowance for absorption the discrepancies in intensities of transmission curves did not exceed 5%, while, without correction for absorption such discrepancies for platy samples attained 150%. The results for the averaging of reflections equivalent within orthorhombic Laue symmetry are listed in Table 2. About 10 different space symmetry groups were suggested in the literature for orthorhombic La-phases with isomorphous Sr impurities Abma, Ammm, Pccn etc. [Shapligin *et al.*,1979; Kajitani *et al.*, 1987; Onada *et al.*,1987]. That is why we undertook a careful analysis of the systematic absences. It was important to make a correct allowance for twinning. Systematic absences of the general hkl type reflections which pointed out to centering of type A Bravais lattice were established only for three samples. They all have a low Sr content, one is $(La_{0.97}Sr_{0.03})_2CuO_{4-\delta}$ and two are of the same chemical composition $(La_{0.96}Sr_{0.04})_2CuO_{4-\delta}$, but they have different Tc of 10K and 15K. One of these crystals (Tc = 10K) was studied immediately after it had been grown, the second one (Tc = 15K) was first annealed in oxygen. There were absences of reflections of (2h,2k,2l+1) and (2h+1, 2k+1, 2l) types, in which two overlapping reflections (hkl) and (khl) from different twin domains vanish. The reflections (2h+1, 2k, 2l), (2h, 2k+1,2l) and (2h+1, 2k, 2l+1) were used to check independently the twin volume ratios in the studied samples.

As for the crystals with a relatively high strontium content, $(La_{0.88}Sr_{0.12})_2CuO_{3.92}$ [Simonov *et al.*,1988a] and $(La_{0.90}Sr_{0.10})_2CuO_{4-\delta}$, their intensity arrays did not contain systematic absences among general type reflections. Bravais lattices of these crystals are primitive. To illustrate this, several of the strongest reflections are listed in Table 2 that upset the centering of Bravais lattices of single crystals $(La_{0.88}Sr_{0.12})_2CuO_{3.92}$ and $(La_{0.90}Sr_{0.10})_2CuO_{4-\delta}$ and show centering of A lattice of $(La_{0.97}Sr_{0.03})_2CuO_{4-\delta}$ and $(La_{0.96}Sr_{0.04})_2CuO_{4-\delta}$ crystals. A further analysis of absences of reflections in crystals with A lattice and allowance for twinning led to two possible symmetry groups: centrosymmetric Abma and acentric Ab2a. We failed to find deviations from the centrosymmetric variant of the structural model for crystals. A further refinement and analysis of the atomic structures of these crystals was performed within the framework of centrosymmetric symmetry space group Abma. In the case of crystals with primitive Bravais lattice, the analysis of additional absences with account of twinning yielded two mostly probable space groups Pbma and Pbmn, which are centrosymmetric subgroups of the symmetry group Abma. Independent refinement of atomic structural models within the framework of these two symmetry groups and the analysis of appropriate difference density maps suggests space group Pbma for the studied crystals of $(La_{0.88}Sr_{0.12})_2CuO_{3.92}$ and $(La_{0.90}Sr_{0.10})_2CuO_{4-\delta}$.

The starting model for structure refinement of crystals

with a low strontium content (space group Abma) was taken
from [Jorgenson *et al.*,1987], reporting the structure of the
stoichiometric phase for La_2CuO_4. The full-matrix refinement

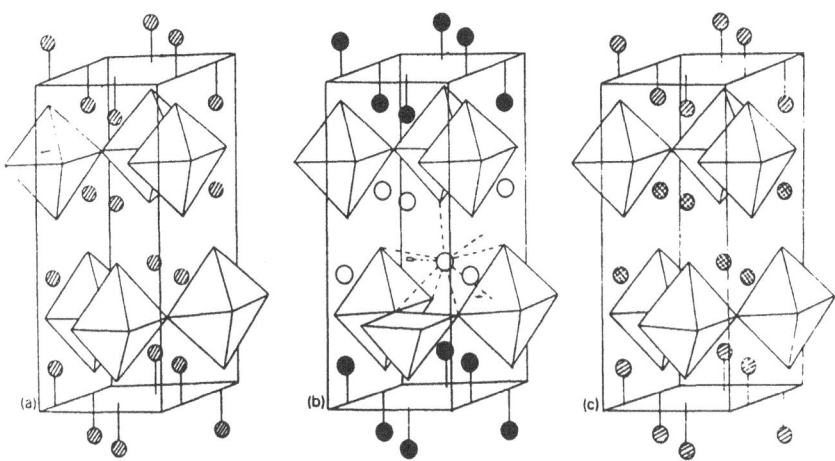

Fig.5. Atomic models of crystals:
  a) $(La_{0.97}Sr_{0.03})_2CuO_{4-\delta}$-even Sr distribution over
     La sites,
  b) $La(La_{0.76}Sr_{0.24})CuO_{3.92}$ - fully ordered Sr
     distribution,
  c) $(La_{0.94}Sr_{0.06})(La_{0.86}Sr_{0.24})CuO_{4-\delta}$ - partially
     ordered Sr distribution.

by least-squares method was made using the PROMETHEUS
program system [Zucker *et al.*,1983], adapted on a NORD
100/560 computer [Muradyan *et al.*,1985]. The chemical
composition was refined from X-ray diffraction data and
composition is in good agreement with the results of X-ray
microanalysis. It should be noted that there is only one
independent La-site in the framework of the space group
Abma, i.e. Sr atoms are evenly distributed over all La
sites, following from symmetry (Fig.5a). Checking of the
site occupancies for Cu, O(1) and O(2) atoms yielded 100%
within the accuracy of our measurements . However, we
suppose that there are oxygen vacancies in the non-annealed
crystals which influence thermal atomic parameters and
vanish completely or partially after annealing in oxygen.
Final R-factors between experimental and calculated moduli
of structure amplitudes for all the three samples are given
in Table 3. Their values testify to the reliability of the

structure parameters which are listed in Table 3. Experimental data from $(La_{0.88}Sr_{0.12})_2CuO_{3.92}$ and $(La_{0.90}Sr_{0.10})_2CuO_{4-\delta}$ suggested two possible symmetry space groups, either Pbma or Pbmn. The La atoms within any of these symmetry groups occupy two crystallographically independent positions. At the first stage of structure refinement all the reflections breaking A centering of the Bravais lattice were excluded. Structural models were refined within sp.gr.Abma, where La atoms occupied one crystallographically independent position. Such refinement of the averaged structure allowed evaluation of the mean Sr content in samples and a rough establishment of the deviation of (La,Sr) atoms from their symmetrical sites. Then reflections violating A centering of the Bravais lattice were included into the structure refinement. The second stage of structure refinement was carried out with a simultaneous analysis of difference electron density maps. A detailed analysis of experimental data showed that the violation of A centering and appearance of additional reflections is first of all due to different probabilities of isomorphous occupation of two crystallographically independent La sites by Sr atoms. Additional contribution to the violation of A centering is made by different displacements of independent (La.Sr) atoms from their pseudotetragonal sites. It is such displacements of basis atoms that are responsible for orthorhombic structural distortions.

The results of structural studies for $(La_{0.88}Sr_{0.12})_2CuO_{3.92}$ sample showed that all the Sr atoms statistically occupy only one lanthanum site. The second lanthanum position in this case is occupied only by La atoms. The structural chemical formula of the sample should then be written as $La(La_{0.76}Sr_{0.24})CuO_{3.92}$. Fig.5b shows a model of this structure which can be split into layers -La-La-Cu-(La, Sr)-(La,Sr)-Cu-, perpendicular to the c axis. It should be noted that we have reliably determined a deficiency of O(22) basis atoms in the structure. These atoms are positioned in layers occupied by La and Sr atoms. Naturally, oxygen losses just in these two layers are due to concentration of divalent Sr atoms in them. Final R-factors between experimental and calculated structure amplitudes are R=2.21 % and Rw =2.32%. Atomic coordinates and other structural parameters of the crystals are presented in Table 3. Results of structural studies of single crystals of $(La_{0.90}Sr_{0.10})_2CuO_{4-\delta}$ are close to the above results. The main difference is the fact that there is incomplete agreement in strontium ordering in this sample. Sr atoms replace La in both of the crystallographically independent sites, though with different probability. Crystallographic

Table 3

Coordinates (x,y,z), effective parameters of thermal motion (B) and occupancy coefficients (q) of basis atoms of the structure of $(La_{1-x}Sr_x)_2CuO_{4-\delta}$.

parameter		Sample				
		(10)	(9)	(8)	(3)	(4)
La(1)	x	0.0058/1 *)	0.0043/2	0.0036/1	0.0030/2	0.0032/3
	y	0.25	0.25	0.25	0.25	0.25
	z	0.11108/5	0.11048/8	0.11102/2	0.11100/2	0.11109/3
	B	0.73/2	0.39/2	0.45/1	0.43/2)	0.46
(La)	q	1.00/2	0.94/2	0.97/1	0.97	0.97
(Sr)	q	0.0	0.06	0.03	0.03	0.03
La(2)	x	0.0041/1	0.0038/3			
	y	0.75	0.75			
	z	0.61119/5	0.61157/8			
	B	0.70/2	0.57/2			
(La)	q	0.76/2	0.86/2			
(Sr)	q	0.24	0.14			
Cu	x	0.0000/5	0.0000/7	0.00	0.0	0.0
	y	0.25	0.25	0.25	0.25	0.25
	z	0.7530/4	0.7501/2	0.75	0.75	0.75
	B	0.72/2	0.52/2	0.38/1	0.36/1	
O(11)	x	0.25	0.25	0.25	0.25	0.25
	y	0.50	0.50	0.50	0.50	0.50
	z	0.756/1	0.756/2	0.7541/9	0.754/2	0.756/8
	B	0.9/1	0.6/2	0.58/6	0.57/7	0.64/9
O(12)	x	0.25	0.25			
	y	0.00	0.00			
	z	0.255/1	0.255/2			
	B	0.9/1)	0.7/2			
O(21)	x	-0.024/2	-0.024/5	-0.017/2	-0.014/3	-0.015/4
	y	0.25	0.25	0.25	0.25	0.25
	z	0.9337/8	0.933/1	0.9336/3	0.9336/3	0.9342/4
	B	1.3/2)	1.1/3)	1.12/8	1.26/9	1.14/9
O(22)	x	-0.025/2	-0.020/5			
	y	0.75	0.75			
	z	0.4345/8	0.434/1			
	B	1.2/2	0.9/2			
	q	0.92/3	1.00/4			
R		0.022	0.025	0.015	0.014	0.017
R_w		0.023	0.024	0.019	0.019	0.022

*) Standard deviation follows the slash.

difference between the two La sites in this structure is realized in the following way: the first La site is 6% statistically occupied by Sr atoms, while the second La site is 14% Sr occupied (Fig. 5c). Thus, the structural formula for this compound should be written as $(La_{0.94}Sr_{0.06})(La_{0.86}Sr_{0.14})CuO_{4-\delta}$. This structural result can be interpreted using two different models. In the first case the single crystal remains uniform in composition and structure. In this model there are two La sites per unit cell, which are characterized by different probability of isomorphous replacement of Sr in them (Fig.5c). The second possible model for interpretation of the obtained structural data is a crystal built up of blocks. Such blocks are uniform in composition and structure, however, the blocks differ in the local chemical composition and Sr distribution. For instance, there can be blocks with complete Sr ordering, whose structure is similar to $La(La_{0.76}Sr_{0.24})CuO_{3.92}$ (Fig.5b). Blocks of other types can have an even distribution of strontium over all crystallographic La sites, similar to crystals with a low Sr content (Fig.5a). Mean chemical composition of the sample is controlled by volume ratio of blocks with different local composition. We failed to determine oxygen defects in $(La_{0.94}Sr_{0.06})(La_{0.86}Sr_{0.14})CuO_{4-\delta}$; only a slight increase of the thermal motion parameter of an appropriate O atom was found.

In crystals with equal Sr content, but differing in its distribution over La sites, we found various orthorhombic distortions. For instance, different degrees of deviations from the tetragonal symmetry were found in two crystals of the same chemical composition $(La_{0.935}Sr_{0.065})_2CuO_{4-\delta}$. Even Sr distribution over all crystallographic La sites reduced orthorhombic structure distortions. Ordering in Sr distribution enhances such structural distortions.

The main interatomic distances for all of the studied samples are listed in Table 4. Mean La-O bonds in $[LaO_9]$-polyhedra are reduced with an increase in Sr content at a particular crystallographic site. This fact agrees with the ionic radii $r(La^{3+})= 1.14\text{Å}$ and $r(Sr^{2+})=1.12\text{Å}$. Vertically positioned O(21) and O(22) atoms have the shortest La-O distances as compared to the anionic environment of La atoms. These O atoms are the farthest from Cu in Cu-octahedra. These distances for all the structures lie within $2.33-2.35\text{Å}$ that is significantly lower than the sum La^{3+} and O^{-2} ionic radii, being 2.46Å. If the structure of La-phases preserved tetragonal symmetry and La and O(21) atoms occupied appropriate symmetrical sites on the same

Table 4

Main interatomic distances in $(La_{1-x}Sr_x)_2CuO_{4-\delta}$ structures.

Distances (Å)	Sample				
	(10)	(9)	(8)	(3)	(4)
La(1)-O(11)	2.593(8)	2.598(7)	2.613(7)	2.608(3)	2.590(2)x2
-O(11)			2.667(8)	2.665(6)	2.683(4)x2
-O(12)	2.662(9)	2.675(8)			
-O(21)	2.590(9)	2.594(8)	2.639(8)	2.656(8)	2.651(7)
-O(21)	2.735(3)	2.737(2)	2.742(1)	2.739(6)	2.741(6)x2
-O(21)	2.902(9)	2.891(8)	2.855(8)	2.834(7)	2.842(8)
-O(21)	2.341(4)	2.346(3)	2.349(3)	2.345(4)	2.337(5)
La(2)-O(12)	2.597(8)	2.596(7)			
-O(12)	2.666(9)	2.676(8)			
-O(22)	2.596(9)	2.624(8)			
-O(22)	2.738(3)	2.742(2)			
-O(22)	2.901(9)	2.873(8)			
-O(22)	2.332(4)	2.346(3)			
Cu -O(11)	1.892(1)	1.895(1)	1.896(1)	1.894(2)x4	1.895(1)x4
-O(12)	1.895(1)	1.895(1)			
-O(21)	2.383(4)	2.416(3)	2.429(3)	2.425(4)x2	2.433(3)x2
-O(22)	2.473(4)	2.431(3)			

vertical axis, these La-O distances would be even shorter. La displacement along the x axis by $+\Delta x(0.031-0.023\text{Å})$ and O atoms towards the opposite side by $-\Delta x(0.107-0.134\text{Å})$ permits an increase of the distance between them and removal of stresses in the structure. Thus, the reason for orthorhombic structural distortions in $(La,Sr)_2CuO_{4-\delta}$ structures is a tendency of La, O(21) and O(22) atoms to be positioned on the bonds allowed from the point of view of crystal chemistry by moving along the x-axis. This conclusion agrees with the known facts of tetragonalization of La-phases doped with isomorphous Sr. However, this is true only in case of even distribution of Sr atoms over all La sites. When Sr is fully ordered one of the sites turns out to be fully occupied by La atoms which are responsible for the appropriate orthorhombic structural distortions. That is why the crystals with a high Sr content $(La_{0.88}Sr_{0.12})_2CuO_{3.92}$ and $(La_{0.90}Sr_{0.10})_2CuO_{4-\delta}$ have a higher extent of orthorhombic distortions. An even distribution of strontium atoms over La sites, as it took place in one of the studied crystals of $(La_{0.94}Sr_{0.06})_2CuO_{4-\delta}$ resulted in the fact that our experimental measurements did not detect orthorhombic structural distortions. In another samples of the same chemical composition but with ordered Sr distribution such distortions were revealed (Table 1).

Cu atoms in all the samples occupy octahedra distorted by Jahn-Teller effect. Mean Cu-O distances are practically equal in different samples. However, $Cu-O(21)$ and $Cu-O(22)$ distances differ significantly: 2.383A and 2.473A. The latter refers to O(22) atoms characterized by noticeably defective site occupancies. Some Cu atoms remain in semioctahedra instead of octahedra due to these oxygen vacancies. In this case Cu atoms go out of the former equatorial plane of two O(11) and two O(12) atoms and approach O(21) atom entering a five coordinate polyhedron. As a result, an average Cu site within the crystal which we determine is displaced towards O(21) atom. A similar tendency although less vivid is observed also in crystals of $(La_{0.90}Sr_{0.10})_2CuO_{4-\delta}$, which is another evidence of probable oxygen vacancies in O(22) site in this compound as well. However, the accuracy of our diffraction experiment does not allow us to determine these vacancies of O(22) atoms.

The analysis of the above literature references first of all pertaining to structural investigations of ceramic samples of La- phases suggests that an increase in isomorphous strontium content stabilizes the tetragonal phase of the structure and lowers the temperature of its transition to the orthorhombic phase. The results of our X-ray structural studies of single crystals demonstrate that such a conclusion is true only in the case of an even Sr distribution over all La sites in a particular structure. The main novelty of our results consists in the fact that, depending on the technique of growing, Sr atoms are distributed differently over La sites. This distribution can be equally probable over all La sites, or Sr atoms can be partially or fully ordered. If the ordering is full, all Sr atoms are distributed only over half of all the La sites. The second half of the sites is 100% La occupied.

When Sr content is low, as is shown in Table 1, an even Sr distribution within the structure is most probable. An even Sr distribution in $(La_{1-x}Sr_x)_2CuO_{4-\delta}$ is retained up to x=0.065. Provided Sr content exceeds this value, the ordering is more probable. It goes without saying that the above limit is not fixed and can vary depending on the technology of preparing crystals and their thermal treatment. Ordering in Sr distribution is accompanied by greater orthorhombic structural distortions. It should be noted that these distortions are due not only to Sr content but also to the character of Sr distribution over La sites. In fact it is revealed in the changing of the probability of the occupation of the two crystallographically independent lanthanum sites by Sr. It is due to this reason that orthorhombic distortions are considerable in $La(La_{0.76}Sr_{0.24})CuO_{3.92}$ crystals where Sr is fully ordered. The analysis of orthorhombic distortions in $(La_{0.935}Sr_{0.065})_2CuO_{4-\delta}$ and $(La_{0.90}Sr_{0.10})_2CuO_{4-\delta}$ shows that they are greater in the former crystal, although mean Sr concentration in it is lower. This is due to the fact that ordering of the Sr atoms in the first crystal is higher than in the second one. We stress once again that the limiting Sr concentration, beyond which it is inevitably ordered, depends on the growing technique and the thermal treatment (annealing, quenching, oxygen pressure etc.). That is why Sr ordering or even distribution can occur over a wide range of Sr contents [Simonov et al.,1990].

From the point of view of the presence of superconductivity and the superconducting transition temperature, Sr ordering over La site plays a negative role. An even Sr atom distribution over all La sites provides for an increase in their concentration when all or almost all oxygen atoms are retained in the structure. Full Sr ordering increases Sr concentration in certain layers of the structure by a factor of 2 as compared to average Sr concentration in the whole sample. It is this doubled concentration that stimulates the formation of oxygen vacancies in the same layers of the structure. Some oxygen losses are a direct cause of a decrease the superconducting transition temperature or a complete loss of superconductivity. A difference in the technologies for obtaining ceramic samples and single crystals leads to different degrees of Sr ordering in them. Then as a rule, the ordering degree is higher in single crystals, oxygen losses are more probable and a lowering of Tc is inevitable. This accounts for a higher Tc in ceramic samples as compared to single crystals when Sr content is the same.

The most reliable parameter of crystals of La-phases that correlates with Tc is the extent of orthorhombic distortions. It depends on Sr content in the sample and on the character of Sr distribution over La sites.

3. Atomic Structure and T_c of Y-phases of High-Temperature Superconductors with Various Oxygen Content.

The second superconducting compound discovered was $YBa_2Cu_3O_7$ with T_c = 93 K, which exceeds the nitrogen boiling temperature and puts all the technical applications of superconductivity to a qualitatively new level. Depending on the conditions of synthesis and treatment of samples,

Y-phases with different oxygen contents and T_c were obtained. In such cases the maximum transition temperature

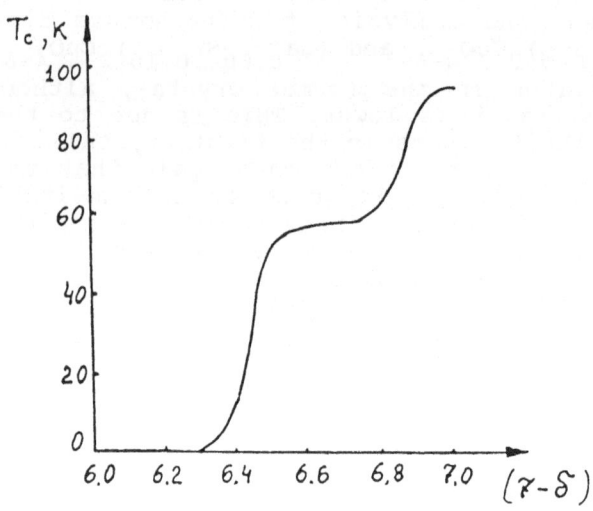

Fig.6. Dependence of T_c on oxygen content in $YBa_2Cu_3O_{7-\delta}$.

T_c = 93 K was observed in orthorhombic samples with oxygen content close to 7 atoms per chemical formula, Fig.6. Tetragonal $YBa_2Cu_3O_6$ crystals do not undergo a superconducting phase transition. A phase with the composition $YBa_2Cu_3O_{6.5}$ with T_c of the order of 60 K was found using electron microdiffraction technique. Fig.7 shows the atomic structures of two orthorhombic and a tetragonal phase of the compositions, respectively, $YBa_2Cu_3O_7$, $YBa_2Cu_3O_{6.5}$ and $YBa_2Cu_3O_6$. Of special interest are Y-phases with intermediate oxygen content. In this case the question of oxygen atom arrangement in the appropriate crystal is the main one.

Accurate structural studies of $YBa_2Cu_3O_{7-\delta}$ single crystals with O content ranging from 6.24 to 6.97 per unit cell have been carried out. The oxygen content was varied by dosage annealing in an argon atmosphere . All the studied samples were, in fact, twins with (1 1 0) and (1 $\bar{1}$ 0) twinning planes. As a result of twinning, each reflection is split into four components, except reflections of the 001 type which are not influenced by twinning, and reflections of the hhl type, in which the components overlap regularly.

Fig. 8 shows a part of reciprocal space at a fixed l. The orientations of the four twin partners are given.

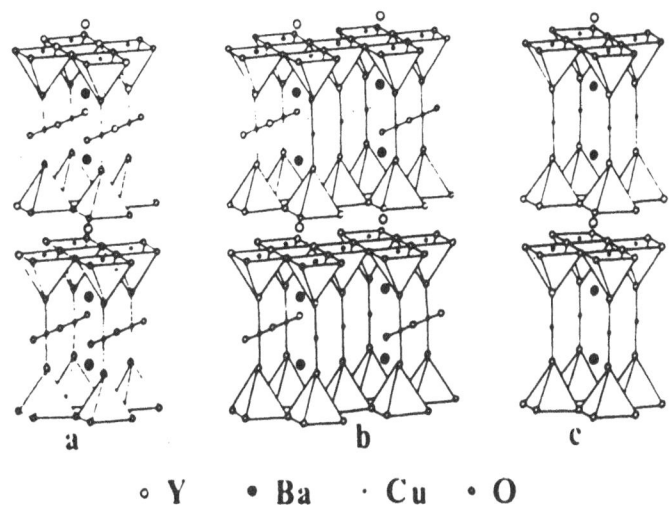

∘ Y   • Ba   · Cu   • O

Fig.7.   Crystal structures of compounds $YBa_2Cu_3O_{7-\delta}$ with various oxygen content:

a) orthorhombic phase I $YBa_2Cu_3O_7$ with Tc = 93K,

b) orthorhombic phase II $YBa_2Cu_3O_{6.5}$ with Tc = 60K,

c) tetragonal phase $YBa_2Cu_3O_6$ which does not undergo a superconducting phase transition.

The two-dimensional scanning technique was applied for experimental determination of the twin volume ratios. The two- dimensional scanning of hk0 reflections was carried out for such values of the ψ angle, when the vector c^* was located in the equatorial plane of the goniometer. This allowed minimization of X-ray absorption in the sample. Fig.9 shows intensity distributions of the reflections 200, 330 and $3\bar{3}0$, in the projection of the Eyler angles of the goniometer. The resolution of the components over ϑ is 0.1° due to a difference between the a and b parameters.

Resolution over χ is about 0.8° due to the fact that the a and b directions of different components do not coinside in the space. In order to improve resolution over ϑ one should essentially reduce the detector aperture, that leads to a decrease of the intensity observed and less reliable data. Therefore, the most informative reflections about twin volume ratios are not the axial h00 ones but diagonal hh0 reflections. Single components in them have the same ϑ

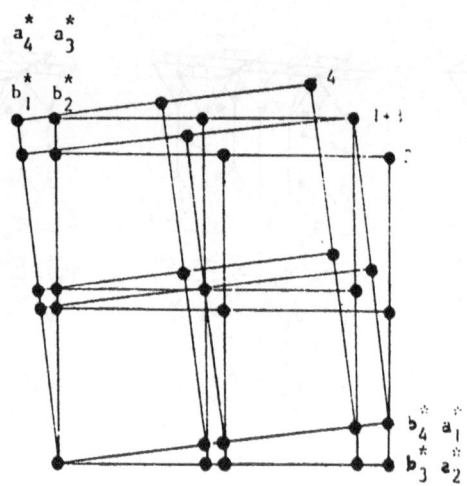

Fig.8. Mutual orientation of the four twin partners of $YBa_2Cu_3O_{7-\delta}$ crystals.

values, which ensures scanning with larger apertures. The good resolutions of the partners over χ is achieved due to reduction of the detector window height without a significant loss in intensity.

The distribution of intensity of the reflections 330 and $3\bar{3}0$ in the projectiopn of the angles ω and χ at a fixed ϑ allows determination of the volume ratios for the four domains of the twin from the relation of integrated intensities of site components. The integrated intensities of the central components, which are a sum of two domains, show the ratio of the crystal volumes in which different twinning planes 110 and $1\bar{1}0$ take part. The data array obtained from twinned samples indicates that the volume ratios of the domains related by a separate twinning plane is kept close to unity, while the ratio of the volumes in which different planes take part can be significantly different. For instance, there were such crystals studied in which the volumes of such domains exhibited a 20-fold difference.

In the crystal with 6.95 oxygen atoms per unit cell the reflections from different twin domains are practically fully resolved,as shown in Fig.10. Integral intensities of these reflections are readily measured because they do not overlap, however, they can be easily approximated by measuring distances and relative intensities from twin

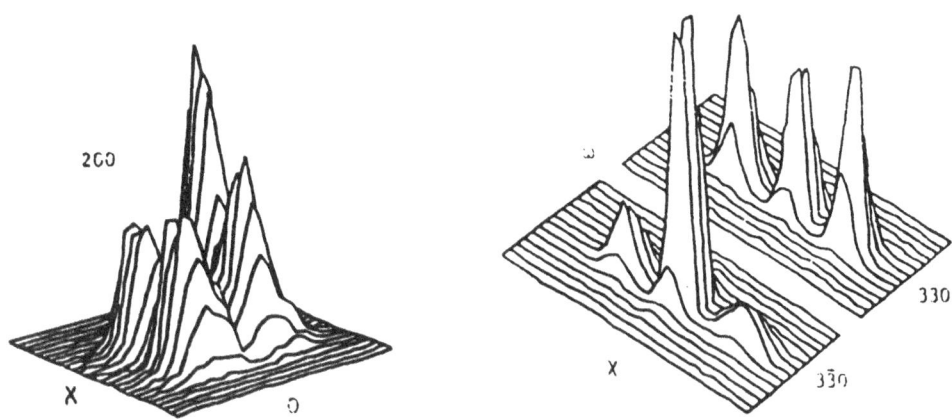

Fig.9. Two-dimensional intensity distribution of 200, 330

and 330 reflections twinned crystal of $YBa_2Cu_3O_{7-\delta}$.

domains. 4 0 0 and 0 4 0 reflection profiles for crystals with 6.46 and especially 6.24 oxygen atoms cannot be interpreted so easily by superimposing on one another the reflections from the orthorombic twin domains. These complex profiles could be interpreted only by adding the scattering from regions of local tetragonal symmetry. Thus, it was found that there is different order in oxygen atom distribution in the sample which we regarded as a single crystal. These orders result in the appearance of three types of regions rather large in size, that have different local symmetry. Two types are the orthorombic twin domains, and the third one is a tetragonal symmetry region. These regions are characterized by different unit cell parameters. Such a difference is best manifestad in a sample with 6.24 oxygen atoms. As is well known from the literature, there is a correlation between oxygen content in a sample of 1-2-3 phase and the period c of this phase. Thus, we can assume that the regions differing in symmetry unit cell parameters have different composition (namely, oxygen content). We have determined mean oxygen content in the sample from statistical O atoms site occupancies. According to the unit cell volume of tetragonal symmetry regions, the composition of these regions can be considered as $YBa_2Cu_3O_6$ from crystal

chemical considerations. The volume ratio of tetragonal and orthorombic regions is calculated from the reflection intensities. These data taken as a whole are sufficient to determine the local chemical composition of orthorombic symmetry regions. In our case for a crystal with a mean

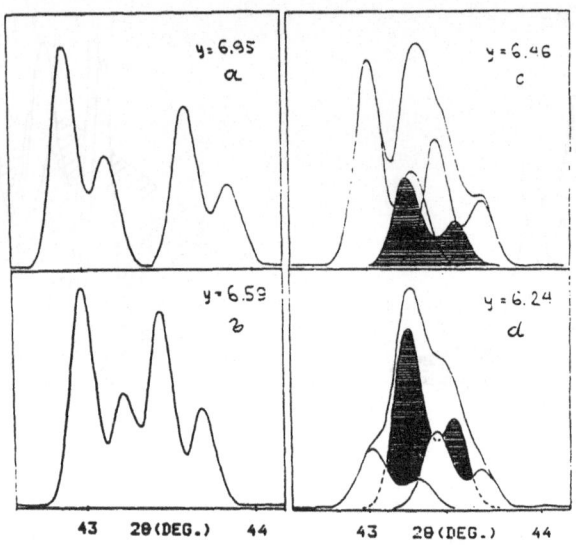

Fig.10.(400) and (040) reflection profiles of $YBa_2Cu_3O_{7-\delta}$
crystals with various oxygen content :
a) 6.95; b) 6.59; c) 6.46; d)6.24. Components
of the scattering by tetragonal regions are shaded.

oxygen content of 6.24 the estimation of the local chemical
composition of orthorombic regions yields 6.56 oxygen atoms
per unit cell. Within the accuracy of our X-ray data this
value is in good agreement with the composition of so-called
orthorombic phase II $YBa_2Cu_3O_{6.5}$, to which a superconducting
phase transition at T_c= 60 K is related.

Thus, on the whole, for the thermodynamically
equilibrium crystals of Y-phases the structures of compounds
with different oxygen content can be regarded as follows. A
strictly tetragonal phase does not undergo a superconducting
phase transition and has the chemical composition $YBa_2Cu_3O_6$.
If such a crystal is enriched with oxygen,
-Cu-O-Cu-O-chains appear in the plane of Cu atoms. Each such
chain alternates with a -Cu-Cu- chain. Thus, an orthorombic
phase II with T_c=60K is formed, with the local chemical
composition $YBa_2Cu_3O_{6.5}$. With an increase in oxygen content
the relative value of this phase as compared to the initial
tetragonal phase $YBa_2Cu_3O_6$ is also increased. The crystal

with the mean total oxygen content 6.5 atoms per unit cell, is fully composed of orthorhombic phase II and undergoes a superconducting phase transition at T_c=60K. A further increase in oxygen content results in the appearance of an orthorhombic phase I of $YBa_2Cu_3O_7$ with T_c=93K. Superconductivity (resistance fall) in the entire sample takes place when regions of orthorhombic phases of $YBa_2Cu_3O_{7-\delta}$ become connected. The connection of structure blocks with different oxygen content is shown in Fig.11.Our patterns of equlibrium atomic structure of $YBa_2Cu_3O_{7-\delta}$ crystals with different oxygen content [Molchanov et al.,1989] were later confirmed by electron microscopy studies of Y-phases [Hiroi et al.,1989].

In the above scheme of crystal structure of $YBa_2Cu_3O_{7-\delta}$ single crystals with various oxygen content there is no place for superconducting phases with tetragonal symmetry. The analysis of the literature data [Arabi et al.,1992; Kajitani et al.,1988; Miceli et al.,1988] on such phases has shown that they ehxibit either Cu atom deficiency or replacement of some copper atoms by Al, Co or Fe atoms. We showed way back in 1988 in our paper [Simonov et al.,1988b] that the tetragonal symmetry of the superconducting phase of $YBa_2Cu_{2.862}O_{6.62}$ is statistical. Its local symmetry remains orthorhombic. The regions of orthorhombic ordering are the blocks which are ordinary for twinned $YBa_2Cu_3O_{7-\delta}$ crystals and are rotated relative to one another exactly by $90°$. In the above considered structure of $YBa_2Cu_{2.862}O_{6.62}$, in the z = 0 plane there are 86 Cu(1) atoms and 14 vacancies per 100 unit cells. The same plane contains 70 O(1) atoms.Each of the Cu(1) atoms is bound to two O(3) atoms, which lie along the z axis above and below the corresponding Cu(1) atom. The ratio of the number of Cu(1) and O(1) atoms unavoidably leads to the result that some of the Cu(1) atoms remain in dumbbells, while the coordination of the rest is made up to squares of O(1) atoms. Estimates give 17% Cu(1) atoms in dumbbells and 69% in squares. If a CuO_4 square is not linked by a common oxygen vertex with another square, then it must unavoidably make contact with a vacancy of a Cu(1) atom. The CuO_4 squares cannot be disconnected, because there are too few copper vacancies in the structure for this. The squares must be joined in chains of the form

$$O(3) \qquad\qquad O(3)$$
$$|\qquad\qquad\quad |$$
$$-\,O(1)\,-\,Cu(1)\,-\,O(1)\,-\,Cu(1)\,-\,O(1)\,-$$
$$|\qquad\qquad\quad |$$
$$O(3) \qquad\qquad O(3)$$

which are cut off at vacancies of Cu(1) atoms . If the vacancies of Cu(1) atoms lie only at the ends of the chains and all of the oxygens in the plane z=0 are in the chains, then we can estimate the mean length of such chains. However, the copper vacancies can form natural accummulations and then the mean length of the chains increases. The chains in the structure form regions of orthorhombic order. Within

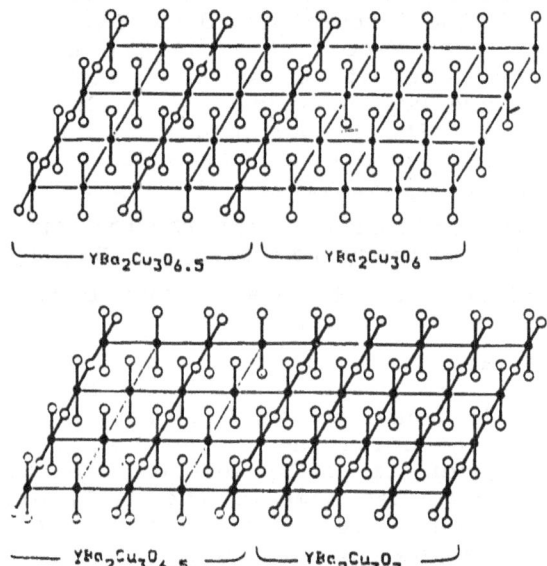

Fig.11.a) Connection of $YBa_2Cu_3O_6$ and $YBa_2Cu_3O_{6.5}$ blocks in structures containing 6.0 - 6.5 oxygen atoms per unit cell,

b) connection of $YBa_2Cu_3O_{6.5}$ and $YBa_2Cu_3O_7$ blocks in structures with oxygen content 6.5 - 7.0 atoms per unit cell.

these regions, which exist in a tetragonal superconducting crystal, the atomic structure completely repeats the structure of the oxygen ordered orthorhombic superconducting phase of the composition $YBa_2Cu_3O_{7-\delta}$.

The above regions alternate in orientations turned through $90°$ in the crystal. This circumstance leads to the mean tetragonal symmetry over the crystal. Fig.12 shows the section of the plane z=0 of the tetragonal unit cell of the investigated single crystal with the statistical population of the positions by Cu(1) and O(1) atoms. The scheme of regions of the orthorhombic ordering is shown in Fig.12b. The atoms O(3) lying above and below the copper are not shown in Fig.12b. The atoms O(1) together with O(3) form

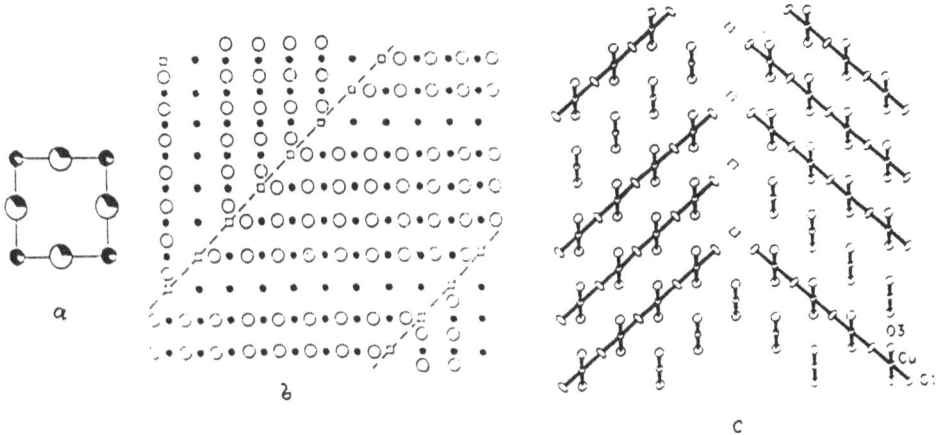

Fig.12. Section z = 0 of the unit cell of structure of $YBa_2Cu_{2.862}O_{6.62}$:
   a) statistical populations of sites by Cu(1) and O(1) (large circles) atoms,
   b) scheme of packing of orthorhombic blocks in a "tetragonal" crystal,
   c) concentration of vacancies at Cu(1) sites at the twin boundary.

squares about the Cu(1) atoms (Fig.12c). Chains of squares are formed by sharing of the O(1) atoms. For an exact imitation of tetragonal symmetry the twin boundaries should be formed by Cu(1) vacancies or the atoms that replace copper and thus do not exhibit the Jan Teller effect.

For independent verification of the existence of regions of orthorhombic ordering in the tetragonal single crystal, we used analysis of the profiles of reflections. The presence of blocks must lead to broadening of the reflections. We investigated the profiles of reflections of various types. The measurements were made in a two-crystal diffractometer in the scheme of symmetrical Bragg-Laue diffraction. For monochromatization and collimation of AgK_α radiation we used the (220) reflection from a germanium crystal, the monochromated beam divergence being about 10". The investigated tetragonal single crystal measuring 0.310 x 0.212 x 0.009 mm^3 was described in detail in [Simonov *et*

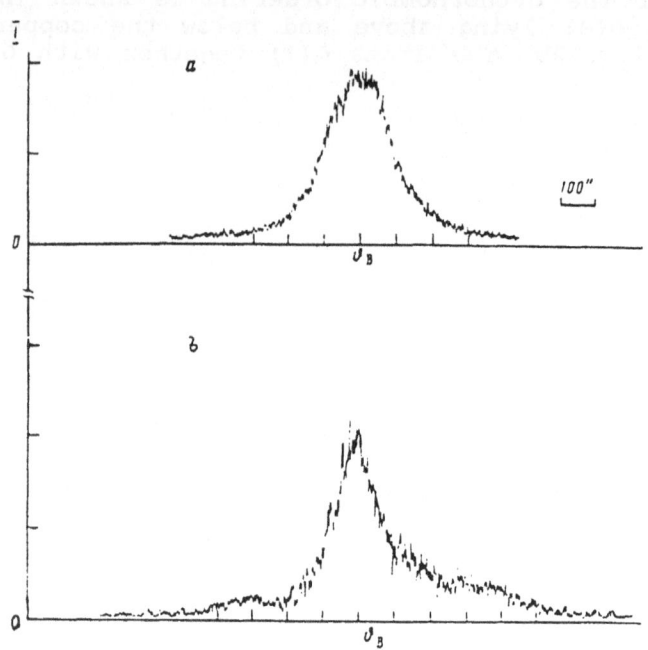

Fig.13.Profiles of X-ray diffraction reflections of a
"tetragonal" superconducting single crystal of
$YBa_2Cu_{2.862}O_{6.62}$:
a) R(200); b) R(220). Primary beam divergence is 10".

al.,1988b]. In Fig.13 we give the most strongly differing profiles. The cross section of the reciprocal lattice points (200) and (220) were obtained with $\vartheta/2\vartheta$ scanning. The profile of the (200) reflection is broadened, and the (220) reflection is seen to possess satellites. The values of the extinction lengths of the (200) and (220) reflections enable us to interpret diffraction in the framework of the kinematical theory. Estimation of the mean size of the blocks (superperiod) from the positions of the satellites gives 160±40Å. The positions of the satellites of the reflection of the type (hh0) indicate that the regions of orthorhombic order in the tetragonal crystal are packed according to the scheme in Fig 12, i.e., their boundaries are parallel to the (110) planes. The chains extend at 45° to these boundaries and their mean length is estimated to be about 60 links. The presence of regions of orthorhombic order in superconducting tetragonal single crystals enables us to draw the following conclusions about the correlation of the structure with the superconductivity for the whole set of Y-phases.

The structural elements determining the superconductivity are the regions of orthorhombic symmetry, which are ordered with respect to the oxygen atoms. The local chemical composition of the most stable phases is $YBa_2Cu_3O_7$ and $YBa_2Cu_3O_{6.5}$. The so-called tetragonal phases of the composition $YBa_2Cu_3O_{7-\delta}$ consist of blocks of orthorhombic symmetry which are rotated relative to one another exactly through 90° and they do not exhibit the usual twin reflection splitting in the diffraction pattern. The above reported [Simonov et al.,1988b] structure of $YBa_2Cu_{3-x}O_{7-y}$ crystals with various oxygen contents is independently confirmed in the course of studies of these compounds by NMR methods [Lutgemeier et al.,1992].

4. Atomic Structure of $Tl_2Ba_2CaCu_2O_8$ Above and Below the Superconducting Phase Transition.

The family of Tl-phases is the largest one among high-temperature superconductors. It can be divided into two classes, with the general chemical formulae $TlBa_2Ca_{n-1}Cu_nO_{2n+3}$ and $Tl_2Ba_2Ca_{n-1}Cu_nO_{2n+4}$. The main structural difference between these two classes is that in the first case there are single TlO layers, while in the second one there are double $(TlO)_2$ layers. The chemical composition of all these compounds can be expressed by a common formula $Tl_mBa_2Ca_{n-1}Cu_nO_{1n+m+2}$, where $m = 1,2$ and $n = 1,2,3,4$. A conditional notation m2(n-1)n was adopted for Tl-phases which indicates the number of appropriate cations in the chemical formula. Fig.14 presents schematically the atomic structures of various Tl-phases. The easiest feature to fix which identifies compounds of Tl-phases is the c period of the unit cell. The c periods are 9.69, 12.73, 15.87, 19.10, 23.15, 29.39, 36.26, 42.00 Å and correspond, respectively, to compounds 1201, 1212, 1223, 1234, 2201, 2212, 2223, 2234. In this case the a and b periods practically coincide in all the structures and are about 3.85 Å.

A specific feature of superconducting Tl-phases is the fact that they are nonstoichiometric in cations. This influences the number of current carriers in the compound and determines the T_c. For instance, in the 1212 compound

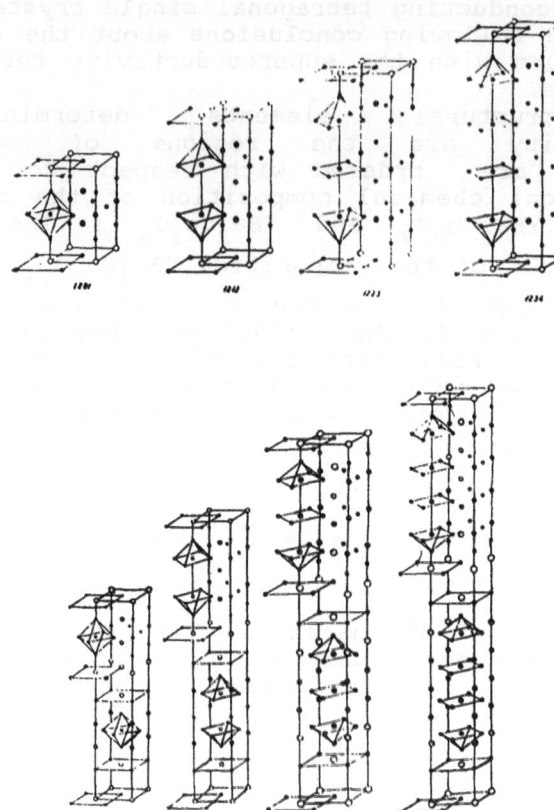

Fig.14. Crystal structures of Tl-phases:
   a) $TlBa_2Ca_{n-1}Cu_nO_{2n+3}$: 1201, 1212, 1223, 1234;
   b) $Tl_2Ba_2Ca_{n-1}Cu_nO_{2n+4}$: 2201, 2212, 2223, 2234.

some Ca cations are replaced by Tl cations. In 2212 crystals some Tl atoms are replaced by copper and, like in the case of 1212 Tl atoms partially replace Ca atoms. Similar replacements were found in 1223 and 2223 compounds. Isomorphous cationic replacements in Tl-phases by cations of higher or lower valencies enable variations of the oxidation degree of Cu atoms , which is responsible for superconductivity and the value of T_c.

    Let us consider an X-ray structural study of a single

crystal thallium phase $Tl_2Ba_2CaCu_2O_8$ (2212) Tc=110K at three temperatures before the phase transition (T=290K, 160K, 130K) and at two temperatures in the superconducting state (T=90K, 60K). A preliminary communication about results of structural studies at T= 290K, 160K and 60K has been published [Blomberg et al.,1992]. The choice of $Tl_2Ba_2CaCu_2O_8$ as an object for investigation resulted from a high T_c=110K and a five-fold coordination of Cu atoms, similar to $YBa_2Cu_3O_7$. In contrast to the latter, however, $Tl_2Ba_2CaCu_2O_8$ crystals are not twins, which ensures the accuracy and reliability of the structure parameters obtained from diffraction data.

The technique of growing $Tl_2Ba_2CaCu_2O_8$ samples was reported in [Muradyan et al.,1991]. X-ray diffraction data were collected from a round sample 0.226mm in diameter, on a Huber 5042 four-circle diffractometer (graphite monochromated, MoK_α-radiation, $\vartheta/2\vartheta$ scan, $sin\vartheta/\lambda \leq 1.22 A^{-1}$),with an attached Displex 202 helium cooling device, close type,with two cooling cycles(Department of Physics, University of Helsinki, Finland). The sample was mounted on a quartz thread using bee wax which prevented any influence of stress on the sample at low temperatures. Unit cell parameters of the tetragonal lattice were found by the least squares method from 18-20 Friedel reflection pairs located in the region $42^o \leq 2\vartheta \leq 56^o$. The intensities of test reflections in experiments at 296K and 160K varied within ±1%. The experiment at 60K was conducted in two stages, six days each, between these periods the sample was kept at room temperature. The total changes of test reflection intensities in this case attained ±4%.

The authors of [Parkin et al.,1988a; Parkin et al., 1988b; Begers et al.,1989] carried out electron diffraction studies of Tl-superconductors and found that almost all the phases exhibited modulated structures, which are manifested in satellite reflections. We attempted to find satellite reflections in the X-ray diffraction patterns of our single crystals at room temperature. However, scrupulous scanning of the volume of reciprocal space, limited by planes (3 0 0), (0 3 0) and (0 0 4) and step scan 0.01, did not yield positive results. This agrees with conclusions given in [Subramanian et al.,1988], where the authors studied single crystals of 1223 and 2212 using synchrotron radiation at 125K; however, they did not find satellite reflections either.

The systematic absences of reflections with h+k+l=2n+1 and no other regular absences in all our arrays are consistent with the space group I4/mmm . The main crystal data on the sample studied as well as characteristics of the diffraction experiments are listed in Table 5.

The structure model was refined by the least squares method using the PROMETHEUS [Zucker et al.,1983] and JANA [Petricek et al.,1988] program systems. In the starting model for the refinement, the atoms were given positions at the ideal sites of the space group I4/mmm according to the data reported in [Onada et al.,1989; Gao et al.,1989; Cox et al.,1988], except for the O(31) atom, which was shifted from the symmetric site (0.5,0.5,z) to the site (0.5+δ, 0.5+δ, z). Refinement of site occupancy coefficients showed that, besides the Ca atoms, heavier Tl(2) atoms occupy the (0,0,0) site. The relation Ca: Tl in that isomorphous mixture was 7:1, approximately and varied within three standard deviations in the course of refinement from the data obtained at different temperatures.

In this model anisotropic amplitudes of Tl(1) atom, however, thermal motion in the ab plane had anomalously large values, whose physical meaning for such heavy atoms was doubtful. In addition, difference electron density maps contained residual peaks which showed displacement of Tl(1) atoms from the symmetric (0.5, 0.5, z) site to the (0.5+δ, 0.5+δ, z) site. Refinement of the structural model with displaced Tl(1) atoms was made by step scan [Muradyan et al.,1989] due to a strong correlation (0.9) between Tl(1) atom displacement and its anisotropic thermal parameter B11. The coordinates of the Tl(1) atom were changed with the step x=y=0.001 and all of the other structure parameters were refined for each value, when the dependence of R-factors on Tl(1) atom position was plotted, the minimum R-factors formed a rather wide plateau, whose center was taken as the solution.

We had carried out an earlier refinement of the structure from diffraction data,which was obtained from the same crystal at room temperature using AgK_α-radiation [Muradyan et al.,1991]. The difference electron density map, from which Tl(1) atom at a shifted site was subtracted, showed a residual density peak at the site (0.5,0.5,z). The analysis of this peak and its environment suggested that some Tl(1) atoms at this site are isomorphously replaced by Cu(2) atoms, which are located at the site (0.5,0.5,z) and not shifted from it [Muradyan et al.,1991]. Difference electron density maps that were calculated in that work from the arrays obtained at various temperatures, showed residual peaks as well. According to the literature, there were suppositions that vacancies at the Tl(1) atoms sites could be occupied by Ca atoms, however, we do not think it is possible. This point is discussed in detail in [Muradyan et al.,1991]. The occupancy of the site (0.5+δ, 0.5+δ, z) by Tl(1) atoms, determined from experimental data for different temperatures, in our case, changed by a value exceeding

Table 5. Crystallographic data for $Tl_2Ba_2CaCu_2O_8$ at different temperatures

Temperature		60K	90K	130K	160K	296K
a,Å		3.8486(3)	3.8490(3)	3.8498(3)	3.8501(3)	3.8542(3)
c,Å		29.232(2)	29.241(2)	29.249(2)	29.261(2)	29.317(2)
Tl1	q	0.902(6)	0.888(6)	0.881(6)	0.886(6)	0.887(5)
	x,y	0.522(3)	0.518(6)	0.524(2)	0.525(2)	0.518(8)
	z	0.21372(3)	0.21372(3)	0.21370(3)	0.21369(3)	0.21381(3)
Cu2	q	0.098	0.112	0.119	0.114	0.113
	x,y	0.5	0.5	0.5	0.5	0.5
	z	0.2122(4)	0.2117(4)	0.2120(4)	0.2122(3)	0.2108(6)
Ba	x,y	0	0	0	0	0
	z	0.12148(2)	0.12143(1)	0.12147(2)	0.12146(2)	0.12162(2)
Ca	q	0.874(3)	0.874(3)	0.870(3)	0.872(3)	0.880(3)
	x,y,z	0	0	0	0	0
Tl2	q	0.126(3)	0.126(3)	0.130(3)	0.128(3)	0.120(3)
	x,y,z	0	0	0	0	0
Cu1	x,y	0.5	0.5	0.5	0.5	0.5
	z	0.05392(4)	0.05393(3)	0.05394(4)	0.05394(3)	0.05398(3)
O1	x	0	0	0	0	0
	y	0.5	0.5	0.5	0.5	0.5
	z	0.0524(1)	0.0526(1)	0.0525(1)	0.0525(1)	0.0525(1)
O2	x,y	0.5	0.5	0.5	0.5	0.5
	z	0.1457(2)	0.1453(2)	0.1454(2)	0.1456(2)	0.1459(2)
O3	x	0.68(1)	0.672(6)	0.672(7)	0.65(1)	0.656(8)
	y	0.5	0.5	0.5	0.5	0.5
	z	0.281(1)	0.2833(7)	0.2826(9)	0.280(1)	0.282(1)
O31	x,y	0.544(6)	0.551(3)	0.549(4)	0.550(4)	0.546(4)
	z	0.281(1)	0.2810(6)	0.2814(7)	0.2816(7)	0.2804(7)
R ,%		3.01	2.61	2.81	2.72	2.71
R_w,%		3.16	2.78	3.06	2.99	2.66

standard deviations. The reason for that is a strong correlation between Tl(1) atom displacement from its symmetric site, occupancy coefficient, and its parameters of the thermal motion. The true error in the determination of the occupancy parameter exceeds the standard deviation. Since we made all the measurements on one sample, we used an average value of the $(0.5+\delta, 0.5+\delta, z)$ site occupancy by Tl(1) atoms for all the five experiments and an average coefficient of Ca atom replacement by Tl(2) atom at the site (0 0 0) for final calculations. The main interatomic distances in the structure at the five different temperatures are listed in Table 6.

First of all, we shall compare the crystal-chemical state, obtained in this work with those, reported in the literature. There are several publications reporting results of X-ray structural studies of the phase 2212 [Onada et al.,1988; Gao et al.,1989; Cox et al.,1988; Hewat et al.,1988; Sequeira et al.,1988; Morosin et al.,1991].

1) Some Tl atom sites, according to the authors of [Cox et al.,1988; Morosin et al.,1991] are vacant, [Onada et al.,1989] these sites are reported to be isomorphously populated by Ca atoms, and, finally, in [Gao et al.,1989; Hewat et al.,1988] they are occupied by Cu atoms. It is hardly probable, from crystal-chemical considerations, that the Tl atom sites can be isomorphously replaced by Ca atoms. Tl atoms are replaced instead by Cu atoms. In this case, the Tl atoms themselves are shifted from the ideal sites, while the Cu atoms, which replace them, occupy symmetrical sites.

2) The authors of [Subramanian et al.,1988; Gao et al.,1989; Morosin et al.,1991] suggested a model without any Tl atom displacement from its ideal site. In this case there is a significant anisotropy of Tl atom thermal motion, and the rotation ellipsoid of Tl atom thermal motion is squeezed along the c-axis. The crystal-chemical analysis of thallium – oxygen compounds shows that its close oxygen coordination normally contains three pairs of O atoms; the distances for the first pair being $\approx 2.0 \overset{\circ}{A}$, for the second $\approx 2.3 \overset{\circ}{A}$, and for the third $\approx 2.8 \overset{\circ}{A}$. That is why a model with Tl atoms displaced from their symmetrical sites is preferable.

3) The third specific feature of the structure 2212 is unanimously accepted by all of the above authors and is manifested in isomorphous occupation of Ca sites by Tl atoms. The replacement of some Ca atoms by Tl atoms, as well as vacancies at the main sites of Tl atoms or their occupation by Cu atoms perturb the stoichiometry of this compound and make a certain concentration of carriers. Experimental studies of photoelectronic spectra and the Hall effect unambiguously confirm that holes are charge carriers in this compound. The chemical formulae of the single crystal that we studied and refined from the X-ray

Table 6. Selected interatomic distances for $Tl_2Ba_2CaCu_2O_8$ at different temperatures.

r,Å		n	60K	90K	130K	160K	296K
Tl -	O2	x1	1.992(6)	2.003(6)	2.002(6)	1.997(6)	1.993(6)
-	O3,O31	x1	1.96- 2.12(3)	1.98- 2.16(2)	1.98- 2.15(3)	1.99- 2.06(3)	1.96- 2.11(3)
-	O3,O31	x4	2.17- 3.24(3)	2.21- 3.24(3)	2.18- 3.25(2)	2.22- 3.18(3)	2.25- 3.28(4)
Ba -	O1	x4	2.789(2)	2.785(2)	2.788(2)	2.789(2)	2.796(2)
-	O2	x4	2.812(2)	2.810(2)	2.811(2)	2.813(2)	2.817(2)
-	O3,O31	x1	2.87- 2.93(3)	2.86- 2.87(2)	2.85- 2.88(3)	2.85- 2.94(3)	2.88- 2.89(3)
Ca -	O1	x8	2.460(2)	2.464(2)	2.462(2)	2.463(2)	2.466(2)
Cu -	O1	x1	1.9248(2)	1.9249(2)	1.9254(2)	1.9255(2)	1.9276(2)
-	O2	x4	2.683(6)	2.672(6)	2.675(6)	2.682(6)	2.695(6)
Cu - Ca			3.1449(6)	3.1455(5)	3.1464(6)	3.1469(5)	3.1515(5)
Cu - Cu		layer	3.152(2)	3.154(1)	3.155(2)	3.157(1)	3.165(1)
O1 - O1		layer	3.064(4)	3.076(4)	3.071(4)	3.072(4)	3.078(4)

For statistically disordered O3 and O31 atoms the range of distances is given only

diffraction   data   is   as   follows:
$(Tl_{0.92}Cu_{0.08})_2Ba_2(Ca_{0.87}Tl_{0.13})Cu_2O_8$.

4) All of the above authors report$_o$ O(3) atom
displacement from its ideal site by 0.23-0.41A [Onada *et
al.*,1988; Gao *et al.*,1989; Cox *et al.*,1988; Hewat *et
al.*,1988; Sequeira *et al.*,1988]. In this case the
displacement of O(3) atoms correlates with Tl atom
displacement [Muradyan *et al.*,1991].

In the structure model with displaced Tl atoms, they
have the coordination of a distorted octahedron, with the
distance set up to O atoms (2+2+2).As mentioned above, this
is typical of thallium. In this case of Tl and O
displacement, the local symmetry of the structure is lowered
to orthorhombic, and the whole crystal retains
(statistically) the tetragonal symmetry. The latter is due
to a combination of local orthorhombic regions which are
rotated relative to one another by 90°. Formally, the
situation does not differ from usual twinning; however,we
did not observe, reflection splitting (even when we meant to
find it). Tl atom displacement from the symmetric sites was
found by the authors of [Toby *et al.*, 1988] based on the
function of distribution of atomic pairs and calculated
using the Fourier transform of powder neutron diffraction
data.

During the analysis of the temperature dependence of
structure parameters, as reported in the literature, special
attention was given to the atoms which form the CuO_2 layers
and the atoms which are close to them. The authors of [Gao
et al.,1989] studied single crystals of the phase 2212 at
125K, that is slightly higher than the phase transition
temperature Tc=110K. The results of the study showed that no
significant structural changes occur near the phase
transition point,except natural temperature reduction of the
periods and, therefore, interatomic distances. The same
conclusion was made after structural studies of the
superconductor $YBa_2Cu_3O_7$ at temperatures lower and higher
than the phase transition point [Capponi *et al.*,1987; Sato
et al.,1988]. However, as we have already pointed out, the
authors of [Scafer *et al.*,1988] observed anomalous values of
structure parameters during the superconducting phase
transition.

Having compared structural models that were refined
from X-ray diffraction data and obtained at various
temperatures, one can conclude that the superconducting
phase transition is not accompanied by substantial changes
in the structure. The symmetry of the crystal as well as the
metrics of its lattice before and after the superconducting
phase transition are retained.

The temperature dependences of some structural

parameters exhibit anomalous features, however. The temperature dependences of lattice parameters and unit cell volume are plotted in Fig.15. No specific behavior of the temperature dependence of unit cell volume is observed, which means a second-order phase transition. The changes that occur during temperature lowering are the result of natural thermal compression. However, a more detailed analysis of the structural data obtained at different temperatures reveals peculiar features of the behaviour of some structural characteristics [Cox *et al.*,1988; Muradyan

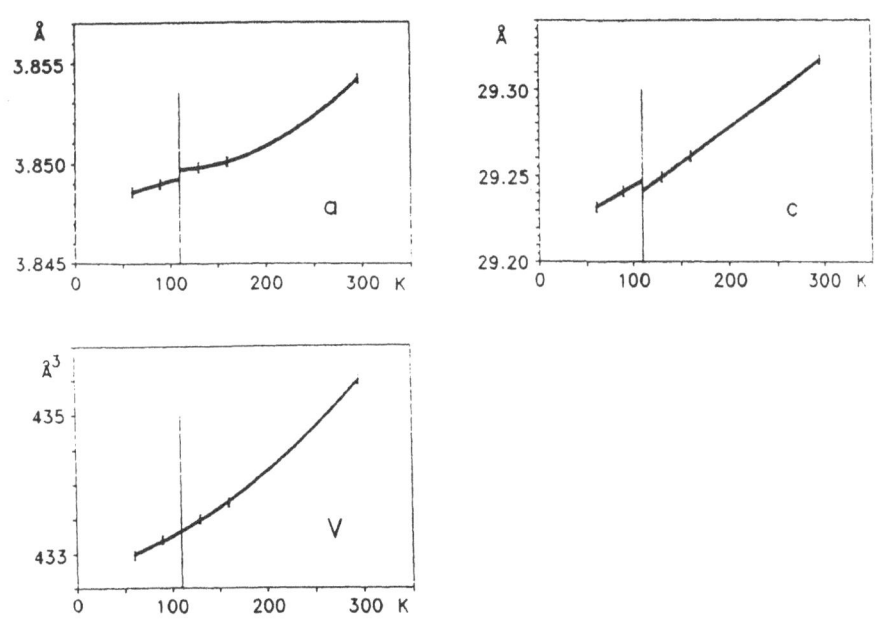

Fig.15.Temperature dependence of lattice parameters and unit cell volume for $Tl_2Ba_2CaCu_2O_8$, Tc=110K;

et al.,1991].

The results of this structural study of single crystals of the 2212 phase at five different temperatures allow a more detailed analysis of such parameters. The analysis of cation-cation distances shows that they are naturally reduced upon temperature lowering below the superconducting phase transition in the studied temperature range. Above the phase transition the behaviour of these distances is slightly changed. The distance between Tl-layers continues to reduce after the phase transition as well. Thallium atoms, however, are shifted from their ideal sites to a larger extent, and Tl-Tl distances are corrected due to that. It is fairly difficult to determine these distances

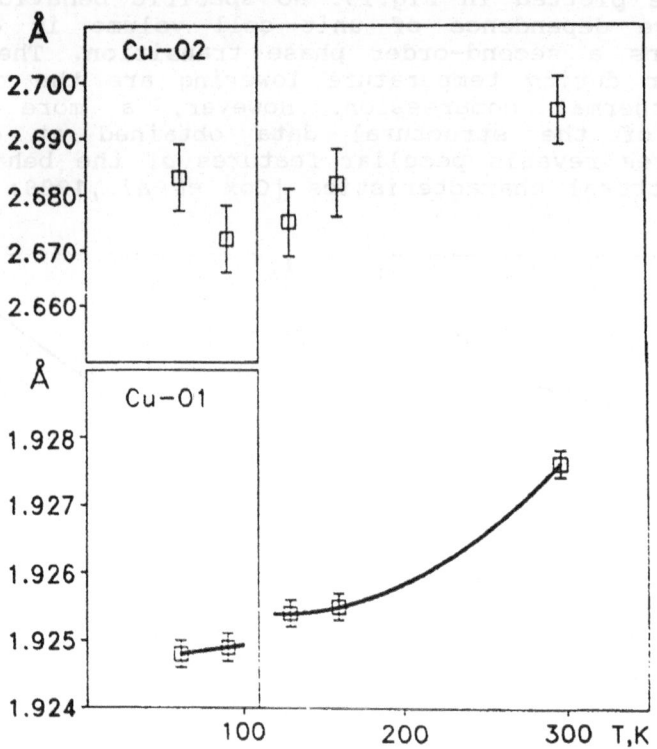

Fig.16. Temperature dependence of interatomic distances
Cu-O(1) and Cu-O(2) in $Tl_2Ba_2CaCu_2O_8$, Tc=110K.

precisely, because the displacement of Tl atoms strongly
correlates with their thermal parameters, and that
sufficiently lowers the accuracy of the determination of
these distances. The distance between thallium layers along
the c-axis does not depend on Tl atom displacement in the
ab-plane and is determined more reliably. Above the phase
transition the behaviour of unit cell parameters is, first
of all, accounted for by the changes in cation-cation
distances as well as those between Tl-layers.
      Below the phase transition most of the cation-anion
distances behave similar to cation-cation distances, i.e.
they are reduced with temperature lowering . The weakest
temperature dependence was found for the Ca-O(1) distances.
The changes of these distances lie within the accuracy of
their determination. The position of the O(1) atom in the
structure is not due to its location in the Ca-polyhedron,
but instead is due to a strong Cu-O(1) bond. The appropriate

Cu-O(1) distance below the phase transition is reduced, yet in the region of the phase transition the temperature dependence of this distance behaves anomalously (Fig.16). During the phase transition this distance is even more reduced, than required by the temperature tendency below the phase transition. No doubt, during the phase transition the Cu-O(1) bond becomes stronger and above the phase transition its ability to reduce is fairly small. These two curves do not meet for independent extrapolation of the temperature dependence of this distance at experimental points below and above the phase transition. The distance between them exceeds by several fold the standard deviation in the determination of this parameter. Further structural studies are required at temperatures, which are closer to the phase transition for a more detailed description of the behaviour of the Cu-O(1) bond.It should be noted that in the phase transition region the Ca-O(1) distance becomes larger within the standard deviation limits. It was shown in [Cox et al.,1988] that the Cu-O(2) distance up to the apical oxygen is extended after the phase transition. These studies confirm anomalous behaviour of the distance Cu-O(2) in the vicinity of the phase transition, although the accuracy of the determination of Cu-O(2) distances is lower than that for Cu-O(1) ones. Temperature dependences in other cation-anion distances exhibit no anomalies.

The CuO_2 planes are the predominant structural units for HTSC of the metaloxide type based on copper. These planes are square nets formed by O atoms. Half of the squares of these nets contain Cu atoms in a staggered arrangement. In the structures of $(La,Sr)_2CuO_4$, $Tl_2Ba_2CuO_6$ and $TlBa_2CuO_5$ oxygen atoms are located above and beneath Cu atoms. Those oxygen atoms complete the squares of the main net to form octahedra. In $YBa_2Cu_3O_7$ and $Tl_2Ba_2CaCu_2O_8$ compounds oxygen atoms are located only on one side of the square and form a tetragonal pyramid around the copper atom; however, in $(Nd,Ce)_2CuO_4$ the copper remains coordinated by a flat oxygen square.

Next, the significance of structural characteristics of CuO_5 groups at the Tc point will be discussed in order to understand the high-temperature superconductivity in $Tl_2Ba_2CaCu_2O_8$. As stated above, a similar effect was observed before in $YBa_2Cu_3O_7$ [Scafer et al.,1988]. Now we shall consider qualitatively some possible causes of an abrupt change in the Cu-O distances in the Tl containing study at the superconduction transition point.

All of the copper-oxide high-temperature superconductors contain plane-square CuO_2 nets as a common structural element. These nets are responsible for the superconductivity effect. They consist of local structural CuO_6, CuO_5 or CuO_4 fragments. Since these structural fragments contain divalent copper ions in a pseudooctahedral site (probably with one CuO_5 or two CuO_4 oxygen vacancies), they can be regarded as Jan-Teller centres. Each CuO_2 net can be interpreted as a cooperative Jan-Teller system which is in a ferrodistortion state with similarly distorted (extended along the c-axis) pseudooctahedral structural fragments. The local symmetry of such fragments is close to tetragonal. This viewpoint is confirmed particularly by the fact that typical Cu-O bond lengths in the main plane are $\approx 1.9 \overset{o}{A}$, while the distances upto the apical oxygen atoms are $2.3-2.6 \overset{o}{A}$. Such distances are typical of all of the other compounds of divalent copper with oxygen ligands, and which contain distorted Jan-Teller pseudooctahedral centres.

   In regular CuO_6 octahedra the crystal field splits a five- fold degenerated 3d-level of the copper ion into the lower three- fold degenerated T2g-level (3dxy, 3dxz and 3dyz -orbitals) and the upper two-fold degenerated Eg-level ($3dx^2-y^2$ and $3dz^2$-orbitals). There are six electrons at the T2g-level and three electrons at the Eg-level. Tetragonal distortion at the Jan-Teller centre (as well as oxygen vacancy in the axial position for the CuO_5 centre) eliminates the degeneracy of the orbital doublet Eg. As a result, the $3dz^2$-orbital of the Cu^{2+}(3d9)-ion is stabilized and filled by two electrons, while the $3dx^2-y^2$-orbital is destabilized and filled by one lone electron. It is known from the Jan-Teller theory [Bersuker et al.,1983] that the symmetry and degree of distortion of the coordination sphere is closely related to occupancies of the Jan- Teller orbitals by electrons, because these orbitals are characterized by anisotropic electron density distribution. For instance, the $3dx^2-y^2$-orbital of copper has four lobes lying in the CuO_2 plane and that are also elongated towards the four close oxygen atoms in the same plane; the electron density of the $3dz^2$-orbital shaped as a dumbell with a central belt is mainly concentrated along the c-axis and directed towards the axial oxygen atoms. For instance, when the $3dz^2$-orbital is populated by two electrons and the $3dx^2-y^2$-orbital is populated only by one electron, the Cu-O distance will be increased along the c-axis due to a stronger repulsion between the $3dz^2$ electrons and the negatively charged axial oxygen atoms. Accordingly the Cu-O distances in the CuO_2 plane will be reduced due to deficiency of the electron density at the $3dx^2-y^2$ and $3dz^2$-orbitals. It is noteworthy that the connection between the population of the $3dx^2-y^2$ and $3dz^2$-orbitals and variations in the Cu-O distances is retained upon the transition from the dielectric copper-oxide systems to the conducting ones in which the populations of these orbitals are no longer integers. In other words, any variation in populations of these orbitals causes structural changes in the close coordination sphere of the Cu ion.

Thus, the effect of reduction of Cu-O bonds in the CuO_2 plane, observed experimentally for $Tl_2Ba_2CaCu_2O_8$ in this work, upon transition from normal to the superconducting phase (and a similar effect observed before for $YBa_2Cu_3O_7$ [Aleksandrov et al.,1992]) can be accounted for by a decrease of the effective occupancy of the $3dx^2-y^2$-orbitals of copper in the CuO_2 plane at $T=T_c$.

Let us consider some additional data in order to relate this decrease with the superconducting phase transition. The experimental data on photoelectron spectra and the Hall's effect indicate the absence of trivalent copper in copper-oxide superconductors. Doping or increase of oxygen content does not lead to a further oxidation of divalent copper, but to the formation of holes in the 2p-shell of oxygen atoms. The corresponding oxygen 2p-zone is not very wide (less than 1 eV). Apparently, this is explained first of all by the fact that due to the alternating character of arrangement of copper and oxygen atoms in the CuO_2 nets the 2p-states at different oxygen atoms in the CuO_2 planes interact not directly, but via the 3d- states of copper. Accordingly, upon hole motion in the CuO_2 plane, the process of quantum jump of the hole from the 2p-orbital of one oxygen atom to a 2p-orbital of the other oxygen atom occurs through an intermediate state of hole migration to the copper atom (trivalent copper, Cu^{3+}(3d8) is formed) and a subsequent movement of the hole to the 2p-orbital of the other oxygen atom, i.e.:

$$O^{1-}-Cu^{2+}-O^{2-}---O^{2-}-Cu^{3+}-O^{2-}---O^{2-}-Cu^{2+} - O^{1-}.$$

Since this virtual process is described by second order perturbation theory, the width of the 2p-zone turns out to be proportional to a squared integral of the overlapping 2p-orbitals of oxygen and 3d-orbitals of copper, $S(2p\ 3d)2$, i.e. quite a small value. Another reason of this band being narrow lies in a strong electron-phonoun interaction of 2p-holes on the oxygen atoms with the lattice, leading to polaronic narrowing of the bond [Aleksandrov *et al.*,1992]. Quite essential, however, is the fact that hole movement along the oxygen band in the CuO_2 plane brings about a decrease of the population of 3d-orbitals of copper, which play the role of virtual intermediate hole acceptors. As is known from the experimental and theoretical data available, and, in particular, from the numerous calculations of the band structure of various HTSC, the 2p orbitals of oxygen and $3dx^2-y^2$, $3dz^2$-orbitals of copper make the largest contribution to the band states near the Fermi level. The role of the $3dx^2-y^2$-orbital turns out to be much more important than that of $3dz^2$-orbital because the former is higher in energy (and consequently is populated only by one electron) and overlaps better with the 2p-orbitals of oxygen, than the latter.

Therefore, any variations of the charge carrier system in the CuO_2 plane, including a superconducting phase transition, should first of all, be reflected in the value of the population of the $3dx^2-y^2$-orbitals of copper in the CuO_2 plane.

Finally, let us consider how the superconductor phase transition in copper-oxide superconductors can lead to a reduction of the effective population of $3dx^2-y^2$-orbitals of copper in the CuO_2 plane. At present, there is no generally accepted viewpoint about the mechanism of superconducting coupling in high temperature superconductors. In particular,it is debated whether the superconducting coupling of charge carriers occurs in the impulse space or the coupling occurs in the real space, with the formation of Bose-particles of the biopolaron type and their subsequent Bose-condensation [Gor'kov *et al.*,1988]. Yet, regardless of a concrete mechanism of superconduction coupling it was established, that the characteristic sizes of these pairs are quite small and make a few Angstroms [Gor'kov *et al.*,1988]. Because this value is much less than that found in ordinary superconductors, the 2p-holes which form the superconductivity pair should be located on oxygen atoms belonging to one or several neighbouring unit cells. Hence, one may suppose that upon the formation of a superconducting condensate below Tc the mean effective concentration of 2p-holes about the given Cu- ion in the CuO_2 plane should increase, as compared to the normal phase. Upon transition from a noncorrelated hole motion at T>Tc to a strongly correlated hole motion at T<Tc, the presence of the 2p- hole

about the given Cu atom should mean a higher probability of a simultaneous presence of another hole near it due to the small size of the superconductor pair. As was pointed out above, the 2p- holes decrease mean occupancies of $3dx^2-y^2$-orbitals first of all, and an increase of the local effective concentration of holes near the copper atom will give rise to a decrease of the total occupancy of the $3dx^2-y^2$-orbital, and consequently, a reduction of Cu-O distances in the CuO_2 plane with a simultaneous increase of the Cu-O distance along the axial c-axis.

Finally, it should be noted that the considered qualitative mechanism can turn out to be universal for various classes of copper-oxygen superconductors, which contain CuO_2 planes. That is why it is very important to carry out direct experimental measurements of electron density on the $3dx^2-y^2$-orbitals of copper using various precise techniques, including construction of deformation electron density maps from diffraction data before and after the superconduction phase transition, so as to reveal a predicted jump of effective population of $3dx^2-y^2$-orbitals at the superconducting phase transition point.

5. Deformation Electron Density in Nd_2CuO_4 Single Crystals.

Expermental determination of valence state of atoms in a crystal, the character of chemical bonding as well as population of atomic orbitals by electrons can be carried out from accurate X-ray diffraction data by way of construction of deformation electron density distribution in the crystal. The deformation electron density $\delta\rho(r)$ is a difference between electron density distribution in the crystal $\rho(r)$ and the sum of electron densities of spherically symmetrical noninteracting atoms which form the crystal:

$$\delta\rho(r) = \rho(r) - \sum_{i=1}^{N} \rho_i(r-r_i).$$

It is rather difficult to construct $\delta\rho(r)$ because reflection intensities should be measured with a high accuracy (of the order of 1-2%). Besides, correction for absorption in the sample, anomalous constituents of absorption, thermal diffuse scattering, exctinction with account of possible anisotropy, atomic thermal motion, including anharmonism, should be made. When the atomic model of the structure is obtained and refined from diffraction data, one should take into account possible correlations between structure parameters. It is most difficult to separate anisotroipy of thermal motion and symmetry in the distribution of valence electrons for the given system of chemical bonding in the sample. The correlation between these charateristics is significant , because atomic thermal motion depends on spatial distribution of chemical bonding.

In order to fulfill all these tasks we have chosen the simplest, from the viewpoint of crystal structure, high-temperature superconductor $(Nd,Ce)_2CuO_4$. The structure of this crystal is defined only by one coordination parameter: z-coordinate of the Nd atom. All the other coordinates were set by elements of crystal symmetry. Of principal importance is the fact that samples of this compound do not tend to twinnning. Below, the first stage of this work is reported. It is concerned with determination of deformation electron density in the starting nonsuperconducting crystal Nd_2CuO_4, grown without doping with Ce. This stage is necessary for elucidation of influence of Ce atoms on the electron structuire of the superconductor.

X-ray diffraction data were obtained from a spherical sample 0.140(5)mm in diameter at room temperature, using AgK_α-radiation. We have measured a total of 5015 reflections for $\sin\vartheta/\lambda \leq 1.30 \overset{\circ}{A}^{-1}$. By averaging the reflections which were equivalent in symmetry an array of 484 independent structure amplitudes was obtained. The structure was refined in the space group I4/mmm, a = 3.9488(3)$\overset{\circ}{A}$, c = 12.1869(9)$\overset{\circ}{A}$ up to final R-factors R_W = 0.82%, R = 0.70%. Details of the refinement are reported in [Makarova et al.,1993].

In the course of the refinement we established violation of stoichiometry of the composition of the single crystal. Its chemical formula, obtained form X-ray diffraction data, can be written as $Nd_{1.890}CuO_{3.85}$. During refinement of the parameters of thermal vibrations of Nd atom the physically meaningful anharmonic constituents characterised by a four-rank tensor were obtained . The atomic model of the Nd_2CuO_4 structure is shown in Fig.17.

Two cuts of the three-dimensional electron density in $Nd_{1.890}CuO_{3.85}$ single crystal are plotted in the same figure. One cut is parallel to the (010) plane and passes through Nd, Cu, O(1) and O(2) atoms. The second cut is perpendicular to the c axis of the crystal and passes through Cu and O(1) atoms.

The highest peaks of deformation electron density are located near the Cu^{2+} atom. They are located beneath and above the Cu atom along the z-axis at a distance of 1.06Å

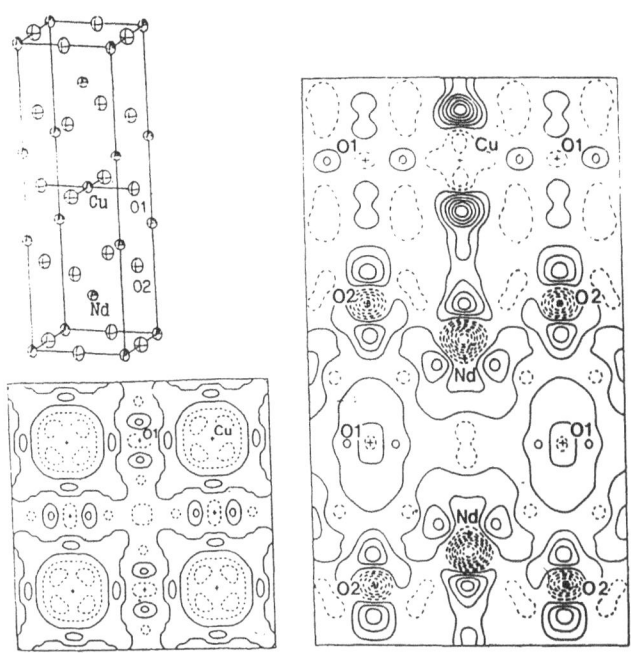

Fig.17. Atomic structure and deformation electron density
sections of a $Nd_{1.890}CuO_{3.85}$ single crystal. The
isoline passes through 0.15 $e \cdot Å^{-3}$.

and their heights attain 0.95 $e \cdot Å^{-3}$. These maxima correspond
to electrons of the orbital $3dz^2$ of the copper atom. The
maxima 0.25 $e \cdot Å^{-3}$ in height are located at the Cu-O(1)
bonds, these maxima correspond to copper orbitals $3dx^2-y^2$.
As for two oxygen atoms, a different character of their
state is evident. The O(1) atoms are involved in covalent
bonds with Cu atoms, while O(2) atoms ehxibit ionic bonding.

A deep minimum $-1.20 e \cdot Å^{-3}$ is observed at a distance of 0.53Å
from the Nd atom. This might me a trace of an electron which
left the 4f shell of the Nd atom. A comparison of the above
mentioned two sections of deformation electron density
reveals their qualitative difference. In the section which
is perpendicular to the c-axis there are neither sharp
minima, nor maxima. This section is characterized by an
essentially uniform density distribution, than the section

parallel to the (010) plane. Deformation electron density maps of the superconductor $(Nd,Ce)_2CuO_4$ obtained at temperatures lower and above the superconducting phase transition are required for further analysis. The reserach in this direction is in progress.

6.References.

Aksenov V.L., Bogolubov N.N., Plakida N.M., editors (1990), World Scientific. Singapore, New Jersey, London, Hong Kong. Progress in High Temperature Superconductivity. Vol.21, International Seminar of High Temperature Superconductivity.

Aleksandrov A.S., Krebs A.S.,(1992), UFN, 162 N5 , 1.

Arabi H., Ciomartan D.A., Clamp M.E., *et al.*,(1992). Physica C 193, p. 90.

Bednorz J.G., Muller K.A.(1986), Z.Physik. B, Condensed Matter, 64, 189-193.

Bersuker I.B., Polinger V.Z.(1983). Vibronic Interaction in moleculas and crystals. Nauka, Moscow.

Beyers R.B., Parkin S.S.P., Lee V.Y. *et al.*,(1989), IBM J. Develop., 33, 228.

Blomberg M.K., Merisalo M.Yu., Molchanov V.N., Tamazyan R.A., Simonov V.I.(1992). Pis'ma Zh.Eksp.Teor.Fiz., 55, N9, 530, in Russian.

Capponi J.J., Chaillout C., Hewat A.W. *et al.*,(1987) Europhys. Lett., 120, 1301.

Cox D.E., Torardi C.C., Subramanian M.A. *et al.*,(1988) Phys.Rev. B 38, 6624.

Gao Y., Li R., Coppens Ph. *et al.*(1989) Acta Cryst., A45 FC11.

Gor'kov L.P., Kopnin N.B.,(1988) UFN, 156 N1 , 117.
Hewat A.W., Hewat E.A., Brynestad J. *et al.*(1988),Physica C, 152 438.

Hiroi Z., Takano M., Bando Y.,(1989) Physica C 158, 269.

Jorgenson J.D., Schutler H.B., Hinks D.G., Capone D.W., Zhang K, Brodsky H., Scalapin D.J., (1987) Phys.Rev.Lett. 58, 1024.

Kajitani T. *et al.*,(1987) Jpn.J.Appl.Phys. Vol. 26, L 1877.

Kajitani T., Kusaba K., Kikuchi M. *et al.*,(1988) Jpn. J. Appl. Phys. Vol. 27, N3, p. L 1877.

Lutgemeier H., Heinman I.,(1992) Proc. Workshop on Phase Separation in Cuprate Superconductors, Ettore Majorana, Erice, Trapani, Italy, May 6 - May 12 .

Makarova I.P., Molchanov V.N., Tamazyan A.A., Simonov V.I., Gamayunov K., Ivanov A.L., Osiko V.V., (1993). Kristallografiya 38, N4, 24, in Russian.

Miceli P.F., Tarascon J.M., Greene L.H., Bardoux P., (1988).Phys. Rev. B , Vol.37, N10, p. 5932.

Molchanov V.N., Muradyan L.A., Simonov V.I. (1989). Pisma v ZHETF, 49, 222, in Russian.

Morosin B., Ginley D.S., Venturini E.H. *et al.*,(1991), Physica C, 172, 413.

Muradyan L.A., Sirota M.I., Makarova I.P., Simonov V.I. (1985), Kristallografiya 30, 258, in Russian.

Muradyan L.A., Radaev S.F., Simonov V.I. (1989). Methods of Structural Analysis, Nauka, Moscow p.5 .

Muradyan L.A., Molchanov V.N., Tamazyan R.A., Simonov V.I., Shibaeva R.P., Korotkov V.E., Kolesnikov N.N., Kulakov M.P., (1991). Superconductivity : Physica, Chemistry, Technology, 4, N2, 277, in Russian.

A.V.Narlikar, editor (1991). Studies of High Temperature Superconductors. Nova Science. NY .

Onada M., Shamoto S., Sato M., Hosoya S.,(1987) Jpn. J. Appl. Phys. 26 , L 363.

Onada M., Kondon S., Fukuda K., Sato M.(1989). Jap. J. Appl. Phys.,27 L 1234.

Parkin S.S.P., Lee V.Y., Nazzal A.I., Savoy R., Beyers R., (1988a). Phys. Rev. Lett., 61, 750.

Parkin S.S.P., Lee V.Y., Nazzal A.I. *et al.*,(1988b). Phys. Rev., B 38, 6531.

Petricek V., Coppens Ph., Becker P.(1988). Acta Cryst., A44, 235.

Putilin S.N., Antipov E.V., Chmaissem O., Marezio M. (1993).

Nature, Vol.362, N 6417, 226.

Sato S., Nakada I., Kohara T., Oda X.,(1988). Acta Cryst. C44, 11.

Scafer W., Jansen E., Will G., Faber J., Veal B.,(1988). Mater. Res. Bull., 23 , 1439.

Sequeira A., Rajagopal H., Gopalakrishan I.K. et al.,(1988). Physica C, 156, 599.

Shapligin I.S., Kakhan B.G., Lazarev V.B.,(1979). Zhurnal Neorganicheskoi Khimii 24 , 1478, in Russian.

R.N.Shelton, W.A.Harrison and N.E.Phillips ,editors (1989), North - Holland. Proceedings of the International Conference on Materials and Mechanisms of Superconductivity High Temperature Superconductors II. Part I and Part II.

Simonov V.I., Muradyan L.A., Tamazyan R.A., Melnikov O.K., Bykov A.B., Vainshtein B.K.,(1988a). Pis'ma Zh. Eksp. Teor. Fiz. 48(5) 290, in Russian.

Simonov V.I., Muradyan L.A., Molchanov V.N., Kov'ev E.K., (1988b). Kristallografiya 33 , 621, in Russian. Simonov V., Muradyan L., et al., (1988). Z. Kristallogr., 1985,P.10, 5.2. TU, 430.

Simonov V.I., Muradyan L.A., Tamazyan R.A., Osiko V.V., Tatarintsev V.M., Gamayunov K.,(1990). Physika C 169, 123-132.

Subramanian N.A., Calabrese J.C., Toradi C.C.,et al.,(1988), Nature, 332 , 420.

Toby B.H., Egami T., Subramanian M.A., et al.,(1988). Phys. Rev. Lett., 61, p.2608. Toby B.H., Egami T., et al., (1990), Phys. Rev. Lett., 64, p.2414

Tokura Y., Takagi H., Uchida S. (1989). Nature, Vol. 337, 345.

Wu M.K., Ashburn J.N., Torng C.J., Hor P.H., Meng R.L., Gao L., Huang Z.J., Wang W.Q., Chu C.W. (1987), Phys. Rev. Lett., 58, 908-911.

Zucker U.H., Perenthaler E., Kuns W.E. et al.,(1983), J. Appl. Cryst., 16, 358.

STRUCTURAL AND PHYSICAL PROPERTIES OF SUPERCONDUCTING La$_2$CuO$_{4+\delta}$

C. CHAILLOUT, M. MAREZIO
Laboratoire de Cristallographie, CNRS-UJF
BP 166
38042 Grenoble cedex 9
France

ABSTRACT. Among high temperature superconducting oxides, La$_2$CuO$_{4+\delta}$ is still one of the most studied compounds as the appearance of superconductivity is not really understood. La$_2$CuO$_4$ can be oxidized either by a treatment at high temperature and under high oxygen pressure, or by using "soft" chemistry methods at room temperature such as electrochemistry. It seems that two regions of extra oxygen concentration have to be considered : first, one with $\delta \leq 0.05$, which corresponds to a superconducting phase with $T_c \approx 32K$; another one with $\delta \geq 0.08$, which corresponds to a superconducting phase with $T_c \approx 45K$. The former is obtained by treating La$_2$CuO$_4$ under high oxygen pressure. In this case, below $\approx 280K$, the excess oxygen atoms migrate in order to form antiferromagnetic La$_2$CuO$_4$ domains and La$_2$CuO$_{4+\delta}$ ones. The migration of the oxygen atoms is accompanied by anomalies in resisitivity and in thermoelectric power. The extra oxygen is found to be in between two LaO layers in a similar position as the oxygen atoms located between two Nd layers in Nd$_2$CuO$_4$ structure. The insertion of extra oxygen causes the displacement of some of the other oxygen atoms. The nature of the bonding between the excess oxygen and the "normal" oxygen atoms of the structure has not yet been completely elucidated.The more oxidized samples ($\delta \geq 0.08$) present a larger orthorhombic distortion than the $\delta \leq 0.05$ samples. It is not clearly established whether or not these $\delta \geq 0.08$ samples present a phase separation at low temperature. The presence of superstucture spots on electron diffraction patterns suggest an ordering of the incorporated oxygen atoms.

1. Preparation, Composition

It is commonly assumed today that a mixed copper oxide becomes superconducting by the introduction of holes in the oxygen sublattice leading to a formal valence state of the Cu cations greater than 2+. For La$_2$CuO$_4$ this can be realized either by doping the La sublattice with a divalent alkali-earth element [1] or by introducing extra oxygen [2] or fluorine [3] or chlorine [3] in the oxygen sublattice.

Beille et al. [4] were the first to report that La$_2$CuO$_4$ becomes superconducting by a treatment under oxygen pressure. Although they only reported the existence of traces of superconductivity their experiment was unconfutable. Subsequently, several other groups [5-8] detected the same behavior for samples of undoped La$_2$CuO$_4$ which had been prepared in oxygen-rich atmospheres. The first reproducible samples with an appreciable Meissner effect were synthesized by Demazeau et al. [9] and by Schirber et al. [10]. The

E. Kaldis (ed.), Materials and Crystallographic Aspects of HTc-Superconductivity, 129–144.
© 1994 *Kluwer Academic Publishers.*

latter authors showed that a heat-treatment at 600° C under an oxygen pressure ranging between 1 and 3 kbar transformed an antiferromagnetic La_2CuO_4 sample into a superconducting one exhibiting 30% of superconducting volume as determined by Meissner effect measurements. It was then that the sample composition became a problem.

Schirber et al. [10] determined the composition of their samples by x-ray microanalysis, iodometric titration, and thermogravimetry. They found that the La/Cu ratio was equal to 2 ; the iodometric titration led to a $\delta=0.032$ oxygen excess, while the thermogravimetry indicated that weight loss corresponded to $\delta=0.13$. This large difference would be explained, according to these authors, if the extra oxygen atoms were incorporated as superoxide O_2^-. This interpretation was corroborated by XPS results obtained by Rogers et al. [11]. On the other hand, Zhou et al. [12] showed that a sample of formula $La_2CuO_{4.05}$ (determined by iodometric titration), heated at 250° C under nitrogen, exhibited a phase transition to an antiferromagnetic state accompanied by an oxygen loss from the entire volume. Moreover, they observed a continuing superficial loss of oxygen around the temperature of the phase transition. This allowed them to conclude that the superoxide identified by XPS was associated with the superficial oxygen, but not to the extra oxygen responsible for the superconductivity.

More recently, Wattiaux et al. [13] showed that it is possible to obtain the superconducting $La_2CuO_{4+\delta}$ by electrochemical oxidation of La_2CuO_4 in an alcali solution (1 N KOH) at room temperature. This method of preparing the superconducting phase has been used by several other groups who utilized either NaOH or KOH [14-16]. By controlling the charge passing through the electrochemical cell it is possible to vary the amount of extra oxygen incorporated into the structure. It seems that the electrochemical method leads to larger amounts of extra oxygen incorporated than that utilizing heat-treatments under high oxygen pressure. Values of $\delta\approx0.09$ and $\delta\approx0.18$ have been reported by Grenier et al. [17] and Chou et al. [16], respectively. On the other hand, Rudolf et al. [18] contest that oxygen be incorporated in the structure by electrochemical process. These authors believe that the starting product is actually an oxyhydroxide containing lanthanum vacancies $La_{2-x}CuO_{4-3x}(OH)_{3x}$ and that the electrochemical oxidation reaction implies a proton loss. The superconducting compound would then contain lanthanum vacancies, $La_{2-x}CuO_4$. Recently Takayama et al. [19] have shown that it is possible to realize the oxygen insertion into La_2CuO_4 by treating the samples in a solution of $KMnO_4$. In this case too, values of $\delta\approx0.09$ can be obtained. These authors believe that the amount of inserted oxygen can be varied by changing the concentration of the $KMnO_4$ solution.

2. Structure

2.1. La_2CuO_4

The structure of La_2CuO_4, of K_2NiO_4 type, belongs to space group Cmca (Fig. 1) with lattice parameters a $\approx$ 5.360 Å, b $\approx$ 13.181 Å, and c $\approx$ 5.393 Å. The atoms are located in the following positions : La in (8f) (0yz), Cu in (4a) (000), O1 in (8f) (0yz), and O2 in (8e) (1/4,y,1/4). The Cu atoms are surrounded by an oxygen octahedron elongated along the b axis. The CuO_{22} planes (formed by the squares perpendicular to the octahedral elongation) are separated by two LaO1 layers.

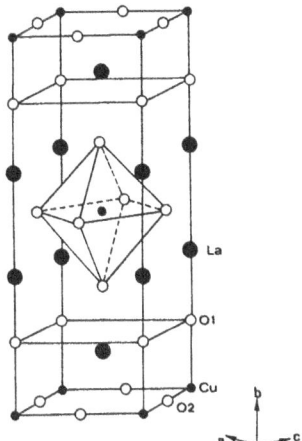

Figure 1 : Three dimensional structural arrangement of La_2CuO_4. One tetragonal unit cell is represented

According to Santoro et al. [20] the structure can be described by the sequence : $(LaO)_o(LaO)_c(CuO_2)_o(LaO)_c(LaO)_o(CuO_2)_c(LaO)_o(LaO)_c$. The o and c indices indicate whether the cation is situated at the origin or at the center of the squares forming the layers. In other words the sequence indicate that each layer is shifted of (1/2,0,1/2) with respect to the one above or below.

The La_2NiO_4 compound which is isostructural with La_2CuO_4 is known to accept a larger concentration of extra oxygen than the latter. Buttrey et al. [21], while investigating the structural behavior of $La_2NiO_{4+\delta}$ samples, suggested that the extra oxygen atoms could be located between two LaO layers in the (8e) (1/4,y,1/4) positions.

Since the symmetry of La_2CuO_4 is slightly orthorhombic, the crystals are in general twinned. The two possible twinning laws are the [010]90°, that is the pseudo four-fold axis along the b axis, and the pseudo diagonal mirror plane (101). For both cases the crystal contains two individuals and this leads to a doubling of most of the reflections, whose indices are hkl and lkh, each corresponding to one individual.

2.2. POWDERS PREPARED UNDER OXYGEN PRESSURE

In 1988 Jorgensen et al. [22] carried out a series of neutron diffraction experiments on samples of $La_2CuO_{4+\delta}$ prepared under different oxygen pressures. These authors showed that below room temperature the samples contained two different, but quite similar orthorhombic phases, one with a stoichiometry very close to La_2CuO_4 and the other with an oxygen stoichiometry greater than 4. The percentage of the superconducting phase increased with increasing oxygen pressure used during the preparation. The analysis of the neutron diffraction diagrams, obtained at different temperatures, showed that the phase separation occurred at about 320 K and was reversible. The corresponding variation of the lattice parameters is shown in Fig. 2.

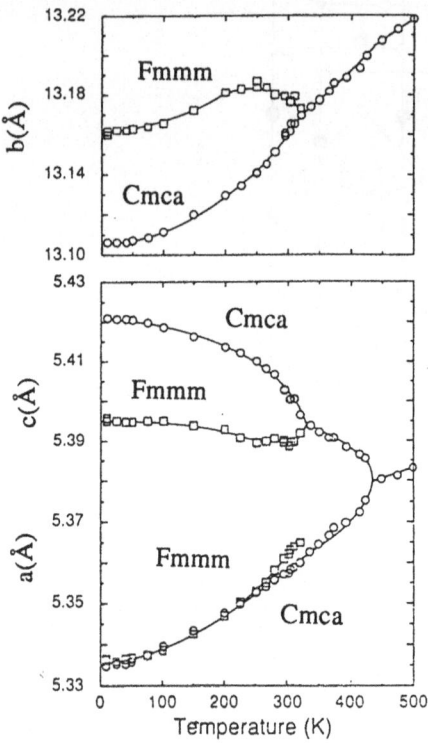

Figure 2 : Variation of the lattice constants vs temperature [22]

Note that the two phase have the same a parameter. The orthorhombic distortion, defined as (a-c), is smaller for the phase richer in oxygen, while the b parameter is larger. Above 320 K the sample contains only one orthorhombic phase, the orthorhombicity decreases with increasing temperature and around 430 K an orthorhombic to tetragonal transition is observed. This temperature, which is definitely smaller than the corresponding transition temperatures (720-800 K) reported for stoichiometric La_2CuO_4, is qualitatively in agreement with the results reported by Johnston et al. [23] who noted a decrease of the transition temperature as y decreases in La_2CuO_{4-y}.

Jorgensen et al. [22] carried out simultaneous refinements of the two structures. The neutron powder diffraction data were analyzed by the Rietveld method. Because of the almost perfect overlapping of the two profiles, it was not possible to locate the extra oxygen atoms present in one of the two phases and to determine unequivocally the respective space group. By comparing the $La_2CuO_{4+\delta}$ system to the one of $La_2NiO_{4+\delta}$,

Jorgensen et al. proposed Cmca for the phase poorer in oxygen and Fmmm for the richer one [22].

The phase separation (occurring below 290 K) and the variation with temperature of the lattice parameters were confirmed by Zolliker et al. [24] by using neutron diffraction and synchrotron x-ray diffraction and a powder of composition $La_2CuO_{4.03}$. These authors observed that below 290 K the amount of the phase rich in oxygen increased with decreasing temperature until 250 K where the two phases were about of the same quantity. No change was observed below 250 K down to 10 K.

Dabrowski et al. [25] reported the results of a neutron diffraction study carried out on powders of composition La_zCuO_4 with z=2.2, 2.0, 1.975, and 1.95. They showed that for z>2 and z<2 the best refinements were obtained by introducing a parasite phase (either La_2O_3 or CuO) whose amount increases, respectively, as |z-2| increases. These results strongly indicate that the La/Cu ratio is indeed equal to 2 at least for the samples prepared under oxygen pressure.

2.3. SINGLE-CRYSTALS

Chaillout et al. [26] carried out neutron diffraction experiments using a single crystal of $La_2CuO_{4+\delta}$ which had undergone a treatment under oxygen pressure. These authors collected the neutron diffracted intensities at 15 K and at room temperature to locate the extra oxygen atoms. Moreover, selected reflections were scanned at different temperatures in order to determine the space group of the two phases [27]. These scans were carried out between 295 K and 118 K by using either x-ray diffraction ($\lambda CuK\alpha$ = 1.5418 Å) or neutron diffraction (λ = 1.26 Å). The phase separation was observed at 273 K. From the behavior of the reflections such as (0,14,0) (see Fig. 3) it can be seen that the contribution of the of the oxygen-poor La_2CuO_4 phase to the total intensity increases from the phase separation temperature to 230 K, after which it remains constant.

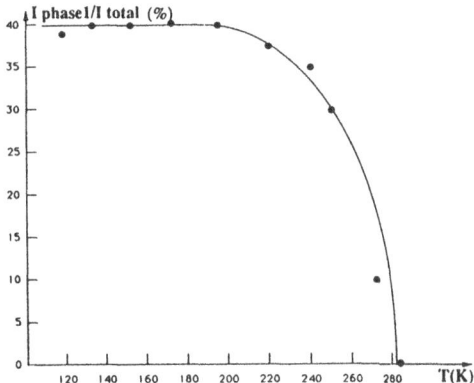

Figure 3 : Variation of the contribution of the oxygen-poor (La_2CuO_4) phase to the total intensity of the (0,14,0) reflection vs temperature

134

The (hkl) reflections with either h = 2n, k = 2n, and l = 2n+1 or with h = 2n+1, k = 2n+1, and l = 2n, which are not affected by twinning and are not allowed by the Fmmm space group, exhibit also well-defined splittings at the phase separation (see Fig. 4). This strongly indicates that both phases of $La_2CuO_{4+\delta}$ do not belong to the Fmmm space group. It must be pointed out that for these reflections the intensity due to the La_2CuO_4 phase is much stronger than that due to the $La_2CuO_{4+\delta}$ phase.

Diffracted intensities were collected at room temperature and at 15 K by using a short wavelength (λ=0.48417 Å) so that the contributions due to the different individuals forming the twin would be as close as possible in reciprocal space. Under these conditions it would not be difficult to scan the total intensity. The structural refinement at room temperature was carried out in the Cmca space group and the atoms placed as cited in § 2.1.. As found in $La_2NiO_{4+\delta}$, the extra oxygen was placed in (1/4,y,1/4) with y≈0.25. It showed that the La/Cu ratio was equal to 2 and that the O1 sites were partially occupied. In fact, about 5% of the O1 atoms are displaced of ≈0.75 Å. The chemical formula obtained from the refinement was $La_2CuO_{4.032}$.

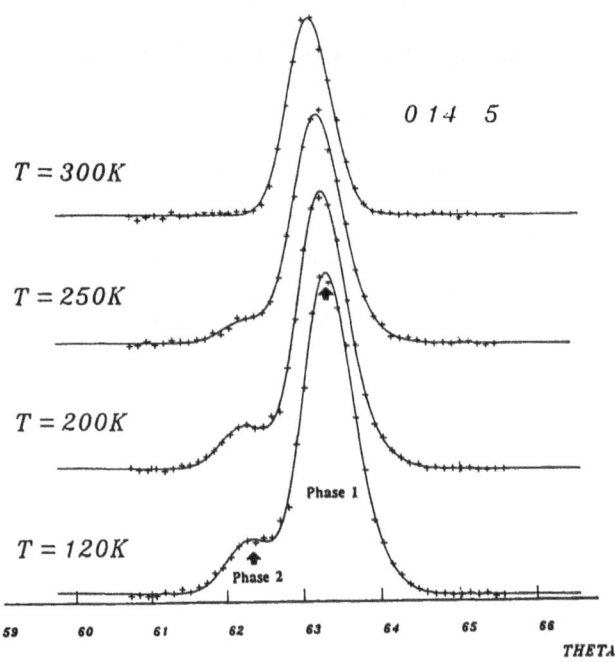

Figure 4 : ω-2θ scan of (0,14 ,5) reflection (neutron diffraction) between 300K and 120K showing the sudden phase separation. Crosses represent experimental points.

The refinements based on the low temperature data with both structures, La_2CuO_4 and $La_2CuO_{4+\delta}$, described in the Cmca space group were carried out simultaneously. The refined volume corresponding to the $La_2CuO_{4+\delta}$ phase represents 70% of the total crystal volume. This is in good agreement with the superconducting volume estimated from ac susceptibility measurements. The stoichiometries of the two phases are La_2CuO_4 and $La_2CuO_{4.048}$, respectively. The results of these refinements and the relevant interatomic distances [26] are reported in Tables I and II.

Table I
Positional, Thermal, and Occupancy Parameters for La_2CuO_4 and $La_2CuO_{4+\delta}$

	PHASE 1	PHASE 2
Y La	0.3616(2)	0.36088(6)
Z La	0.0096(5)	0.0053(2)
Y O1	0.1843(2)	0.1823(1)
Z O1	- 0.048(1)	- 0.0261(5)
Y O2	- 0.0112(5)	- 0.0031(2)
X O3		0.033(5)
Y O3		0.185(2)
Z O3		0.105(4)
Y O4		0.242(4)
P O1	0.99(1)	0.922(8)
P O2	0.99(1)	0.999(5)
P O3		0.040(4)
P O4		0.024(4)
U11 La	- 0.002(1)	0.011(1)
U22 La	0.0056(7)	0.0018(2)
U33 La	0.004(1)	0.0008(7)
U23 La	0.0004(4)	- 0.0005(4)
U11 Cu	0.001(2)	0.005(2)
U22 Cu	0.0043(9)	0.0051(4)
U33 Cu	0.001(2)	- 0.000(1)
U23 Cu	- 0.0002(6)	- 0.0003(6)
U11 O1	0.005(1)	0.015(2)
U22 O1	0.0018(8)	0.0051(4)
U33 O1	0.006(2)	0.0087(7)
U23 O1	- 0.0003(8)	- 0.0000(8)
U11 O2	0.001(1)	0.011(1)
U22 O2	0.003(1)	0.0127(5)
U33 O2	0.003(1)	- 0.000(1)
U13 O2	0.0008(6)	0.0006(4)
U O3		0.005 fixed
U O4		0.005 fixed
R_W = 4.45%	R_{uw} = 3.10%	χ^2 = 4.18

Figure 5 illustrates the modifications undergone by the structure with the introduction of the extra oxygen atoms O4. These are located between two LaO1 layers which are stacked as in a NaCl structure. The two layers form a face-sharing cube layer whose eight corners are occupied by 4 La and 4 O1, each set of atoms arranged as a tetrahedron. The O4 atoms are located at the center of these cubes with an occupancy factor equal to 2.4%. Each time that an O4 position is occupied, the coordination of the 4 La atoms increase from 9 to 10, but their positions are not affected. On the other hand, the introduction of the O4 induces the displacement of some of the first nearest O1 toward the O3 sites. According to the refinement, the ratio of O3/O4 is equal 3.3 with an estimated standard deviation of 0.6. Taking into account this value and the possible O1-O4 and O3-O1 distances, given in Table II, different models can be proposed. To choose among these models one should take into consideration whether or not the oxygen octahedra surrounding the Cu cations, are to be considered as "rigid" entities (see Fig. 6).

Table II : Selected Interatomic Distances

	PHASE 1		PHASE 2	
Cu—O1	2.437(1)	x 2	2.401(1)	x 2
Cu—O2	1.9057(2)	x 4	1.9005(1)	x 4
Cu—O3			2.50(3)	x 2
La—O1	2.352(2)	x 1	2.354(2)	x 1
	2.463(3)	x 1	2.592(3)	x 1
	2.750(4)	x 2	2.7369(3)	x 2
	3.070(3)	x 1	2.924(3)	x 1
La—O2	2.559(2)	x 2	2.624(2)	x 2
	2.710(2)	x 2	2.651(2)	x 2
La—O3			2.38(3)	x 2
			2.25(2)	x 2
			2.64(3)	x 2
			2.97(3)	x 2
			3.30(2)	x 2
La—O4			2.35(3)	x 2
			2.44(3)	x 2
O1—O2	3.096(3)	x 2	3.032(3)	x 2
	3.091(3)	x 2	3.092(3)	x 2
O2—O2	2.675(2)	x 2	2.675(2)	x 2
	2.715(2)	x 2	2.700(2)	x 2
O1—O4			2.06(3)	x 2
			2.15(2)	x 2
O2—O3			2.77(3)	x 2
			2.94(3)	x 2
			3.34(3)	x 2
			3.48(3)	x 2
O3—O4			1.59(4)	x 2
			1.86(3)	x 2
			2.44(3)	x 2
			2.62(3)	x 2

The model corresponding to the determined by R O3/O4 = 3.3(6), consists of the presence of one O4 which displaces three near-neighbor O1 towards the positions O3, forcing one of the four O4—O distances to be 1.59(4) Å (indicated as short bond in fig. 6). This value is close to the O—O distances with peroxide character (1.48 Å). If R O3/O4 were 2, only two O1 atoms would be displaced towards the O3 positions. Two cases exist : if such displacements are localized, these two atoms would be those which would increase their distance from O4 (see Fig. 6). Thus, the formation of the short O3—O4 distance would not occur. On the other hand, one could equally select a short distance together with the apically opposite O3 as the appropriate model. Within a few estimated standard deviations of R O3/O4 ≥4, models would employ rigid octahedra. For example, for R O3/O4 = 4 a model can be envisaged in which only two O1 nearest neighbors of O4 would be displaced towards the O3 positions, the other two displaced O1 being those apically opposite. This case would be in a way equivalent to that corresponding to R O3/O4 = 2 as only two O3 would be nearest neighbor to O4 and, therefore, the formation of the short bond would not occur. For R O3/O4 = 6 it would be like the case given above for R O3/O4 = 3, but employing rigid octahedra. More complex models involving close or adjacent O4 defects are also possible. For example, a rigid octahedron model can be proposed for R O3/O4 = 3, if one assumes that two extra oxygen atoms occupy two O4 adjacent positions. These two O4 would share one O3 which would necessarily form at least one short bond with one of these two O4. There would be one short bond for every two O4. It can be seen that the structural results lead to several models, corresponding to the formation of either one short bond with a peroxide character or of all normal distances. Unfortunately, the precision of the ratio O3/O4 does not allow us to draw any definite conclusion.

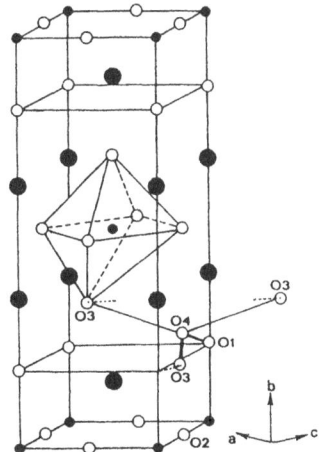

Figure 5 : A schematic representation of the structure of La₂CuO₄.₀₄₈

138

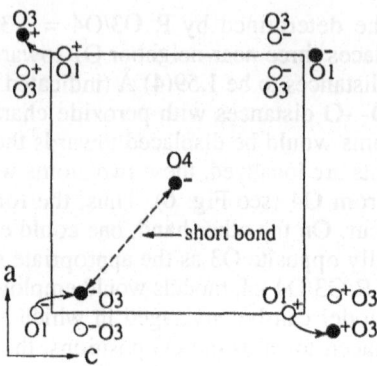

Figure 6 : A projection of the section containing O1, O3, and O4 atoms on the ac plane (only a/2 by c/2 shown ; - and + superscripts indicate atoms below or above the y = 1/4 plane, respectively). This model corresponds to R O3/O4 = 3.

2.4. POWDERS PREPARED BY ELECTROCHEMISTRY

The first x-ray diffraction studies of $La_2CuO_{4.09}$ powders prepared by electrochemical procedures [17], revealed that the orhorhombic distortion of these samples is larger than that of La_2CuO_4. This is exactly the opposite of what happens for the samples prepared under oxygen pressure for which $\delta \approx 0.03$. For the latter samples, in fact, the orthorhombic distortion is smaller than the undoped one. Subsequent studies [16, 28] showed that there are actually two distinct domains of oxygen stoichiometry. For $\delta \leq 0.05$ there would exist a superconducting phase ($T_c \approx 32$ K) for which the orthorhombic distortion would decrease with increasing δ ; on the contrary, for the highly oxidized samples ($T_c \approx 45$ K) the orthorhombic distortion would increase with increasing δ. This behavior has been explained independently by two groups [16, 28]. For low values of δ, the insertion of oxygen atoms into the strucure of La_2CuO_4 would induce an increase of the La-O distances. On the other hand, because of the hole-doping on the CuO_2 planes, the Cu-O distances decrease. The consequence of this situation is that the strain existing between the LaO and CuO_2 planes decreases. Since the orthorhombicity depends upon these strains, it decreases with increasing oxygen insertion. Such an analysis is in agreement with that proposed by Chaillout et al. [26, 29] based on the single-crystal $La_2CuO_{4.032}$ data.

For the high oxygen doping, Chou et al. [16] and Grenier et al. [28] suggest that the additional oxygen atoms could order on a sublattice of interstitial sites and the charge transfer toward the CuO_2 planes would no longer take place. The strain between LaO and CuO_2 planes would no longer decrease and the oxygen insertion would induce an increase of the orthorhombic distortion. By theoretical calculations Loktev et al. [30] showed that a superstrucure should exist for $La_2CuO_{4.125}$.

The structural characterization performed for the highly oxidized samples is not as detailed as that carried out for the low oxidized ones. In particular, no low-temperature studies have been performed in order to show whether or not the phase separation occurs for the high concentrations of extra oxygen atoms.

3. Physical properties

The superconducting transition temperatures, measured either by a.c. susceptibility or by resistivity, of samples prepared under high oxygen pressure (powders or single crystals) range usually between 35K and 40K. For powders with different δ values, Dabrowski et al. [25] observed an anomaly between 150K and 300K in the variation versus temperature of the resistivity. This anomaly does not occur in samples treated under N_2 which do not present the phase separation at low temperature. This behavior suggests that the anomaly in the resistivity is related to the phase separation.

Later, Hundley et al. [31] measured the in-plane (parallel to the CuO_2 plane) and out-of-plane (perpendicular to the CuO_2 plane) resistivity, the thermoelectric power and the magnetic susceptibility for superconducting single crystals of $La_2CuO_{4+\delta}$ samples prepared under oxygen pressure. Distinct anomalies were observed for all three quantities between 200K and 280K, which confirms the results of Dabrowski et al. [25]. The in-plane resistivity shows a large ($\approx 30\%$) and abrupt change beginning at 280K, whereas the out-of-plane resistivity shows only a tiny drop upon cooling at the same temperature. It seems from these results that the transport properties along the b axis are only slightly influenced by the phase separation. The thermoelectric power is positive at all temperatures, which is indicative of a hole-like transport which is corroborated by the Hall-effect measurements [32]. The thermoelectric power drops significantly at the phase separation temperature and its variation, as well as that of the in-plane resistivity, exhibit clear signs of hysteresis upon warming and cooling. This behavoir indicates that the phase separation process is weakly first order.

More recently, Ryder et al. [33] studied the transport and magnetic properties of powder samples of $La_2CuO_{4+\delta}$. They confirmed the presence of a bump in the resistivity between 200K and 240K and reported that the samples, which had been quenched between room temperature and 4.2K, had a broader superconducting transition than those which had been slowly-cooled. In agreement with the results of Hundley et al. [31], they observed hysteresis between 200K and 280K in the variation of the resistivity of slowly-cooled samples.

All the transport data can be consistently explained by the migration of the excess oxygen atoms below 280K and the formation of regions of samples which are metallic (and ultimately superconducting) and regions that are highly insulating and magnetic.

For $La_2CuO_{4+\delta}$ powders prepared under oxygen pressure, Oda et al. [34] observed three distinct T_C values depending on the cooling process of the samples. On quenched samples one T_C onset was observed at 38K, in agreement with the previously mentioned results ; but the samples which had been either slowly-cooled or quenched from below 200K, presented two T_C onsets at 48K and 42K.

Superconducting transition temperatures higher than 40K have also been reported in the case of samples prepared by electrochemistry or by treatment in $KMnO_4$ (44K [13], 45K [14], 40K [35], 40K [19]). As already mentioned in § II4, Chou et al. [16] associate an onset of $\approx$45K to a highly oxidized phase with $\delta \geq 0.08$. The variations vs temperature of the resistivity and of the thermoelectric power measured on powders oxidized by electrochemistry [17] do not exhibit the anomalies observed in the case of samples prepared under oxygen pressure.

Different mechanisms have been proposed in order to explain the phase separation process and the migration of the oxygen atoms. Some authors [31] believe that magnetic fluctuations involving Cu moments have to be taken into consideration ; others [33] reject the idea of a magnetic origin. The question is still open.

4. Spectroscopy

As soon as superconductivity was discovered in $La_2CuO_{4+\delta}$, the question of the nature of the excess oxygen arose : O^{2-}, O^-, or O_2^-? It is indeed important to know whether or not the occurrence of superconductivity in this compound is related to the presence of a peroxide or superoxide bond. We have seen that a definitive answer cannot be given from structural studies.

The first XPS experiments carried out by Rogers et al. [11] led to the conclusion that the excess oxygen was incorporated in the structure as a superoxide ion, O_2^-. This interpretation was contested by Zhou et al. [12] who thought that the superoxide species identified by Rogers et al. were associated with superficial oxygen. The polemics about this question went on with the studies of Strongin et al. [36] who explained that the peak observed by Rogers et al. at 532 ev can be associated more satisfactorily with a peroxide species than to a superoxide one. These authors note that a peak has been observed at the same energy by Buttrey et al. [21] in $La_2NiO_{4+\delta}$ and has been assigned to peroxide. The interpretation of Strongin et al. is still contested by Shinn et al. [37].

Besides, Mc Carty et al. [38] studied $La_2CuO_{4+\delta}$ single crystals by Raman scattering. They observed that a phonon peak is present at 630 cm^{-1} in the oxidized compound but not in La_2CuO_4. They concluded that, if this peak results from a local vibration mode of a dioxygen species in the structure, then the observed frequency is consistent with a peroxide bond and not with a superoxide one.

The presence of a peroxide bond between the interstitial oxygen and an oxygen atom of the lattice is also corroborated by the reports of Shinn et al. [39] who studied the thermal desorption of isotopically-labelled oxygen from $La_2CuO_{4+\delta}$ single crystals prepared under high oxygen presuure. They argue that up to 50% of the lattice ions can be exchanged with interstitial oxygen during the high pressure annealing and that the activation energy for this exchange is apparently less than the activation energy for oxygen diffusion process through the bulk.

After the announcement of the possible oxidation of La_2CuO_4 by electrochemistry [13], the question of the presence of OH^- ions arose. However, the chemical analyses and thermogravimetric measurements of Grenier et al. [17] proved that the intercalated species are not OH^-. This was also corroborated by the absence of an OH^- band in the infra-red spectra and of protons in NMR spectra of these samples [40].

The last point concerns the NMR and NQR experiments of Ueda et al. [41] (powder samples) and Hammel et al. [42,43] (single crystals) who studied the signals from Cu and La. They confirm the phase separation occurring in $La_2CuO_{4+\delta}$ into an antiferromagnetic phase and a paramagnetic one which becomes superconducting at low temperature. In the superconducting phase, two peaks are present in the La spectrum, attributed to two distinct La sites in the structure. These results are in agreement with the structural ones [26,29] which indicate that some of the La atoms are coordinated to nine oxygen atoms while others are coordinated to ten.

5. Electron microscopy

The position occupied by the excess oxygen in the structure of $La_2CuO_{4+\delta}$ has been determined by single crystal neutron diffraction studies. However, no information has been obtained concerning the possible ordering of these atoms below the phase separation temperature. Electron diffraction and microscopy are more appropriate techniques to handle this problem. Ryder et al. [33] have performed such experiments on $La_2CuO_{4+\delta}$ powders prepared under high oxygen pressure. When cooling from room temperature to 100K, they observed the development of a strong contrast between white and black fringes inside each orthorhombic twin domain. The fringes periodicity is about 300Å. From one domain to the adjacent one the orientation of the fringes changes and forms an herringbone arrangement. These authors associate this contrast with the distortion of the cations sublattice created by the incorporation and the ordering of oxygen atoms.
For samples oxidized at room temperature (by electrochemistry or oxidation in $KMnO_4$), Takayama et al. [19] and Weill et al. [44] have observed superstructure spots, which suggests that such preparation methods allow the ordering of the incorporated oxygen atoms. This is in agreement with the suggestion by Chou et al. [16] and Grenier et al. [28] (see § II). However, no structural model has been proposed so far.

6. Conclusion

Among high temperature superconducting oxides, $La_2CuO_{4+\delta}$ is still one of the most studied compounds as the appearance of superconductivity is not really understood. The possibility to oxidize La_2CuO_4 at room temperature by using soft chemistry methods has given a revival of interest for the compound since larger δ values can be reached.
It seems that two regions have to be considered depending on the extra oxygen concentration : first, one with $\delta \leq 0.05$, which corresponds to a superconducting phase with $T_c \approx 32K$; another one with $\delta \geq 0.08$, which corresponds to a superconducting phase with $T_c \approx 45K$. The former, obtained as early as 1987 by treating La_2CuO_4 under high oxygen pressure has been extensively studied from the structural and physical properties point of view. Below $\approx 280K$, the excess oxygen atoms migrate in order to form antiferromagnetic La_2CuO_4 domains and $La_2CuO_{4+\delta}$ ones which become superconducting below $\approx 32K$. The migration of the oxygen atoms is accompanied by anomalies in the resisitivity and in the thermoelectric power. The nature of the driving force of this mechanism is not yet clearly understood. From neutron diffraction studies the position of the excess oxygen atoms has been determined as well as the structural distortions related to the oxygen incorporation. The nature of the bonding between the excess oxygen and the "normal" oxygen atoms of the structure has not yet been elucidated. The structural analysis does not allow one to conclude definitively between a "normal" and a peroxide bond and controversy still exits concerning the interpretation of the spectroscopy results.

The more oxidized samples, recently obtained and only as powders, are not so well characterized. For example, it is not clearly established whether or not these samples present a phase separation at low temperature. In case of a negative answer, one could wonder whether the ordering of the excess oxygen suggested by the superstructures spots may interfere with its diffusion. This ordering has to be confirmed and studied in detail. It would be also worthwhile to study samples which have been slowly-cooled below 200K and which exhibit a superconducting critical temperature of 48K. One could determine if these samples are similars, from the structural point of view, to the ones prepared by electrochemistry, i.e. at room temperature. All these conjectures and speculations make $La_2CuO_{4+\delta}$ still an interesting compound.

References

[1]   J.B. Bednorz, K.A. Müller, Z. Phys. B, 64, 189, (1986)
[2]   A. Tressaud, B. Chevalier, B. Lepine, K. Amine, L. Lozano, E. Marquestaut,
        J. Etourneau, Eur. J. Solid State Inorg. Chem., 309, 309, (1990)
[3]   A. Tressaud, C. Robin, B. Chevalier, L. Lozano, J. Etourneau,
        Physica C, 177, 330, (1991)
[4]   J. Beille, R. Cabanel, C. Chaillout, B. Chevalier, G. Demazeau, F. Deslandes,
        J. Etourneau, P. Lejay, C. Michel, J. Provost, B. Raveau, A. Sulpice,
        J.L. Tholence, R. Tournier,
        C.R.Acad. Sc. Paris, t. 304, série II n°18, 1097, (1987)
[5]   P.M. Grant, S.S.P. Parkin, V.Y. Lee, E.M. Engler, M.L. Ramirez, G. Lim,
        R.D. Jacowitz, Phys. Rev. Lett., 58, 2482, (1987)
[6]   K. Sckizawa, Y. Takano, S. Takigami, T. Inaba,
        Jpn. J. Appl. Phys., 26, L840, (1987)
[7]   S.A. Shaheen , N. Jisrawi, Y.H. Lee, Z. Zhang, M. Croft, W.L. Mc Lean,
        H. Zhen, L. Rebelsky, S. Horn, Phys. Rev. B, 36, 7214, (1987)
[8]   J.E. Schirber, J.F. Kwak, E.L. Venturini, B. Morosin, D.S. Ginley, W.S. Fu,
        R.J. Baughman, Proc. Boston Fall Meeting (1987)
[9]   G. Demazeau, F. Tresse, Th. Plante, B. Chevalier, J. Etourneau, C. Michel,
        M. Hervieu, B. Raveau, P. Lejay, A. Sulpice, R. Tournier,
        Physica C, 153-155, 824, (1988)
[10]   J.E. Schirber, B. Morosin, R.M. Merill, P.F. Hlava, E.L. Venturini,
        J.F. Kwak, P.J. Nigrey, R.J. Baughman, D.S. Ginley,
        Physica C, 152, 121, (1988)
[11]   J.W. Rogers Jr, N.D. Shinn, J.E. Schirber, E.L. Venturini, D.S. Ginley,
        B. Morosin, Phys Rev. B, 38, 5021, (1988)
[12]   J. Zhou, S. Sinha, J.B. Goodenough, Phys. Rev. B, 39, 12331, (1989)
[13]   A. Wattiaux, J. C. Park, J.C. Grenier, M. Pouchard,
        C.R.Acad. Sc. Paris, t. 310, série II , 1047, (1990)
[14]   J.C. Bennett, M. Olfert, G.A. Scholz, F.W. Boswell,
        Phys. Rev. B, 44, 2727, (1991)
[15]   R. Suryanarayanan, O. Gorochov, M.S.R. Rao, L. Ouhammou, W. Paulus,
        G. Heger, Physica C, 185-189, 573, (1991)
[16]   F.C. Chou, J.H. Cho, D.C. Johnston, Physica C, 197, 303, (1992)
[17]   J.C. Grenier, A. Wattiaux, N. Lagueyte, J.C. Park, E. Marquestaut,
        J. Etourneau, M. Pouchard, Physica C, 173, 139, (1991)
[18]   P. Rudolf, W. Paulus, R. Schöllhorn, Adv. Mater., 3 (9), 438, (1991)
[19]   E. Takayama-Muromachi, T. Sasaki, Y. Matsui, Physica C, 207, 97, (1993)
[20]   D.J. Buttrey, P. Ganguly, J.M. Honig, C.N.R. Rao, R.R. Schartman,
        G.N. Subbanna, J. Solid State Chem., 74, 233, (1988)
[21]   A. Santoro, F. Beech, M. Marezio, R.J. Cava, Physica C, 156, 693, (1988)
[22]   J.D. Jorgensen, B. Dabrowski, S. Pei, D.G. Hinks, L. Soderholm,
        B. Morosin, J.E. Schirber, E.L. Venturini, D.S. Ginley,
        Phys. Rev. B, 38, 11337, (1988)
[23]   D.C. Johnston, J.P. Stokes, D.P. Goshorn, J.T. Lewandowski,
        Phys. Rev. B, 36, 4007, (1987)

144

[24]   P. Zolliker, D.E. Cox, J.B. Parise, E.M. Mc Carron III, W.E. Farneth,
Phys. Rev. B, 42, 6332, (1990)

[25]   B. Dabrowski, D.G. Hinks, J.D. Jorgensen, D.R. Richards,
Materials Research Society Symposium Proceedings, vol. 156, 69, (1989)

[26]   C. Chaillout, J. Chenavas, S.W. Cheong, Z. Fisk, M. Marezio, B. Morosin,
J.E. Schirber, Physica C, 170, 87, (1990)

[27]   C. Chaillout, P. Bordet, J. Chenavas, S.W. Cheong, Z. Fisk, M. Marezio,
B. Morosin, J.E. Schirber, Proc. Boston Fall Meeting , vol. 169, (1989)

[28]   J.C. Grenier, N. Lagueyte, A. Wattiaux, J.P. Doumerc, P. Dordor,
J. Etourneau, M. Pouchard, Physica C, 202, 209, (1992)

[29]   C. Chaillout, S.W. Cheong, Z. Fisk, M.S. Lehmann, M. Marezio,
B. Morosin, J.E. Schirber, Physica C, 158, 183, (1989)

[30]   V.M. Loktev, H.M. Tatarenko, Phys. Stat. Sol. (b), 166, 191, (1991)

[31]   M.F. Hundley, J.D. Thompson, S.W. Cheong, Z. Fisk, J.E. Schirber,
Phys. Rev. B, 41, 4062, (1990)

[32]   H. Kitazawa, K. Katsumata, Physica C, 185-189, 1255, (1991)

[33]   J. Ryder, P.A. Midgley, R. Exley, R.J. Beynon, D.L. Yates, J.A. Wilson,
Physica C, 173, 9, (1991)

[34]   Y. Oda, A. Sumiyama, T. Kohara, M. Yamada, K. Asayama, S. Kashiwai,
M. Motoyama, Physica C, 185-189, 917, (1991)

[35]   G. Rajaram, R. Sryanarayanan, N. LeNagard, O. Gorochov, L. Ouhammou,
Physica C, 199, 139, (1992)

[36]   M. Strongin, S.L. Qiu, J. Chen, C.L. Lin, E.M. Mc Carron,
Phys. Rev. B, 41, 7238, (1990)

[37]   N.D. Shinn, J.W. Rogers Jr., J.E. Schirber, Phys. Rev. B, 41, 7241, (1990)

[38]   K.F. Mc Carty, J.E. Schirber, S.W. Cheong, Z. Fisk,
Phys. Rev. B, 43, 7883, (1991)

[39]   N.D. Shinn, M. E. Bartram, J.E. Schirber, J.W. Rogers Jr, D.L. Overmyer,
Z. Fisk, S.W. Cheong, Physica C, 179, 303, (1991)

[40]   J. Grenier, A. Wattiaux, J.P. Doumerc, P. Dordor, L. Fournes,
J.P. Chaminade, M. Pouchard, J. Solid State Chem., 96, 20, (1992)

[41]   K. Ueda, T. Sugata, Y. Kohori, T. Kohara, Y. Oda, M. Yamada,
Solid State Comm., 73, 49, (1990)

[42]   P.C. Hammel, E.T. Ahrens, A.P. Reyes, R.H. Heffner, P.C. Canfield,
S.W. Cheong, Z. Fisk, J.E. Schirber, Physica C, 185-189, 1095, (1991)

[43]   P.C. Hammel, A.P. Reyes, Z. Fisk, M. Takigawa, J.D. Thompson,
R.H. Heffner, S.W. Cheong, J.E. Schirber, Phys. Rev. B, 42, 6781, (1990)

[44]   N. Lagueyte, F. Weill, A. Wattiaux, J.C. Grenier, submitted

ELECTRON ENERGIES IN OXIDES

JOHN B. GOODENOUGH
Center for Materials Science and Engineering, ETC 9.102
University of Texas at Austin, Austin, TX 78712-1084

ABSTRACT. Construction of outer-electron energies in oxides is reviewed within the context of an ionic model. The $5s$ and $6s$ conduction bands of Group-B metals are distinguished from the s bands of other main-group oxides. Localized $4f^n$ configuration energies are located relative to the edges of the broad bands in rare-earth oxides, and some physical properties imparted by $4f$ electrons in mixed-valent or intermediate-valent oxides are indicated. Conditions for localized *versus* itinerant d electrons and for ionic *versus* covalent bonding in the transition-metal oxides are stressed in addition to the location of the Fermi energy relative to the edges of the broad s and p bands.

1. INTRODUCTION

In a conventional superconductor, superconductivity is due to a condensation of electron pairs from a normal metallic state; the pairs are coupled by a retarded electron-phonon interaction and have a coherence length $\xi \sim 1000$ Å. The superconductive pairs, known as Cooper pairs, are itinerant electrons of wave vector **k** and **-k**, of spin s = 1/2 and -1/2; the characteristic energy of the coupling $k\theta_D$ is determined by the Debye temperature θ_D. Condensation opens up an energy gap Δ at the Fermi surface on cooling through a critical temperature T_c; strong overlap of the superconductive pairs results in a second-order transition at T_c that is well-described by mean-field theory with an order parameter Δ/Δ_0, where Δ_0 is the gap energy at T = 0 K [1]. It follows that

•Any other phenomenon that introduces an energy gap at the Fermi energy E_F competes with superconductivity; *superconductivity can be viewed as nature's last resort to stabilize occupied states at the expense of the empty states of a normal metal.*
•Any phenomenon that removes the spin degeneracy of the metallic state (*e.g.* ferromagnetism or paramagnetic impurities that introduce an internal magnetic field at neighboring metallic atoms) suppresses the formation of Cooper pairs.

If we are to understand superconductivity in oxides, it is important to ask ourselves some fundamental questions. In this lecture I address two such questions:

•Where can we expect to find metallic oxides?
•What mechanisms compete with superconductivity in oxides?

My second lecture explores the question

•What is unusual about the normal state of the copper-oxide superconductors?

E. Kaldis (ed.), Materials and Crystallographic Aspects of HTc-Superconductivity, 145–160.
© 1994 *Kluwer Academic Publishers.*

146

My final lecture reviews some aspects of the structural architecture of the copper-oxide superconductors in order to examine

•What distinguishes the n-type from the p-type copper-oxide superconductors?

2. PRELIMINARIES

It is useful to consider initially three quite different classes of metal oxides: the main-group oxides, the d-block transition-metal oxides, and the rare-earth oxides [2]. The main-group elements have only s and p electrons in partially occupied outer shells, and only these valence electrons are active in bonding. A partially filled $4f$ shell on a rare-earth atom or ion forms a localized $4f^n$ configuration having an energy E_n that is separated from the $4f^{n+1}$ and $4f^{n-1}$ configuration energies E_{n+1} and E_{n-1} by a large, on-site electron-electron coulomb energy $U \sim 10$ eV. The outer $4f$ electrons of a rare-earth atom do not contribute significantly to the bonding. A partially filled d shell of a d-block transition-metal atom participates significantly in the metal-oxygen (M-O) bonding; this participation gives rise to the crystal-field splitting and the formation of crystal-field wave functions. However, whether the electrons of metal-atom d parentage remain in localized d^n configurations or are delocalized in a particular oxide depends on the magnitude of the intraatomic electron-electron coulomb energy U separating configurations E_n and E_{n+1} relative to the interatomic metal-metal (M-M) and/or metal-oxygen-metal (M-O-M) interactions as well as the location of E_F relative to the top of the O-2p bands.

A measure of the strength of the interatomic M-M or M-O-M interactions is the bandwidth W; and in tight-binding theory

$$W \approx 2 zb \tag{1}$$

where b is a spin-independent resonance integral for like nearest-neighbor atoms:

$$b_{ij} \equiv (\psi_i, H'\psi_j) \approx \varepsilon (\psi_i, \psi_j) \tag{2}$$

H' is the perturbation of the potential at $\mathbf{R}_j$ due to the presence of a metal atom at $\mathbf{R}_i$ and ε is a one-electron energy. Where E_F lies above the top of the O-2p bands, the M-O bonding can be treated as ionic, and the ψ_i, ψ_j are the antibonding crystal-field wavefunctions at metal-atom position $\mathbf{R}_i$ and $\mathbf{R}_j$. From (2) we may obtain a qualitative estimate of W from the overlap integral (ψ_i, ψ_j). The d^n configurations remain localized for W < U; they are delocalized for W > U. Where a W $\approx$ U occurs, electron-lattice interactions may stabilize a static charge-density wave (CDW) or electron-electron interactions a spin-density wave (SDW); a CDW and/or a SDW change the translational symmetry of the lattice so as to open up an energy gap at the Fermi energy E_F, or at least over a portion of the Fermi surface. Therefore the CDW and the SDW generally compete with superconductivity.

Table 1 contrasts several insulating oxides with some isostructural metallic oxides that contain a metal atom in a single valence state. It should be noted that the insulator SnO_2 and the metal PbO_2 are both main-group oxides, the ferromagnetic insulator EuO and the antiferromagnetic metal GdO are both rare-earth oxides, and the d-block transition-metal oxides range from antiferromagnetic insulators (MnO and $LaFeO_3$) to compounds exhibiting first-order insulator-metal transitions (VO_2 and $NdNiO_3$) to metals with and without a superconductive transition at lowest temperatures (TiO and $LaNiO_3$).

The spinel structure lends itself to the possibility of mixed valency. Mn_3O_4, for example, contains Mn^{2+} ions on tetrahedral sites and Mn^{3+} ions on octahedral sites; it is a ferrimagnetic

Table 1 : INSULATING *vs* METALLIC OXIDES
Single-Valent Compounds

Insulators		Metals	
MgO, EuO, MnO		GdO, TiO	(rocksalt)
SnO_2	VO_2	PbO_2	(rutile)
$LaFeO_3$	$NdNiO_3$	$LaNiO_3$	(perovskite)

Mixed Valent Spinels

$Mn^{2+}[Mn_2^{3+}]O_4$

$Li[Mn^{4+}Mn^{3+}]O_4$

$Fe^{3+}[Fe^{3+}Fe^{2+}]O_4$

$Li[V_2^{3.5+}]O_4$

$Li[Ti_2^{3.5+}]O_4$

Monoxides with the rocksalt structure

MgO	=	insulator ($E_g > 6$ eV)
EuO	=	ferromagnetic ($T_C = 69$ K, $\mu_{Eu} = 7\,\mu_B$) semiconductor ($E_g = 1.1$ eV)
MnO	=	antiferromagnetic ($T_N = 118$ K, $\mu_{Mn} = 5\,\mu_B$) insulator ($E_g = 3.2$ eV)
GdO	=	antiferromagnetic ($\mu_{Gd} = 7\,\mu_B$) metal
TiO	=	superconductor ($T_C = 1$ K)

Dioxides with rutile structure

SnO_2	=	insulator ($E_g = 3$ eV)
VO_2	=	insulator-metal transition $T_t = 67\ ^\circ C$
PbO_2	=	·metal

Perovskites

$LaFeO_3$	=	antiferromagnetic ($T_N = 750$ K) insulator
$NdNiO_3$	=	first-order antiferromagnetic-insulator to metallic Pauli paramagnet at $T_t = 201$ K

Spinels

$Mn^{2+}[Mn_2^{3+}]O_4$	=	tetragonal (c/a > 1), complex-spin-configuration ferrimagnetic insulator
$Li[Mn^{3+}Mn^{4+}]O_4$	=	small-polaron conductor
$Fe^{3+}[Fe^{2+}Fe^{3+}]O_4$	=	ferrimagnet, semiconductor-semiconductor transition at $T_V = 119$ K
$Li[V_2^{3.5+}]O_4$	=	metallic, Curie-Weiss paramagnet
$Li[Ti_2^{3.5+}]O_4$	=	superconductor ($T_C = 13$ K)

insulator with localized d^5 and d^4 configurations. E_F lies in an energy gap between the occupied d^5 configuration on tetrahedral-site Mn^{2+} and an empty d^5 configuration at the octahedral-site Mn^{3+} ions. A cooperative Jahn-Teller distortion associated with the Mn^{3+} ions distorts the crystal to tetragonal ($c/a > 1$) symmetry.

In the remaining spinels of Table 1, equal concentrations of two different valance states occupy energetically equivalent sites. This situation places E_F at the standard redox potential for the given redox couple. The critical competition in this case is the time $\tau_h \approx \hbar/W$ for an electron transfer from a d^{n+1} to a d^n configuration relative to the period ω_R^{-1} of the local lattice deformation that would trap the mobile electron at sites of larger M-O bond length. Electrons "dressed" in a local lattice deformation are called *small polarons*. Small polarons have a diffusional motion; their mobility contains a motional enthalpy, which results in a semiconductor temperature dependence of the conductivity. In Table 1 we see an evolution from a small-polaron semiconductor $Li[Mn_2]O_4$ with $\tau_h > \omega_R^{-1}$ to a superconductor $Li[Ti_2]O_4$ with $\tau_h < \omega_R^{-1}$. Both local moments, as in Fe_3O_4 and $Li[V_2]O_4$, and the lattice deformations associated with small-polaron formation compete with superconductivity.

3. MAIN-GROUP OXIDES

The outer s and p electrons have relatively large radial extensions of their wavefunctions beyond the core electrons, which makes them particularly active in bonding; in solids they are generally described by band theory ($W > U$). Metal oxides have a strong ionic component to their bonding; the antibonding bands of s and p parentage have a primarily metal-atom character and the bonding band a primarily oxygen character as illustrated schematically in Fig. 1. Since the metal and oxygen atoms are distinguishable, an oxide is a non-Bravais lattice; the bottom of the antibonding s band generally lies above the top of the bonding p bands by an energy gap E_g.

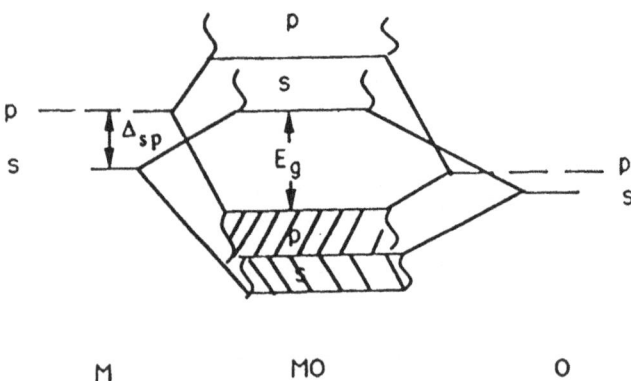

FIG. 1. Outer s and p electrons in main-group oxides

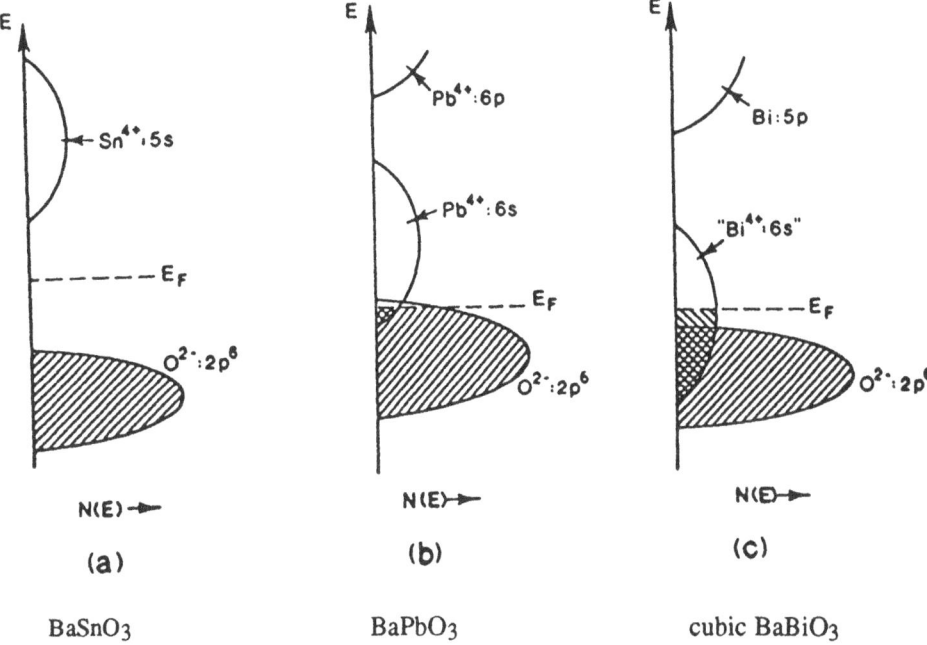

FIG 2.   Effect of a larger Δ_{sp} in heavy Group-B metal oxides.

FIG. 3.   Oxygen displacement in distortion from cubic to monoclinic BaBiO$_3$.

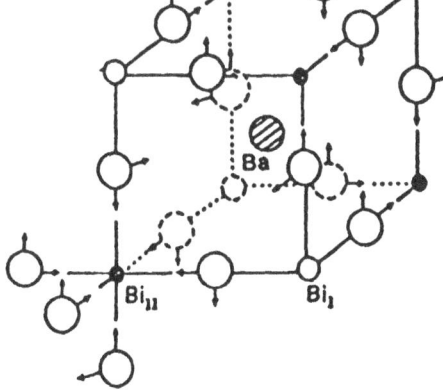

The splitting Δ_{sp} between s and p states of a given atomic species reflects the different screening parameters σ_l in the effective nuclear charge $(Z - \sigma_l)e$ seen by electrons of different quantum number l for the same quantum number n. Δ_{sp} increases with the number of screening core electrons and hence with the mass of the atom. In most main-group oxides, the bottom of the s band remains too high in energy to be accessible to electrons; attempts to dope n-type an oxide like MgO or Al$_2$O$_3$ lead to the formation of native defects. However, the heavy Group-B metals have a large Δ_{sp}, which allows $5s$ or $6s$ electrons to be introduced into their oxides. In SnO$_2$, for example, the Sn-$5s$ band lies about 3.0 eV above the top of the O-$2p$ bands, and this compound may be doped n-type to become a metal transparent to visible light. In PbO$_2$, the $6s$ band overlaps the top of the O-$2p$ bands, which is why it is metallic. Fig. 2 illustrates the analogous situation in the perovskites BaSnO$_3$ and BaPbO$_3$. However it is more common to encounter $5s$ or $6s$ electrons as highly polarizable $5s^2$ or $6s^2$ lone pairs at, for example, Sn^{2+} or Pb^{2+} or Bi^{3+} ions. In fact, partially filled $5s$ and $6s$ bands at heavy Group-B metals tend to be unstable relative to a disproportionation reaction that stabilizes $5s^2$ or $6s^2$ cores at metal atoms of longer M-O bond lengths and no $5s$ or $6s$ electrons at metal atoms with shorter M-O bond lengths. The perovskite BaBiO$_3$ provides an illustration of such a reaction.

The half-filled Bi-$6s$ band shown for cubic BaBiO$_3$ in Fig. 2(c) is unstable with respect to the monoclinic distortion illustrated in Fig. 3. There are two components to the distortion: a displacement of the oxygen atoms away from the Bi$_I$ atoms towards the Bi$_{II}$ atoms and a cooperative tilting of the Bi$_{II}$O$_6$ octahedra. The first component changes the translational symmetry of the crystal so as to halve the volume of the first Brillouin Zone, thereby opening an energy gap E_g at E_F. The distortion stabilizes occupied states at the expense of empty states to transform the compound from a metal to a semiconductor. In the limit

$$2 \, \text{Bi (IV)} \rightarrow \text{B}_I \, \text{(III)} + \text{Bi}_{II} \, \text{(V)} \tag{3}$$

the Bi$_I$ contain a spin-paired $6s^2$ core and the Bi$_{II}$ have no $6s$ electrons. In such an internal redox reaction, the electrostatic energy U lost to create a pair of $6s^2$ electrons at a Bi$_I$ must be more than compensated by the added covalent bonding at the Bi$_{II}$, so such a disproportionation is called a "negative-U" charge-density wave (CDW). In fact, the opening of a gap at E_F does not require the transfer of an integral electron charge $-e$ from the Bi$_{II}$ to the Bi$_I$; the measured charge transferred in this case is about half an electron charge:

$$2 \, \text{Bi (IV)} \rightarrow \text{Bi}_I \, \text{(IV} - \delta) + \text{Bi}_{II}(\text{IV} + \delta) \tag{3'}$$

with $\delta \approx 0.5$ [3]. Formation of the CDW suppresses superconductivity.

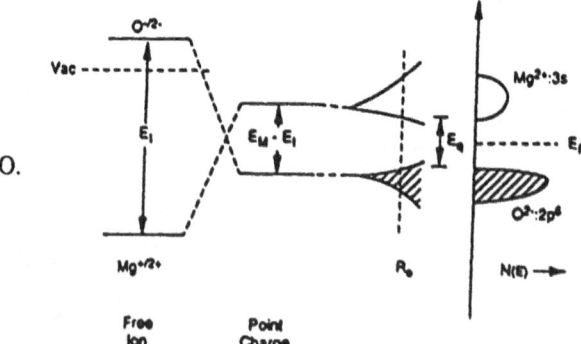

FIG. 4. Ionic model for MgO.

An alternate approach to the construction of the broad s-p bands of a main-group oxide is the ionic model illustrated for MgO in Fig. 4. This model begins with the energy E_I to remove the last electron from a metal atom and place it on an O^- ion at infinite separation so as to convert it to an O^{2-} ion. The free-ion redox energy O^-/O^{2-} lies above the vacuum level (Vac); the O^- ion has a negative electron affinity. Next, the ions are assembled on the lattice positions of an MgO crystal; if they can be considered point charges, we may calculate an electrostatic lattice energy

$$E_M = N\Sigma_j (\pm)q^2/r_j$$

which is called the Madelung energy. For MgO, $q = \pm 2e$ is the magnitude of the point charge. Such an ionic model requires an $E_M > E_I$ in order to stabilize O^{2-} ions. Introduction of covalence into this model via second-order perturbation theory admixes O-$2p$ character into the antibonding conduction band and Mg-$3s$, $3p$ character into the bonding O-$2p$ bands. Although back transfer of charge lowers the effective magnitude of q and hence E_M, the quantum-mechanical repulsion between bonding and antibonding states compensates for the reduction in E_M - E_I, and calculation of the binding energy of the solid from a point-charge model gives a surprisingly good result. In such a model, the O-$2p$ and Mg-$3s$ levels are broadened into energy bands by O-O and Mg-Mg interactions.

In summary, the main-group oxides are mostly insulators with a large energy gap E_g between the bottom of a metal-s conduction band and the top of the O-$2p$ bands. However, the $5s$ bands of a Group-B metal such as tin are energetically accessible; SnO_2, for example, can be doped so as to make it an n-type metal. The 6s bands of the Group-B metals Tl, Pb, and Bi may be sufficiently stable for an overlap of the O-$2p$ and M-6s bands as in PbO_2 and $BaPbO_3$. However, a near-half-band occupancy of a $6s$ band may lead to a disproportionation reaction that introduces an energy gap at E_F, thus suppressing metallic conductivity and any possibility of a superconductive state. Superconductivity is found in the perovskite systems $BaBi_{1-x}Pb_xO_3$ and $Ba_{1-x}K_xBiO_3$ only where the 6s bands are sufficiently depleted of electrons that the CDW state is suppressed, see Fig. 5.

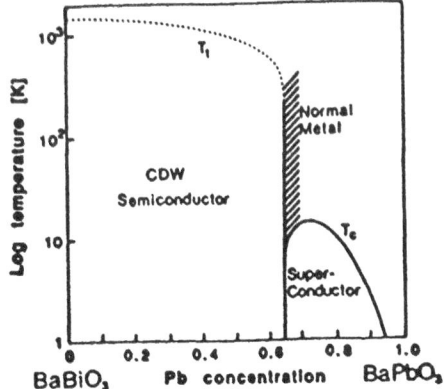 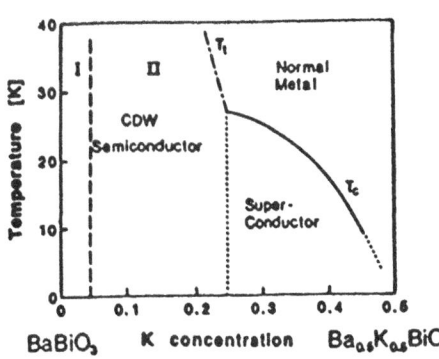

FIG. 5.   Preliminary phase diagrams for
the systems $BaBi_{1-x}Pb_xO_3$ and $Ba_{1-x}K_xBiO_3$

4. RARE-EARTH OXIDES

The energies of the $5d$ and $6s$ states of the rare-earth oxides are similar, and the $5d$ orbitals have sufficient radial extension to act like the valence $6s$ and $6p$ orbitals; therefore the rare-earth $5d$ orbitals form antibonding conduction bands. Moreover, stronger covalent M-O mixing with the $6s$ bands leaves the bottom of the conduction band with primarily $5d$ character, particularly where there is M-M bonding across a shared octahedral-site edge as in the rocksalt structure. In fact, the bottom of the $5d$ band is energetically accessible to electrons in the heavier rare-earth monoxides, and metallic conductivity occurs where E_F falls in the $5d$ conduction band. We illustrate this phenomenon with EuO and GdO.

The Hamiltonian for a localized $4f^n$ configuration in an oxide is given by

$$H = H_0 + V_{el} + \Delta_{LS} + \Delta_c + H_z \qquad (4)$$

where the first three terms correspond to the free-ion Hamiltonian: H_0 is the spherical approximation and V_{el} is the electron-electron electrostatic term responsible for separating the E_n and E_{n+1} configurations by an energy U, which is reduced in an oxide by covalent screening to a $U \sim 10$ eV $> E_g$; Δ_{LS} is the spin-orbit-coupling energy responsible for multiplet splitting. The fact that Δ_{LS} is larger than the crystal-field splitting gives an atomic moment

$$\mu_J = gJ\mu_B \text{ with } g = 1 + \frac{J(J+1) + S(S+1) - L(L+1)}{2J(J+1)}$$

as on the free ion. The Zeeman splitting due to internal molecular fields is only found below a long-range magnetic-ordering temperature. With $U > E_g$, the position of the $4f^n$ - configuration energy E_n relative to the edge of the conduction band or to E_F is critical.

The half-filled $4f^7$ configuration at a Gd^{3+} or a Eu^{2+} ion is separated by a particularly large energy U from the $4f^8$ configuration. The $4f^7$ configuration at a Gd^{3+} sees a larger effective nuclear charge than that at a Eu^{2+} ion, so the $4f^7$ electrons are more tightly bound and lie at a significantly lower energy in GdO than in EuO. Nevertheless, the Gd $4f^8$ configuration lies above the bottom of the $5d$ band, so that GdO has one electron per molecule in the $5d$ conduction bands as illustrated in Fig. 6. With a $Gd:4f^7 5d^1$ configuration, GdO is metallic, and the $\mu_{Gd} = 7 \mu_B$ atomic moments are coupled via the $5d$ electrons by indirect exchange; the atomic moments magnetize locally the Gd-$5d$ electrons, which removes their spin degeneracy and suppresses superconductive pair formation. On the other hand, Gd_2O_3 is an insulator, and the Gd^{4+}: $4f^6$ state is energetically inaccessible.

In EuO, the $Eu^{2+}:4f^7$ level lies 1.1 eV below the bottom of the Eu - $5d$ conduction band; therefore EuO is a semiconductor [4]. Oxidation of EuO to $Eu_{1-\delta}O$ lowers E_F into the $Eu^{2+}:4f^7$ level. The $4f^7$ electrons are sufficiently localized that the lattice has time to relax and trap a $4f^6$ configuration at a Eu^{3+} ion having shorter Eu-O bonds. The small-polaron states are lifted above E_F by the reorganization energy in exact analogy with a redox couple in solution; however, the small-polaron hole can move diffusively to make $Eu_{1-\delta}O$ a p-type semiconductor. Reduction of EuO to $EuO_{1-\delta}$, on the other hand, creates oxygen vacancies that act as two-electron traps; the two-electron trap energy lies just below the bottom of the $5d$ conduction band, and a single-electron trap energy lies below it by a coulomb energy U_t. On cooling through the ferromagnetic Curie temperature T_c, the $5d$ states with spin parallel to the spin of the $4f^7$ configurations are stabilized relative to the states of antiparallel spin, and the trap state of antiparallel spin is raised above the bottom of the $5d$ band of parallel spin. Transfer of trapped electrons to the $5d$ band transforms $EuO_{1-\delta}$ from a semiconductor for $T > T_c$ to a metal for $T < T_c$; the conductivity

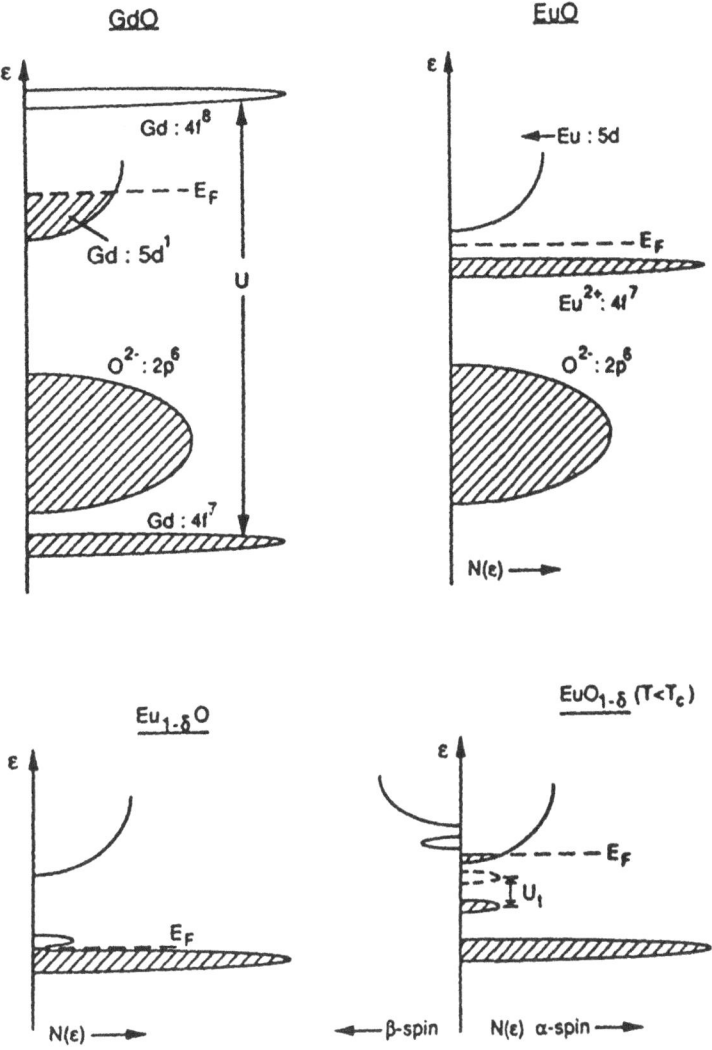

FIG. 6.   Placement of $4f^n$ levels relative to the edges of the
Ln-$5d$ and O-$2p$ bands in GdO and EuO.

changes by six orders of magnitude. However, removal of the spin degeneracy suppresses superconductive pair formation.

Under a pressure of 300 kbar, the $5d$ band is broadened sufficiently to overlap the $4f^7$ level in EuO. A first-order phase change with decrease in volume is associated with a transfer of about 0.5 electrons/Eu to the $5d$ band; hybridization of the $4f$ and $5d$ orbitals near E_F splits the $4f^7$ energy E_7 to stabilize half of the $4f^7$ configurations above E_F and half below. EuO is then said to be in an *intermediate-valance state* with valence fluctuations (VF) between the configurations $4f^6 5d^1$ and $4f^7 5d^0$ at the Eu atoms.

5. TRANSITION-METAL (d-BLOCK) OXIDES

The main-group oxides have only s and p valence electrons; for these electrons, interatomic interactions dominate the intraatomic interactions ($W >> U$), and conventional band theory is applicable. The rare-earth oxides contain, in addition, $5d$ bands having a $W_{5d} > U$ and localized $4f^n$ configurations in which the intraatomic interactions dominate the interatomic interactions ($W_{4f} << U_{4f}$). The outer d electrons of the d-block transition-metal oxides are of intermediate character. In some oxides the d^n configurations remain localized ($W_d < U_d$) and in others the d electrons form itinerant-electron energy bands ($W_d > U_d$). In every case, the M-O bonding is important; it gives rise to crystal-field splittings that are larger than the multiplet splitting Δ_{LS}, which introduces a *quenching* of the orbital angular momentum; the cubic component of the crystalline fields may even compete with the intraatomic energy U to stabilize a *low-spin*, localized d^n configuration. Therefore, it is customary to consider first the M-O interactions and their contributions to the crystal-field splitting of the d wavefunctions. For this purpose, we consider the angular dependences of the d wavefunctions as these determine the symmetry arguments for the M-O interactions.

The atomic d wavefunctions f have a z component of the orbital angular momentum defined by

$$L_z f = -i\hbar \partial f / \partial \varphi = m\hbar f \tag{6}$$

with azimuthal quantum number $m = 0, \pm 1, \pm 2$. Therefore we write

$$f_0 = [(z^2 - x^2) + (z^2 - y^2)]/r^2 = 3\cos^2\theta - 1 \qquad m = 0$$
$$f_{\pm 1} = 2 (yz \pm izx) /r^2 = \sin 2\theta \exp (\pm i \phi) \qquad m = \pm 1 \tag{7}$$
$$f_{\pm 2} = [(x^2 - y^2) \pm ixy] /r^2 = \sin^2\theta \exp (\pm i 2\phi) \qquad m = \pm 2$$

Provided the ionic condition $E_M - E_I > 0$ is fulfilled, back transfer of charge from the oxide ions to the *empty* d orbitals can be treated in second-order perturbation theory. This virtual back transfer of charge mixes into the atomic wavefunctions f some O-$2p$, $2s$ character having the same symmetry. With $E_M - E_I > 0$, the states f form antibonding M-O admixtures that destabilize the f orbitals by an amount

$$\Delta\varepsilon = | b^{ca} |^2 /\Delta E \tag{8}$$

where $\Delta E = E_M - E_I$ for the O-$2p$ orbitals and $\Delta E = \Delta E_s > E_M - E_I$ for the O - $2s$ orbitals is the energy required for the back transfer of an electron. The matrix element $b^{ca} \approx \varepsilon(f, \phi)$ is proportional to the resonance integral associated with an overlap of the interacting cation (f) and anion (ϕ) wavefunctions.

Fig. 7 illustrates the crystal-field splitting for a $^2D(d^1)$ configuration at an octahedral site. In cubic symmetry, the xy, yz, zx orbitals remain degenerate and overlap only with O-$2p_\pi$ orbitals;

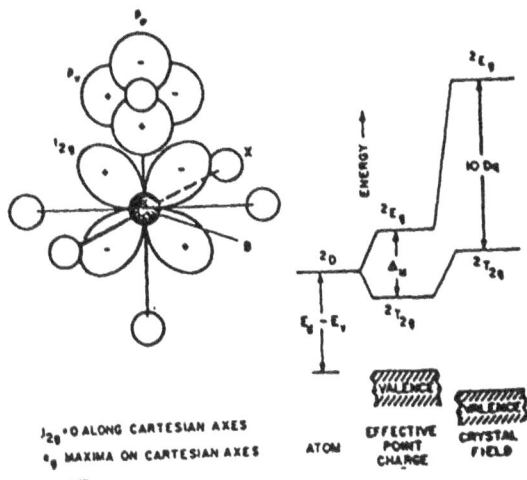

FIG. 7 Crystal-field splitting of a ^2D configuration in an octahedral-site crystalline field.

integration over all space makes vanish the overlap with O-$2p_\sigma$ and O-$2s$ orbitals. The $(x^2 - y^2)$ and $[(z^2 - x^2) + (z^2 - y^2)]$ orbitals form a second degenerate set that overlap the O-$2p_\sigma$ and O-$2s$ orbitals, but not the O-$2p_\pi$. We designate (Mulliken rotation) the orbitals of the threefold degenerate set by t and those of the twofold-degenerate set by e. The corresponding crystal-field wavefunctions are

$$\psi_e = N_t (f_t - \lambda_\pi \varphi_\pi) \tag{9}$$
$$\psi_e = N_e (f_e - \lambda_\sigma \varphi_\sigma - \lambda_s \varphi_s)$$

where the φ are linear combinations of O-$2p_\pi$, $2p_\sigma$, or $2s$ wavefunctions at the six neighboring oxygen atoms that have the same symmetry as the f orbital with which they mix. The

$$\lambda \equiv b^{ca}/\Delta E \tag{10}$$

are the covalent-mixing parameters. Because the overlap integrals (f, φ) in b^{ca} are larger for σ than for π bonding, the $^2E_g(e^1)$ state is raised higher in energy than the $^2T_{2g}(t^1)$ state according to Equation (8). The total cubic-field splitting is

$$\Delta_c = \Delta_M + (\lambda_\sigma^2 - \lambda_\pi^2) (E_M - E_I) + \lambda_s^2 \Delta E_s \tag{11}$$

where Δ_M is a small electrostatic component of uncertain sign. The dominant contribution to the crystal-field splitting comes from the covalent component of the M-O bond.

Two other features should be noted. First, splitting of the x^2-y^2 and xy orbitals quenches the orbital angular momentum associated with $f_{\pm 2}$; the two e orbitals have $m = 0$ and the three t orbitals have $m = 0, \pm 1$. It is customary to express the magnetic moment imparted by a d^n configuration as $\mu_A = g S \mu_B$ and to determine g experimentally. An effective $g \approx 2$ holds only where the orbital angular momentum is completely quenched. Second, the splitting of localized-electron configurations E_n and E_{n+1} becomes

$$U_{eff} = U + \begin{cases} 0 \\ \Delta_c \\ \Delta_{ex} \end{cases} \tag{12}$$

where an intraatomic coulomb energy Δ_{ex} must be added to U where the next electron is the first to require double occupancy of a crystal-field orbital, *e.g.* on going from $4f^7$ to $4f^8$ or $3d^5$ to $3d^6$, and a Δ_c must be added where the next electron is the first to occupy a crystal-field orbital of different symmetry, *e.g.* on going from t^3e^0 to t^3e^1. On the other hand, covalent mixing spreads the wavefunctions out over the neighboring oxygen atoms, which reduces U and Δ_{ex}. Where the covalent mixing is stronger, a $\Delta_c > \Delta_{ex}$ may stabilize a *low-spin* state. For example, the energy separations of localized d^3 and d^4 configurations in an octahedral site are $U + \Delta_c$ for t^3e^0 to t^3e^1 and $U + \Delta_{ex}$ for t^3e^0 to t^4e^0; the high-spin t^3e^1 (S = 2) configuration is more stable if $\Delta_c < \Delta_{ex}$, but the low-spin t^4e^0 (S = 1) configuration is more stable if $\Delta_c > \Delta_{ex}$. The $4d^n$ and $5d^n$ configurations are low-spin in oxides; the $3d^n$ configurations are high-spin where $E_M - E_I$ is relatively large (> 2eV) and low-spin where $E_M - E_I$ is smaller.

Finally, whether the d electrons form localized d^n configurations or become delocalized depends on the strength of the M-M and/or M-O-M interactions, which give a bandwidth W, relative to the intraatomic energy U_{eff}. For $W_d < U_{eff}$, the electrons are localized and the M-M and/or M-O-M interactions give the spin-spin interactions responsible for magnetic order. For $W_d \geq U_{eff}$, electron-electron interactions are taken into account by a renormalization procedure that transforms the electrons into *quasiparticles*; the quasi-particles of this Fermi liquid have momentum vectors **k**, an energy dispersion, and a well-defined Fermi surface in momentum space like the Fermi gas of conventional band theory. Moreover, in the absence of a CDW state, the electron-lattice interactions may be treated in perturbation theory (Migdal's theorem). The transition from localized to itinerant electrons is probably not smooth in single-valent oxides. For example, $NdNiO_3$ undergoes a first-order transition from an antiferromagnetic insulator to a Pauli paramagnetic metal near 200 K; the volume is larger in the antiferromagnetic state where the Ni-$3d$ holes are localized [5].

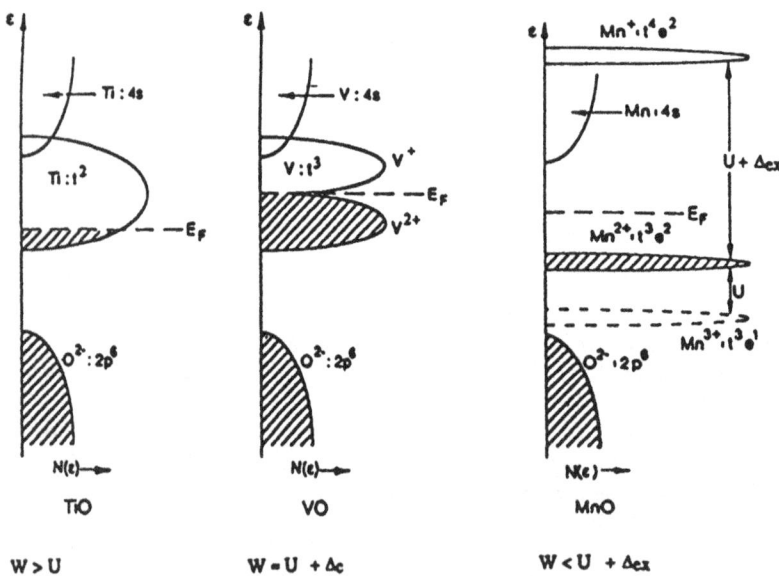

FIG. 8. Schematic electron energies for
(a) TiO, (b) VO, and (c) MnO

Fig. 8 compares the electron energy diagrams for the superconductor TiO, the semi-metal VO, and the antiferromagnetic insulator MnO. In TiO and VO, the M-O interactions split the $3d$ bands into bands of t and e parentage; the M-M interactions broaden the t states into a band of width $W_t > U_{eff} = U_t$ in TiO and $W_t \approx U_{eff} = U_t + \Delta_{ex}$ in VO where $\Delta_c > \Delta_{ex}$. TiO is a superconductor ($T_c = 1$ K) [6] and VO changes from p-type to n-type conductivity on passing from $V_{1-\delta}O$ to $VO_{1-\delta}$ [7]. The empty bands of e parentage, not shown, have a width determined by the M-O-M interactions.

A localized Mn : $3d^5$ configuration in MnO lies in the middle of the gap separating the Mn-4s and O-2p bands; it gives a high-spin (t^3e^2) manganese atomic moment $\mu_{Mn} \approx 5 \mu_B$. A slight oxidation of MnO lowers E_F toward the top of the d^5 level, but small-polaron formation lifts the empty d^4 configuration about 0.3 eV above the occupied d^5 configurations, and $Mn_{1-\delta}O$ is a p-type semiconductor with E_F in a Mn^{3+}/Mn^{2+} redox couple. The Mn^+ : $3d^6$ configuration is not energetically accessible because $U_{eff} = U + \Delta_{ex}$ is large; but the Mn^{3+} : $3d^4$ configuration lies above the top of the O-2p bands because $U_{eff} = U_e$ for the e electrons is greatly reduced by the σ-bond covalency. Consequently the formal valence states Mn^{2+}, Mn^{3+}, and Mn^{4+} are available in oxides. Moreover, placement of the empty d^4 configuration over 2 eV above the top of the O-2p bands makes the high-spin t^3e^1 configuration more stable at a Mn^{3+} ion. The empty d^4 configuration at a Mn^{4+} ion lies far enough above the top of the O-2p bands that an ionic model still applies to this oxidation state. Finally, both Mn-Mn and Mn-O-Mn interactions remain sufficiently weak that a bandwidth $W \leq U_{eff}$ leaves the Mn(III): d^4 configuration localized. The spin-spin interactions between localized t^3e^2, t^3e^1, t^3e^0 configurations at Mn^{2+}, Mn^{3+}, and Mn^{4+} ions can be treated by superexchange perturbation theory.

Fig. 9 shows an idealized phase diagram for a half-filled band. The parameter b is the operative resonance integral. In the rocksalt structure, the M-M interactions across shared octahedral-site edges give a b_t for the t orbitals that decreases exponentially with increasing M-M separation R; the M-O-M interactions give a $b_e \sim \varepsilon_\sigma (\lambda_\sigma^2 + \lambda_s^2)$ for the e orbitals. In the domain $b < b_g$ applicable to MnO, a $U_d > W_d$ opens an energy gap between d^n and d^{n+1} configurations. For smaller b/U, the interatomic superexchange interactions are manifest in an antiferromagnetic ordering temperature $T_N \sim b^2/U$, which increases unambiguously with increasing b as long as the superexchange perturbation expansion remains valid; a T_N that decreases with increasing b signals a change to itinerant-electron antiferromagnetism. In MnO, the magnetic order below T_N shows that the Mn-O-Mn interactions are stronger than the Mn-Mn interactions.

For $b > b_g$, a $W > U$ results in a half-filled band of itinerant electrons. In VO, the half-filled t band has $W \approx U_{eff}$. The nature of the cross-over from localized-electron antiferromagnetism to itinerant-electron behavior with superconductivity below a critical temperature T_c has proven difficult to explore experimentally in single-valent systems because of first-order transitions to CDW/SDW states as in VO_2. However, the mixed-valent system $La_{1-x}Sr_xTiO_3$ appears to be an antiferromagnetic semiconductor for stoichiometric $LaTiO_3$ and to be metallic for $0.05 \leq x < 1$ with evidence of increasing electron-electron interactions as x decreases to x = 0.05 [8]. The first-order magnetic-nonmagnetic transition in $NdNiO_3$ could be due to a breakdown of the ionic model.

Therefore we come, finally, to the question of what happens if the ionic model breaks down. For illustration, consider the nickel oxides. As shown in Fig. 10, NiO has an empty Ni(I) : $3d^9$ level lying just below the bottom of the Ni-4s conduction band. In a few special oxides, it is

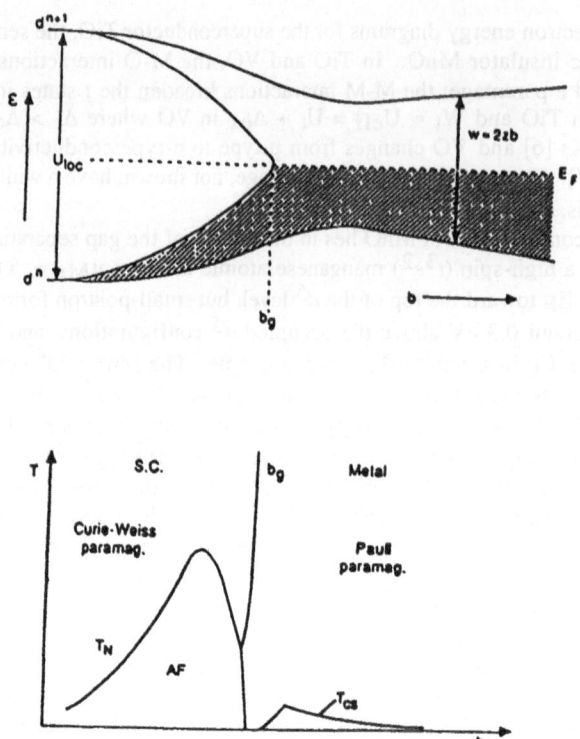

FIG 9. Idealized splitting of a half-filled band by the on-site electrostatic energy U and the associated phase diagram. A first-order phase change may occur as b increases toward b_g from a localized-electron antiferromagnetic phase to an itinerant-electron antiferromagnetic SDW.

possible to stabilize the $Ni^+ : 3d^9$ state [9]. A charge-transfer gap $\Delta \approx 3.0$ eV appears to be smaller than the $U_{eff} = U + \Delta_c$ for octahedral-site $Ni^{2+} : t^6 e^2$, and it becomes uncertain whether an E_M-$E_I > 0$ exists to permit discussion of the valence state Ni^{3+} within an ionic model. Although the antibonding states of e parentage must lie at the top of the O-$2p$ bands, it is not clear whether the equilibrium reaction

$$Ni^{3+}O^{2-} = Ni^{2+}O^-$$

is biased to the left or to the right. An ionic model requires a bias to the left; it has a larger Ni-O equilibrium bond length than the covalent bond represented by a bias to the right. The antibonding states have a larger O-$2p$ than Ni-$3d$ character in the covalent Ni-O bond. In such a case, the character of the Ni-O bond in a Ni(III) oxide may depend upon the magnitude of E_M and also subtle changes in oxygen polarizability with changing countercation in a mixed-metal oxide. Although electrochemical and thermodynamic data indicate an E_M-$E_I > 0$ for the $Ni^{2+} : 3d^8$ configuration, *i.e.* a bias to the left in the equilibrium reaction, an E_M-E_I could still be small enough for the perturbation expansion of the ionic model to break down.

159

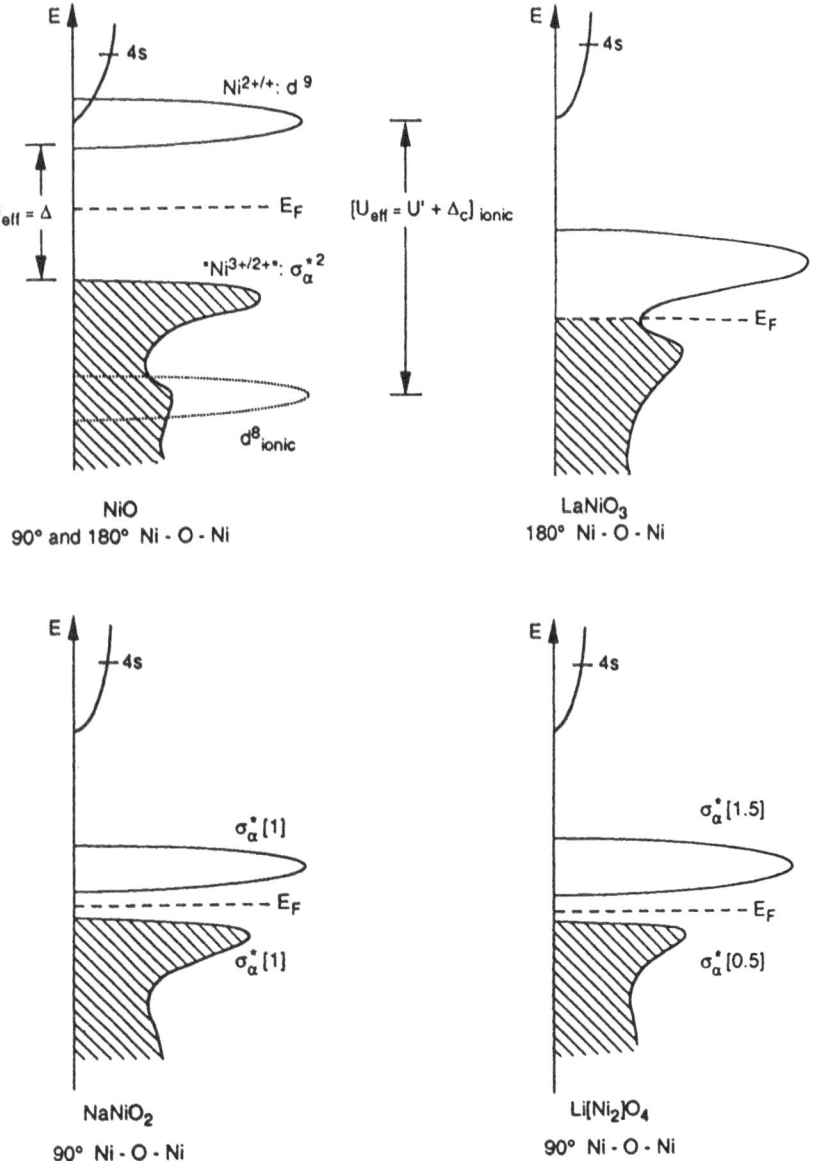

FIG. 10. Schematic electron energies for several nickel oxides.

Interestingly, the magnitude of the Ni-O-Ni interaction decreases from its maximum value at a bond angle of 180° to zero at a bond angle of 90°. Metallic $LaNiO_3$ has a bond angle approaching 180°; the bond angle is smaller in $NdNiO_3$ and it is near 90° in metamagnetic $NaNiO_2$. The layered oxide $NaNiO_2$ has ferromagnetic planes of edge-shared NiO_6 octahedra [10]; it is an insulator because the orthogonal e orbitals give zero bandwidth even though the Ni-O interactions are strong. The mixed-valent spinel $Li[Ni_2]O_4$ also has 90° Ni-O-Ni interactions and therefore remains a magnetic insulator even though the $Ni^{4+}O^{2-} = Ni^{3+}O^-$ reaction is biased to the right [11]. Metallic $LaNiO_3$ is not a superconductor; the bands remain sufficiently narrow that strong ferromagnetic correlations between electrons in the quarter-filled σ^* band of e parentage suppress superconductivity.

In the next lecture, we consider the problem of ionic versus covalent bonding in a mixed-valent copper oxide with nearly 180° Cu-O-Cu bonding.

ACKNOWLEDGMENTS

Financial support from the Robert A. Welch Foundation, Houston, TX, from Texas Advanced Research Program, and from the National Science Foundation is gratefully acknowledged.

REFERENCES

1. J. Bardeen, L. N. Cooper, and J. R. Schriefter, Phys. Rev. **108**, 1175 (1957).
2. J. B. Goodenough,, Prog. Solid State chem. **5**, 141 (1971).
3. C. Chaillout, A. Santoro, J. P. Remeika, A. S. Cooper, G. P. Espinosa, and M. Marezio, Solid State Chem. **65**, 1363 (1988).
4. M. R. Oliver, J. O. Dimmock, A. L. McWhorter, and T. B. Reed, Phys. Rev. B **5**, 1078 (1972).
5. J. L. Garcia-Munoz, J. Rodriguez-Carvajal, P. Lacorre, and J. B. Torrance, Phys. Rev. B **46**, 4414 (1992)
6. M. D. Banus, Mater. Res. Bull. **3**, 723 (1968).
7. M. D. Banus and T. B. Reed, in *The Chemistry of Extended Defects in Non-Metallic Solids*, L. Eyring and M. O'Keeffe, eds, (North Holland, Amsterdam, 1970) p. 488.
8. Y. Tokura, Y. Taguchi, Y. Okada, Y. Fujishima, T. Arima, K. Kumagai, and Y. Iye, Phys. Rev. Lett. **70**, 2126 (1993).
9. Ph. Lacorre, J. Solid State Chem. **97**, 495 (1992).
10. P. F. Bongers and U. Enz, Solid State Commun. **4**, 153 (1966).
11. M. G. S. R. Thomas, W. I. F. David, J. B. Goodenough, and P. Groves, Mater. Res. Bull. **20**, 1137 (1985).

THE SYSTEM La$_{2-x}$Sr$_x$CuO$_4$

JOHN B. GOODENOUGH
Center for Materials Science and Engineering, ETC 9.102
University of Texas at Austin,, Austin, TX 78712-1084

ABSTRACT. A complex phase diagram for the strontium-doped 214 lanthanum-copper-oxide superconductive system is systematically developed and interpreted. It is argued that below 300 K the superconductive phase is thermodynamically distinguishable from both the antiferromagnetic parent compound and the overdoped metallic phase. The mobile holes introduced by oxidation of the copper-oxide sheets form unconventional polarons characterized by covalent bonding within a polaron and ionic bonding without; molecular-orbital (MO) states inside a polaron are coupled vibronically to localized copper states without. Below 300 K, the polaron gas is unstable relative to the formation of a distinguishable normal state consisting of coupled polarons having extended vibronic states with a Fermi surface near the locus predicted by band theory, but with a band splitting at the half-band position.

1. INTRODUCTION

The superconductive system La$_{2-x}$Sr$_x$CuO$_4$, $0 \leq x \leq 0.34$, has been extensively studied both because of the relative simplicity of its crystal structure and because it can be prepared over the entire compositional range in which superconductivity is found. The parent compound La$_2$CuO$_4$ is an antiferromagnetic insulator, the overdoped samples with $x > 0.28$ are metallic with no apparent transition to superconductivity at lowest temperatures. The superconductive compositional range with maximum critical temperature T$_C$ is confined to the narrow compositional range $0.10 < x < 0.22$; I shall argue that this compositional range represents, in the temperature interval T < 300 K, a unique phase that is thermodynamically distinguishable from the antiferromagnetic phase of the parent compound on the underdoped side and from the metallic phase on the overdoped side. Moreover, in this system $x = p$ gives unambiguously the number p of holes per Cu atom in the CuO$_2$ sheets; it appears that the maximum values of T$_C$ for all the p-type superconductors are found in the range $0.10 < p < 0.22$ holes per Cu atom in the superconductive CuO$_2$ sheets. I shall also argue that what makes the superconductive phase distinguishable is a condensation of an intermediate-size-polaron gas below 300 K to a polaron liquid in which *extended vibronic states* are formed that have a momentum vector **k**, an energy dispersion, and a Fermi surface locus in momentum space near that predicted by Fermi-liquid theory (FLT). However, molecular-orbital coupling within a polaron introduces an energy gap at the half-band position separating bonding from antibonding states within a polaron. This unusual electron-phonon coupling allows the system to accommodate to a coexistence of ionic and covalent Cu-O bonding.

E. Kaldis (ed.), Materials and Crystallographic Aspects of HTc-Superconductivity, 161–174.
© 1994 *Kluwer Academic Publishers.*

162

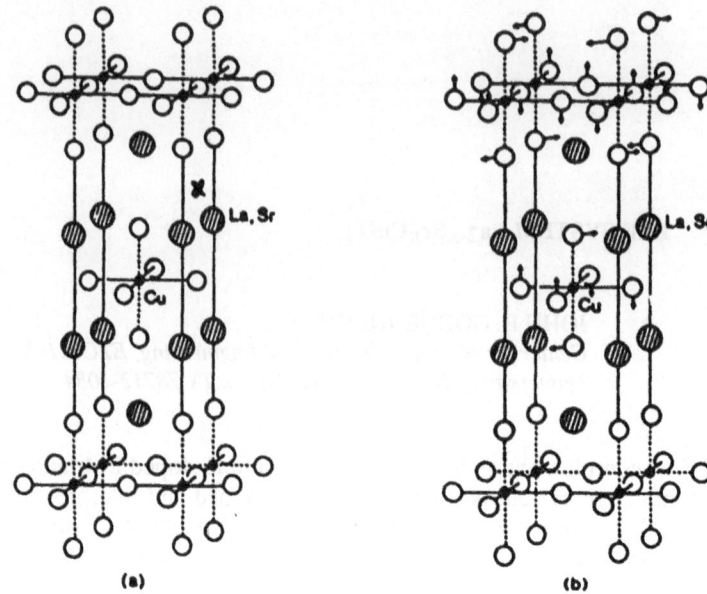

FIG. 1. The structure of La$_2$CuO$_4$: (a) tetragonal T > T$_t$ phase and (b) cooperative CuO$_6$ rotation (arrows) in orthorhombic T < T$_t$ phase. In (a), x marks position of interstitial oxygen in La$_2$CuO$_{4+\delta}$.

2. THE La$_{2-x}$Sr$_x$CuO$_4$ PHASE DIAGRAM

At high temperatures, stoichiometric La$_2$CuO$_4$ has the tetragonal structure of Fig. 1; it consists of an intergrowth of CuO$_2$ planes containing 180° Cu-O-Cu bonds alternating with (001) rock-salt bilayers of LaO. We represent the structure as

| LaO LaO | CuO$_2$.| LaO LaO | CuO$_2$ | (1)

where the vertical lines represent the interfaces between intergrowth layers on traversing the c-axis. In this structure, the bond-length mismatch across an interface is expressed by a tolerance factor

t = (La-O)/$\sqrt{2}$ (Cu-O)   (2)

where La-O and Cu-O are the equilibrium bond lengths. Because the La-O bond has the larger thermal expansion, t decreases with decreasing temperature; at room temperature, a t < 1 can be calculated from the sums of the ionic radii. A t < 1 places the CuO$_2$ sheets under compression and the LaO·LaO layers under tension; below a transition temperature T$_t$ ≈ 540 K, a cooperative rotation of the CuO$_6$ octahedra about [110] axes (arrows in Fig. 1) distorts the crystal from tetragonal to orthorhombic symmetry so as to relieve the internal stresses. The rotation bends the Cu-O-Cu bonds from 180° and creates some shorter La-O bonds.

The internal stresses are also relieved by an ordering of the 3d hole on the Cu^{2+} : 3d^9 configuration into an x^2 - y^2 orbital (x and y axes directed towards nearest-neighbor Cu atoms of tetragonal cell); removal of one electron/Cu from the antibonding $x^2 - y^2$ orbitals reduces the equilibrium Cu-O bond length in the CuO_2 planes, and a large tetragonal (c/a > 1) distortion of the CuO_6 octahedra is found.

Substitution of a larger Sr^{2+} ion for a La^{3+} in $La_{2-x}Sr_xCuO_4$ increases the tolerance factor t not only by increasing the mean equilibrium (La,Sr)-O bond length, but also by removing x antibonding electrons/Cu more from the CuO_2 sheets. Consequently T_t decreases monotonically with increasing x, vanishing near x = 0.22, as shown in the preliminary phase diagram of Fig. 2 [1].

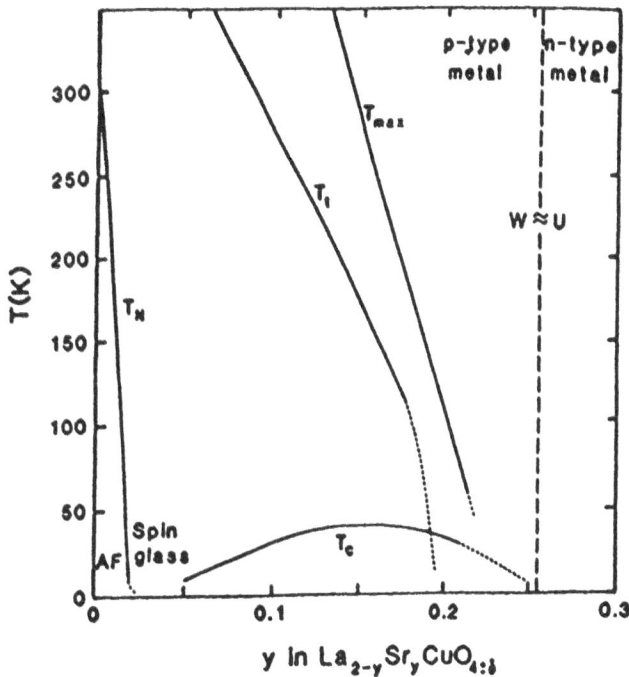

FIG. 2. Preliminary phase diagram for $La_{2-x}Sr_xCuO_4$, after ref [1].
T_N = Néel temperature
T_t = orthorhombic-tetragonal transition temperature
T_m = temperature of maximum magnetic susceptibility
T_c = superconductive critical temperature.

The parent compound is an antiferromagnetic insulator with a Néel temperature $T_N \approx 320$ K that decreases sensitively with up-take of oxygen in air to form $La_2CuO_{4+\delta}$. The Cu-O-Cu interactions within a CuO_2 sheet are much stronger than the Cu-O-O-Cu interactions between sheets, so short-range magnetic order within a sheet sets in below a temperature $T_{max} > T_N$, where T_{max} is the temperature at which the magnetic susceptibility shows a broad maximum. In two-dimensional systems, long-range magnetic order below T_N commonly occurs at a significantly lower temperature than T_{max}. With the introduction of Sr^{2+}, T_N shows a remarkably sharp decrease with hole concentration, dropping to 10 K by x = 0.02. What was originally interpreted to be spin-glass behavior at low temperatures in the range $0.02 \leq x \leq 0.08$ has been reinterpreted by Cho *et al* [2], on the basis of ^{139}La nuclear quadrupole resonance (NQR), to be quasi-static fluctuations of locally ordered antiferromagnetic domains. The decrease in T_{max} with increasing x, originally reported by Torrance *et al* [1], is no less significant. T_{max} decreases smoothly to zero near x = 0.22; at higher Sr concentrations, a Pauli paramagnetism indicates a transition from the superconductive phase to a conventional metal. The vanishing of T_{max} represents a collapse of the on-site electron-electron electrostatic energy U in the metallic phase.

The transport properties of the system $La_{2-x}Sr_xCuO_4$ divide the phase diagram of Fig. 2 at compositions near x = 0.10 and x = 0.22 [3,4]. However, the division at x = 0.10 is blurred by a structural anomaly occurring at x = 1/8. In the isostructural system $La_{2-x}Ba_xCuO_4$, a transition to a low-temperature tetragonal (LTT) phase at compositions near x = 1/8 suppresses superconductivity; at this composition the holes can be ordered by a static CDW that splits the $x^2 - y^2$ conduction band at the Fermi energy. In the system $La_{2-x}Sr_xCuO_4$, competition between the static CDW and superconductivity reduces T_c in compositions near x = 1/8; but T_c is not completely suppressed.

In the compositional range $0 < x \leq 0.10$, the resistivity varies nearly as $\rho \sim T$ in the range $T_\rho < T < T_l$. Above T_l, the ρ vs T curve bends toward the temperature axis ($d\rho/dT$ decreases with increasing T), and below T_ρ the resistivity increases with decreasing temperature; the resistance drops precipitously to zero below a critical temperature $T_c < T_\rho$ in the range $0.05 \leq x \leq 0.10$. In the superconductive compositions $0.10 < x < 0.22$, a $\rho \sim T$ is found for all temperatures $T > T_c$ below the decomposition temperature of the solid. In the compositional range x > 0.22, the critical temperature T_c drops to zero by x = 0.28, and above T_c the ρ vs T curve exhibits a power-law behavior; such behavior is predicted from FLT for a correlated-electron metal [3,5].

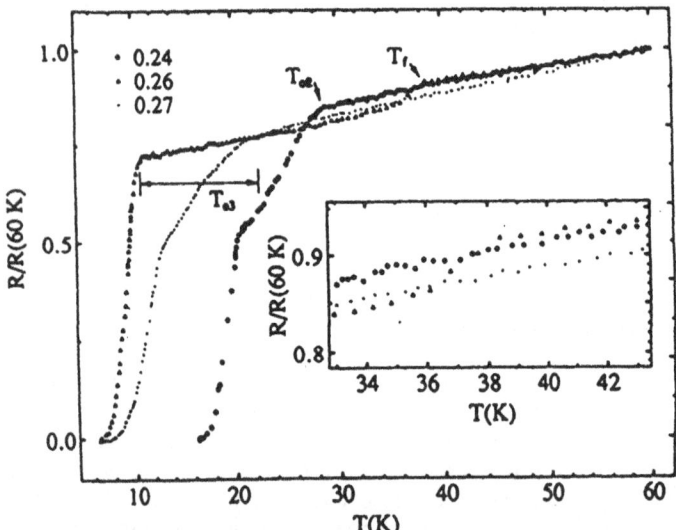

FIG. 3. Normalized resistance *versus* temperature for three different values of x in the system $La_{2-x}Sr_xCuO_4$.

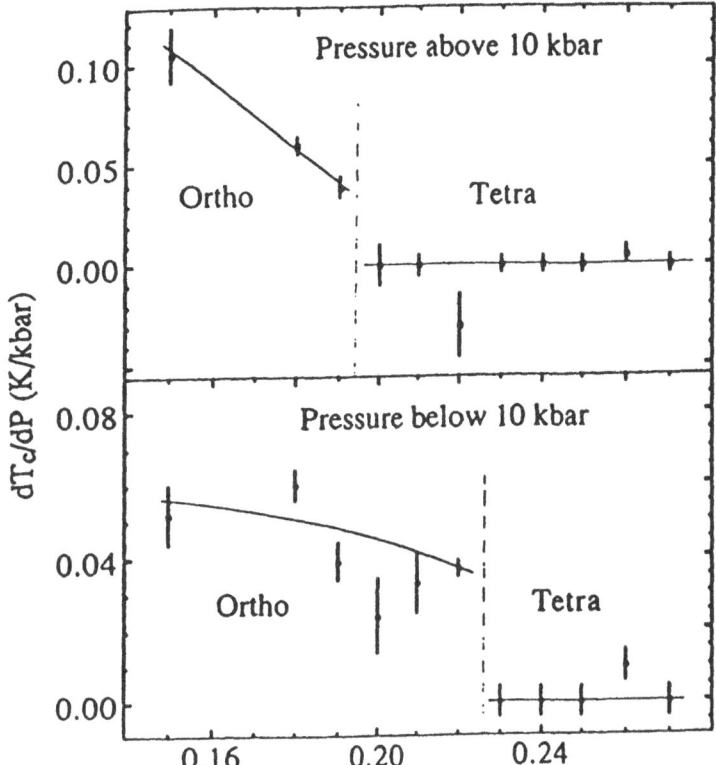

FIG. 4. Variation with x of the pressure dependence of T_c for
$P > 10$ kbar and $P < 10$ kbar in the system $La_{2-x}Sr_xCuO_4$.

Returning to the superconductive critical temperature T_c, it is observed to increase nearly linearly with x in the range $0.05 \leq x \leq 0.10$; it exhibits a definite dip in the neighborhood of $x = 1/8$, but otherwise it exhibits a broad maximum in the onset temperature $T_0(max) \approx 34$ K near $x = 0.15$ in the interval $0.15 \leq x \leq 0.20$. We find a $T_f = T_0$ (max) over the entire superconductive compositional range $0.05 < x < 0.30$ even where a $T_0 < T_f$ is found. The resistance versus temperature for $x = 0.27$ in Fig. 3 exhibits a clear separation of T_f and T_c. The anomaly at T_f does not appear to be due to the presence of a second superconductive phase. By monitoring the pressure dependence of T_c, Fig. 4, it is possible to show that superconductivity extends into the tetragonal phases where $dT_c/dP = 0$; in the orthorhombic phase a $dT_c/dP > 0$ is found [4,6]. It follows that the distortion to orthorhombic symmetry is not critical for stabilization of the superconductive phase.

Fig. 4 also shows that pressure stabilizes the tetragonal phase relative to the orthorhombic phase. According to Equation (2) and our interpretation of the driving force for the orthorhombic distortion, this finding means that the Cu-O bond is more compressible than the La-O bond. This deduction is contrary to all experience of the relative compressibilities of the La-O and M-O bonds in transition-metal $LaMO_3$ oxides having a $t > 1$ [7]; it indicates that there is something unusual about the high compressibility of the Cu-O bond in the copper-oxide superconductors. Independent measurements also show that the Cu-O bond has indeed an exceptionally high compressibility [8]. This observation turns out to be a necessary condition for a vibronic state in which ionic and covalent Cu-O bonding coexist and fluctuate.

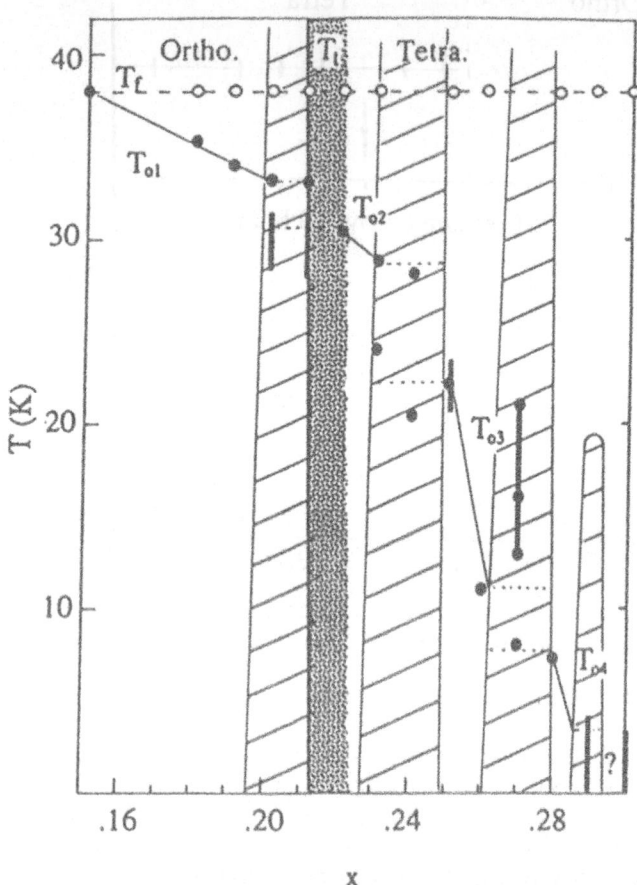

FIG. 5. Low-temperature phase diagram of $La_{2-x}Sr_xCuO_4$ for $0.15 \leq x \leq 0.30$.

Fig. 5 shows how the superconductive onset temperature decreased with increasing x in the range $x \geq 0.15$ for a set of polycrystalline samples all similarly prepared below $1150°$. Oxygen stoichiometry and homogeneity were obtained, but the resolution of our x-ray diffractometer was not adequate to guarantee complete La and Sr homogeneity. In the tetragonal phase, T_o decreases rapidly from 30 K at $x = 0.21$ to below 4 K at $x = 0.29$. Surprisingly, it appears to exhibit a sequence of superconductive phases of narrow compositional range; the appearance of such a sequence is consistent with the proposition that the superconductive phase is thermodynamically distinguishable from the overdoped n-type metallic phase stable for x > 0.29, but it could also reflect some sample inhomogeneity.

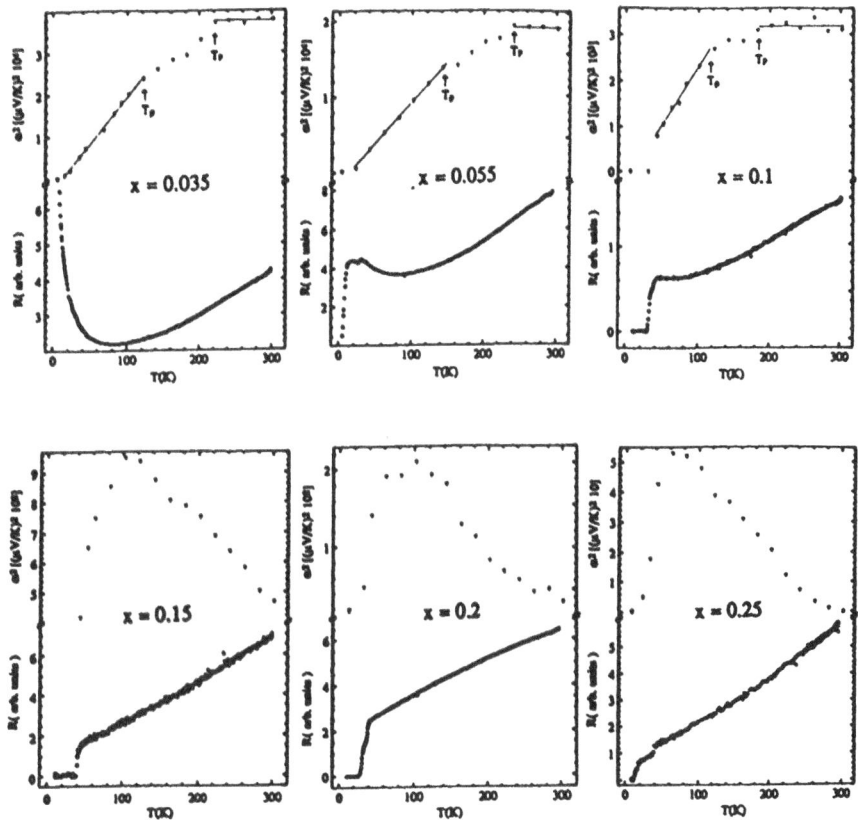

FIG. 6. Temperature dependence of the square of the Seebeck coefficient (α^2) and the resistance R for several values of x in the system $La_{2-x}Sr_xCuO_4$.

As Fig 6 shows, the temperature dependence of the Seeback coefficient α exhibits a remarkable change not only on passing from x < 0.10 to x > 0.10, but also on passing from $T > T_l$ to $T < T_l \approx 320$ K. At temperatures $T > T_l$, there is a complete solid solution that is thermodynamically stable for $0 \leq x \leq 0.34$, and a large temperature-independent α decreases in magnitude over the range $0 \leq x < 0.10$; in the range $0.10 \leq x < 0.22$ a small slope $d\alpha/dT$ increases as α decreases further with increasing x. Below T_l, a temperature-independent α persists down to a temperature T_F in the range $0 < x \leq 0.10$ whereas in the range $0.10 < x < 0.22$ the temperature dependence of α increases abruptly on cooling through T_l; α passes through a broad maximum with decreasing T before it vanishes abruptly at T_c. Near x = 0.26, α becomes small and, like the Hall effect, shows a change of sign from p-type to n-type conduction.

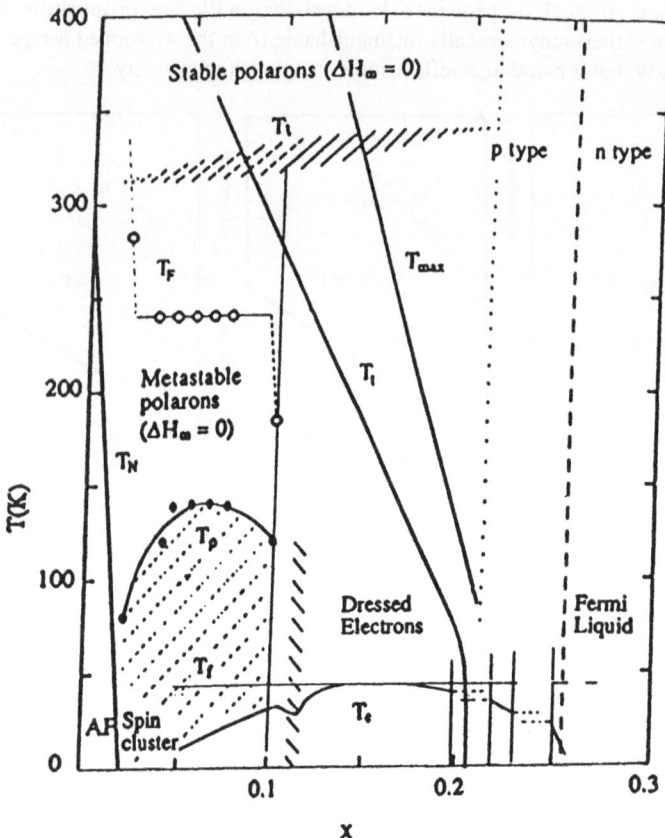

FIG. 7. Phase diagram for the system $La_{2-x}Sr_xCuO_4$, $0 \leq x \leq 0.30$.

Fig. 7 shows a more complete phase diagram; it includes the compositional dependencies of T_ρ and T_F as well as T_l and T_f. In the interval $T_\rho < T < T_F$, α changes by less than 60 $\mu V/K$, which is consistent with a loss of the spin-degeneracy factor β in the statistical term of the Seebeck coefficient

$$\alpha \approx \frac{k}{e} \ln [\beta(N_{eff}-p)/p] \qquad (3)$$

where $N_{eff}/N \leq 1$ is the effective fraction of Cu sites available to a mobile hole. Below T_F the antiferromagnetic spin fluctuations associated with the short-range magnetic order below T_{max} become so slow that the mobile hole sees an ordered array of spins. The resistivity exhibits no anomaly at T_F [9].

On the other hand, α changes both at T_F and T_ρ. Below T_ρ an $\alpha^2 \sim T$ dependence is observed. The changes in ρ and α at T_ρ are similar to those observed in the system $La_2CuO_{4+\delta}$, where a spinodal phase segregation occurs below 300 K in the range $0 < \delta < 0.05$ [10]. A spinodal decomposition below $T_l \approx 300$ K is possible in this system because the interstitial oxygen are mobile down to 220 K, which allows a classical segregation into oxygen-rich and oxygen-poor regions in the metastable temperature range $T < T_l$. In the system $La_2CuO_{4+\delta}$, an ordering of the interstitial oxygen for $\delta \approx 0.05$ may drive the phase segregation; but the driving force may also be enhanced by a metastability of the uncoupled polarons. The compositional range $0 < \delta < 0.05$ corresponds to a hole concentration per Cu atom in the CuO_2 planes of $0 < p < 0.10$ and hence to $0 < x < 0.10$ in the system $La_{2-x}Sr_xCuO_4$. Since no ions are mobile below T_l in $La_{2-x}Sr_xCuO_4$, the system remains metastable. However, the variations in ρ and α below T_ρ suggest a phase segregation is occurring below T_ρ; at these temperatures in this system a phase segregation can only be expressed by atomic displacements in lieu of atomic diffusion. We are accustomed to observing atomic displacements stabilizing a static CDW; in this case no static CDW is observed, which leads to the *inference* that cooperative atomic displacements are introducing a phase segregation that expresses itself as dynamic charge fluctuations. This interpretation is consistent with the deduction [2] that quasistatic antiferromagnetic domains are stabilized at low temperatures in the range $0.02 \leq x \leq 0.08$.

The unusually rich phase diagram of Fig. 7 would therefore appear to indicate that below T_l in the narrow compositional range $0.10 < x < 0.22$ a distinguishable thermodynamic state is stabilized and that the high-T_c superconductive phenomenon is associated with this unique phase. Before we go further with the interpretation of Fig. 7 and a characterization of the superconductive phase, it is useful to return to a consideration of the electronic energies in the system $La_{2-x}Sr_xCuO_4$.

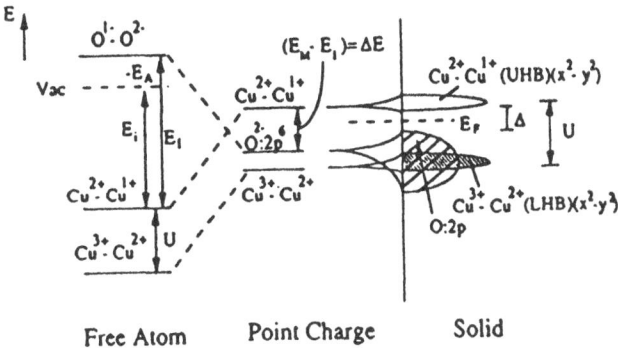

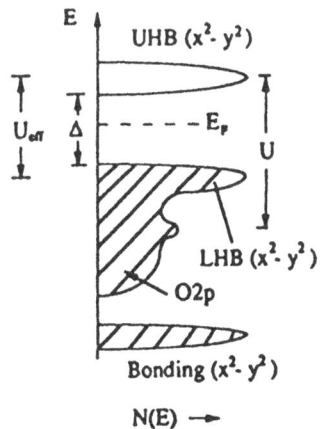

FIG. 8. Electron energies for La_2CuO_4: (a) ionic model and (b) after hybridization

3. ELECTRON ENERGIES

The parent compound La_2CuO_4 may be described with an ionic model for the Cu-O bonding. As described in Lecture I, this model requires an $E_M - E_I > 0$ for the Cu^{2+}/Cu^+ redox couple, which places an *empty* $Cu:3d^{10}$ level in the energy gap between the O-2p and Cu-4s bands as depicted in Fig. 8. The tetragonal ($c/a > 1$) distortion of the CuO_6 octahedra places the 3d hole at a formally Cu^{2+} ion in the antibonding $x^2 - y^2$ orbital, and optical spectroscopy measures a charge-transfer gap $\Delta = 2.0$ eV. The fact that La_2CuO_4 is an antiferromagnetic insulator means that the antibonding $x^2 - y^2$ band is split by an energy U into an upper and a lower Hubbard band. From photoelectron spectroscopy, $U \approx 6 - 8$ eV is estimated [11], which would place the lower Hubbard band (LHB) below the top of the O-2p bands. However, hybridization of the overlapping O-2p band and the LHB creates antibonding $x^2 - y^2$ states at the top of the O-2p bands to make $\Delta \approx U_{eff}$. What remains ambiguous is the sign of $E_M - E_I$ in the ionic model; with $E_M \leq E_I$, an ionic model cannot be appropriate for the states of the LHB. It follows that holes introduced into the LHB by oxidation of the CuO_2 sheets (1) occupy two-dimensional (2D) $x^2 - y^2$ bands and (2) introduce covalent Cu-O bonding coexisting with ionic Cu-O bonding.

Transport measurements on single crystals in the system $La_{2-x}Sr_xCuO_4$ by, for example, Nakamura and Uchida [12] confirm that the conductivity is strongly anisotropic with $\rho_c \gg \rho_{a,b}$; moreover, in the normal state of the superconductive phase ($T_c < T < T_l$ and $0.10 \leq x \leq 0.20$) the anisotropy of both ρ and α is enhanced by over an order of magnitude compared to band-theory predictions. In the normal state of the superconductive phase, the principal action is clearly within a two-dimensional $x^2 - y^2$ band.

4. UNCONVENTIONAL POLARONS AND VIBRONIC DISPERSION

A temperature-independent α at $T > T_l$ is suggestive of a polaron model with α described by Equation (3); however, the temperature dependence of ρ shows that the hole mobility is not activated. Moreover, the magnitude of α is too low for a small-polaron model. If the principal contribution to α comes from the statistical term Equation (3), as opposed to the transport term, the magnitude of the polaron must encompass 4 to 5 Cu atoms.

These considerations have led to the postulate [13] of an *unconventional polaron* of intermediate size within which a mobile hole becomes trapped by cooperative atomic displacements. What is unconventional about the polaron is the existence of covalent Cu-O bonds within the polaron and ionic Cu-O bonds without. Within a polaron, the covalent bonds are described by non-magnetic molecular-orbital (MO) cluster orbitals, and a *vibronic coupling* of the MO states to the vibrational modes that define the size of the polaron allows the polarons to be *mobile with no motional enthalpy* ($\Delta H_m = 0$) in their diffusional mobility at $T > T_l$. Non-magnetic (small U_{eff}) MO states within a polaron of intermediate size accounts well qualitatively for the rapid decrease with x in T_N. The presence of c-axis oxygen permits some mixing of apical $O-p_z$ and $Cu-(3z^2 - r^2)$ character into the MOs, and hole transfer to the oxygen allows dynamic O-O bonding also in these MO clusters.

This postulate is based on the observation of an unusually large compressibility of the Cu-O bond together with the coexistence of covalent and ionic Cu-O bonding. Covalent Cu-O bonding has a shorter equilibrium Cu-O bond length than ionic bonding, which makes the transition from ionic to covalent bonding first-order. Moreover, a double-well potential for the Cu-O bond would make an ionic Cu-O bond highly compressible. The coexistence of ionic and covalent bonding can only be accommodated by a phase segregation into antiferromagnetic ionic domains and covalent domains within the unconventional polarons in which holes are trapped. Vibronic coupling that allows these domains to be mobile can give rise at high temperatures $(T > T_l)$ to mobile intermediate-size polarons; at lower temperatures $(T < T_l)$ coupling between polarons in the superconductive phase $0.10 < x < 0.22$ may cause the polaron gas to condense into a polaron liquid. In the compositional range $0 < x < 0.10$, condensation into a polaron liquid would cause a phase segregation (at $T < T_\rho$) into quasistatic antiferromagnetic regions and polaron-rich regions in which the polarons are coupled as in the superconductive phase. However, we cannot rule out the alternative interpretation that the apparent phase segregation represents a condensation into bipolarons [14].

Fig. 9(a) shows the flat dispersion curves expected for the two-dimensional UHB and LHB of the parent compound. Fig. 9(b) illustrates the change in these dispersion curves at $T > T_l$ on doping with holes. Creation of MO states within a polaron implies a transfer of spectral weight from the Hubbard bands to the polaron state that is greater than the hole concentration, and this transfer has been observed with both optical and photoelectron spectroscopies [15,16]. The bonding and antibonding MO states are split by a finite energy gap Δ_{MO}, and trapping of the hole at the top of the bonding MO states pins the Fermi energy E_F. The polarons of a polaron gas move diffusively and randomly at temperatures $T > T_l$, so the MO states transferred from the LHB and the UHB would have no dispersion.

In the compositional range $0.10 \leq x \leq 0.22$, condensation of the polarons below T_l into a *polaron liquid*, i.e. from uncoupled to coupled polarons, would lead to *extended vibronic states* in which electronic states are coupled to phonons rather than to local cooperative vibrational modes. However, retention of identifiable polarons would retain the molecular feature of a finite splitting at the half-band position of Cu-O-Cu bonding and antibonding x^2-y^2 states as illustrated in Fig. 9(c). The vibronic states have a narrower dispersion than that calculated for a Fermi liquid, and angle-resolved photoelectron spectroscopy now indicates an extremely flat dispersion near E_F [17]. A flat dispersion curve of this shape was used by Kaiser [18] to obtain a shape of the α vs T curve like that observed in the range $0.10 < x < 0.22$ for $T < T_l$. Even more significant is the collapse of U_{eff} for the extended vibronic states, which gives a well-defined Fermi surface with a locus near that predicted by FLT. Observation of a Fermi surface at this locus has been a signal success of FLT; however, FLT cannot then account for the observation of p-type conduction. On the other hand, a gap at the half-band position due to MO formation with the polarons allows for a continuation of p-type conduction; n-type conduction is only established where the polaron density would be too high to sustain polaron formation. With the suppression of polarons, the system transforms to an n-type metal as indicated in Fig. 9(d). The hole-poor regions separating the coupled polarons contain ionic Cu-O bonds, so states associated with these regions would continue to give some spectral weight at the center positions of the original UHB and LHB in Fig. 9(d).

This model accounts well for the high anisotropy of the transport properties in the temperature range $T_c < T < T_l$. Independent evidence for the condensation of a polaron gas to a polaron liquid comes also from ^{63}Cu $1/T_1$ data obtained by Imai et al [19] with nuclear quadrupole resonance. At temperatures $T > T_l$, the nearest-neighbor exchange interaction is almost independent of doping; the nuclear spin-lattice relaxation rate $1/T_1$ is dominated by short-range order, and the data indicate that the holes doped into the CuO_2 sheets do not change significantly the short-range order, which is precisely what is predicted with a polaron model. On the other hand, the d-electron spins were found to undergo a cross-over to a more itinerant regime on cooling through T_l. A polaron, whether formed by spin-spin exchange or by electron-phonon interactions, would be "dressed" in antiferromagnetic spin fluctuations associated with hole-poor regions. The fact that the ^{63}Cu $1/T_1$ data exhibit a magnetic-field dependence only for $x < 0.04$ that the driging force for coupling the polarons lies primarily in the electron-phonon interactions;

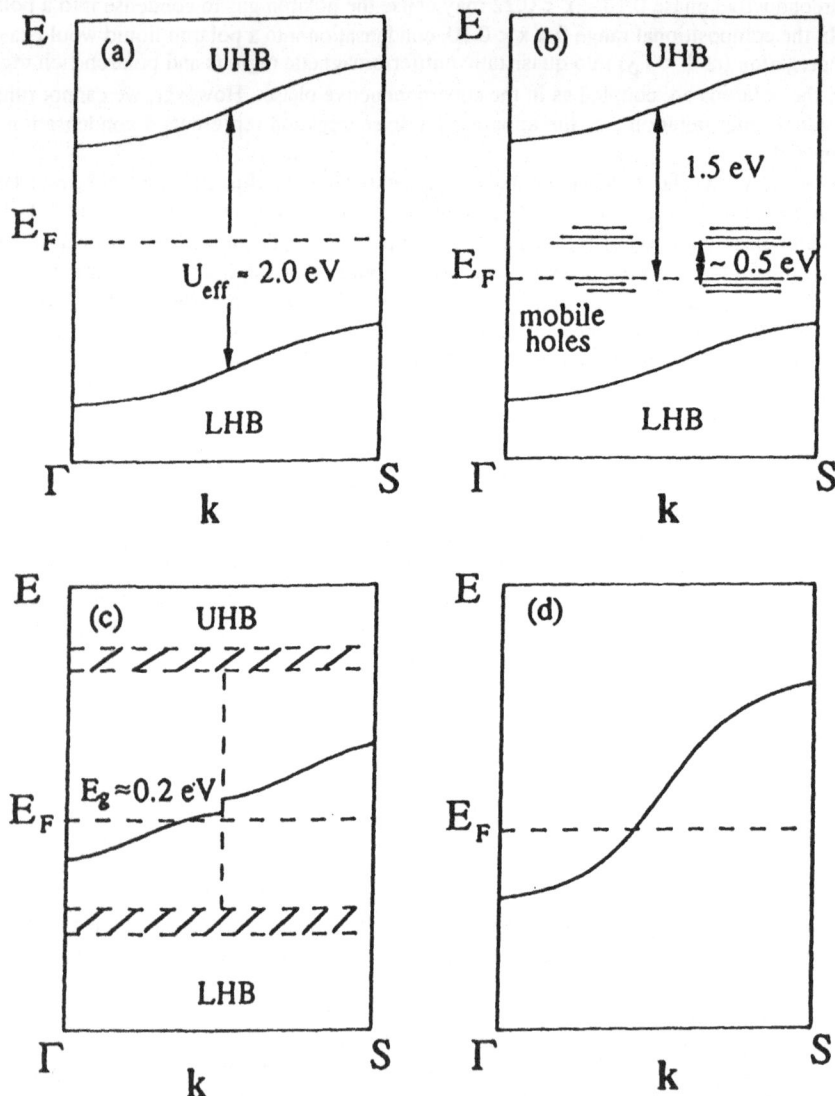

FIG. 9. Schematic room-temperature dispersion curves for La$_{2-x}$Sr$_x$CuO$_4$: (a) x = 0, (b) 0 < x < 0.10, (c) 0.10 ≤ x ≤ 0.22, and (d) x > 0.28.

but the spin fluctuations should nevertheless reflect the localized *versus* itinerant character of any vibronic states.

No attempt is made here to identify the attractive forces responsible for the stabilization of vibronic Cooper pairs below T_c. Coupling between adjacent CuO_2 planes could play a decisive role; in the $La_{2-x}Sr_xCuO_4$ system, the apical oxygen would be expected to be a principal mediator of such a coupling. However, we note that the model contains two features that were used to construct a correlation-bag model of high-T_c superconductivity, *viz* strong electron-phonon coupling and a collapse of the U parameter on going from a region of ionic bonding to one of covalent bonding [20]. These features would replace the Debye energy $k\theta_D$ of BCS theory by a characteristic energy of the form $(\varepsilon_F \Delta U)^{1/2}$, where ΔU is the change in U_{eff} on passing from an ionic to a covalent region of Cu-O bonding and ε_F is the Fermi energy measured from the top of the lower half-band.

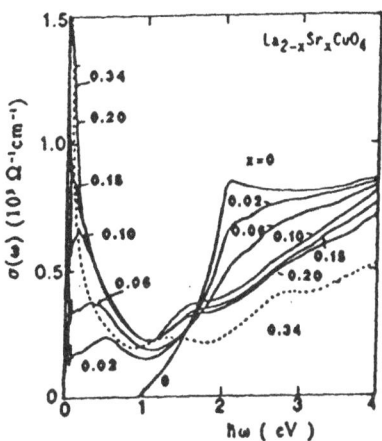

FIG. 10. Optical conductivity $\sigma(\omega)$ obtained from Kramers-Kronig transformation of the $E \perp c$-axis reflectivity spectra for various compositions x of single-crystal $La_{2-x}Sr_xCuO_4$ from ref. [21].

5. OPTICAL DATA

In closing, it is instructive to compare the optical-conductivity data of Fig. 10 taken from Uchida *et al* [21] with the predictions from Fig. 9. The 2 eV gap found for La_2CuO_4 remains throughout the compositional range $0 \leq x \leq 0.20$, but with a spectral weight that decreases rapidly with increasing x. A mid-IR peak at 0.5 eV appears immediately with doping; in Fig 9(b) it would correspond to the excitation of an electron from a bonding to an antibonding MO state. However, in the range $0.10 \leq x \leq 0.22$ the mid-IR peak is shifted to about 0.2 eV and decreases as x increases as expected for a ΔMO that is smaller for coupled polarons and collapses at the transition to a Fermi liquid at higher values of x. The appearance of a Drude component for x > 0.10 signals the stabilization of extended states within the polaron liquid, Fig. 9(c), as does the appearance of a lower intensity charge-transfer gap shifted to 1.5 eV. The 1.5 eV excitations from the MO bonding states to the UHB appear where the MO states are transformed to band states with a meaningful **k** vector.

ACKNOWLEDGMENTS

Financial support from the Robert A. Welch Foundation, Houston, TX, from Texas Advanced Research Program, and from the National Science Foundation is gratefully acknowledged.

REFERENCES

1. J. B. Torrance, A. Bezinge, A. I. Nazzal, T. C. Huang, S. S. Parkin, D. T. Keane, S. J. LaPlaca, P. M. Horn, and G. A. Held, Phys. Rev. B **40**, 8872 (1989).
2. J. H. Cho, F. Borsa, D. C. Johnston, and D. R. Torgeson, Phys. Rev. B **46**, 3179 (1992).
3. J. B. Goodenough, J.-S. Zhou, and J. Chan, in *Lattice Effects in High-T_cSuperconductors*, T. Egami *et al*, eds (World Scientific, Singapore, 1993) p. 486.
4. J.-S. Zhou, H. Chen, and J. B. Goodenough (unpublished).
5. H. Takagi, R. J. Cava, M. Marezio, B. Batlogg, J. T. Krajewski, W. F. Peck, Jr., P. Bordet, and D. E. Cox, Phys. Rev. Lett. **68**, 3777 (1992).
6. N. Yamada and M. Ido, Physica C **203**, 240 (1992).
7. J. B. Goodenough, J. A. Kafalas, and J. M. Longo, in *Preparative Methods in Solid State Chemistry*, P. Hagenmuller, ed. (Academic Press, New York, 1972) p. 74.
8. M. Gupta and R. P. Gupta, Physica C **173**, 381 (1991).
9. J. B. Goodenough, J.-S. Zhou, and K. Allan, J. Mater. Chem. **1**, 715 (1991).
10. J.-C. Grenier, N. Lagueyte, A. Wattiaux, J.-P. Doumerc, P. Dordor, J. Etourneau, M. Pouchard, J. B. Goodenough, and J.-S. Zhou, Physica C **202**, 209 (1992).
11. Z.-X. Shen, J. W. Allen, J. J. Yeh, J.-S. Kang, W. Ellis, W. Spicer, J. Lindau, M. B. Maple, Y. D. Yalichaouch, M. S. Torikachvidi, J. Z. Sun, and T. H. Geballe, Phys. Rev. B **36**, 8414 (1987).
12. Y. Nakamura and S. Uchida, Phys. Rev. B **47**, 8369 (1993).
13. J. B. Goodenough and J.-S. Zhou (unpublished).
14. W. Y. Liang, Proc. SPIE **1362**, Pt 1, 127 (1991).
15. M. Eskes and G. A. Sawatzky, Phys. Rev. Lett. **61**, 1415 (1988).
16. J. W. Allen, C. G. Olson, M. B. Maple, J.-S. Kang, L. Z. Liu, J.-H. Park, R. O. Anderson, W. P. Ellis, J. T. Markert, Y. Dalichaouch, and R. Liu, Phys. Rev. Lett. **64**, 595 (1990).
17. W. Spicer, private communication.
18. A. B. Kaiser, Phys. Rev. B **29**, 7088 (1984).
19. T. Iwai, C. P. Slichter, K. Yoshimura, and K. Kosuge, Phys. Rev. Lett. **70**, 1002 (1993).
20. J. B. Goodenough and J.-S. Zhou, Phys. Rev. B **42**, 4276 (1990).
21. S. Uchida, T. Ido, H. Takagi, T. Arima, Y. Tokura, and S. Tajima, Phys. Rev. B **43**, 7942 (1991).

THE n-TYPE COPPER OXIDE SUPERCONDUCTORS

JOHN B. GOODENOUGH
Center for Materials Science and Engineering, ETC 9.102
University of Texas at Austin, Austin, TX 78712-1084

ABSTRACT. The basic intergrowth principles of the crystal architecture of the copper-oxide superconductors is first reviewed. It is then argued that n-type superconductivity requires not only planes of copper in corner-shared square-coplanar coordination, but also internal stresses that stretch a copper-oxygen bond length to 1.95 Å or longer in the parent oxide. Two types of structures have been found to satisfy those requirements. The synthesis of n-type superconductors in one of these structure types requires either high-pressure synthesis or epitaxial growth of films. In the other structure, compounds with rare-earth atoms of smaller size cannot be made superconductive because the copper-oxygen bond lengths in the parent compounds are too short.

1. STRUCTURAL CONSIDERATION

In Lecture II, the unusual normal state occurring between T_c and $T_l \approx 300$ K in the system $La_{2-x}Sr_xCuO_4$ was found to be two-dimensional; the metallic conductivity is confined to the CuO_2 sheets. Moreover, the common feature in all the copper-oxide superconductors is the existence of CuO_2 sheets. It is therefore convenient to describe the crystal architecture of the copper-oxide superconductors as an intergrowth of alternating "superconductive" and "non-superconductive" layers. However, it should be noted that below T_c the superconductive phenomenon itself is three-dimensional. Moreover, stabilization of an infinite n-type superconductive layer has been demonstrated.

The superconductive layers have a fixed oxygen content and contain one or more CuO_2 sheets; the top and bottom faces of a superconductive layer are CuO_2 sheets. The sheets consist of a square array of Cu atoms bridged by 180° Cu-O-Cu bonds; a cooperative bending of the Cu-O-Cu bonds from 180° may distort the square Cu array to a lower symmetry. Where there is more than one CuO_2 sheet in a superconductive layer, the sheets are separated by planes of Y^{3+}, Ln^{3+}, Ca^{2+}, and/or Sr^{2+} ions each coordinated by eight oxygen atoms, four from each neighboring CuO_2 sheet. Complete substitution of Sr^{2+} ions into these planes is facilitated by high-pressure synthesis or in thin films grown epitaxially on a single-crystal substrate of slightly larger lattice parameter.

The non-superconductive layers always contain a minimum of two sheets, a top and a bottom face. These two faces are the same for any given non-superconductive layer, but they may be either a rocksalt (001) AO sheet or a plane of cations like the planes between CuO_2 sheets within a superconductive layer. The cation planes contain no oxygen with which to coordinate a Cu of a neighboring CuO_2 sheet; the oxygen of an AO sheet coordinate a Cu atom. Therefore, where a single CuO_2 sheet interfaces AO sheets on opposite sides, the Cu have sixfold octahedral oxygen

175

E. Kaldis (ed.), Materials and Crystallographic Aspects of HTc-Superconductivity, 175–186.
© 1994 *Kluwer Academic Publishers.*

coordination with the two apical oxygen belonging to the two neighboring AO sheets. Where a CuO_2 sheet interfaces an AO sheet on only one side, the Cu are in fivefold square-pyramidal coordination; and a CuO_2 plane between two cation planes has only four square-coplanar coordination. *Although superconductivity is found for all three types of CuO_2 sheets, it occurs only where essentially all the Cu of any given sheet have the same oxygen coordination: sixfold, fivefold, or fourfold.* A change in oxygen coordination at Cu atoms within a CuO_2 sheet perturbs the periodic potential within a CuO_2 sheet, which suppresses the formation of extended wavefunctions. Therefore the observation that superconductivity is suppressed by a perturbation of the periodic potential supports Cooper-pair formation below T_C as against a Bose condensation of bipolarons.

To date, n-type superconductivity has only been found where the Cu atoms have fourfold oxygen coordination. The Madelung energy E_M at a Cu site increases with the oxygen coordination, and *a smaller E_M appears to be critical for stabilization of n-type superconductivity.*

A great variety of structural elements of variable anion and/or cation content may be stabilized between the two outer faces of a non-superconductive layer. This architecture permits either oxidation or reduction of the CuO_2 sheets without changing the oxygen coordination at a Cu atom. It also allows for the establishment of internal redox processes between the superconductive and non-superconductive layers. Where charge transfer occurs between the superconductive and non -superconductive layers, the non-superconductive layers have been referred to as "charge reservoirs" for the superconductive layers. It is striking that (a) the presence of a charge reservoir does not suppress superconductivity; it may in fact enhance coupling of superconductive pairs across a non-superconductive layer, and (b) charge transfer tends to stabilize an optimum charge-carrier density for superconductivity in the superconductive layers, which is consistent with the hypothesis that the superconductive phase is a distinguishable phase of narrow charge-carrier range within the superconductively active CuO_2 sheets.

Finally, it is noted that the cations neighboring a superconductively active sheet may have mixed formal valences 3+ and 2+ or 3+ and 4+; variation of the electronegativity of the larger cations over a limited range does not perturb the periodic potential in the CuO_2 sheets sufficiently to suppress the high-T_C phenomenon. Interestingly, superconductivity is not suppressed even by the introduction of rare-earth Ln^{3+} ions that carry a localized magnetic moment $\mu_J = gJ\mu_B$ associated with a $4f^n$ configuration unless the $4f^n$ energy E_n lies close enough to E_F for significant hybridization to take place between the occupied $4f$ and empty x^2-y^2 orbitals, as occurs for the $Pr^{3+} : 4f^2$ configuration in $Y_{1-x}Pr_xBa_2Cu_3O_{6+x}$. However, substitution of other cations for Cu, whether aliovalent or isovalent, in a superconductively active CuO_2 sheet sharply suppresses superconductivity.

2. SIMPLEST ILLUSTRATIONS

In a p-type superconductor, the simplest non-superconductive layer is a double rocksalt layer and the simplest superconductive layer is a single CuO_2 sheet; the La_2CuO_4 structure of Fig. 1(a) represents, therefore, the simplest architecture for a p-type superconductor. Substitution of Sr for La or insertion of interstitial oxygen between the rocksalt bilayers allows oxidation of the CuO_2 sheets without changing the sixfold oxygen coordination at a Cu atom, and both $La_{2-x}Sr_xCuO_4$ and $La_2CuO_{4+\delta}$ contain a compositional range exhibiting high-T_C p-type superconductivity.

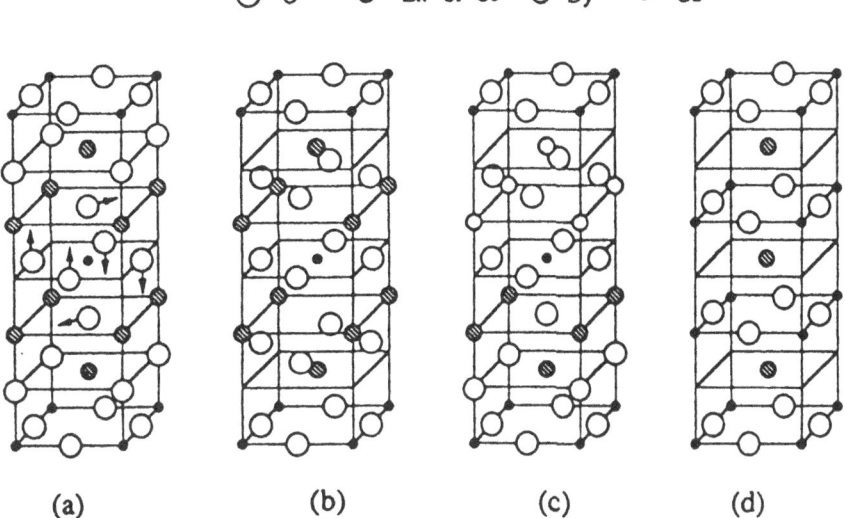

○ O   ◉ Ln Sr Ce   ○ Dy   • Cu

      (a)   (b)   (c)   (d)

FIG. 1. (a) The T/O structure of La_2CuO_4, (b) the T' structure of Nd_2CuO_4, (c) the T* hybrid structure, and (d) the infinite-layer $ACuO_2$ structure.

In an n-type superconductor, the CuO_2 planes interface only all-cation planes. This arrangement is possible in an infinite superconductive layer, Fig. 1(d), or in an intergrowth structure where a plane or layer of negative charge binds the top and bottom faces of cation planes. The simplest intervening layer is a plane of anions occupying the tetrahedral interstices of the two outer planes of cations; this situation is illustrated by the $Nd-O_2-Nd$ non-superconductive layer found in the tetragonal T' structure of Nd_2CuO_4 shown in Fig. 1(b). Substitution of Ce for Nd in $Nd_{2-x}Ce_xCuO_4$ or of F for O in the non-superconductive layer dopes the compound n-type without changing the fourfold oxygen coordination at all the Cu atoms. However, in order to ensure that there are no apical oxygen present at Cu atoms, it appears to be necessary to reduce such an n-type compound slightly.

The tetragonal T* structure of $LaDyCuO_4$ shown in Fig 1(c) represents a hybrid containing alternately the non-superconductive rocksalt bilayer of Fig. 1(a) and the fluorite layer of Fig. 1(b). With this arrangement, the Cu of the single CuO_2 sheets become fivefold coordinated, and it has only been possible to oxidize the CuO_2 sheets in this structure. In this hybrid structure, the smaller Dy^{3+} ions occupy preferentially the fluorite layers and the larger La^{3+} ions the rocksalt layers. This ordering minimizes the bond-length mismatch between the superconductive and non-superconductive layers. Bond-length mismatch plays a key role in determining which structure is stabilized and whether a given structure can be doped p-type or n-type.

3. BOND-LENGTH MISMATCH

As pointed out in Lecture II, a measure of the bond-length mismatch between the CuO_2 sheets and the $LaO \cdot LaO$ rocksalt bilayers in La_2CuO_4 is the tolerance factor

$$t \equiv La\text{-}O/\sqrt{2}\,(Cu\text{-}O) \tag{1}$$

where La-O and Cu-O represent equilibrium bond lengths for a given temperature and pressure. At atmospheric pressure, the room-temperature bond lengths are given by the sums of the empirical ionic radii. A larger thermal expansion of the La-O bond makes a $t < 1$ decrease with decreasing temperature; on lowering the temperature, increasingly the CuO_2 sheets are under compression and the $LaO \cdot LaO$ bilayers are under tension. The resulting internal stresses have five consequences:

• The hole in the 3d shell of a Cu^{2+} : $3d^9$ configuration is ordered into an antibonding $x^2\text{-}y^2$ orbital, which makes for a large tetragonal ($c/a > 1$) distortion of the CuO_6 octahedra.

• On cooling in air in the tetragonal phase, interstitial oxygen atoms are increasingly introduced between the two LaO sheets of the bilayers; a $\delta \approx 0.02$ is found at room temperature in $La_2CuO_{4+\delta}$. The interstitial oxygen atoms partially relieve the tensile stress on the bilayers, and transfer of additional electrons from the antibonding $x^2\text{-}y^2$ orbitals to form interstitial O^{2-} ions partially relieves the compressive stress on the CuO_2 sheets.

• A cooperative buckling of the Cu-O-Cu bonds below an orthorhombic-tetragonal transition temperature T_t relieves further the compressive stress on the CuO_2 sheets.

• Substitution of a larger Sr^{2+} ion for La^{3+} relieves the tension on the bilayers and also removes another antibonding $x^2\text{-}y^2$ electron to relieve the compressive stress on the CuO_2 sheets; therefore the system $La_{2-x}Sr_xCu_{4+\delta}$ is stable, and both T_t and δ decrease with increasing x on firing in air. On the other hand, substitution of a smaller Ce^{4+} ion for La^{3+} would increase the tension on the bilayers and also add an antibonding electron to the CuO_2 sheets to increase their compressive stress; an n-type superconductor $La_{2-x}Ce_xCuO_4$ cannot be made by normal synthetic techniques.

• Substitution of a smaller Ln^{3+} ion for La^{3+} reduces $t < 1$ sufficiently to induce a transition from the T/O phase of Fig. 1(a) to the T' phase of Fig. 1(b). Fig. 2 illustrates the phase diagram obtained for the system $La_{2-y}Nd_yCuO_4$ at different firing temperatures [1]. Starting with the oxalate precursors allows synthesis at temperatures as low as 500° C; by increasing the firing temperature, it is possible to increase the critical value of y at which the transition from the T/O to the T' phase occurs. The calculated value of the room-temperature tolerance factor decreases with increasing y, but t increases with temperature. Therefore, if the transition occurs at a critical value of $t = t_c$, then the plot of Fig. 2 shows how t increases with temperature. Extrapolation to lower firing temperatures shows that synthesis of La_2CuO_4 below 400° C should yield a stable T' phase for this compound.

Displacement of the apical oxygen atoms of La_2CuO_4 to the interstitial oxygen positions yields the T' structure, and repulsion between O^{2-} ions in the plane of the interstitial sites inverts the stress field: in the T' structure the fluorite $Ln\text{-}O_2\text{-}Ln$ layers are under compression and the CuO_2 planes are under tension. The T' phases Ln_2CuO_4 (Ln = Pr, $\cdots$,Sm) have been made n-type superconductors by substitution of quadrivalent Ce^{4+} [2] or Th^{4+} [3] for Ln^{3+}, or by substitution of F^- for O^{2-} ions in the fluorite layer [4]. It has not been possible to substitute larger Sr^{2+} or Ca^{2+} ions for Ln^{3+} to dope them p-type. Co-doping of Sr^{2+} and Ce^{4+} produces the T^* structure [5].

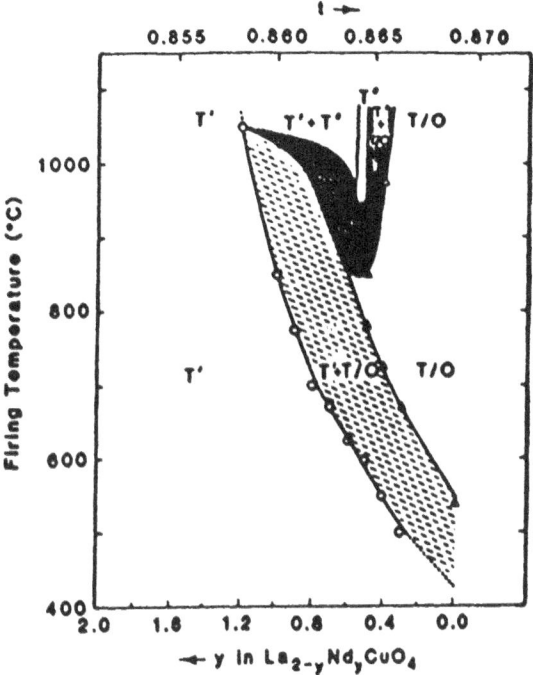

FIG. 2. The stability ranges of the T' *versus* T/O structures in the system La$_{2-y}$Nd$_y$CuO$_4$ for samples prepared at different firing temperatures. The room-temperature tolerance factor for different values of y are also shown. An ordered T" phase is stabilized at higher firing temperatures. From ref. [1].

4. THE n-TYPE SUPERCONDUCTORS WITH T' STRUCTURE

The parent T' phases Ln$_2$CuO$_4$ have been prepared at atmospheric pressure for Ln = Pr, $\cdots$, Gd; partial substitution of Dy^{3+} for a Ln^{3+} ion is able to reduce the Cu-O bond length in a CuO$_2$ plane to about 1.94 Å. Smaller Ln^{3+} ions have been introduced by high-pressure synthesis [6].

The parent compounds contain only Cu^{2+} and are antiferromagnetic insulators like the parent La$_2$CuO$_4$ compound of the T/O structure. Moreover, in each T' system Ln$_{2-x}$M$_x$CuO$_4$ (M = Ce or Th), n-type superconductivity is found in only a narrow compositional range in the interval $0.15 \leq x \leq 0.22$, and at larger x a normal metallic state persists to lowest temperatures with the n-type conductivity changing to p-type with increasing x > 0.22. Thus the n-type superconductors appear to be n-type analogues of the p-type superconductors with a distinguishable thermodynamic superconductive phase occurring in a narrow compositional range corresponding to a charge-carrier concentration $0.10 \leq n < 0.20$ in the CuO$_2$ sheets as in the p-type superconductors. The principal difference between the n-type and p-type superconductors appears only in the underdoped compositional range $0 < n < 0.15$.

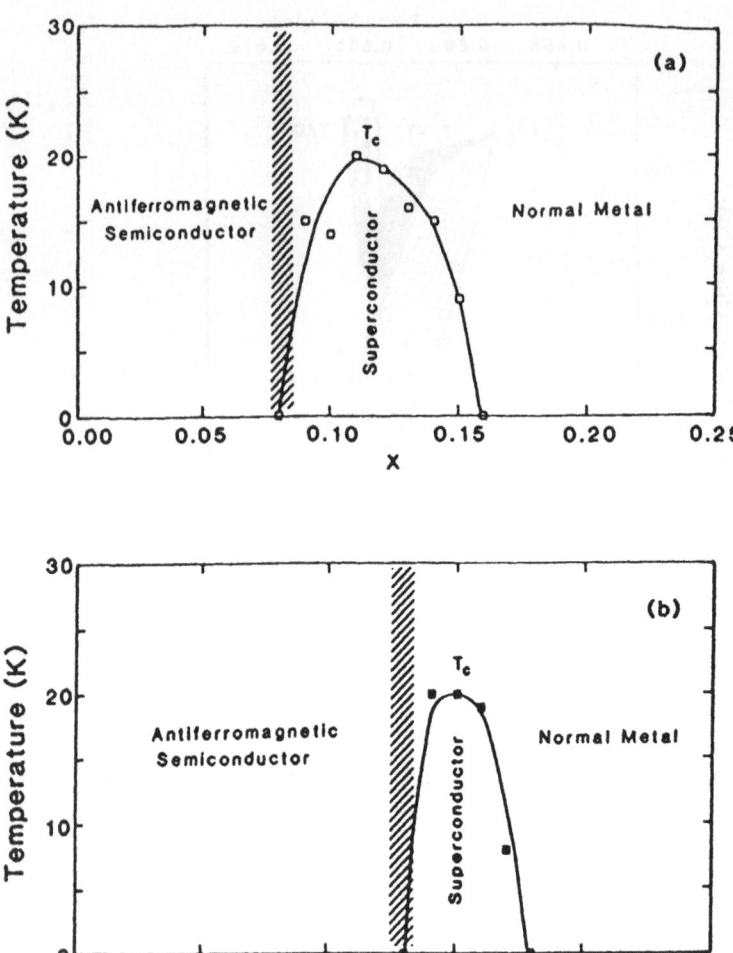

FIG. 3. Variations of T_c with x for (a) $LaNd_{1-x}Ce_xCuO_4$ and (b) $Nd_{2-x}Ce_xCuO_4$. Shaded areas refer to two-phase regions. After ref. [7].

The charge carriers in the underdoped T' systems are not unconventional polarons; they act like small polarons up to a critical concentration $n_c = x_c$ where there is a first-order transition from an antiferromagnetic phase to a superconductive phase; however, the appearance of superconductivity requires synthesis or subsequent annealing in N_2 atmosphere or under pressure so as to ensure removal of apical oxygen from all the Cu atoms. Charge fluctuations do not appear at low temperatures in the antiferromagnetic-semiconductor phase with $x < x_c$. Moreover, x_c increases and the solubility limit x_l of Ce decreases as the Cu-O bond length of the parent phase decreases. Fig. 3 compares the phase diagram for $LaNd_{1-x}Ce_xCuO_4$ and $Nd_{2-x}Ce_xCuO_4$; in the former $x_c = 0.10$ and $x_l = 0.23$ while in the latter $x_c = 0.14$ and $x_l = 0.20$ [7]. For Ln = Gd, an $x_c > x_l = 0.15$ occurs, so no superconductive phase is stabilized. The decrease in x_l with

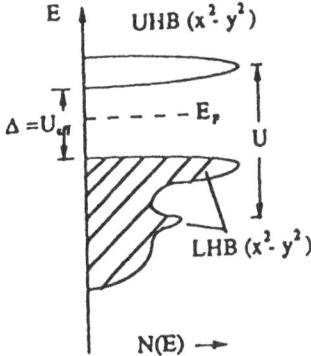

FIG. 4. Density of electronic states associated with CuO_2 planes in T' Nd_2CuO_4.

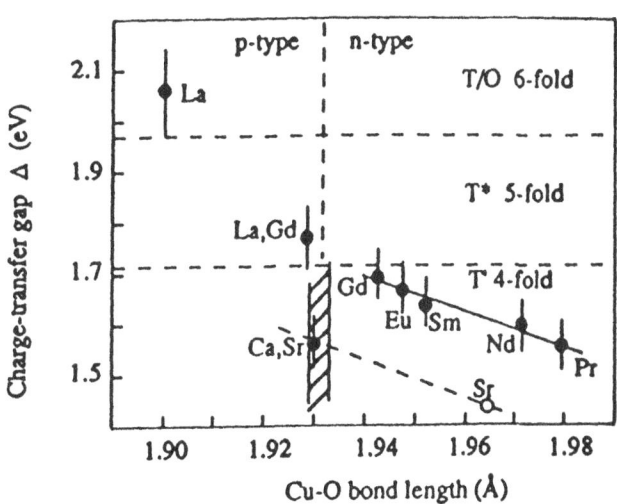

FIG. 5. Charge-transfer gap *versus* Cu-O bond length for Ln_2CuO_4 oxides, after ref. [9].

decreasing size of the Ln^{3+} ion appears to reflect the decreasing tensile strain on the CuO_2 sheets in the parent oxides; the Cu-O bond length decreases from over 1.98 Å in $LaNdCuO_4$ to about 1.94 Å in Gd_2CuO_4. Although the transition from superconductor to normal metal increases with x_c, nevertheless the width Δx of the superconductive range decreases from $\Delta x = 0.06$ for $Ln = La_{0.5}Nd_{0.5}$ to $\Delta x = 0.04$ for $Ln = Nd$ and can be expected to have an upper limit near $x = 0.20$. The increase in x_c with decreasing Cu-O bond length in the parent oxide provides a clue as to how the observation of a high-T_c phenomenon in both n-type and p-type copper oxides might be reconciled with the mechanistic concepts discussed in Lecture II [8].

Like La_2CuO_4, the parent T'-tetragonal Ln_2CuO_4 compounds have electron energies like those shown in Fig. 4; however, the lower oxygen coordination at the Cu atoms reduces the charge-transfer energy $\Delta = E_M-E_I$ from the 2.0 eV in La_2CuO_4 to 1.5 eV in Pr_2CuO_4. Moreover, Fig. 5 shows that the gap Δ varies sensitively with the Cu-O bond length in the CuO_2 planes of the T' structure [9]. Doping the CuO_2 planes n-type introduces small polarons as is typical for a mixed-valent ionic compound; the Néel temperature T_N decreases only slowly with electron concentration typical of a simple dilution with non-magnetic Cu^+ ions. However, as x increases through x_c in a T' $Ln_{2-x}Ce_xCuO_4$ system, there is a first-order isostructural phase change to a superconductive phase. In the superconductive phase, there is a transfer of spectral weight from the Hubbard bands to a narrow mid-gap band with a gap at the half-band position that is analogous to the situation in the p-type superconductive phase. Since doping introduces antibonding electrons into the x^2-y^2 orbitals, it expands the Cu-O bond length and lowers E_M. We may imagine that at a critical Cu-O bond length, which occurs at a larger x_c the smaller the Cu-O bond length of the parent compound, there is a first-order phase change to a system with unconventional, intermediate-size polarons. Such a phase change requires a breakdown of the ionic model, *i.e.* too small a Δ, only within the unconventional polaron. Covalent Cu-O bonding introduces MO states within the polaron; however, ionic Cu-O bonding is retained outside the polaron.

Implicit in this argument are two assumptions: (1) the x^2-y^2 states of the UHB have an overlap integral (f, φ) that is less sensitive to the Cu-O bond length than is the Madelung energy E_M and (2) E_M decreases with the net charge on the Cu atom. With these assumptions, a failure of the ionic model must occur first at the relaxed Cu-O bonds dressing a small polaron, and the transition to covalent Cu-O bonding within a small cluster can be expected to remain confined to a polaron of intermediate size. Once the unconventional polarons of intermediate size are formed, the analogy with p-type superconductivity is established.

The qualitative predictions that follow from this model are quite compatible with the experimental data. Since the transition at $x = x_c$ occurs only at polaron concentrations high enough for condensation below $T_l \approx 300$ K to a polaron liquid, the transition at $x = x_c$ is directly from an antiferromagnetic to a superconductive state. However, if the Cu-O bond in the parent compound is too small, as occurs in Gd_2CuO_4, doping cannot reduce E_M sufficiently rapidly with x to give $x_c < x_l$, and no superconductive phase is stabilized. Moreover, since n-type superconductivity requires a small E_M, *its appearance is restricted not only to compounds with Cu in fourfold, square-coplanar oxygen coordination, but also to those where the Cu-O bonds of the CuO_2 planes are so stretched by internal stresses in the parent compound that n-type doping to a concentration $x_l > x_c$ is possible.*

5. INFINITE-LAYER SUPERCONDUCTORS

The infinite-layer structure of Fig. 1(d) was first stabilized in the compound $Ca_{0.86}Sr_{0.14}CuO_2$ by Siegrist et al [10]. However, synthesis at atmospheric pressure allows little variation of the Ca/Sr ratio; in $SrCuO_2$, for example, the larger Sr^{2+} ion breaks apart the CuO_2 planes, and CuO_2 double chains are formed instead, Fig. 6. Moreover, attempts to dope $Ca_{0.86}Sr_{0.14}CuO_2$ sufficiently to make it either an n-type or a p-type superconductor were also unsuccessful. As can be seen from Fig. 5, a Cu-O bond length of only 1.93 Å in the parent compound is too small to allow n-type doping to an $x_l > x_c$ even though a charge-transfer gap Δ = 1.55 eV is comparable to that in the n-type parent compounds Nd_2CuO_4 and Pr_2CuO_4. On the other hand, Takano et al [11], showed that with synthesis under a hydrostatic pressure of 25 kbar, a high-pressure $SrCuO_2$ with the infinite-layer structure can be stabilized at atmospheric pressure and room temperature; moreover, in this phase the Cu-O bond length is stretched to 1.965 Å. The high-pressure $SrCuO_2$ is therefore a parent compound that satisfies both criteria for n-type superconductivity: Cu in square-coplanar coordination and a Cu-O bond stretched to longer than 1.95 Å.

Realization that high-pressure $SrCuO_2$ satisfies both criteria led Smith et al [12], to prepare under pressure infinite-layer n-type superconductors in two systems, $Sr_{1-y}Nd_yCuO_2$ and $Sr_{1-y}Pr_yCuO_2$. Er et al [13] quickly confirmed this finding by demonstrating n-type superconductivity in the system $Sr_{1-y}La_yCuO_2$. Fig. 7 shows the expected increase in Cu-O bond length, $a/2$, with n-type doping in the system $Sr_{1-y}Nd_yCuO_2$. A $y \geq 0.14$ for the superconduction phase is indicated by the susceptibility data of Fig. 8.

Later Azuma et al [14] reported superconductivity at 110 K in a high-pressure $(Sr_{0.7}Ca_{0.3})_{0.9}CuO_2$ phase that they claimed to be a p-type superconductor having the infinite-layer structure of Fig. 1(d). However, the samples were never single-phase, so it remains unclear whether a p-type superconductor with only square-coplanar oxygen coordination has been prepared. Adachi et al [15] provide evidence that the superconductive phase has a different structure and intergrows with the infinite-layer structure. Hiroi et al [16] also find superconductivity associated with defect intergrowth layers in the system $(Ca_{0.3}Sr_{0.7})_{0.95}CuO_{2-z}$. High-resolution electron microscopy reveals an intergrowth of defect layers, Fig. 9, which they postulate to be $A_{1-\alpha}|CuO_{2-\delta}|A_{1-\alpha}$ layers with A = Ca, Sr. Treatment in an oxidizing atmosphere gave a $T_c \approx 100$ K, in a reducing atmosphere a $T_c \approx 40$ K, and in an inert atmosphere no superconductivity. They interpreted this result to mean that oxygen can be inserted reversibly into the $CuO_{2-\delta}$ planes and that the oxidized samples are p-type, the reduced samples n-type. This speculation was not confirmed with transport measurements. From a theoretical point of view, there is no à priori reason to exclude the possibility of p-type superconductivity in planes of four-coordinated copper. However, identification of superconductivity with defect layers having Cu in different oxygen coordination must be suspect; these defect layers may serve as charge reservoirs that dope the neighboring CuO_2 layers p-type or n-type. The experimental situation with respect to these compounds must be considered still unresolved.

Superconductivity has also been observed in infinite-layer thin films grown epitaxially on single-crystal $SrTiO_3$ substrates: Terada et al [17], Kanai et al [18], or Sugii et al [19], for example. Stabilization of the infinite-layer structure is due to the bonding forces across the substrate-film interface. Here also extended defects plague the experimentalist, but the technique may allow a study of the variation of T_c with film thickness and hence with the number of like CuO_2 planes within a superconductive layer.

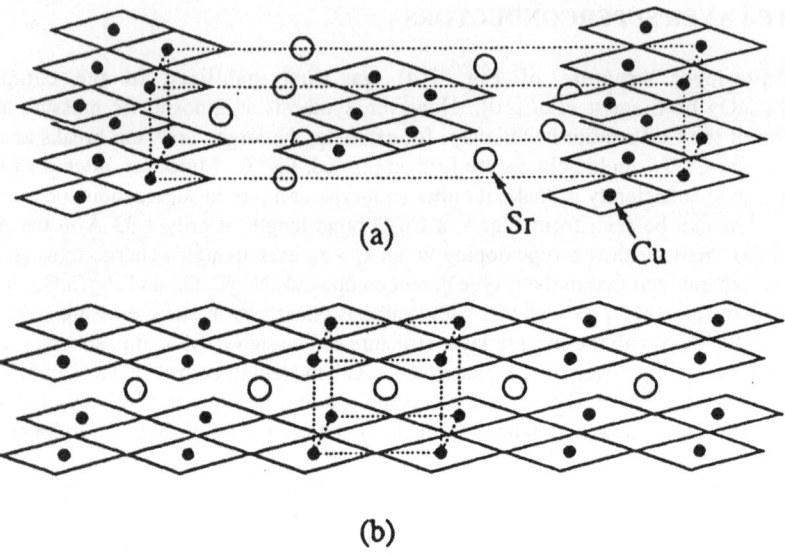

FIG. 6. Comparison of the (a) atmospheric-pressure and (b) high-pressure forms of $SrCuO_2$.

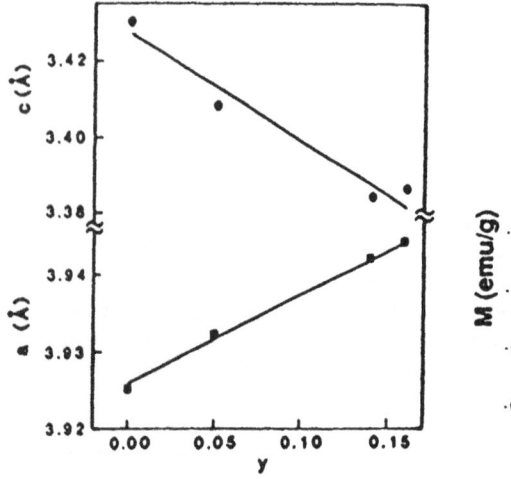

FIG. 7. Variation of lattice parameters with Nd concentration in the infinite-layer system $Sr_{1-y}Nd_yCuO_2$..

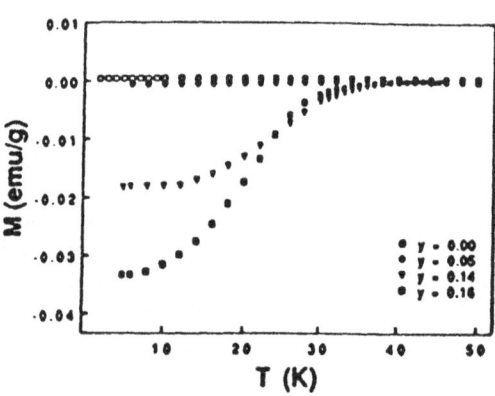

FIG. 8. A. C. magnetic susceptibility *versus* temperature for the infinite-layer system $Sr_{1-y}Nd_yCuO_2$.

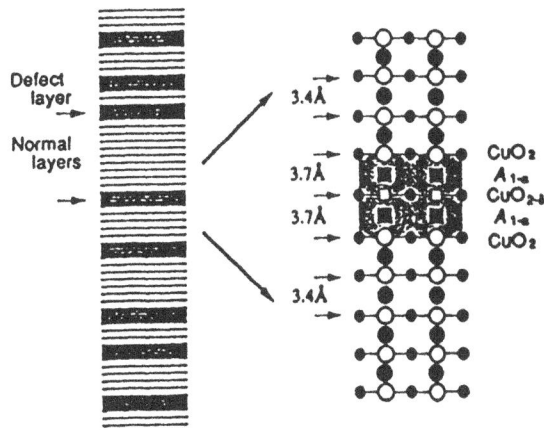

FIG. 9. Schematic representation of the intergrown
infinite-layer structure of $A_{1-x}CuO_{2-z}$, from
ref. [16].

The importance of coupling between CuO_2 sheets is an issue in need of experimental resolution. One approach has been to vary the number of CuO_2 sheets in a superconductive layer. In the p-type intergrowth structures, this approach has been frustrated by the fact that the outer CuO_2 sheets of a p-type superconductive layer have Cu in fivefold oxygen coordination whereas the inner CuO_2 sheets have Cu in fourfold coordination; this configuration tends to trap the holes in the outer CuO_2 sheets as these have Cu with a larger Madelung energy E_M.

Pressure studies provide an alternate strategy. However, where the non-superconductive layers provide a charge reservoir, studies of the pressure dependence of the superconductive critical temperature may be dominated by charge transfer between the CuO_2 sheets and the reservoir. In the Ln_2CuO_4 and $Sr_{2-y}Ln_yCuO_2$ systems discussed in these lectures, there is no charge reservoir, and a study of the pressure dependence of T_c, *i.e.* of dT_c/dP, can provide some insight into the relative influences of intraplane *versus* interplane bond-length changes.

In the p-type system $La_{2-x}Sr_xCuO_4$, a $dT_c/dP > 0$ distinguishes the superconductivity of the orthorhombic phase from that of the tetragonal phase where $dT_c/dP = 0$, see Lecture II. In the n-type systems, there is no bending of the Cu-O-Cu bonds, so dT_c/dP provides a more straightforward measure of the influence of bond-length changes. In n-type $Sm_{1-x}Ce_xCuO_{4-y}$, a $dT_c/dP < 0$ was interpreted by Beille *et al* [20] to signal a decrease in T_c with decreasing Cu-O bond length that was more important than any increase in T_c due to enhanced interlayer coupling. In $Sr_{0.84}Nd_{0.16}CuO_2$, which has an interlayer spacing of only *ca.* 3.8 Å as against 6 Å in $Sm_{2-x}Ce_xCuO_{4-y}$, the pressure-induced changes in interlayer coupling can be more important, and a $dT_c/dP = 0.06 \pm 0.02$ K/kbar has been interpreted to mean they have become dominant [21].

ACKNOWLEDGMENTS

Financial support from the Robert A. Welch Foundation, Houston, TX, from Texas Advanced Research Program, and from the National Science Foundation is gratefully acknowledged.

REFERENCES

1. A. Manthiram and J. B. Goodenough, J. Solid State Chem **92**, 231 (1991).
2. Y. Tokura, H. Tagaki, and S. Uchida, Nature **337**, 345 (1989); Phys. Rev. Lett. **62**, 1197 (1989).
3. J. T. Markert and M. B. Maple, Solid State Commun. **70**, 145 (1989); Physica C **158**, 178 (1989)
4. A. C. W. P. James, S. M. Zahurak, and D. W. Murphy, Nature **338**, 240 (1989).
5. H. Sawa, S. Suzuki, M. Watanahe, J. Akimitsu, H. Matsubaro, H. Watabe, S. Uchida, K. Kokusho, H. Alana, F. Izumi, and E. Takayamo-Muromachi, Nature **337** (1989).
6. H. Okada. M. Takano, and Y. Takeda, Phys. Rev. B **42**, 6813 (1990).
7. A. Manthiram, J. Solid State Chem. **100**, 383 (1992).
8. J. B. Goodenough, J.-S. Zhou, and J. Chan, in *Lattice Effects in High-T_c Superconductors*, T. Egami *et al.* eds. (World Scientific, Singapore, 1993) p. 486.
9. T. Arima, K. Kikuchi, M. Kasuya, S. Koshihara, Y. Tokura, T. Ido, and S. Uchida, Phys. Rev. B **44**, 917 (1991).
10. T. Siegrist, S. M. Zahurak, D. W. Murphy, and R. S. Roth, Nature **334**, 231 (1988).
11. M. Takano, Y. Takeda, H. Okada, M. Miyamoto, and T. Kusako, Physica C **159**, 375 (1989).
12. M. G. Smith, A. Manthiram, J.-S. Zhou, J. B. Goodenough, and J. T. Markert, Nature **351**, 549 (1991).
13. G. Er, Y. Miyamoto, F. Kanamaru, and S. Kikkawa, Physica C **181**, 206 (1991).
14. M. Azuma, Z. Hiroi, M. Takano, Y. Bando, and Y. Takeda, Nature **356**, 775 (1992).
15. S. Adachi, H. Yamauchi, S. Tanaka, and N. Môri, Phisica C **208**, 227 (1993).
16. Z. Hiroi, M. Azuma, M. Takano, and Y. Takeda, Physica C **208**, 286 (1993).
17. N. Terada, G. Zongenalis, M. Jo, M. Hirabayashi, K. Kaneko, H. Ihara, Physica C **185 - 189**, 2019 (1991).
18. M. Kanai, T. Kawai, and S. Kawai, Appl. Phys. Lett. **58**, 771 (1991).
19. N. Sugii, K. Kubo, M. Ichikawa, K. Yamamoto, H. Yamauchi, and S. Tanaka, Jpn. J. Appl. Phys. **31**, L1024 (1992).
20. J. Beille, A. Gerber, T. Grenet, M. Cyrot, J. T. Markert, E. A. Early, and M. B. Maple, J. Less-Common Met. **164 - 165**, 800 (1990).
21. C. L. Wooten, B. Hoan, J. T. Markert, M. G. Smith, A. Manthiram, J. Zhou, and J. B. Goodenough, Physica C **192**, 13 (1992).

CHARGE DYNAMICS IN HIGH T_c COPPER OXIDES

Shin-ichi UCHIDA

The University of Tokyo
Yayoi 2-11-16, Bunkyo-ku, Tokyo 113, Japan

ABSTRACT. A review is given of recent progress in the experimental study of high-T_c superconductors aiming to understand the unique charge dynamics in-and out-of-the doped CuO_2 plane.

1. Introduction

During the last three years no truly new cuprate superconductors have been discovered, but experimental progress made in this period might surpass that in the previous three years of high-T_c discovery, 1986-1989. This owes to improved sample quality and the availability of single crystals of various materials. Systematic study has become possible covering a wide variety of cuprate families as well as a wide composition or carrier density range. The progress has deepened our understanding of the high-T_c cuprates and provided constraints to plausible theoretical models.

In this article emphasis is given on the origin of the linear temperature dependence of the resistivity that every high-T_c cuprate shows in the normal state. Apart from the unprecedentedly high-T_c values, the most remarkable property which characterizes the normal state is the in-plane resistivity proportional to temperature from above T_c up to 1000 K or higher. The finding of the T-linear resistivity by Bednorz and Müller[1] and by the present author's group[2] was striking, since none of the oxides with the layered K_2NiF_4 structure had shown a truly metallic behavior at all temperatures[3]. Now it has been shown that the T-linearity over a wide temperature range is observed only near the optimal dopant concentration at which T_c shows a maximum. Thus, to pursue the origin of the T-linear resistivity is expected to be an effective access to the mechanism of high-T_c superconductivity. The following sections in this article are organized to clarify unique and novel electronic structure of the doped CuO_2 plane, the common structural unit of all the known high-T_c materials, and very characteristic spin or charge fluctuations occuring there which would give rise to the T-linear resistivity as well as high T_c.

2. Electronic Structure of Doped CuO_2 Plane

2.1 SPECTROSCOPIC STUDIES

The optical spectrum of the undoped compound such as La_2CuO_4 and Nd_2CuO_4 is characterized by the O2p-Cu3d charge-transfer (CT) excitation with a threshold

187

E. Kaldis (ed.), Materials and Crystallographic Aspects of HTc-Superconductivity, 187–201.
© 1994 *Kluwer Academic Publishers.*

energy in the 1.5—2.0 eV range[4]. Doping causes a drastic change in the optical spectrum. The CT excitation is rapidly suppressed and instead an edge shows up around 1.0 eV in the reflectivity spectrum, indicating that the spectrum is dominated by low-energy excitations in the energy range lower than 1.0 eV[5].

The optical conductivity spectrum demonstrates that the conductivity above 1.5 eV in the case of La214 decreases while increases below 1.5 eV as doping proceeds, so that the spectral weight of the CT excitaiton is transfered to low-energy excitations(Fig. 1). This is commonly observed in N-type and other P-type cuprate systems[5, 6]. The implication of this spectral change would be that the states, O2p and Cu3d, initially separated by a CT energy gap due to large U, are redistributed on doping to form new states in the original CT gap region[7]. These 'in-gap' states would be reminiscent of p-d hybridized band, as the spectrum is eventually transformed into a Drude-type one in the overdoped region.

Formation of the 'in-gap' states on doping has also been suggested by electron-energy-loss (EELS)[8] and X-ray absorption (XAS)[9] from the core O1s level. In particular, the EELS result on a series of La214 has clearly demonstrated that upon doping new states are formed in the gap region along with the reduction of the lowest empty states, $i.e.$, the upper Hubbard band, of La_2CuO_4.

2.2 ITINERANCY OF THE STATES —FERMI SURFACE AND ELECTRICAL CONDUCTION

The reconstructed states have predominantly in-plane, $O2p_{x,y}$ and $Cu3d_{x^2-y^2}$ chatacter as evidenced from EELS and XAS results[8, 9]. Angle-resolved photoemission spectroscopy (ARPES) shows that these states have momentum dispersion, forming a band of about 1 eV width[10-12]. The low-energy spectrum is thus determined by the intraband excitations in this band. The evolution of Drude-like optical spectrum in the metallic phase indicates that the states near the Fermi level become highly itinerant and the energy range of the itinerant states increases as doping proceeds, finally extending over the whole band in the overdoped region[5]. The development of itinerant states is reflected in the x-dependences of Hall coefficient (R_H)[13] and in-plane resistivity (ρ_{ab})[14]. For $La_{2-x}Sr_xCuO_4$ it has been demonstrated that R_H^{-1} and ρ_{ab}^{-1} increases at a rate much faster than x, suggesting that, in addition to O2p states, initially localized Cu3d states become progressively delocalized.

For the metallic phase of $La_{2-x}Sr_xCuO_4$, the evolution of the resistivity is shown in Fig. 2. One should note that the T-linear resistivity over a wide temperature range is observed in a narrow compositional range ($x \sim 0.20$)[14]. The in-plane resistivity for the overdoped material ($x = 0.30$) shows a superlinear temperature dependence ($\rho_{ab} \sim T^\alpha$ with $\alpha > 1$), whereas underdoped materials ($x < 0.15$) exhibit a complicated T-dependence with an inflexion at 300-400 K. One should also note that the magnitude of the in-plane resistivity is nearly the same for other materials with $\delta \sim 0.2$ where the T-linear dependence is observed over a wide T-range[15]. The results for representative systems are shown in Fig. 3.

2.3 TWO-DIMENSTIONAL ELECTRONIC STATE

Studies of transport and various spectroscopic properties have revealed that the in-gap states and quasiparticles there have two-dimensional character. The temperature

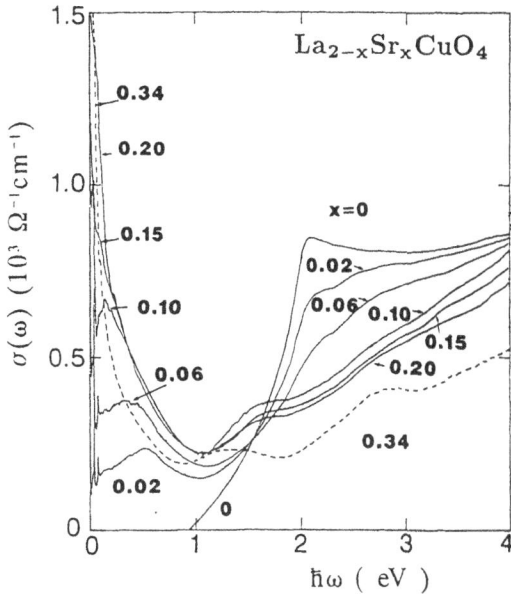

Fig.1 Evolution of the optucal conductivity spectrum of $La_{2-x}Sr_xCuO_4$ with doping.

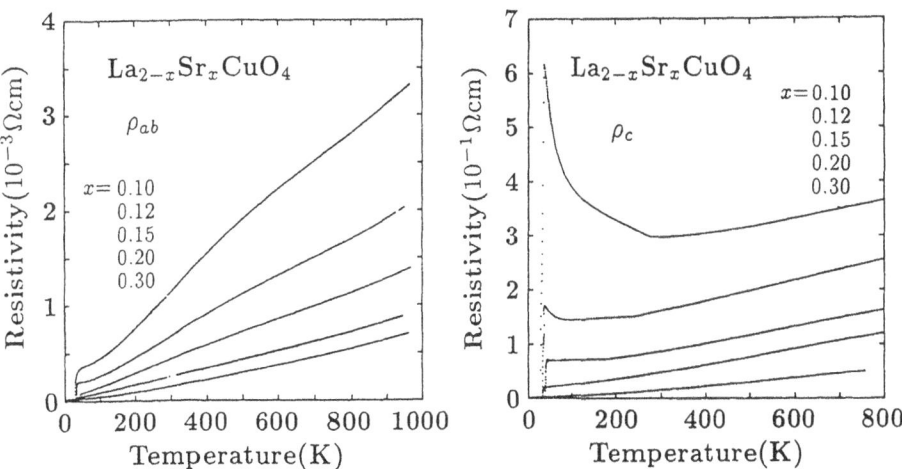

Fig.2 Temperature dependences of the in-plane (ρ_{ab}) and out-of-plane (ρ_c) resistivity measured for metallic $La_{2-x}Sr_xCuO_4$ single crysrals.

dependence of the out-of-plane resistivity (ρ_c) is shown in Fig. 2 together with $\rho_{ab}(T)$ for various compositions of $La_{2-x}Sr_xCuO_4$[14]. The values of ρ_c are always by two orders of magnitude a larger than those of ρ_{ab} but decrease rapidly with x as in the case of ρ_{ab}. For $x < 0.20$ a kink is seen in the $\rho_c(T)$ curve at temperature which coincides with the structural phase transition from HTT to LTO[16]. This temperature T_0 decreases with increasing x and a kink is not clearly seen for $x = 0.20$. The temperature dependence of ρ_c apparently changes at T_0, as most clearly seen for $x = 0.10$; ρ_c is metallic for $T > T_0$ but non-metallic for $T < T_0$. Thus, in the orthorhombic phase, where ρ_{ab} is metallic but ρ_c is non-metallic, the conduction mechanism seems completely different between the two directions. This gives evidence for a two-dimensional metallic state in the superconducting compositional region.

In the case of overdoped compound ρ_c shows a metallic conduction similar to ρ_{ab}. This is demonstrated in Fig. 4 where the anisotropic resistivity ratio ρ_c/ρ_{ab} is plotted as a function of T. The anisotropy of overdoped $x = 0.30$ samples is almost constant (~ 100) over the whole temperature range, giving an evidence for the same conduction mechanism in all directions and thus for a three-dimensional metal. The same seems true for the superconducting compositions in the high temperature region where ρ_c/ρ_{ab} is nearly T-independent and depends little on the composition x in contrast to strongly x — and T-dependent anisotropy in the low temperature region. In this regard, Fig. 4 demonstrates a crossover from three-dimensional to two-dimensional metal as temperature is lowered.

The results should be compared with the band-structure calculations. The band theory predicts an anisotropic three-dimensional metal for any material and for any [17]. Assuming the same scattering process in all directions, the anisotropic resistivity is estimated to be $\rho_c/\rho_{ab} = 28$ for La214 with $x = 0.15$. This value is not very dependent on x. In the actual material the anisotropy is larger by more than an order of magnitude in the superconducting regime. This is also the case with Y123 where the experimental value $\rho_c/\rho_{ab} \sim 40$ should be compared with $\rho_c/\rho_{ab} = 7$ in the band calculation[17]. The results suggest that a certain mechanism is working to confine electrons in the CuO_2 plane more tightly than expected by the band theory.

Recent c-axis polarized spectroscopic studied have presented evidence for the anisotropic electronic states[18]. The c-axis reflectivity spectra of $La_{2-x}Sr_xCuO_4$ ($x = 0.15$) are shown in Fig. 5. The features above 3 eV correspond to the interband trasitions involving the states in the LaO layers and they are observed also in the $a - b$ plane spectrum. By contrast, the low-energy spectrum, involving excitations within or between CuO_2 planes is very anisotropic. While the $a - b$ plane spectrum is dominated by large electronic contribution, the c-axis spectrum is typical of semiconductors, being dominated by two major optical phonon bands even for heavily doped compounds. The electronic contribution is seen only as a small background absorption in the optical conductivity spectrum. The electronic contribution becomes significant in the overdoped region in conjuction with the result of anisotropic resistivity. The much depressed c-axis conductivity over a wide energy range gives another evidence for two-dimensional character of the electronic structure in the high-T_c regime.

Though polarized XAS experiments have confirmed predominant in-plane character of the in-gap states, recent experiments on $La_{2-x}Sr_xCuO_4$ have suggested a small contribution from apical O2p_z state increasing with increase of x[19, 20]. The result can be taken as evidence for increased three-dimensionality with increasing x and be related to the observed metallic c-axis conductivity at higher dopant concentrations.

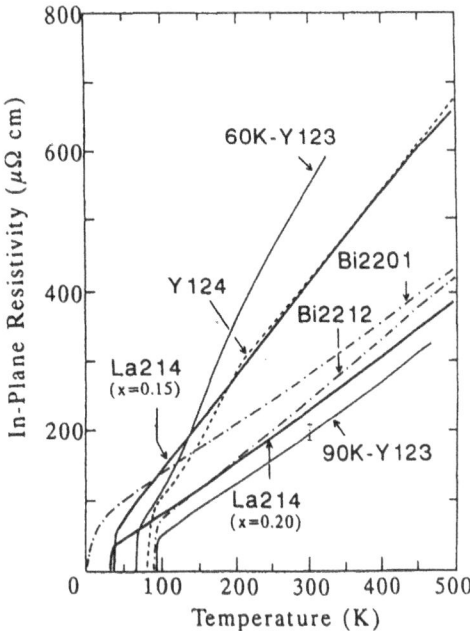

Fig.3 Temperature dependence of in-plane resistivity for various cuprates near optimum compositions.

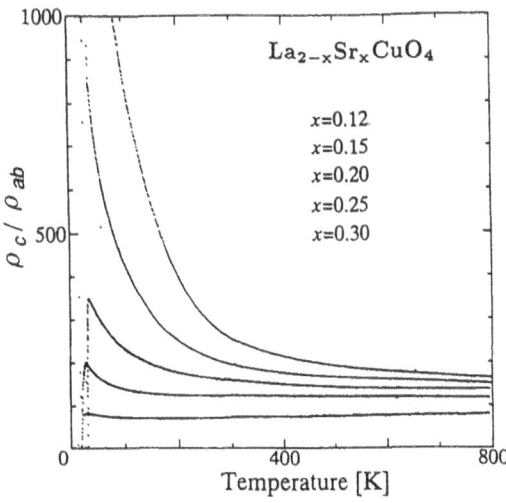

Fig.4 Temperature variations of anisotropic resistivity ratio ρ_c/ρ_{ab} of metallic $La_{2-x}Sr_xCuO_4$.

3. Charge and Spin Excitations in the Doped CuO_2 Plane

3.1 CHARGE FLUNCTUATIONS AND CARRIER DYNAMICS

As discussed in 3.2, beyond the exsistence of the "FS", a fundamental question concerning the metallic state in the high-T_c regime is how well defined is the quasi-particle peak in the vicinity of the Fermi energy. PES on Bi2212 indicated that the quasiparticle peak width or lifetime Γ broadens linearly, $\Gamma \sim \mid E - E_F \mid$[10], on moving away from E_F. The optical conductivity spectrum of Y123, when analysed in terms of Drude carriers with ω-dependent scattering time[21, 22], $\sigma(\omega) = (\omega_p^2/4\pi)(m/m^*(\omega))\tau(\omega)/[1 - i\omega\tau(\omega)]$, gives the scattering rate or lifetime broadening of carriers τ^{-1} apparently increasing linearly with ω. In ordinary Fermi liquid, $\Gamma \sim \mid E - E_F \mid^2$ and $\tau^{-1} \sim \omega^2$ are expected. Actually, τ^{-1} of the overdoped compound exhibits a superlinear ω dependence in the low ω-range[23].

Though the basic physics is not clear, it is tempting to consider that the above ω-dependent scattering may be related to the T-dependence of electrical resistivity; the ω-linear scattering to the T-linear resistivity at the optimal composition and quadratic ones in the overdoped region. One may asume that $\tau^{-1} < \omega + \pi T$ (or max $(\omega, \pi T)$), analogous to $\tau^{-1} \propto \omega^2 + (\pi T)^2$ found for heavy Fermions[24, 25]. The latter relation is expected for electron-electron scattering in the framework of a Fermi liquid picture[26]. The apparent ω- and T-linear dependence of τ^{-1} in the high-T_c regime is assumed to arise from a liquid at marginal edge between Fermi liquid and non-Fermi liquid[27].

There existis an alternative view on the optical spectrum where one assumes that the optical conductivity spectrum is decomposed into two components: a narrow Drude one peaked at $\omega = 0$ and a broad band contribution in the mid-infrared region, so-called mid-IR band[28] (Fig. 6). The two-component spectrum is obvious for underdoped material, e.g., $x < 0.10$ for $La_{2-x}Sr_xCuO_4$. However, as doping proceeds, the spectral weights of both components increase and simultaneously the peak of the mid-IR band shifts to lower energies which makes the spectrum of apparently single component for $x > 0.15$.

The appearance of mid-IR band is not unique to the doped copper oxides but rather universally observed in the spectra of the systems with strong electron correlation. The extensive optical study on 3d-TM oxide series has made it clear that the presence of mid-IR band is correlated with the presence of short-range AF order and thus local spins on 3d-TM sites even in the metallic compounds[29].

In the subsequent section it will be argued that, though the long-range AF order of localized Cu spins is destroyed upon doping, AF correlation within CuO_2 plane persists well into the superconducting phase, up to 0.15 or higher in the case of La-cuprate and up to fully oxygenated Y123. This indicates that the Cu spins play a vital role and thus favors the decomposition of the spectrum into Drude and mid-IR band likely has signifincance even in the high-T_c regime.

On the basis of two-componet, Drude+Mid-IR, analysis the Drude part is directly related to the dc resistivity. The in-plane resistivity is factorized as $\rho_{ab}^{-1} = (\omega_{pD}^2/4\pi)\,\tau$, where ω_{pD}^2 corresponds to the spectral weigh of the drude part and τ^{-1} to the width of the Drude peak. For optimally doped, i.e. $\delta \sim 0.2$, compounds τ^{-1} is shown to be proportional to temperature T, $\tau^{-1} = 2\pi\lambda k_B T$ where λ stands for a coupling constant between carriers and scatterers. Combination of resistivities in Fig.3 and

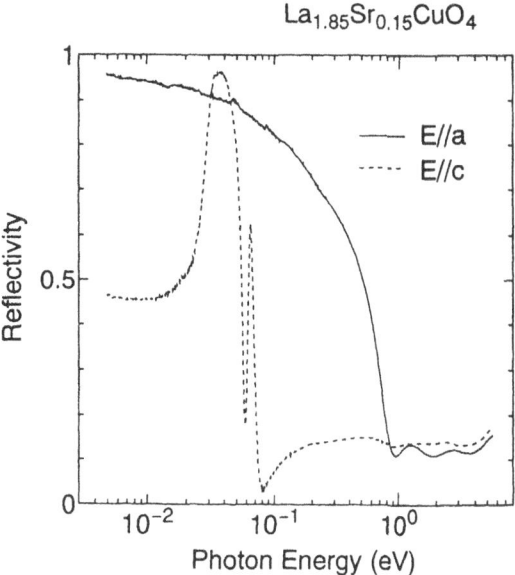

Fig.5 C-axis reflectivity spectrum of $La_{1.85}Sr_{0.15}CuO_4$ measured at room temperature. Also shown is the a-axis spectrum of the same crystal.

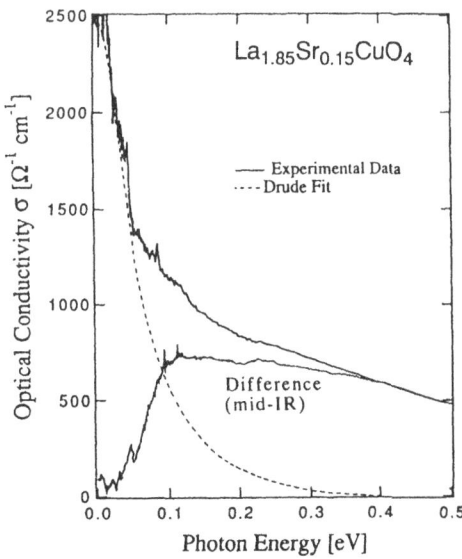

Fig.6 Optical conductivity spectrum of $La_{2-x}Sr_xCuO_4$ in the normal state, decomposed into Drude and Mid-IR part.

the estimated values of ω_{pD}[30], yields surprisingly similar values of τ for optimally doped materials. This is equivalent to suppose nearly identical coupling constant $\lambda \sim 0.3$ among various cuprates with $\delta \sim 0.2$. A direct but rough estimate of τ from the width of the Drude spectral peak also yields the same value $\lambda \sim 0.3$ for so-far measured cuprates[30]. The small λ is consistent with the T- linear resistivity persisting to very high temperatures.

3.2 SPIN FLUCTUATIONS

Spin fluctuations have been extensively investigated by inelastic neutron scattering (INS) and NMR/NQR experiments. Systematic neutron experiments have so far been done mainly on $YBa_2Cu_3O_{6+x}$ system[31, 32]. The Cu spins order antiferromagnetically in the undoped compound and spin excitation is a spin-wave of two-dimensional character with unprecedentedly high spin-wave velocity corresponding to extremely large superexchange coupling $J \sim 0.15$ eV. It has been shown that the antiferromagnetic correlations or antiferromagnetic spin fluctuations persist to the superconducting phase at least to $x = 0.69$, 60 K-Y123. A broad peak is observed in the imaginary part of the dynamical spin susceptibility Im $\chi(q,\omega)$ at $q = Q$, the AF wavevector. Since the peak broadens and the spectral weight shifts to higher energies with increasing hole density, the AF spin correlations are difficult to be seen in more doped 90 K-Y123 ($x \sim 1.0$) by inelastic neutron scattering[31, 32].

NMR experiments provide informaiton on spin fluctuations Im $\chi(q,\omega)/\omega$ at very low ω through the nuclear relaxation rate $(1/T_1)$. The results are in good agreement with Im χ at much higher ω from INS and suggestive of dominant contribution from Im $\chi(Q,\omega)$[33, 34]. $1/T_1T$ due to Im $\chi(Q,\omega)$ turns out to exhibit a T-dependence following a Curie-Weiss law, not a Korringa type, at higher temperatures which has been confirmed also in $La_{2-x}Sr_xCuO_4$ for $x \leq 0.15$[35]. The NMR result reveals a similar behavior in $1/T_1T$ also in 90 K-Y123 that arises from a broad peak of Im χ at $q = Q$ and gives an evidence that the AF spin correlations persist even in this optimal but slightly overdoped region.

As we have seen in Fig. 2 for La214, the T-linear resistivity over a wide temperature range is observed in a narrow compositional range around the optimum doping. In the underdoped region the in-plane resistivity apparently deviates downward from the T-linear extrapolation in the low temperature region. The deviation from the T-linearity is systematically seen in the case of $YBa_2Cu_3O_{6+x}$[36] (Fig. 7). For this system, particularly for 60 K-123, it has been strongly suggested by the T-dependence of NMR-$(T_1T)^{-1}$ and by INS that a gap opens in the spin excitation spectrum [spin gap] at temperature ($\sim 180K$) well above superconducting T_c. It is found that the deviation from the T-linear dependence in resistivity starts at the temperature where the spin gap starts to open. This temperature decreases, while the magnitude of the spin gap increases, as the oxygen content increases. The spin gap seems to merge into the superconducting gap at $x \sim 1$, that is, a gap opens simultaneously in spin and charge excitations[31, 32]. This gives an evidence for an intimate relation between in-plane charge transport and spin fluctuations. The opening of spin gap would lead to a suppression of the carrier scattering and thus to a reduction of resistivity. Therefore, the T-linear resistivity in the optimally doped materials likely arises form a coupling with a zero-gap spectrum of spin fluctuations. As discussed before, the coupling strength λ, estimated from the coefficient of T-linear resistivity, is about 0.3 independent of material.

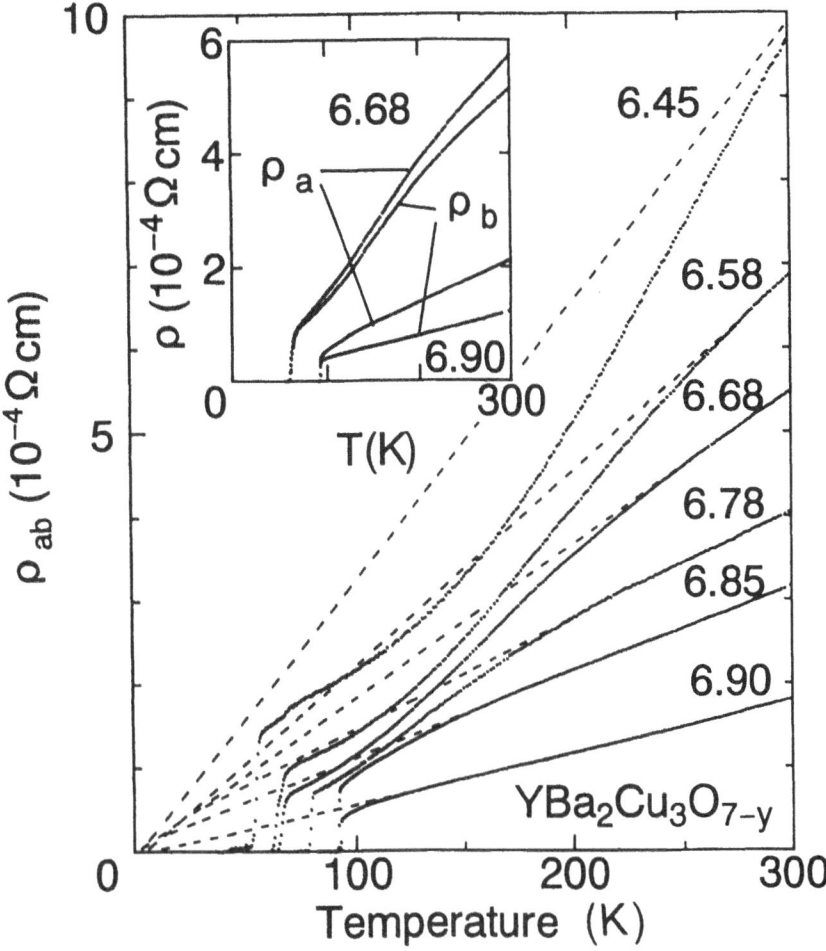

Fig.7 Temperature dependence of the in-plane resistivity of Y123 with various oxygen contents. The inset shows the resistivities for two in-plane directions ($b \parallel$ CuO chains) measured on two twin-free single crystals (T_c=60K and 90K), which gives evidence that ρ_{ab} of twinned crystals is determined by contribution from the CuO_2 planes.

The presence of spin gap in the underdoped region has also been suggested for other systems, $YBa_2Cu_4O_8$[37] and Bi2212[38], and similar non-linear T dependence has been observed in the in-plane resistivity. However, the spin gap has not been identified for underdoped La214 ($x \lesssim 0.15$),though the resistivity shows non-T-linear behavior as shown in Fig.2. The evolution of $\rho_{ab}(T)$ with x is certainly different from that in Y123, $e.g.$ the onset of the steeper decrease in ρ_{ab} takes place always at the same temperature. Aithough the reason for the different behavior in La214 is not clear at present, it is tempting to ascribe this to the different spectrum of spin excitations between Y123 and La214[39, 40].

4. Magnetic Penetration Depth

4.1 MAGNETIC PENETRATION DEPTH AND QUASIPARTICLE LIFETIME

Though it appears certain that the energy scale for pairing in the CuO_2 plane is fairly large, it is still an open question whether or not a BCS gap is observed or a BCS gap has significance in the high-T_c superconductivity. On the other hand, the spectral weight which is transferred or redistributed below T_c seems more relevant to the spectroscopic responses.

In the case of the IR response the spectral weight is transferred to a δ-function at $\omega = 0$. The intensity of the δ-function, equal to the missing spectral weight, is related to the magnetic penetration depth $\lambda = 1/2\pi\omega_{ps}$. This quantity is determined by measurements of muon-spin-rotation (μSR) or surface impedance in the microwave and millimeter wavelength region.

At finite temperatures, λ^{-1} or superconducting carrier density is reduced due to quasiparticle excitations across a superconducting gap. The optical responses, both IR absorption and microwave surface impedance, are then determined by density and scattering rate of the quasiparticles. The quasiparticle lifetime in the conventional superconductors is presumably equal to that of normal-state electrons and independent of temperature. Though the conductivity at finite frequencies below $2\Delta_s$ is completely suppressed at $T \ll T_c$, the quasiparticles contribute to conductivity for $\omega << 2\Delta_s$ at finite temperatures. The temperature dependence of the quasiparticle conductivity normally exhibits a so-called coherence peak just below T_c.

High-T_c cuprates do not show such a conductivity coherence peak[41] in coincidence with the absence of a coherence peak in the T-dependence of NMR relaxation rate T_1^{-1} [41]. For 90 K-material a broad peak is observed in the T-dependence of conductivity at low frequencies ($\omega < 10$ cm^{-1}) which might have an origin different from the coherence effect[42-44]. A possible interpretation might be that the quasiparticle lifetime is rapidly enhanced below T_c ($\tau \sim T^{-\alpha}$ with $\alpha > 1$) which competes with the decrease in the quasiparticle density making a conducivity peak at a certain temperature (Fig. 8). The suppression of quasiparticle scattering may be caused by opening a gap in the spectrum of some entity that scatters charge carriers. This would be analogous to the reduction of resistivity due to the opening of a spin gap observed in 60 K-material. Thus, it suggests that the entity is the spin system where a gap opens accompanied with the opening of a superconducting gap below T_c.

4.2 ANISOTROPY

Recently, it becomes possible to perform reliable and accurate optical measurements on the c-axis (perpendicular to the CuO_2 planes) properties owing to the progress in synthesizing large single crystals with sufficient quality. As we have seen in the previous section, CuO_2 planes are almost isolated electronically each other in the normal state and would be nearly so in the superconducing state. In this case the $a - b$ plane gap would be only one energy scale in the problem, that is, pairing occurs only within a CuO_2 plane and thus anisotropic gap, pairing between planes, does not make sense[45].

A systematic c-polarized optical experiment has recently been done on $La_{2-x}Sr_x$ CuO_4 where large single crystals are now available over a wide x range[46]. In contrast to the $a - b$ plane spectrum, the c-axis spectrum is dominated by optical phonons even at overdoped compositions, indicating a very small electronic contribution. The electronic contriction is not of a Drude form but is a broad and featureless continuum extending up to about 0.1 eV in the normal state.

A dramatic change takes place in the reflectivity spectrum when the sample goes into the superconducting state (Fig. 9). A sharp edge appears at frequency below the lowest phonon band where any feature is not seen in the normal state. The edge position shifts to higher frequencies and the edge becomes sharper as temperature decreases. The energy of the edge at lowest temperatures is found to increase with increasing x, so it does not scale with T_c. Moreover, the edge energy in units of $k_B T_c$ is appreciably lower than 3.5, $\hbar\omega/k_B T_c \sim 1.7$ for $x = 0.15$, and no feature is identified at frequencies higher than $\hbar\omega/k_B T_c = 3.5$. It turns out that the edge corresponds to the zero-crossing of the real part of the dielectric function, not to the zero of the conductivity. From these facts it is concluded that the origin of the edge is not a superconducting gap but is a screened plasma edge of the carriers condensed into the superconducting state which form a δ-function peak at $\omega = 0$ in the optical conductivity spectrum[46]. This can happen only when an extraordinary condition is realized, that is, when the gap $2\Delta_s$ is larger than the screened plasma frequency $\omega'_{ps} = \omega_{ps}/\sqrt{\epsilon_\infty}$ with ϵ_∞ being the dielectric constant contributed by high-energy electronic excitations and low-energy optical phonons.

Thus, one cannot see the superconducting gap in the c-axis optical spectrum, either. The result is seggestive of a single and large energy scale for gap excitations in every direction, implying that the pairing predominaritly occurs within a CuO_2 plane. Then, it might be likely that the supercurrent along the c-axis is carried by in-plane pairs via interplane Josephson coupling. Several attempts to confirm the Josephson coupling have been made on Bi2212 which has largest anisotropy among the cuprates[47, 48]. The experiments made on films on a tilt $SrTiO_3$ substrate or thin single crystals showed Josephson-like current-voltage characteristics without obvious excess currents, but there are controversial results concerning ac or dc Josephson effects.

5. Summary

The spin and charge fluctuations in the daped CuO_2 plane and interplay between them have become progressively clear owing to systematic experiments on well-characterized samples. NMR and inelastic neutron scattering have revealed that the AF spin correlations persist well into the high-T_c regime. The coupling between

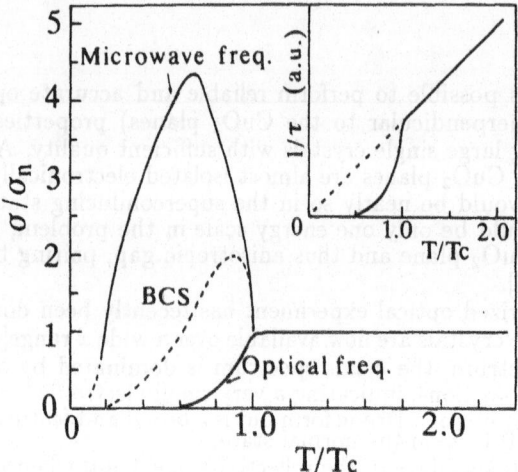

Fig.8 Temperature dependence of in-plane conductivity of 90 K-Y123 at frequencies in the microwave and optical regions. The BCS coherence peak observed in ordinary superconductors is indicated by the dashed curve. The inset shows the quasiparticle scattering rate below T_c deduced from the conductivity at microwave frequency.

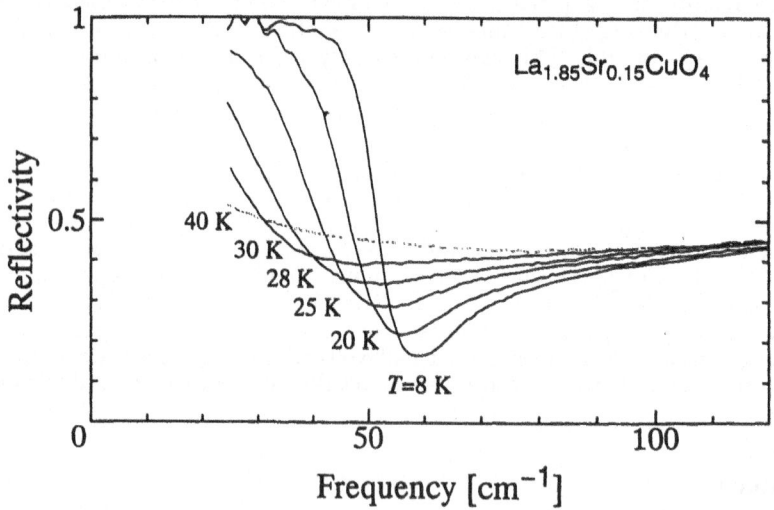

Fig.9 C-axis reflectivity spectrum of $La_{2-x}Sr_xCuO_4$ ($x = 0.15$) measured above and below T_c.

charge and spin fluctuations gives rise to the T-linear resistivity in the normal state of the optimally doped compounds. The coupling strength is found to be nearly the same among various cuprates, indicating the same electronic circumstances in the CuO_2 plane.

The electrons or quasiparticles in the CuO_2 plane show unusual properties. The detailed analyses on the electrical and magnetic properties indicate that the electrons which play a dominant role in the magnetic properties as "localized" spins are the same as the "itinerant" charge carriers — it is not the case that charge carriers are the $O2p$ holes whereas magnetism arises from $Cu3d$ spins. The unusual metallic state is highlighted by the opening of spin gap in underdoped Y123 (and probably in Bi2212) at temperature T_s well above T_c, which obviously affect the charge transport below T_s. This suggests that a gap exists only in the spin-degree of freedom between T_c and T_s, suggestive of nearly independent behavior of spin and charge on the same electron[49, 50] analogous to the spin-charge separation in the Luttinger liquid in one dimension.

The results of the out-of-plane transport and optical properties have yielded an evidence that the electronic state and quasiparticle dynamics are two-dimensional in the normal state. The electrons (holes) are confined within the CuO_2 plane much more strongly than expected by the band theory and coherent charge transport between planes is severely inhibited. The out-of-plane properties are fairly well explained by the gauge theory of the uniform RVB state which also presumes the spin-charge separation in the CuO_2 palne[51].

At the superconducting transition, redistribution of the states takes place over a fairly large energy scale. Though a well-defined superconducting energy gap is not observed nor the symmetry of the gap, either s- or d-wave pairing, is still controversial, there is a large characteristic energy scale in both in-plane and out-of-plane directions. There appears to be only one energy scale in the superconducting state of these anisotropic superconductors which is an evidence for predominant in-plane pairing and weak coupling between CuO_2 planes. The large anisotropy in the normal state is found to be retained in the superconducting state.

Acknowledgment

The author gratefully acknowledges his laboratory members, K. Takenaka, K. Tamasaku, Y. Nakamura, T. Ido, S. Ishibashi, T. Ito, H. Eisaki, H. Sato and H. Takagi for their enthusiasm in performing experiments. Many of their experimental results are shown in this article. He was benefitted by the discussions with Prof. N. Nagaosa at University of Tokyo.

References

1) J. G. Bednorz and K. A. Müller: Z. Phys. B64 (1986) 189.

2) S. Uchida, H. Takagi, K. Kitazawa and S. Tanaka: Jpn. J. Appl. Phys. 26 (1987) L151.

3) C. N. R. Rao and P. Ganguly: in *The Metallic and Nonmetallic States of Matter*, eds. P. P. Edwards and C. N. R. Rao (Taylor&Francis, 1985) p.329.

4) Y. Tokura *et al.*: Phys. Rev. B41 (1990) 11657.

5) S. Uchida *et al.*: Phys. Rev. B43 (1991) 7942.

6) S. L. Cooper *et al.*: Phys. Rev. B41 (1990) 11605, T. Arima, Y. Tokura, T. Ido and S. Uhcida: Physica C185-189 (1991) 1021.

7) S. Maekawa, Y. Ohta and T. Tohyama: *Physics of High-Temperature Superconductors*, eds., S. Maekawa amd M. Sato (Springer Verlag, 1991) p.29.

8) H. Romberg, M. Alexander, N. Nücker, P. Adelman and J. Fink: Phys. Rev. B32 (1990) 8868.

9) C. T. Chen *et al.*: Phys. Rev. Lett. 66 (1991) 104.

10) C. G. Olson *et al.*: Phys. Rev. B42 (1990) 381.

11) T. Takahashi *et al.*: Nature 334 (1988) 691.

12) R. Manzke, T. Buslaps, R. Claessen and J. Fink: Europhys. Lett. 9 (1989) 477.

13) H. Takagi, T. Ido, S. Ishibashi, M. Uota and S. Uchida: Phys. Rev. B40 (1989) 2254.

14) Y. Nakamura and S. Uchida: Phys. Rev. B47 (1993) 8369, H. Takagi *et al.*: Phys. Rev. Lett. 69 (1992) 2975.

15) B. Batlogg: Physics Today 44, No.6 (1991) 44.

16) S. Kambe *et al.*: Physica C160 (1989) 243, T. Kimura *et al.*: Physica C192 (1992) 247.

17) W. E. Pickett: Rev. Mod. Physics 61 (1989) 433.

18) S. Uchida: J. Phys. Chem. Solids 53 (1992) 1603.

19) C. T. Chen *et al.*: Phys. Rev. Lett. 68 (1992) 2543.

20) E. Pellegrin *et al.*: Phys. Rev. B47 (1993) 3354.

21) Z. Schlesinger *et al.*: Phys. Rev. B41 (1990) 11237.

22) L. D. Rotter *et al.*: Phys. Rev. Lett. 67 (1991) 2741.

23) S. Uchida: Physica C185-189 (1991) 28.

24) B.C.Webb, A. J. Sievers and T. Mihalisin: Phys. Rev. Lett. 57 (1986) 1951.

25) K. Miyake, T. Matsuura and C. M. Varma: Solid State Commun. 71 (1989) 1149.

26) P. E. Sulewski *et al.*: Phys. Rev. B38 (1988) 5338.

27) C. M. Varma, P. B. Littlewood, S. Schmitt-Rink, E. Abrahams and A. E. Ruckenstein: Phys. Rev. Lett. 63 (1989) 1996.

28) T. Timusk and D. B. Tanner: in *Physical Properties of High Temperature Superconductors I*, ed. D. M. Ginsberg (World Scientific, 1989) p.339.

29) S. Uchida, H. Eisaki and S. Tajima: Physica B186-188 (1993) 975.

30) D. B. Tanner and T. Timusk: in *Physical Properties of High Temperature Superconductors III*, ed. D. M. Ginsberg (World Scientific, 1992) p.363.

31) M. Sato: JJAP Series 7, *Mechanisms of Superconductivity* (1992) p.167, and M. Sato *et al.*: J. Phys. Soc. Jpn. 62 (1993) 263.

32) J. Rossat-Mignod *et al.*: Physica B180&181 (1992) 383.

33) H. Yasuoka: *Strong Correlation and Superconductivity*, eds. H. Fukuyama, S. Maekawa and A. P. Malozemoff (Springer-Verlag, 1989) p.254.

34) R. E. Walstedt and W. W. Warren, Jr.: Science 248 (1990) 1082.

35) For a recent view, Y. Kitaoka, K. Ishida, S. Ohsugi, K. Fujiwara and K. Asayama: Physica C185-189 (1991) 98, and Appl. Magn. Reson. 3 (1992) 549.

36) T. Ito, K. Takenaka and S. Uchida: Phys. Rev. Lett. 70 (1993) 3995.

37) T. Machi *et al.*: Physica C185-189 (1991) 1147.

38) R. E. Walstedt, R. E. Bell and D. B. Mitzi: Phys. Rev. B44 (1991) 7760.

39) S. -W. Cheong *et al.*: Phys. Rev. Lett. 67 (1991) 1791.

40) T. Tanamoto, H. Kohno and H. Fukuyama: J. Phys. Soc. Jpn. 61 (1992) 1886.

41) R. T. Collins *et al.*: Phys. Rev. B43 (1991) 8701.

42) M. C. Nuss, P. M. Mankiewich, M. L. O'Malley, E. H. Westerwick and P. B. Littlewood: Phys. Rev. Lett. 66 (1991) 3305.

43) D. A. Bonn, P. Dosanjh, R. Liang and W. N. Hardy: Phys. Rev. Lett. 68 (1992) 2390.

44) T. Shibauchi, A. Maeda, H. Kitano, T. Honda and K. Uchinokura: Physica C203 (1992) 315.

45) R. A. Klemm and S. H. Liu: Phys. Rev. B44 (1991) 7526.

46) K. Tamasaku, Y. Nakamura and S. Uchida: Phys. Rev. Lett. 69 (1992) 1455.

47) R. Kleiner, F. Steinmeyer, G. Kunkel and P. Müller: Phys. Rev. Lett. 68 (1992) 2394.

48) J. S. Tsai and J. Fujita: unpublished.

49) P. W. Anderson: Physica C185-189 (1991) 11.

50) N. Nagaosa and P. A. Lee: Phys. Rev. Lett. 64 (1990) 2450 and Phys. Rev. B43 (1991) 1233.

51) N. Nagaosa: J. Phys. Chem. Solids 53 (1992) 1493.

LATTICE INSTABILITIES AND SUPERCONDUCTIVITY IN La-214 COMPOUNDS

Y. MAENO
Department of Physics
Faculty of Science
Hiroshima University
Higashi-Hiroshima 724
Japan

ABSTRACT. Studies of interplay between structural instabilities and superconductivity in La-214 compounds revealed a novel nature of electronic bands near the Fermi level.
In $La_{2-x}Ba_xCuO_4$ (LBCO) a transition to a low-temperature tetragonal phase and the disappearance of bulk superconductivity occur only at the carrier concentration $p\sim1/8$ per copper. We will review the physical properties of carrier and ion-size controlled systems based on LBCO.
Ultrasonic measurements on $La_{2-x}Sr_xCuO_4$ (LSCO) single crystals show that only a particular transverse elastic constant softens below about 50 K and that this softening is due to the growth of structural fluctuation in the normal state. This behavior suggests the presence of a narrow electronic band, which couples with a particular in-plane strain.
We argue that these phenomena are closely related because both the structural change in LBCO and the applied shear strain in LSCO lift the degeneracy of in-plane oxygen sites. They indicate the importance of strong coupling between the normal-state electronic system and the lattice by Peierls-type mechanism.
We also review recent results on the question whether orthorhombic crystal symmetry is necessary for superconductivity in LSCO.

1. Introduction

La-214 compounds constitute a model system of high-T_c copper oxides, including the first high-T_c superconductor discovered by Bednorz and Müller [1], $La_{2-x}Ba_xCuO_4$ (LBCO). In these compounds, a variety of structural instabilities characteristic of perovskite-related K_2NiF_4 structure are present. The purpose of this paper is to discuss the importance of interplay between the lattice instabilities and superconductivity in these compounds.

Physical properties of LBCO is depicted in the phase diagram of Fig. 1. The parent compound, La_2CuO_4 with $x=0$ is an insulator with its Cu spins ordering antiferro-magnetically below $T_N\sim310$ K. The insulating nature of the parent compound is a consequence of a charge-transfer gap of $\Delta=2$ eV formed by strong correlations among the $3d$ electrons. The Néel temperature T_N decreases quickly upon doping, and disappears above $x=0.02$ [2].

It seems true in nearly all high-T_c superconductors that if the hole concentration p can be varied, the variation of the superconducting transition temperature T_c with p follows a bell-shaped curve. In the "lightly-doped" region, T_c increases with p and gradually reaches a maximum value, specific to a given structure. When p is further increased

E. Kaldis (ed.), Materials and Crystallographic Aspects of HTc-Superconductivity, 203–222.
© 1994 *Kluwer Academic Publishers.*

204

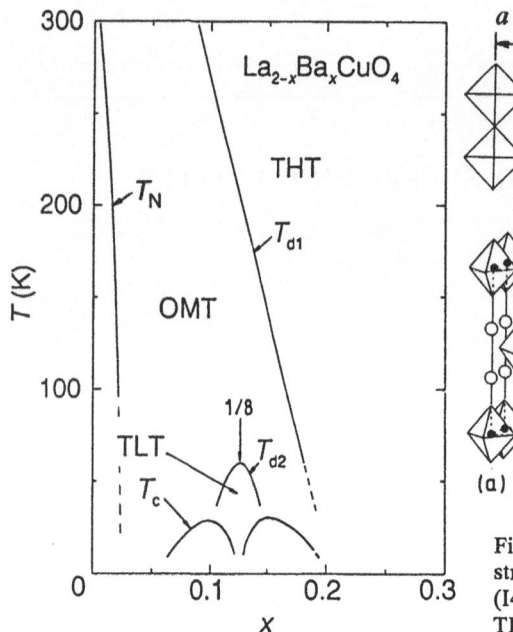

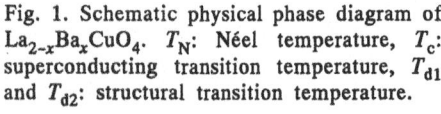

Fig. 1. Schematic physical phase diagram of $La_{2-x}Ba_xCuO_4$. T_N: Néel temperature, T_c: superconducting transition temperature, T_{d1} and T_{d2}: structural transition temperature.

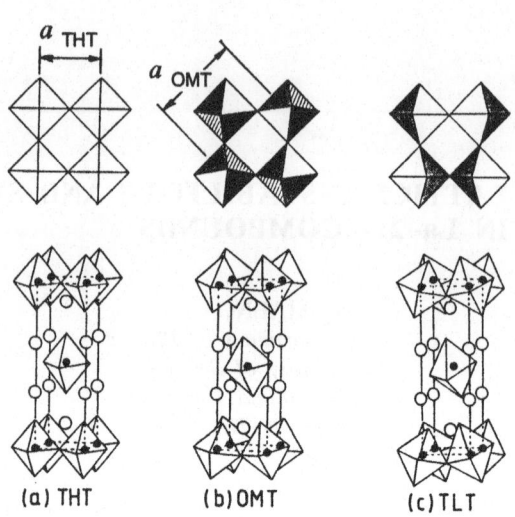

(a) THT   (b) OMT   (c) TLT

Fig. 2. Schematic representation of the crystal structures of $La_{2-x}Ba_xCuO_4$. (a) THT phase (I4/mmm), (b) OMT phase (Bmab) and (c) TLT phase ($P4_2/ncm$). Closed and open circles represent Cu and La(Ba) atoms, respectively. Oxygen atoms are located at each apex of octahedra drawn by solid lines. The upper figures show the top view of the structures, clarifying how CuO_6 octahedra are tilted.

beyond such optimum p into the "heavily–doped" region, T_c decreases with p and finally the system becomes a normal–Fermi liquid in the "overdoped" region.

For LBCO, $p=x$ as long as the amount of the oxygen vacancies is negligible. As soon as the metallic conductivity appears with increasing x, the system exhibits superconductivity at low temperatures. For LBCO the variation of T_c with x is quite unique, as first pointed out by Moodenbaugh et al. [3]: as shown in Fig. 1, T_c attains two maxima of about 30 K at $x \approx 0.10$ and 0.15. Between these x, T_c sharply drops in a very narrow range of x around 0.125, where bulk superconductivity is lost.

The structural phase boundaries are also schematically shown in Fig. 1. The parent compound La_2CuO_4 has a tetragonal phase at high temperatures (the THT phase*, space group I4/mmm) above 530 K [4], but transforms to a less symmetric orthorhombic phase

---

*Both abbreviations THT and HTT for the tetragonal phase at high temperatures have been widely used at least since mid 1970's in the study of organic K_2NiF_4-type compounds. In light of the recently–found orthorhombic phase at low temperatures in Nd–substituted $La_{2-x}Sr_xCuO_4$, we much prefer to place O (for orthorhombic, or T for tetragonal) in front and designate Bmab and Pccn phases as OMT (midtemperature orthorhombic) and OLT (low–temperature orthorhombic), respectively.

at midtemperatures (the OMT phase, Bmab) on cooling. The phase boundary T_{d1} of the THT–OMT transition, which is of the second order, decreases almost linearly with x at the initial rate of $dT_{d1}/dx = -2.8 \times 10^3$ K [5] and disappears at $x \sim 0.2$.

Corresponding to the disappearance of bulk superconductivity at $x \sim 1/8$, the structural phase changes abruptly from the OMT phase to a low–temperature tetragonal (TLT, space group $P4_2/ncm$) phase on cooling below $T_{d2} \sim 60$ K [6–8]. As in Fig. 1, T_{d2} sharply peaks at $x \sim 1/8$. In $La_{2-x}Sr_xCuO_4$ (LSCO) a slight depression of T_c is also observed at $x \sim 0.12$, but the OMT phase is known to be stable to the lowest temperatures by diffraction measurements.

These structural phases are characterized by the difference in the ordering of tilted CuO_6 octahedra, as schematically represented in Fig. 2. The THT–OMT transition follows soft–mode behavior involving zone–boundary phonons, as found in neutron scattering measurements [9]. The displacements associated with these modes consist principally of tilting of CuO_6 octahedra about the axes along the <110> direction of the THT phase, resulting in the tilting represented in Fig. 2 (b). This transition is accompanied by the doubling of the unit–cell volume with a_{OMT}, $b_{OMT} \sim \sqrt{2} \cdot a_{THT}$ and $c_{OMT} \sim c_{THT}$. In the OMT–TLT transition, the direction of the tilt changes by $45°$ as shown in Fig. 2 (c). The tilt pattern is rotated by $90°$ in the adjacent CuO_2 layers, so that the total symmetry becomes tetragonal. The principal axes and the unit–cell volume are essentially unchanged in the OMT–TLT transition.

In the following, we focus on three aspects of the interplay between lattice instability and superconductivity in the La–214 compounds. We first discuss the problems involved in the OMT–TLT transition. In particular, we review how the low–temperature instabilities are modified in carrier and ion–size controlled systems, obtained by divalent, trivalent and tetravalent substitutions at the La–site in LBCO. We will discuss the mechanism responsible for the strong correlation between the lattice transformation and the suppression of superconductivity. In Sec. 3, we will address a controversial issue of possible relationship between the THT–OMT transition and the disappearance of superconductivity in the overdoped region. In Sec. 4, we will discuss the results of ultrasonic measurements on LSCO, which show that only the elastic constant of the in–plane shear mode softens below about 50 K and that this softening is due to the growth of structural fluctuation in the normal state. We argue that this behavior is closely related to the 1/8 problem in LBCO because both the structural change in LBCO and the applied shear strain in LSCO lift the degeneracy of in–plane oxygen sites. They clarify the importance of strong coupling between the normal–state electronic system and the lattice by Peierls–type mechanism.

2. The Low–Temperature Tetragonal Phase and Superconductivity

2.1. PROPERTIES OF THE LOW–TEMPERATURE TETRAGONAL PHASE

Properties of the tetragonal phase at low temperatures (the TLT phase) in LBCO and related compounds have been reviewed by several authors [10–13].

A sequence of structural transitions from the THT to OMT and from the OMT and TLT phases by varying the temperature is clearly seen by X–ray powder diffraction. Figure 3 represents a portion of the diffraction spectra of LBCO with $x=0.125$ at selected temperatures. The (220) reflection of the THT phase at room temperature, shown in Fig. 3(a), splits into (040) and (400) reflection peaks of the OMT phase below $T_{d1} =$

210 K as in Fig. 3(b). On further cooling, a third peak appears between the two peaks and coexists with these OMT peaks, as in Fig. 3(c). The central peak is the (400) reflection of the TLT phase and its intensity grows at the expense of those of the OMT peaks at low temperatures, as in Fig. 3(d). This behavior indicates that the OMT–TLT transition is of the first order.

The temperature dependence of the lattice parameters are plotted in Fig. 4. The orthorhombic splitting emerges continuously on cooling across the THT–OMT transition, as it is expected from the symmetry consideration of these structures. For the coexisting OMT and TLT phases, the volume fraction of each phase may be estimated from the ratio of the integrated diffraction intensities of the peaks shown in Fig. 3. After fitting the data by multiple gaussian peaks, we obtained the temperature dependence of the integrated intensity ratios shown in Fig. 5.

Changes in the microstructures across the structural transitions have been studied in detail by electron microscopy, diffraction and dark–field imaging by Chen *et al.*[14]. They found that in LBCO with $x=0.125$ the microdomain structures arising from the twin boundaries in the OMT phase transforms into antiphase domain boundaries in the TLT phase on cooling, without changing their physical locations.

The THT–OMT transition has a profound effect on the transport properties of LSCO. For single crystals of LSCO with $x=0.12$ and 0.14, Kambe *et al.* [15] argued based on their measurements of resistivity that the THT phase above T_{d1} is treated as a highly anisotropic but three–dimensional metal: the OMT phase, on the other hand, should rather

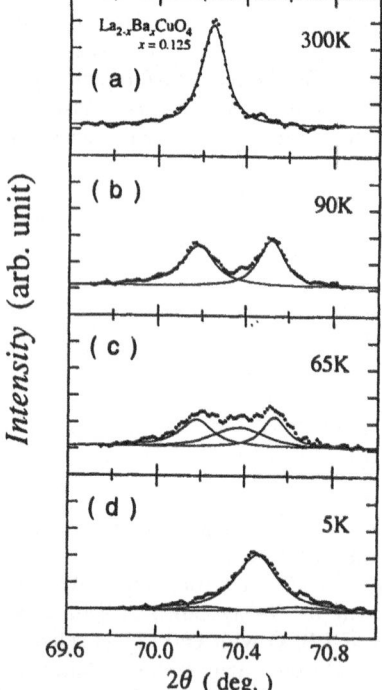

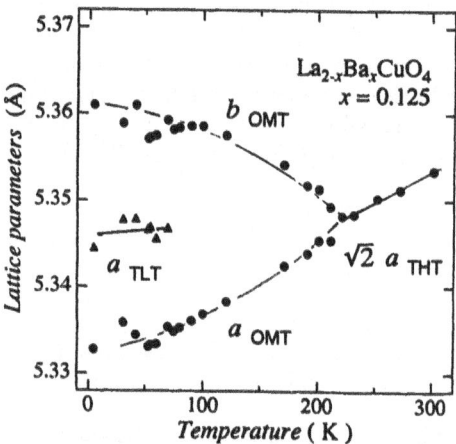

Fig. 4. (Above) Temperature dependence of the lattice parameters for $La_{2-x}Ba_xCuO_4$ with $x=0.125$.

Fig. 3. (Left) A portion of powder X-ray spectra for $La_{2-x}Ba_xCuO_4$ with $x=0.125$ at selected temperatures.

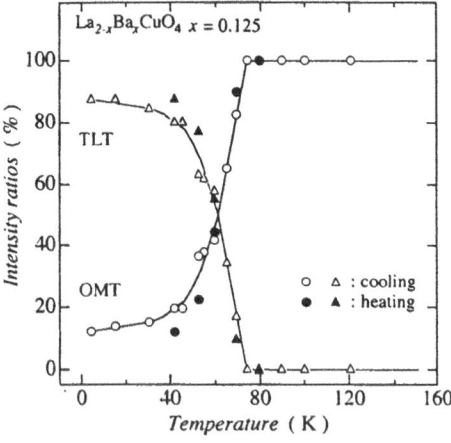

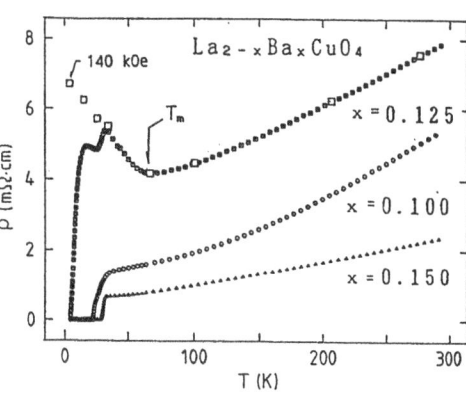

Fig. 5. Variations of the integrated intensity ratios of the X-ray peaks of the OMT and TLT phases with temperature for $La_{2-x}Ba_xCuO_4$ with $x=0.125$.

Fig. 6. Electrical resistivity of $La_{2-x}Ba_xCuO_4$ with various values of x. open squares shows the data for $x=0.125$ under a magnetic field of 14 T.

be regarded as a two-dimensional metal, because the electrical conduction along the c-axis becomes semiconductor-like. The structural transition has little effect on the conduction in the ab-plane. For LBCO, we are not aware of any corresponding study on single crystals.

In sharp contrast to the THT-OMT transition, a dramatic change occurs in the in-plane conductivity at the OMT-TLT transition. The temperature dependence of the resistivity of LBCO is shown in Fig. 6. Only with $x\approx0.125$, the resistivity takes a minimum at ~ 60 K and the conduction becomes non-metallic below this temperature. These results are for polycrystalline samples, but quite likely reflecting the change in the in-plane conductivity. One should be very cautious, however, in identifying the minimum in the resistivity to the structural transition temperature. In LBCO the correspondence seems quite good, but in $La_{2-x}Ba_{x-y}Sr_yCuO_4$ [5,16] or $La_{2-y-x}Nd_yBa_xCuO_4$ [17], for example, the minimum in the resistivity is always found at about 60 K, whereas the structural transition temperature is either suppressed or enhanced. The resistivity minimum in these cases probably reflects localization of carriers caused by reduction of the carrier density and by the presence of microdomains in the TLT phase. For $x\approx0.125$, a double-step transition as in Fig. 6 is typically observed, but the superconductivity is filamentary because Meissner signal becomes very small.

Other quantities, such as paramagnetic susceptibility in the normal state, Seebeck coefficient and Hall coefficient all exhibit characteristic changes [18]. The Seebeck coefficient becomes negative below about 50 K only near $x\sim1/8$. This behavior is interpreted as due to the formation of fine structure in the electronic density of states near the Fermi level in the TLT phase. The enhancement of the thermal conductivity just below T_{d2} suggests that the phonon mean fee path is enhanced because of the decrease in the number of conduction electrons that scatter phonons [19].

Magnetic ordering of the Cu spins has been found at $x\sim0.12$ only for LBCO by the measurements of nuclear Schottky term in the specific heat below 1 K [20], by muon-spin-rotation (μSR) measurements [21,22], and by NMR measurements [22]. Because the observed Néel temperature $T_N=38$ K [22] is substantially lower than $T_{d2}\sim60$ K, it is unlikely that the magnetic interaction is the driving mechanism of the OMT–TLT transition.

The OMT–TLT transition temperature T_{d2} of LBCO is strongly suppressed by pressure. According to the neutron diffraction study by Katano et al.[23], T_{d2} of LBCO with $x = 0.125$ disappears at 0.7 GPa. However, Yamada and Ido [24] found that even when the formation of the TLT phase is strongly suppressed by the pressure up to $P=2.0$ GPa, T_c determined from the ac susceptibility recovers only up to 10 K. Therefore, they argued that there exists another origin of the depression of T_c in addition to the structural transition.

Another curious property of La–214 compounds at $x\sim0.12$ is an extraordinary enhancement of the oxygen isotope effect in both LBCO and LSCO [25]. The coefficient of oxygen isotope effect, α_O, increases with x and reaches a value as large as 0.8 at $x=0.12$, then sharply drops below 0.2 for x greater than 0.14. The behavior of α_O seems quantitatively similar in both LBCO and LSCO. To account for this unusual isotope effect, a model has been proposed which assumes the presence of 2–dimensional van Hove singularities in the electronic density of states [26].

We will next consider how the stability of the TLT phase is modified by various atomic substitutions at the La site of LBCO.

2.2. IMPORTANCE OF THE CARRIER CONCENTRATION: $La_{2-y-x}Th_yBa_xCuO_4$

The electronic and structural instabilities described above occur at the doping of 1/8. But this condition either means the Ba content x being 1/16 of the La–site or the carrier concentration p being equal to 1/8 per Cu ion. In order to clarify this point, various attempts have been made to control the carrier density in LBCO by means other than the amount of divalent ions [27–29]. It turned out that thorium is an ideal substitute for this purpose [30] because it is stable in the Th^{4+} ($5f^0$) state in 9 coordination and the size of the ion is effectively comparable to that of La^{3+}.

Polycrystalline samples of $La_{2-y-x}Th_yBa_xCuO_4$ (LTBCO) were prepared by solid–state reaction and formation of the desired compound has been confirmed by a wavelength-dispersive electron–probe microanalyzer (EPMA). By comparing the values determined by iodine titration, we conclude that the change in oxygen content by Th substitution is insignificant, and Th^{4+} ion acts to compensate the hole carrier introduced by Ba^{2+}; thus $p\sim x-y$.

Figure 7 compares the x–dependence of T_c for LBCO and LTBCO with $y=0.020$. It is clear that the center of sharp depression of T_c shifts from $x\approx0.125$ for LBCO to $x\approx0.145$ for LTBCO. In both cases, the minimum in T_c occurs at $p\sim x-y\approx1/8$. The temperature dependence of field–cooled magnetization is shown in Fig. 8. In LTBCO with $y=0.020$, a strong Meissner signal is recovered for $x=0.125$. For $x=0.145$, the onset of superconductivity remains sharp and a substantial Meissner volume fraction is retained in spite of the remarkable depression of T_c. By increasing x further, T_c recovers again to about 30 K. Correspondingly, the distinctive minimum in resistivity ρ at about 60 K, reproducibly present in LBCO with $x=0.125$, now appears at ~ 50 K only for $x\sim0.145$.

The variation of the lattice parameters with temperature is plotted in Fig. 9. LTBCO

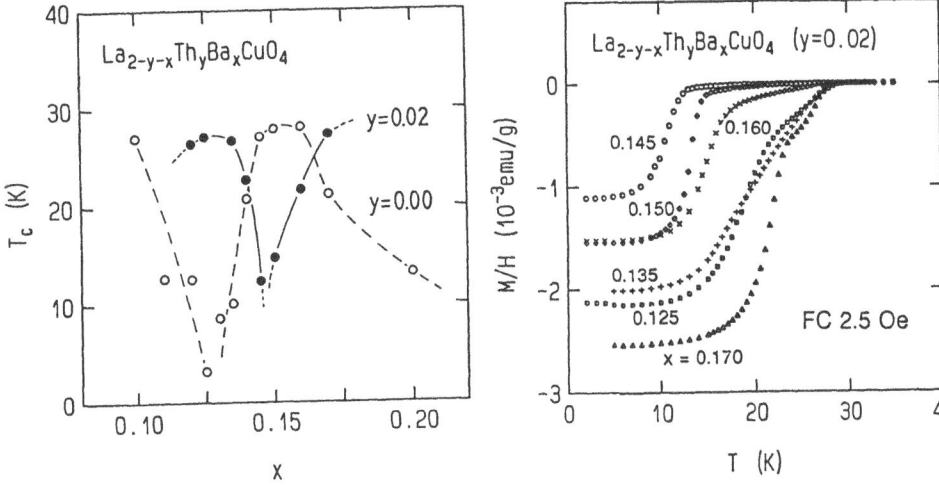

Fig. 7. The variation of T_c with x for $La_{2-x}Ba_xCuO_4$ (O) and $La_{2-y-x}Th_yBa_xCuO_4$ with $y=0.020$ (●). T_c is defined at the temperature of 1 % volume fraction of perfect diamagnetism. In both systems, the anomalous suppression of T_c is centered at $p \approx x - y \approx 0.125$.

Fig. 8. Magnetization of $La_{2-y-x}Th_yBa_xCuO_4$ with $y=0.020$. For $x=0.145$ a strong suppression of T_c is evident, but the onset of superconductivity remains sharp and a substantial Meissner volume fraction is retained.

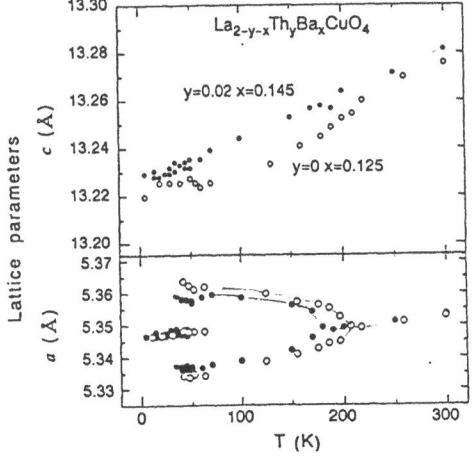

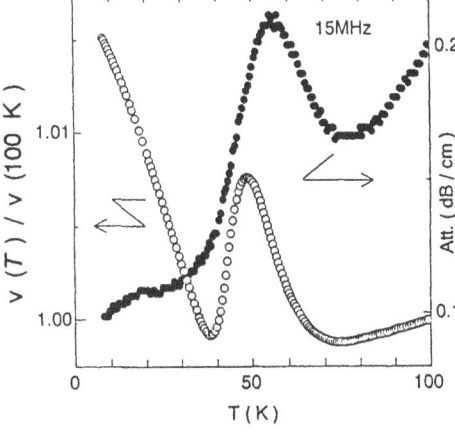

Fig. 9. Temperature dependence of the lattice parameters of LBCO with $x=0.125$ and LTBCO with $y=0.020$ and $x=0.145$. Although both samples exhibit minimum in T_c with respect to varied x, the higher structural-transition temperature $T_{d1}=180$ K of LTBCO is as much as 25 K lower than that for Th-free LBCO.

Fig. 10. Temperature dependence of the longitudinal ultrasound velocity v normalized by the value at 100 K and the attenuation coefficient α of a polycrystalline sample of $La_{2-y-x}Th_yBa_xCuO_4$ with $y=0.020$ and $x=0.145$.

with most reduced T_c exhibits $T_{d1}=180\pm5$ K whereas LBCO with most reduced T_c shows $T_{d1}=205\pm5$ K. This difference is substantial and excludes any interpretation of the behavior of LTBCO in terms of those of Th–free LBCO. With $y = 0.020$, $T_{d2}=38$ K for $x=0.145$ if defined at the TLT–OMT intensity ratio of 50 % in the X–ray diffraction spectra. The intensity ratio reaches 90 % at 32 K. The TLT intensity never reaches 80 % for other values of x. This is a clear direct evidence that the structural anomaly at low temperatures is also controlled by the condition $x-y\approx1/8$.

Similar results are obtained for a series of samples with a fixed value of x at 0.145 and with varied y from 0 to 0.040. Suppression in T_c and maximum in T_{d2} occur sharply around $x-y\approx0.125$.

Even if each Th^{4+} ion were paired with Ba^{2+} ion on a neighboring site, such a pair would not compensate for the local ionic distortion because Ba^{2+} has a commandingly greater ionic radius in comparison with matching La^{3+} and Th^{4+}. Therefore we conclude that the hole concentration $p=1/8$ is the essential condition which governs the low–temperature instabilities in LBCO.

Judging from the volume fractions of the Meissner signal and of the TLT phase, we infer that the bulk superconductivity with reduced T_c is sustained even in the low–temperature phase in Th–substituted LBCO. It is not possible to conclude, however, whether truly tetragonal domains with $P4_2/ncm$ symmetry sustain bulk superconductivity.

The longitudinal ultrasound velocity of LBCO increases with the stiffening of the lattice in the TLT phase and continues to increase at least down to 1.8 K. In LTBCO with $y=0.020$ and $x=0.145$, in contrast, a large softening is observed well below T_{d2} and hardening starts again at lower temperatures, as shown by open circles in Fig. 10 [10]. Since we were not able to detect any corresponding change by X–ray diffraction, we do not have a definitive explanation for the second anomaly of the ultrasound velocity, although it is plausible that the second lattice instability is present in LTBCO.

2.3. DIVALENT SUBSTITUTIONS AT THE La-SITE: $La_{2-x}Ba_{x-y}M_yCuO_4$ (M = Sr and Ca)

The size of the cations in the La–site is no doubt an important parameter, especially to characterize the strength of the instabilities. In $La_{2-x}Sr_xCuO_4$(LSCO) weak depression of T_c has also been observed at $x\approx0.12$ but the structural transition to the TLT phase does not occur. Study of low–temperature structural transition [5] indicated that in $La_{2-x}Ba_{x-y}Sr_yCuO_4$ with $x=0.125$, T_{d2} decreases almost linearly with y for $0\leq y\leq0.05$, but disappears rather abruptly beyond $y\sim0.07$. Therefore it is not likely that the TLT phase exists in bulk in LSCO. Lattice stiffening and a small attenuation peak at $T\approx10$ K in LSCO with $x\approx0.12$, observed by ultrasound [31], may correspond to the development of short–range formation of the TLT–type buckling of the CuO_2 planes.

In an effort to quantitatively describe the variations of T_{d2} with different divalent–ion substitutions at the La–site, we studied alloyed compounds $La_{2-x}Ba_{x-y}M_yCuO_4$ with M = Sr (LBSCO) and M = Ca (LBCCO) with $x=0.125$ (fixed) and $0\leq y\leq0.125$ [16].

Variations of T_c and T_{d2} in both Ca^{2+}– and Sr^{2+}–substituted systems are plotted in Fig. 11 as functions of the averaged effective ionic radius at the La^{3+}–site, r_{ave}. The effective size is given by subtracting 0.1 Å from the ionic radius of nine–coordinated M^{2+} ion [32] as an empirical correction for the valence change [10,27]. The effective ionic radii used are 1.37 Å, 1.21 Å and 1.08 Å, respectively for Ba^{2+}, Sr^{2+} and Ca^{2+}. This correction, however, does not alter the qualitative interpretations discussed here. With decreasing r_{ave}, T_{d2} decreases almost linearly at first, but it disappears rather

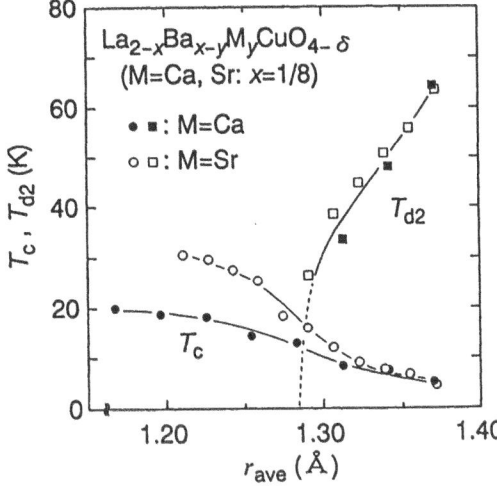

Fig. 11. The variations of T_c and T_{d2} plotted against the average ionic size of divalent ions at the La^{3+}-site in $La_{2-x}Ba_{x-y}M_yCuO_4$ (M = Ca, Sr) with $x=0.125$. T_{d2} was determined as the peak temperature of the small anomaly in the specific heat; T_c was determined from the electrical resistivity.

abruptly near $r_{ave} \sim 1.28$ Å. Correspondingly, T_c recovers rapidly below this r_{ave}. As reported previously [33] T_c of $La_{2-x}Ca_xCuO_4$ is limited to about 20 K, because of greater amount of oxygen vacancies. Aside from this difference in the highest attainable T_c, the variations of both T_c and T_{d2} in both systems are characterized well by r_{ave}. The disappearance of the TLT phase is correlated with the recovery of T_c, and vice versa, the emergence of superconductivity seems to suppress T_{d2}.

The lattice parameters a and c at room temperature in both systems decrease with decreasing r_{ave}. The variations of c are quite well scaled by r_{ave}. Those of a, however, are scaled well by the amount of doping, y. Therefore, the variations of T_{d2} may also be characterized by the lattice parameter c. The corresponding c-dependence of T_{d2} for large r_{ave} is approximately $dT_{d2}/d(c/c_0) = 2.1 \times 10^4$ K. Here, c_0 is the value of c at $y=0$.

Let us compare this variation of T_{d2} with that caused by lattice compression under hydrostatic pressure, because both the chemical substitution with smaller ions and application of hydrostatic pressure lead to the contraction of the lattice parameters. The compressibility along the c-axis for $y=0$ at room temperature is $\kappa_c = -2.4 \times 10^{-3}$ GPa^{-1}. Thus the variation of T_{d2} caused by chemical substitution is interpreted as caused by the application of "chemical pressure": $dT_{d2}/dP_{chem} = dT_{d2}/d(c/c_0) \times \kappa_c = -(50 \pm 7)$ K/GPa. According to the thermal expansion experiments under hydrostatic pressure [24], the variation of T_{d2} is estimated as $dT_{d2}/dP \sim -40$ K/GPa. This value corresponds well to the chemical pressure effect. The increase of chemical pressure, as well as hydrostatic pressure, suppresses the structural transition.

2.4. TRIVALENT SUBSTITUTIONS AT THE La-SITE: $La_{2-x-y}Nd_yBa_xCuO_4$

Partial substitution of Nd^{3+} and other smaller rare-earth trivalent ions for La^{3+} induces low-temperature structural transitions even in LSCO, as first reported by Büchner et al. [34] and Crawford et al. [35]. In these systems, T_{d2} is enhanced even though the average ionic radius at the La-site is reduced. Various other new features are reported on the low temperature properties of $La_{2-x-y}Nd_ySr_xCuO_4$ (LNSCO).

(1) The TLT phase is stable in a wide range of x, not restricted to a narrow region near 1/8. But the minimum in T_c still occurs near $x=1/8$ [35].

(2) Between the OMT and TLT phases, another orthorhombic phase (the OLT phase, space group is probably Pccn) with much reduced orthorhombic deformation exists. The OMT–OLT transition is either of the second order or of the first order, depending on the values of x and y [35]. The OLT–TLT transition is of the second order.

(3) Bulk superconductivity and the TLT phase does not seem to coexist for $x=0.15$. For x substantially away from 0.15, they do not exclude each other: bulk superconductivity is sustained even in the TLT phase [36].

(4) The pressure dependence of the OMT–TLT transition temperature, T_{d2}, is positive at least for $x=0.15$ [36], contrary to that in LBCO.

Discontinuous changes in various transport coefficients are reported on single crystals of LNSCO with $y(Nd)\sim0.4$ by Nakamura et $al.$ [37]. Crystals with both $x=0.12$ and $x=0.20$ exhibit the structural transition at $T_{d2}=70\sim80$ K. However, a remarkable difference is that superconductivity is lost for $x=0,12$, whereas a sharp diamagnetic transition is observed at $T_c=17$ K for $x=0.20$. Moreover, for $x=0.12$ the in–plane resistivity ρ_{ab} exhibits a jump at T_{d2} and becomes non–metallic below T_{d2}, whereas for $x=0.20$ ρ_{ab} does not show any change at T_{d2}.

We investigated the structural and transport properties of $La_{2-y-x}Nd_yBa_xCuO_4$ (LNBCO) to clarify the contrasting effects of partial substitutions by smaller divalent and smaller trivalent ions at the La site [17]. Powder X–ray diffraction study revealed that LNBCO with $x=0.125$ and with any values of y undergoes the same sequence of structural transition as LBCO with $x=0.125$. We did not detect any indication of the complicated sequence of transitions as reported by Hara et $al.$ [38].

Figure 12 shows the y dependence of T_{d1} and T_{d2}. The latter is defined at the temperature where the integrated intensities for the OMT and TLT phases equal. The vertical bars represent the temperature range in which the diffraction peaks of both the OMT and TLT phases coexist.

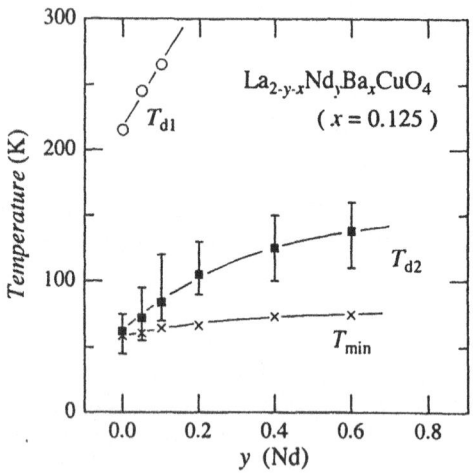

Fig. 12. Variations of the structural transition temperatures T_{d1} and T_{d2} with y in $La_{2-y-x}Nd_yBa_xCuO_4$ (LNBCO) for $x=0.125$. T_{min} is the temperature at which the resistivity takes a minimum value. Bulk superconductivity was not observed in these samples.

The variation of T_{d1} with y is expressed approximately by $T_{d1} = 220 + 500\ y$ K. This is very similar to the result for LNSCO with $x=0.150$; $T_{d1} = 180 + 550\ y$ K [39]. This implies that the Nd substitution plays the same role in stabilization of the OMT phase in both systems.

Substantial increase of T_{d2} also occurs in LNBCO and LNSCO. However, the amplitude of changes in the transport coefficients at T_{d2} are different between the two systems. For LNSCO with $x=0.12$ and $y=0.40$, a clear jump in the resistivity is reported in both single–crystalline and sintered samples [37, 39]. On the other hand, we only observed a distinctive resistivity minimum at $T_{min}=60{\sim}70$ K, not corresponding to T_{d2}. This difference is probably because of the increased randomness in LNBCO caused by greater ionic–size mismatch at the La site, which makes the effects at T_{d2} unclear in LNBCO.

2.5. DRIVING MECHANISMS OF THE INSTABILITIES AT $p \sim 1/8$

Properties of Th–substituted LBCO revealed that change in the normal–state, the low–temperature lattice instability, as well as the sharp suppression of T_c, are governed by the specific value of $p{\sim}1/8$ for the density of carriers per Cu. Therefore, the low–temperature structural transition is likely driven by a mechanism intimately related to the conduction–electron system.

Let us consider again the different tilting of the CuO_6 octahedra depicted in Fig. 2. In the THT and OMT phases all the in–plane oxygen sites are equivalent. In contrast, there are two distinctive sites in the TLT phase: half of the oxygen ions reside precisely in the plane while the other half slightly above or below the plane. If the electronic band near the Fermi level E_F splits in the presence of the distinctive sites of oxygen, a Peierls–type mechanism may trigger a structural transition to the TLT phase [40–42].

Since the critical value of p for the instability is very close to 1/8, it is tempting to consider that a band instability commensurate with the lattice triggers the lattice deformation to lift this degeneracy. We speculate that the states in the charge–transfer gap ("in–gap states") around the Fermi level, generated by hybridization of O $2p$ and Cu $3d$ orbits develop a new gap structure with reduced density of states (DOS) at E_F in the TLT phase by this deformation and lower the energy of the system. We believe that this reduction in the density of states is the main reason for the strong suppression of T_c, as well as for the change in the normal–state properties.

It is not clear at present whether the TLT phase induced by Nd–substitution for La has any direct relevance to the $p=1/8$ problem, because the presence of the TLT phase is by no means restricted to a narrow region around $p=1/8$. In this connection, it is worth recalling that insulating La_2CoO_4 [43] and La_2NiO_4 [44] also exhibit cascade structural transitions from the THT to OMT and from the OMT to TLT phases, indicating that La–214 compounds with K_2NiF_4 structure are quite susceptible to the transition to the TLT phase.

Once the transition to the TLT phase occurs, the band originating from the in–plane oxygen orbitals would develop an additional structure. A remarkable difference of superconducting properties in LNSCO crystals with different x [37] described in Sec. 2.4 is quite suggestive. If E_F is located near the peak of DOS for $x=0.12$, the structural transition would have a substantial effect on the electronic properties. In contrast, if E_F is away from the DOS peak for $x=0.20$, the modification of the electronic DOS would have a much reduced effect on superconductivity.

3. The High–Temperature Tetragonal Phase and Superconductivity

In La–214 compounds, superconductivity mainly occurs in the OMT phase. A question of possible relation between deviation from tetragonal crystal symmetry and super-conductivity has been actively investigated in $La_{2-x}Sr_xCuO_4$. According to the phase diagram of LSCO by Torrance *et al.*[45], superconductivity persists up to the carrier concentration $p\sim0.32$ for samples with no oxygen vacancies. From various structural studies mentioned in Sec. 1, the THT (Tetragonal at High–Temperatures) phase is stable to the lowest temperature for $x>0.21$. Therefore the THT phase does also seem to support superconductivity. To clarify this issue, however, we need a more detailed and systematic study to overcome ambiguities due to broadening of the superconducting transition and the question of sample homogeneity at high doping.

Application of pressure P dramatically depresses the THT–OMT structural boundary, T_{d1}. Kim, Moret *et al.*[46,47] studied the effect of quasi-hydrostatic pressure on T_{d1} of LSCO single–crystals using high–pressure, low–temperature X–ray diffraction. They described that their measurements were not done under fully hydrostatic pressure, because the twin boundaries present in the OMT phase were formed predominantly in one preferred direction. As in Fig. 13, they found that for $x=0.12$, T_{d1} decreases from 205 ± 10 K at the initial rate of $dT_{d1}/dP=-100$ K/GPa up to 1.2 GPa, then drops more rapidly and vanishes near 1.5 GPa. In other words, at pressures as much as 2 GPa, the THT phase is stable to the lowest temperatures for most of the range of x where superconductivity is expected; see Fig. 1. By comparing the results of structural study with those of transport measurements under pressure (references cited by Kim and Moret [46]), they argued that both the OMT and THT phases are superconductive. They also suggested that dT_c/dP is smaller in the THT phase.

These expectations have been confirmed by a recent systematic study by Yamada and Ido [24]. They measured ac susceptibility, electrical resistivity, and thermal expansion coefficient of polycrystalline LSCO samples in a pressure cell with P up to 2 GPa.

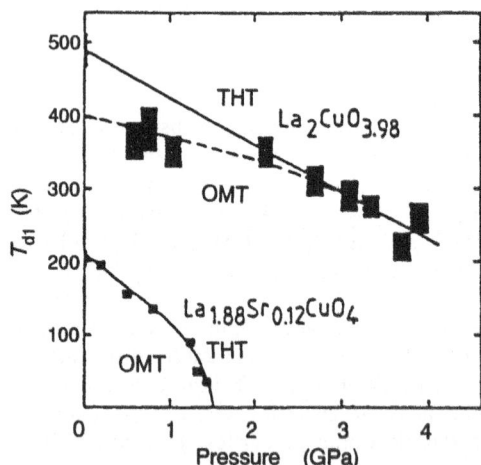

Fig. 13. Pressure dependence of the tetragonal (THT)–orthorhombic (OMT) transition temperature T_{d1}. Pressure stabilizes the THT phase (Kim and Moret [46]).

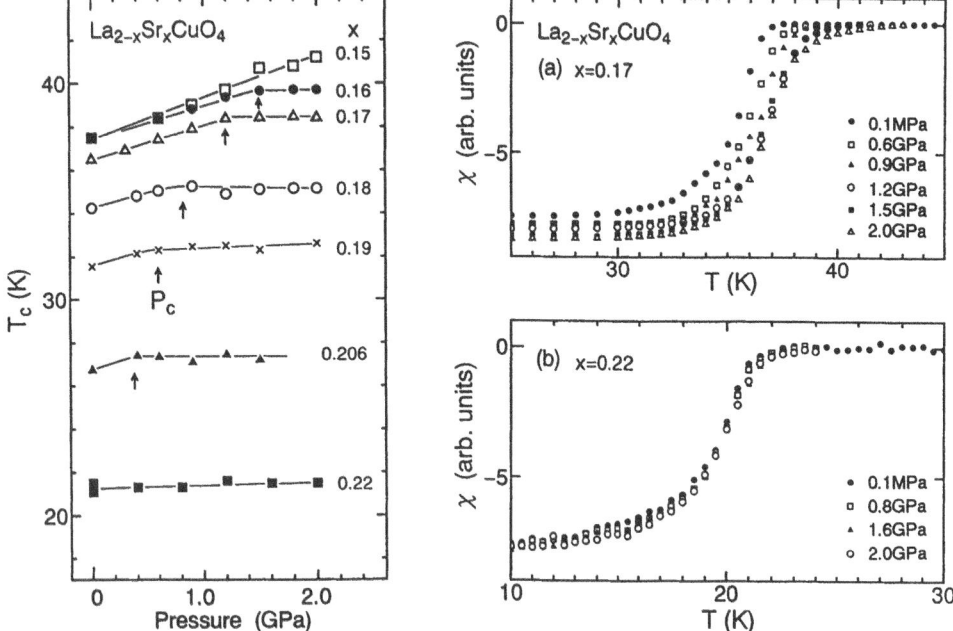

Fig. 14. Pressure dependence of the super-conducting transition temperature T_c of La$_{2-x}$Sr$_x$CuO$_4$ with x=0.15–0.22. The arrow indicates the critical pressure P_c above which T_c is almost independent of pressure (Yamada and Ido [24]).

Fig. 15. The diamagnetic transition curve of superconductivity at various pressures for the samples of La$_{2-x}$Sr$_x$CuO$_4$ with x=0.17 (a) and x=0.22 (b) (Yamada and Ido [24]).

Figure 14 summarizes pressure dependence of T_c for various x. The superconducting transition temperature T_c was determined from ac susceptibility with a magnetic field of 0.01 Oe and a frequency of 90 Hz, as shown in Fig. 15 for selected values of x. A clear change in the slope of a T_c vs. P curve occurs at P_c, which decreases with increasing x. For each x, the value of P_c agrees well with the THT–OMT phase boundaries determined from the pressure dependence of the anomaly in thermal expansion coefficient α measured by a strain–gauge method. This means that dT_c/dP for the OMT phase (1.9 K/GPa for x=0.15 and 1.6 K/GPa for x=0.17, estimated from the original graph by Yamada and Ido [24]) is much greater than that for the pressure–stabilized THT phase (~ 0.01 to 0.2 K/GPa). The data also suggest that the variation of T_c with P is nearly continuous across the structural boundary. Moreover, the susceptibility in Fig. 15(a) clearly shows that the Meissner fraction does not change as the structure is altered from orthorhombic to tetragonal by pressure, because for x=0.170 the structure remains tetragonal for pressure above 1.2 GPa. It therefore seems evident that the THT phase stabilized by pressure sustains bulk superconductivity.

Figure 15(b) indicates that the bulk superconductivity is maintained in a LSCO sample with x=0.22, which remains in the THT phase to the lowest temperature even at ambient pressure. The smallness of the pressure dependence of T_c (~0.01 K/GPa) as shown in

Fig. 14 is also consistent with the values for the pressure–induced THT samples with different x.

Recently, Takagi *et al.*[48] reported structural and superconducting properties of polycrystalline LSCO annealed in oxygen at 800°C for up to one month, and cooled slowly to room temperature at the rate of 10°C/hr. The results of high–resolution X–ray diffraction show that the solid solution exists at least up to $x\sim0.3$. Therefore in LSCO, phase separation does not occur at least up to $x\sim0.3$ in carefully prepared samples. The orthorhombic splitting of diffraction peaks at 30 K decreases continuously with increasing x and disappears above $x=0.22$.

They observed a significant drop of the Meissner signal at the doping corresponding to the OMT–THT boundary. On the basis of this observation, they claimed that the disappearance of superconductivity on overdoping occurs near $x\sim0.2$, which is substantially less than the value widely believed, and this disappearance is associated with a structural phase transformation from the OMT to THT phase.

Their claim of incompatibility of superconductivity in the THT phase apparently contradicts with the results of Yamada and Ido [24]. Results for pressure–induced THT phase seem persuasive, but the results at ambient pressure may require cross checks of their results, especially because annealing procedures were different between the two investigations and because Meissner signal from unpowdered ceramic samples [48] are affected largely by flux penetration in the grain boundaries and flux pinning. Kitazawa [49] suggested that for the latter reasons, zero–field–cooled shielding measurements will give much better estimates of the volume fraction of superconductivity.

Most of the high–T_c copper oxides have orthorhombic structures at low temperatures. However, a simple correlation between the orthorhombic deformation of the crystal symmetry and high–T_c superconductivity does not hold true in general. A counter example is $La_{2-x}Sr_xCaCu_2O_6$, for which the tetragonal symmetry has been confirmed down to 10 K by neutron diffraction powder profile refinement [50]. The THT phase of LSCO under pressure can now be added to the list of counter examples.

We have already noted in Sec. 1 that the THT–OMT structural transition affects only the out–of–plane conductivity but not the in–plane conductivity [15]. The conduction along the c–axis changes from metallic to semiconductor–like as the temperature in decreased. This indicates modification of the electronic coupling between the planes. The persistence of superconductivity in the THT phase suggests that this change of out–of–plane conduction may not be a crucial factor for superconductivity.

A more relevant and important question may concern the local symmetry or the dynamic symmetry of the lattice, in case they are different from the static crystal symmetry. Studies of local symmetry by neutron pair distribution function analysis revealed short-range buckling–type distortion to occur near T_c on $Tl_2Ba_2CaCu_2O_8$ [51] and $YBa_2Cu_3O_7$ [52]. It is hoped that such detailed symmetry of the THT phase of LSCO be clarified in future study.

4. Interplay between Lattice Instabilities and Superconductivity in $La_{2-x}Sr_xCuO_4$

Ultrasonic measurements on superconductors have been used to study the gap properties and the strain dependence of superconducting transition temperature. Measurements of ultrasound attenuation coefficients provides information of the gap structure, complemental to the nuclear relaxation rate, $1/T_1$. Whereas measurements of ultrasound velocities or of ultrasound velocities or elastic constants will give thermodynamic properties including the strain dependence of the superconducting condensation energy.

In this section, we will present the results of recent high–resolution measurements of the elastic constants of LSCO single crystals by Nohara et al.[53], which clarified a novel lattice instabilities at low temperatures competing with superconductivity.

In that study the elastic constants for various ultrasonic modes were measured in magnetic fields up to 14 T applied both parallel and perpendicular to the Cu–O layers. They are longitudinal C_{11} (with sound propagation direction k // [100] and displacement direction u//[100]), C_{33} (with k//u//[001]), transverse $(C_{11}-C_{12})/2$ (k//[110] and u//[1$\bar{1}$0]), C_{44} (k//[100] and u//[001], or k//[110] and u//[001]) and C_{66} (k//[100] and u//[010]). Here the elastic constants and corresponding strains are defined using crystalline coordinates in the THT phase.

The crystals used in the measurements were grown by a traveling–solvent floating–zone method [54]. The Sr concentration of x=0.138±0.03 was determined by electron–probe microanalysis (EPMA). The superconducting transition temperature was T_c=36 K with a transition width of 4 K, as determined magnetically. Sound velocity was measured by using a phase–comparison method [55] with a frequency of 30 to 45 MHz and the relative resolution better than 10^{-6}. The elastic constant was calculated from the relation $C = \rho v^2$ using the room–temperature value of mass density $\rho = 6.97$ g/cm^3. No correction for contraction of the sample with temperature was made since it is expected to be 1 part in 10^{-3} over the entire temperature interval studied.

The THT–OMT structural transition temperature is T_{d1}=210 K for x=0.14, according to the analysis of the temperature dependence of C_{66}, which shows by as much as 70 % softening toward T_{d1} in the THT phase [56,57]. At temperatures way below T_{d1}, only $(C_{11}-C_{12})/2$ exhibits remarkable softening (reduction in the elastic constant). The elastic constant $(C_{11}-C_{12})/2$ is a susceptibility to a shearing strain $\varepsilon_{xx}-\varepsilon_{yy}$, by which the [100]–axis (x) is stretched and [010] axis (y) is shrunk and vice versa. It is worth noting that the elastic $(C_{11}-C_{12})/2$ mode is a pure shear mode with B_{1g} symmetry in both the THT and OMT structure.

Let us now look into the results for $(C_{11}-C_{12})/2$ mode in more detail. The variations of $(C_{11}-C_{12})/2$ with temperature are presented in Fig. 16. The elastic constant $(C_{11}-C_{12})/2$ hardens with decreasing the temperature from 300 K as it does for an ordinary solid. A slight change is seen at T_{d1}, which is not shown here [56]. The most remarkable feature of the zero–field measurements is that $(C_{11}-C_{12})/2$ starts to soften at a temperature around 50 K, which is substantially higher than T_c, but turns to a rapid hardening just below T_c. This is markedly different from the behavior of all other elastic constants, which do not show any corresponding anomaly above T_c [56,57]. The softening of $(C_{11}-C_{12})/2$ can neither be explained by mean–field expectations at the super-conducting transition nor by superconducting fluctuations [58]. Therefore the softening is attributed to an intrinsic property of the lattice in the normal state.

This interpretation is confirmed by the results under magnetic fields, which reduce T_c and retain the normal state to lower temperatures. Figure 16 contains the results for H//[001]$\perp u$. It is important to note first that in this configuration, T_c is substantially reduced by H. The transition temperature $T_c(H)$ decreases with increasing H, as indicated by arrows in Fig. 16: it is 32 K at 0 T, 30 K at 1 T, 26 K at 5 T, and 14 K at 14 T. These were determined from the temperature dependence of C_{33}, which exhibits a discontinuity at T_c, corresponding to the well known specific–heat jump. As the temperature is decreased, the elastic constant $(C_{11}-C_{12})/2$ continues to soften as long as

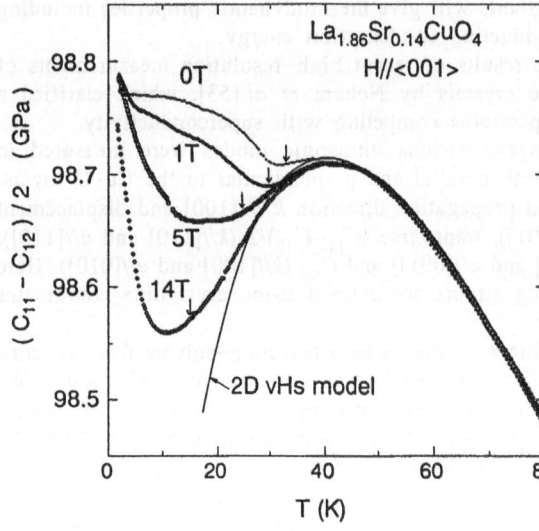

Fig. 16. Temperature dependence of the transverse elastic constant $(C_{11}-C_{12})/2$ in single–crystalline $La_{2-x}Sr_xCuO_4$ with $x=0.14$ under magnetioc fields. The field direction is $H \parallel c \perp u$. Arrows indicate superconducting transition temperature $T_c(H)$ for each magnetic field. The solid line is a fit by a two–dimensional van Hove singularity model (Nohara et al. [53]).

the system remains in the normal state. On further cooling, a recovery hardening occurs just below T_c. Now, the sound wave in this configuration acts to shear the flux–line lattice (FLL) below an irreversibility temperature T^* and a shear modulus of FLL should be superposed on the elastic constant of the crystal lattice. However, the magnitude of the shear modulus of the FLL is known to be too small to account for the recovery hardening below $T_c(H)$ [59].

The softening of $(C_{11}-C_{12})/2$ below 50 K indicates a structural instability, namely a growth of the structural fluctuation, in response to the shearing strain $\varepsilon_{xx}-\varepsilon_{yy}$. This strain displaces the in–plane oxygen ions in such a way to produce two distinctive sites for them. Therefore for this particular strain, the degeneracy of the oxygen sites is lifted and the electronic bands originating from the oxygen orbitals would split. In this case the instability can be regarded as the Peierls–type instability and is essentially the same as the one associated with the OMT–TLT transition in LBCO described in Sec. 2. The softening of $(C_{11}-C_{12})/2$ follows naturally from this coupling between the electronic states and $\varepsilon_{xx}-\varepsilon_{yy}$ as described below.

Consider a doubly degenerate electronic band with the density of states (DOS) $N(E)$, for which the electronic level splits into $E_k = E_{k0} \pm d(\varepsilon_{xx}-\varepsilon_{yy})$ in response to the shearing strain. Here, E_k^0 is an electronic energy without the strain and d is the coupling energy between the electrons and strain. In this model, the deviation of the elastic constant from the background $C^0(T)$, the value in the absence of the coupling, is given by the following formula [60];

$$\Delta(C_{11} - C_{12})/2 = -2\,d^2\langle N\rangle_T \tag{1}$$

where k_B is Boltzmann constant. Here $<N>_T$ is the "thermal average" of DOS around the Fermi level E_F at temperature T given below, in which f is the Fermi distribution function.

$$\langle N \rangle_T = - \int_0^\infty N(E) \frac{\partial f}{\partial E} \, dE. \tag{2}$$

As evident from Eqs. (1) and (2), the elastic constant should exhibit a large softening if (1) the coupling d is large for the particular strain under consideration and (2) the band width is narrow enough to give a significant temperature dependence of $\langle N \rangle_T$. By assuming a simple degenerate rectangular band with the width $2W$ and the DOS N, a fit to the data results in $2W = 80$ K and $2d^2N = 1.9 \times 10^8$ J/m^3. The width seems much too small for the band formed by many–body correlations arising from p or d orbitals. Two–dimensional van Hove singularity (vHs) with $N(E) = N_0 \ln (W/|E - E_F|)$, which has been assumed to explain high-T_c superconductivity on a two–dimensional square lattice, effectively has a narrow band width because of its logarithmical diverging DOS. A solid line in Fig. 16 is a fit using this vHs model with $W = 5000$ K. The fit seems satisfactory, indicating a presence of a narrow band. But again, we do not know of reliable explanation for the preservation of such a sharp peak in the presence of many–body correlations.

Now let us consider the origin of the hardening of $(C_{11} - C_{12})/2$ below $T(H)$ in Fig. 16. In the superconducting state, $(C_{11} - C_{12})/2$ consists of at least three components: contributions from superconductivity itself $\Delta C^S(T)$, from flux–line lattice $\Delta C^F(T)$, and from crystalline lattice $C^L(T)$. Since $\Delta C^S \propto T_c \Delta c_p$ [61], this contribution should be reduced by H. As we noted before, the FLL contribution ΔC^F is known to be negligible in this configuration of the field and the displacement. In the normal state, $C^L(T)$ is the observed $(C_{11} - C_{12})/2$, which softens by the lattice instability. If this instability persists in the superconducting state, $(C_{11} - C_{12})/2$ would continue to soften, or at least retain the value at $T_c(H)$. In this case, $C^L(T) + \Delta C^S(T)$ can not explain the observed $(C_{11} - C_{12})/2$, which tends to the same value independent of H at the lowest temperature. Therefore the observed feature can be explained only when the instability disappears in the superconducting state and $C^L(T)$ is restored to the background value $C^0(T)$.

In LSCO we believe that the total energy may be reduced more by forming a superconducting gap rather by forming a gap structure induced by the shearing strain. In this case the lattice instability of the OMT structure disappears as superconductivity emerges.

In concluding this section, we found that the transverse elastic constant $(C_{11} - C_{12})/2$ of LSCO with $x = 0.14$ exhibits a remarkable softening starting from about 50 K and a rapid hardening below $T_c(H)$. This softening clearly indicates the existence of a structural instability of the OMT phase in the normal state. We propose that a Peierls–type instability caused by coupling between the in–plane shearing strain $\varepsilon_{xx} - \varepsilon_{yy}$ and the electronic states is responsible for the softening of $(C_{11} - C_{12})/2$. The rapid hardening of $(C_{11} - C_{12})/2$ in the superconducting state is most likely a signature of the disappearance of the structural instability of the OMT phase. Our observation clarifies a strong interplay between lattice instability and high-T_c superconductivity in LSCO.

5. Conclusion

We have discussed three topics concerning the interplay between the lattice instability and high-T_c superconductivity in La–214 compounds. It is now clear that in these compounds there exist lattice instabilities which are driven by the coupling with the electronic system and compete with superconductivity.

In LBCO a structural transition to the TLT phase occurs at a particular hole concentration $p \approx 1/8$ per copper. A Peierls-type mechanism has been proposed in which the displacement of the in-plane oxygen atoms in the TLT phase results in the formation of a gap structure near the Fermi level and lower the energy of the system.

Partial substitution of smaller trivalent ions for La^{3+} in both LBCO and LSCO stabilizes the TLT phase. We argued that the principal mechanism in this case may be quite different from that of the "1/8 instability", and the displacements of apical oxygen, rather than the in-plane oxygen, play an important role. Although the TLT phase appears in a wide range of p in Nd-substituted LSCO, the center of suppression in T_c remains at $p \approx 1/8$. This behavior implies that the Fermi level is located near the peak of the density of states for $p \approx 1/8$.

Ultrasonic study revealed that in LSCO with $x=0.14$ a growth of structural fluctuation exists in the normal state and that this instability disappears with the emergence of superconductivity. The ultrasonic study provides novel information of the nature of the bands responsible for the lattice instability and superconductivity. The electronic bands are considerably sharp and peaked near the Fermi level. Moreover, they are restricted to couple with the in-plane, shearing strain $\varepsilon_{xx}-\varepsilon_{yy}$ with B_{1g} symmetry. Such detailed information is no doubt important in understanding the mechanism of the high-T_c superconductivity.

We argue that these phenomena are intimately related because both the structural change to the TLT phase in LBCO and the applied shear strain in LSCO lift the degeneracy of the in-plane oxygen sites. In this sense, these lattice instabilities are qualitatively different from the THT-OMT structural transition, because the latter does not alter the degeneracy of the in-plane oxygen sites.

It is not clear at this point whether the interaction between the lattice and electronic systems has any active role in the pairing mechanism of high-T_c superconductivity. Nevertheless the importance of the strong coupling between the lattice and normal-state electronic system should be more emphasized. After all it is crucially important to clarify the unusual normal state properties, such as electronic states as well as phonons, in order to understand the mechanism of high-T_c superconductivity.

Acknowledgements

The author is grateful to his colleagues, especially to Professor T. Fujita, Dr. T. Suzuki, Dr. F. Nakamura and Mr. M. Nohara for their valuable suggestions and comments. He is thankful to Mr. Y. Tanaka and Mr. K. Yoshida for preparing some of the figures for this article. He also acknowledges valuable comments from Professors M. Ido, H.Takagi, K. Kitazawa, K. A. Müller, M. Sato, Y. Koike, K. Nasu and A. Fujimori.

References

1. J.G.Bednorz and K.A.Müller, Z. Phys. B **64** (1986) 189.
2. T.Fujita, Y.Aoki, Y.Maeno, J.Sakurai, H.Fukuba and H.Fujii, Jpn. J. Appl. Phys. **26** (1987) L368.
3. A.R.Moodenbaugh, Y.Xu, M.Suenaga, T.J.Folkerts and R.N.Shelton, Phys. Rev. B **38** (1988) 4596.
4. P.Ganguly ang C.N.R.Rao, Mater. Res. Bull. **8** (1973) 408.
5. Y.Maeno, A.Odagawa, N.Kakehi, T.Suzuki and T.Fujita, Physica C **173** (1991) 322.

6. J.D.Axe, D.E.Cox, K.Mohanty, H.Moudden, A.R.Moodenbaugh, Y.Xu and T.R.Thurston, IBM J. Res. Dev. **33** (1989) 382.
7. T.Suzuki and T.Fujita, J. Phys. Soc. Jpn. **58** (1989) 1883; Physica C **159** (1989) 111.
8. J.D.Axe, A.H.Moudden, D.Hohlwein, D.E.Cox, K.M.Mohanty, A.R.Moodenbaugh and Y.Xu, Phys. Lett. **62** (1989) 2751.
9. R.J.Birgeneau, *et al.*, Phys. Rev. Lett. **59** (1987) 1329.
10. Y.Maeno, F.Nakamura, T.Suzuki and T.Fujita, JJAP Series **7** (1992) 91.
11. J.D.Axe, in *Proceedings of Lattice Effects in High-T_c Superconductors*, eds. Y.Bar-Yam, T.Egami, J.Mustre-de-Leon and A.R.Bishop, (World Scientific, 1992), p.517.
12. M.K.Crawford, W.E.Farneth, R.L.Harlow, E.M.McCarron, R.Miao, H.Chou and Q.Huang, in *Proceedings of Lattice Effects in High-T_c Superconductors*, eds. Y.Bar-Yam, T.Egami, J.Mustre-de-Leon and A.R.Bishop, (World Scientific, 1992), p.531.
13. Y.Maeno, N.Kakehi, Y.Tanaka, T.Tomita, F.Nakamura and T.Fujita, in *Proceedings of Lattice Effects in High-T_c Superconductors*, eds. Y.Bar-Yam, T.Egami, J.Mustre-de-Leon and A.R.Bishop, (World Scientific, 1992), p.542.
14. C.H.Chen, S.-W.Cheong, D.J.Werder and H.Takagi, Physica C **206** (1993) 183.
15. S.Kambe, K.Kitazawa, M.Naito, A.Fukuoka, I.Tanaka and H.Kojima, Physica C **160** (1989) 35.
16. K.Yoshida, F.Nakamura, Y.Tanaka, Y.Maeno, T.Kagayama, G.Oomi and T.Fujita, submitted to Proc. LT20 (1993).
17. Y.Tanaka, Y.Maeno, F.Nakamura and T.Fujita, submitted to Proc. LT20 (1993).
18. M.Sera, Y.Ando, S.Kondoh, K.Fukuda, M.Sato, I.Watanabe, S.Nakashima and K.Kumagai, Solid State Commun. **69** (1989) 851.
19. M.Sera, S.Shamoto and M.Sato, Solid State Commun. **74** (1990) 951; M.Sato, M.Sera, S.Shamoto, S.Yamagata and K.Kakurai: Physica C **185~189** (1991) 905.
20. N.Wada, S.Ohsawa, Y.Nakanura and K.Kumagai, Physica B **165&166** (1990) 1345.
21. G.M.Luke, L.P.Le, B.J.Sternlieb, W.D.Wu, Y.J.Uemura, S.Ishibashi and S.Uchida, Physica C **185-189** (1992) 1175.
22. K.Kumagai, I.Watanabe, K.Kawano, H.Matoba, K.Nishiyama, K.Nagamine, N.Wada, M.Okaji and K.Nara, Physica C **185-198** (1991) 913.
23. S.Katano, private communication.
24. N.Yamada and M.Ido, Physica C **203** (1992) 240.
25. M.K.Crawford, W.E.Farneth, E.M.McCarron, III, R.L.Harlow and A.H.Moudden, Science **250** (1990) 1390.
26. C.C.Tsuei, D.M.Newns, C.C.Chi and P.C.Pattnaik, Phys. Rev. Lett. **65** (1990) 2724.
27. S.Shamoto, M.Sera and M.Sato, Solid State Commun. **76** (1990) 923.
28. Y.Maeno, A.Odagawa, N.Kakehi, T.Fujita, K.Matsukuma, T.Ekino and H.Fujii, in Springer Proceedings in Physics **60**, *The Physics and Chemistry of Oxide Superconductors*, eds. Y.Iye and H.Yasuoka (Springer, Heidelberg, 1992) p.201.
29. Y.Koike, T.Kawaguchi, N.Watanabe, T.Noji and Y.Saito, Solid State Commun. **79** (1991) 155.
30. Y.Maeno, N.Kakehi, M.Kato and T.Fujita, Phys. Rev. B **44** (1991) 7753.
31. T.Fukase, T.Nomoto, T.Hanaguri, T.Goto and Y.Koike, Physica B **165&166** (1990) 1289.
32. R.D.Shannon, Acta. Cryst. A **32** (1976) 751.
33. K.Oh-ishi, M.Kikuchi, Y.Syono, N.Kobayashi, T.Sasaoka, T.Matsuhira, Y.Muto and H.Yamauchi, Jpn. J. Appl. Phys. **27** (1988) L1449.
34. B.Büchner, M.Braden, M.Cramm, W.Schlabitz, W.Schnelle, O.Hoffels, W.Braunisch, R.Müller, G.Heger and D.Wohlleben, Physica C **185-189** (1991) 903.
35. M.K.Crawford, R.L.Harlow, E.M.McCarron, W.E.Farneth, J.D.Axe, H.Chou and Q.Huang, Phys. Rev. B **44** (1991) 7749.
36. B.Büchner, M.Cramm, M.Braden, W.Braunisch, O.Hoffels, W.Schnelle, R.Müller, A.Freimuth, W.Schlabitz, G.Heger, D.I.Khomskii and D.Wohlleben, Europhys. Lett **21** (1993) 953.
37. Y.Nakamura and S.Uchida, Phys. Rev. B **46** (1992) 5841.

222

38. N.Hara, T.Kamiyama, I.Kamata, I.Nakahara, H.Hayakawa, E.Akiba and H.Asano, Solid State Commun. **82** (1992) 975.
39. Büchner *et al.*, in *Proc. Physics and Material Science of HTSC II*, ed. R.Kossowsky (Kluwer Publ., 1991).
40. R.S.Markiewicz, J.Phys.: Condens. Matter **2** (1990) 6223; Physica C **193** (1992) 323.
41. S.Barišić and J.Zelenko, Solid State Commun. **74** (1990) 367.
42. W.E.Pickett, R.E.Cohen and H.Krakauer, Phys. Rev. Lett. **67** (1991) 228.
43. K.Yamada *et al*, Phys. Rev. **B 39** (1989) 2336.
44. G.H.Lander, P.J.Brown, J.Spalek and J.M.Honig, Phys. Rev. **B 40** (1989) 4463.
45. J.B.Torrance, Y.Tokura, A.I.Nazzal, A.Bezinge, T.C.Huang and S.S.P.Parkin, Phys. Rev. Lett. **61** (1988) 1127.
46. H.J.Kim and R.Moret, Physica C **156** (1988) 363.
47. R.Moret, J.P.Pouget, C.Noguera and G.Collin, Physica C **153–155** (1988) 968.
48. H.Takagi, R.J.Cava, M.Marezio, B.Batlogg, J.J.Krajewski, W.F.Peck, Jr., P.Bordet and D.E.Cox, Phys. Rev. Lett. **68** (1992) 3777; in *Proceedings of Lattice Effects in High-T_c Superconductors*, eds. Y.Bar–Yam, T.Egami, J.Mustre–de–Leon and A.R.Bishop, (World Scientific, 1992), p.548.
49. K.Kitazawa, private communication.
50. R.J.Cava, A.Santoro, J.J.Krajewski, R.M.Fleming, J.V.Waszczak, W.F.Peck, Jr. and P.Marsh, Physica C **172** (1990) 138.
51. B.H.Toby, T.Egami, J.D.Jorgensen and M.A.Subramanian, Phys. Rev. Lett. **64** (1990) 2414.
52. M.Arai, K.Yamada, Y.Hidaka, S.Itoh, Z.A.Bowden, A.D.Taylor and Y.Endo, Phys. Rev. Lett. **69** (1992) 359.
53. M.Nohara, T.Suzuki, Y.Maeno, T.Fujita, I.Tnaka and H.Kojima, to be published in Phys. Rev. Lett. (1993).
54. I.Tanaka and H.Kojima, Nature **337** (1989) 21.
55. T.Goto *et al.*, the Bulletin of the Research Institute for Scientific Measurement, Vol. **38** (Tohoku University, Sendai, Japan, 1989) p. 65.
56. T.Suzuki, T.Fujita, M.Nohara, Y.Maeno, I.Tanaka and H.Kojima, in *Mechanisms of Superconductivity*, JJAP Series 7, ed. Y.Muto (Jpn. J. Appl. Phys. Publication Office, 1992) p.219.
57. M.Nohara T.Suzuki, Y.Maeno, T.Fujita, I.Tanaka and H.Kojima, Physica C **185–189** (1991) 1397.
58. A.J.Mills and K.M.Rabe, Phys. Rev. **B 38** (1988) 8908.
59. P.Lemmens P.Fröning, S.Ewert, J.Pankert, G.Marbach and A.Comberg, Physica C **174** (1991) 289.
60. T.J.Moran and B.Luthi, Phys. Rev **187** (1969) 710.
61. B.Luthi, J. Magn. Magn. Mat. **52** (1985) 70.
62. L.R.Testardi, Phys. Rev. **B 12** (1975) 3849.

Part II

Physics of HTSC

PROBING CRYSTALLOGRAPHIC AND MATERIALS PROPERTIES OF Y-Ba-Cu-O SUPERCONDUCTORS BY NMR AND NQR

D. BRINKMANN
Physik-Institut
University of Zürich
Schönberggasse 9
CH-8001 Zürich
Switzerland

ABSTRACT. The lecture presents examples of Nuclear Magnetic Resonance (NMR) and Nuclear Quadrupole Resonance (NQR) studies performed in compounds of the Y-Ba-Cu-O family in order to show the kind of information these microscopic techniques can provide with respect to structural and electronic properties. Topics which are addressed are the following: calculation of electric field gradients; valences and bonding and temperature and pressure effects; oxygen distribution; doping with cations; Knight shifts and relaxation times and their relations to hole concentration, one-component susceptibilities, coupling of planes, spin-gap and BCS orbital-pairing.

1. Introduction

Nuclear Magnetic Resonance (NMR) played an important role in understanding classical superconductors such as aluminum when the Bardeen-Cooper-Schrieffer (BCS) theory provided the first microscopic explanation of this extraordinary state of matter. Thus it was logical that soon after the discovery of the new high-temperature superconductors (HTSC) by G. Bednorz and K.A. Müller, NMR and Nuclear Quadrupole Resonance (NQR) were among the leading techniques to explore these fascinating and puzzling materials.

The power of NMR and NQR in the studies of various aspects of superconductivity lies in the fact that these methods can obtain information on static and dynamic properties on an atomic scale. How this has been accomplished for "classical" superconductors has been reviewed in [1]. For the HTSC an immense amount of detailed information on structural, electronic and superconducting aspects has been collected in recent years. Studies of the most intensively investigated structure, $YB_2Cu_3O_7$ (1-2-3-7 for short), were reviewed in [2]; work on these and many other compounds is described in [3]. A summary of reports of the present activities in many leading NMR/NQR laboratories in the field of HTSC research has recently been published [4]. A review of the present status of NMR/NQR research in HTSC will appear shortly [5].

In this lecture we will show how NMR and NQR contributed to the understanding of many features and properties of the HTSC. The emphasis will be on crystallographic and materials properties, *i.e.*, we will present investigations concerned with electric field gradients, valences and bonding and the effect of temperature and pressure, the oxygen distribution and cation doping effects. Many of these examples are from our laboratory where we have been particularly interested in materials aspects of the HTSC. Since the borderline between materials aspects and issues which are more concerned with the mechanism of superconductivity is not clear cut, we will also include studies of Knight shifts and relaxation times. The latter list will be by no means exhaustive.

E. Kaldis (ed.), Materials and Crystallographic Aspects of HTc-Superconductivity, 225–248.
© 1994 *Kluwer Academic Publishers.*

For the benefit of the non-expert in the field of NMR/NQR we will start with a short summary of the basic notions and principles of these techniques. For a comprehensive and detailed treatment of the topics we recommend Ref. [6].

2. Basic Principles of NMR and NQR

NMR is concerned with resonance phenomena of nuclear magnetic moments, μ, in all three phases of matter. Nearly all elements of the periodic table possess at least one isotope with a non-zero nuclear spin, $\hbar \vec{I}$. This is the prerequisite for the existence of μ. Both quantities are coupled by the relation $\vec{\mu} = \gamma \hbar \vec{I}$. γ, the gyromagnetic ratio, is the characteristic property of each magnetic nucleus. Its value, for instance, for protons is 42.577 MHz/T.

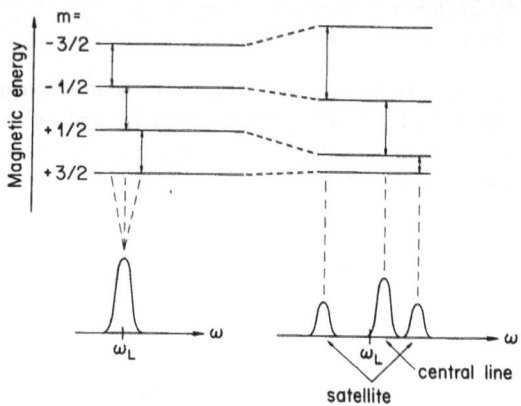

Fig. 1. Energy level scheme for a spin $I = 3/2$ without (left) and with (right) quadrupolar interaction. The arrows indicate transitions induced by a radio frequency field. The bottom shows the resulting NMR spectrum.

In an external static magnetic field, B_0, the magnetic nuclei are described by the Zeeman Hamiltonian: $\mathcal{H}_Z = -\vec{\mu} \vec{B}_0$. The magnetic energies, $E = \gamma \hbar B_0 m$ (m = magnetic quantum number), are split into $2I$ values as shown in Fig. 1 (left hand side) for $I = 3/2$ nuclei such as the copper isotopes ^{63}Cu and ^{65}Cu. To perform an NMR experiment, a small oscillating magnetic field, B_1, is applied perpendicular to B_0. This field which has an angular frequency ω_0 in the range of radio frequencies, then induces transitions between the nuclear levels. The energy balance yields the resonance condition $\omega_0 = \gamma B_0$. In a modern NMR pulse experiment, a short radio-frequency pulse at or near ω_0 is applied causing the spin system to respond by a free-induction decay (FID) signal which then is Fourier transformed to yield the NMR signal in the frequency domain.

About one third of all magnetic nuclei have a spin greater than 1/2 and they therefore possess another very important property, the electric quadrupole moment eQ where e is the elementary charge. The quadrupole moment is a measure of the nucleus' deviation from spherical symmetry. The quadrupole moment interacts with inhomogeneous electric field gradients (EFG) present at the nuclear site thus creating the quadrupole Hamiltonian, $\mathcal{H}_Q$, in addition to $\mathcal{H}_Z$. $\mathcal{H}_Q$ is given by

$$\mathcal{H}_Q = \frac{eQV_{zz}}{4I(2I-1)}[3I_z^2 - I(I+1) + \eta(I_x^2 - I_y^2)] \tag{1}$$

Here, V_{ii} are the principal components of the EFG given in a system x, y, z defined such that $|V_{zz}| \geq |V_{yy}| \geq |V_{xx}|$. η is the asymmetry parameter defined as $\eta = (V_{xx} - V_{yy})/V_{zz}$. The parameter eQV_{zz}/h is called the quadrupole coupling constant.

In many cases we have $\mathcal{H}_Q \ll \mathcal{H}_Z$ and perturbation theory may be used to calculate the energy levels as shown on the right hand side of Fig. 1. The NMR experiment now yields three signals, the central signal corresponding to the transition $m = +1/2 \leftrightarrow -1/2$ and two satellite signals arising from the $m = \pm 1/2 \leftrightarrow \pm 3/2$ transitions. In a single crystal, the pattern of the resonance lines changes with the orientation of the EFG with respect to B_0 thus allowing to determine the EFG tensor including its orientation with respect to the crystallographic axes.

Let us take an example! In the case of Cu and O NMR in the HTSC, the spectrum is a superposition of a quadrupolar pattern and a frequency shift, the Knight shift, to higher frequencies. The Knight shift (which is a tensor) is defined as

$$K = \frac{\Delta\nu}{\nu_0} \tag{2}$$

$\Delta\nu$ represents the displacement of the NMR frequency in a metal with respect to the frequency ν_0 in a non-metallic reference compound. For ^{17}O having a spin $I = 5/2$ and in the presence of a strong field, there are five transition frequencies given by

$$\nu_{(-\frac{1}{2},+\frac{1}{2})} = (1 + K_{cc})\nu_0 + 800/81(\nu_{aa} - \nu_{bb})^2/\nu_0 \tag{3}$$

$$\nu_{(\pm\frac{1}{2},\pm\frac{3}{2})} = (1 + K_{cc})\nu_0 \pm 20/3\nu_{cc} + 500/81(\nu_{aa} - \nu_{bb})^2/\nu_0 \tag{4}$$

$$\nu_{(\pm\frac{5}{2},\pm\frac{3}{2})} = (1 + K_{cc})\nu_0 \pm 40/3\nu_{cc} - 400/81(\nu_{aa} - \nu_{bb})^2/\nu_0 \tag{5}$$

where $\nu_0 = \gamma_n B_0$, with γ_n the ^{17}O gyromagnetic ratio, and $\nu_{\alpha\alpha} = (|eQ|/h)V_{\alpha\alpha}$, with $\nu_{\alpha\alpha}$ being the α component of the quadrupole coupling tensor and B_0 is along the crystallographic c direction. Similar equations are valid for $B_0 \parallel a, b$.

Fig. 2 gives an example, the ^{17}O NMR spectra obtained from an oriented powder sample of the double-chain compound 1-2-4-8 for two different field orientations [7]. Since a sufficiently large single crystal of high quality was not available an "oriented powder" sample was used where the crystallographic c-axes of all the crystallites are parallel to each other. The unit cell of 1-2-4-8 contains eight oxygen atoms, located on four crystallographically inequivalent sites: four, O(2) and O(3), in the CuO_2 planes, two, O(4), in the double Cu–O chains and two, O(1), in the bridging sites between the chains and the planes. According to eq.(3)-(5) each oxygen site yields a set of five frequencies with four unknown parameters. Thus it was possible to assign the lines of the $B_0 \parallel c$ spectrum to four different ^{17}O magnetic shift and quadrupole coupling tensors. Each set of lines is denoted by a capital letter with indices (0,1 and 2), referring to the central line, the inner and outer satellite, respectively. The *exact* values for the shift and quadrupole coupling tensors were determined by fitting the experimental positions of the resonance lines to theoretical values which were obtained from the diagonalisation of the total Hamiltonian.

For large quadrupolar interactions as in the case of Cu in the HTSC, it is advantageous to employ the opposite condition, $\mathcal{H}_Z \ll \mathcal{H}_Q$, or even reduce B_0 to zero. This is NQR! A

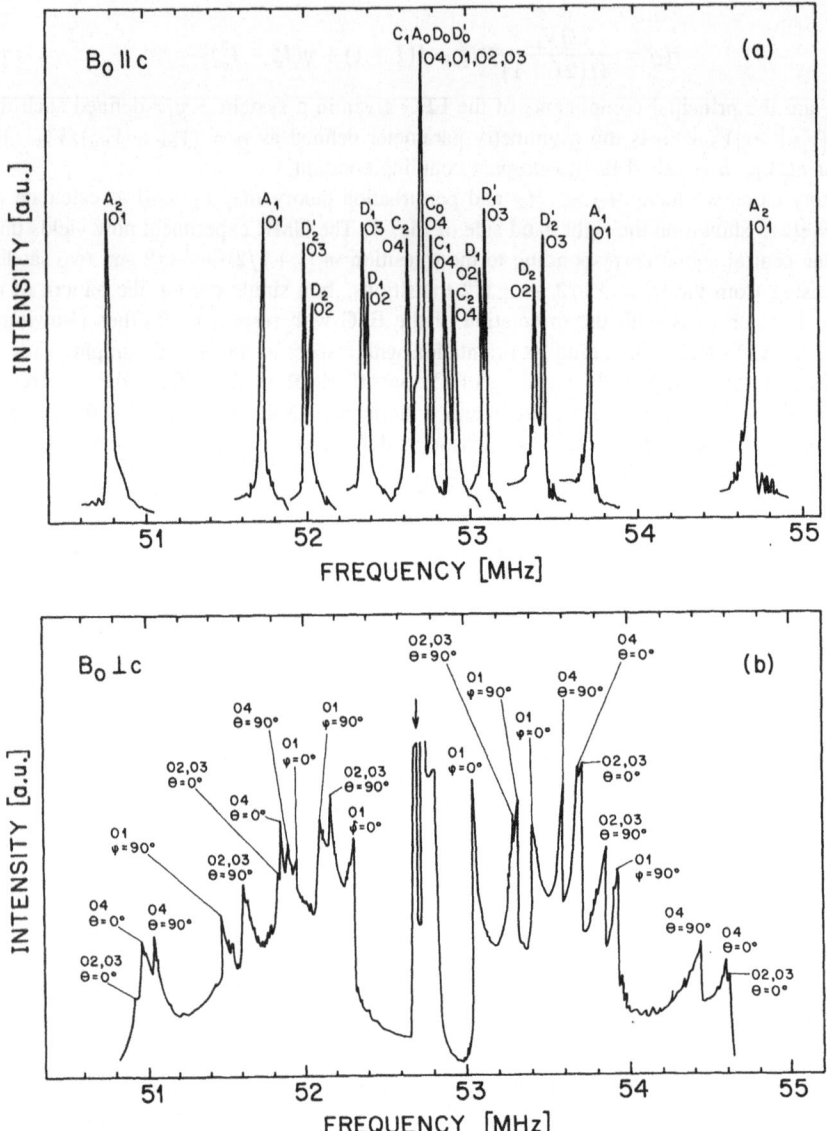

Fig. 2. ^{17}O NMR spectrum at 100 K with B_0 = 9.129 T along the c axis (a) and perpendicular to the c axis (b). Each set of lines is denoted by a capital letter with indices (0,1 and 2) referring to the central line, the inner and outer satellite, respectively. The intensities of different lines are not to scale. The arrow in (b) denotes the central transitions. θ and ϕ are the polar and azimuth angle, respectively, of the direction of B_0 in the EFG's principal axes system.

magnetic oscillating field can induce magnetic dipole transitions between the energy levels of $\mathcal{H}_Q$. In the case of $I = 3/2$, we are mostly interested in, $\mathcal{H}_Q$ contains two degenerate levels, $\pm$ 3/2 and $\pm$ 1/2, and hence only one NQR frequency:

$$\nu_Q = \frac{eQV_{zz}}{2h}\sqrt{1 + \frac{1}{3}\eta^2} \tag{6}$$

As an example, Fig. 3 shows the Cu NQR spectrum from 1-2-3-7 [8] which we will discuss in Chapter 3. In order to determine the complete EFG tensor either an NMR experiment or an NQR experiment in a low magnetic field are necessary.

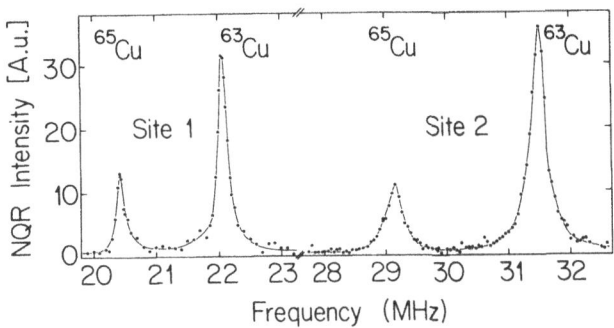

Fig. 3. Cu NQR spectrum from $YBa_2Cu_3O_7$

We conclude our survey of NMR/NQR notions by introducing the relaxation times. We will only be concerned with the spin-lattice relaxation time T_1. It is the time constant for the spin system to approach thermal equilibrium with its surrounding, the so-called "lattice". The nuclei are acted on by internal fluctuating magnetic fields and EFG arising from other atoms, ions or conduction electrons. If the Fourier spectrum of the fluctuations contains frequencies of appreciable amplitude in the neighborhood of an NMR or NQR transition frequency, transitions in the spin system are induced, hence, the populations of the various nuclear energy levels are changed and the spin system may attain equilibrium.

It turned out that nearly all mechanisms which are responsible for spin-lattice relaxation in the HTSC are based on electron-nuclear interactions. The only exception seems to be the Ba relaxation in 1-2-4-8 [9,10] which is due to a two-phonon Raman process, *i.e.* a coupling between phonons and the Ba quadrupole moment. Thus, for Cu, O and Y the relaxation is described by the Moriya formula [11] which relates the relaxation rate to the hyperfine fields and the dynamical electron susceptibility. We write this formula in the following way:

$$^i\left(\frac{1}{T_1}\right)_\alpha = \frac{\gamma_i^2 k_B T}{2\mu_B^2} \sum_{\vec{q},\alpha\neq\alpha'} |^i A_{\alpha'}(\vec{q})|^2 \frac{\chi''_{\alpha'\alpha'}(\vec{q},\omega_i)}{\omega_i} \tag{7}$$

Here, i denotes the nuclear species and ω_i is its NMR frequency. The magnetic field B_0 is oriented along the α direction which coincides with one of the crystallographic axes (a, b, c). $\chi''_{\alpha'\alpha'}(\vec{q},\omega_i)$ is the imaginary (dissipative) part of the complex electron susceptibility which depends on the excitation wave vector $\vec{q}$ and the frequency. $A_{\alpha'}(\vec{q})$ are formfactors related to the hyperfine fields. Eqn 7 demonstrates that the spin-lattice relaxation provides a tool to probe the average over the *q-dependent* electron susceptibility while the Knight shift yields information about the *static* susceptibility.

3. Interpretation of Electric Field Gradients

Our first topic we will address is the EFG which is an important *static* property of any solid since it depends sensitively on the charge distribution around the nucleus in question. The EFG thus reflects structural and crystallographic aspects of the compound and provides information about ionicity and bonding.

The Cu NQR spectrum obtained from the single-chain compound 1-2-3-6+x (where x is very close to 1 [8]) shown in Fig. 3 and the similar Cu NQR spectrum from the double-chain compound 1-2-4-8 [12] contain four signals each. This proves immediately the presence of two inequivalent Cu sites in the unit cell, the chain and plane site Cu(1) and Cu(2), respectively. The EFG at these sites differ by their magnitude.

The total number of signals is four because of the two copper isotopes, ^{63}Cu and ^{65}Cu, which differ by their quadrupole moment. According to eqn 6 the ratio of the NQR frequencies of the two isotopes (measured at the same site) is equal to the ratio of their quadrupole moments (1.081). We have confirmed this experimentally for both Cu sites in 1-2-3-7 and 1-2-4-8 and for Cu(1) in the paramagnetic phase of the semiconducting structure 1-2-3-6. This proves that no internal magnetic field due to any type of magnetic ordering is present at these Cu sites (in the temperature range we have studied). Such a field would split or shift the NQR lines. For the Cu(2) signal in the anti-ferromagnetic phase of 1-2-3-6 a shift has been measured which allows to determine the internal magnetic field due to antiferromagnetism.

Table 1.

Quadrupole coupling constant C and asymmetry parameter η at the ^{63}Cu sites in $YBa_2Cu_3O_7$ and $YBa_2Cu_4O_8$.

Substance	Cu1	Cu2
$YBa_2Cu_3O_7$ powder (Reanalysis of data of Ref.[13], 300 K)	C=38.9 ± 0.5 MHz η =0.95 ± 0.05	C=62.40 ± 0.05 MHz η =0.04 ± 0.04
$YBa_2Cu_3O_7$ single crystal, ([14], 100 K)	C=38.54 ± 0.02 MHz η=0.984 ± 0.005	C=62.38 ± 0.02 MHz η =0.012 ± 0.003
$YBa_2Cu_4O_8$ oriented powder (Ref. [15], 150 K)	C = 35.101 ± 0.01 MHz η = 0.951 ± 0.005	C = 59.42 ± 0.02 MHz η = 0.015 ± 0.005

In Table 1 we have summarized the Cu(1) and Cu(2) EFG coupling constants and asymmetry parameters for 1-2-3-7 and 1-2-4-8. The powder and single-crystal data for 1-2-3-7 agree within the error limits. The similarity of the two structures shows up in small differences between corresponding parameters of the EFG tensors. The coupling constants C at both sites are slightly smaller in 1-2-4-8. The nearly axial symmetry of the EFG at the Cu(2) site (η being close to zero) in both compounds reflects the small deviation of this site from tetragonal symmetry. On the other side, the large difference of the values of η for the Cu(1) and Cu(2) sites in each compound is striking. A value of η close to 1 must reflect the peculiar electronic structure at the Cu(1) site. Comparing the single crystal data for 1-2-3-7 with those for 1-2-4-8, we notice a slight decrease of asymmetry in the Cu(1) EFG which might arise from the structural changes in 1-2-4-8 introduced by the second copper chain.

How can the EFG's be related to charge distributions and the electronic structure of the respective compound? Among others one motivation for measuring the dependence of the NQR frequencies on both temperature and pressure (to be discussed later) was to have as much experimental information about the EFG as possible in order to put constraints on proposed calculations.

The simplest calculation of the EFG is by means of the point-charge model. The EFG at the nuclear site is determined by two contributions: one arises from the point charges on neighboring ions, the second from incomplete electronic shells of the ion under consideration (valence contribution). Thus the total EFG may be written as

$$V_{zz} = (1 - \gamma_\infty)V_{zz}^o + (1 - R)V_{zz}^{val} \tag{8}$$

Here, V_{zz}^o and V_{zz}^{val} are the EFG produced by neighboring ions and incomplete shells, respectively, and γ_∞ and R are the respective Sternheimer factors. The valence contribution originating from the holes in the open Cu $3d$ shell can be written in terms of the number of holes in the different d orbitals as follows:

$$V_{zz}^{val} = A < r^{-3} >_{3d} [n_h(3d_{3z^2-r^2}) - n_h(3d_{x^2-y^2}) - n_h(3d_{xy}) + \frac{1}{2}n_h(3d_{xz}) - \frac{1}{2}n_h(3d_{yz})] \tag{9}$$

where A is a constant. This way one can estimate the hole concentration from the change of ν_Q.

Among the more reliable or nearly *ab initio* calculations we mention those by the groups around T.P. Das employing Hartree-Fock cluster investigations [16,17] and K.H. Schwarz. The latter group has developed a method to compute EFG's from a full potential linearized augmented plane wave (LAPW) band structure calculation [18]. It should be emphasized that in these calculations there is no need for Sternheimer factors.

Results of an application to 1-2-3-7, 1-2-3-6 and 1-2-3-6.5 [19] and to 1-2-4-8 [20] have been reported. The good agreement with experimental values for 1-2-3-7, 1-2-3-6 and 1-2-3-6.5 (except for the planar Cu(2) site!) was already discussed in [19]. In the meantime, the EFG's in 1-2-4-8 for both Cu sites [12,15] , all four O sites [7] and the Ba site [21] have been measured. Together with the recent experimental result for the Ba site in 1-2-3-7 [22], it is now established that the LAPW method within the local density approximation adequately describes the charge distributions for Cu(1), all O and Ba sites in 1-2-3-7 and 1-2-4-8.

The theory reveals [20] that the Cu(1) EFG is dominated by the valence contribution, that the Ba EFG originates from the interaction with the oxygen neighbors and that by going from 1-2-3-7 to 1-2-4-8 one expects to see effects in the chain O position. In particular, the difference between theoretical and experimental oxygen EFG's is about 12 % for all sites.

For the apex O(1) and the planar O(2) and O(3) sites, the corresponding EFG tensors of 1-2-3-7 and 1-2-4-8 differ by only about 10 %. However, a pronounced difference between both systems is observed for the chain O(4) site, whose asymmetry parameter changes from 0.48 in 1-2-3-7 to 0.868 in 1-2-4-8. Thus, the charge distribution in both structures is quite similar except for the chain oxygen site where the doubling of the chain affects the asymmetry of the charge distribution and causes a charge transfer from O(2) and O(3) to Cu(1) [20]. This exemplifies the interplay between the chains and the planes in these structures.

However, for the Cu(2) site in all these structures the theoretical values are too low, in case of 1-2-4-8 by a factor of about 3. The origin of this discrepancy probably is that the local density approximation not fully accounts for exchange-correlation effects between Cu $3d$ states [19]. In other words: the EFG calculations confirm the exceptional role played by the planar Cu and that the CuO_2 planes are mainly responsible for superconductivity.

4. Temperature and Pressure Dependence of NQR Frequencies

The dependence of the NQR frequency ν_Q on temperature, T, and pressure, p, can provide detailed information on structural and electronic changes of the compound since the EFG tensor is very sensitive to small atomic displacements or modifications of charge distributions. In Fig. 4 we have plotted $\nu_Q(T)$ for both Cu sites in a 1-2-3-7 sample with oxygen deficiency $x = 0.018 \pm 0.001$ [23,24]. Our data, at least those for Cu(1(, do not provide evidence for any major structural changes at T_c. The change of slope in the $\nu_Q - vs - T$ dependence at T_c and at 240 K is attributed to effects such as small changes of inter-atomic distances, Cu-O bond corrugations, valence state changes due to a minute electron transfer between orbitals and lattice vibrations. Perhaps Jahn-Teller excitations should be taken into account as a conceivable origin of distinct changes of ν_Q.

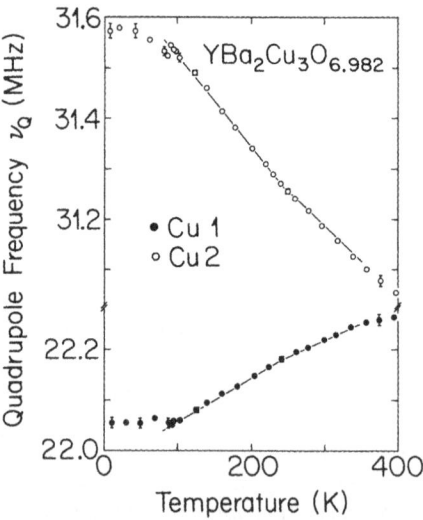

Fig. 4. Temperature dependence of the [63]Cu NQR frequencies at both Cu sites in $YBa_2Cu_3O_{6.982}$. The curves are guides to the eye.

On the other side, Riesemeier *et al.* [25,26] who have measured with high precision ν_Q around T_c, reported a *discontinuous* change of ν_Q at T_c (Fig. 5). Within a small temperature range ν_Q steeply decreases at T_c, but increases again in the superconducting state. In addition, a change in slope of the $\nu_Q - vs - T$ curve at T_c was observed, in agreement with our findings. The authors believe that the discontinuity at T_c is indicative for a ferroelectric or anti-ferroelectric transition as suggested also by ultrasonic investigations. So far, the discrepancy with our data at T_c could not be resolved.

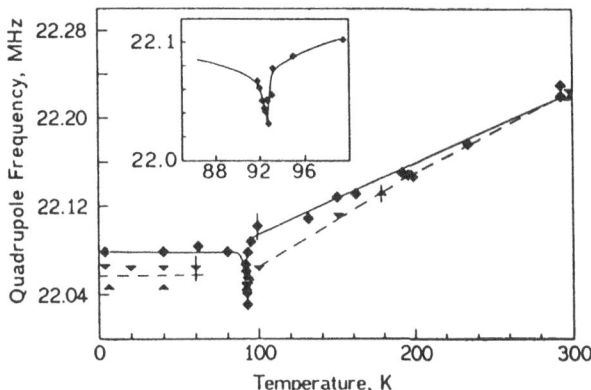

Fig. 5. Temperature dependence of the ^{63}Cu NQR frequency at the Cu(1) site in different $YBa_2Cu_3O_{7-\delta}$ powder samples. From [26].

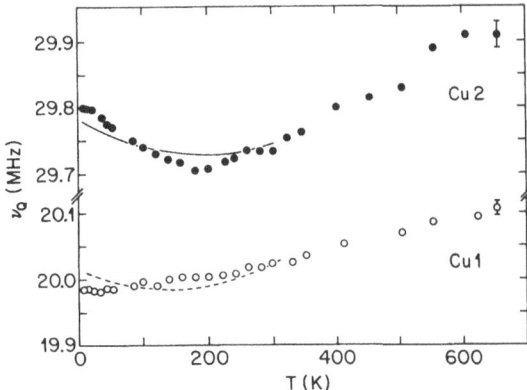

Fig. 6. Temperature dependence of the ^{63}Cu NQR frequencies at both Cu sites in $YBa_2Cu_4O_{8.04}$. The fit curves are explained in the text.

In 1-2-4-8, the $\nu_Q(T)$ curves of ^{63}Cu for both Cu sites [15] exhibit even more subtle changes (Fig. 6) which are outside the experimental errors and the data above 300 K behave quite differently from those of Cu(2) in 1-2-3-7. The high precision of this data allow to discern even smallest changes of ν_Q with temperature, the relative changes over the whole temperature range are below 0.7 % at both sites. Since the EFG calculations based on the LAPW method are reliable within only 10 % we have performed a point charge model calculation of the temperature dependent EFG thereby taking into account a valence contribution originating from the holes in the open Cu $3d$ shell. We have used the published values for the atomic positions at 300 K [27] and the temperature dependent lattice parameters [28]. The solid and dashed lines in Fig. 6 are fits to the data using the following ionic charges: Y^{+3}, Ba^{+2}, $Cu(1,2)^{+2}$, $O(1)^{-2}$, $O(2,3)^{-1.917}$ and $O(4)^{-1.667}$, a Sternheimer factor of -11.8 and a 0.8 occupancy of the Cu $3d_{x^2-y^2}$ orbital. No further fit parameters are used.

The fit accounts for the overall behavior of the data. That means that the temperature dependence of ν_Q is mainly caused by changes of the lattice parameters; there is no evidence for a change of the electronic structure. In other words: the degree of localisation of the hole in the Cu $3d$ orbital does not change between 6 and 750 K. That one may have some confidence in these calculations follows from the fact that the calculated asymmetry parameters η are in excellent agreement with data we obtained from high field NMR measurements [15,12]. A similar calculation for 1-2-3-6 and 1-2-3-7 has also given remarkable agreement with the experimental data.

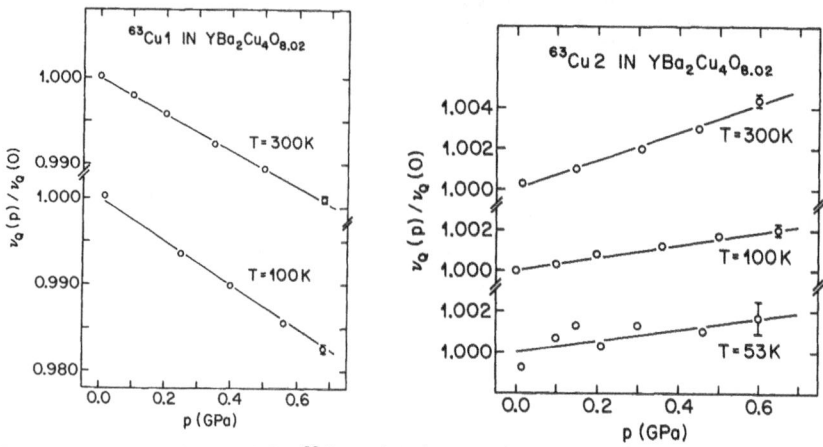

Fig. 7. Pressure dependence of the ^{63}Cu NQR frequencies at both Cu sites in $YBa_2Cu_4O_{8.02}$. The fit curves are explained in the text.

Let us now turn to the pressure dependence of the NQR frequency. Fig. 7 presents the pressure dependence of the ^{63}Cu ν_Q for Cu(1) and Cu(2) in 1-2-4-8 at different temperatures [15]. In both cases the pressure dependence of ν_Q is linear as expected for small volume changes. The slope $d\nu_Q/dp$ is only weakly temperature dependent. To explain the data we have again employed the point charge model (same values for Sternheimer factors) and have used pressure dependent structure data [27]. We obtain good agreement with the experimental data if a charge transfer from the chains to the planes is assumed. The effective charges are now -2-(1/12)-(1/2)(dx/dp)p for O(2,3) ions and -2-(1/3)+(dx/dp)p for O(4) ions, where dx/dp is the number of charge carriers (cc) transferred per Cu(2) and per 1 GPa; it is used as a free parameter. Good agreement with the data is obtained for dx/dp = 0.01 cc/GPa which is compatible with the pressure dependence of the electric conductivity.

Since $d\nu_Q/dp$ is nearly equal for 1-2-4-8 and 1-2-3-7 [29] one infers the pressure induced chain-plane charge transfer to be equal for both compounds. This contrasts with the large difference in dT_c/dp in both compounds where the 1-2-4-8 value of 5.7 K/GPa exceeds that of 1-2-3-7 by a factor of 10. A similar difference is observed for 1-2-3-6.5 and 1-2-3-7. The charge carriers in 1-2-3-6.5 are nearly confined to the planes as can be deduced from the extreme anisotropy of the conductivity. The anisotropy of conductivity in 1-2-3-6.5 and the anisotropy of the upper critical fields in 1-2-4-8 suggest that these two compounds have more two-dimensional ($2d$) character than 1-2-3-7. In an extremely $2d$ material, T_c sensitively depends on the coupling between the planes. Therefore one would expect by applying pressure and hence increasing interplane coupling, a much stronger effect in the compounds with more $2d$ character.

5. NQR Linewidth and Oxygen Ordering

The large NMR linewidths found in HTSC have been an obstacle for many detailed investigations. Do these linewidths arise from imperfections of the structure such as oxygen disorder and are they an intrinsic property of these materials or are they the result of "imperfect" sample preparation? To answer this question we have measured the linewidth $\Delta\nu$ of the Cu NQR signals in carefully prepared 1-2-3-(7-δ) samples for oxygen concentrations ranging from about 6.940 to 6.983 with uncertainties of ±0.001 [30]. Fig. 8 shows that $\Delta\nu$ of the Cu(1) signal increases linearly with oxygen deficiency δ. Extrapolation of the data to $\delta = 0$ leaves a "rest" linewidth of about 40 kHz whose origin is due to other effects than oxygen deficiency, for instance crystal structure defects. The intrinsic linewidth arising from spin-spin relaxation is only about 10 kHz.

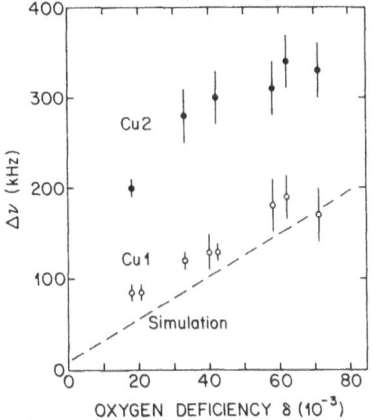

Fig. 8. Linewidth $\Delta\nu$ ("full width at half height") of the ^{63}Cu NQR signals of both Cu sites in $YBa_2Cu_3O_{7-\delta}$ as a function of oxygen deficiency. The dashed line is a result of a simulation discussed in the text.

To our knowledge these are the smallest Cu NQR linewidths measured so far in powder samples of HTSC. For the Cu(2) linewidth, the $\Delta\nu$ vs. δ dependence is less clear and more experimental data at small δ are required.

We have calculated the effect of the EFG of the oxygen vacancies on $\Delta\nu$ by a computer simulation assuming a random distribution of oxygen vacancies on the chain 0(4) sublattice and using the point charge model [30]. The simulated $\Delta\nu$ increases linearly with δ (see Fig. 8) and is practically the same for Cu(1) and Cu(2). This result agrees quite well with the experimental data for the Cu(1) signals. That the Cu(2) signals are about twice as broad as the Cu(1) signals may be traced back to the fact that an oxygen vacancy wipes out the contribution of its next-neighbor Cu(1) nucleus to the Cu(1) signal while the effect on the Cu(2) - a charge redistribution around Cu(2) neighbors - is less severe leading to a detectable broadening of the signal.

In 1-2-4-8 the linewidths at 350 K are about 60 KHz for Cu(1) and 170 kHz for Cu(2) [12]. They seem to be compatible with fully occupied oxygen sites as seen in neutron diffraction [31].

6. Cation Doped Y-Ba-Cu-O Material

Among the many activities to "tailor" HTSC for a particular purpose is doping in order to increase T_c. For instance, Miyatake et al. [32] could enhance T_c of 1-2-4-8 from 82 K to 90 K by doping with 10 % Ca. For the understanding of the enhancement it is crucial to know which crystallographic site the Ca^{+2} ion is occupying and whether the substitution changes the hole concentration.

Because of the controversy in this issue we investigated this problem by measuring the Cu NQR spectrum at 100 K in 1-2-4-8 doped with 10 % Ca ($T_c = 91$ K) [33]. As can be seen in Fig. 9, the NQR lines of the plane and chain Cu sites appear at the same frequencies ν_Q as in the pure compound, however, both lines are broadened by static defects. Within the linewidths no temperature dependence of ν_Q was observed and no indication for magnetic ordering at the Cu sites was found.

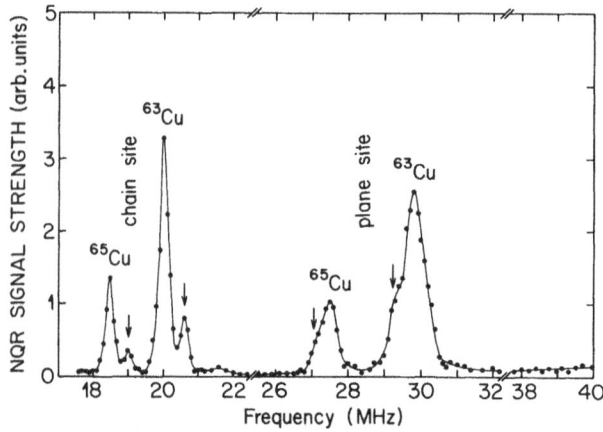

Fig. 9. The Cu NQR spectrum in $YBa_{1.9}Ca_{0.1}Cu_4O_8$ at 100 K. The arrows indicate the resonance lines that have not been detected in pure $YBa_2Cu_4O_8$.

At 20.600 MHz, near to the Cu(1) line, a resonance line has been detected which is absent in the spectrum of pure 1-2-4-8. Because of its similarities with the Cu(1) spectrum we have identified the 20.600 MHz peak as a Cu(1) resonance line arising from those Cu(1) nuclei which are nearest neighbors of a lattice defect. A careful discussion of the signal intensities has revealed that the defect is a Ca ion occupying a Ba site. We thus concluded that at least a major fraction of Ca ions occupies Ba sites.

The next question is whether this type of substitution changes the hole concentration. If we assume that the lattice contribution to the EFG at Cu(1) and Cu(2) sites is not changed by Ca doping we conclude that according to eqn 9 the hole concentration does not change. In particular, it is known that ν_Q of the plane Cu(2) nuclei is shifted towards higher values with increasing hole concentration. Since ν_Q of Cu(2) in Ca doped 1-2-4-8 remains in the limit of the linewidth unchanged with respect to pure 1-2-4-8, the increase of the number of holes is estimated to be less than 0.013 per Cu. Last not least, as a second surprising result we found that the spin-lattice relaxation time T_1 of both the Cu(1) and Cu(2) nuclei in 1-2-4-8 are not changed by 10 % Ca doping [34]. Since it is well known that T_1 is related to the hole concentration this result supports our conclusion that according to NMR/NQR data, Ca doping induces only a negligible change of the hole concentration. It therefore seems that the increase of T_c by Ca doping must be explained by another mechanism.

The opposite conclusion concerning this point has been reached by Machi *et al.* [35]. They claim that is possible to separate in the relaxation data the contribution arising from the Brillouin zone corner which reflects the anti-ferromagnetic (AF) spin fluctuations. A reduction of this component due to Ca doping then suggests an increase of the hole concentration. The authors, however, do not comment on the additional Cu(1) line we have observed. Our NMR study of the [43]Ca nucleus itself in Ca doped 1-2-4-8 being in progress, hopefully can decide between the two possibilities.

A somewhat opposite situation occurs by doping fully oxygenated 1-2-3-7 by Zn or Al. Zn is thought to occupy primarily Cu(2) sites and produces a dramatic *decrease* in T_c [36] whereas Al ions go mainly into Cu(1) sites and have very little effect on T_c [37]. The effect of the dopants could be studied via [89]Y NMR. Balakrishnan *et al.* [38] noticed that Zn causes a temperature dependent broadening of the line and an increase of T_1^{-1} while Al produces only a temperature independent broadening and has no appreciable effect on T_1. Alloul *et al.* [39] also observed the broadening effect of Zn which was taken as evidence for the generation of local magnetic moments on or near the Cu(2) sites [38,39] since Zn unlike Al produces a large increase in Curie-like magnetism and therefore should affect the [89]Y NMR.

Warren Jr. *et al.* [40] and Walstedt and Warren Jr. [41] pursued the question whether NMR can tell the difference between 1-2-3-7 samples whose T_c has been reduced either by Zn doping or by removing oxygen. Zn doping causes, among others, an increase of the Cu(2) linewidth which is believed to arise from a Ruderman-Kittel-Kasuya-Yosida (RKKY) effect, *i.e.* an indirect exchange interaction of local moments at Cu(2) sites via conduction electrons. Dipolar broadening by magnetic impurities cannot account for the magnitude of the effect. The interpretation is consistent with the previous analyses. As to relaxation, Zn doping leads to an appreciable decrease of the [63]Cu spin-lattice relaxation rates and this for both sites. This contrasts with the increase of the [89]Y relaxation rate [38]. The different behavior is attributed [41] to a breaking of the magnetic symmetry around the Y site which then allows the AF fluctuations to enhance the Y relaxation rate.

7. Knight Shift Investigations

So far we have reviewed NMR/NQR studies which have a strong crystallographic and materials science aspect. We will now turn to topics which are primarily concerned with electronic aspects

238

and ultimately address questions about the mechanism of superconductivity.

Among the important NMR parameters in HTSC studies which reflect *static* properties is the Knight shift, $K(T)$, defined in eqn 2. In the field of HTSC NMR it is customary to define $K(T)$ as that part of the total shift that arises from the interaction between the nuclear spin and the *spin* paramagnetism of the conduction electrons.

The importance of the Knight shift stems from the fact that it is proportional to the electron spin susceptibility χ_S and the hyperfine interaction $\vec{A}$ of the nucleus. Thus, K may be expressed as

$$K(T) = \frac{1}{g\mu_B} \sum_i A_i \chi_S \qquad (10)$$

where g is the Landé factor and μ_B the Bohr magneton. For a classical superconductor with s-wave spatial symmetry (as in the BCS theory), the Knight shift is constant in the normal state, declines in the superconducting state due to spin-singlet pairing and approaches zero with zero slope [42]. In several HTSC, however, K starts to decrease far above T_c and the shift is anisotropic.

To explain these facts and other features, Mila and Rice [43] proposed the following model of the hyperfine interactions. For the Cu(2) nucleus, there is a contribution $\vec{A}$ from the interaction with an electron of the on-site $3d_{x^2-y^2}$ orbital and a transferred interaction $\vec{B}$ with electrons of the $3d_{x^2-y^2}$ orbitals of the four neighboring copper ions. For the planar oxygen nuclei, there is a transferred hyperfine coupling $\vec{C}$. Due to the weak hybridization of Cu-4s and O-2s states with Cu-d and O-p states, a hyperfine coupling between Cu and O may be anticipated.

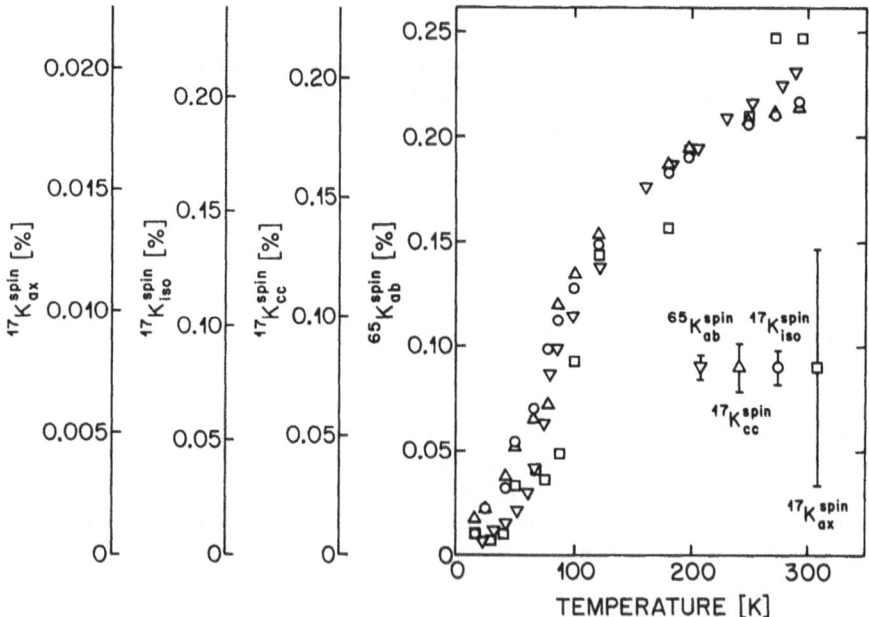

Fig. 10. Temperature dependence of various Knight shift components at Cu and O sites in $YBa_2Cu_4O_8$.

Fig. 10 shows the temperature dependence of various planar Cu and O Knight shift components in 1-2-4-8 [44-46]. All components exhibit the same temperature dependence with a decline which starts already well above T_c. Together with very similar results for 1-2-3-6.63 [47], these facts support the "single spin fluid" model which states that the Cu-d holes and the doped holes (which mainly go into O-$2p$ states) have one spin degree of freedom. Similar ^{17}O data for 1-2-4-8 have been reported by Machi *et al.* [48] and similar Cu Knight shift data were obtained in $Bi_2Sr_2CaCu_2O_8$ [49]. Also strong evidence for the single spin fluid model has been provided by ^{89}Y and ^{205}Tl Knight shift investigations, see [50,51].

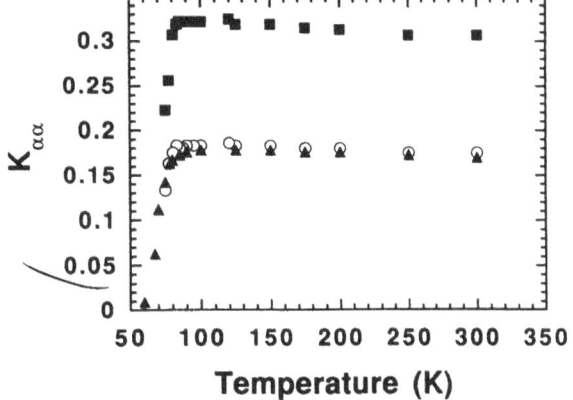

Fig. 11. Temperature dependence of the principal components of the shift tensor at O(2,3) sites in $YBa_{1.92}Sr_{0.08}Cu_3O_7$. ■ K_{ZZ} ($B_0 \parallel$ Cu-O-Cu bond); ○ K_{YY} ($B_0 \perp$ Cu-O-Cu bond and in ab plane); ▲ K_{XX} ($B_0 \parallel$ c-axis). From [52].

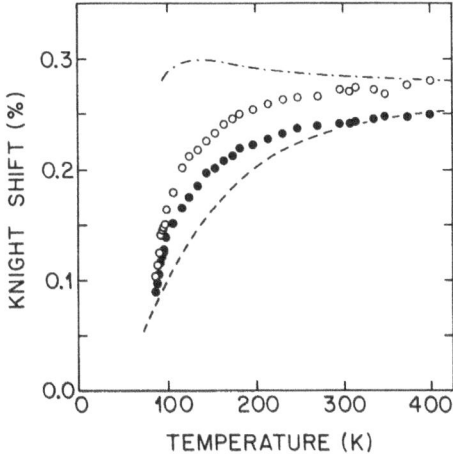

Fig. 12. Temperature dependence of the Knight shift of the two planar Cu sites Cu(2) (1-2-3-7 block, open circles) and Cu(3) (1-2-4-8 block, full circles) in $Y_2Ba_4Cu_7O_{15}$. From [53]. The dashed-dotted and the dashed curves are the shifts of the "mother" substances $YBa_2Cu_3O_7$ and $YBa_2Cu_4O_8$, respectively.

In over-doped samples, the Knight shift behavior is quite different. The group around C. Berthier in Grenoble has performed detailed studies of the spin susceptibility as probed at Cu(2) and O(2,3) sites in 1-2-3-6+x *single* crystals for the under-doped and over-doped regime (for a summary see [52]). As an example, Fig. 11 shows the temperature dependence of the principal components of the $K_{\alpha\alpha}$ tensor at the O(2,3) sites in $YBa_{1.92}Sr_{0.08}Cu_3O_7$. The slight increase of $K_{\alpha\alpha}$ with decreasing temperature is now recognized as the signature of slightly over-doped samples.

A particular interesting Y-Ba-Cu-O structure is the "mixed-chain" compound $Y_2Ba_4Cu_7O_{15}$ (2-4-7-15) which consists (along the c axis) of a sequence of alternating one-chain 1-2-3-7 and double-chain 1-2-4-8 blocks. The structure thus offers the unique possibility to study the coupling between *inequivalent* CuO_2 planes. Fig. 12 shows our results for the Knight shift at the two inequivalent Cu sites: Cu(2) in the 1-2-3-7 block and Cu(3) in the 1-2-4-8 block [53]. Apparently the coupling between the two inequivalent layers establishes a common (static) spin susceptibility. So to speak, the Cu(3)O$_2$ plane "imposes" its behavior with respect to the susceptibility on the Cu(2)O$_2$ plane.

8. Spin-Lattice Relaxation Studies

The spin-lattice relaxation study of superconducting aluminum [54] yielded two important observations: the exponential dependence of the rate on inverse temperature and the occurrence of the so-called Hebel-Slichter coherence peak [1] just below T_c. These were the first direct proofs of the basic correctness of the BCS theory. In the HTSC so far neither of these features were observed. Instead, in most HTSC the relaxation behavior below T_c follows a power law $1/T_1 \propto T^n$ with n between 3 and 4. If one plots for planar Cu sites in various HTSC, the normalized relaxation rates, *i.e.* the rate at temperature T, $1/T_1$, divided by the rate at T_c, $(1/T_1)_{T_c}$, as a function of T/T_c, the data fall on a "universal" curve. This is a remarkable result in view of the differences in structure, composition, doping and T_c values.

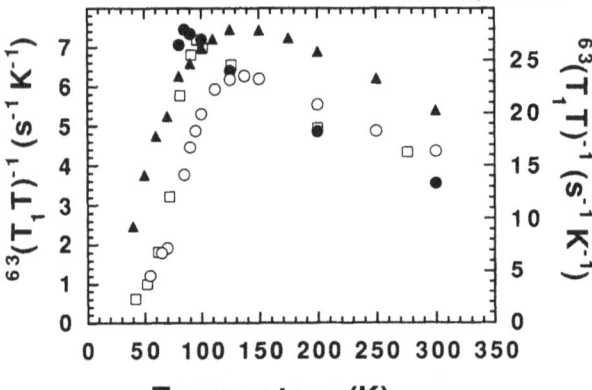

Temperature (K)

Fig. 13. Temperature dependence of $^{63}(T_1T)^{-1}$ for Cu(2) sites. Left-hand scale: □ 1-2-3-7 (oriented powder); ○ $YBa_{1.93}Sr_{0.07}Cu_3O_{6.92}$ (single crystal, $B_0 \parallel c$); ▲ $YBa_2Cu_3O_{6.52}$ (single crystal, $B_0 \parallel c$). Right-hand scale: ● $YBa_{1.92}Sr_{0.08}Cu_3O_7$ (single crystal, $B_0 \parallel a, b$ plane. From [52].

Here we will summarize some experimental facts about relaxation studies and we will focus on the normal conducting phase. In Fig. 13 Cu data from single crystal studies of Berthier's group are compared [55]. For the over-doped compound 1-2-3-7, $^{63}(T_1T)^{-1}$ increases when the temperature is lowered and then drops *at T_c*. For the two other samples with oxygen content $O_{6.52}$ and $O_{6.92}$ which are under-doped, $^{63}(T_1T)^{-1}$ reaches a maximum already *above T_c* and then steadily decreases. While for the $O_{6.92}$ sample the decrease of $^{63}(T_1T)^{-1}$ becomes faster just below T_c, the $O_{6.52}$ sample exhibits a more "gentle" decrease.

For the planar oxygens in these samples, it was found that $^{17}(T_1T)^{-1}$ in the O_7 sample increases when the temperature is lowered, similar to the behavior of the Cu isotope although less pronounced. Below T_c, $^{17}(T_1T)^{-1}$ decreases sharply as found by many groups. On the other hand, the $O_{6.52}$ sample exhibits a continuous decrease of $^{17}(T_1T)^{-1}$ with decreasing temperature.

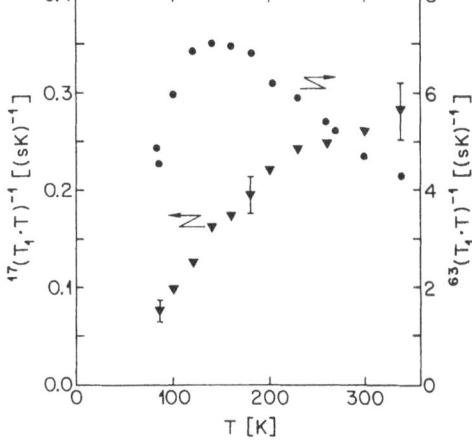

Fig. 14. Temperature dependence of $(T_1T)^{-1}$ for the planar Cu and O in $YBa_2Cu_4O_8$.

In 1-2-4-8 we have detected for planar Cu and O a behavior [9] similar to that of the oxygen-deficient 1-2-3-6+x. $^{63}(T_1T)^{-1}$ reaches a maximum around 130 K and rapidly decreases towards lower temperatures (Fig. 14). On the other side, $^{17}(T_1T)^{-1}$ displays a monotonous decrease with lowering temperature. This again may be taken as signature of under-doped material.

We conclude the listing of Y-Ba-Cu-O studies by presenting the planar Cu relaxation data (Fig. 15) for the "mixed-chain" compound 2-4-7-15 [9,53]. As may be seen from the insert, the two relaxation rates are proportional to each other in the normal phase thus demonstrating that not only the static electron susceptibilities (as exemplified by the Knight shift) but also the *dynamic* susceptibilities are governed by the same temperature dependence. Obviously the two planes must be strongly coupled. At T_c, however, the coupling breaks down.

For the interpretation of relaxation data it advantageous to have as many experimental constraints as possible. Two such crucial parameters are the orientation in the external magnetic field, B_0, and the strength of this field. Let us denote by W_α the spin-lattice relaxation rate at a planar site in an Y-Ba-Cu-O sample where $\alpha = a, b, c$ specifies the orientation of B_0. Then,

242

$r = W_{ab}/W_c$ is the relaxation-rate anisotropy at this site.

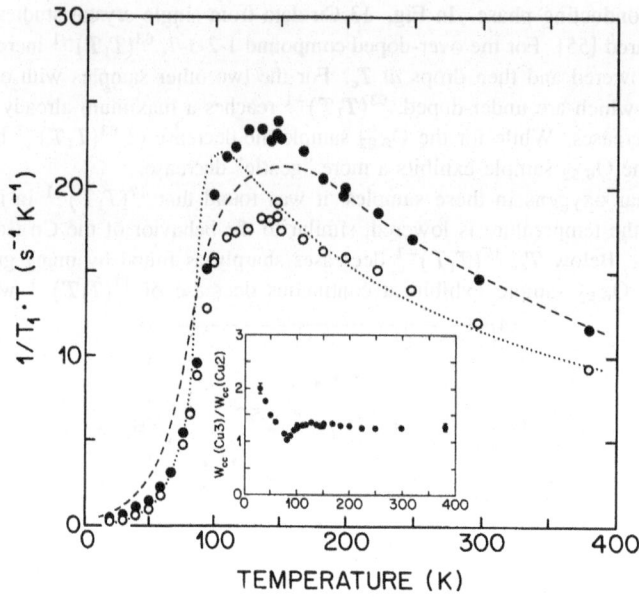

Fig. 15. Temperature dependence of $^{63}(T_1T)^{-1}$ for the two planar Cu sites Cu(2) (1-2-3-7 block, open circles) and Cu(3) (1-2-4-8 block, full circles) in $Y_2Ba_4Cu_7O_{15}$. The dotted and dashed lines denote the relaxation rates in the "mother" substances $YBa_2Cu_3O_7$ and $YBa_2Cu_4O_8$, respectively. The insert shows the temperature dependence of the ratio of the Cu(3) and Cu(2) relaxation rates, W_{cc}.

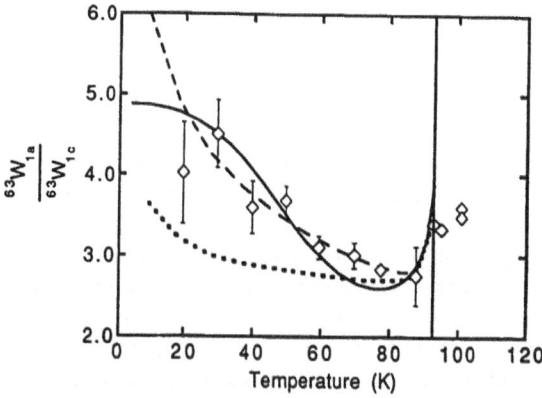

Fig. 16. Anisotropy of the Cu(2) relaxation rate in $YBa_2Cu_3O_7$. The various lines are discussed in Chapter 9. From [57].

The Urbana and Los Alamos NMR groups were the first to report systematic investigations of the quantity r. ^{63}r of Cu(2) is independent of temperature and equal to 3.74 in the normal phase of 1-2-3-7 but undergoes a rapid change just below T_c [56]. Since the change of r due to a magnetic field could be a secondary effect caused for example by fluxoid cores or by T_c suppression in a magnetic field, the high field experiments should be extended to low fields where the field induced anisotropy effects may be neglected. Fig. 16 shows the temperature dependence of ^{63}r in an oriented powder sample of 1-2-3-7 [57]. The various lines are theoretical fits based on a BCS spin-singlet, orbital-d-wave pairing state to be discussed in the next Chapter.

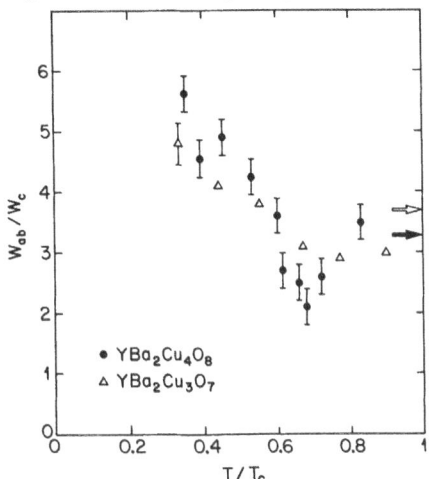

Fig. 17. Cu(2) relaxation rate anisotropy r vs. the reduced temperature T/T_c in a weak magnetic field: ● $YBa_2Cu_4O_8$ in $B_0 = 0.58$ T (from [59]); △ $YBa_2Cu_3O_7$ in $B_0 = 0.44$ T (from [58]). The arrows denote the respective r values above T_c.

Takigawa et al. [58] independently measured the temperature dependent Cu(2) anisotropy below T_c in 1-2-3-7 using also low magnetic fields. Their results are given in Fig. 17 where we will compare them with our data for 1-2-4-8 [59]. Below T_c, both the high and low field r values for 1-2-4-8 decrease with the low field r passing through a pronounced minimum at 55 K. The low field r values for 1-2-4-8 and 1-2-3-7 are quite similar. However, the upturn of r with decreasing temperature is much more pronounced in 1-2-4-8 than in 1-2-3-7. Furthermore, the 1-2-4-8 data exhibit just below T_c a much softer decrease of r; the derivative dr/dT is almost zero at T_c. It remains to be shown whether the different behaviour of the two structures can be reproduced by recent models based on d-wave pairing BCS theory to be discussed in the next Chapter.

A summary of the temperature dependence of the Cu relaxation rates in 1-2-4-8 for various fields is given in Fig. 18 [59]. It was found that W_c depends stronger on field than W_{ab}. The field dependence increases with decreasing temperature. Down to about $T = 0.7\,T_c$, an applied field slightly enhances W_{ab} as observed already in 1-2-3-7 [57] and also recently reported by Borsa et al. [60]. However, below 0.7 T_c the enhancement gives way to a suppression that becomes more evident at lower temperatures; this contrasts with the behavior of W_{ab} in 1-2-3-7. The opposite response of W_{ab} and W_c to the application of a magnetic field seems to rule out the possibility that fluxoid cores or thermally activated fluxoid motion cause the field dependence as it has been discussed for 1-2-3-7 [57]. To account for the opposite response of W_{ab} and W_c an unexpected field related breaking of the spin-rotation invariance in the superconducting state has to be considered. Finally, we want to stress the fact that in 1-2-4-8 the high field W_c is larger than the NQR rate from the lowest temperatures used in our experiment, up to $T_c + 13$ K (see Fig. 18). This is in contrast with a recent observation by Borsa *et al.* [60] who found in the temperature region just above T_c the opposite behaviour for 1-2-3-7 and $La_{1.85}Sr_{0.15}CuO_4$.

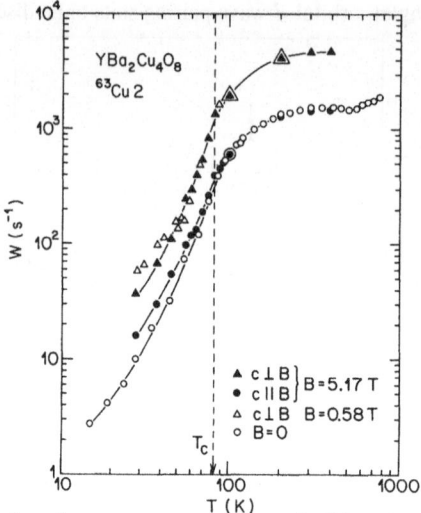

Fig. 18. Cu(2) spin-lattice relaxation rates *vs* temperature in $YBa_2Cu_4O_8$ for different magnetic fields and orientations. The triangles are for $B_0 \perp c$ and the circles for $B_0 \parallel c$.

9. Interpretation of Relaxation and Knight Shift

It is now widely believed that magnetic spin-lattice relaxation in several HTSC imply the existence of AF fluctuations and correlations. Inelastic neutron scattering experiments in 1-2-3-6+x (see, *e.g.*, [61-64]) have shown how these dynamic spin correlations survive once doping has destroyed the long-range magnetic ordering. The fluctuations peak near $\vec{Q} = (\pi/a, \pi/a)$ which is the AF ordering point in $\vec{q}$-space. Given these AF fluctuations, the relaxation rates of Cu, O and Y nuclei will follow the Moriya expression (eqn 7).

Based on these facts, Millis, Monien and Pines have put forward a phenomenological model (for short MMP) [65] which together with its variants [66,67], has quantitatively described the normal state relaxation of Cu, O and Y in 1-2-3-7, 1-2-3-6.63 and $La_{1.85}Sr_{0.15}CuO_4$. The main features of the MMP model are as follows. The hyperfine interactions between the electron spin and the nuclei are given by the Mila-Rice Hamiltonian (see Chapter 7). The electron spins interact anti-ferromagnetically with a finite temperature-dependent correlation length $\xi(T)$ which increases with decreasing temperature. The electron susceptibility, $\chi(\vec{q}, \omega)$, treated in a mean field approximation, consists of two parts: a quasi-particle (normal Fermi-liquid-like) contribution and a second one arising from AF correlations.

The MMP model gives a quantitative account of relaxation data at Cu, O and Y sites in 1-2-3-7 [65] provided that ξ^2 has approximately a Curie-Weiss like temperature dependence. Also the Cu(2) relaxation data in 1-2-4-8 could be fitted with a similar ξ [15], However, since neutron scattering investigations [62] concluded that χ is temperature independent the applicability of the MMP model has been questioned. Instead, several groups have suggested, inspired by neutron studies, that the development of a (pseudo) spin gap, Δ, in the magnetic excitation spectrum of the AF fluctuations is responsible for the temperature variation of the relaxation rate, at least in the normal conducting state.

Without concerning for detailed theoretical justification, the relaxation rate is written as (compare for instance [69,70])

$$1/T_1 \propto T^{-\alpha}[1 - \tanh^2(\frac{\Delta}{2T})] \tag{11}$$

with α values around 1.5. Fig. 19 shows a fit of eqn 11 to the Cu relaxation rates we measured in 2-4-7-15 [53]. However, it is noteworthy that a completely different approach based on a charge excitation in a two component Hubbard description [71] derives for the relaxation rate an expression mathematically identical to eqn 11 with the exception that Δ is replaced by the chemical potential. This confirms that the issue of NMR relaxation in the HTSC is not yet settled.

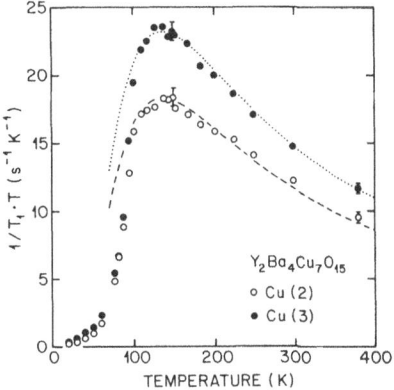

Fig. 19. Fit of eqn 11 to the Cu(2) and Cu(3) relaxation rates in $Y_2Ba_4Cu_7O_{15}$. The parameters of both fits are $\Delta = 240$ K and $\alpha = 1.25$.

This conclusion also applies to the superconducting state. Bulut and Scalapino [72,73], Lu [74] and Lu and Pines [75] showed that the Knight shift and relaxation data below T_c can be understood within a BCS framework which includes AF correlations, spin-singlet states but orbital-d-wave pairing. Fig. 16 shows such fits to the relaxation rate anisotropy which are based on a BCS spin-singlet, orbital-d-wave pairing state. As to the results of such fits to 1-2-3-7 data, we quote a summary from Ref. [76]: "(i) The Knight shifts of the Cu(2) show that the gap opens quickly as one lowers T below T_c. An unfortunate consequence is that in the low temperature region there is not enough precision to decide whether or not the orbital pairing is s- or d-wave. (ii) The Cu(1) Knight shifts have a less rapid drop as one cools below T_c. Near T_c, the data can fit either orbital s- or d-wave pairing, but at low T the experimental points are all fit by a d-wave calculation, and do not quite fit the s-wave result. (iii) The anisotropy in the spin-lattice relaxation rate of the Cu(2) nuclei is well explained by orbital d-wave pairing, and not by orbital *s*-waves." It remains to be shown how our Knight shift and relaxation data for 1-2-4-8 [53] fit into this context.

10. Conclusion

The examples we have presented have shown both the power and the diversity of the application of NMR-NQR techniques to a variety of topics in studies of high-temperature superconductors. Information on static and dynamic phenomena can be obtained on the atomic scale. Although we could cover only a limited range of the present activities, we have tried to demonstrate how these results of NMR-NQR investigations present significant contributions to the understanding of various phenomena in this class of new materials. We are convinced that NMR-NQR has not yet come to an end in this field of research.

References

[1] MacLaughlin DE (1976) in: Solid State Physics, vol. 31. Academic Press, New York, p. 1
[2] Pennigton CH, Slichter CP (1990) in: Ginsberg DM (ed) *Physical Properties of High-Temperature Superconductors*, vol. II. World Scientific, New Jersey
[3] Kuzmany H, Mehring M, Fink J (eds) (1990) *Electronic Properties of High-T_c Superconductors and Related Compounds*, Springer Series in Solid-State Sciences 99, Springer, Berlin
[4] Mehring M (guest ed) (1992) Appl. Magn. Reson. 3
[5] Brinkmann D, Mali M (1993) in: Blümich B (guest ed) *Solid State NMR*, NMR - Basic Principles and Progress, vol. 31, Springer, invited paper, to be published
[6] CP Slichter (1990) *Principles of Magnetic Resonance*, Springer, Berlin
[7] Mangelschots I, Mali M, Roos J, Brinkmann D, Rusiecki S, Karpinski J, Kaldis E (1992) Physica C 194: 277
[8] Mali M, Roos J, Brinkmann D (1988) Physica C 153-155: 737
[9] Mangelschots I, Mali M, Roos J, Stern R, Bankay M, Lombardi A, Brinkmann D (1993) in: Müller KA (ed) *Phase Separation in Cuprate Superconductors*, World Scientific, in press
[10] Lombardi A, Mali M, Roos J, Brinkmann D, Yakubowskii A: to be published
[11] Moriya T (1963) J. Phys. Soc. Jpn 18: 516
[12] Zimmermann H, Mali M, Brinkmann D, Karpinski J, Kaldis E, Rusiecki S (1989) Physica C 159: 681
[13] Mali M, Brinkmann D, Pauli L, Roos J, Zimmermann H, Hulliger J (1987) Phys. Lett. A 124: 112
[14] Pennington CH, Durand DJ, Slichter CP, Rice JP, Bukowski, Ginsberg DM (1989) Phys. Rev. B 39: 274
[15] Zimmermann H (1991) Ph.D. Thesis, University of Zürich
[16] Sahoo N, Markert S, Das TP, Nagamine K (1990) Phys. Rev. B 41: 220
[17] Sulaiman SB, Sahoo N, Das TP, Donizelli O, Torikai E, Nagamine K (1991) Phys. Rev. B 44: 7028
[18] Blaha P, Schwarz K, Herzig P (1985) Phys. Rev. Lett. 54: 1192
[19] Schwarz K, Ambrosch-Draxl C, Blaha P (1980) Phys. Rev. B 42: 2051
[20] Ambrosch-Draxl C, Blaha P, Schwarz K (1991) Phys. Rev. B 44: 5141
[21] Lombardi A, Mali M, Roos J, Brinkmann D, Yakubowskii A: to be published

[22] Yakubowskii A, Egorov A, Lütgemeier H (1992) Appl. Magn. Reson. 3: 665

[23] Brinkmann D (1990) in: Kuzmany H, Mehring M, Fink J (eds) *Electronic Properties of High-T_c Superconductors and Related Compounds*, Springer Series in Solid-State Sciences 99, p. 195, Springer, Berlin

[24] Mangelschots I, Zimmermann H, Meister HP, Mali M, Roos J, Brinkmann D, Kaldis E, Karpinski J, Rusiecki S (1990) in: Mehring M, von Schütz JU, Wolf HC (eds) 25 Congress AMPERE, Extended Abstracts, p. 280, Springer, Berlin

[25] Riesemeier H, Grabow C, Scheidt EW, Müller V, Lüders K, Riegel D (1987) Solid State Commun. 64: 309

[26] Riesemeier H, Kamphausen H, Scheidt EW, Stadermann G, Lüders K, Müller V (1990) in: Kuzmany H, Mehring M, Fink J (eds) *Electronic Properties of High-T_c Superconductors and Related Compounds*, Springer Series in Solid-State Sciences 99, p. 225, Springer, Berlin

[27] Yamada Y, Jorgensen JD, Pei S, Lightfoot P, Kodama Y, Matsumoto T, Izumi F (1991) Physica C 173: 185

[28] Alexandrov OV (1991) private communication

[29] Müller K, Mali M, Roos J, Brinkmann D (1989) Physica C 162 - 164: 173

[30] Schiefer H, Mali M, Roos J, Zimmermann H, Brinkmann D, Karpinski J, Kaldis E, Rusiecki S (1989) Physica C 159: 681

[31] Kaldis E, Fischer P, Hewat AW, Hewat EA, Karpinski J, Rusiecki S (1989) Physica C 159: 668

[32] Miyatake T, Gotoh S, Koshizuka N, Tanaka S (1989) Nature 341: 41

[33] Mangelschots I, Mali M, Roos J, Zimmermann H, Brinkmann D, Rusiecki S, Karpinski J, Kaldis E, Jilek E (1990) Physica C 172: 57

[34] Mangelschots I, Mali M, Roos J, Zimmermann H, Brinkmann D, Karpinski J, Kaldis E, Rusiecki S (1990) J. Less-Common Metals 164 & 165: 78

[35] Machi T, Tomeno I, Miyatake T, Tai K, Koshizuka N, Tanaka S, Yasuoka H (1991) Physica C 185 - 189: 1147

[36] Borges HA, Wells GL, Cheong SW, Kwok RS, Thompson JD, Fisk Z, Smith JL (1987) Physica B 148: 411

[37] Siegrist T, Sunshine S, Murphy D, Cava RJ, Zahurak SM (1987) Phys. Rev. B 35: 7137

[38] Balakrishnan G, Caves LJW, Dupree R, McPaul D, Smith ME (1989) Physica C 161: 9

[39] Alloul H, Mendels P, Casalta H, Marucco JF, Arabski J (1991) Phys. Rev. Lett. 67: 3140

[40] Warren Jr. WW, Walstedt RE, Bell RF, Schneemeyer LF, Waszczak J, Dupree R: preprint

[41] Walstedt RE, Warren Jr. WW (1992) Appl. Magn. Reson 3: 469

[42] Yosida K (1958) Phys. Rev. 110: 769

[43] Mila F, Rice T.M (1989) Physica C157: 561

[44] Zimmermann H, Mali M, Mangelschots I, Roos J, Pauli L, Brinkmann D, Karpinski J, Rusiecki S, Kaldis E (1990) J. Less-Common Metals 164&165: 138

[45] Mangelschots I, Mali M, Roos J, Brinkmann D, Rusiecki S, Karpinski J, Kaldis E (1992) Physica C 194: 277

[46] Bankay M, Mangelschots I, Mali M, Roos J, Brinkmann D, Kaldis E, Karpinski J (1993) Spring Meeting of Swiss Physical Society and to be published

[47] Takigawa M, Reyes AP, Hammel PC, Thompson JD, Heffner RH, Fisk Z, Ott KC (1991) Phys. Rev. B 43: 247

[48] Machi T, Tomeno I, Miyatake T, Koshizuka N, Tanaka S, Imai T, Yasuoka H (1990) Physica C 173: 32

[49] Walstedt RE, Bell RF, Mitzi DB (1991) Phys. Rev. B 44: 7760

[50] Alloul H, Mendels P, Collin G, Monod P (1988) Phys. Rev. Lett. 61: 746

[51] Alloul H, Ohno T, Mendels P (1989) Phys. Rev. Lett. 63: 1700

[52] Berthier C, Berthier Y, Butaud P, Clark WG, Gillet JA, Horvatič M, Ségransan, Henry JY (1992) Appl. Magn. Reson. 3: 449

[53] Stern R, Mali M, Mangelschots, Roos J, Brinkmann D, Heinmaa I, Genoud JY, Graf T, Muller J (1993) to be published

[54] Hebel LC, Slichter CP (1957) Phys. Rev. 107: 901

[55] Berthier C, Berthier Y, Butaud P, Clark WG, Gillet JA, Horvatič M, Ségransan, Henry JY (1992) Appl. Magn. Reson. 3: 449

[56] Barrett SE, Martindale JA, Durand DJ, Pennington CH, Slichter CP, Friedmann TA, Rice JP, Ginsberg DM (1991) Phys. Rev. Lett 66: 108

[57] Martindale JA, Barrett SE, Klug CA, O'Hara KE, DeSoto SM, Slichter CP, Friedmann TA, Ginsberg DM (1992) Phys. Rev. Lett 68: 702

[58] Takigawa M, Smith JL, Hults WL (1991) Phys. Rev. B 44: 7764

[59] Bankay M, Mali M, Roos J, Mangelschots I, Brinkmann D (1992) Phys. Rev. B 46: 11228

[60] Borsa F, Rigamonti A, Corti M, Ziolo J, Hyun O, Torgeson DR (1992) Phys. Rev. Lett. 68: 698

[61] Rossad-Mignod J, Regnault LP, Vettier C, Burlet P, Henry JY, Lapertod G (1991) Physica B 169: 58

[62] Rossad-Mignod J, Regnault LP, Vettier C, Bourges P, Burlet P, Bossy J, Henry JY, Lapertod G (1991) Physica B 185 - 189: 86

[63] Bourges P, Gehring PM, Hennion B, Moudden AH, Tranquada JM, Shirane G, Shamoto S, Sato M (1991) Phys. Rev. B 43: 8690

[64] Tranquada JM, Gehring PM, Shirane G, Shamoto S, Sato M (1992) Phys. Rev. B 46: 5561

[65] Millis AJ, Monien M, Pines D (1990) Phys. Rev. B 42: 167

[66] Monien M, Pines D, Takigawa M (1991) Phys. Rev. B 43: 258

[67] Monien M, Monthoux P, Pines D (1991) Phys. Rev. B 43: 275

[68] Brinkmann D (1991) Z. Naturforsch 47 a: 1

[69] Takigawa M (1992) Appl. Magn. Reson. 3: 495

[70] Mehring M (1992) Appl. Magn. Reson. 3: 383

[71] Eremin MV, Markendorf R, Zavidonov AY, Brinkmann D, Bankay M, Mali M, Mangelschots I, Roos J, Stern R (1993) submitted to Solid State Commun.

[72] Bulut N, Scalapino DJ (1992) Phys. Rev. Lett. 68: 706

[73] Bulut N, Scalapino DJ (1992 Phys. Rev. B 45: 2371

[74] Lu JP (1992) Modern Physics Letters B 6: 547

[75] Lu JP, Pines D (1992) private communication to C. P. Slichter

[76] Slichter CP, Barrett SE, Martindale JA, Durand DJ, Pennington CH, Klug CA, O'Hara KE, DeSoto SM, Imai T, Rice JP, Friedmann TA, Ginsberg DM (1992) Appl. Magn. Reson. 3: 423

INFRARED PROPERTIES OF SELECTED HIGH T_c SUPERCONDUCTORS.

Z. SCHLESINGER, R. T. COLLINS, L. D. ROTTER, F. HOLTZBERG, C. FEILD,
IBM Watson Research Center, Yorktown Heights, New York 10598
U. WELP(1,2), G. W. CRABTREE(2), J. Z. LIU* AND Y. FANG(2)
1.Science and Technology Center for Superconductivity, 2.Materials Science Division, Argonne National Lab, Argonne, Illinois 60439
*Physics Dept., University of California, Davis, California, 95616

ABSTRACT. We wish to investigate to what extent the infrared properties of cuprates are similar to and different from that of ordinary metals and superconductors. We therefore examine and discuss the infrared reflectivity and conductivity in the normal and superconducting states for various optimally doped and under-doped cuprates. We focus on the conductivity of the CuO_2 planes in compounds with high T_c, bulk superconductivity. There we find a very simple but unusual normal state phenomenology, as well as evidence for a transition to superconductivity that is very different from the conventional gap opening transition of BCS theory.

1. Introduction

The infrared region of the electromagnetic spectrum covers the range corresponding to about 5 to 20,000 K. Since this is the region in which the characteristic energy scales relevant to high T_c superconductors are expected to lie, infrared can be expected to provide a very useful probe of fundamental properties of the cuprate, bismuthate, organic and C_{60} based compounds. As mentioned in my talk, many groups throughout the world have worked on the the study of infrared and optical properties; references to that work can be found in references[1-3].

The basis for understanding the response of a solid to infrared or optical radiation, lies in the equations of motion for an electron in an oscillating electric (and magnetic) field in the presence of damping (scattering of electrons)[4-6]. The applied fields induce currents, and the relationship between the field and current is expressed in terms of a frequency dependent conductivity, $\sigma(\omega)$ ($J(\omega) = \sigma(\omega)E(\omega)$). (For strongly interacting systems including cuprates, heavy fermion metals and doped C_{60}, the mean free path is shorter than both the infrared wavelength and penetration depth, in which case infrared measurements probe a local (q = 0) response.) In the infrared, the conductivity, $\sigma(\omega)$, has both real and imaginary parts, corresponding to the in-phase and out-of-phase response to the applied field. Infrared conductivity can be simply viewed as an extension of ordinary d.c. conductivity (1/resistivity) to finite frequency.

The conductivity can be divided into two parts, one referred to as interband, the other intraband. The former is associated with transitions from one band to another, distinct band, i.e., transitions from occupied bands below the Fermi level, E_f, to unoccupied bands above E_f. These interband transitions occur in all materials, including insulators, semiconductors, and metals, since there are always

249

E. Kaldis (ed.), Materials and Crystallographic Aspects of HTc-Superconductivity, 249–263.

250

many filled bands below E_f and many empty bands above E_f. Intraband conductivity, on the other hand, involves an excitation right at the Fermi surface within a single band, and occurs only in compounds which have a partially filled band, i.e. metals. The interband excitations are typically important at high frequency, while the intraband response occurs at low frequency. It is the intraband part of $\sigma(\omega)$, which is centered at $\omega=0$, that is directly connected to the d.c. conductivity and to superconductivity.

2. Normal State

The simplest form for the intraband part of the conductivity is the Drude form[4-6], where the conductivity is characterized by a damping (scattering rate), an effective mass, and a carrier density, which are independent of frequency[7]. In this case the real part of the infrared conductivity falls like $1/\omega^2$ for frequencies larger than the scattering rate.

In Fig. 1 the reflectivity and the real part of the conductivity, $\sigma_1(\omega)$, are shown for ordinary copper[6]. The reflectivity of Cu is very high in the infrared, where the conductivity (fig. 1b) indeed follows the $1/\omega^2$ form of the Drude model. The

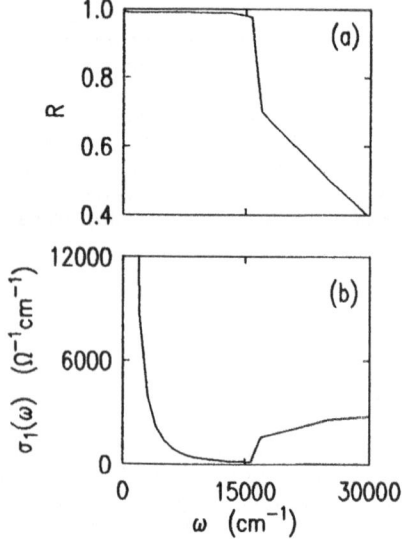

FIGURE 1
Reflectivity and conductivity for elemental copper (Cu). The onset for interband absorption occurs at about 16,000 cm^{-1}. Below about 16,000 cm^{-1} the conductivity is intraband, and follows the conventional $1/\omega^2$ form expected for an ordinary metal. In this region, the reflectivity is very high ($\sim$99%).

reflectivity begins to fall near 16,000 cm^{-1} (red) due to the onset of interband absorption at this energy. Cu thus does not reflect so well in the green-blue-violet part of the electromagnetic spectrum, which accounts for its reddish-yellow color (16,000 cm^{-1}=red). Below the threshold for interband transitions ($\sim$16,000 cm^{-1}) the conductivity is intraband conductivity. This conductivity (fig 1b) rises to a maximum value of order several hundred thousand at $\omega=0$, corresponding to the room temperature resistivity of roughly 3 $\mu\Omega$ – cm for Cu at 300 K. Above the scattering rate of roughly 300 cm^{-1}, $\sigma_1(\omega)$ falls as a function of frequency like 1/ω^2. This is the conventional behavior expected for a metal[6].

The reflectivity and conductivity for a cuprate superconductor in the normal state are shown in figure 2. Here the separation of the conductivity into distinct inter and intraband parts is less clear than in Cu (fig 1), as discussed particularly in[8]. We assume that in the region where $\sigma_1(\omega)$ drops smoothly and monotonically, e.g., $\omega \lesssim 4000$ cm^{-1}, the conductivity is primarily intraband. The really unusual aspect of these data is that there is no frequency range where the

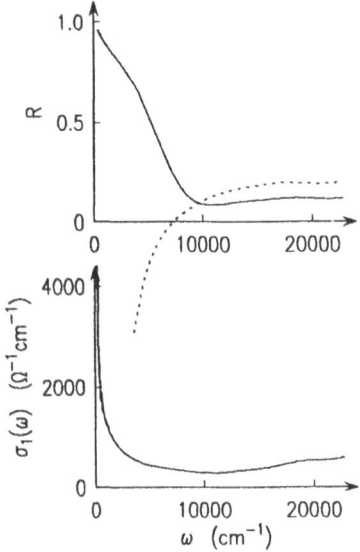

FIGURE 2

Room temperature reflectivity, conductivity and $\varepsilon_1(\omega)$ for the CuO_2 planes in the cuprate superconductor $YBa_2Cu_3O_7$ ($T_c \simeq 92$ K). (Measured with the electric field polarized along the $\hat{a}$ direction.) These spectra typical of cuprates in the optimal or moderately under-doped region. (The dotted curve is $\varepsilon_1(\omega)$, which crosses zero at about 7500 cm^{-1} and has a high frequency value of about 3.5 at 20,000 cm^{-1}.) $\sigma_1(\omega)$ (solid line) has a simple, but very unusual form, since it decreases like 1/ω at low frequency instead of the conventional 1/ω^2.

conductivity falls like $1/\omega^2$, but, instead, the conductivity falls like $1/\omega$ up to $\sim$8000 cm^{-1} (Actually the exponent may be closer to 0.8 if we wish to be very precise.) This simple and essential difference between the cuprates and ordinary metals turns out to be very important for a theoretical understanding of the normal state. Its implications are difficult to fully understand, and have been talked about in terms of Luttinger liquid theory, marginal Fermi liquid theory, strong scattering of charge by spin fluctuations, and many other approaches. This leads into a realm of great subtlety and difficulty, however, we should not let it distract us from the inherent simplicity of the phenomenology of the normal state: the conductivity in the infrared falls like $1/\omega$[9], which is much slower than the conventional $1/\omega^2$ form.

Having made the discussion of the normal state infrared conductivity very simple, let us digress now into some discussion of the complexity that can arise in actually measuring the conductivity. This discussion is not necessary for understanding the essential physics of the cuprates, but instead is relevant to the historical development of our understanding, and to certain material related complexities. Complexities in the study of infrared properties arise primarily from two sources. One is the inherent anisotropy of the cuprates. For all cuprates the conductivity is much lower in the $\hat{c}$ direction (perpendicular to the CuO_2 planes), than it is in the a-b plane. For most cuprates the conductivity is also different along the $\hat{a}$ and b directions in the plane. The study of samples with unresolved anisotropies, such as twinned $YBa_2Cu_3O_{7-x}$, or the extreme case of polycrystalline samples, has, historically, lead to confusion and ambiguity in the interpretation of infrared data. In figure 2 we have shown the reflectivity and conductivity measured with the infrared electric field polarized parallel to the $\hat{a}$ axis. With this polarization the electric field is perpendicular to the chains in $YBa_2Cu_3O_7$ and one probes only the response of the CuO_2 planes[3,9]. One can also measure the reflectivity along the b direction, in which case one probes both chain and plane response[1-3,8-11]. This makes the conductivity more complicated and confusing. We focus in fig. 2 and in this paper on the response (conductivity) of the CuO_2 planes because they are the essential ingredient for superconductivity.

The second source of complexity, has to do with the study of the infrared conductivity as a function of doping, particularly in the low doping region[12,13]. The conductivity evolves in complex ways as one dopes the half-filled CuO_2 band, and this evolution has been the subject of much study. At low doping strong localization effects (disorder) tend to suppress the d.c. and very low frequency part of the infrared conductivity. The issue of the relevant frequency scales and where and how the insulator-metal transition occurs as a function of doping remains unresolved.

Leaving these complex issues aside, we note again that for superconductors doped high enough to have T_c's of $\sim$50 K or above, the high temperature conductivity (normal state) has essentially the form shown in figure 2. Details of the changes in the normal state conductivity with T_c in the optimal and underdoped regimes are discussed in references [3,10,11]. The essential result is that in the infrared the normal state conductivity decreases much more slowly as a function of frequency, than the usual (Drude) $1/\omega^2$ form. Particularly for the optimally doped compounds, the phenomenology of the normal state can be summarized quite simply, as illustrated in fig 3. We know from d.c. resistivity measurements that $\sigma_{1n}(\omega)$ is proportional to 1/T at $\omega = 0$, while the infrared measurements show that $\sigma_{1n}(\omega)$ is proportional to $1/\omega$ (for $\omega \gtrsim 3kT$).

As described in references [3,9], the assumption that the low frequency conductivity is primarily intraband, leads naturally to the association of the slowly falling nature of the conductivity with a scattering rate that increases with frequency. This implied frequency dependence of the scattering rate is directly analogous to the temperature dependence of the scattering rate, which leads to the temperature dependence of the d.c. resistivity. As the doping level (and T_c) are changed, the resistivity changes in detail (the surprisingly linear dependence is observed near the optimum T_c's), however, a strong temperature dependence is always found. Similarly in the infrared conductivity the details of the frequency dependence change, while the data reflect a strong frequency dependence for $\tau^{-1}(\omega)$ over a wide range of doping[10,11]. Aside from effects of disorder (localization), which influence primarily the d.c. conductivity, a fundamental relationship between the frequency and temperature dependence of the scattering rate (which are both related to the imaginary part of the self-energy) is expected to be maintained. A possible origin for the frequency and temperature dependence observed, may lie in the physics of charge-spin interactions and the incumbent frequency and temperature dependent renormalizations.

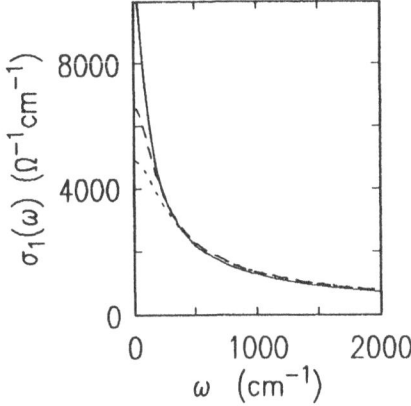

FIGURE 3

Schematic summary of the normal state conductivity in a cuprate superconductor. The dotted curve shows the conductivity at a high temperature ($\sim$300 K); the dashed curve shows $\sigma_1(\omega)$ at an intermediate temperature (200 K); and the solid curve shows $\sigma_1(\omega)$ just above T_c ($\sim$100 K). At $\omega = 0$ the value of the conductivity goes like 1/T, reflecting the linear T dependence of the d.c. resistivity. The temperature dependence is strong for $\omega \lesssim 3kT/\hbar$ Above this frequency the conductivity falls with the highly unusual form, $\sigma_1(\omega) \sim 1/\omega$.

Another approach has been to fit the conductivity using a conventional Drude term and a sequence of Lorentz oscillators. This can be used to generate a reasonable fit, however, many free parameters are required, and the lowest oscillator frequency is typically around 300 cm^{-1}. Since this is too low to be plausibly associated with an interband transition, it is not clear what the physical significance of the individual terms are. Another problem with this approach, which is discussed in ref.[3,8], is that it provides no fundamantal connection between the temperature dependence of the d.c. conductivity and the frequency dependence of the infrared conductivity. The more generally accepted view is that the conductivity below at least 1000 or 2000 cm^{-1} is all intraband, and that the "Drude" part of the conductivity is unconventional, as discussed above.

3. Superconducting State

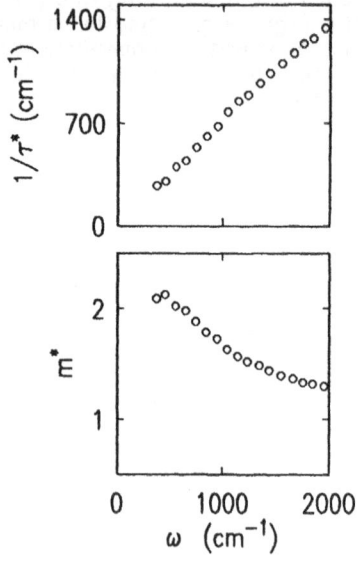

FIGURE 4

Scattering rate vs frequency as inferred from the unusually slow falling form ($\sim 1/\omega$) of the infrared conductivity. Part b shows the incumbent frequency dependent enhancement of the carrier effective mass. (data from YBa$_2$Cu$_3$O$_7$). Essentially the same frequency dependence can be inferred from Bi$_2$Sr$_2$CaCu$_2$O$_{8-y}$ and La$_{2-x}$Sr$_x$CuO$_4$ data. The linear dependence of the scattering rate on frequency is directly related to the linear temperature dependence of the d.c. resistivity.

Let us now consider the superconducting state. There the conductivity and reflectivity change dramatically in frequency range up to roughly 500 cm^{-1} for the cuprates. We will first provide some introduction to the changes in the infrared expected for ordinary (BCS) superconductors.

In the BCS model, a gap opens at the Fermi level in the density of states. At low (0) temperature the density of states is zero for an energy range Δ above and below E_f. The missing density of states from the gap region "piles up" just above and below the gap, forming the famous BCS density of states peaks[15]. With this density of states gap, the intraband conductivity will be zero up to the energy 2Δ, which is the lowest energy to excite an electron from the filled levels below E_f to the empty levels above E_f. This excitation corresponds to breaking a superconducting pair, and is thus referred to as the pair excitation energy or threshold.

In the BCS theory the gap becomes smaller with increasing temperature (due to effects of thermal quasiparticles). This reduction of the gap becomes significant above about $0.6T_c$, and the gap collapses to zero as T reaches T_c. In the BCS theory, the creation of pairs, and the condensation of those pairs to form a superfluid are precisely coincidental. More generally, one can have preformed pairs for which the onset of phase coherence occurs at one temperature, while the pairs may begin forming at another, higher temperature. In this case the transition to the superfluid state is referred to as Bose condensation, while in the BCS case the transition is described as a Fermi surface instability. The usual example for Bose condensation is He4, where the superfluid transition occurs near 2 K. Roughly speaking, the pair excitation energy would correspond to the ionization energy to remove one electron from the He4. This energy is so large that we tend not to think of it when we consider the superfluid transition of He4. He4 and BCS superconductivity thus provide two extreme cases: one where the pairs form at an energy which is essentially infinitely larger than the superfluid transition temperature, T_c; and the other, a mean field transition in which there is no pairing at all above T_c. As we discuss below, some of the data from cuprates can be interpreted in terms of an in-between scenario, in which pair correlations persist into the normal state, but with pairing energy which is not so extremely large compared to T_c as one finds in He4. We will discuss first some general aspects of gap measurement in superconductors, and then proceed to examine the unusual aspects of the optimally-doped and under-doped cuprates.

Figure 5 shows measurements of the ratio of the reflectivity in the superconducting state to that in the normal state for a variety of superconducting compounds[16-19]. In each case the reflectivity is enhanced at low frequency in the superconducting state (due to the reduced value of the conductivity in the gap region.) One can infer a superconducting gap (or, more precisely, a characteristic energy) from the frequency range over which the enhancement occurs. For all the compounds shown in figure 5 except $YBa_2Cu_3O_7$, one obtains an energy scale of about 3 to 4 kT_c from the reflectivity enhancement. This is the expected range for the mean field BCS theory. $YBa_2Cu_3O_7$ is unique in that the energy scale inferred from the infrared reflectivity enhancement is $\sim 8kT_c$. Given the context that for all previously studied superconductors the energy gap has fallen in the range from 3.5 to 4.6kT_c, this result from $YBa_2Cu_3O_7$ is quite unusual.

The energy gap can be studied more rigorously with infrared measurements by looking at the conductivity as a function of frequency, as we did in the normal

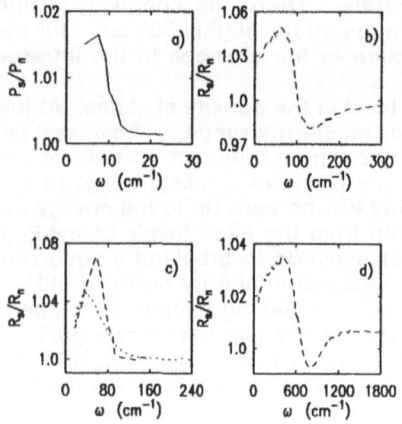

FIGURE 5

For four types of superconductors,(a) Sn (conventional), (b) $Ba_{0.6}K_{0.4}BiO_3$ (bismuthate), (c) Rb_3C_{60} (Fulleride) and (d) $YBa_2Cu_3O_7$ (cuprate), we show reflectivity in the superconducting state divided by that in the normal state[16-19,1]. The enhancement of the reflectivity at low frequency is generally associated with the development of an energy gap or a region of reduced loss below a characteristic pair breaking energy.

state. Figure 6 shows a comparison of the infrared evidence for an energy gap in Pb ($T_c \sim$ 7 K) (from Palmer and Tinkham), and for $YBa_2Cu_3O_7$ (untwinned, probing only the CuO_2 planes)[9,21]. For Pb, the measurement shows that at least 90 % of the conductivity below the pair-excitation threshold of about 22 cm^{-1} has disappeared in the superconducting state. In the BCS theory, there is no absorption below this energy, and we assume therefore that $\sigma_1(\omega)$ is 0 (at T = 0) below this gap energy of 22 cm^{-1}. (Tunneling measurements of Pb also show close to zero density of states in this energy range.) For $YBa_2Cu_3O_7$, the infrared measurements (fig 6b) show that at least 80 % of the conductivity below 500 cm^{-1} disappears in the superconducting state. Whether there is any intrinsic absorption below this frequency is not settled by the experiment, and as yet there is no theory for the gap in cuprates. With experimental uncertainties related both to the measurements and to the quality of the samples and their surfaces, it is quite possible that most or all of the conductivity below 500 cm^{-1} in $YBa_2Cu_3O_7$ is extrinsic or phonon related. In any case, it is clearly appropriate to associate the rapid rise of $\sigma_1(\omega)$ at 500 cm^{-1} with a pair excitation threshold, i.e., the energy to break apart the superconducting pairs. Again we emphasize that this energy, 500 cm$^{-1} \simeq$ 60 meV $\simeq$ 700 K, is unusually large relative to T_c (corresponding to $2\Delta/kT_c \simeq 8$).

One can speculate about why the gap might be so large compared to T_c. One possibility, in principle, would be extremely strong coupling, however, this can be basically ruled out by a variety of experiments. A more generic approach is to postulate that there is a mean field T_c of order 200 K for the bi-layer cuprates (i.e., about 1/3.5 of the gap temperature), and that something is suppressing T_c. The idea here is that based on the size of the gap at low temperature, T_c would be expected to be about 200 K (using $T_c \simeq 2\Delta/3.5k$), thus a lower T_c could mean that something is happening to make the material go normal at a lower temperature than expected based on mean field theory. Although the coherence lengths is short, the discrepancy between the hypothetical mean field T_c and the real T_c is probably too large to be accounted for by orinary fluctuation effects. One possibility is to invoke a temperature dependent pair-breaking scattering, which becomes strong as $T \rightarrow T_c$. This could arise from a charge-spin interaction, which is weak at low temperature when the spin spectrum is gapped, but is strong in the

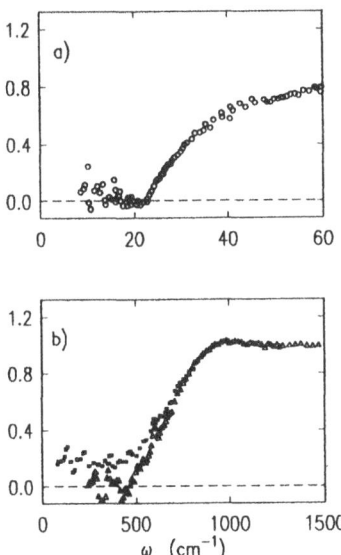

FIGURE 6

For the conventional superconductor, Pb, and for $YBa_2Cu_3O_7$, we show the conductivity as a function of frequency at low temperature normalized by the conductivity in the normal state. For $YBa_2Cu_3O_7$ data from both reflectivity[9] and absorptivity[21] measurements are shown. The threshold at about 22 cm^{-1} for Pb (a) is interpreted as transition across the energy gap due to the breaking of a superconducting pair. A similar, although somewhat sharper, threshold seen in $YBa_2Cu_3O_7$ (b) at 500 cm^{-1} is also usually interpreted as an energy gap or superconducting pair-excitation energy. From the experiments one one cannot tell whether the conductivity is exactly zero below the excitation threshold. Differences in the results of reflectivity and absorptivity measurements may be associated with a flux layer on the surface or other sample related issues.

normal state. (Recall that we talked about spin scattering possibly being responsible for the unusual frequency dependence of the normal state intraband conductivity.) This description is probably too superficial, however, these may be interesting things to wonder about even if the physics may not be fully understood. These speculations also suggest that the nature of the transition from the normal state may be unusual, or at least very different from the mean field (gap-opening) transition of BCS theory. Let us now return to the description of the phenomenology, with a look at the temperature dependence of the infrared conductivity in the vicinity of the normal to superconducting state transition.

In figure 7a, we show the temperature dependent development of the infrared conductivity near and below T_c for the optimally doped ($T_c \sim 92$ K) $YBa_2Cu_3O_7$. In the normal state we see the conductivity falling approximately like $1/\omega$, as discussed in the last section. In the superconducting state, we see the gradual disappearance of the of the conductivity in the range $\omega \lesssim 500$ cm^{-1} as the temperature is reduced. One unusual aspect of these data is that one sees no evidence for the gap getting smaller close to T_c. This is generally a difficult thing to see, because one has to get get very close to T_c to see a significant reduction, however, with these data one can see a conductivity depression even at T$=$90 K (0.98 T_c), and the energy scale is not noticeable smaller than at 20 K. This is very different from the expectation from conventional BCS theory, where the superconducting transition involves the opening of an energy gap, i.e., the gap is 0 at T_c and then grows to its full value as temperature is lowered.

Thus far, we have seen two unusual aspects to the gap; both its size and its temperature dependence are very different from BCS theory expectations. It has been suggested that the 500 cm^{-1} feature is too unusual, and that it is not directly related to the superconductivity or to a superconducting gap. Possible alternative interpretations are discussed in reference[3]. We feel, however, that the comparison of figure 6, and the temperature dependence of figure 7, provide too strong a case to seriously consider not associating this feature with the superconducting state, presumably as a pair excitation threshold or gap. The issue of whether the conductivity is exactly zero up to 500 cm^{-1} is one that cannot be resolved by experiment. Independent of this question, one can confidently say that 500 cm^{-1} is the characteristic energy for the CuO_2 plane superconducting state. Functionally this means that it is the energy that is related to the (a-b plane) coherence length by a characteristic (Fermi) velocity, and that it is related to a peak in the density of states at half this energy below the Fermi level.

Let us digress for a moment to look at the relationship between the infrared conductivity and the penetration depth, or superfluid density. We have talked about the infrared conductivity disappearing at low frequency in the superconducting state. Actually there is a sum rule on the conductivity, so it cannot really vanish, but must go somewhere. Where it goes is into a Dirac-δ function at $\omega = 0$ which represents the dissipationless response of the superconducting condensate. (A δ-function is an infinitely sharp peak, hence $\sigma_1(\omega)$ is infinite at $\omega = 0$.) The superfluid density is directly proportional to the area of $\sigma_1(\omega)$ that goes into the δ-function (the "missing area"). The penetration depth is inversely proportional to the square root of this "missing area". Normally one measures penetration depth directly, using the low frequency methods discussed by Professor Keller at this

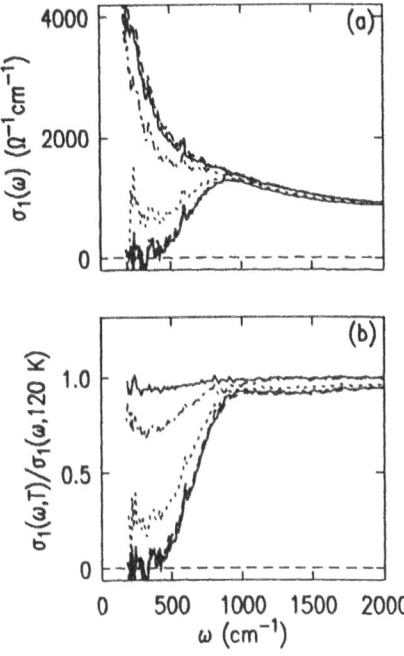

FIGURE 7
Temperature dependence of the infrared conductivity of YBa$_2$Cu$_3$O$_7$ (T$_c$≃92 K) near and below T$_c$. Part a) shows the absolute conductivity, part b) shows each spectrum normalized by the conductivity at 120 K.

School. One can, however, as a consistency check, calculate the penetration depth from the area missing from $\sigma_1(\omega)$ in the infrared. The result is about 170 nm at low temperature, which agrees reasonably well with more direct measurements of λ discussed by Prof. Keller. The temperature dependence of the penetration depth and the infrared conductivity are also intimately related. For T between about 50 and 90 K, the infrared conductivity fills in the gap region. This means that there is less missing conductivity going into the δ-function, and hence a longer penetration depth. The temperature dependence of the infrared conductivity and the penetration depth, which grow as T approaches T$_c$, are actually in reasonably good correspondence[---]. These observations indicate, however, that there is no simple relation between the temperature dependence of the penetration depth and the size of the gap, as there is in BCS theory. Instead the T dependence of λ reflects a filling-in of the gap (which can be described as a growth of a imaginary part of Δ), and a phenomenology that is very different from that of BCS theory.

Thus far we have simplified the discussion of the energy gap by focussing only on the CuO_2 plane response and avoiding the complexities associated with anisotropy. In general, the response along the low conductivity, c-axis direction can be different from that of the a-b plane[1], and the energy scales inferred from changes below T_c tend to be lower. The chains (along the b direction) also provide added complexity, and any gap development in the chains seems to occur at much lower frequency than for the planes[8]. Presumably the CuO_2 planes are of primary agent for superconductivity and the whatever modest changes occur on the chains are induced and of only peripheral importance.

Leaving these more technical aspects aside, the frequency and temperature dependence of the in-plane infrared conductivity can be summarized very simply and succinctly for the 90 K (optimally doped) superconductors. The essential aspects of the normal state are embodied in figure 4, where the d.c. conductivity is proportional to $1/T$, while the infrared conductivity falls like $1/\omega$. In the superconducting state (figure 7) (most of) the conductivity below 500 cm^{-1} disappears at low temperature, going into the δ-function at $\omega = 0$; the temperature dependence is unusual in that the gap does not appear to get smaller as $T \rightarrow T_c$. In figure 7b, the strongest evidence evidence for the gap remaining large close to T_c comes from the spectrum at 90 K, where in the conventional scenario one would expect the gap to be very small because it is so close to T_c. Actually it is surprising that we can see anything at all related to the gap in this spectrum, taken only 2 degrees below T_c. (The temperature measurement is accurate to about 1 K.) In fact, if we look at spectra at 95 K or even 100 K, both temperatures in the normal state, one can still see a subtle remnant of a gap like feature[1]. Such a persistence of a gap-like feature into the normal state is quite surprising, however, the effects are subtle and the reader would be well justified in being skeptical if this were the only evidence. What we find, however, is that this is a foreshadowing of an aspect of the data which becomes much more extreme in the underdoped (reduced T_c) samples.

For the underdoped $YBa_2Cu_3O_{7-x}$ gap-like feature seems to persist well above T_c. This has been discussed, for example, by Cooper et al.[22], Orenstein et al.[2], and Rotter et al.[10]. The latter work is the most clear because it involves untwinned crystals, in which ambiguities associated with the confusion of chain and CuO_2 plane related aspects of the data can be resolved. Figure 8 shows the conductivity as a function of frequency and temperature for three untwinned crystals, beginning with an optimally doped sample($T_c \simeq 92$ K), and moving to progressively more underdoped ($T_c \simeq 80$ and 56 K) crystals. For each sample, the dotted curve shows the temperature closest to T_c. One can see that as the crystals become more underdoped, the persistence of a gap-like feature in the infrared conductivity above T_c becomes more clear. For example, for the $T_c \simeq 56$ K crystal, the feature in $\sigma_1(\omega)$ at ~ 500 cm^{-1} does not disappear at T_c, but instead fades away gradually with increasing temperature in the normal state.

The temperature dependence of the infrared data shown in figure 8 is quite unusual, and is strikingly similar to that of the NMR relaxation rates[10], which also show a relatively sharp transition for optimally doped samples and a much more gradual temperature dependence extending into the normal state for the under-doped crystals. This relationship, which is discussed in reference[--], suggests that there may be some fundamental significance in these anomalous data, and that it is not just a reflection of systematic error or ambiguous interpretation associated with a particular measurement technique. Resistivity measurements also seem to show a closely related change in the their temperature dependence with doping. That is, for the optimally doped samples the resistivity divided by T is essentially constant from high temperature down to about 10 K above T_c, while for underdoped crystals[23] this quantity begins to decrease well above T_c ($\sim$150 K). Taken together, the infrared and d.c. data can be interpreted in terms of a narrowing of the low frequency peak in $\sigma_1(\omega)$ due to a reduction in scattering.

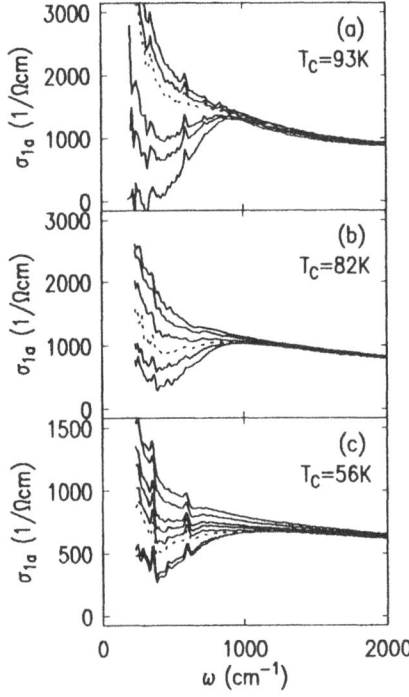

FIGURE 8
Temperature dependence of the infrared conductivity is show for an optimally doped crystal (a), and two under-doped crystals, (b) and (c). These YBa$_2$Cu$_3$O$_{7-x}$ crystals are untwinned and have T_c's of 92, 80 and 56 K, respectively. In each case the dotted curve shows the spectrum at the temperature closest to T_c.

262

For the under-doped compounds, the NMR, infrared and d.c. data have all been discussed in terms of the opening of a spin gap above T_c. For NMR the association is fairly direct since NMR probes the spin relaxation rate, which can be expected to decrease when a spin gap opens. For the charge related probes one can adopt the view that some sort of charge-spin interaction is primarily responsible for establishing the d.c. resistivity and for the frequency dependence of the infrared conductivity. The opening of a spin gap will thus lead to a reduction of scattering (at low frequency). This will allow the low frequency (d.c.) conductivity to increase, while at the same time causing a reduction in the conductivity at higher frequency (up to ~the spin gap energy), due to the reduction of inelastic scattering. Thus a "spin gap scenario" allows at least a hand-waving sort of understanding of the data. An actual theory remains to be worked out, and may be difficult because the spin and charge degrees of freedom are not neither completely distinct (as the phonon and charge degrees of freedom are in the Eliashberg theory), nor do they seem to be exactly the same for the cuprates. The experimental observations seem to be very consistent with the Fukuyama-Nagosa-Lee picture in which spin gap develops above T_c, especially in the reduced T_c compounds.

To summarize, for the $T_c \sim 90$ K superconductors the conductivity in the normal state falls like $1/\omega$; while in the superconducting state most ($\gtrsim 80$ %) of the conductivity below ~ 500 cm^{-1} disappears (into the superconducting δ-function). These data are unusual in the the rate of decrease in the normal state is much slower than the conventional $1/\omega^2$, and the energy scale for the superconducting state is about $8kT_c$, instead of the conventional 3.5-$4kT_c$. The temperature dependence in the superconducting state is unusual in that the gap does not seem to diminish in magnitude as $T \rightarrow \& Tc$, but instead "fills-in from below", which indicates a growth of the imaginary part of Δ rather the a collapse of the real part, as expected in BCS theory. A further unusual aspect is that a remnant of the gap seems to persist slightly (~ 10 K) above T_c. It is reasonable to presume that the unconventional size, temperature dependence and persistence above T_c are interrelated.

For underdoped crystals the persistence of a gap-related feature to temperatures well above T_c becomes more striking and dramatic. Infrared, NMR and d.c. resistivity all seem to follow the same trend in the changes in temperature dependence with doping, from a relatively sharp transition for optimally doped crystals, to a much more gradual transition with strong suggestions of pairing tendencies well above T_c for underdoped compounds. Taken together, the data suggest a phenomenology for the cuprate superconductors that is significantly different from BCS theory, and a transition that may lie somewhere between the extremes of the BCS Fermi surface instability and the Bose condensation of preformed pairs.

Acknowledgements: The authors acknowledge valuable conversations with P. W. Anderson, E. Abrahams, B. Bucher, H. Fukuyama, M. P. A. Fisher, P. A. Lee, P. B. Littlewood, T. M. Rice, D. J. Scalapino, C. P. Slichter, and C. M. Varma

References:

1. Z. Schlesinger et al., Phys. Rev. B41, 11237 (1990)

2. J. Orenstein et al., Phys. Rev. B42, 6342 (1990)

3. Z. Schlesinger et al., Physica C185-189, 57 (1991)

4. N. W. Ashcroft and N. E. Mermin, Solid State Physics, Holt, Rinehart and Winston, New York, 1976

5. J. M. Ziman, Principles of the Theory of Solids, Cambridge University Press, London, 1964

6. Frederick Wooten, Optical Properties of Solids, Academic Press, San Francisco, 1972

7. In general, inelastic scattering can make both the scattering rate and the carrier effective mass dependent on frequency. This can be treated in the context of a generalized Drude formula, in which τ and m^* are functions of ω (and are related by a Kramers-Kronig relation[6]). Effects of strong disorder (strong localization) can cause the intraband conductivity to be very small or zero at low frequency, and to initially increase with frequency. These effects, which are most important in low or moderately doped samples where the screening is poor, lead to a more complex form for $\sigma(\omega)$ which is not describable in terms of any Drude-type model.

8. Z. Schlesinger et al., in Physics of High Temperature Superconductors (Solid State Sciences 106), Editors S. Maekawa and M. Sato, Springer-Verlag, Berlin, 1992

9. Z. Schlesinger et al., Phys. Rev. Lett.65, 801 (1990)

10. L. D. Rotter et al. Phys. Rev. Lett.67, 2741 (1991)

11. S. L. Cooper et al., Phys. Rev. B45, 2549 (1992)

12. S. Uchida, Physica C185-189, 28 (1991)

13. S. Uchida et al., Phys. Rev. B43, 7942 (1991)

14. K. Kamaras et al., Phys. Rev. Lett.64, 84 (1990), and Erratum, ibid, 1962 (1990)

15. M. Tinkham, Introduction to Superconductivity, Robert E. Krieger publishing company, Malabar, Florida (and McGraw Hill), 1975

16. P. L. Richards and M. Tinkham, Phys. Rev.119, 575 (1960)

17. Z. Schlesinger et al., Phys. Rev. B40, 6862 (1989)

18. L. D. Rotter et al., Nature355, 532 (1992)

19. L. Degiorgi et al., Phys. Rev. Lett.69, 2987 (1992)

20. L. H. Palmer and M. Tinkham, Phys. Rev.165, 588 (1968)

21. T. Pham et al., Phys. Rev. B44, 5377 (1991)

22. S. L. Cooper et al., Phys. Rev. B40, 11358 (1989)

23. B. Bucher et al., Phys. Rev. Lett.70, 2012 (1993)

References:

1. Z. Schlesinger et al., Phys. Rev. B41, 11237 (1990).
2. Orenstein et al., Phys. Rev. B42, 6342 (1990).
3. Z. Reihmeyer et al., Physica C185-189, 67 (1991).
4. N. W. Ashcroft and N. D. Mermin, Solid State Physics, Holt, Rinehart and Winston, New York, 1976.
5. J. M. Ziman, Principles of the Theory of Solids, Cambridge University Press, Cambridge, 1964.
6. Frederick Wooten, Optical Properties of Solids, Academic Press, San Francisco, 1972.

PROBING HIGH-TEMPERATURE SUPERCONDUCTIVITY WITH POSITIVE MUONS

H. KELLER
Physik-Institut der Universität Zürich
Schönberggasse 9
CH-8001 Zürich
Switzerland

ABSTRACT. A short review on some recent muon-spin rotation (μSR) investigations on cuprate superconductors is presented. The possibilities and limitations of the μSR technique for studying the microscopic magnetic properties of these materials are briefly discussed. In the superconducting state, the local magnetic flux distribution $p(B)$ can be measured in the bulk of the sample, yielding valuable information on the complex vortex structure in these layered materials. In the mixed state the magnetic penetration depth can be deduced from the measured $p(B)$. Recent measurements of the angular-dependent $p(B)$ in single-crystal $YBa_2Cu_3O_{7-\delta}$ and $Bi_2Sr_2CaCu_2O_{8+\delta}$ and of the magnetic penetration depth in various cuprates are reported. A 'universal' parabolic relationship between the transition temperature T_c and the zero-temperature μSR relaxation rate (condensate density) is discussed, implying common trends in the pressure dependence of T_c and in the isotope effect for a particular class of doped cuprates. These empirical relations have important consequences for the microscopic theory.

1. Introduction

The basic pairing mechanism responsible for superconductivity in the perovskite-based copper oxides and the origin of the high transition temperatures remains elusive in spite of the fact that many novel models have been proposed [1]. Only a detailed knowledge of the normal-state and superconducting properties of the CuO_2-based systems can be helpful in discussing possible pairing mechanisms of superconductivity and to test theoretical models.

Over the past fifteen years the muon-spin rotation (μSR) technique has been developed into a powerful and well recognized tool for investigating a large variety of problems in condensed matter physics [2,3]. Extensive μSR investigations on cuprate and related systems have demonstrated that this rather exotic method yields important information on the microscopic magnetic properties of these novel materials, which are not easily obtained with standard experimental techniques [3-11]. In contrast to decoration experiments, NMR and ESR which probe local fields near the surface of a superconductor, μSR provides a sensitive *microscopic* probe (spin-polarized positive muons μ^+) of local magnetic fields in the *bulk* of a superconducting specimen. The local fields can be observed by means of muon-spin precession or relaxation. In this respect μSR has many similarities to NMR and ESR. In particular, μSR allows a direct measurement of the magnetic flux

E. Kaldis (ed.), Materials and Crystallographic Aspects of HTc-Superconductivity, 265–288.
© 1994 *Kluwer Academic Publishers.*

distribution in the mixed state of a type-II superconductor from which the magnetic penetration depth λ - one of the fundamental length of a superconductor - can be extracted. Since λ is related to the superconducting order parameter, the temperature dependence of λ is a potential probe of the pairing state (symmetry of the wave function, coupling strength). Moreover, μSR can also be used to investigate the complex vortex structure and vortex dynamics in the highly anisotropic layered cuprate systems on a *microscopic* scale. As a result of this rapidly growing field, an enormous number of papers on this subject has appeared in the literature [3-11]. Here a representative, but very incomplete selection of some typical examples of recent μSR experiments in cuprate superconductors is presented, demonstrating the potential and unique applications of this technique [12-20].

The organization of the paper is as follows: In Section 2 the basic principles of the μSR technique, insofar as is relevant to the interpretation of the following experiments, are briefly described. A short description of the magnetic flux distribution in the mixed state of an extreme type-II superconductor (isotropic and anisotropic case) and its relation to the magnetic penetration depth is given in Section 3. In Section 4 a systematic μSR investigation of the magnetic penetration depth in various cuprate superconductors is presented [12-17]. In addition, recent μSR measurements of the angular-dependent local magnetic flux distribution in single-crystal $YBa_2Cu_3O_{7-\delta}$ and $Bi_2Sr_2CaCu_2O_{8+\delta}$ are reported, revealing some interesting results concerning the flux-line structure in these materials [18-20]. 'Universal' relations and trends in extreme type-II superconductors are discussed in Section 5. Uemura *et al.* [21-23] found empirically a remarkable relationship between the transition temperature T_c and the zero-temperature μSR relaxation rate ('Uemura plot'). Motivated by this relationship we recently proposed a simple 'universal' scaling ansatz which connects the rescaled transition temperature with the rescaled condensate density [24,25]. This ansatz implies important consequences for the pressure dependence of T_c and the oxygen isotope effect in the cuprates and provides essential constraints on the microscopic theory [24-26]. The summary follows in Section 6.

2. Basic Principles of the μSR Technique

The positive muon is a lepton with a mass of $m_\mu \simeq 207 m_e$ (electron mass m_e), spin $S = 1/2$ and a finite lifetime of $\tau_\mu \simeq 2.2$ μs. For convenience, some of the fundamental properties of the positive muon are summarized in Table 1. From a particle physics point of view, the μ^+ may be considered as a heavy positron (e^+). In a solid, however, positive muons behave more like light unstable protons ($m_\mu \approx m_p/9$). Due to its positive charge, the muon is repelled by the nuclei in a solid and therefore goes preferentially to interstitial regions in the lattice and serves, due to its magnetic moment of $\mu_\mu \simeq \mu_B/207$, as a *microscopic* magnetic probe of the local magnetic field B_μ seen by the muon.

The basic principle of a standard transverse-field μSR experiment is illustrated schematically in Fig. 1a. Nowadays highly spin-polarized muon beams are available which are polarized either parallel (antiparallel) or perpendicular to the muon momentum. Here we discuss the situation for μ^+ with their spins antiparallel to their momentum. In a standard μSR experiment, spin-polarized μ^+ are implanted into a sample, which is situated in an external magnetic field B_{ext} perpendicular to the initial muon-spin polarization $P_\mu(0)$. During the short thermalization process ($< 10^{-10}$ s) in the sample the muons are not depolarized and subsequently undergo Larmor precession in the local magnetic field B_μ (at the muon site) with a precession frequency $\omega_\mu = \gamma_\mu B_\mu$, where γ_μ

Table 1: Some properties of the positive muon.

Property	Value
Rest mass m_μ	105.658 MeV/c^2
	206.768 m_e
	0.1124 m_p
Charge q	$+e$
Spin S	1/2
Magnetic moment μ_μ	$4.836 \times 10^{-3} \mu_B$
	3.183 μ_p
Gyromagnetic ratio $\gamma_\mu/2\pi$	135.5387 MHz/T
Lifetime τ_μ	2.197 μs

is the gyromagnetic ratio of the muon (see Table 1). The implanted muons decay after a mean lifetime of 2.2 μs by emitting a positron (e^+) and two neutrinos ($\nu_e, \bar{\nu}_\mu$) according to the decay scheme indicated in Fig. 1b. Because of parity violation in this decay process, the emission probability of the decay positrons with respect to the muon-spin direction is anisotropic (see Fig. 1b) and therefore allows a direct observation of the muon-spin precession by detecting the number of decay positrons as a function of time after a muon has stopped in the sample. This can be realized by using an arrangement of particle detectors (M, E_1, and E_2) as shown schematically in Fig. 1a. Before entering the sample, a muon is registered by a muon detector (M) and starts a clock (t_{start}). The muon stops in the sample and starts to precess in the local magnetic field B_μ at the position of the muon. After a certain time (mean lifetime 2.2 μs) the muon decays. When the corresponding decay positron is detected by one of the two positron detectors (E_1 or E_2), the clock is stopped (t_{stop}). The μSR signal is then obtained by counting the number of decay positrons N_{e^+} as a function of the time difference $t = t_{stop} - t_{start}$ and accumulating these events in a memory. As an example, a typical μSR time spectrum of copper taken in a magnetic field of 50.6 mT is displayed in Fig. 1a. One can clearly observe oscillations due to muon-spin precession superimposed on the exponential muon decay. In general, the observed μSR time spectrum may be written in the following form [2]:

$$N_{e^+}(\theta, t) = N_o \exp(-t/\tau_\mu)[1 + A\,R(t)\cos(\omega_\mu t + \theta)] + b, \qquad (1)$$

where N_0 is a normalization constant, τ_μ is the muon lifetime, A is the precession amplitude, $\omega_\mu = \gamma_\mu B_\mu$ is the muon-spin precession frequency, θ is the initial phase, and b is a time-independent background. For the positron counters E_1 and E_2 the initial phase is $\theta = 0$ and π, respectively. The relaxation function $R(t)$ describes the damping of the precession signal. In the special case of a Gaussian distribution of static local fields $p(B_\mu)$ the relaxation function has the simple form [2]:

$$R(t) = \exp(-\sigma^2 t^2). \qquad (2)$$

Here σ is the muon-spin relaxation rate which is related to the second moment of $p(B_\mu)$ according to the expression:

$$\langle \Delta B_\mu^2 \rangle = \langle B_\mu^2 \rangle - \langle B_\mu \rangle^2 = 2\sigma^2/\gamma_\mu^2, \qquad (3)$$

MUON SPIN ROTATION (μSR) TECHNIQUE

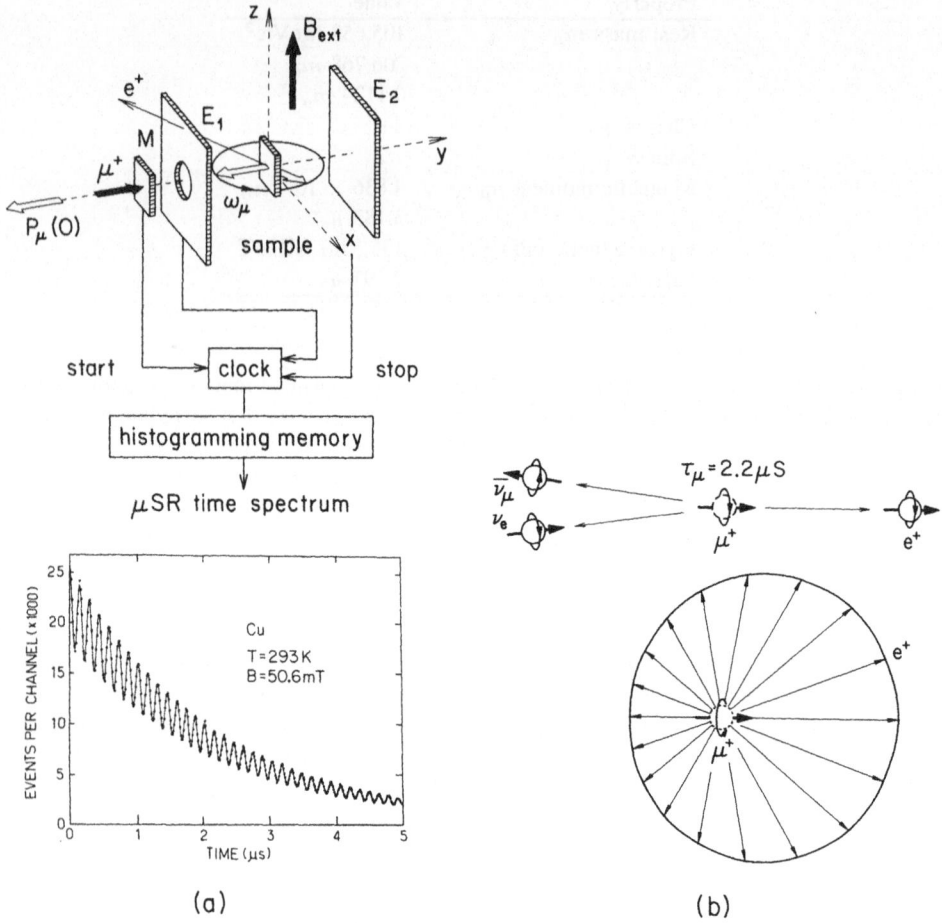

(a) (b)

Figure 1: (a) Schematic diagram of a time-differential, transverse-field μSR apparatus. The muon detector (M) registers the incoming muons (μ^+) with their spins pointing perpendicular to the external magnetic field (B_{ext}). The positron detectors (E_1 and E_2) are used to count the decay positrons (e^+) arising from muons stopped in the sample. The lower part of the figure shows a typical μSR time spectrum of copper. (b) Muon-decay scheme and angular-distribution pattern (emission probability) of the decay positrons with respect to the muon-spin direction. Note that the decay positrons are preferentially emitted along the muon-spin direction.

where $\langle B_\mu \rangle = \omega_\mu / \gamma_\mu$ is the first moment of $p(B_\mu)$. Note that in (magnetic) solids (ferromagnets, superconductors) the local field B_μ seen by the muon is in general not identical with the external field B_{ext}.

Most of the μSR experiments in high-T_c systems are performed with low-energy muons ($\approx$ 4 MeV) with their spins pointing perpendicular to the muon momentum. In this case the external magnetic field B_{ext} is directed along the muon momentum, in contrast to the situation shown in Fig. 1a. These so-called spin-rotated 'surface' muons are ideal for μSR studies in superconductors for the following reasons: (i) the muon beam is focussed by B_{ext} on the sample (low background signal from muons stopping in the cryostat windows and target holder), (ii) relatively small samples (typically 1 cm^2 $\times$ 1 mm) can be used, and (iii) the unwanted background signal arising from beam positrons is small. Beside the standard transverse-field μSR technique described here, other techniques such as zero-field μSR (no external magnetic field) and longitudinal-field μSR (external magnetic field parallel to the initial muon-spin polarization) have been used for investigating specific problems of high-T_c systems [1-10]. For a detailed description of the μSR technique and its applications in solid state physics we refer to Ref. [2].

3. Magnetic Flux Distribution and Magnetic Penetration Depth

There are two fundamental lengths that characterize the physical properties of a superconductor: the coherence length ξ and the magnetic penetration depth λ. For isotropic 'dirty' type-II superconductors, the zero-temperature limit of λ is generally given by [27]

$$1/\lambda^2(0) = [1 + \xi/\ell]^{-1}[\mu_0 e^2 n_s(0)/m^*], \tag{4}$$

where ξ is the coherence length, ℓ is the mean free path, m^* is the effective mass of the superconducting carriers, and $n_s(0)$ is the single-particle superfluid density at zero temperature. In the clean limit ($\xi/\ell \ll 1$) the magnetic penetration depth is simply determined by the London penetration depth $\lambda_L(0)$ [28]:

$$1/\lambda^2(0) \simeq 1/\lambda_L^2(0) = \mu_0 e^2 n_s(0)/m^*. \tag{5}$$

It was argued by Uemura *et al.* [23] that the cuprate superconductors due to their extremely short coherence length are 'clean' superconductors ($\xi/\ell \ll 1$), and therefore one expects the London formula in Eq. (5) to be approximately valid for these materials.

Since the superfluid density is directly related to the superconducting order parameter, the temperature dependence of λ provides essential information on the type of pairing mechanism (symmetry of the wave function, coupling strength) [28-30]. In particular, at low temperature $\lambda(T)$ is a potential probe of the pairing state in superconductors [30]. For conventional s-wave superconductors $\lambda(T)$ is generally described by the *empirical* Gorter-Casimir expression of the two-fluid (TF) model [28]:

$$\lambda^2(0)/\lambda^2(T) = 1 - (T/T_c)^4. \tag{6}$$

For extreme type-II superconductors ($\lambda/\xi \gg 1$), such as the cuprates, Eq. (6) represents a good approximation of $\lambda(T)$ for an s-wave superconductor with strong coupling [29].

The study of the magnetic flux distribution $p(B)$ in the mixed state of a type-II superconductor has been the goal of several μSR investigations, long before the discovery of high-temperature superconductivity. Here one should mention the impressive μSR investigations of $p(B)$ in type-II niobium by Herlach *et al.* [31]. But only recently in connection with high-temperature supercon-ductivity, it was realized that μSR also provides a powerful tool for determining the magnetic penetration depth λ in the *bulk* of a type-II superconductor. In the mixed state, an external magnetic field $B_{ext} > B_{c_1}$ (lower critical field) penetrates the superconducting specimen in the form of a regular vortex lattice, each vortex having an elementary flux quantum $\Phi_0 = h/2e$. This gives rise to a broad distribution $p(B)$ of local fields (at the muon site). The width (second moment) of $p(B)$ is determined by λ. The magnetic flux distribution $p(B)$ in an extreme type-II superconductor ($\lambda/\xi \gg 1$) may be calculated using the simple London-theory approach [32,9]. As a result, for an isotropic superconductor the second moment of $p(B)$ for a perfect triangular vortex lattice is given by [32,9]

$$\langle \Delta B^2 \rangle = 0.00371 \Phi_0^2 \lambda^{-4}. \tag{7}$$

This equation is only valid for high magnetic fields ($B_{ext} > 2B_{c_1}$), where $\langle \Delta B^2 \rangle$ is independent of B_{ext}.

In the case of a uniaxial superconductor, such as the layered cuprate systems, one has to consider two principal-axis effective masses m_{ab}^* and m_c^*, associated with supercurrents flowing in the ab-planes (CuO_2-planes) and along the c-axis (perpendicular to the CuO_2-planes), respectively [33,34,9]. The London formula defined in Eq. (5) then has the more general form [33,34]:

$$1/\lambda_{ab,c}^2 = \mu_0 e^2 n_s/m_{ab,c}^*, \tag{8}$$

where λ_{ab} and λ_c are the corresponding principal-axis penetration depths. For a uniaxial super-conductor the second moment of $p(B)$ depends on the orientation of the external magnetic field B_{ext} with respect to the symmetry axis (crystallographic c-axis). For an arbitrary angle θ between B_{ext} and the c-axis, the angular dependence of the second moment of $p(B)$ for a superconductor with uniaxial mass anisotropy is given within the London approximation by [18-20,9]

$$\langle \Delta B^2 \rangle(\theta) = \langle \Delta B^2 \rangle(\theta = 0)[\gamma^{-2} \sin^2 \theta + \cos^2 \theta], \tag{9}$$

where

$$\langle \Delta B^2 \rangle(\theta = 0) = 0.00371 \Phi_0^2 \lambda_{ab}^{-4} \tag{10}$$

is the second moment for $B_{ext} \parallel c$, and

$$\gamma = (m_c^*/m_{ab}^*)^{1/2} = \lambda_c/\lambda_{ab} \tag{11}$$

is the anisotropy ratio.

It is evident that the application of Eqs. (7) and (9-11) requires that a stable and regular vortex lattice is actually formed in the sample. Although the rather unconventional magnetic $(B - T)$ phase diagram of the cuprate superconductors is not fully understood at present [35], there is experimental evidence from Bitter decoration experiments [36] and recent neutron diffraction experiments [37,38] on $YBa_2Cu_3O_{7-\delta}$ that at low temperatures a short-range ordered vortex lattice does indeed exist.

In cuprate superconductors pinning is generally present and leads to a distortion of the perfect vortex lattice. As a consequence, random vortex pinning increases the magnetic field variation far above the ideal value for a perfect lattice given in Eq. (7) [32,9]. It has been suggested [32,39] that the effect of lattice disorder due to random pinning and thermal fluctuations can be treated by using a convolution of the ideal local magnetic flux distribution $p_{ideal}(B)$ with an appropriate broadening function $b(B)$, which in the case of a vortex glass or amorphous state may be approximated by a Gaussian function. For instance, this procedure was successfully applied in order to study the local magnetic flux distribution in the pinned vortex state of $YBa_2Cu_3O_7$ powder [39]. In highly anisotropic layered superconductors, such as the cuprates, the vortex structure is considered to be quite different from that in weakly anisotropic systems. It was proposed that in this case the flux-line structure is formed of stacks of two-dimensional (2D) 'pancake' vortices confined to the weakly-coupled CuO_2-layers [40]. In contrast to less anisotropic systems with stiff flux lines, random pinning of 2D 'pancake' vortices in the layers may drastically reduce the magnetic field variation $\langle \Delta B^2 \rangle^{1/2}$ *below* the intrinsic value for the perfect triangular lattice [Eq. (7)] [41]. This is due to the fact that strong pinning forces may prevent interlayer alignment of the 'pancake' vortices, giving rise to partial destructive interference of the magnetic field modulation in different layers [41,42]. Indeed, an extremely narrow and almost symmetric field distribution was recently observed in highly anisotropic single-crystal $Bi_2Sr_2CaCu_2O_{8+\delta}$ by μSR at low temperatures and high magnetic fields, consistent with a layered vortex structure controlled by random pinning forces within individual layers [42].

In conclusion, the evaluation of the intrinsic (due to a perfect vortex lattice) local magnetic field distribution $p(B)$ from measured μSR time spectra in cuprate superconductors is in general a difficult task, since random pinning, thermal fluctuations and demagnetization effects (in poly-crystals) lead to an unwanted distortion of the spectra. Although powerful methods have been developed in order to solve this problem, they are normally not easy to use. However, if one is mainly interested in the second moment of $p(B)$ and not necessarily in the exact shape of $p(B)$, it was demonstrated [13] that in a sintered sample a simple Gaussian distribution of local fields [cf. Eq. (2)] represents a good approximation of $p(B)$ and yields a resonable estimate of the magnetic field variance $\langle \Delta B^2 \rangle^{1/2}$ in terms of the μSR relaxation rate σ [Eq. (3)]. In this paper we use the Gaussian approximation of $p(B)$ which has the advantage that no uncontrolled systematic errors are introduced in the data evaluation procedure, so that results from different samples and workers can easily be compared.

4. Measurements of the Magnetic Penetration Depth and Magnetic Anisotropy

The μSR technique has been widely used to extract the magnetic penetration depth from the measured local magnetic field distribution in high-T_c superconductors [1,3-11]. In order to determine the principal-axis penetration depths λ_{ab} and λ_c by means of μSR rather large (typically $1 \text{ cm}^2 \times 1 \text{ mm}$) single crystals are required. However, big single crystals of high quality are not readily available. Therefore, most μSR experiments are performed on powder samples which are available in large sizes and of high quality. In a polycrystalline sample one measures an effective penetration depth (average over all orientations) which for highly anisotropic superconductors ($\gamma \geq 5$) is proportional to the in-plane penetration depth λ_{ab} [34,9,13]. Here we present a systematic μSR study of $\lambda_{ab}(T)$ in various cuprate superconductors: $YBa_2Cu_3O_x$ ($6.5 < x < 7.0$) (123), $YBa_2Cu_4O_8$ (124), and $Bi_2Sr_2CaCu_2O_{8+\delta}$ (2212) [12-17]. The samples were high-quality sintered disks. In addition, low-temperature measurements of the angular-dependent magnetic flux distribution in the mixed state of single-crystal $YBa_2Cu_3O_{7-\delta}$ and $Bi_2Sr_2CaCu_2O_{8+\delta}$ are discussed [18-20].

In order to ensure that a 'regular' vortex lattice is established in the mixed state, all μSR

272

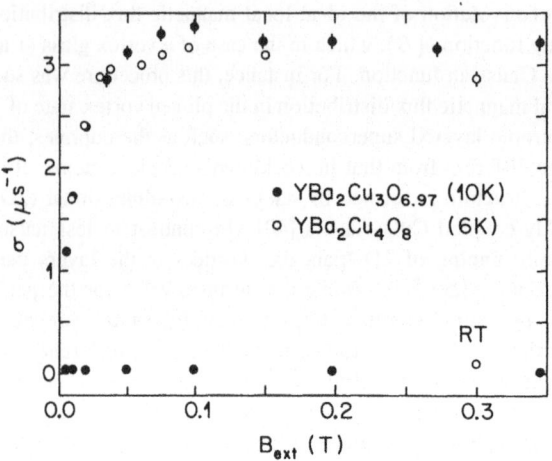

Figure 2: Low-temperatue relaxation rate σ as a function of the external magnetic field B_{ext} (FC) for sintered YBa$_2$Cu$_3$O$_{6.97}$ (solid circles) and YBa$_2$Cu$_4$O$_8$ (open circles). For comparison the room temperature data (RT) are also shown.

experiments were performed after field-cooling (FC) the samples slowly in a high external magnetic field ($B_{ext} > 2B_{c_1} \approx 100$ mT) [13]. The μSR spectra were analyzed assuming a Gaussian distribution of local magnetic fields [Eqs. (1-3)], yielding the relaxation rate σ which is a measure of the second moment of $p(B)$ (see Sec. 3). In order to estimate a value for λ from the relaxation rate σ one has to be certain that the applied field is high enough for Eq. (7) to be valid. Figure 2 shows the field dependence of σ for sintered YBa$_2$Cu$_3$O$_{6.97}$ and YBa$_2$Cu$_4$O$_8$, measured at 10 K and 6 K, respectively. In this experiment each data point was obtained after FC the sample from above T_c. For both samples $\sigma(T)$ increases until reaching a maximum, then decreases slightly and becomes nearly field independent, as expected for a vortex lattice with local or noncollective pinning [32,39]. The maximum of $\sigma(B)$ at lower fields can be explained by a distortion of the flux lattice due to strong nonlocal pinning. Assuming that a short-range order triangular vortex lattice is indeed established in a sintered sample, the powder average of the in-plane penetration depth $\langle \lambda_{ab} \rangle$ can be extracted from the measured relaxation rate σ by combining Eqs. (3) and (7) [13]:

$$\langle \lambda_{ab} \rangle = (0.0431 \gamma_\mu \Phi_0 / \sigma)^{1/2}. \tag{12}$$

λ_{ab} can be determined from $\langle \lambda_{ab} \rangle$ by correcting for the effective mass anisotropy m_c^* / m_{ab}^* [34,9]. For practical use λ_{ab} can be extracted directly from the relaxation rate σ measured in a polycrystalline sample by using the simple relation:

$$\lambda_{ab} \, [\text{nm}] \simeq 224 / \sqrt{\sigma \, [\mu s^{-1}]}. \tag{13}$$

For a single-crystal sample the numerical factor in front of Eq. (13) has to be replaced by 275. Estimated values of $\lambda_{ab}(0)$ from μSR data for some cuprate superconductors are listed in Table 2. In general, the values of the penetration depth obtained with μSR are in fair agreement with those determined using other experimental techniques [11,13,38,43,44]. As an example,

Table 2: Values of T_c, $\sigma(0)$ and $\lambda_{ab}(0)$ for various cuprate superconductors. (After Ref. [16].)

	T_c [K]	$\sigma(0)$ [μs^{-1}]	$\lambda_{ab}(0)$ [nm]
$YBa_2Cu_3O_{6.970}$	89.5(1)	3.06(2)	130(10)
$YBa_2Cu_3O_{6.556}$	59.4(2)	0.993(4)	220(20)
$YBa_2Cu_3O_{6.516}$	57.7(2)	0.777(3)	250(20)
$YBa_2Cu_4O_8$	80.2(1)	1.93(1)	160(15)
$Bi_2Sr_2CaCu_2O_{8+\delta}$	86.4(2)	1.18(1)	210(15)

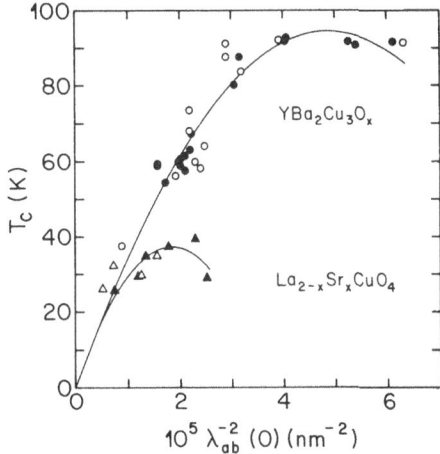

Figure 3: T_c versus $\lambda_{ab}^{-2}(0)$ for $YBa_2Cu_3O_x$ and $La_{2-x}Sr_xCuO_4$. $YBa_2Cu_3O_x$: μSR data [12-17] (solid circles), magnetic torque data [43] (open circles); $La_{2-x}Sr_xCuO_4$: μSR data [22] (solid triangles), magnetization data [44] (open triangles). The solid curves represent fits to the μSR data using Eq. (17). See text for a detailed explanation.

Fig. 3 shows a plot of T_c versus μSR estimates of $\lambda_{ab}^{-2}(0)$ for $YBa_2Cu_3O_x$ ($6.5 < x < 7.0$) [12-17] and $La_{2-x}Sr_xCuO_4$ (various x) [22]. For comparison, experimental values of $\lambda_{ab}(0)$ determined from magnetic torque [43] and from magnetization [44] measurements for $YBa_2Cu_3O_x$ and $La_{2-x}Sr_xCuO_4$, respectively, are also included in Fig. 3. Note that for a particular family of cuprates λ_{ab} increases with reduced T_c. In addition, it was found from magnetic torque measurements in oxygen deficient $YBa_2Cu_3O_x$ that the increase of λ_{ab} is accompanied by a substantial increase of the effective mass ratio $\gamma^2 = m_c^*/m_{ab}^*$ as the oxygen concentration is reduced [43]. It is evident from Fig. 3 that μSR provides values of $\lambda_{ab}(0)$ that are consistent with those obtained with macroscopic techniques. At this point we should note that values of the penetration depth measured by μSR should be considered as lower limits of the intrinsic values. As discussed in Sec. 3, random pinning, thermal fluctuations and demagnetization effects lead to a broadening of the measured field distribution, giving rise to a systematic error in the in-plane

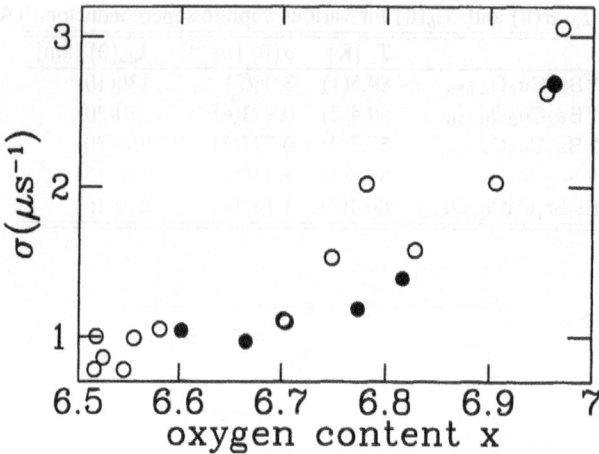

Figure 4: Relaxation rate σ as a function of oxygen content x for sintered $YBa_2Cu_3O_x$, measured in an external field of 350 mT (FC) at 10 K. Open circles indicate data points from Refs. [14,15] and black dots represent recent measurements from Ref. [17].

penetration depth deduced from Eq. (12) or (13). However, a misinterpretation of $\langle \Delta B^2 \rangle$ of say 25 % would according to Eq. (7) affect λ_{ab} by only 7 %, so that the systematic errors in μSR estimates of λ_{ab} are expected to be rather small (< 15 %). Moreover, Harshman et al. [39] analyzed the μSR lineshape in the mixed state of $YBa_2Cu_3O_7$ powder in detail, including a critical-state model for flux trapping and penetration-depth anisotropy in the analysis. As a result, the best estimate of the in-plane penetration depth was found to be $\lambda_{ab}(0) = 130(10)$ nm, in excellent agreement with the value for $YBa_2Cu_3O_{6.97}$ obtained from a simple Gaussian fit (see Table 2). This clearly demonstrates that a Gaussian approximation of local fields may indeed provide reasonable estimates of λ_{ab}.

It is well known that the normal state and superconducting properties of $YBa_2Cu_3O_x$ sensitively depend on the oxygen content x [45,46]. For instance, the superconducting transition temperature T_c increases with increasing x, exhibiting two characteristic plateaus at 60 K and 90 K. The oxygen content x determines the average occupancy of oxygen sites in the whole system (CuO_2-planes and CuO-chains) and therefore controls the density of the holes in the CuO_2-planes which are responsible for superconductivity in these materials. Therefore, a systematic investigation of the magnetic penetration depth in oxygen deficient $YBa_2Cu_3O_x$ is of great interest for studying the influence of x on the superconducting condensate density $\rho_s = n_s/m^*$ in the CuO_2-planes. μSR experiments on a series of oxygen deficient samples ($6.5 < x < 7.0$) revealed some remarkable results [14-17]. Figure 4 shows the low-temperature relaxation rate σ as a function of oxygen content x after field-cooling to 10 K in a magnetic field of 350 mT. The sudden increase of $\sigma(x)$ between $x = 6.7$ and 6.8 which was observed in earlier measurements [14,15] could not be reproduced in recent experiments [17]. The reason for this discrepancy is still unclear, but may be well due to sample homogeneity and preparation. It is interesting to note that around $x \simeq 6.75$ an anomalous behavior of the pressure dependence of T_c was reported

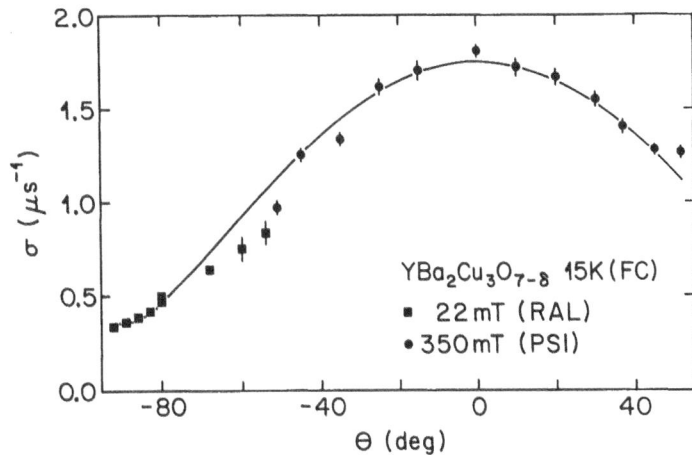

Figure 5: Low-temperature relaxation rate σ vs. angle θ between the external magnetic field and the c-axis for single-crystal $YBa_2Cu_3O_{7-\delta}$. The experiments at 22 mT and 350 mT were performed at the Rutherford Appleton Laboratory (RAL) and at the Paul Scherrer Institute (PSI), respectively. The solid line is a fit to the data to the 3D anisotropic London expression [Eq. (9)] with $\gamma = 5$. (After Ref. [19].)

by Bucher *et al.* [47], the pressure coefficient dT_c/dp showing a distinct maximum (see Fig. 10) in coincidence with the sudden increase of $\sigma(x)$ (Fig. 4). This maximum can be related to the oxygen content x, where T_c changes substantially from the 60 K to the 90 K plateau. Furthermore, the behavior of $\sigma(x)$ as well as the pressure dependence of T_c appear to be correlated with structural properties (oxygen ordering [45], anomaly in the lattice constant c [46]). Because σ is proportional to the condensate density ρ_s, the behavior of $\sigma(x)$ may be associated with a positive charge transfer from the CuO-chains to the CuO_2-planes [45,48,46], occurring rather monotonically (except around $x \simeq 6.75$) with increasing x. Indeed, $\sigma(x)$ shows interesting similarities to the x-dependence of the bond valence sum around the CuO_2-plane copper [45], reflecting the actual charge sitting on that particular copper atom. This may have some important consequences for theoretical considerations, since it has been suggested that charge-transfer excitations and phonons play an essential role in high-temperature superconductivity [49,50].

Recent work on highly anisotropic cuprate superconductors has suggested that in these systems due to their extremely short coherence length and due to their weakly-coupled layered structure an unconventional 'vortex lattice' is formed in the mixed state which substantially deviates from the classical Abrikosov lattice [1,35,40] (see Sec. 3). Among the cuprates $Bi_2Sr_2CaCu_2O_{8+\delta}$, having an extremely large anisotropy ($\gamma > 150$ [51]), is a potential candidate to test predictions of unconventional vortex lattices, such as the 'pancake' model [40]. In order to study the anisotropic properties of the vortex structure in layered superconductors the local magnetic field distribution in the mixed state of single-crystal $YBa_2Cu_3O_{7-\delta}$ [18,19] and $Bi_2Sr_2CaCu_2O_{8+\delta}$ [20] was investigated as a function of the angle θ between the external magnetic field and the c-axis. The samples were mosaics of carefully aligned single crystals. The low-temperature angular dependence of

276

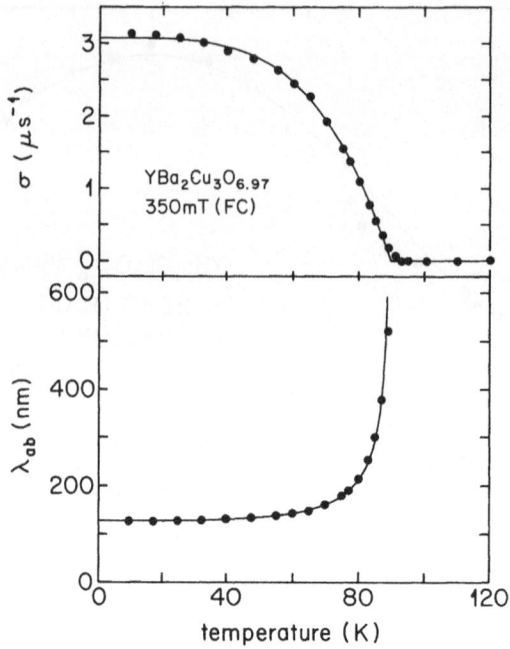

Figure 6: Relaxation rate σ (top) and in-plane penetration depth λ_{ab} (bottom) as a function of temperature for sintered $YBa_2Cu_3O_{6.97}$ measured in a field of 350 mT (FC). Solid curves correspond to fits to the two-fluid model [Eq. (6)]. (After Ref. [13].)

the relaxation rate σ for single-crystal $YBa_2Cu_3O_{7-\delta}$ is shown in Fig. 5. For each individual data point the sample was slowly field-cooled to 15 K from above T_c in order to establish a stable flux-line configuration at low temperature. The observed angular dependence of $\sigma \propto \langle \Delta B^2 \rangle^{1/2}$ is well described by Eq. (9) given for a 3D anisotropic Ginzburg-Landau-type (AGL) model with a uniaxial mass anisotropy of $\gamma = (m_c^*/m_{ab}^*)^{1/2} \simeq 5$, as indicated by the solid curve in Fig. 5. This value of γ is in fair agreement with those obtained from Bitter decoration experiments [36] and other experiments summarized in Ref. [13]. However, very recent measurements of $\sigma(\theta)$ at low temperatures for single-crystal $Bi_2Sr_2CaCu_2O_{8+\delta}$ revealed marked deviations from the 3D AGL behavior [20]. In magnetic fields of 400 mT and 22.6 mT the relaxation rate σ at 10 K showed an unexpectedly weak variation with angle in the range $0° < \theta < 55°$, followed by a rapid decay at higher angles. Indeed, a comparison of the data for $\theta < 55°$ with the expression predicted for the 3D AGL model [Eq. (9)] leads to an unrealistic low value of $\gamma \simeq 1.1$, in spite of the fact that this system is extremely anisotropic ($\gamma > 150$ [51]). This suggests that a 3D AGL model with stiff flux lines, which describes well the angular dependence of σ in $YBa_2Cu_3O_{7-\delta}$ (Fig. 5), does not provide an appropriate description of the vortex structure in the mixed state of $Bi_2Sr_2CaCu_2O_{8+\delta}$. In conclusion, this unusual behavior of $\sigma(\theta)$ in $Bi_2Sr_2CaCu_2O_{8+\delta}$ must reflect the as yet undetermined vortex structure of this strongly anisotropic layered superconductor, which, according to various models (such as the 2D 'pancake' model [40]), may be complicated

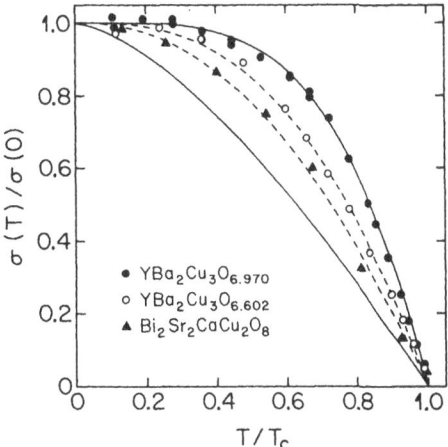

Figure 7: Normalized relaxation rate $\sigma(T)/\sigma(0)$ versus reduced temperature T/T_c for YBa$_2$Cu$_3$O$_{6.970}$, YBa$_2$Cu$_3$O$_{6.602}$, and Bi$_2$Sr$_2$CaCu$_2$O$_{8+\delta}$ measured in a field of 350 mT (FC). The solid lines correspond to Eq. (14) with $n = 3/2$ and $n = 4$, respectively. The dashed lines represent fits to Eq. (14), not shown for YBa$_2$Cu$_3$O$_{6.970}$, where the fit ($n = 3.89$) almost coincides with the solid curve for $n = 4$.

[20]. For a more detailed discussion of these interesting results we refer to Ref. [20].

In sintered samples of highly anisotropic layered superconductors the normalized relaxation rate $\sigma(T)/\sigma(0) = \lambda_{ab}^2(0)/\lambda_{ab}^2(T)$ reflects the temperature dependence of the in-plane magnetic penetration depth λ_{ab} [12-17]. As an example, Fig. 6 shows $\sigma(T)$ and the corresponding $\lambda_{ab}(T)$ of YBa$_2$Cu$_3$O$_{6.97}$, after subtraction of a small background contribution from the copper nuclear moments. As indicated by the solid line $\sigma(T) \propto \lambda_{ab}^{-2}(T)$ is very well described by the empirical formula of the two-fluid (TF) model [cf. Eq. (6)]. This finding is consistent with conventional s-wave pairing and suggests strong coupling [29]. Moreover, the temperature behavior of σ is in agreement with the prediction of a phenomenological, marginal-Fermi-liquid model [52] with a superconducting gap to T_c ratio of $2\Delta(0)/k_BT_c \simeq 5$, which is well in excess of the BCS weak-coupling ratio of 3.52 [28]. In general, however, not for all cuprate superconductors $\sigma(T)$ exhibits TF behavior (strong coupling) [16]. For instance, in YBa$_2$Cu$_3$O$_x$ with low x, YBa$_2$Cu$_4$O$_8$ and Bi$_2$Sr$_2$CaCu$_2$O$_{8+\delta}$ the temperature dependence of σ is in better agreement with weak-coupling BCS theory [16]. This observation suggests that in YBa$_2$Cu$_3$O$_x$ the coupling strength becomes weaker with reduced x. Why some cuprates exhibit weak-coupling (BCS) and other more strong-coupling (TF) behavior is not understood. However, one may certainly not conclude that cuprate systems which show a BCS-like temperature dependence of the magnetic penetration depth are indeed conventional BCS superconductors.

In order to investigate the peculiar behavior of $\lambda_{ab}(T)$ in various cuprate systems in a more systematic way, we analyzed our μSR data in terms of the more general empirical expression:

$$\sigma(T)/\sigma(0) = \lambda_{ab}^2(0)/\lambda_{ab}^2(T) = 1 - (T/T_c)^n, \qquad (14)$$

where $3/2 \leq n \leq 4$. $n = 4$ corresponds to the TF model [Eq. (6)] and $n = 3/2$ to the ideal

278

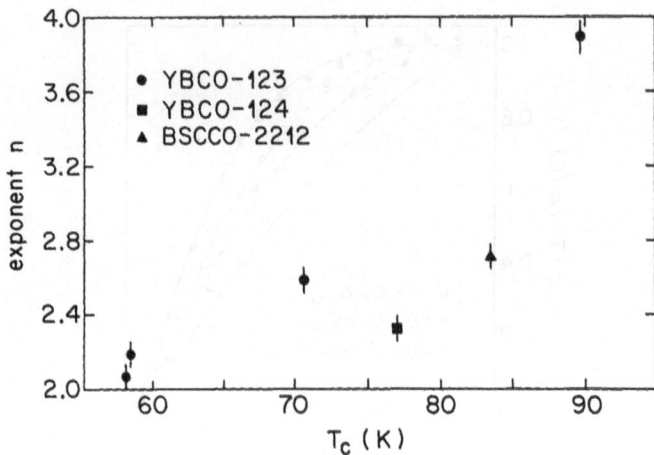

Figure 8: Exponent n versus T_c for various cuprate superconductors as obtained from a fit of the measured relaxation rate $\sigma(T)$ to Eq. (14).

charged Bose gas [53]. Figure 7 shows the measured $\sigma(T)/\sigma(0)$ versus T/T_c for $YBa_2Cu_3O_x$ ($x = 6.970, 6.602$) and $Bi_2Sr_2CaCu_2O_{8+\delta}$, as well as the calculated curves for $n = 3/2$ and $n = 4$ for comparison. It is seen that $\sigma(T)$ is rather well described by Eq. (14) and that the exponent n decreases with reduced T_c. This is evident in Fig. 8 where we plotted n versus T_c for various cuprate superconductors. The exponent n decreases with reduced T_c from $n \simeq 4$ for $T_c \approx 90$ K and approaches $n \approx 2$ around $T_c \approx 60$ K. However, recent small-angle neutron [38], microwave [54] and μSR [54] work on single-crystal $YBa_2Cu_3O_{x-\delta}$ with $T_c \approx 93$ K show a temperature dependence of λ_{ab}^{-2} with an exponent between $n = 2$ (neutron scattering, microwave studies) and $n = 2.37$ (μSR). This variation of λ_{ab} with temperature is contrary to the conventional s-wave two-fluid behavior observed in highly oxygenated sintered samples with $n \simeq 4$.

In conclusion, the observed trends in the temperature dependence of the magnetic penetration depth in doped cuprates are consistent with the behavior of the 3d interacting Bose gas and therefore seem to favor a pairing mechanism involving Bose condensation of preformed weakly charged and interacting pairs (see contribution of Schneider [26]).

5. 'Universal' Relations and Trends in Extreme Type-II Superconductors

Detailed μSR investigations of the magnetic penetration depth in various classes of cuprate superconductors were carried out by several groups [1,3-11]. In particular, Uemura et al. [21-23] performed a systematic μSR study of the relaxation rate σ in a large number of doped cuprates and related high-T_c superconductors at low temperatures. They found a remarkable empirical relation between T_c and $\sigma(T \to 0) \propto \rho_s(0) = n_s(0)/m_{ab}^*$ that seems to be 'universal' for many cuprate high-T_c systems ('Uemura plot'). This 'universal' relation $T_c(\sigma)$ has the following interesting features: (i) with increasing carrier doping, T_c initially increases almost linearly ($T_c \propto \rho_s(0)$), then saturates, and finally is suppressed for high carrier doping, (ii) this general trend is observed in single layer $La_{2-x}Sr_xCuO_4$ (214), double layer $YBa_2Cu_3O_x$ (123) and $Bi_2Sr_2CaCu_2O_8$ (2212), triple layer $Bi_2Sr_2Ca_2Cu_3O_{10}$ (2223), and other related extreme type-II superconductors, and (iii) the initial slope of the linear increase of $T_c(\sigma)$ appears to be almost the same for different families, however, the saturation and suppression of T_c starts at different values of σ for each family. The 'Uemura plot' is of considerable interest for testing theoretical models of high-temperature superconductivity, since a satisfactory theory should give a plausible explanation of the 'universal' behavior of $T_c(\sigma)$. For instance, the relationship between T_c and $\rho_s(0)$ points to a non-retarded pairing interaction, or equivalently to a high-energy scale for the coupling between the superconducting carriers, and therefore cannot be explained within the framework of weak-coupling BCS theory [21,22].

Extensive investigations of the pressure (p) and oxygen isotope mass (m) dependence of T_c in a large number of various families of cuprate superconductors have indicated that the pressure coefficient

$$\alpha = \frac{d \ln T_c}{dp} \tag{15}$$

and the isotope effect coefficient

$$\beta = -\frac{d \ln T_c}{d \ln m} \tag{16}$$

exhibit a generic behavior for several classes of doped cuprates as a function of T_c [24-26]. Both $\alpha(T_c)$ and $\beta(T_c)$ have two nearly symmetric branches, one with positive and the other with negative values for the coefficients. The two branches merge at the maximum T_c^m, where both coefficients nearly vanish, and the magnitude of both coefficients increases with decreasing T_c. These trends seem to be a common feature of many classes of doped cuprates and related systems, sharing in addition the rather unique relation between T_c and the low-temperature μSR relaxation rate $\sigma(0) \propto \lambda_{ab}^{-2}(0) \propto \rho_s(0)$, which for a particular family is characterized by a maximum (optimum) value of the transition temperature T_c^m and the corresponding optimum zero-temperature relaxation rate $\sigma^m(0)$ (see Fig. 3). We recently proposed that $T_c(\sigma)$ close to the optimum condensate density $\rho_s^m(0) \propto \sigma^m(0)$ can be approximated by a simple parabolic 'universal' scaling ansatz of the form [24,25]:

$$\overline{T_c} = 2\,\overline{\sigma}\,(1 - \overline{\sigma}/2), \tag{17}$$

where $\overline{T_c} = T_c/T_c^m$ and $\overline{\sigma} = \sigma(0)/\sigma^m(0)$ are the reduced temperature and relaxation rate, respectively. As expected, T_c vanishes for $\sigma(0) = 0$. Since $\sigma(0) \propto n_s(0)/m_{ab}^*$, this implies a vanishing superconducting carrier density $n_s(0)$ and/or an infinite effective mass m_{ab}^*. According to Eq. (17), T_c versus $\sigma(0)$ should fall on a single parabola for a particular family, characterized by T_c^m and $\sigma^m(0)$. As an example, in Fig. 3 we plotted T_c versus $\lambda_{ab}^{-2}(0) \propto \sigma(0)$ for $YBa_2Cu_3O_x$ and $La_{2-x}Sr_xCuO_4$. Indeed, the μSR data are consistent with the parabolic behavior (17), as indicated by the solid curves. To provide evidence that this behavior is not just an artifact of the μSR method, sample preparation and homogeneity, we also included in Fig. 3 estimated values of $\lambda_{ab}(0)$ from magnetic torque [43] and magnetization [44] measurements. The excellent agreement between the two data sets obtained with quite different experimental techniques and samples clearly suggests that the parabolic behavior of T_c versus $\sigma(0) \propto \rho_s(0)$ is really an intrinsic property. The scaled plot of $\overline{T_c}$ versus $\overline{\sigma}$ for a large number of cuprates and Chevrel phase superconductors [24,25] is displayed in Fig. 9, showing a remarkable consistency with the parabolic scaling ansatz (17). Moreover, it was found that the optimum values T_c^m and $\sigma^m(0)$ obtained for each family of doped systems are nearly proportional to each other [25,26].

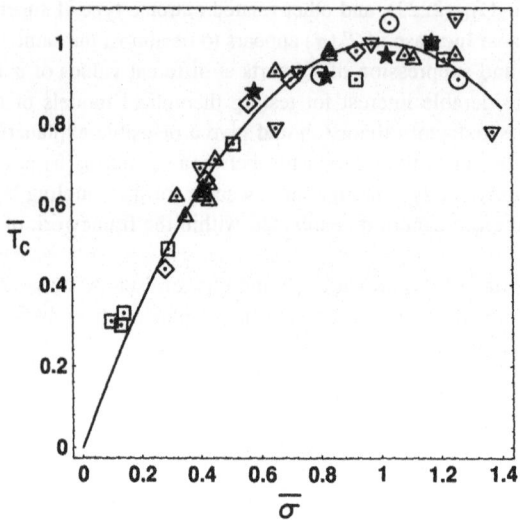

Figure 9: Reduced transition temperature $\overline{T}_c$ versus reduced relaxation rate $\overline{\sigma}$ for various extreme type-II superconductors: $\triangle$: $YBa_2Cu_3O_x$ ($6.5 < x < 7$), $\diamond$: $Y_{1-x}Pr_xBa_2Cu_3O_{6.97}$ ($0.05 \leq x \leq 0.4$), $\triangledown$: $La_{2-x}Sr_xCuO_4$ (various x), $*$: $Bi_2Sr_2Ca_{1-x}Y_xCu_2O_{8+\delta}$ ($0 \leq x \leq 0.45$), $\odot$: $Tl_2Ba_2Ca_2Cu_3O_{10}$, $Tl_{0.5}Pb_{0.5}Sr_2Ca_2Cu_3O_9$, $Bi_{2-x}Pb_xSr_2Ca_2Cu_3O_{10}$, $\square$: $LaMo_6Se_8$, $SnMo_6S_8$, $PbMo_6S_8$, $SnMo_6S_4Se_4$, $SnMo_6S_7Se$, $SnMo_6S_1Se_7$, $PbMo_6S_4Se_4$. The solid curve corresponds to Eq. (17). (After Refs. [24,25], where all the details are given.)

As shown in Fig. 9, the majority of the data points are concentrated on the left branch of the parabola (underdoped region: $\overline{\sigma} \leq 1$). The physical meaning of the right branch of the parabola (overdoped region: $\overline{\sigma} \geq 1$) is still unclear. One should keep in mind that, although Eq. (17) provides a reasonable empirical description of the behavior of $T_c(\sigma)$ in the present case, it should not be pushed too far. In this context it is interesting to note that the dependence of T_c on the hole concentration p in $La_{2-x}Sr_xCuO_4$, $YBa_2Cu_3O_x$ and other compounds also seem to exhibit a parabolic behavior as proposed in Eq. (17) [55]. The hole concentration can be measured either directly as in $La_{2-x}Sr_xCuO_4$ or estimated from bond-valence-sum calculations as in $YBa_2Cu_3O_x$. This implies that the condensate density $\rho_s(0) = n_s(0)/m^*_{ab}$ is a linear function of the hole concentration p [26]. Recently, a similar universal relationship between T_c and the hole content p in p-type cuprate superconductors was reported [56]. However, in this case the universal curve $\overline{T}_c = T_c/T_c^m$ versus p is characterized by a plateau, rather than a parabola, with sharp bends at both sides.

The parabolic ansatz (17) can now be used to predict trends in the pressure (α) and isotope mass (β) coefficients defined in Eqs. (15) and (16) [24-26]. Both pressure [57] and oxygen isotope effect [50] are considered to be crucial in understanding the microscopic mechanism responsible for high-temperature superconductivity. For example, the pressure coefficient α can be expressed in terms of Eq. (17). In principle all quantities $\sigma(0)$, $\sigma^m(0)$, and T_c^m in Eq. (17) depend on pressure p. However, it is an empirical fact that for most classes of cuprate high-T_c systems

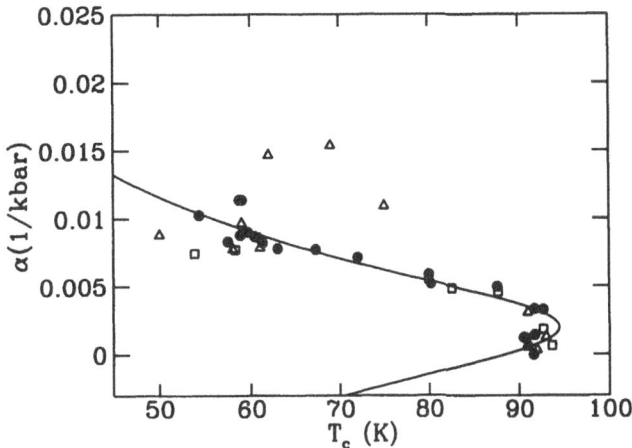

Figure 10: Pressure coefficient α versus T_c for oxygen deficient $YBa_2Cu_3O_x$. Triangles: pressure data from Bucher *et al.* [47], squares: pressure data from Almasan *et al.* [58]. Solid circles represent values of α as deduced from μSR data [14-17] using Eq. (19) with $\alpha_m = 0.002$ kbar^{-1} and $\alpha_0 = 0.0037$ kbar^{-1}. The solid curve corresponds to the predicted $\alpha(T_c)$ in Eq. (18). (From Ref. [17].)

dT_c^m/dp and in turn $d\sigma^m(0)/dp$ are very small [57], so that the dominant contribution to the pressure coefficient α arises from the induced change of the condensate density $\sigma(0) \propto \rho_s(0)$. Thus, using Eqs. (15) and (17), α may be expressed by these simple relations [24-26]:

$$\alpha = \frac{1}{T_c} \frac{dT_c}{dp} \simeq \alpha_m \pm 2 \frac{\sqrt{1 - \overline{T_c}}}{\overline{T_c}} \alpha_0 , \tag{18}$$

or

$$\alpha = \frac{1}{T_c} \frac{dT_c}{dp} \simeq \alpha_m + \frac{1 - \overline{\sigma}}{\overline{\sigma}(1 - \overline{\sigma}/2)} \alpha_0 , \tag{19}$$

with

$$\alpha_m = \frac{1}{T_c^m} \frac{dT_c^m}{dp} , \tag{20}$$

and

$$\alpha_0 = \frac{1}{\sigma^m(0)} \frac{d\sigma(0)}{dp} . \tag{21}$$

α_0 is a material specific constant which may also depend on doping, and α_m accounts for a possible small pressure dependence at the optimum transition temperature T_c^m. For compounds with T_c close to T_c^m one expects α_0 to be nearly constant, and α_m is in general quite small [57]. The pressure dependence of T_c in a large number of cuprates was investigated in detail by several workers [57]. As an example, we present here some pressure data for oxygen deficient $YBa_2Cu_3O_x$ taken from Refs. [47,58]. The resulting plot of the pressure coefficient α versus T_c is depicted in Fig. 10. The magnitude of α increases rather monotonically with reduced T_c (or

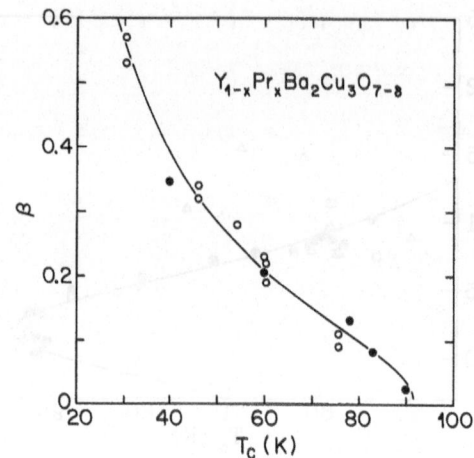

Figure 11: Isotope effect coefficient β versus T_c for $Y_{1-x}Pr_xBa_2Cu_3O_{7-\delta}$. Open circles: direct measurements of Franck *et al.* [59]. Solid circles represent values of β as deduced from μSR data of Seaman *et al.* [60] using Eqs. (19), (22) and (23) with $\beta_m = 0.007$ and $\beta_0 = 0.113$. The solid curve corresponds to the predicted $\beta(T_c)$ in Eq. (18). (From Ref. [17].)

oxygen content x). The anomaly in α discussed below clearly shows up as a broad maximum at $T_c \simeq 70$ K ($x \simeq 6.75$). According to Eqs. (18-21) α can be expressed in terms of the measured relaxation rate $\sigma(x)$ (see Fig. 4) and $\sigma^m(0) = 2.44$ μs^{-1} (Ref. [25]) with only two adjustable parameters α_m and α_0. As shown in Fig. 10, a fair agreement between the pressure and μSR data was found with $\alpha_m = 0.002$ kbar^{-1} and $\alpha_0 = 0.0037$ kbar^{-1}. The solid curve represents the predicted $\alpha(T_c)$ given in Eq. (18). However, the μSR data do not reproduce the broad maximum in α observed in the pressure data at $T_c \simeq 70$ K ($x \simeq 6.75$) [47]. The reason for this discrepancy between the two data sets is not yet understood. The most plausible explanation is that the parameter $\alpha_0 \propto d\sigma(0)/dp$ is changing substantially in the region of x, where T_c changes from the 60 K to the 90 K plateau. Identical expressions to those in Eqs. (18-21) can also be derived for the isotope mass coefficient β [24-26]. In particular, β can be expressed in terms of the μSR relaxation rate σ by replacing α by β in Eq. (19) with

$$\beta_m = -\frac{m}{T_c^m}\frac{dT_c^m}{dm} , \qquad (22)$$

and

$$\beta_0 = -\frac{m}{\sigma^m(0)}\frac{d\sigma(0)}{dm} . \qquad (23)$$

In analogy to the pressure coefficient α, the experimental data for various cuprate systems are in fair accordance with the predicted behavior $\beta(T_c)$ [24-26]. As an example, in Fig. 11 we plot β for $Y_{1-x}Pr_xBa_2Cu_3O_{7-\delta}$ versus T_c from the work of Franck *et al.* [59]. For comparison we include the corresponding μSR estimates of β as obtained from the μSR data of Seaman *et al.* [60] using Eqs. (19), (22) and (23) with $\beta_m = 0.007$ and $\beta_0 = 0.113$. The curve corresponds to

Eq. (18) with the quoted values for β_m and β_0. The agreement between the direct measurements and the μSR estimates of β is remarkable.

In conclusion, the parabolic scaling ansatz (17) provides a natural explanation of the common 'universal' trends in $\alpha(T_c)$ and $\beta(T_c)$ observed in a large number of cuprate and related superconductors [24-26], strongly suggesting that the condensate density is a key property of high-temperature superconductivity which controls the transition temperature T_c. A detailed discussion of these interesting empirical relations is given in the contribution of Schneider [26].

6. Summary

The μSR technique is a powerful and unique tool for investigating the *local* magnetic flux distribution $p(B)$ (at the positive muon) in the *bulk* of a superconducting specimen (no surface effect). In particular, in the mixed state of a type-II superconductor, the magnetic penetration depth λ can be determined directly from the second moment of $p(B)$. A precise knowledge of the temperature dependence of λ is of fundamental importance, since $\lambda(T)$ essentially reflects the type of the pairing mechanism (symmetry of the wave function and coupling strength). Moreover, an investigation of $p(B)$ as a function of the angle between the external magnetic field and the c-axis provides important information on the complex vortex structure in the highly anisotropic layered cuprates.

The in-plane penetration depth $\lambda_{ab}(0)$ was determined in various families of cuprates using μSR. The measured values of $\lambda_{ab}(0)$ are consistent with those obtained with other experimental techniques. For $YBa_2Cu_3O_x$ with high oxygen content x the temperature dependence of λ_{ab} was found to be well described by the empirical two-fluid model. This behavior is consistent with conventional s-wave pairing and suggests evidence for strong coupling. However, for compounds with low x the temperature behavior of λ_{ab} is in better agreement with weak-coupling BCS theory.

A systematic investigation of the low-temperature penetration depth in $YBa_2Cu_3O_x$ as a function of oxygen content x is of fundamental importance for understanding the relationship between oxygen doping and the condensate density $\rho_s(0) = n_s(0)/m_{ab}^*$ in the CuO_2-planes. The low-temperature relaxation rate σ was found to increase rather monotonically upon doping, indicative of a positive charge transfer taking place continuously from the CuO-chains to the CuO_2-planes with increasing x [45,48,46]. In fact, $\sigma(x)$ exhibits a quite similar behavior than the bond valence sum around the CuO_2-plane copper [45]. The relaxation rate σ and the pressure dependence of T_c in oxygen deficient $YBa_2Cu_3O_x$ appear to be correlated with structural parameters (e.g. lattice constant c), suggesting that some structural and superconducting (condensate density) properties are closely related to each other and are very sensitive to small changes of the oxygen content x [46].

Detailed studies of the local magnetic flux distribution $p(B)$ as function of the angle between the external magnetic field and the c-axis in single-crystal $YBa_2Cu_3O_{7-\delta}$ and $Bi_2Sr_2CaCu_2O_{8+\delta}$ revealed some remarkable results. In YBCO the low-temperature angular dependence of the second moment of $p(B)$ is well described by a 3D anisotropic Ginzburg-Landau-type (AGL) model with uniaxial effective mass anisotropy, yielding $m_c^*/m_{ab}^* \simeq 25$. However, in BSCCO substantial deviations from the 3D AGL model prediction were observed in the angular dependent relaxation rate in the mixed state, indicative of an unconventional vortex structure in this highly anisotropic layered superconductor.

Uemura *et al.* [21-23] found empirically an interesting relation between the superconducting transition temperature T_c and the low-temperature relaxation rate $\sigma(0) \propto n_s/m_{ab}^*$ (condensate density) which seems to be observed 'universally' for various families of cuprate and related superconductors ('Uemura plot'). Motivated by the Uemura plot we recently proposed [24,25] a simple parabolic scaling ansatz [Eq. (17)] which relates the rescaled transition temperature $\overline{T_c}$ 'universally' to the rescaled relaxation rate $\overline{\sigma}$: $\overline{T_c} = 2\overline{\sigma}(1 - \overline{\sigma}/2)$. μSR data of a large number of cuprate and Chevrel-phase superconductors exhibit a remarkable consistency with this scaling ansatz. Moreover, the proposed ansatz implies 'universal' trends in the T_c-dependence of the pressure (α) and isotope mass (β) coefficients. Indeed, experimental results for various classes of cuprate systems are in fair agreement with the predicted common behavior of $\alpha(T_c)$ and $\beta(T_c)$. These findings strongly suggest that the condensate density, which essentially determines T_c, plays a crucial role in high-temperature superconductivity [25,26].

The temperature dependence of the penetration depth λ was measured in a series of cuprates and analyzed in terms of the power law given in Eq. (14). As a result, for all compounds investigated $\lambda(T)$ lies between the limits of the two-fluid model ($n = 4$) and the ideal charged Bose gas ($n = 3/2$). More specifically, for compounds with high T_c the temperature dependence of λ is very well described by the two-fluid model, whereas for compounds with reduced T_c the behavior of $\lambda(T)$ changes gradually with decreasing T_c towards the limit of the ideal charged Bose gas.

As discussed in the contribution of Schneider [26], these trends in the temperature dependence of the penetration depth, as well as the empirical 'universal' scaling relation between T_c and the low-temperature relaxation rate (condensate density) together with the common trends in the pressure and isotope effect coefficients suggest Bose condensation of weakly charged bound pairs as a possible scenario of high-temperature superconductivity.

Acknowledgments

I would like to thank E. Kaldis and L. Riva di Sanseverino (Lodovico) for organizing this International School of Crystallography "Materials and Crystallographic Aspects of High-T_c Superconductivity" and for the opportunity to contribute. The kind collaboration with K. Conder, R. Cubitt, E.M. Forgan, E. Kaldis, J. Karpinski, P.H. Kes, C. Krüger, S.L. Lee, T.W. Li, A.A. Menovsky, S. Rusiecki, I.M. Savić, T. Schneider, H. Simmler, B. Stäuble-Pümpin, Z. Tarnawski, M. Warden, P. Wenk, D. Zech and P. Zimmermann, who participated in this work, is gratefully acknowledged. I also wish to thank K.A. Müller for many helpful and stimulating discussions and his great interest in our work. This work was supported by the Swiss National Science Foundation and partly by a special grant from the British-Swiss Joint Research Programme.

References

[1] See, e.g., *Proceedings of the International Conference on Materials and Mechanisms of Superconductivity, High Temperature Superconductors III*, Kanazawa, Japan, Physica C **185-189** (1991).

[2] See, e.g., A. Schenck, *Muon Spin Rotation Spectroscopy: Principles and Applications in Solid State Physics* (Adam Hilger, Bristol 1985).

[3] See, e.g., *Proceedings of the 5th International Conference on Muon Spin Rotation, Relaxation and Resonance*, Oxford, U.K., Hyperfine Interactions **63-65** (1990).

[4] See, e.g., Y.J. Uemura *et al.* , J. Phys. (Paris) Colloq **49**, C8-2087 (1988).

[5] See, e.g., H. Keller, IBM J. Res. Develop. **33**, 314 (1989), and references therein.

[6] See, e.g., H. Keller, in *Earlier and Recent Aspects of Superconductivity*, ed. by J.G. Bednorz and K.A. Müller, Springer Series in Solid-State Sciences, Vol. **90** (Springer-Verlag Berlin, Heidelberg 1990), pp. 222-239, and references therein.

[7] See, e.g., Y.J. Uemura *et al.* , in *Proceedings of NATO Advanced Research Workshop on 'Dynamics of Magnetic Fluctuations in High Temperature Superconductors'*, Crete, Greece, 1989 (Plenum, 1990).

[8] See, e.g., V.J. Emery, Hyperfine Interactions **63**, 13 (1990).

[9] See, e.g., Yu. M. Belousov, V.N. Gorbunov, V.P. Smilga, and V.I. Fesenko, Sov. Phys. Usp. **33**, 911 (1990), and references therein; V.I. Fesenko, V. N. Gorbunov, and V.P. Smilga, Physica C **176**, 551 (1991).

[10] See, e.g., H. Keller, in *Exotic Atoms in Condensed Matter*, ed. H. Schneuwly and G. Benedek, Springer Proceedings in Physics, Vol. **59** (Springer-Verlag, Berlin, Heidelberg 1992), pp. 191-212, and references therein.

[11] See, e.g., D.R. Harshman and A.P. Mills, Phys. Rev. B **45**, 10684 (1992), and references therein.

[12] B. Pümpin, H. Keller, W. Kündig, W. Odermatt, I.M. Savić, J.W. Schneider, H. Simmler, P. Zimmermann, J.G. Bednorz, Y. Maeno, K.A. Müller, C. Rossel, E. Kaldis, S. Rusiecki, W. Assmus, and J. Kowalewski, Physica C **162-164**, 151 (1989).

[13] B. Pümpin, H. Keller, W. Kündig, W. Odermatt, I.M. Savić, J.W. Schneider, H. Simmler, P. Zimmermann, E. Kaldis, S. Rusiecki, Y. Maeno, and C. Rossel, Phys. Rev. B **42**, 8019 (1990).

[14] B. Pümpin, H. Keller, W. Kündig, I.M. Savić, J.W. Schneider, H. Simmler, P. Zimmermann, E. Kaldis, S. Rusiecki, and C. Rossel, Hyperfine Interactions **63**, 25 (1990).

[15] B. Pümpin, H. Keller, W. Kündig, I.M. Savić, J.W. Schneider, H. Simmler, P. Zimmermann, E. Kaldis, S. Rusiecki, C. Rossel, and E.M. Forgan, Journal of the Less-Common Metals **164 & 165**, 994 (1990).

286

[16] H. Keller, W. Kündig, I.M. Savić, H. Simmler B. Stäuble-Pümpin, M. Warden, D. Zech
P. Zimmermann, E. Kaldis, J. Karpinski, S. Rusiecki, J.H. Brewer, T.M. Riseman,
J.W. Schneider, Y. Maeno, and C. Rossel, Physica C **185-189**, 1089 (1991).

[17] P. Zimmermann et al. , private communication.

[18] E.M. Forgan, S.L. Lee, S. Sutton, F. Wellhofer, J.S. Abell, C.E. Gough, S.F.J. Cox, and
C.A. Scott, Supercond. Sci. Technol. **3**, 217 (1990).

[19] E.M. Forgan, S.L. Lee, S. Sutton, J.S. Abell, S.F.J. Cox, C.A. Scott, H. Keller, B. Pümpin,
J.W. Schneider, H. Simmler, P. Zimmermann, and I.M. Savić, Hyperfine Interactions **63**,
71 (1990).

[20] R. Cubitt, E.M. Forgan, M. Warden, S.L. Lee, P. Zimmermann, H. Keller, I.M. Savić, P.
Wenk, D. Zech, P.H. Kes, T.W. Li, A.A. Menovsky, and Z. Tarnawski, Physica C (1993),
in press.

[21] Y.J. Uemura, V.J. Emery, A.R. Moodenbaugh, M. Suenaga, D.C. Johnston, A.J. Jacobson,
J.T. Lewandowski, J.H. Brewer, R.F. Kiefl, S.R. Kreitzman, G.M. Luke, T. Riseman, C.E.
Stronach, W.J. Kossler, J.R. Kempton, X.H. Yu, D. Opie, and H.E. Schone, Phys. Rev. B
38, 909 (1988).

[22] Y.J. Uemura, G.M. Luke, B.J. Sternlieb, J.H. Brewer, J.F. Carolan, W.N. Hardy, R. Kadono,
J.R. Kempton, R.F. Kiefl, S.R. Kreitzman, P. Mulhern, T.M. Riseman, D.Ll. Williams, B.X.
Yang, S. Uchida, H. Tagaki, J. Gopalakrishnan, A.W. Sleight, M.A. Subramanian, C.L.
Chien, M.Z. Cieplak, Gang Xiao, V.Y. Lee, B.W. Statt, C.E. Stronach, W.J. Kossler, and
X.H. Yu, Phys. Rev. Lett. **62**, 2317 (1989).

[23] Y.J. Uemura, L.P. Le, G.M. Luke, B.J. Sternlieb, W.D. Wu, J.H. Brewer, T.M. Riseman,
C.L. Seaman, M.B. Maple, M. Ishikawa, D.G. Hinks, J.D. Jorgensen, G. Saito, and H.
Yamochi, Phys. Rev. Lett. **66**, 2665 (1991).

[24] T. Schneider and H. Keller, Phys. Rev. Lett. **69**, 3374 (1992).

[25] T. Schneider and H. Keller, Physica C **207**, 366 (1993).

[26] T. Schneider, this volume.

[27] P.G. de Gennes, *Superconductivity of Metals and Alloys* (Benjamin, New York, 1966).

[28] M. Tinkham, *Introduction to Superconductivity* (McGraw-Hill, New York 1975).

[29] J. Rammer, Europhys. Lett. **5**, 77 (1988).

[30] J. Annett, N. Goldenfeld, and S.R. Renn, Phys. Rev. B **43**, 2778 (1991).

[31] D. Herlach, G. Majer, J. Major, J. Rosenkranz, M. Schmolz, W. Schwarz, A. Seeger, W.
Templ, E.H. Brandt, U. Essmann, K. Fürderer, and M. Gladisch, Hyperfine Interactions **63**,
41 (1990).

[32] E.H. Brandt, Phys. Rev. B **37**, 2349 (1988); J. Low. Temp. Phys. **73**, 335 (1988).

[33] S.L. Thiemann, Z. Radović, and V.G. Kogan, Phys. Rev. B **39**, 11406 (1989).

[34] W. Barford and J.M.F. Gunn, Physica C **156**, 515 (1988).

[35] See, e.g., D.A. Huse, M.P.A. Fisher, and D.S. Fisher, Nature **358**, 553 (1992), and references therein.

[36] G.J. Dolan, G.V. Chandrashekhar, T.R. Dinger, C. Feild, and F. Holtzberg, Phys. Rev. Lett. **62**, 827 (1989); G.J. Dolan, F. Holtzberg, C. Feild, and T.R. Dinger, Phys. Rev. Lett. **62**, 2184 (1989).

[37] E.M. Forgan, D. McK. Paul, H.A. Mook, P.A. Timmins, H. Keller, S. Sutton, and J.S. Abell, Nature **343**, 735 (1990).

[38] M. Yethiraj, H.A. Mook, G.D. Wignall, R. Cubitt, E.M. Forgan, D.M. Paul, and T. Amstrong, Phys. Rev. Lett. **70**, 857 (1993).

[39] D.R. Harshman, A.T. Fiory, and R.J. Cava, Phys. Rev. Lett. **66**, 3313 (1991).

[40] J.R. Clem, Phys. Rev. B **43**, 7837 (1991); L.N. Bulaevskii, M. Ledvij, and V.G. Kogan, Phys. Rev. **46**, 366 (1992).

[41] E.H. Brandt, Phys. Rev. Lett. **66**, 3213 (1991).

[42] D.R. Harshman, E.H. Brandt, A.T. Fiory, M. Inui, D.B. Mitzi, L.F. Schneemeyer, and J.V. Waszczak, Phys. Rev. B **47**, 2905 (1993).

[43] B. Janossy, D. Prost, S. Pekker, and L. Fruchter, Physica C **181**, 51 (1991).

[44] Q. Li, M. Suenaga, T. Kimura, and K. Kishio, Phys. Rev. B **47**, 2854 (1993).

[45] R.J. Cava, A.W. Hewat, E.A Hewat, B. Batlogg, M. Marezio, K.M. Rabe, J.J. Krajewski, W.F. Peck Jr., and L.W. Rupp Jr., Physica C **165**, 419 (1990).

[46] S. Rusiecki, B. Bucher, E. Kaldis, E. Jilek, J. Karpinski, C. Rossel, B. Pümpin, H. Keller, W. Kündig, T. Krekels, and G. van Tendeloo, Journal of the Less-Common Metals **164 & 165**, 31 (1990).

[47] B. Bucher, E. Kaldis, S. Rusiecki, and P. Wachter, XIII AIRAPT Conference, Bangalore (India), 1991; Proceedings (Oxford, New Delhi 1992).

[48] E. Kaldis, P. Fischer, A.W. Hewat, E.A. Hewat, J. Karpinski, and S. Rusiecki, Physica C **159**, 668 (1989).

[49] A.R. Bishop, R.L. Martin, K.A. Müller, and Z. Tesanovic, Z. Phys. B - Condensed Matter **76**, 17 (1989).

[50] K.A. Müller, Z. Phys. B - Condensed Matter **80**, 193 (1990).

[51] J.C Martinez, S.H. Brongersma, A. Koshelev, B. Ivlev, P.H. Kes, R.P. Griessen, D.G. de Groot, Z. Tarnavski, and A.A. Menovsky, Phys. Rev. Lett. **69**, 2276 (1992), and references therein.

288

[52] E.J. Nicol and J.P. Carbotte, Phys. Rev. B **43**, 1158 (1991).

[53] M.R. Schafroth, Phys. Rev. **100**, 463 (1955).

[54] D.A. Bonn, Ruixing Liang, T.M. Riseman, D.J. Baar, D.C. Morgan, Kuan Zhang, P. Dosanjh, T.L. Duty, A. MacFarlane, G.D. Morris, J.H. Brewer, W.N. Hardy, C. Kallin, and A.J. Berlinsky, Phys. Rev. B **47**, 11314 (1993).

[55] J.L Tallon and N.E. Flower, Physica C **204**, 237 (1993).

[56] H. Zhang and H. Sato, Phys. Rev. Lett. **70**, 1697 (1993).

[57] See, e.g., N. Mori, in *The Physics and Chemistry of Oxide Superconductors*, ed. by Y. Iye and H. Yasuoka, Springer Proceedings in Physics, Vol. **60** (Springer-Verlag, Berlin, Heidelberg 1992) (pp. 191-196), and references therein.

[58] C.C. Almasan, S.H. Han, B.W. Lee, L.M. Paulius, M.B. Maple, B.W. Veal, J.W. Downey, A.P. Paulikas, Z. Fisk, and J.E. Schirber, Phys. Rev. Lett. **69**, 680 (1992).

[59] J.P. Franck, J. Jung, M. A-K. Mohamed, S. Gygax, and G.I. Sproule, Phys. Rev. B **44**, 5318 (1991).

[60] C.L. Seaman, J.J. Neumeier, M.B. Maple, L.P. Le, G.M. Luke, B.J. Sternlieb, Y.J. Uemura, J.H. Brewer, R. Kadono, R.F. Kiefl, S.R. Kreitzman, and T.M. Riseman, Phys. Rev. B **42**, 6801 (1990).

EXTREME TYPE II SUPERCONDUCTORS: UNIVERSAL PROPERTIES AND TRENDS

T. SCHNEIDER
IBM Research Division,
Zurich Research Laboratory,
8803 Rüschlikon, Switzerland

ABSTRACT. We review and analyze experimental results of the specific heat, the muon-spin resonance (μSR) relaxation rate and the fluctuation contributions to the magnetization and dc conductivity of extreme type II superconductors in the context of critical phenomena. Our estimates of critical exponents and amplitudes, the measured scaling behavior, the consistencies with the universal relations between the critical amplitudes and the fall of the transition temperature (T_c) with reduced thickness, which corresponds to a dimensional crossover from 3d to 2d xy-critical behavior, provide considerable evidence for a three-dimensional xy-critical point. The estimated volume of the critical correlation length amplitudes turns out to be comparable to that in superfluid helium and is several orders of magnitude smaller than in BCS superconductors. Moreover, motivated by the Uemura plot, which relates the measured T_c to the zero temperature μSR relaxation rate, and by the hyperuniversal relation between T_c and critical amplitudes of the London penetration depth and phase correlation length, we explore the implications of a simple "universal" scaling ansatz, where the plot of the rescaled transition temperature versus rescaled μSR relaxation rate should fall on a single parabola. Our analysis of the μSR data reveals excellent agreement with this scaling ansatz for a large class of cuprate and Chevrel-phase superconductors. The resulting dependence of T_c on the zero-temperature magnetic penetration depth is then used to explore universal trends in the pressure (α) and isotope (β) coefficients. In good agreement with experiment we find that α and β fall onto common $T_c - \alpha$ and $T_c - \beta$ regions, respectively, forming two branches, one for systems with positive and the other for compounds with negative pressure or isotope coefficient. The two branches merge at the maximum T_c where the coefficients vanish and the magnitude of the coefficients increases with reduced T_c. The "universal" scaling ansatz also implies that the critical amplitudes of the penetration depth and phase correlation length are related to the zero temperature magnetic penetration depth. Empirically, this quantity is also related to the hole concentration. Finally we discuss the temperature dependence of the penetration depth. μSR measurements indicate that for various cuprates the temperature dependence appears to be bounded by the behavior of the two-fluid model and the dilute Bose gas. High-T_c compounds turn out to be closer to two-fluid behavior, while the behavior of systems with reduced T_c points to a crossover to the dilute gas. Combined with the "condensate" dependence of T_c these trends suggest that Bose condensation of weakly charged pairs is the mechanism that drives the transition.

1. Introduction

An essential and unifying property of superconductivity is the phase transition from the normal to the superconducting state. Although interest has focused on the microscopic mechanism of superconductivity in the cuprates and their peculiar normal state properties, their thermodynamic and transport properties appear to be far more distinct from those of conventional superconductors. A crucial difference between the

E. Kaldis (ed.), Materials and Crystallographic Aspects of HTc-Superconductivity, 289–310.
© 1994 IBM.

cuprates and standard type II superconductors is that the latter exhibit a gradual decrease of their electrical resistance with temperature in the presence of a magnetic field. This phenomenon is attributed to the interplay of thermal fluctuations and disorder, and it raises the fundamental question of whether the cuprates and, more generally, extreme type II superconductors have zero resistance when cooled in a magnetic field [1]. Another essential difference is the drastic enhancement of thermal fluctuations, resulting from a combination of not unrelated factors: (i) higher transition temperature, (ii) short correlation length which characterizes the size of the bound electron pairs, (iii) large magnetic penetration depth reflecting weak supercurrents because of the relatively low density of mobile electron pairs, and (iv) pronounced anisotropy. Indeed, estimates of the critical regime, that is the temperature interval around T_c where the standard mean-field treatment fails and critical fluctuations dominate, is experimentally accessible [2], while in conventional superconductors (with the exception of ultrathin films and thin wires), the fluctuation dominated regime is largely an academic issue. Although these issues, associated with strong thermal fluctuations, disorder and their interplay, can be addressed entirely within the context of the Ginzburg-Landau theory, it will take more than the standard mean-field treatment to resolve them.

Here, we analyze and review the evidence and implications of critical fluctuations in extreme type II superconductors with short correlation lengths. Recent measurements of the specific heat [3,4], the fluctuation contributions to the magnetization [4], susceptibility [5] and dc conductivity [6] revealed remarkable evidence for three-dimensional (3d) xy-critical behavior in superconducting YBa$_2$Cu$_3$O$_{7-x}$. It is important to recognize that the critical fluctuations include more than just the Gaussian ones which have been studied rather extensively in the context of superconductivity for thermodynamic and transport properties [7]. Indeed, close to T_c, critical fluctuations invalidate mean-field treatments, even with the inclusion of Gaussian fluctuations, and lead to singular behavior in the thermodynamic and transport quantities of interest. This singular behavior is usually characterized by critical amplitudes and exponents. A central concept in the modern theory of critical phenomena is the universality of critical exponents, critical amplitude ratios and critical amplitude relations [8,9]. Within the universality class, the lattice structure, coupling constants, etc. may vary, but such variations only affect the nonuniversal factors [8,9]. Thus the concept of universality greatly reduces the variety of different types of critical behavior by dividing them into universality classes.

In extreme type II superconductors the complex and scalar order parameter Ψ is supposed to have stronger thermal fluctuations than the magnetic field. Thus, by approaching T_c in zero external magnetic field, these fluctuations are expected to alter the mean-field into critical 3d xy-behavior. In other words, the crossover upon approaching T_c is initially to the critical regime of a weakly charged superfluid where the fluctuations in Ψ are nearly those of an uncharged superfluid, like He^4 or the xy-model [10,11]. In this regime the fluctuation behavior is dominated by a critical point belonging to the 3d xy-universality class.

Following Refs. [11] and [12], the paper is structured as follows: We begin in Sect. II with a sketch of the 3d xy-critical point behavior, extended to extreme type II superconductors with uniaxial effective mass anisotropy. This behavior includes critical exponents and amplitudes, and the universal relations between the critical exponents, the critical amplitudes and between the critical amplitudes and the transition temperature. Indeed, the critical amplitudes of the specific heat and the correlation lengths are not mutually independent and the transition temperature can be expressed in terms of the amplitudes of the London penetration depth and the phase correlation length. Along with the critical exponents, these relations are universal for the 3d *xy*-critical point. Moreover, we sketch the universal scaling functions of the specific heat, the fluctuation contribution to the magnetization, magnetic susceptibility and dc conductivity. Rather accurate estimates of the critical exponents, universal critical amplitude ratios and combinations for the 3d *xy*-critical point are available from renormalization group methods [9,13] and are in agreement with experiment [9,14], in particular for the transition to the superfluid state in superfluid helium. For systems that are finite in one direction, this scenario also implies a dimensional crossover from 3d to 2d *xy*-critical behavior and a fall of the transition temperature from the bulk value to $T_c = T_{KT}$ with reduced thickness. Indeed, the phase transition in any *xy*-system, finite in one direction and infinite in the two remaining ones, will in principle be a Kosterlitz-Thouless (KT) transition with no long-range order even below T_c [15,16]. The KT transition is due to vortex unbinding which occurs with increasing temperature in strictly 2d systems. The essential KT predictions are also expected to hold for films, provided the temperature is sufficiently close to the transition temperature $T_c = T_{KT}$. Here we sketch how the fall of T_c with reduced thickness and the associated crossover from 3d to 2d *xy*-critical behavior can be described by crossover scaling, in close analogy to superfluid helium films [16].

It should be kept in mind, however, that the phase transition in superfluid helium is particularly suited to test critical point universality. Owing to the absence of sample inhomogeneities and strains, there is a sharp transition and T_c can be approached very closely. Continuous phase transitions in solids, including the one to the superconducting state, are blurred by internal strains, sample inhomogeneities and finite size effects that arise from the finite extent of the crystallites. Moreover, the singular part of the measured quantity is usually accompanied by a vast background. In view of this, it appears to be difficult to examine the consistency with a particular critical point just from experimental estimates of critical exponents. By including the hyperuniversal relations for the critical amplitudes and the universal scaling functions, however, the experimental evidence for a particular critical point can be considerably strengthened.

In Sect. 3 we present, review and analyze recent specific heat [3,4], magnetization [4], susceptibility [5], μSR relaxation rate [17-20] and dc conductivity [6] measurements for cuprate superconductors to assess the consistency with 3d *xy*-critical point behavior. The temperature dependence of the zero field specific heat in $DyBa_2Cu_3O_{7-x}$ [3] appears to be fully consistent with a 3d *xy*-critical point. Further evidence is provided by the universal scaling property of the specific heat and magnetization in an applied magnetic field [4] and by the fluctuation contribution to the susceptibility in the zero field limit [5]. The hyperuniversal relations involving the critical amplitudes of the phase correlation lengths, the specific heat, the London penetration depth and the transition temperature are then used to check the consistency of the estimates for the critical amplitudes and to account for the decrease observed in the heat-capacity anomaly of oxygen-deficient $YBa_2Cu_3O_{7-x}$ [21]. The critical amplitude of the London penetration depth will be estimated from the measured μSR relaxation rates [17-20].

Finally we turn to the fluctuation contribution of the dc conductivity. Invoking dynamic scaling and purely relaxational dynamics (Model A) [22] the singular behavior turns out to be proportional to the correlation length. Again, the measured temperature dependence in sintered $YBa_2Cu_3O_{6.96}$ [6] turns out to be remarkably consistent with a 3d *xy*-critical point. Thus, the estimates for the critical exponents, critical amplitudes, the consistency with the universal relations and the universal scaling behavior point clearly to a 3d *xy*-critical point. Although this seems to be unrelated at first glance, T_c was found to drop with reduced thickness of ultrathin slabs of YBCO that are M unit cells thick, accompanied with KT behavior in the temperature dependence of the resistance and in the current-voltage characteristics [23-25]. Here we concentrate on the fall of T_c with reduced thickness. In agreement with crossover scaling this fall appears to be fully determined by the temperature dependence of the bulk condensate density [15].

In Sect. 4 we turn to the universal trends. Motivated by the Uemura plot [23], which relates T_c and the zero temperature value of the μSR relaxation rate, as well as by the hyperuniversal relation between T_c and the critical amplitudes of the London penetration depth and the phase correlation length, we explore a simple parabolic scaling ansatz [12,26]. T_c is expressed in units of the maximum value T_c^m for a particular family of compounds, and the μSR relaxation rate $\sigma(0)$ is measured in units of its value $\sigma^m(0)$ at the maximum T_c. In this scaled form the μSR data should fall on a single parabola. According to our analysis of the μSR data for several families of cuprate and Chevrel phase superconductors, there is remarkable evidence for this simple scaling behavior. Moreover, our estimates for T_c^m and $\sigma^m(0)$ strongly suggest that these quantities are not independent, but linearly related. Noting that $\sigma(0)$ is proportional to the zero temperature value of the condensate density, it appears natural that the "universal" scaling property of T_c implies universal trends in the pressure and isotope dependence of T_c, measured in terms of the pressure (α) and isotope (β) coefficients. Indeed, the resulting trends are in good agreement with experiment [27-34]. α and β fall onto common $T_c - \alpha$ and $T_c - \beta$ regions, respectively, forming two branches, one for systems with positive and the other for compounds with negative pressure or isotope coefficient. The two branches merge at the maximum T_c where the coefficients vanish and the magnitude of the coefficients increases with reduced T_c. Both pressure [27] and isotope effect [34] are considered to be instrumental to the understanding of the mechanism responsible for superconductivity. The "universal" scaling ansatz also implies that the critical amplitudes of the penetration depth and phase correlation length are related to the zero temperature condensate density. Moreover, we explore the relation between zero temperature condensate density and hole concentration. Finally, we discuss the temperature dependence of the penetration depth. μSR measurements indicate that for various cuprates the temperature dependence for $0 < T < T_c$ appears to be bounded by that of the two-fluid model and dilute Bose gas, respectively. High-T_c compounds turn out to be close to the two-fluid behavior, while in compounds with reduced T_c, a crossover to the dilute Bose gas appears to occur. Combined with the estimates of the asymptotic $T \to 0$ behavior and the dependence of T_c on "condensate density" these trends point to Bose condensation of interacting and weakly charged bound pairs as the mechanism driving the transition. In any case, our analysis and in particular the strong evidence for a 3d interaction driving the transition places crucial constraints on the microscopic theory.

2. Universal and Critical Properties of Extreme Type II Superconductors

Critical scaling behavior occurs in all continuous phase transitions and can be attributed to the existence of a characteristic length which diverges at the critical temperature T_c [8,9]. The closely related concept of universality greatly reduces the variety of different types of critical behavior by dividing systems into universality classes. Characteristics of a universality class are the space dimension d of the system, the dimensionality of the order parameter and the range of the interaction driving the transition. In addition, all systems within a given universality class have the same critical exponents, scaling functions and, along with the relations between exponents, corresponding universal relations between critical amplitudes [8,9].

Extreme type II superconductors with short correlation lengths are supposed to belong to the 3d xy-universality class with the complex scalar order parameter

$$\Psi(r) = |\Psi(r)|\,e^{i\varphi(r)} = |\Psi(r)|\,(\cos(\varphi(r)) + i\sin(\varphi(r)))\ , \tag{1}$$

In general both the magnitude $|\Psi(r)|$ and the phase $\varphi(r)$ will vary in space but the phase fluctuations dominate the critical behavior in the xy-universality class. In the limit of vanishing applied magnetic field, the specific heat C and the fluctuation contribution to the magnetic susceptibility χ diverge in the vicinity of the transition temperature T_c as [8,11,12]

$$C = B^{\pm} + (A^{\pm}/\alpha)\,|t|^{-\alpha}\ , \qquad \chi(t, H=0) \propto |t|^{-\gamma}\ , \qquad t = 1 - T/T_c\ . \tag{2}$$

$A^{\pm}$ is the amplitude of the specific heat singularity and α the associated critical exponent. The index $+$ refers to $T > T_c$ and $-$ to $T < T_c$. In the presence of a magnetic field, the specific heat difference $\Delta C(T, H) = C(t, H=0) - C(t,H)$ and the magnetization m scale as [11,12]

$$\Delta C(t, H_{\perp})H_{\perp}^{\alpha/2\nu} = C(x)\ , \qquad m/H^{1/2} = M(x)\ , \qquad x = t/H_{\perp}^{1/2\nu}\ . \tag{3}$$

$M(x)$ and $C(x)$ are universal scaling functions.

In Table I we list theoretical estimates for critical exponents and universal amplitude relations for the 3d xy-critical point. Only two exponents are needed to determine the others. Moreover, due to the universal combinations of amplitudes, only two are needed to fix the others. In particular, the ratio A^+/A^- and the amplitude combinations [11,12]

$$R_{\varphi}^3 = A^-(\xi_{\|0}^{\varphi})^2 = \xi_{\perp 0}^{\varphi}\ , \qquad R_{\xi}^3 = A^+(\xi_{\|0}^+)^2\xi_{\perp 0}^+ \tag{4}$$

adopt universal values. The phase correlation length is associated with the fluctuations in the imaginary part of the complex order parameter ξ^{φ}, while the correlation length ξ probes the fluctuations of the real part. ξ^{φ} is defined as being below T_c and ξ^+ as being above T_c. Here we assumed uniaxial mass anisotropy in the Ginzburg-Landau functional. The correlation lengths and the London penetration depth perpendicular and parallel to the layers are related by [11,12]

$$\frac{\xi_{\|}^{\varphi}}{\xi_{\perp}^{\varphi}} = \left(\frac{\lambda_{\|}}{\lambda_{\perp}}\right)^2 = \frac{M_{\|}}{M_{\perp}}\ , \qquad \frac{\xi_{\|0}^{\varphi}}{\xi_{\|0}^+} = \sqrt{\frac{M_{\|}}{M_{\perp}}}\,\frac{\xi_{\perp 0}^{\varphi}}{\xi_{\perp 0}^+}\ , \qquad \frac{\xi_{\|}^+}{\xi_{\perp}^+} = \sqrt{\frac{M_{\perp}}{M_{\|}}} \tag{5}$$

where the London penetration depth is given by

$$1/\lambda_{\|,\perp}^2 = \frac{4\pi n_s e^2}{M_{\|,\perp}c^2}\ , \tag{6}$$

where n_s denotes the condensate density and $M_{\|,\perp}$ the effective Ginzburg-Landau mass. Thus, $\|$ denotes the behavior parallel and $\perp$ perpendicular to the layers. From the universal relation and Table I it is clear that the volume of the correlation amplitudes is fully determined by the specific heat amplitude, expressed as inverse volume.

Relations (4) are a consequence of hyperuniversality, where $dv = 2 - \alpha$ [9]. For the 3d xy-critical point hyperuniversality also implies that [9,11,12]

$$k_B T_c = \frac{\hbar^2}{4M_\parallel} n_{s0} \xi^\varphi_{\parallel 0} = \frac{\hbar^2}{4M_\perp} n_{s0} \xi^\varphi_{\perp 0} = \frac{\hbar^2 c^2}{16\pi e^2} \frac{\xi^\varphi_{\parallel,\perp 0}}{\lambda^2_{\parallel,\perp 0}} \quad , \tag{7}$$

where $n_s = n_{s0}|t|^{-\nu}$. To obtain the latter term we used the London expression (6). Thus, the transition temperature, the amplitudes of the London penetration depth and phase correlation length are not mutually independent.

Table I: Estimates for critical exponents and universal quantities for the 3d xy-critical point [9,13]

Critical exponents	Universal quantities
$\nu \approx 0.669$	$A^+/A^- \approx 1.05$
$\alpha = 2 - 3\nu \approx -0.007$	$R_\xi \approx 0.3$
	$R_\varphi \approx 0.8$

To summarize, only two exponents are needed to determine the others. Moreover, due to the universal combinations of amplitudes, only two are needed to fix the other amplitudes. The universality class to which a given system belongs is thus not only characterized by its critical exponents but also by the relations between the critical amplitudes. Since universal amplitude combinations vary more widely than the corresponding exponents, the amplitude relations are very useful to determine or eliminate a given universality class. .

In this context it is important to recognize that in a superconductor the order parameter is coupled to the vector potential A. Indeed, the "boson" responsible for superfluidity carries a nonzero charge ($\Phi_0 = hc/2e$) in addition to its mass, and the charge couples the order parameter to the electromagnetic field. For extreme type II superconductors, however, where

$$\kappa_{\parallel,\perp} = \frac{\lambda_{\parallel,\perp}}{\xi_{\parallel,\perp}} \tag{8}$$

is supposed to be large, the resulting effective dimensionless charge $\tilde{e}_{\parallel,\perp} = 1/\kappa_{\parallel,\perp}$ [10] is small and the screening is weak. In this case the thermodynamic properties associated with the fluctuations of the order parameter are essentially those of an uncharged superfluid or xy-model. It is important to recognize, however, that in the 3d xy-critical regime this behavior is restricted by

$$\tilde{e} = \frac{1}{\kappa_{\parallel,\perp}} = \frac{\xi_{\parallel,\perp 0}|t|^{-\nu}}{\lambda_{\parallel,\perp 0}|t|^{-\nu/2}} = \frac{\xi_{\parallel,\perp 0}}{\lambda_{\parallel,\perp 0}} t^{-\nu/2} \ll 1 \quad . \tag{9}$$

Thus, as one approaches T_c the charge becomes relevant. When $\tilde{e}$ becomes of order unity, where $t = t^*$, the system is expected to cross over into another critical regime, that of a charged superfluid. For the typical values $\xi_{\|0} \approx 10\,A$ and $\lambda_{\|0} \approx 1000\,A$ we obtain $t^* \approx 10^{-6}$ corresponding to a correlation length of $\xi_\| \approx 10^5\,A$, which appears to be inaccessible due to sample inhomogeneities. Thus the crossover upon approaching T_c is essentially due to the critical regime of a weakly charged superfluid where the fluctuations in Ψ are close to those of an uncharged superfluid, like He^4 or the xy-model.

For bulk systems that become finite in one direction a dimensional crossover from 3d to 2d critical behavior will occur. In the xy-universality class the 2d transition is of KT type, with no long-range order below the transition temperature T_c. The KT transition is due to vortex unbinding which occurs with increasing temperature in strictly 2d systems. Nevertheless the essential KT predictions are also expected to hold for slabs, provided one is sufficiently close to the transition temperature T_{KT} [16]. Clearly, the lateral dimension must always be large compared to the thickness. Finite size scaling then describes the crossover from 3d to 2d behavior in terms of a crossover function, which connects the aerial condensate density n_{s2} with the bulk counterpart n_s [16]:

$$n_{s2} = n_s d\Phi\left(\frac{\xi_\perp^\varphi}{d} \right) . \tag{10}$$

$\Phi(x)$ is a universal crossover function and d the slab thickness. Clearly $\Phi(x) \to 1$ for $x \to 0$. In terms of the phase correlation length one requires

$$\xi_\perp^\varphi(T_{KT}) = dx_c \tag{11}$$

at the KT transition of a slab of thickness d in order for the finite system to be two-dimensional. The term x_c is of order unity and depends on the boundary conditions, but is supposed to be independent of thickness d. Combining this with the universal expression for the KT transition temperature, we obtain [15]

$$T_{KT} = \frac{\pi}{8} \frac{\hbar^2}{M_\|} n_{s2}(T_{KT}) = \frac{\pi}{8} \frac{\hbar^2}{M_\|} n_s(T_{KT}) d\Phi(x_c) , \qquad \Phi(x_c) = \frac{2}{\pi} x_s . \tag{12}$$

Invoking the hyperuniversal expression for the bulk transition temperature [15] we obtain

$$\frac{T_{KT}(d)}{T_c} = \frac{n_s(T_{KT})}{n_{s0}} \frac{d}{\xi_\|^\varphi} 0\,x_c , \qquad x_c = \frac{\xi_\perp^\varphi(T_{KT})}{d} . \tag{13}$$

Thus, the fall of the transition temperature with reduced thickness appears to be fully determined by the temperature dependence of the bulk condensate density. It is important to emphasize that this scenario only applies if the physics remains unchanged for thinner slabs.

Up to now we assumed in the bulk system three-dimensionality that $\xi_\|^\varphi \gg c$ for $T < T_c$ and $\xi_\perp \gg c$ for $T > T_c$, where c denotes the layer spacing. In the opposite limit $\xi_\|^\varphi < c$ ($T < T_c$) and $\xi_\perp < c$ ($T > T_c$) the system corresponds to nearly independent quasi-2d superconductors, each undergoing a KT phase transition. In this context it is important to note that in zero magnetic field and for $T < T_c$ the phase correlation length, which increases with rising anisotropy (Eq. 5), sets the scale. It is instructive to estimate the temperature regime where 2d fluctuation behavior can occur. Taking $YBa_2Cu_3O_7$ as an example we obtain with the estimates for the correlation length amplitudes listed in Table II ($\xi_\perp = \xi_{\perp 0} t^{-\nu} \approx 1.2 t^{-2/3} = 12$ A) $t = (T - T_c)/T_c \geq 0.03$ for $T > T_c$, while for $T < T_c$ there is no 2d fluctuation regime because $\xi_\parallel^\varphi \approx 83$ A $< c = 12$ A.

3. Experimental Evidence for Critical 3d xy-Behavior and 3d-2d xy-Crossover

3.1 SPECIFIC HEAT

First we consider the critical behavior in zero magnetic field. Due to the small value of the critical exponent α (Table I) the specific heat expression (2) can be written as

$$C^\pm \approx \frac{A^\pm}{\alpha} + B^\pm - A^\pm \ln|t| . \tag{14}$$

$B^\pm$ accounts for the background. As an example we reproduced in Fig. 1 the specific heat measurements of polycrystalline $DyBa_2Cu_3O_{7-x}$ [3]. In the temperature range $6 > -\ln|t| > 4$ the measured specific heat, plotted versus $\ln|t|$, is seen to fall on nearly parallel branches with finite slope. The upper branch corresponds to $T < T_c$ and the lower one to $T > T_c$ (Fig. 1b) so that $B^- + A^-/\alpha > B^+ + A^+/\alpha$. This reflects the mean field behavior, accounted for here as background. Apparently, these measurements are fully consistent with a 3d xy-critical point and provide an estimate for the critical amplitude $A^+ \approx A^-$. In Table II we list recent estimates for $1/A^+ \approx 1/A^-$ and the effective mass ratio for cuprate superconductors and superfluid helium [11]. Using the hyperuniversal amplitude relations (4) with Eq. (5), the amplitudes of the correlation

---

Table II: Measured values of T_c [K] and $M_\perp/M_\parallel$ experimental estimates for $A^+ \approx A^-$ after Ref. [11] in units of A^{-3}. The amplitudes of the correlation lengths and the associated volume where derived from Eqs. (4) and (5) with R_φ and R_ξ taken from Table I.

	T_c	$\dfrac{M_\perp}{M_\parallel}$	$\dfrac{1/A^+\approx}{1/A}$	$\dfrac{(\xi_{\parallel 0}^\varphi)^2}{\xi_{\perp 0}^\varphi}$	$\xi_{\parallel 0}^\varphi$	$\xi_{\perp 0}^\varphi$	$\xi_{\parallel 0}^+$	$\xi_{\perp 0}^+$
$DyBa_2Cu_3O_7$	91.4	[25]	1130	579	2.9	71	5.3	1.1
$YBa_2Cu_3O_7$	91.7	25	1790	916	3.3	83	6.2	1.2
He^4	2.18		600	307	6.7		2.5	

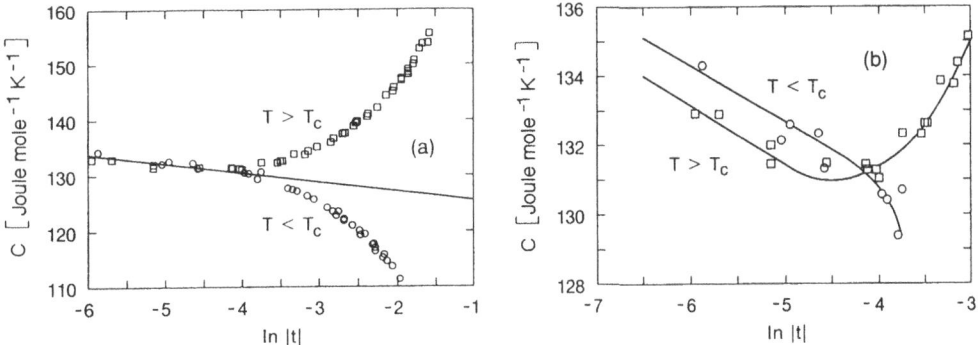

Fig. 1: Specific heat versus $\ln |t|$ of $DyBa_2Cu_3O_{7-x}$ after Ref. 3. [From Ref. 11].

lengths and the associated volume are then readily calculated. These quantities are also included in Table II. Surprisingly enough, the correlation volume $(\xi_{\parallel 0}^{\varphi})^2 \xi_{\perp 0}^{\varphi}$, corresponding roughly to the volume of a bound pair of electrons, are comparable to the volume of the unit cell and to that of He^4, but are several orders of magnitude smaller than in conventional superconductors [12].

Further evidence for critical 3d xy-behavior has been provided in terms of the data collapse of the temperature and field-dependent specific heat [4]. Indeed, the specific heat data scaled according to Eq. (3) should fall on a single curve given by $\Delta C(t, H_\perp) H_\perp^{\alpha/2\nu} = C(x)$ where $x = t/H_\perp^{1/2\nu}$. The results of Salamon et al. [4] for $YBa_2Cu_3O_{7-x}$ are reproduced in Fig. 2 for the critical exponents listed in Table I. The collapsing of the data is excellent, except in the vicinity of T_c where rounding effects appear.

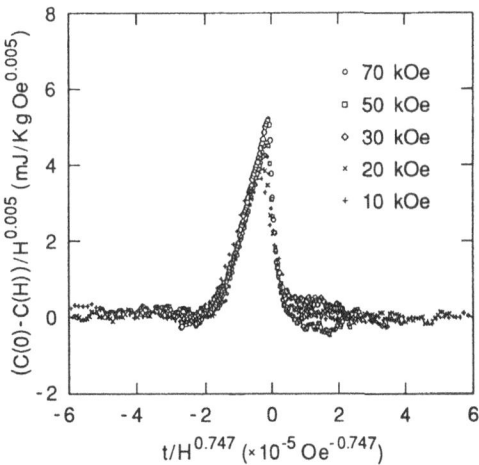

Fig. 2: Scaled heat capacity data for $YBa_2Cu_3O_{7-x}$ after Salamon et al. [4].

3.2 MAGNETIZATION AND SUSCEPTIBILITY

In the zero field limit the contribution of critical fluctuations to the susceptibility is given by $\chi(t, H = 0) \propto |t|^{-\nu}$ (Eq. 1). Figure 3 shows the fluctuation contribution to the susceptibility in sintered $YBa_2Cu_3O_{6.96}$ measured just above T_c [5]. In the reduced temperature range $10^{-4} < t < 10^{-2}$, an analysis of the data in terms of Eq. (32) yields $\nu = 0.66(5)$ and $T_c = 92.14(4)$ K, in agreement with the theoretical value for the 3d xy-critical point (Table I). In a finite magnetic field, the magnetization should scale according to $m/H^{1/2} = M(x)$, where $x = t/H^{1/2\nu}$ (Eq. 3). The scaled experimental data of Salomon et al. [4] for single-crystal $YBa_2Cu_3O_{7-x}$ are reproduced in Fig. 4 with the critical exponent ν appropriate for the 3d xy-critical point (Table I). The collapsing of the data is impressive.

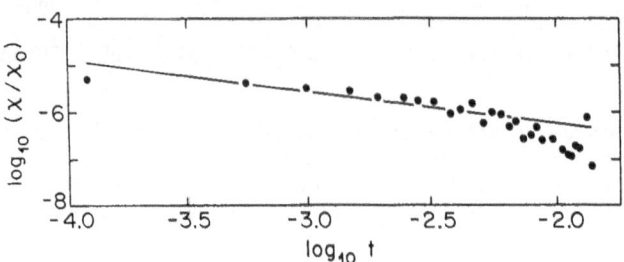

Fig. 3: Fluctuation contribution to the susceptibility $\log(x/x_0)$ versus $\log|t|$ for sintered $YBa_2Cu_3O_{6.96}$ [5]. According to the power law (2), the slope of the straight line determines the critical exponent $\nu = 0.66(5)$.

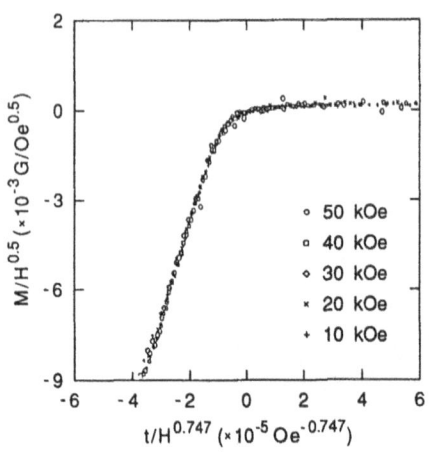

Fig. 4: Scaled magnetization data for $YBa_2Cu_3O_{7-x}$ after Salomon et al. [4].

3.3 THE HYPERUNIVERSAL RELATION FOR THE TRANSITION TEMPERATURE

The hyperuniversal relation (7) reveals that the transition temperature and the amplitudes of the London penetration depth and phase correlation length are not mutually independent. Because $\lambda(t)$ and T_c in turn $\lambda_{\parallel, \perp 0}$ can be measured experimentally, this relation provides an independent estimate of the phase correlation length amplitude. Indeed, the values listed in Table II have been derived from the specific heat amplitude A^-, the ratio of the effective masses $M_\perp/M_\parallel$ and the hyperuniversal relation (4). Moreover, Eq. (7) provides a basis for the correlation observed between T_c and the μSR relaxation rate, which is simply related to the penetration depth [17,35],

$$\sigma = \left(\frac{2754B}{\lambda_\parallel} \right)^2. \tag{15}$$

λ is measured in A and σ in μsec^{-1}. In sintered materials, we have $B = 0.813$ and in single crystals, $B = 1$ [17]. In Table III we list μSR data and estimates for the amplitudes of penetration depth λ and phase correlation length ξ, as derived from Eqs. (7) and (15). Comparing the two estimates for $\xi_{\parallel 0}^\varphi$ in YBCO with nearly optimum T_c, one determined from the hyperuniversal relation (4) the experimental specific heat amplitude A and effective mass anisotropy (Table II) and the other from the hyperuniversal relation (7) in terms of measured T_c and $\lambda_{\parallel 0}$ (Table III), we observe satisfactory agreement, which reveals the consistency of the independent estimates with the 3d xy-critical point.

Table III: μSR estimates [14-17] for the zero temperature values and the critical amplitudes of the μSR relaxation rate $\sigma_\parallel(0)$, the penetration depth $\lambda_\parallel(0)$ and its critical amplitude $\lambda_{\parallel 0}$ derived from the experimental data in the vicinity of T_c by assuming the power law $1/\lambda_\parallel^2 = 1/\lambda_{\parallel 0}^2 |t|^\nu$ with $\nu = 2/3$. The corresponding value of $\xi_{\parallel 0}^\varphi$ was evaluated with Eq. (7).

	T_c	$\sigma(0)$	σ_0	$\lambda_\parallel(0)$	$\lambda_{\parallel 0}$	$\xi_{\parallel 0}^\varphi$
YBa$_2$Cu$_3$O$_{6.97}$	91.6	3.06	4.6	1280	1293	2.5
YBa$_2$Cu$_3$O$_{6.556}$	58.8	1.00	1.4	2239	1892	3.4
YBa$_2$Cu$_4$O$_8$	80.2	1.93	2.0	1612	1583	3.2
Bi$_2$Sr$_2$CaCu$_2$O$_8$	86.4	1.18	1.3	2061	1964	5.3

Rewriting the hyperuniversal relation (4) in the form

$$A^- = \frac{M_\parallel}{M_\perp} \frac{R_\varphi^3}{(\xi_{\parallel 0}^\varphi)^3} \tag{16}$$

we find that the ξ values listed in Table III might be used to explore trends in the amplitude of the specific heat singularity. Indeed the amplitude of the specific heat singularity is seen to decrease with increasing effective mass anisotropy and phase correlation length amplitude. From specific heat measurements on YBa$_2$Cu$_3$O$_{7-x}$ it is known that A^- decreases with reduced transition temperature [21]. According to Table III and Eq. (16), a considerable reduction is indeed expected, because the phase correlation length amplitude increases with reduced oxygen content. A further reduction results from the increase of the effective mass anisotropy with reduced oxygen content [35]. Noting that the amplitude of the phase correlation length in YBCO-124 and BISCO-2212 is considerably larger than that of YBCO with maximum T_c (Table III) and taking the more pronounced effective mass anisotropy of these compounds for granted, one expects in these compounds a considerably reduced specific heat anomaly.

The hyperuniversal relations (4) and (7) might also be combined to eliminate the amplitude of the phase correlation length in terms of directly measurable quantities. This yields

$$k_B T_c = R_\varphi \, \frac{\hbar^2 c^2}{16\pi e^2} \, \frac{1}{\lambda_\parallel^2} \left(\frac{1}{A^-} \, \frac{M_\parallel}{M_\perp} \right)^{1/3} \qquad (17)$$

and reveals that T_c is determined by the amplitudes of specific heat and London penetration depth and the effective mass anisotropy.

3.4 DIMENSIONAL CROSSOVER AND FALL OF T_c WITH REDUCED THICKNESS

Recently it became possible to fabricate thin YBCO films, M unit cells thick, as sandwiches or superlattices with thick PrPCO slabs in between. PrPCO is supposed to be insulating. This experiments provide clear evidence for the fall of T_c with reduced thickness of the superconducting slabs [23-25]. Within the xy-universality class this phenomenon can be understood in terms of a dimensional crossover from 3d to 2d behavior. According to the crossover expression (13), this can be rewritten in the form

$$\frac{T_{KT}(d)}{T_c M} = \frac{n_s(T_{KT})}{n_{s0}} \, \frac{c}{\xi_{\parallel 0}^\varphi} \, x_c \, , \qquad x_c = \frac{\xi_\perp^\varphi(T_{KT})}{Mc} \, , \qquad (18)$$

where the slab thickness d was replaced by $d = Mc$. M denotes the number of unit cells and c the lattice spacing perpendicular to the layers. In Fig. 5 we plotted the available estimates for the zero resistance transition temperature for various YBCO slabs M unit cells thick in terms of $T_{KT}(M)/(T_c M)$ versus $T_{KT}(M)/T_c$ [23-25]. According to the dimensional crossover scaling expression (18), the experimental points should follow the temperature dependence of the condensate bulk density n_s. In YBCO this quantity

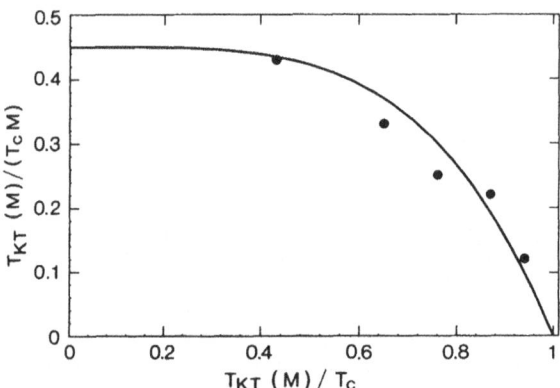

Fig. 5: $T_{KT}(M)/(T_c M)$ versus $T_{KT}(M)/T_c$ for YBCO slabs $M = 1,2,3,4,8$ unit cells thick. The temperature dependence resulting from the two-fluid model for the bulk condensate density is included for comparison [From Ref. 15].

is known to follow the two-fluid model reasonably well, with

$$n_s(t) = n_s(0)\left(1 - \left(\frac{T}{T_c}\right)^4\right), \qquad n_{s0} = 4n_s(0) . \tag{19}$$

The resulting behavior is shown in Fig. 5 with $x_c c/(4\xi_{\parallel 0}^0 \approx 0.45$.) For $c = 12 A$ and $\xi_{\parallel 0}^0 \approx 2.5 A$ (Table II), we arrive at $x_c \approx 0.38$, while the experimental data for He^4 films gives $x_c \approx 0.3$ [15]. Thus, the near coincidence of the x_c values of the YBCO and He^4 films strongly suggests that the quality of the films is comparable and that the main contribution to the observed fall of T_c in the YBCO slabs must be attributed to dimensional crossover.

3.5 CONDUCTIVITY

Finally we turn to conductivity. In mean-field theory it jumps discontinuously from the normal state value above T_c to infinity below T_c. Fluctuations of the order parameter modify this result. According to the theory of dynamic critical phenomena, the characteristic order parameter relaxation time diverges as $\tau \propto \xi^z$ where z is the dynamic critical exponent. Thus, below T_c the low-frequency conductivity is expected to scale as [10,11]

$$\sigma(\omega) = n_s F^-(\omega \xi^z) \propto \xi^{2-d+z} F^{\pm}(\omega \xi^z) . \tag{20}$$

The second term is valid above T_c as well and F is a universal scaling function. In the critical regime the dc conductivity is then given by

$$\sigma(0) \propto \xi^{2-d+z} . \tag{21}$$

In an uncharged superfluid, like He^4, model-E dynamics would apply [22]. In anisotropic charged superconductors, however, there are plasmons even for long wavelengths. Due to collisions with phonons and other excitations, the charge current is not expected to be conserved. This implies a finite lifetime of the plasmons even at long wavelengths. On these grounds, hydrodynamic modes are not expected to occur in superconductors, and model-A dynamics (which has no conserved quantities) should apply when $z = 2$. In this case we ultimately obtain for $d = 3$

$$\sigma(0) \propto |t|^{-\nu} . \tag{22}$$

In Fig. 6 we show the conductivity data for sintered in YBa$_3$Cu$_3$O$_{6.96}$ terms of $\log(\Delta\sigma/\sigma_0)$ versus $\log|t|$ [6]. Here $\Delta\sigma = \sigma - \sigma_B$ is the fluctuation contribution to the conductivity, σ_B denotes the temperature-dependent background and $\sigma_0 = \sigma(150\ K)$. The straight line in Fig. 6 corresponds to a fit of the power law (22) to the data, yielding $\nu = 0.67(4)$ and $T_c = 92.44(5)\ K$ in the reduced temperature range $6 \times 10^{-3} < t < 10^{-2}$ in excellent agreement with the susceptibility measurements shown in Fig. 3. The resulting estimate for ν is again in remarkable agreement with the theoretical value for the 3d xy-critical point (Table I). The deviations appearing close to T_c are most likely due to a finite size effect. Indeed, by approaching T_c, the correlation increases and at some temperature $T > T_c$ reaches the linear dimension of the grains.

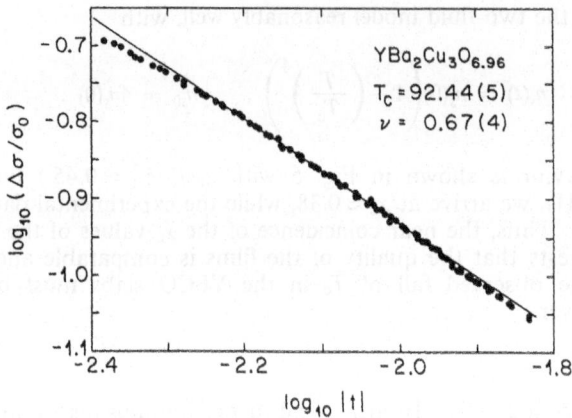

Fig. 6: Fluctuation contribution to the conductivity for $\log(\Delta\sigma/\sigma_0)$ versus $\log|t|$ for sintered $YBa_2Cu_3O_{6.96}$ [6]. According to Eq. (22), the slope of the straight line determines the critical exponent $\nu = 0.67(4)$.

To summarize this section, it is fair to say that the estimates for the critical exponents, the scaling behavior of the magnetization and the specific heat, the consistency of the critical amplitudes with the hyperuniversal relations, the experimental evidence for the resulting trends and the fall of T_c with reduced slab thickness (which corresponds to a 3d-2d xy-crossover) provide considerable evidence for xy-critical point behavior. Moreover, the cuprate superconductors considered here are very far from the BCS limit, because the volume of the correlation length amplitudes is comparable to that of superfluid helium and several orders of magnitude smaller than in conventional superconductors (Table II).

4. Universal Trends

μSR measurements strongly suggest that many extreme type II superconductors share the unique property that their transition temperature T_c is closely related to the zero temperature value of the μSR relaxation rate $\sigma(0)$ [26] which is proportional to the square of the inverse zero temperature penetration depth (Eq. 15). Moreover, extended studies of the pressure [27-30,38,39] and isotope mass [31-34,40,41] dependence of T_c in various cuprates, measured in terms of the coefficients

$$\alpha = \frac{1}{T_c}\frac{dT_c}{dP} \quad \text{and} \quad \beta = -\frac{d\ln T_c}{d\ln m} \quad , \tag{23}$$

revealed remarkable generic trends. Indeed, for various compounds, α and β fall into common $T_c - \alpha$ and $T_c - \beta$ regions, respectively, forming two branches, one for systems with positive and the other for compounds with negative pressure or isotope coefficient. The two branches merge at the maximum T_c where the coefficients vanish and the magnitude of the coefficients α and β decreases with raising T_c. From Fig. 7 it is seen that these trends appear to be a common feature of a large class of doped extreme type II superconductors which share a rather unique dependence of the

303

transition temperature on the inverse zero temperature penetration depth squared $(\sigma(0) \propto 1/\lambda_{\parallel}^2(0) \propto n_s/M_{\parallel})$. Indeed the assumption of a parabolic maximum $\sigma(0)$ yields the simple scaling form

$$\overline{T}_c = 2\overline{\sigma}(1 - \overline{\sigma}/2),$$
$$\overline{T}_c = T_c/T_c^m,$$
$$\overline{\sigma} = \sigma(0)/\sigma^m(0) .$$
(24)

Thus plotting T_c versus $\sigma(0)$, we find that a particular family forms a unique parabolic branch characterized by T_c^m and $\sigma^m(0)$. As expected, T_c vanishes at $\sigma(0) = 0$, implying a vanishing condensate density or an infinite effective mass $(\sigma(0) \propto n_s(0)/M_{\parallel})$. The plot T_c versus $\sigma(0)$ is shown in Fig. 8 for various cuprate and Chevrel phase superconductors. For comparison we included the parabolas resulting from ansatz (25), which are supposed to characterize the families. The scaled version of the data is depicted in Fig. 9, which reveals remarkable consistency with the universal parabolic behavior of ansatz (25). From Fig. 10, showing T_c^m versus $\sigma^m(0)$, it is seen that these two quantities are nearly proportional to each other.

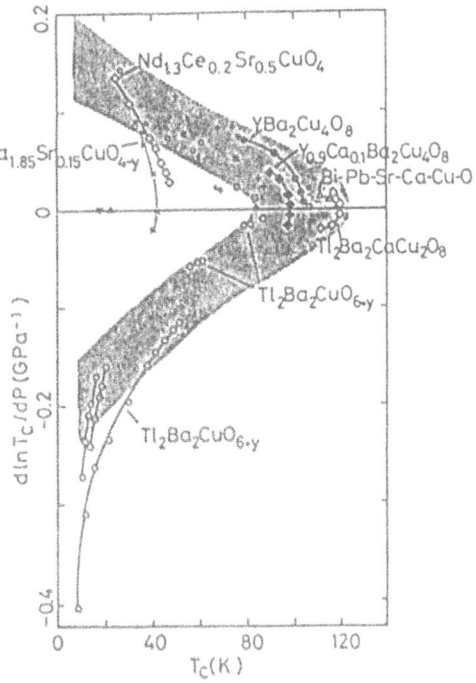

Fig. 7: Pressure coefficient α versus T_c for various cuprate superconductors. [From Ref. 27].

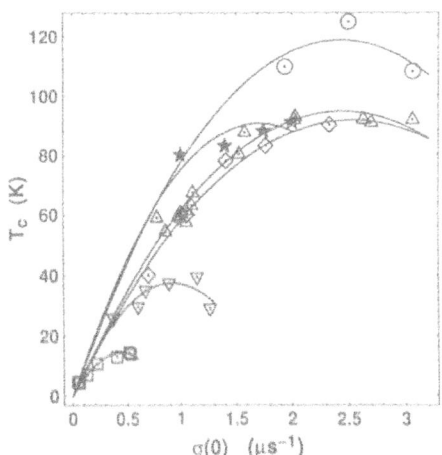

Fig. 8: T_c versus $\sigma(0)$. The parabolas are fits to Eq. (24), yielding the T_c^m and $\sigma^m(0)$ shown in Fig. 10. [From Ref. 26].
⊚ $Tl_2Ba_2Ca_2Cu_3O_{10}$,
   $Tl_{0.5}Pb_{0.5}Sr_2Ca_2Cu_3O_9$,
   $Bi_{2-x}Pb_xSr_2Ca_2Cu_3O_{10}$
◇ $Y_{1-x}Pr_xBa_2Cu_3O_{6.97}$
▲ $YBa_2Cu_3O_x$
▽ $La_{2-x}Sr_xCuO_4$
★ $Bi_2Sr_2Ca_{1-x}Y_xCu_2O_{8+\delta}$
⊡ $SnMo_6S_8$, $PbMo_6S_8$,
   $SnMo_6S_4Sc_4$, $SnMo_6S_7Se$,
   $SnMo_6S_1Se_7$, $PbMo_6S_4Se_4$.

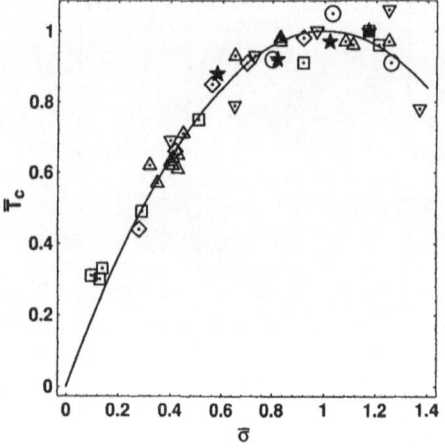

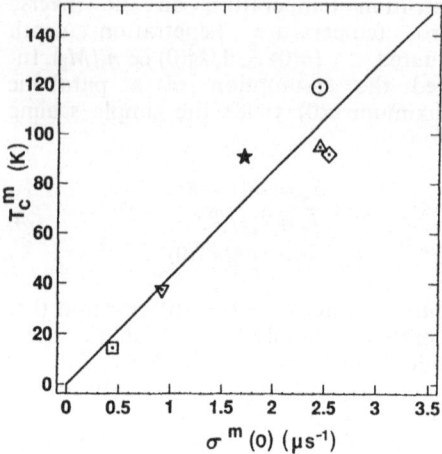

Fig. 9: $\overline{T}_c$ versus $\overline{\sigma}$ for the data shown in Fig. 8. The parabola corresponds to Eq. (24).

Fig. 10: T_c^m versus $\sigma^m(0)$. [From Ref. 12].

To check this scenario further, we consider the pressure and isotope effect coefficients. Adopting Eq. (25), we find that pressure and isotope mass can enter only via $\sigma(0)$, $\sigma^m(0)$ and T_c^m. In principle all these quantities depend on pressure. In view of the empirical fact that in most compounds (Fig. 7) dT_c^m/dP and in turn $d\sigma^m(0)/dP$ are very small and assuming the same behavior for dT_c^m/dm, the dominant contribution to the pressure and isotope coefficients arises from the induced change of the "condensate density," $\sigma(0) \propto n_s(0)/M_{\parallel}$. Thus, using Eqs. (24) and (25) the pressure and isotope effect coefficients are given by

$$\alpha = \frac{1}{T_c} \frac{dT_c}{dP}\bigg|_{P=0} = \frac{1}{T_c} \frac{dT_c}{d\sigma(0)} \alpha_{HP} = \pm 2 \frac{\sqrt{1-\overline{T}_c}}{\overline{T}_c} \alpha_{HP}, \quad \alpha_{HP} = \frac{1}{\sigma^m(0)} \frac{d\sigma(0)}{dP}\bigg|_{P=0} \quad (25)$$

and

$$\beta = -\frac{m}{T_c} \frac{dT_c}{dm} = \frac{1}{T_c} \frac{dT_c}{d\sigma(0)} \beta_{HM} = \pm 2 \frac{\sqrt{1-\overline{T}_c}}{\overline{T}_c} \beta_{HM}, \quad \beta_{HM} = -\frac{m}{\sigma^m}(0) \frac{d\sigma(0)}{dm}. \quad (26)$$

The indices $+$ and $-$ refer to $\sigma(0) < \sigma^m(0)$ and $\sigma(0) > \sigma^m(0)$, respectively. Because the "condensate density" ($\sigma(0) \propto n_s(0)/M_{\parallel}$) increases with reduced volume, the coefficients α_{HP} and β_{HM} are expected to be positive and nearly constant for T_c close to T_c^m. In view of this, the plot pressure and isotope effect coefficients versus $\overline{T}_c = T_c/T_c^m$ are expected to exhibit identical trends. In the limit $\overline{T}_c^m \to 1$ the coefficients vanish, their magnitude increases with reduced $\overline{T}_c^m$ and there are two symmetric branches, one for positive and the other for negative values of the coefficients.

In Fig. 11 we show α plotted versus $\overline{T}_c = T_c/T_c^m$ for YBa$_2$Cu$_3$O$_x$ ($6.35 \leq x \leq 7$) [30]. In the high-$\overline{T}_c$ regime and for $\alpha_{HP} \approx 0.0035$ (kbar)$^{-1}$, the data agree well with Eq. (25). However, α_{HP} is not expected to remain constant over the full $\overline{T}_c$ range and it

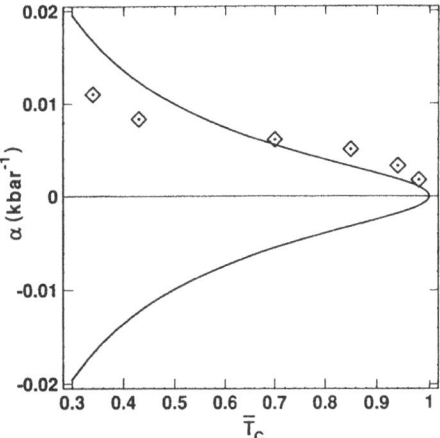

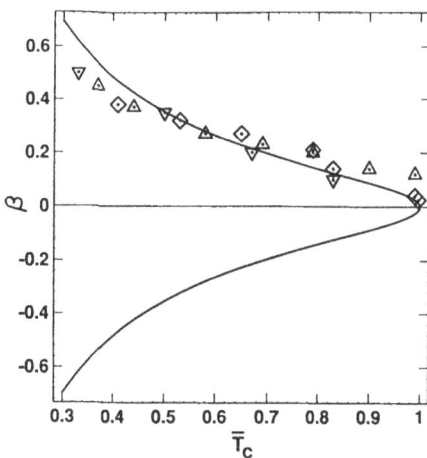

Fig. 11: Pressure coefficient α versus $\overline{T} = T_c/T_c^m$ for $YBa_2Cu_3O_x$ ($6.35 \leq x \lesssim 7$) taken from Ref. 30. The line corresponds to Eq. (25) with $\alpha_{HP} = 0.0035\,kbar^{-1}$. [From Ref. 26].

Fig. 12: Isotope effect coefficient β versus $\overline{T} = T_c/T_c^m$ for:

△ $La_{1.85}Sr_{0.15}Cu_{1-x}Ni_xO_4$ [33]
◇ $YBa_{2-x}La_xCu_3O_z$ [31]
▽ $Y_{1-x}Pr_xCa_yBa_2Cu_3O_7$ [31].

The line corresponds to Eq. (27) with $\beta_{HM} = 0.125$. [From Ref. 26].

might adopt family or compound specific values. In view of this, universal behavior is restricted to the trends implied by $\alpha \propto \pm (1 - \overline{T}_c)^{1/2}/\overline{T}_c$ that have been observed in a large class of cuprate superconductors (Fig. 7). Clearly, if the pressure or isotope mass dependence of T_c^m and $\sigma^m(0)$ is appreciable, the contribution from T_c^m will shift the minimum of α and β to a finite value and the term arising from $\sigma^m(0)$ might change the sign of the coefficients. This behavior appears to occur in the pressure coefficient of $La_{2-x}Sr_xCuO_4$ [42].

According to Eqs. (25) and (26), identical behavior is expected for the coefficient β of the isotope effect. In Fig. 12 we show β versus $\overline{T}_c$ for $\beta_{Hm} = 0.125$. This value fits the higher $\overline{T}_c$ range of the experimental data for the oxygen isotope coefficient in $La_{1.85}Sr_{0.15}Cu_{1-x}Ni_xO_4$ [33], $YBa_{2-x}La_xCu_3O_z$ [31] and $Y_{1-x}Pr_xCa_yBa_2Cu_3O_7$ [32] quite well. Analogous to α_{HP}, however, β_{Hm} is not expected to remain constant over the full $\overline{T}_c$ range and might vary from system to system. Thus, the universal behavior of the isotope effect coefficient is restricted to trends implied by the $\overline{T}_c$-dependent part of Eq. (26), namely vanishing β in the limit $\overline{T}_c \to 1$, increasing magnitude of β with reduced $\overline{T}_c$ and the appearance of two symmetric branches, one for positive and the other for negative β values. Indeed, negative β [40] and negative α values (Fig. 7) have been observed in various compounds. As noted above in the context of the pressure coefficient, deviations from this behavior are expected if doping modifies the characteristic parameters T_c^m and $\sigma^m(0)$.

Additional information on the behavior of the amplitude of the phase correlation length is obtained by combining the hyperuniversal relation (17) with ansatz (24), which yields

$$\overline{T}_c = A \frac{\sigma_0}{\sigma^m(0)} \frac{\sigma^m(0)}{T_c^m} \xi_{\|0}^{\varphi} = 2\overline{\sigma}\left(1 - \frac{\overline{\sigma}}{2}\right) \tag{27}$$

where

$$A = \frac{\hbar^2 c^2}{16\pi k_B e^2 (2754B)^2} \quad , \quad \sigma_0 = \frac{(2754B)}{\lambda_{\|0}^2} \quad . \tag{28}$$

Noting that $\sigma^m(0)/T_c^m$ (Fig. 9) and $\sigma_0/\sigma^m(0)$ (Table III) are nearly constant, it follows that

$$\overline{T}_c \approx y(2-y) \;, \qquad y = \frac{\overline{T}_c}{\overline{\sigma}} \propto \xi_{\|0}^{\varphi} \;. \tag{29}$$

The resulting behavior is shown in Fig. 13, where we included the experimental data shown in Fig. 9 for comparison. Apparently, there is reasonable agreement with the parabolic behavior in (29). Accordingly, at $y = 1$, the amplitude of the phase correlation length adopts the optimum value $\xi_{\|0}^{\varphi m}$. In the limit $y \rightarrow 0$, where $\sigma(0) = 2\sigma^m(0)$, it vanishes and superconductivity is suppressed at any finite temperature (Fig. 9). For $y \rightarrow 2$, however, the amplitude of the phase correlation length remains finite and superconductivity disappears because $\sigma(0) \propto 1/\lambda_{\|}^2 \propto n_s/M_{\|}$ vanishes.

In the context of the hyperuniversal relation (7), expressing the transition temperature in terms of the amplitudes of penetration depth and phase correlation length, we recover ansatz (24) by setting

$$T_c \propto \sigma_0 \xi_{\|0}^{\varphi} \propto \frac{\xi_{\|0}^{\varphi}}{\lambda_{\|0}^2} \propto 2\overline{\sigma}\left(1 - \frac{\overline{\sigma}}{2}\right) T_c^m \;. \tag{30}$$

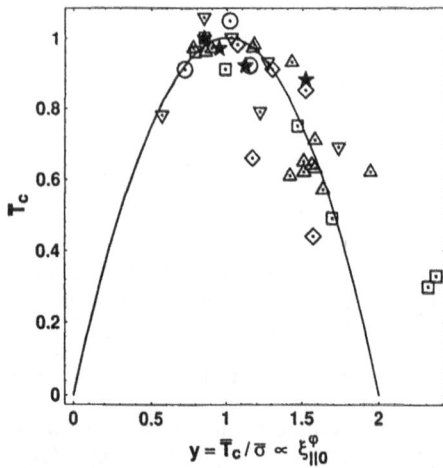

Fig. 13: $\overline{T}_c$ versus $y = \overline{T}_c/\overline{\sigma} \propto \xi_{\|0}^{\varphi}$ for the data shown in Fig. 8. The parabola corresponds to Eq. (29).

Assuming in addition, as proposed in Eq. (26), that

$$\xi_{\|0}^{\varphi} \propto \frac{\overline{T}_c}{\overline{\sigma}} \;, \tag{31}$$

we obtain

$$T_c^m \propto \sigma^m(0) \frac{\sigma_0}{\sigma(0)} \approx \sigma^m(0) \;, \tag{32}$$

which is consistent with experiment (Fig. 10). Thus, ansatz (24) implies an explicit relation (Eq. 30) between the critical amplitudes $\lambda_{\|0}$ and $\xi_{\|0}^{\varphi}$ and the zero temperature penetration depth ($\sigma(0) \propto 1/\lambda_{\|}^2 \propto n_s/M_{\|}$). A model where this connection between critical and zero temperature behavior can be verified is the ideal charged Bose gas. It does not belong, however, to the 3d xy-universality class. In this model: $T_c \propto \xi^{\varphi}_{\parallel 0}/(\lambda_{\parallel 0})^2 \propto n^{2/3}$, where n is the zero temperature boson density [43].

Another property revealing universal trends is the hole concentration dependence on T_c. For most superconducting cuprates the doped hole concentration p can be estimated by chemical means using titration, thermal gravimetry under hydrogen reduction or other techniques [44,45]. This is not always possible, as for $YBa_2Cu_3O_x$, where the charge is distributed between both chains and planes in a complicated manner. Nevertheless, a method which shows considerable promise for estimating the hole concentration is the use of bond valence sums (BVS) determined from interatomic bond lengths [46]. The dependence of T_c on hole concentration p in $YBa_2Cu_3O_x$ as determined from BVS [46] and compared with that in $La_{2-x}Sr_xCu_3O_4$ obtained from direct estimates [44] is shown in Fig. 14. The parabolic curve corresponds to [46],

$$\overline{T}_c = \frac{T_c}{T^m_c} = 1 - \beta(p - p_0)^2 , \tag{33}$$

with

$$p_0 \approx 0.16 , \qquad \beta \approx 82.6 . \tag{34}$$

Analogous behavior was found in $Bi_2Sr_2CaCu_2O_{8+\delta}$ [45]. It is believed that this relation is rather universal for hole doped superconductors. Combining Eqs. (24) and (33) we obtain a connection between the zero temperature condensate density and the hole concentration, namely

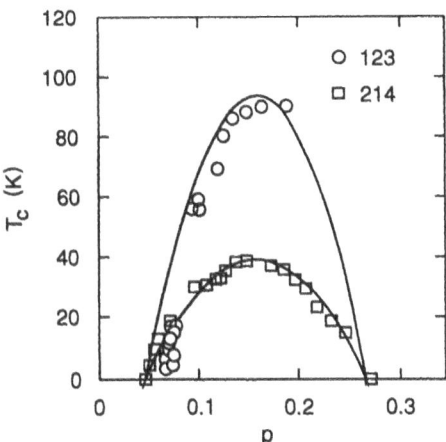

Fig. 14: T_c versus hole concentration p for $La_{2-x}Sr_xCu_2O_4$ and $YBa_2Cu_3O_x$, where the hole concentration for 1-2-3 is estimated at BVS [From Ref. 46]. The full lines correspond to Eq. (33).

$$\overline{\sigma} = \frac{\sigma(0)}{\sigma^m(0)} = 1 \pm \sqrt{\beta} \, |p - p_0| , \tag{35}$$

where $+$ and $-$ denote $\sigma(0) > \sigma^m(0)$ and $\sigma(0) < \sigma^m(0)$ respectively. Thus, the condensate density depends linearly on the hole concentration with $\overline{\sigma} = 1$ at $p = p_0$, $\overline{\sigma} = 0$ at $p = p_0 - \beta^{-1/2}$ and $\overline{\sigma} = 2$ at $p = p_0 + \beta^{-1/2}$. This relation between the condensate density and hole concentration might be used to express α_{HP} and β_{HM} in the pressure and isotope coefficients in terms of the hole concentration. An essential feature of Eqs. (24) and (35) is the disappearance of finite temperature superconductivity at and above the hole concentration $p = p_0 + \beta^{-1/2}$ (Fig. 14), corresponding to the maximum condensate density $\sigma(0) = 2\sigma^m(0)$ (Fig. 9).

Finally we turn to the temperature dependence of the penetration depth. In the intermediate regime

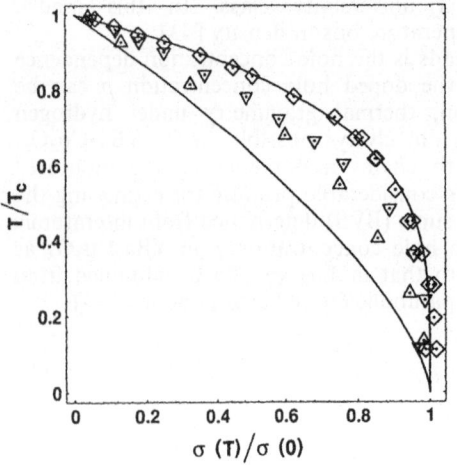

Fig. 15: μSR estimates $\sigma(T)/\sigma(0)$ versus T/T_c [17-20,35]
$\Diamond$ YBa$_2$Cu$_3$O$_{6.97}$
$\triangle$ YBa$_2$Cu$_3$O$_{6.602}$
$\triangledown$ BiSr$_2$CaCu$_2$O$_8$
The full lines corresponds to Eq. (36) with $n = 3/2$ and 4 respectively.

Fig. 16: n versus $\overline{T}_c$ for various cuprates obtained from a fit of Eq. (36) to the measured temperature dependence of the μSR relaxational rate [17-20,35].

$0 < T < T_c$ it is well described by the ansatz

$$\sigma(T) = \sigma(0)\left(1 - \left(\frac{T}{T_c}\right)^n\right) , \qquad (36)$$

where $n = 4$ corresponds to the two-fluid model and $n = 3/2$ to the dilute charged Bose gas [43]. Clearly, sufficiently close to T_c, $1/\lambda_{\parallel}^2(T) = 1/\lambda_{\parallel 0}^2 (1 - T/T_c)^\nu$ with $\nu \simeq 2/3$ for the 3d xy-critical point. In Fig. 15 we show μSR estimates for $\sigma(T)/\sigma(0)$ plotted versus T/T_c for YBa$_2$Cu$_3$O$_{6.97}$, YBa$_2$Cu$_3$O$_{6.602}$ and BiSr$_2$CaCu$_2$O$_8$. For comparison we included the temperature dependence given by Eq. (36) with $n = 3/2$ and $n = 4$. Noting that T_c decreases from 91.6 K (YBa$_2$Cu$_3$O$_{6.97}$) to 83 K (BiSr$_2$CaCu$_2$O$_8$) and 57.5 K (YBa$_2$Cu$_3$O$_{6.602}$), it is seen that n decreases with reduced T_c. To substantiate this trend we analyzed the μSR relaxation rate of various cuprates in the intermediate temperature regime according to Eq. (36). The resulting plot of n versus $\overline{T}_c = T_c/T_c^m$ is shown in Fig. 16. Consistent with the trend seen in Fig. 15, n decreases with reduced $\overline{T}_c$. Sufficiently close to T_c, however, critical 3d xy-behavior sets in, while in the $T \to 0$ limit the available experimental data favors the power law, $((\lambda_{\parallel}(T) - \lambda_{\parallel}(0))/\lambda_{\parallel}(0))^2 \propto (T/T_c)^4$ [47]. This behavior is reminiscent of the neutral 3d Bose gas with repulsive interaction; namely, T^4 in the $T \to 0$ limit and 3d xy-critical behavior close to T_c while the temperature dependence in the intermediate depends on the boson density and reflects the location of the crossover from the dilute to the dense condensate limit in terms of the exponent n (Eq. 36); $n \to 3/2$ in the dilute and $n \to 4$ in the dense Bose system [48]. Moreover, in the dilute Bose system T_c rises with increasing condensate density, saturates and falls in the dense regime due to the repulsive boson interaction, in remarkable agreement with ansatz (25). This analogy to the neutral 3d Bose gas with repulsive interaction is completely valid if the effective mass $M_{\parallel}$ entering $\sigma(0) \propto 1/\lambda_{\parallel}^2 \propto n_s(0)/M_{\parallel}$ is nearly constant. In this case, the empirical trends in temperature dependence of the penetration depth, together with the experimental evidence for the universal trends emerging from ansatz (25), favor Bose condensation of weakly charged and interacting pairs as the mechanism that drives the transition from the normal to the superconducting state.

In summary, the simple scaling ansatz (25), which expresses T_c in terms of the zero temperature "condensate density" $\sigma(0) \propto 1/\lambda_{\parallel}^2 \propto n_s(0)/M_{\parallel}$, accounts rather quantitatively for the empirical trends in various families of extreme type II superconductors. This is particularly true for the $\sigma(0)$ dependence of T_c and the T_c dependence of both pressure and isotope coefficients. In view of the fact that these trends are not restricted to the cuprates, but include the Chevrel phase superconductors as well (Fig. 8), the scenario outlined here, which connects the 3d xy-critical behavior with the zero temperature "condensate density," the universal trends in the pressure and isotope coefficients as well as the trends in the temperature dependence of the "condensate density" might well be generic for extreme type II superconductors with short correlation length amplitudes. Indeed previous experiments [29] on the pressure dependence of T_c in Chevrel phase compounds support this expectation. Moreover, the trends in the temperature dependence of the penetration depth, including the asymptotic low temperature behavior, suggest Bose condensation of weakly charged and interacting pairs as the mechanism that drives the transition. Indeed, consistent with ansatz (25) and analogous to the dilute Bose gas, T_c is expected to rise with increasing condensate density, while in the dense regime a nonlocal repulsive interaction can account for the saturation and the fall of T_c.

It should be kept in mind, however, that our analysis relies on the assumption that the fraction of the sample that is superconducting does not change upon oxygen removal or atomic substitution. Otherwise a decreasing specific heat singularity [21] or μSR relaxation rate would simply reflect the reduced superconducting fraction.

I wish to thank G. Bednorz, H. Keller and K.A. Müller for suggestions and stimulating discussions.

References

[1]   D.A. Huse, M.P.A. Fisher and D.S. Fisher, Nature, 358, 553 (1992)
[2]   C.J. Lobb, Phys. Rev. B 36, 3930 (1987)
[3]   A. Kozlowski, Z. Tarnawski, A. Kolodziejczyk, J. Chmist, T. Scizor and R. Zalecki, Physica C 184, 113 (1991)
[4]   M.R. Salamon, J. Shi, N. Overend and M.H. Howson, to appear in Phys. Rev. B (1993)
[5]   D. Zech et al. (private communication)
[6]   P. Wenk, Diploma Thesis, University of Zurich (1993)
[7]   W.J. Skocpol and M. Tinkham, Rep. Prog. Phys. 38, 1049 (1975)
[8]   M.E. Fisher, Rev. Mod. Phys. 46, 597 (1974)
[9]   V. Privman, P.C. Hohenberg and A. Aharony, in Phase Transitions and Critical Phenomena, Vol.14, edited C. Domb and J.L. Lebowitz (Academic Press, 1991)
[10] D.S. Fisher and M.P.A. Fisher, Phys. Rev. B 43, 130 (1991)
[11] T. Schneider and D. Ariosa, Z. Phys. B 89, 267 (1992)
[12] T. Schneider and H. Keller, Physica C, 207, 366 (1993)
[13] J.C. LeGuillou and J. Zinn-Justin, J. de Phys. 46, L137 (1985)
[14] G. Ahlers, in Quantum Liquids, edited by J. Ruvalds and T. Regge (North Holland, Amsterdam, 1978), p.1

310

[15] T. Schneider, Z. Phys. B 88, 249 (1992)
[16] V. Ambegaokar, B.I. Halperin, D.R. Nelson and E.D.D. Siggia, Phys. Rev. B 21, 1806 (1980)
[17] B. Pumpin et al., Phys. Rev. B 42, 8019 (1990)
[18] B. Pumpin et al., J. Less-Common Metals, 164&165, 994 (1990)
[19] B. Pumpin et al., Hyperfine Interactions 63, 25 (1990)
[20] H. Keller et al., Physica C 185-189, 1089 (1991)
[21] K. Ghiron, M.R. Salamon, B.W. Veal, A.P. Paulikas and J.W. Downey, Phys. Rev. B 46, 5837 (1992)
[22] P.C. Hohenberg and B.I. Halperin, Rev. Mod. Phys. 49, 435 (1977)
[23] J.M. Triscone et al., Phys. Rev. Lett. 64, 804 (1990)
[24] Q. Li et al., Phys. Rev. Lett. 64, 3086 (1990)
[25] D.H. Lowndes et al., Phys. Rev. Lett. 65, 1160 (1990)
[26] T. Schneider and H. Keller, Phys. Rev. Lett. 69, 3374 (1993)
[27] H. Takahashi and N. Môri, The Physics and Chemistry of Oxide Superconductors, Springer Proceedings in Physics, Vol. 60, edited by Y. Iye and H. Yasuoka, (Springer Verlag, Berlin, 1992) p. 237.
[28] J.G. Lin, K. Matsuishi, Y.Q. Wang, Y.Y. Xue, P.H. Hor and C.W. Chu, Physica C 175, 627 (1991)
[29] R.N. Shelton, A.C. Lawson and D.C. Johnston, Mat. Res. Bull. 10, 297 (1975)
[30] C.C. Almasan et al., Phys. Rev. Lett. 69, 680 (1992); S. Rusieki et al., J. Less-Common Metals, 164, 31 (1990)
[31] H.J. Bornemann and D.E. Morris, Phys. Rev. B 44, 5322 (1991)
[32] J.P. Franck et al., Physica C 162-164, 733 (1989), Physica C 185-189, 1379 (1991), Phys. Rev. B 44, 5318 (1991)
[33] N. Babushkina et al., Physica C 185-189, 901 (1991)
[34] K.A. Müller, Z. Phys. B 70, 193 (1990)
[35] H. Keller, this volume
[36] Y.J. Uemura et al., Physica C 162-164, 857 (1989)
[37] Y.J. Uemura, Phys. Rev. Lett. 66, 2665 (1991)
[38] C. Murayama et al., Physica C 183 (1991)
[39] N. Mori et al., J. Jap. Phys. Soc. 59, 3839 (1990)
[40] H.J. Bornemann, D.E. Morris and H.B. Liu, Physica C 182, 132 (1991); H.J. Bornemann et al., preprint
[41] M.K. Crawford et al., Phys. Rev. B 41, 282 (1990)
[42] N. Tanahashi et al., Jpn. J. Appl. Phys., 28, L762 (1989)
[43] M.R. Schafroth, Phys. Rev. 100, 463 (1955)
[44] J.B. Torrance et al., Phys. Rev. Lett. 61, 1127 (1990)
[45] W.A. Groen et al., Physica C 165, 55 (1990)
[46] J.L. Tallon and N.E. Flower, Physica C 204, 237 (1993)
[47] J.F. Annett and N. Goldenfeld, J. Low Temp. Phys. 89, 197 (1992)
[48] P.B. Weichman, Phys. Rev. B 38, 8739 (1988)

IR-Excited Raman Spectroscopy on HT_c Superconductors

G.RUANI[§]
Istituto di spettroscopia Molecolare- C.N.R.
via de'Castagnoli, 1
40126 Bologna - Italy

I. Introduction

The Raman scattering study of high critical temperature superconductors, shortly after their discovery, has played a significant role in characterizing materials and in investigating such low energy excitations as phonons, electronic or magnetic transitions and their interactions. In spite of some limitation of Raman scattering (the first order transitions are limited to the Γ point of the Brillouin zone) compared with other spectroscopic techniques, like neutron scattering, the high accuracy and resolution that can be achieved allow detailed study of the lineshape of the phonon bands, their interaction with low energy electronic excitations, and also the detection of some minor changes that can be determined by doping or temperature changes. The strong intensity dependence of the Raman scattering on laser excitation frequency can provide information on the nature and the interband transitions and/or, because of theresonantly enhanced scattering, on the microstructure of the sample. In this sense it is very important to study the Raman scattering in as wide range of excitation energy as possible. In this paper we report the Raman scattering in $MBa_2Cu_3O_{6+x}$ exciting at 1.16 eV and compare the results, obtained at different oxygen content, with different oxygen isotopes and different annealing procedures with those obtained by visible excitation. In particular we will discuss the peculiar behaviour of the apex O(4) oxygen atom in this system, showing the existence of an anharmonic potential associated to this atom, the evidence for the existence of

[§] work done in collaboration with M. Muccini, C. Taliani and
V.M. Burlakov, V.N. Denisov, A.G. Mal'shukov, E. Schönner (chapter III)
V.N. Denisov, P. Radaelli, C. U. Segre (chapter IV);
K. Conder, E. Kaldis, H. Keller, K.A. Müller, D. Zech (chapter V);
V.M. Burlakov, V.N. Denisov, A. Erb, H.P. Geserich (chapter VI)

E. Kaldis (ed.), Materials and Crystallographic Aspects of HTc-Superconductivity, 311–329.
© 1994 *Kluwer Academic Publishers.*

electronic levels, related to the O(4), at the Fermi energy and the relevance of these observations on the formation of the superconducting state.

II. IR excited FT-Raman spectroscopy.

The Raman scattering cross section is proportional to ω_s^4 where ω_s is the frequency of the scattered light; this means that the scattering efficiency for a near infrared excitation light (1.16 eV) is approximately 20 times smaller than for a visible excitation light. Infrared excited Raman spectra were performed in a back-scattering configuration by means of an IFS 88 Bruker FT spectrometer equipped with the FT-Raman module at 4 cm^{-1} resolution. A diode pumped c.w. Nd-YAG laser is used for excitation (1.16 eV). A cooled Ge detector allowed us to collect the scattered radiation in a wide spectral range around the excitation energy (from 100 to 3300 cm^{-1} in Stokes side and from 100 to 2100 cm^{-1} in the anti-Stokes one). In order to avoid heating and possible damaging effects due to absorption of the incident laser radiation, we kept the laser power as low as possible (less than 75 mW within a spot of about 0.3 mm in diameter). The temperature is estimated from the Stokes/anti-Stokes ratio of the observed Raman bands (duly corrected for the response of the detector at both frequencies).

III. $YBa_2Cu_3O_{6+x}$ (x<2): IR-Raman scattering at low oxygen concentration.

The study of the resonant Raman scattering at different excitation frequencies allows clarification of important feature of the electronic band structure and electron-phonon coupling of solids. The Raman excitation profiles in $YBa_2Cu_3O_{6+x}$ were measured in the region of $1.8 < \hbar\omega_L < 2.7$ eV by E.T. Heyen et al.[1]. Good agreement was found with Raman intensities calculated in the local density approximation[2] for x = 1, indicating that electronic states of $YBa_2Cu_3O_7$ can be treated, in first approximation, within the mean-field approach as for ordinary metals. However, electronic correlations are strong in the tetragonal semiconducting phase[3]. So far there is no reliable quantitative description of the metal-insulator transition in the high-T_c oxides. Moreover the origin of the insulating gap and the nature of the electronic states, filling the gap upon doping, are not clear. In this context, measurements of Raman spectra of the semiconducting phase with laser excitation at photon energy lower than the charge transfer gap ($\approx$ 1.7 eV) can provide useful information. The first Raman spectra of $YBa_2Cu_3O_{6+x}$ ceramics (x = 0.15; 0.3 and 0.85) exciting at 1.16 eV were reported in Ref. 4 and 5. The main spectral feature was a strong band at 505 cm^{-1} with overtones at 1010 cm^{-1} and 1515 cm^{-1} for x = 0.15.

This observation suggested the resonant nature of the Raman scattering at 1.16 eV as well as a strong electron-phonon coupling.

By site selective oxygen isotope substitution in the semiconducting $YBa_2Cu_3O_{6+x}$ we have unambiguously assigned the strong resonant Raman band at about 505 cm^{-1} to the A_g mode of the apex oxygen in the proximity of short chain segment. The relevant energy shift of this mode with respect to the unperturbed $YBa_2Cu_3O_6$ structure is a clear indication of the strong interaction of the apex oxygen atoms with electronic excitation.

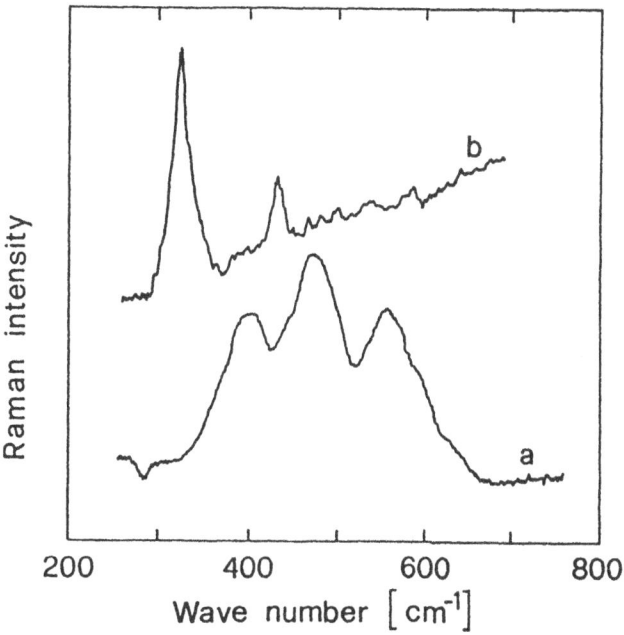

Fig. 1 Raman spectra of $YBa_2Cu_3{}^{18}O_{6.0}$ obtained (a) by exciting at 1.16 eV and (b) by exciting at 2.41 eV

The behaviour of the Raman scattering of semiconducting $YBa_2Cu_3O_{6+x}$ samples is strongly dependent on laser excitation energy. As an example IR (1.16 eV) and visible (2.41 eV) excited Raman spectra of $YBa_2Cu_3{}^{18}O_6$ are reported in Fig. 1. The IR excited Raman spectrum (Fig. 1a) shows three strong bands at 387, 459 and 552 cm^{-1} which are absent in the visible excited spectrum (Fig. 1b). At the same time for $\hbar\omega_L = 1.16$ eV no signs are detected of the 320cm^{-1} band (340 cm^{-1} in $YBa_2Cu_3{}^{16}O_6$), which on the contrary dominates the Raman spectrum obtained with visible excitation. The latter band is usually present in Raman spectra of ceramics and crystals in the wide range of excitation energies from 1.8 eV to 2.7 eV[1]. Such behaviour shows that the IR excited Raman scattering in these systems is characterized by peculiar resonant conditions.

The oxygen doping dependence of the visible excited Raman scattering of $YBa_2Cu_3O_{6+x}$ is well known[6]; there is only a minor shift of the bands (less then 2 cm^{-1} at room temperature) and a small redistribution of their intensities by increasing the oxygen content from $x = 0$ to $x = 0.3$ (the same behaviour has been confirmed by us in these samples). On the contrary, the IR excited Raman scattering exhibits a strong dependence of the spectral features on oxygen content (see Fig. 2 a,b). Also the strong bands clearly observed for $\hbar\omega_L = 1.16$ eV, despite the fourth power dependence on frequency of the Raman cross-section, indicate their resonance enhanced character suggesting the presence of an electronic dipole transition deep inside the semiconducting charge transfer gap.

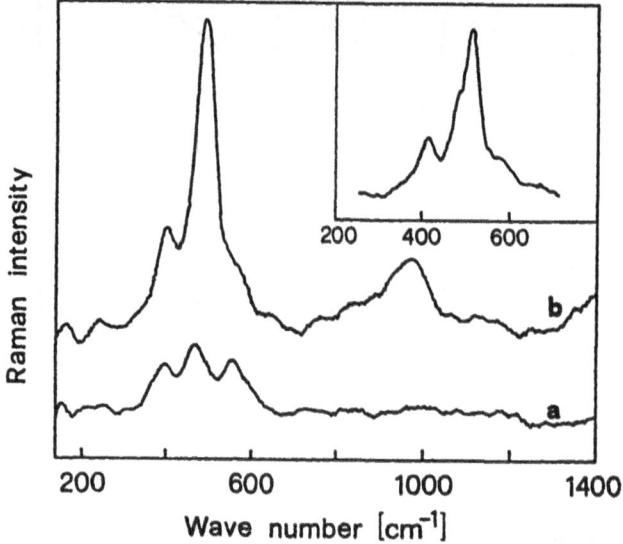

Fig. 2. FT-Raman spectra obtained by exciting at 1.16 eV of: (a) $YBa_2Cu_3{}^{18}O_{6.0}$ and (b) $YBa_2Cu_3{}^{18}O_{6.2}$. The spectrum of $YBa_2Cu_3{}^{16}O_{6.05}$ is shown in the inset.

Upon doping a new resonant band appears at 477 cm^{-1} (in O^{18} substituted samples). For $x = 0.2$ the peak at 477 cm^{-1} is the strongest (Fig. 2b), and its overtones are clearly seen. Inset of Fig. 2 represents the spectrum of $YBa_2Cu_3{}^{16}O_{6.05}$. All frequencies are shifted with respect to the upper spectrum due to the isotope substitution of oxygen. The peak at 507 cm^{-1} (477 cm^{-1} for ^{18}O) and its overtones are weaker. We can conclude that the new doping induced band at 507 cm^{-1} (477 cm^{-1} in ^{18}O substituted samples) is induced by the oxygen occupation of the O(1) chain sites in the basal plane (see Fig. 3). In some samples it's possible to observe overtones associated to this band up to the limit of our dector working range as is shown in Fig. 4 for $YBa_2Cu_3{}^{16}O_{6.2}$.

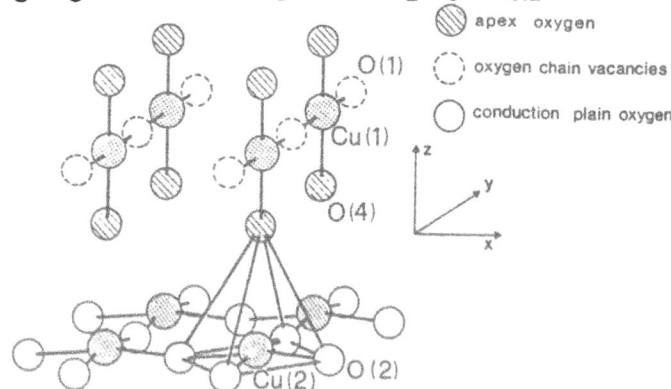

Fig. 3. Schematic representation of the crystal structure of conduction and basal planes in $YBa_2Cu_3O_{6+x}$.

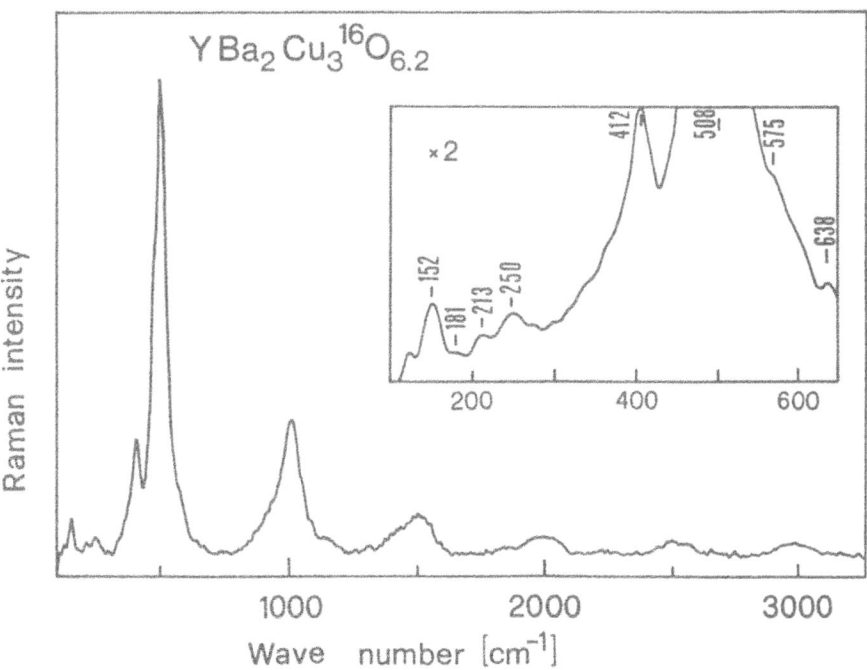

Fig. 4. FT-Raman spectrum exciting at 1.16 eV of $YBa_2Cu_3{}^{16}O_{6.2}$ (the spectrum is not corrected for the detector response in order to emphasize the presence of overtones up to the 6th order). In the inset is shown a blow-up of the 1st order phonons region.

The frequency of the peak at 507 cm^{-1} is very close to the frequency of the A_g phonon mode of the O(4) atoms in the fully doped (x=1) system, and it can be attributed to vibration of the O(4) atoms in microscopic domains of the orthorhombic phase. In order to confirm this assignment we have performed site selective isotope substitution of oxygen ^{16}O in a $YBa_2Cu_3{}^{18}O_6$ sample by means of low temperature annealing in air. Pristine $YBa_2Cu_3{}^{18}O_6$ IR excited Raman spectrum is shown in Fig. 5a. Fig. 5b shows the spectrum of the same sample after 30 min. annealing at 200°C in air. As expected, the spectrum changes with the oxygen doping and a new peak at 477 cm^{-1} (ω_2*) accompanied by a weak overtone ($2\omega_2*$) appears. At this annealing temperature, oxygen ^{16}O (from the atmosphere) diffuses within the sample only in the basal plane but cannot jump in other oxygen sites occupied by ^{18}O atoms[7]. We notice that this new ω_2* band at 477 cm^{-1} has the same frequency as the main band of the spectrum of the fully ^{18}O substituted sample $YBa_2Cu_3{}^{18}O_{6+x}$ (Fig. 2b). This is indeed a confirmation of our assignment. In fact, chain sites are occupied after annealing by ^{16}O and therefore the doping induced band cannot be assigned to vibrations of oxygen atoms on chain sites because in this case we should have observed the band characteristic of ^{16}O at 507 cm^{-1}. A second heat treatment of the sample is performed at 300°C. At this temperature O atoms in the O(1) sites can exchange with the nearest neighbour apex O(4) sites. At the same time, as reported in Ref. 7 and 8, the temperature is not so high as to allow oxygen in CuO_2 planes to diffuse out of the plane. After 60 minute of annealing at this temperature a dramatic change in the Raman spectrum is observed (Fig.5c). A shift towards higher frequencies of the bands at 385cm^{-1} (ω_1), 477cm^{-1} (ω_2*) and the overtone of the latter band ($2\omega_2*$) is clearly seen together with the increase in intensity of ω_2* and $2\omega_2*$. The similarities between the spectra of the annealed sample in Fig. 5c with the one of $YBa_2Cu_3{}^{16}O_{6.2}$ in the inset of Fig. 2, is quite evident; the frequencies of the modes in the two spectra do not correspond exactly because of a not complete isotope substitution of ^{18}O on apical positions.

The spectrum in Fig. 5c looks like that of Fig. 2b approximately shifted by 20 cm^{-1} towards higher energy. However, the band ω_3 in Fig. 5a that is clearly seen as a shoulder at 552 cm^{-1} in Fig. 2b is not present in Fig. 5c. If this band is related to vibrations of oxygen atoms on apex sites one should expect that the band at 575 cm^{-1}, which is seen in the spectrum of $YBa_2Cu_3{}^{16}O_{6.05}$ (see inset of Fig. 2) should be present also in Fig. 5c, at least as a shoulder, similar to Fig. 3. The absence of this feature indicates its relevance to vibrations of oxygen atoms in CuO_2 layers which were not isotopically substituted during the annealing of the sample. Thus we conclude that all oxygen vibrational bands that appear in the IR excited Raman spectra of $YBa_2Cu_3O_{6+x}$, with the exception of the highest frequency 575 cm^{-1} peak (in ^{18}O samples), are associated with phonons whose eigenvectors are mainly associated to the apex oxygen atoms displacement.

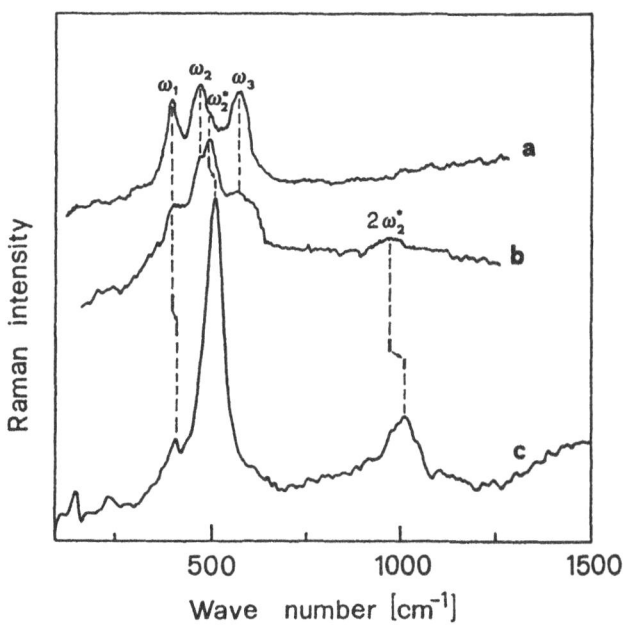

Fig. 5. FT-Raman spectra exciting at 1.16 eV of: (a) $YBa_2Cu_3{}^{18}O_{6.0}$, (b) the same sample after 30 minutes of annealing at 200°C in air and (c) the same sample after a second annealing of 60 minutes at 300°C in air.

The resonance nature of the scattering suggests that the resonance electronic transition near 1 eV must involve 2p-3d electronic orbitals of the O(4)-Cu(1)-O(4) complex which, for x = 0, seems to be weakly interacting with the CuO_2 planes due to the increased distance between planes and apex oxygen atoms.

Site selective substitution of oxygen isotope in the $YBa_2Cu_3O_{6+x}$ system therefore allows assignment of the bands observed in the IR excited Raman spectra. It is now widely recognized that the ω_2 band in the semiconducting samples (x≈0) is assigned to a totally symmetric (A_g) axial vibration of the apex oxygen[1]. We assign the ω_2* peak appearing upon slight oxygen doping to an A_g vibration of apex oxygen atoms adjacent to short chain segments.

The assignment of the ω_1 and ω_3 bands is not so straightforward. These bands were not present in the visible excited Raman spectra. This means that these bands are very weak or not at all Raman active when out of resonance.

The best candidate for ω_1 (at 412 cm^{-1} in ^{16}O samples) is the Raman forbidden IR E_u(LO) phonon calculated[9], at 414 cm^{-1} and found in reflectance measurements[10, 11] near 420 cm^{-1}. According to calculations[9] this mode involves vibrations of the apex oxygen atoms parallel to the CuO$_2$ planes. It is reasonable to assume that the LO component of the E_u mode becomes Raman active near resonance due to the Frölich electron-phonon mechanism[12] through the quadrupole term. Other E_u(LO) modes, rather weak but reproducible in all samples, are also seen in the IR excited Raman spectra. In the spectrum shown in Fig. 3 (inset) at least three other peaks at 181 cm^{-1}, 250 cm^{-1} and 638 cm^{-1} can be assigned to E_u(LO) modes which were observed in infrared reflectance measurements[10] at 199 cm^{-1}, 266 cm^{-1}, 637 cm^{-1} respectively. Confirmation of this assignment is given by the observation of the same four modes in YBa$_2$Cu$_3$O$_6$ when the red (676.4 nm) excitation light is used[13].

We envisage at least two possible assignments for the ω_3 band at 575 cm^{-1}. Isotope substitution, as was discussed above, shows that this band is related to the vibrations of oxygen atoms O(2) in the CuO$_2$ planes. Calculations[9] predict only one Raman active mode close to 575 cm^{-1} in YBa$_2$Cu$_3$O$_6$ which is assigned to the E_g vibration at 577 cm^{-1} of the O(2) oxygen atoms. Another possibility for the assignment of the 575 cm^{-1} band is a defect induced local vibration found in reflectance measurements (see Ref. 10 and 11), respectively at 595 cm^{-1} and 588 cm^{-1}. Comparing calculated eigenvector of this mode (502 cm^{-1}) and the one of the E_u(LO) (414 cm^{-1}) vibration[6] one can see that both have a similar mixed character. Therefore, its increased Raman intensity for 1.16 eV excitation may be caused by the admixture of the resonantly enhanced E_u(LO) mode near 412 cm^{-1}. Raman measurements on single crystals in zx, zy geometries may allow a choice to be made between these two possible assignments.

IV. ErBa$_2$Cu$_3$O$_{6+x}$: 0.1≤x≤0.9 (the IIT Chicago samples)

The ErBa$_2$Cu$_3$O$_{6+x}$ (x=0.9) were prepared by the standard solid state technique. The vacuum annealing of the oxygen reach material allows the oxygen deficient (x=0.1) one to be obtained. The intermediate concentrations, 0.1≤x≤0.9, were obtained by annealing the oxygen rich and the oxygen deficient materials mixed in appropriate proportions for two weeks at 400°C in sealed ampullas which were then slowly cooled over a period of two months[14].

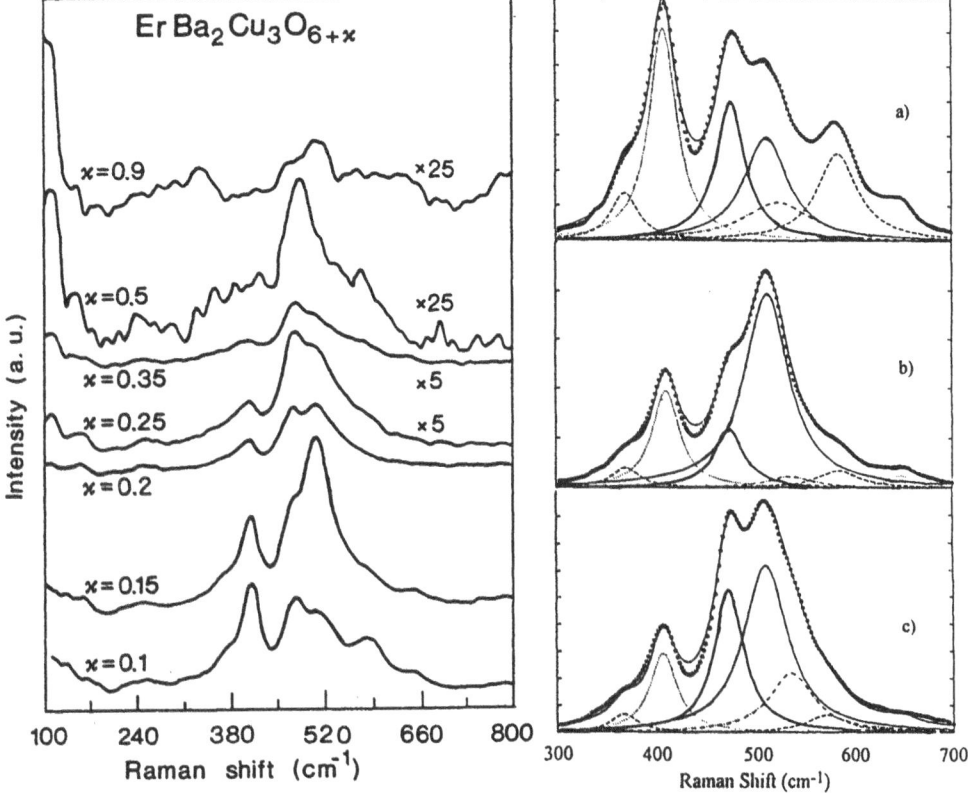

Fig. 6. Room temperature Raman spectra of ceramic ErBa$_2$Cu$_3$O$_{6+x}$ with 0.1≤x≤0.9.

Fig. 7. Room temperature Raman spectra of ceramic ErBa$_2$Cu$_3$O$_{6+x}$ with x=0.1 (a), 0.15 (b) and 0.2 (c). The curve fitting has been executed using the same set of Lorentzians and only the intensities as free parameters.

As in the case of previous samples, at low oxygen concentration the IR excited Raman of ErBa$_2$Cu$_3$O$_{6+x}$ spectra are characterized by the structure at 510 cm^{-1} (see Fig. 6 and 7). In this case the resonant behaviour with a sharp intragap band is more evident. In the sample with lower oxygen content (x=0.1) the ω_2^* 510 cm^{-1} band appears as a shoulder of the ω_2 475 cm^{-1} band, while the most intense band is the ω_1 LO phonon at 408 cm^{-1}. As it is obtained by the fitting with a summation of Lorentzians of the spectra (see Fig. 7), at x=0.15 the intensity of the ω_2 remains practically constant, the ω_1 slightly decreases and ω_2^* becomes the most pronounced one. At x=0.2, the ω_2 band is still of the same intensity while both the ω_1 and ω_2^* decrease in intensity.

At higher oxygen content the appearance of macroscopic orthorhombic domains is observed[14]; at x=0.25 the total scattering intensity is strongly reduced (Fig. 6) because of the appearance of metallic islands associated with the orthorhombic domains and ω_1 and ω_2^* intensities are further reduced relatively to ω_2. The profile of the Raman spectra does not change till x=0.4. The Raman spectra of samples with $0.42 \leq x \leq 0.7$ show only a single band associated with the apex oxygen O(4) at about 480 cm^{-1}, above a broad "electronic" background, (see Fig. 6) that shifts towards higher energy with the increase of oxygen content. At x=0.75 a band at about 340 cm^{-1} (associated to the antisymmetric plane-dimpling mode of oxygen O(2)/O(3)) start to be visible in the spectra and, at x=0.9, its intensity becomes comparable with the apex oxygen mode (see Fig. 7) whose energy at this oxygen concentration is comparable with the one indicated in the low oxygen concentration samples as ω_2^*. The relative intensity behaviour of these two A_g modes observed in IR excited Raman is the opposite of that observed by visible excitation[1-3]. As mentioned before, the resonant Raman conditions of these two modes are very different. Also LMTO calculations[1] show that the Raman efficiency of the A_g O(2)/O(3) mode is reduced only by a factor of 4 while the Raman efficiency of the A_g O(4) is reduced more than two orders of magnitude when the excitation energy is changed from 2.5 eV to 1 eV.

These results are in agreement with the recent NQR measurements by R. De Renzi et al.[15] on the same samples. They have observed that only a small number of well defined frequency lines, whose intensities are x dependent, can be detected in all the samples. Such a result in not consistent with a macroscopic picture obtained by X-ray or neutron diffraction techniques that observe a continuous change of the lattice parameter with oxygen content, but indicates that the crystal structure of 1-2-3 systems obtained using diffraction techniques only an average of four coexistent phases: pure tetragonal (no oxygen in the chain sites), pure orthorhombic (ORTHO I), isolated chains and alternate empty and full chains (ORTHO II). The Raman scattering with excitation at 1.16 eV, because of its peculiar resonant condition in these compounds, allows the presence of these phases to be distinguished.

V. ^{18}O Metallic $YBa_2Cu_3O_{6+x}$ Substituted Samples (the Zürich samples)

Metallic $YBa_2Cu_3O_{6+x}$ samples with different oxygen content ($0.48 < x < 0.974$) both with ^{16}O and with 96% of substituted ^{18}O have been prepared by using high purity

Y, Ba and Cu metals and ^{16}O or ^{18}O oxygen as starting materials. The oxidation of Y, Ba and Cu and the following sintering of the .oxides have been described elsewhere[7].

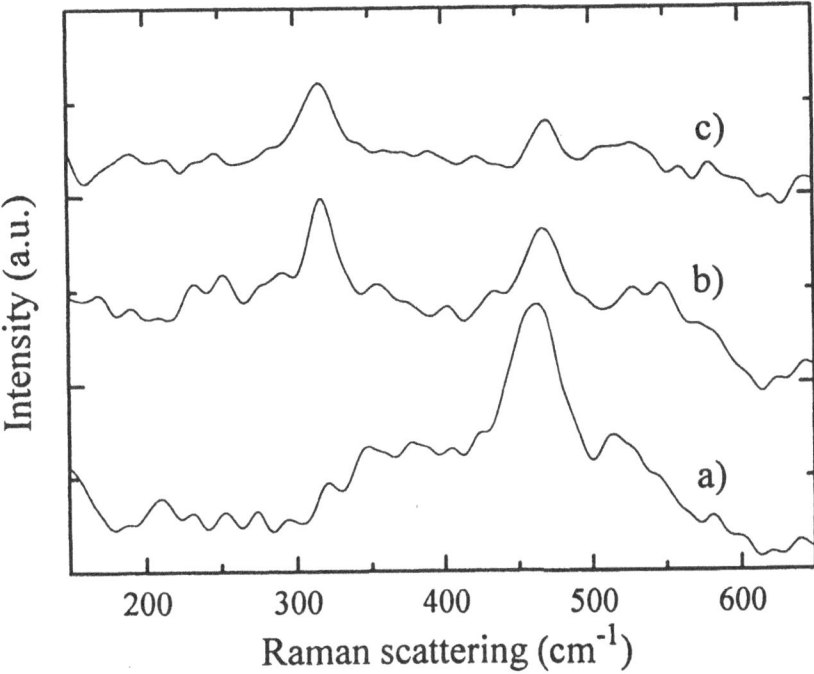

Fig. 8. RT Raman spectra of ceramic $YBa_2Cu_3{}^{18}O_{6+x}$ with x=0.52 (a), 0.85 (b) and 0.953 (c).

As mentioned before, the IR excited Raman spectrum of $MBa_2Cu_3O_{6+x}$ in the metallic phase (see Fig. 6 for x ~ 0.8) is characterized mainly by the presence of two phonon bands above a background between 250 - 600 cm^{-1} . The band at lower energy, around 340 cm^{-1} in ^{16}O samples, is associated with the A_g out of phase displacement along c of the O(2)/O(3) atoms; the second band, around 490 cm^{-1} in ^{16}O samples, is the A_g O(4) phonon mode. The relative intensities of these two bands change with oxygen content: at x < 0.6 only the A_g O(4) phonon mode is strong enough to be detected while at x ≈ 1 the dominant feature of the spectrum is the A_g O(2)/O(3) phonon mode (see Fig. 8); this behaviour, opposite to what is observed by visible excited Raman[1-3], is related to the very different resonant condition between 1.16 eV and ~ 2.5 eV photon energy and to the change of low energy electronic states induced by doping.

322

The oxygen isotope substitution is observed in Raman by the shift of the phonon modes whose eigenvectors are represented by displacements of oxygen atoms. As is shown in Fig 9 for the $YBa_2Cu_3O_{6+x}$ samples with $x \approx 0.8$, the almost complete (96%) homogeneous oxygen isotope substitution of ^{16}O with ^{18}O determines a systematic shift of all the phonon features of the spectrum[16].

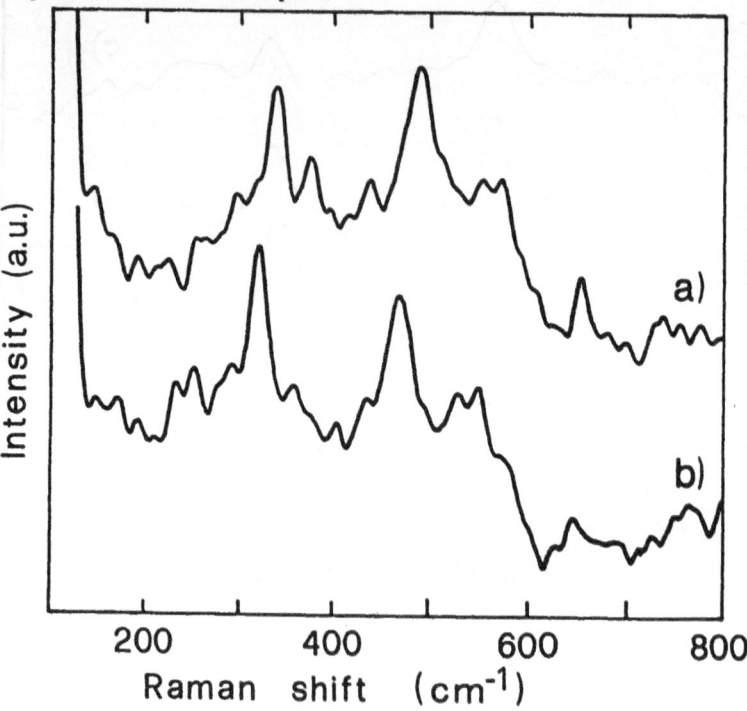

Fig. 9. RT Raman spectra of ceramic $YBa_2Cu_3{}^{18}O_{6.8}$ (a) and $YBa_2Cu_3{}^{16}O_{6.8}$ (b).

The peak position of the A_g O(2)/(3) Raman band for different samples is reported in Fig. 10 for oxygen ^{16}O (full squares) and for ^{18}O (open dots) as a function of oxygen content. The full triangles represent the "calculated" frequency of the same band for a sample with 96% of ^{18}O. The isotope shift of the energy relative to this phonon, and also of the Ag O(4) one, has been calculated by several methods, always within the harmonic aproximation. The results are not strongly dependent on the method used (see Ref. 17 and references therein). To obtain what is indicated as the "calculated" frequency of ^{18}O samples, we have multiplied the experimental Raman frequency of the $YBa_2Cu_3{}^{16}O_{6+x}$ of the same mode for a proportional factor relative to the frequency isotope shift obtained from the "shell model" calculations[18], taking into account that our ^{18}O samples are not completely isotope substituted. The Raman frequency obtained in this way coincides, in the case of the A_g O(2)/(3), with the frequency observed experimentally for the ^{18}O substituted samples. This confirms that the harmonic aproximation is valid in the case of oxygen atoms in the CuO_2 planes.

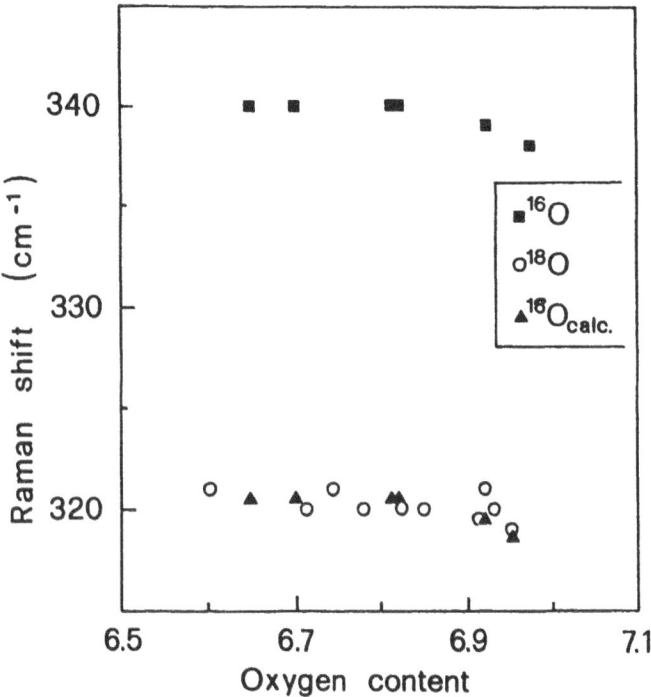

Fig. 10. Frequency of the Raman A_g mode associated to the out of phase motion of O(2)/O(3) oxygen atoms along the c axis; full squares: experimental value of $YBa_2Cu_3^{16}O_{6+x}$ samples; open dots: experimental values of $YBa_2Cu_3^{18}O_{6+x}$ samples; and full triangle: computed value of 96% of ^{18}O substituted samples from the $YBa_2Cu_3^{16}O_{6+x}$ experimental value isotopically shifted using the harmonic approximation. The lines are guide to the eyes.

In Fig. 11 the Raman shift of the A_g O(4) mode versus the oxygen content is reported. In this case the value of the computed isotopic shift ($\Delta\omega/\omega|_{calc.}=5.6\%$) is much larger than the average shift observed experimentally ($<\Delta\omega/\omega|_{exp.}>=4.7\%$). Such a difference is too large to be ascribed to experimental errors in measuring the oxygen content x of the samples or to the evaluation of the peak position.

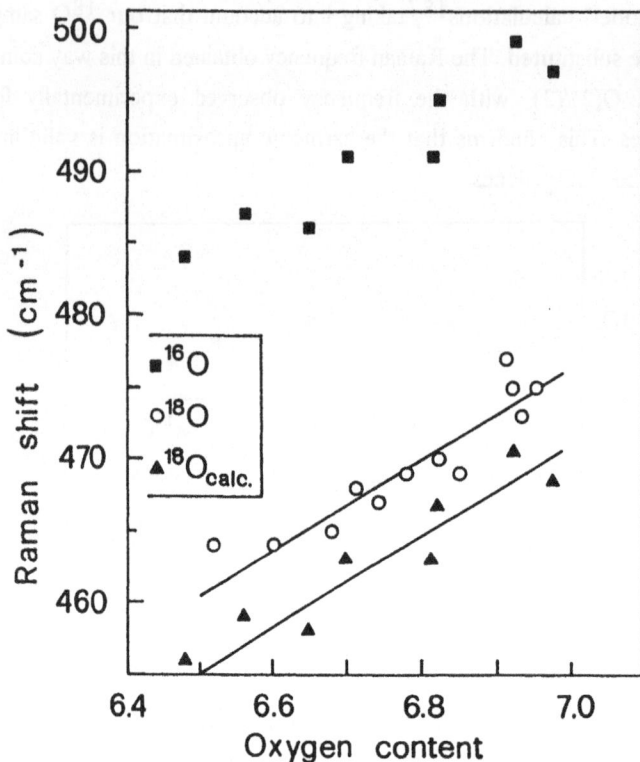

Fig. 11. Frequency of the Raman A_g mode associated to the displacement of the O(4) oxygen atoms along the c axis; open circles: full squares: experimental value of $YBa_2Cu_3{}^{16}O_{6+x}$ samples; open dots: experimental values of $YBa_2Cu_3{}^{18}O_{6+x}$ samples; and full triangle: computed value of 96% of ^{18}O substituted samples from the $YBa_2Cu_3{}^{16}O_{6+x}$ experimental value isotopically shifted using the harmonic approximation. The lines are guide to the eyes.

Similar behaviour of the isotope induced shift of the two Raman bands has been observed by J.C. Irwin et al.[17]. They have observed a larger isotope shift of the A_g O(2)/O(3) band than for the O(4) band. To explain such a difference they suggested the hypothesis of a different substitution efficiency for oxygen in O(2)/O(3) and O(4) sites (higher mobility of the oxygen in the CuO_2 planes than in the BaO planes). As mentioned before, our sample have been prepared from oxides already isotopically enriched with oxygen ^{16}O or ^{18}O, so it is not possible in these conditions to obtain samples with different site occupancy even if different kinds of diffusion processes with different activation energies in the different oxygen sites exist. Moreover, it has been demonstrated by low temperature annealing[7, 19] that the oxygen isotope substitution and therefore the oxygen diffusion proceeds from basal (chain) plane to BaO plane (apex) and then to CuO_2 plane, leading to the conclusion that a higher percentage of isotope substitution in the $O(2)/O(3)$ sites with respect to the $O(4)$ sites is not possible.

The small isotope shift of the A_g phonon energy observed for the apex oxygen is consistent with the existence of an asymmetric anharmonic double well potential. As has been suggested in several papers, the existence of an anharmonic potential related to the apex oxygen can have a strong impact on the superconducting properties of high T_c materials[20-23]. There is evidence from several experiments that suggest the existence of an anharmonic potential for the apex oxygen[24-26]. Moreover, recently J.H. Nickel *et al.* have shown by site selective oxygen isotope substitution in $YBa_2Cu_3O_7$ that apex oxygen $O(4)$ vibrations contribute to the pairing mechanism. Such results are emphasizing the validity of models that describe the Cooper pair formation in the conducting CuO_2 plane as induced by the coupling of correlated electrons to anharmonic vibrations of the $O(4)$ atom, as for example, the model proposed by Bishop et al.[27] in which a strong charge transfer coupled with the apex oxygen vibration along c within a double well anharmonic potential is the source of superconductivity in the close CuO_2 plane of $YBa_2Cu_3O_{6+x}$.

VI. Untwinned 1-2-3 Crystal: Electronic Background (the Karlsruhe sample)

Several Raman studies of twin-free $Yba_2Cu_3O_7$ crystals have been reported[28-35]. These investigations were carried out with visible light excitation and revealed differences in the relative intensities of the A_g phonons depending upon whether the polarization of the incident and scattered light was along the x or y direction. The complete polarization study of Raman active phonons were reported in Ref. 31 and the properties of the background in the Raman spectra of YBCO were investigated in Ref. 33-35.

Here we report a Raman spectra of untwinned $Yba_2Cu_3O_7$ crystal obtained with excitation by laser energy of 1.16 eV ($\lambda=1.06\mu m$).

The single-phase YBCO $T_c \sim 90K$ crystals with dimensions $1 \times 1 \times 0.3$ mm^3 were grown and detwinned according to a method described in detail elsewhere[36]. The characterization of the crystal was made by means of inductive measurements, polarized microscope and reflectivity measurements of plasma edge. Room temperature Raman spectra for three scattering geometries are shown on Fig. 12. The (yy) spectrum (y corresponds to Cu-O chains direction) is rather complicated, it consists of intensive broad maximum at ~400 cm^{-1} serving as a background for relatively sharp and lower intensive phonon lines. There are three well pronounced lines in the (yy) spectrum at frequencies 337, 491, and 575 cm^{-1}. The (xx) spectrum contains only one phonon line at 337. The (zz) spectrum in Fig. 12 contains only one well pronounced line at 498 cm^{-1}. Our measurements in crossed (xy) polarization did not reveal any structure in the spectrum, so all the features in the spectra shown in Fig. 12 should be regarded as having A$_g$ character.

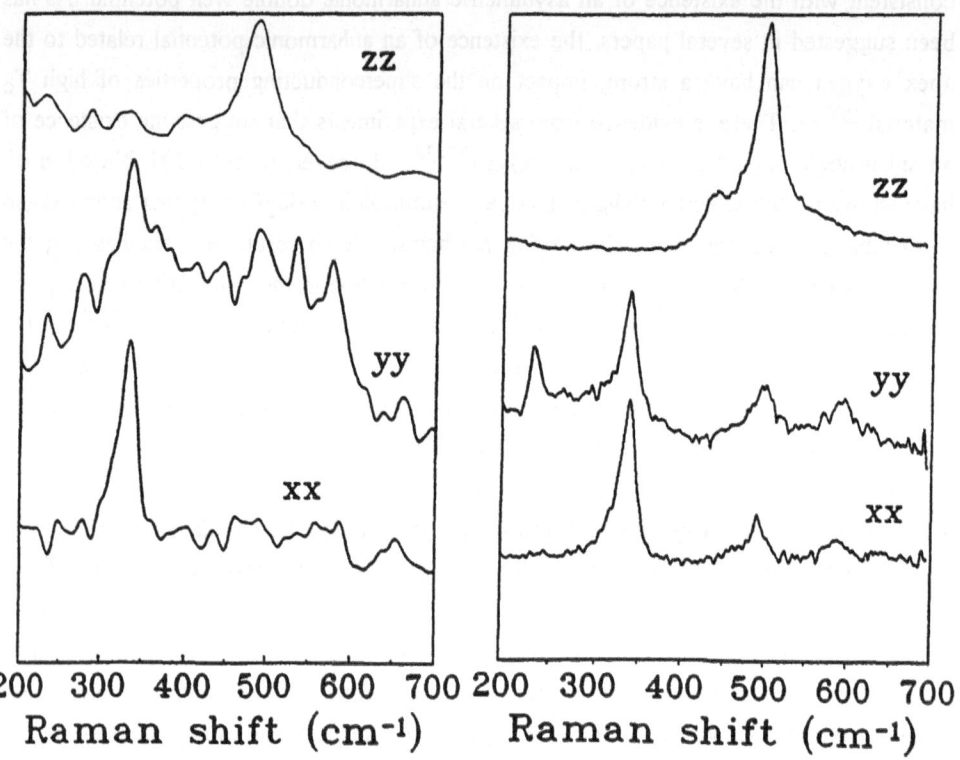

Fig. 12. Raman spectra of YBa$_2$Cu$_3$O$_{7-\delta}$ untwinned single crystals for different scattering geometries exciting at 1.16 eV.

Fig. 13. Raman spectra of YBa$_2$Cu$_3$O$_{7-\delta}$ untwinned single crystals for different scattering geometries exciting at 2.51 eV.

The visible Raman spectra of the same crystal is shown in Fig. 13. The most pronounced difference with respect to the IR excited spectra is the absence of this rather broad maximum in the (yy) spectrum around 400 cm^{-1}. A similar broad feature is observed in the ceramic samples with x > 0.5 (see. Fig. 7-9). The relative intensity with respect to the phonon band and the width of this background depend on oxygen content x: at x ~ 1 it is approximately two times larger than at x ~ 0.5. In Fig. 14 the IR excited Raman spectra of the untwinned single crystal in yy geometry with together those obtained in 1-2-3 ceramic samples with similar oxygen content are reported. It is clearly seen that the broad peak is identical in all the three samples. The nature of this peak is clearly non phononic: in fact the comparison of the Raman spectra of the two ceramic YBCO samples with the same oxygen content but different isotopes shows no detectable shift. If this structure had a phononic origin, its relatively "high" energy would imply the oxygen nature of the mode with the consequent shift related to change in the mass of the isotope.

As mentioned before, such a broad peak is not observed in the visible excited Raman spectrum of the same sample, moreover, nobody has observed by visible Raman excitation any similar structure in 1-2-3 at any oxygen content[37]. On the contrary, recent measurements on ^{18}O substituted samples have shown the existence of some features in the $\varepsilon_2(\omega)$ spectra below 800 cm^{-1} that is independent of the isotope substitution[39]. These structures have been interpreted, by I.I. Mazin *et al.*[40] as related to low energy electronic interband transitions on the base of accurate LDA energy bands calculations. In particular one of these has been assigned to the transition between two bands at the S point in the Brillouin zone around the Fermi level E_F. The Upper curve, the one crossing the E_F has prevalently $O(4)(p_y)$ in character and the lower one mainly $O(4)(p_x)$ in character. In accordance with LDA calculations[40], such a transition should be observed in resonant conditions also by Raman scattering; in particular the authors claim that such a resonant condition would be matched at energy smaller than 1.9 eV or higher than .8 eV. This would explain why this intraband Raman scattering is observed only by IR excited Raman.

In conclusion, the confirmation of the existence of low energy electronic states associated to apex oxygen around the E_F, i.e. belonging neither to CuO_2 planes nor to chains, and the observation, outside the experimental error, of the anharmonicity of the apex oxygen $O(4)$ in YBCO superconducting system, underlines the peculiar role played by this atom in the formation of the superconducting state. Moreover the similarity of the $O(4)$ in YBCO with the anharmonic highly polarizable O^{2-} anion in the highest T_c ferroelectrics strongly supports those theories that consider a possible common origin for ferroelectricity and superconductivity in oxides.

Acknowledgements

I wish to thank Prof. N.M. Plakida and Dr. M. Zoli for helpful discussions. Financial support by National Research Council of Italy, C.N.R. under the Progetto Finalizzato "Superconductive and Cryogenic Technologies" and EC contract SCI*CT91-0751 is acknowledged.

References

1. E.T. Heyen, R. Liu, C. Thomsen and M. Cardona in "Electronic Properties of High-T_c Superconductors and Relative Compounds", edited by H.Kuzmany, M. Mehring and J. Fink, (Springer-Verlag, Berlin Heidelberg 1990), p.324.

2 E.T. Heyen, S.N. Rashkeev, I.I. Mazin, O.K. Andersen, R. Liu, M. Cardona and O. Jepsen, Phys. Rev. Lett. 65, 3048 (1990).

3 H. Eskes, M.B.J. Meinders and G.A.Sawatzky, Phys. Rev. Lett. 67, 1035 (1991).

4 R. Zamboni, G. Ruani, A.J.Pal and C. Taliani, Solid State Commun 70, 813 (1989).

5 G. Ruani, R. Zamboni, C. Taliani, V.N. Denisov, V.M. Burlakov and A.G. Mal'shukov, Physica C 185-189, 963 (1991); V.N. Denisov, C. Taliani, A.G. Mal'shukov, V.M. Burlakov, E. Schönherr and G. Ruani, Phys. Rev. B, submitted.

6. C. Thomsen, R. Liu, M. Bauer, A. Wittlin, L.Genzel, M. Cardona, E. Schönherr, W. Bauhofer and W. König, Solid State Commun. 65, 55 (1988).

7. K.Conder, E. Kaldis, M. Maciejewski, E.F. Steigmeier and K.A. Muller, Physica C, in press.

8. R. Nishitani, N. Hoshida, Y. Sasaki, Y. Nishina, H. Yoshida-Katayama, Y. Okabe and T. Takahashi, Japan. Journal of Appl. Phys. 29, L50 (1990).

9. C. Thomsen, M. Cardona, W. Kress, R. Liu, L. Genzel, M. Bauer and E. Schînherr, Solid State Commun. 65, 1139 (1988).

10. M. Bauer, L.B. Ferreira, L. Genzel, M. Cardona, P. Muragaraj and J. Maier, Solid State Commun. 72, 551 (1989).

11. M.K. Crawford, G. Burns and F. Holtzberg, Solid State Commun. 70, 557 (1989).

12 M. Cardona in "Light Scattering in Solids II", ed. M. Cardona and G. Guntherodt (Springer, 1982), p1.

13 M. Cardona, Physica C 185-189, 65 (1991).

14. P.G. Radaelli, C.U. Segre, D.G. Hinks and J.D. Jorgensen, Phys. Rev. B 45, 4923, 1992.

15. R. Derenzi et al., Conferenza SATT, Riccione, 1993.

16. G. Ruani, C. Taliani, M. Muccini, K. Conder, E. Kaldis, K.A. Müller, H. Keller and D. Zech, in progress.

17 J.C. Irwin, J. Chrzanowski, E. Altendorf, J.P. Franck and J. Jung, J. Mater. Res 5, 2780 (1990); E. Altendorf, J. Chrzanowski, J.C. Irwin and J.P. Franck, Phys. Rev. B 43, 2771 (1991).

18 M. Cardona, R. Liu, C. Thomsen, W. Kress, E. Schönherr, M. Bauer, L. Genzel and W. Konig, Solid State Commun 67, 789 (1988); C.Thomsen, Hj. Mattausch, M. Bauer, W. Bauhofer, R. Liu, L. Genzel and M. Cardona, Solid State Commun. 67,, 1069 (1988)

19 J.H. Nickel, D.E. Morris, J.W. Ager III, Phys. Rev. Lett. 70, 81, (1993).

20 K.A. Müller, **Z. Phys. B 80**, 193 (1990).

21 T. Galbaatar, S.L. Drechsler, N.M. Plakida, G.M. Vujicic, **Physica C 176**, 496 (1991); T. Galbaatar, S.L. Drechsler, N.M. Plakida, G.M. Vujicic, **Physica C 185-189**, 1529 (1991).

22 D. Mihailovic, C.M. Foster, K.F. Voss and N. Herron, **Phys. Rev. B 44**, 237 (1991).

23 M. Frick, I. Morgenstern and W. von der Linden, **Z. Phys. B 82**, 339 (1991).

24 J. Mustre de Leon , S.D. Conradson, A.R. Bishop and I. Batistic, **Phys Rev. Lett. 65**, 4675 (1990).

25 S.D. Conradson, I.D. Raistrick and A.R. Bishop, **Science 248**, 1394 (1990).

26 D. Mihailovic and C.M. Foster, Solid State Commun., 74, 753 (1990).

27 A.R. Bishop, R.L. Martin, K.A. Müller and Z. Tesanovic, **Z. Phys. B 76**, 17 (1989).

28 C. Thomsen, M. Cardona, B. Gegenheimer, R. Liu and A. Simon, **Phys. Rev. B. 37**, 9860 (1988)

29. F. Slakey, S.L. Cooper, M.V. Klein, J.P. Rice and D.M. Ginsberg, **Phys. Rev. B. 39**, 2781 (1989)

30. L.V. Gasparov, V.D: Kulakovskii, O.V. Misochko and V.B. Timofeev, **Phisica C 157**, 341 (1989)

31. V.D. Kulakovskii, O.V. Misochko and V.B. Timofeev, **Sov.Phys.Solid State 31** (9), 1599 (1989)

32. K.F. McCarty, J.Z. Liu, R.N. Shelton and H.B. Radovsky, **Phys. Rev. B 41**, 8792 (1990)

33. F. Slake, M.V. Klein, J.P. Rice and D.M. Ginsberg, **Phys. Rev. B 43**, 3764 (1991)

34. D. Rein, A. Kotz, S.L. Cooper, M.V. Klein, W.C. Lee and D.M. Ginsberg, Proc. Int. Workshop on Electronic Prop. and Mech. of High-T_c Superconductors (IWEPM), Tsukuba, Japan (1991).

35. D. Reznik, S.L. Cooper, M.V. Klein, W.C. Lee, D.M. Ginsberg and S.W.Cheong, in "Electronic Properties of High-Tc Superconductors", edited by H.Kuzmany, M. Mehring and J. Fink, (Springer-Verlag, Berlin Heidelberg 1993), p. 215.

36. A. Zibold, M. Dürrler, H.P. Geserich, A. Erb and G. Müller-Vogt, **Physica C 171**, 151 (1990)

37. A band-like structure around 400 cm^{-1} is observed in Bi based HT$_c$ superconductors in the Raman spectra at T<<T_c. In that case the peack is related to the opening of the superconducting gap and at the consequent redistribution of the electronic states[38]; on the contrary the broad band observed exciting at 1.16 eV is present also at room metereture than is not relate to the superconducting state.

38. R. Nemetschek, T. Staufer, O.V. Misochko, D. Einzel, R. Hackl, P. Müller and K. Andres, in "Electronic Properties of High-Tc Superconductors", edited by H.Kuzmany, M. Mehring and J. Fink, (Springer-Verlag, Berlin Heidelberg 1993), p. 215.

39. A.V.Bazhenov and K.B. Razchikov, **Physica C 192**, 411 (1992).

40. I.I. Mazin, S.N. Rashkeev, A.I. Leichtenstein, and O.K. Andersen, **Phys. Rev. B 46**, 11232 (1992); I.I. Mazin, O. Jepsen, O.K. Andersen, A.I. Leichtenstein, S.N. Rashkeev and Y.A. Uspemskii, **Phys. Rev. B 45**, 5103 (1992).

Classification Physics Abstracts 74.20

Numerical Simulation of

High Temperature Superconductors

I.Morgenstern, J.M.Singer, Th.Hußlein and H.-G.Matuttis

Department of Physics, University of Regensburg,
D-8400 Regensburg, Germany

Abstract

Motivated by experiments on high-T_c superconducting compounds and made possible by the power of present day supercomputers, the quantum Monte Carlo method is used to gain more insight into Hubbard-like many particle models for these high-T_c materials. We report on numerical studies of various high-T_c models, the quantum Monte Carlo method is described in detail. For the simulations we employed predominantly the projector quantum Monte Carlo method (PQMC) to study the ground state properties of the systems. Our calculations are based on a generalized version of this technique allowing us a simultaneous treatment of fermions and dynamical phonons.

A possible pairing mechanism is linked to the Apex oxygen with booster effects originating from the multilayered structure and the presence of a van Hove singularity. Application oriented simulations are carried out for weak links, pinning centers and glassy behavior. The consequences of the short coherence length (depression of the order parameter etc.) are discussed. It is emphasized that the Monte Carlo method has the potential to detect superconducting mechanisms in correlated electron systems, which are not treatable by standard analytic techniques. The Monte Carlo simulations provide an approximation-free approach in contrast to most of the standard analytical methods and give access to informations about systems much larger than those accessible for exact diagonalization algorithms.

E. Kaldis (ed.), Materials and Crystallographic Aspects of HTc-Superconductivity, 331–351.
© 1994 Kluwer Academic Publishers.

1. Introduction

Using the quantum Monte Carlo method described in the next chapter we consider several high-T_c models which are based on the Hubbard model. The general model is defined by the lattice Hamiltonian

$$\mathcal{H} = H_0 + H' + H_\perp + H_H + H_{e-ph} + H_{ph} \quad ,$$

which consists of the following parts:

Inplane particle motion (kinetic term) between nearest neighbours $\langle ii' \rangle$ with hopping element t

$$H_0 = -t \sum_{\substack{\langle ii' \rangle \sigma \\ \text{inplane,nn}}} (c_{i\sigma}^\dagger c_{i'\sigma} + h.c.) \quad .$$

inplane particle motion (kinetic term) between next-nearest neighbours $\langle jj' \rangle$ with hopping element t'
("t-prime term")

$$H' = -t' \sum_{\substack{\langle jj' \rangle \sigma \\ \text{inplane,nnn}}} (c_{j\sigma}^\dagger c_{j'\sigma} + h.c.) \quad .$$

tunneling of particles between layers with interlayer hopping term $t_\perp$
("Double Layer term")

$$H_\perp = -t_\perp \sum_{\substack{(k,k')\sigma \\ \text{interplane,nn}}} (c_{k\sigma}^\dagger c_{k'\sigma} + h.c.) \quad ,$$

Hubbard interaction

$$H_H = U \sum_i n_{i\uparrow} n_{i\downarrow} \quad ,$$

as well as the electron-phonon interaction with coupling strength g

$$H_{e-ph} = g \sum_i \left(\sum_\delta n_{i+\delta} \right) s_i^z \quad .$$

and phonon dynamics with phonon frequency Ω

$$H_{ph} = -\Omega \sum_i (\cos \Phi s_i^z + \sin \Phi s_i^x) \quad .$$

(The last two terms are referred as "Apex-part"). The hopping term t serves in the following description as an energy unit; for a detailed description and motivation of the different parts see reference /1/.

We concentrate on the following special versions of this model:

a) **Apex-Oxygen-Model**

Here we consider single and double layered systems, t' is usually zero. Most of the simulations are carried out for one single layer. The model exhibits several features of the high-T_c compounds.

b) **Double-Layer Hubbard Model**

It consists of two purely electronic Hubbard layers without any coupling to the phonon (i.e. Apex) part; t' is zero, too. The double layer model shows superconductivity for 15% particles, i.e. 85% doping away from half-filling. Until now we have not been able to perform well converged simulations in a more realistic doping range, but first results are very promising for the s-wave as well as for the nodeless d-wave channel.

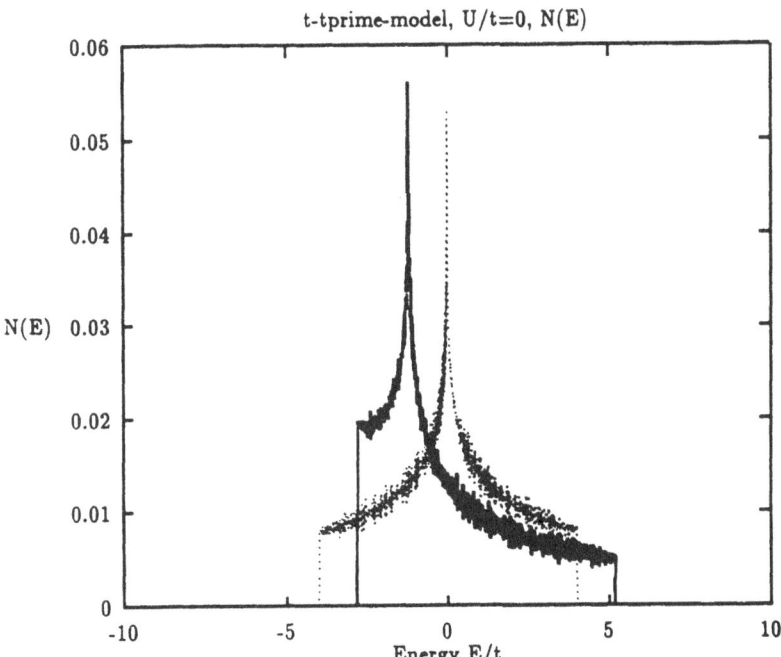

Figure 1 Shifting of the van Hove singularity away from halffilling; the plot shows the energy distribution $N(E)$ versus energy in units of t in the $U = 0$ t-tprime-model for $tprime = -0.3t$ (full curve) and $tprime = 0.0$ (dotted curve).

c) **t-t'-Model**

Here again we have no coupling to the phonon terms, but t' is different from zero allowing hopping to next nearest neighbour sites with a shifting of the van

Hove singularity away from half-filling. The t-t'-model is the simplest model to exhibt a van Hove singularity away from half-filling. So far we could only obtain a "positive vertex" indicating "local correlation (attraction)" between electrons but off-diagonal long range order as an evidence for superconductivity ("saddlepoint pairing", /2/) could not be established satisfactorily due to an enormous consumption of CPU time to reach equlibrium.

2. Numerical Method

The projector quantum Monte Carlo method (PQMC) /3/ uses the projection operator $\exp(-\theta\mathcal{H})$ to obtain the groundstate wave function $|0\rangle$. $\mathcal{H}$ is the Hamiltonian and the projection parameter θ has to be choosen large enough to assure a reliable projection starting with a test wave function $|\psi^T\rangle$ (usually originating from mean field calculations). $|\psi^T\rangle$ should not be orthogonal to $|0\rangle$.

Then we have

$$\exp(-\theta\mathcal{H}) = \exp(-\theta E_0)\left(\langle 0|\psi^T\rangle|0\rangle + \sum_{n>0}\exp(-\theta(E_n - E_0))\langle n|\Psi^T\rangle|n\rangle\right)$$

$$\xrightarrow[\theta \to \infty]{} \quad C \cdot |0\rangle \quad ,$$

where $|n\rangle$ are the eigenfunctions of the Hamiltonian $\mathcal{H}$, E_n denote the appropriate energy eigenvalues.

Therefore we obtain

$$|0\rangle = \lim_{\theta \to \infty} \exp(-\theta\mathcal{H})|\psi^T\rangle \quad .$$

In the simulation we have to take finite values for θ and try to find an appropriate approximation. The value of θ obviously depends on the difference $E_n - E_0$, the energy gap for the lowest states. For the proper evaluation of $\exp(-\theta\mathcal{H})$ we make use of the Trotter-Suzuki-Transformation /4/

$$\exp(A + B) = \lim_{m \to \infty} (\exp(A/m) \cdot \exp(B/m))^m \quad ,$$

where A and B are quantum mechanical operators. We use the transformation in a somewhat different form, which looks in the case of the single layer Apex model like

$$\exp(-\theta\mathcal{H}) = \left[\exp(-\frac{\theta}{m}\mathcal{H})\right]^m$$

$$\simeq \prod_{\nu=1}^{m}\left[\exp(-\frac{\epsilon}{2}H_0)\exp(-\epsilon(H_H + H_{e-ph} + H_{ph})\exp(-\frac{\epsilon}{2}H_0)\right] \quad ,$$

with $\epsilon = \theta/m$. The deviation is of order ϵ^2.

Now we have to apply this product on the trial wave function $|\psi^T\rangle$. The major difficulty here is that the Hilbert space increases exponentially with the system size. To circumvent this we transform our problem into a single particle problem using the discrete Hubbard-Stratonovich transformation /5/. Then we obtain for the interaction part of the Hubbard model $H_H = U \sum_i n_{i\uparrow} n_{i\downarrow}$:

$$e^{-\epsilon H_H} = \prod_i e^{-\epsilon U n_{i\uparrow} n_{i\downarrow}}$$

$$= \prod_i \left[e^{-\frac{\epsilon}{2}\hat{n}_i U} \sum_{\sigma_i = \pm 1} e^{\lambda \sigma \hat{m}_i} \right]$$

$$= \sum_{\{\sigma_i = \pm 1\}} e^{-\frac{\epsilon}{2}\hat{N}U} e^{\lambda \sum_i \sigma_i \hat{m}_i} \quad ,$$

where

$$\cosh(\lambda) = \exp(\frac{\epsilon}{2}|U|)$$

and

$$\hat{n}_i = \left\{ \begin{array}{ll} n_{i\uparrow} + n_{i\downarrow} - 1 & U \leq 0 \\[2mm] n_{i\uparrow} + n_{i\downarrow} & U > 0 \end{array} \right\} \quad , \quad \hat{N} = \sum_i \hat{n}_i$$

$$\hat{m}_i = \left\{ \begin{array}{ll} n_{i\uparrow} + n_{i\downarrow} - 1 & U \leq 0 \\[2mm] n_{i\uparrow} - n_{i\downarrow} & U > 0 \end{array} \right\} \quad ,$$

therefore obtaining a single particle problem. The interaction between up and down spins ($U n_{i\uparrow} n_{i\downarrow}$) is now replaced by an interaction with an Ising field σ_i, but for the price that one has to sum over all possible configurations $\{\sigma_i = \pm 1\}$. This summation of course cannot be carried out and we have to rely on the Monte Carlo procedure to sum over the *important* states (*Monte Carlo importance sampling*). One can show that applying the exponential of a single particle operator on a Slater-determinant results in a Slater-determinant again. Therefore it is possible in our case to propagate a Slater determinant as starting wave function. The Monte Carlo process now performs importance sampling of the appropriate Ising configurations. The necessary statistical weight of a certain configuration is obtained by calculating the value of the resulting determinants, i.e. the product for particles with spin up and spin down respectively.

There a major problem arises, the *minus-sign problem*:

Calculating the determinants we find for $U > 0$ (repulsive interaction !) in certain cases negative values, resulting in a negative transition probability, which is a priori not defined. But using the absoulte value we have to take care of the minus sign in the calculation of the expectation values. We have, for example, for the energy:

$$\langle E \rangle = \frac{E^+ - E^-}{Z^+ - Z^-} \quad,$$

where E^+ and E^- are the energies of "positive" and "negative" Ising configurations. Nevertheless the partition function Z also consists of positive and negative contributions. Using only the absolute values for the transition probabilites we then calculate

$$\langle E_p \rangle = \frac{E^+ - E^-}{Z^+ + Z^-}$$

and the "average sign"

$$\langle sign \rangle = \frac{Z^+ - Z^-}{Z^+ + Z^-} \quad.$$

This leads to the following correction for the expectation value:

$$\langle E \rangle = \frac{\langle E_p \rangle}{\langle sign \rangle} \quad.$$

So far there seems no problem, but the average sign behaves as

$$\langle sign \rangle \propto \exp(-\theta N) \quad,$$

where θ is the projection parameter and N the system size.

Therefore just the systems of interest have the major difficulties. The statistical errors in some case do not allow the calculation of expectation values, but for the PQMC the minus sign problem is not as severe as in the case of the "world line approach" /6/, for instance. System sizes up to 16×16 sites with θ up to 16 have been treated., and especially for certain fillings, so-called closed shell configurations, we have a relative large $\langle sign \rangle$. But we would like to mention that just in the for high-T_c superconductivity interesting regime we encounter for the double layer Hubbard model quite severe problems.

In our opinion a restriction of the Ising configurations to such equivalent to "world-lines" improves the simulation. But this new CPU-time consuming approach has to be checked more precisely against previous results /7/. Nonetheless it is possible to obtain reliable results with PQMC showing superconductivity. For the Apex Oxygen model we considered different values for θ where we have a mixing-in of higher excited states. In the case of superconductivity there are still clear indications of the long-range plateau-regime signaling superconductivity. The plateau for smaller θ's is lower as we mix in more and more of the higher lying non-superconducting states. Due to the fact that we deal with a projector method we do not simulate a partition function which according to Mermin-Wagner, Hohenberg /8/ shows the absence of superconductivity at finite temperatures. Therefore we consider PQMC just for the detection of superconductivity superior to partition function methods. Of course we are not able to detect the critical temperature T_c directly therefore relying on approximate indirect methods.

Here it should be mentioned that the PQMC is able to provide evidence in favour of superconductivity but the opposite is extremely hard to realize. In other words to show the absence of superconductivity with PQMC is in practical work almost impossible. There are too many obvious reasons why one may not obtain evidence for off-diagonal long range order in the simulation. The most common are that the projection parameter θ is too small, that the correlation length of the superconductiong correlation function is two large compared to the system size and therefore the system considered is just too small. A plateau value is a clear indication of superconductivity but it has to stabilize as the system size increases. This can be expected as long as we have relatively short correlation length. One of the major drawbacks of PQMC is that we deal with a finite system with a fixed number of particles involved. The influence of boundaries is much worse than in the classical case and finite size scaling virtually impossible as shown by T.Schneider /9/ and coworkers.

Therefore we would like to conclude that PQMC is a rather complex method to look for superconductivity. But so far it is the only method which provides reliable numerical evidence for superconductivity in certain models. The fact that we did not find superconductivity in the case of the single-layer Hubbard model and the Emery model does not mean that there is definitly no superconductivity but in the light of other sucessful simulations it seems more and more unlikely.

3. Off-Diagonal Long Range Order

According to Yang /10/ the relevant order parameter for the examination of the superconducting properties in a system with fixed particle number is the reduced two-particle density matrix or rather the Cooper pair correlation function (CPCF)

$$\chi_m(il) = \langle 0|c^\dagger_{i+\frac{m}{2},\uparrow}c^\dagger_{i-\frac{m}{2},\downarrow}c_{i+l+\frac{m}{2},\uparrow}c_{i+l-\frac{m}{2},\downarrow}|0\rangle$$

χ measures the CPCF between a Cooper pair of extension m at site i and a second one at distance l. One can define certain Cooper pair symmetries like s-wave, extended-s-wave, d-wave etc. /11/. To exclude the effects of any residual quasiparticle-interaction we focus our interest on the vertex CPCF by substracting the corresponding one-particle contributions /12/. This leads to a measure for the effective interaction between the carrieres within a Cooper pair. A macroscopic quantum state, i.e. superconductivity, is indicated by off-diagonal long range order (ODLRO) /10/, which is present, if the vertex correlation function approaches a finite constant value ('plateau value') for large Cooper pair distances. Kohn and Sherrington /13/ have shown that the presence of ODLRO implies the existence of a Meissner effect and thus superconductivity.

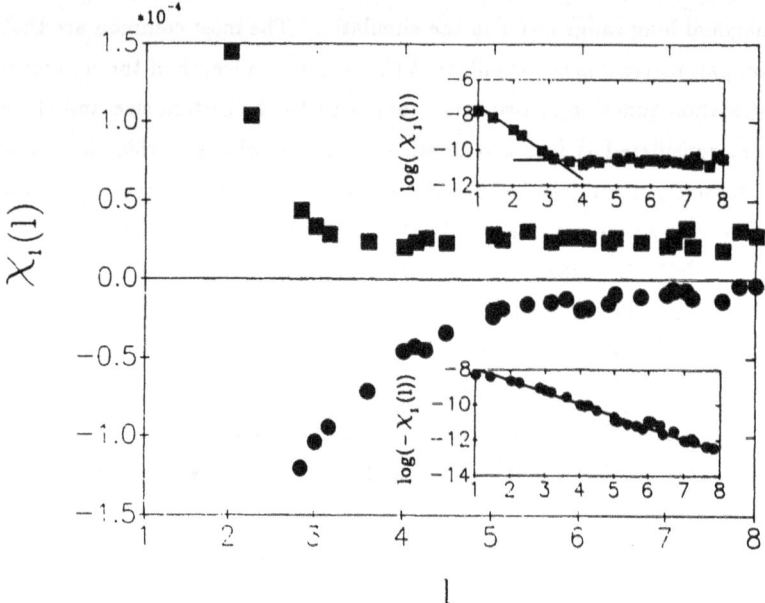

Figure 2 Decay of the nearest-neighbour vertex CPCF (extended-s-wave symmetry) with the Cooper Pair distance l for the electron-phonon model (Apex-Oxygen model, upper part) and the Hubbard model (lower part). Insets: Semilog-arithmic plots. The errorbars are of the size of the symbols. A 16×16 lattice was carried out for $U/t = 6$ and 15% doping. Electron-phonon coupling $g/t = 1$ and phonon frequency $\Omega/t = 1.0$ The plateau is clearly reached in contrast to the purely electronic single layer Hubbard model.

4. High-Temperature Superconductivity

Using the described method we obviously obtain a plateau value in the case of the Apex-Oxygen model /1/. Furthermore there is evidence also for the double layer

Hubbard model /14/.

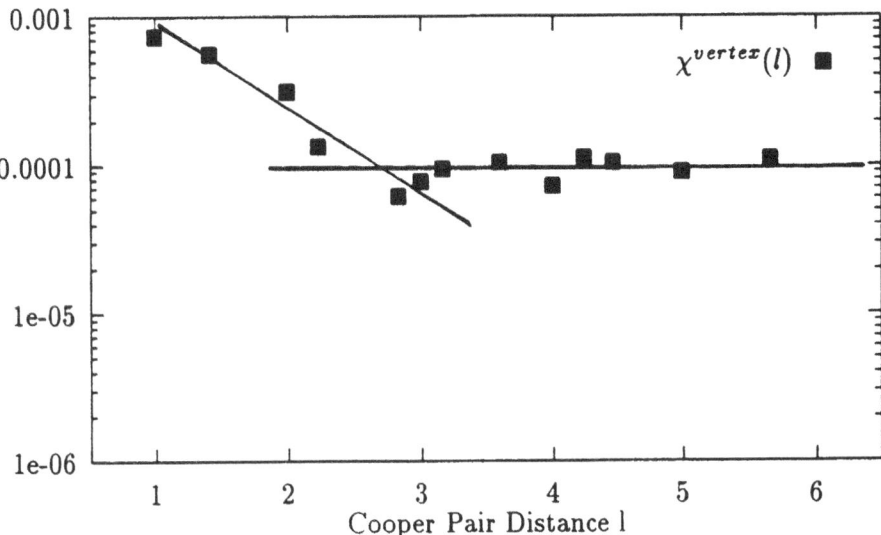

Figure 3 Semilogarithmic plot of the (nodeless d-wave) vertex CPCF versus Cooper pair distance for the purely electronic double layer Hubbard model with size $8 \times 8 \times 2$, $U/t = 8$ and $t_\perp = 0.1t$. Notice that we only have 15% particles (i.e. 85% doping away from halffilling). Here we again find a clear plateau value similiar to the el-ph-system in figure 1.

Recent simulations for the t-t'-model in collaboration with D. M. Newns and P. C. Pattnaik resulted in a positive vertex (i.e. local attraction), but the plateau value could not be established due to much too short simulations /15/. Large scale simulations are currently under way to give clear evidence in favour of the idea of the saddlepoint-pairing mechanism proposed by News and collaborators /2/.

This results in the following possible scenario:

a) Basic mechanism: Apex Oxygen

b) Booster mechanism: Double layer or multi-layer structures

c) Booster mechanism: Van Hove singularity

At the moment it is not clear whether a permutation of these points is also possible. Therefore the generic model for the high-T_c compounds should be a double layer

Emery model plus Apex-Oxygen.

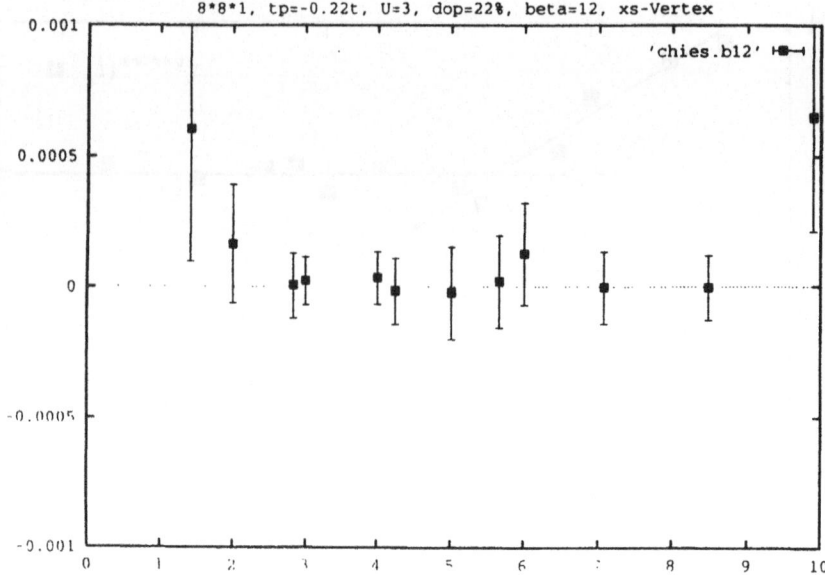

Figure 4 Extended-s-wave vertex CPCF versus Cooper pair distance originating from the purely electronic t-t'-model with system size $8 \times 8 \times 1$, $t' = -0.22t$, $U/t = 3$, $\beta = 12$, 22% doping away from halffilling. The position of the Fermi level E_F corresponds with the mean-field position of the van Hove singularity. The larger Cooper pair distances arise from the periodicity of the system. Here we only find a positive vertex but no direct indication for a plateau. Notice the large error bars due to insufficiently long runs; further simulations are necessary.

The drawback is that at the moment we do not a have sufficient amount of CPU time to carry out significant simulations. But with the introduction and spreading of the new parallel architectures we expect this to be the case in the near future. So far up to this point we still have to rely on simple but accessible models.

5. Experimental Predictions

At this point we are already able to provide experimental predictions to support the above mentioned scenario. Assuming that the Apex oxygen is the major reason for superconductivity we should consider the Isotope effect as a reliable indication of the influence of the Apex oxygen. Therefore a substitution of the apical O^{16} by O^{18}

should give a clear effect. It is also possible to have samples with almost all O^{18}. Just exposing this sample to O^{16} we should first have a substitution in the Apex-regime, possibly controlled by Raman measurements. Changing more and more of the Apex oxygen should give a steady increase of the isotope effect. According to our simulations we can obviously have a positive or even negative isotope effect. The closer the frequency is to the maximum in T_c the smaller the isotope exponent α should be.

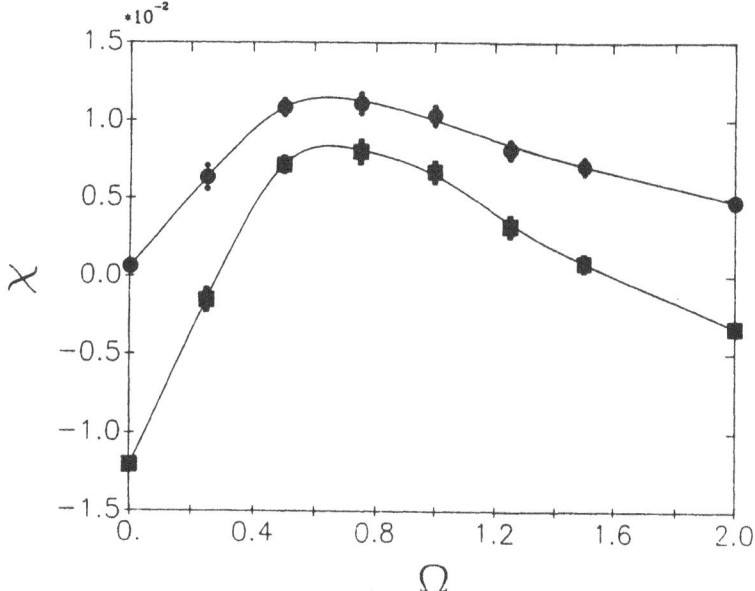

Figure 5 Integrated extended-s-wave vertex CPCF (lower curve) and next-nearest neighbour extended-s-wave vertex CPCF (upper curve) versus TLS frequency Ω/t. Parameters: $8 \times 8 \times 1$ lattice, $g/t = 1.0$, $U/t = 6.0$, 16% doping. Notice the maximum of the CPCF.

Here at this point we would like to comment on the experimental situation concerning the isotope effect. For samples close to maximum T_c the isotope shift exponent α has to be small /2,21,22/. This is actually also the case for the fully O^{18}-subsituted as well as apex site-selective subsituted YBCO-123 samples as reported at this conference by D.Zech and coworkers. So the current status gives no direct support for the apex conjecture, it is only in agreement with it. The crucial test for the apex is therefore a measurement of α in the underdoped (or overdoped) region with smaller T_c. We give a rough estimate for 123 with a T_c in the regime of $40K$ to $50K$: α

342

0.20 − 0.25 at least and about 70 − 80% contribution from the apex vibrations. We consider this experiment as the major test for the apex conjecture. Furthermore we would like to comment that the Apex mechanism cannot be ruled out just by consideration about the absence of the proposed anharmonicity of the Apex motion. Even in the case of a harmonic potential our two-level-system can be explained as a representation of the two lowest states. But in this context we have to point out that our simulations lead to superconductivity only if carried out with an asymmetric double well potential. The symmetric case as proposed by Bishop and coworkers /23/ does not work !. Therefore we propose the possibility of a double well potential with a strong asymmetric component (parameter Φ in the H_{ph}-Term of Hamiltonian (1) !) up to $50 meV$. In particular neutron scattering results should be analyzed along this line. The experimental results of Ruani et al. and Bianconi et al. /24/ presented at this conference agree according to private communication with this conjecture.

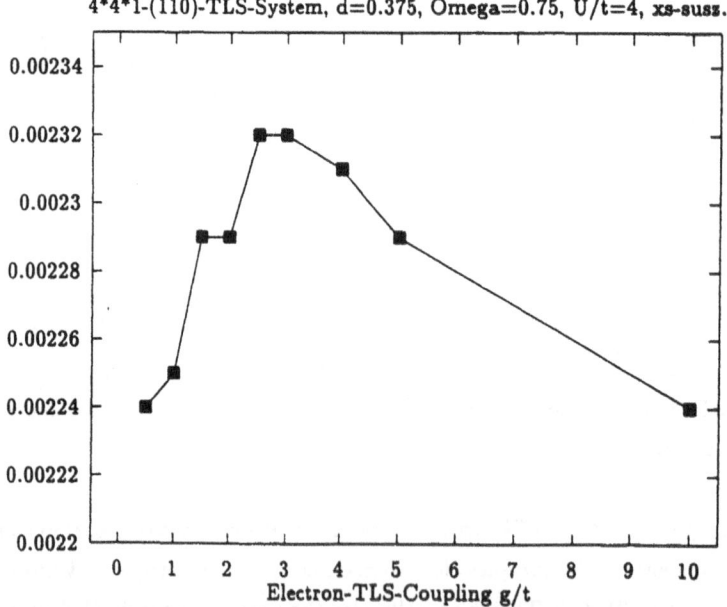

Figure 6 Integrated extended-s-wave vertex CPCF versus electron-TLS-coupling g/t. Parameters: $4 \times 4 \times 1$ lattice, $\Omega/t = 0.75$, $U/t = 4.0$, doping $\delta = 0.375$. Notice the increase of the values with coupling (corresponding to applied uniaxial pressure) up to a maximum. Further pressure should again decrease the suszeptibility, i.e. the critical temperature.

Another crucial role in the coupling to the Apex vibrations plays the coupling constant g. Increasing g with a fixed frequency Ω also leads to a broad peak with a maximum. Assuming now a positive isotope effect for the Apex materials increasing g should first lead to an increase of T_c up to a maximum and then a decrease. In an experiment this should be equivalent to applying unaxial pressure on a high-T_c compound sample. Using now the ($O^{16} \longleftrightarrow O^{18}$)-substituted samples we predict a change in the isotope exponent α with applied uniaxial pressure as long as the change affects the Apex oxygen. The obvious reason is the change of the coupling value g to the Apex system, and applying uniaxial pressure should lead to a decrease in α.

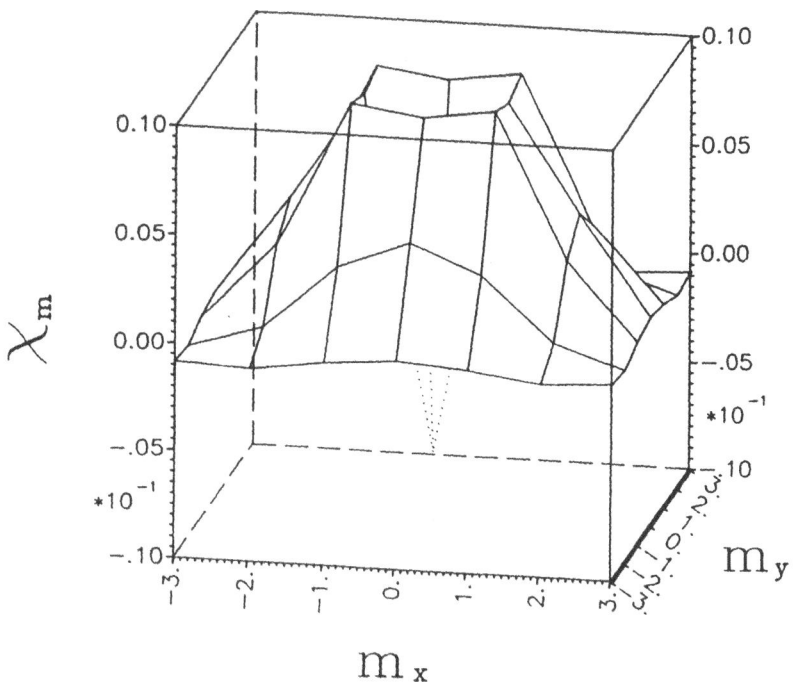

Figure 7   Spatial structure of the Cooper pair ("vulcano structure"). Integrated vertex CPCF as a function of the pair extension $m = (m_x, m_y)$, el-ph-model, parameters as figure 1.

For large enough pressure even a change in the sign of α seems possible. Therefore we would like to propose following experiments: First have a sample with as close as possible to 100% O^{18}. Measure the T_c-curve against applied uniaxial (c-direction !) pressure. Then expose the sample to O^{16}. Raman measurements indicate the amount of O^{16}-substitution in the Apex sites. Measure again T_c versus pressure.

344

Continue exposing the sample more and more to O^{16}. We expect as mentioned above a decrease of α, perhaps even a change of the sign. Increasing concentration of O^{16} in the apical oxygen positions will increase the effect on α, i.e. smaller changes with pressure for smaller concentrations of O^{16}, larger deviations for larger concentrations. The described scenario is according to our simulations only possible if the Apex Oxygen mechanism is the main reason for superconductivity in the high-T_c copper oxide compounds.

6. Applications

Assuming now the essential role of the Apex oxygen we can expand our simulations to some application oriented cases. The most crucial points in increasing the critical currents in ceramics and thin films is the understanding of the *weak link* and *pinning* behaviour of the materials /16/. As a consequence glassy behaviour also in single crystals can be simulated. First we would like to turn to the weak link phenomenon:

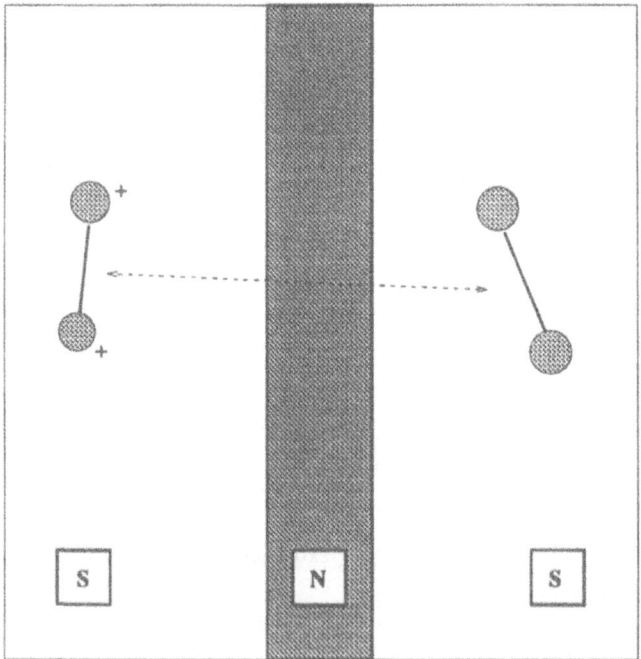

Figure 8   Schematic drawing indicating the measurement of the CPCF over a weak link. The Cooper pair is created on the left superconducting domain (denoted by S) and anihilated on the other side. The middle part denoted by N contains no Apex couplings and is therefor a simple doped Hubbard system.

6.1 Weak Links

In our simulations we assume the Apex-Oxygen model and consider in a first step only single layers. As it turned out including weak links increases the CPU consumption dramatically and we had to restrict ourselves at the moment to quite simple cases. But simulations for more complex systems like double layer weak links, layer-to-layer tunneling etc. are planned under way or planned for the future, preliminary results already exist.

In the case of weak links we just discard the apical oxygen over a certain width of the lattice, therefore dividing the lattice into a superconducting, a non-superconducting and a second superconducting domain (like a SNS junction). Obviously in this case we have to deal with free, non-periodic boundary conditions. Now we measure the correlations between a Cooper pair created in the part left to the link (first superconducting domain) and annihilated on the other side, thus obtaining the superconducting correlations accross the weak link. As expected the correlation decreases exponentially with the width of the link /17/.

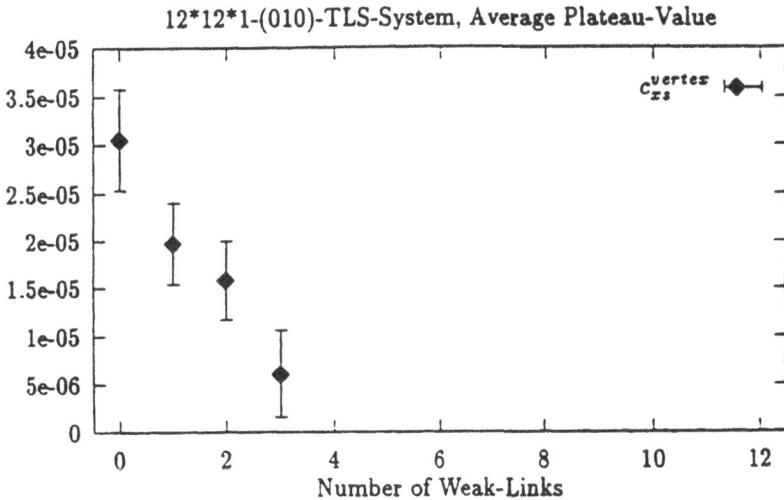

Figure 9 Plateau value of the extend-s-wave vertex CPCF over a weak link; the width of the weak link is measured in numbers of apex-free rows ("Number of Weak Links"). System size $12 \times 12 \times 1$, apex system with parameters similiar to figure 1. Notice the the strong decay of the plateau value.

But according to P.Chaudhari /18/ the decay is one order of magnitude slower in his most recent experiments. This suggests that weak links still can be improved considerably, implying that one should be in the position to fabricate ceramic materials with sufficiently high critical currents. Further aspects to improve critical currents are according to our simulations the influence of the multilayer structure; assuming at least a booster effect originating from double or multilayer systems we argue that close to a weak link the layered structure should be as intact as possible including and most importantly the Apex oxygen structure. A further possibility is the influence of the weak link boundary on the location of the van Hove singularity. A slightly different doping concentration should help at least in a theoretical approach.

To contribute to the theoretical understanding we consider the influence of the weak link on the structure of the Cooper pair crossing the link. Far from the link we have a "vulcano-like" correlation structure for the Cooper pairs, i.e. a ring of maximum correlation between the fermions of a Cooper pair due to the Coulomb repulsion and the short coherence length. The diameter of the vulcano is roughly the coherence length of the pair. Now approaching the link the vulcano-structure starts to "melt", the height of the whole structure decreases and the vulcano flattens out. Reaching the other superconducting domain across the weak link, the influence of the Apex is felt again, the "flattened vulcano" pushed back to its original structure. Increasing the width of the link, the Cooper pair obviously has more time and space to melt. At a certain point, the pair melts completely and recombination of Cooper pairs with now different electrons occurs on the other side. This lowers the correlation across the link structure dramatically. From this point of view it is clear that links which exceed the coherence length of the Cooper pairs considerably are really "bad". For the improvement of weak links we propose:

- Weak link width as small as possible (obviously !).
- Apex oxygen structure intact as cleanly as possible.
- Double or multilayer structure intact as closely as possible.
- Different doping close to the the links (van Hove singularity ! Further simulations are necessary to decide about increasing or decreasing doping).

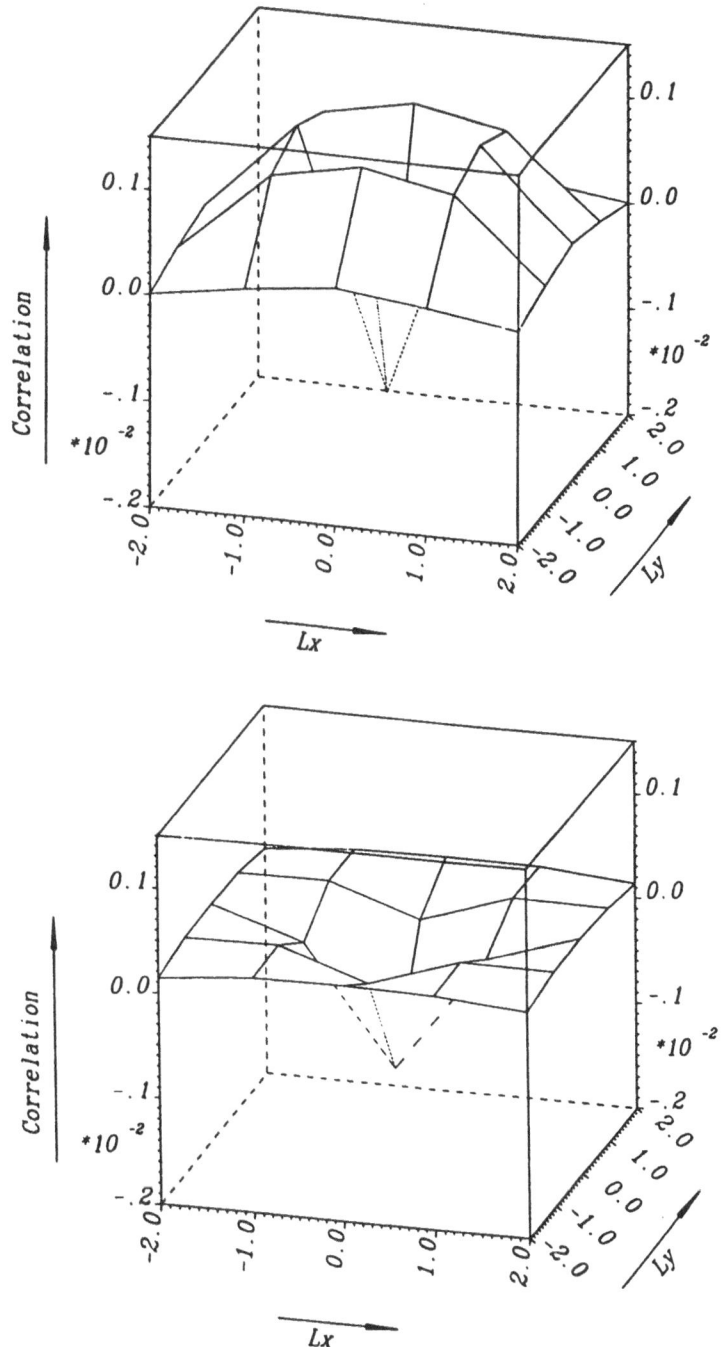

Figure 10 Spatial Cooper pair structure similiar to figure 5 in the weak link system. In the upper figure the values are measured outside the link in the superconducting domain, whereas the measurement in the lower part has taken place in the center of the link. "Melting of the vulcano-structure".

6.2 Pinning Centers

Removing the apical oxygens cancels out the central mechanism for superconductivity. This is essentially true also for a single site. We find a deep drop in the superconducting correlation function implying that already a single site defect in the apical oxygen structure may serve as a pinning center. Larger clusters of several sites show an even more dramatic effect /19/. Thus we propose that removing the Apex oxygen atoms or replacing them for instance with chlorine atoms within a certain concentration regime should lead to reliable pinning centers. The main message is that pinning centers are possible on an atomic scale as long as the planes are still intact.

6.3 Glassy Behaviour

Replacing the apical oxygen randomly in the planes should lead to glassy behaviour even in single crystals. Considering the three-dimensional layered structure and applying a magnetic field in c-direction the flux lines would arrange along the pinning centers. Positional disorder of the pinning centers in the different layers leads to glassy behaviour in the sense of some type of directed polymer glasses. The flux lines experience a different potential at the pinning centers. According to M.P.Fisher a certain relation to quantum Monte Carlo simulations exists. Considering the flux lines as "world-lines" in quantum Monte Carlo simulations the problem can be mapped onto a quantum mechanical XY-model with disorder. Here we would like to mention that the original models for glassy behaviour of ceramics were classical XY-models with disorder /20/. The occurence of glassy behaviour can be used to increase the critical currents by locking in the system into a relatively low and narrow valley in phase space. This can be achieved by changing the concentration of removed Apex oxygens or by a certain amount of texturing. On the other hand glassy behaviour of single crystals or thin films could be related to a ramified Apex structure. Raman measurements could reveal this effect. Furthermore an especially prepared crystal should reveal glassy behaviour, too. This concept opens a large field for numerical simulations based on the quantum Monte Carlo concept.

7. Conclusion

We have demonstrated the potential of the quantum Monte Carlo method. The projector version is in our opinion still the favourite method to detect superconductivity. On the other hand it is extremely hard to show the absence of this effect, thus our conclusion is:

Superconductivity in the Apex-Oxygen model /1/ but most likely not in the single layered Hubbard or Emery model. A booster mechanism is provided by multilayer structures /14/ . The influence of the van Hove singularity could also provide a booster mechanism /2/, although numerical evidence for a purely electronic saddlepoint pairing /15/ mechanism is still quite weak. A scenario for experiments related to the isotope effect is given based on the Apex mechanism. The influence of applying uniaxial pressure is outlined, it is predicted to change the sign of the isotope exponent α in some cases (high enough pressure before destroying the crystal). Application oriented simulations give some indications to improve the weak link behaviour based on the Apex-Oxygen model. Effective pinning centers can be obtained by decoupling the Apex oxygen vibrations, by substituting the apical oxygen atoms with other elements. In this case we predict glassy behaviour also for single crystals in the presence of a magnetic field.

8. Acknowledgement

The authors would like to thank K. A. Mueller, D. M. Newns, P. C. Pattnaik, P. Chaudhari, W. von der Linden, T.Schneider, W. Hanke, R. Hetzel, J. Keller, G. Baskaran and P. W. Anderson for helpful and stimulationg comments and discussions in particular M. Frick and H. de Raedt for their support in developing essential parts of the algorithm as well as Ch. Baur and W. Fettes for their help in performing some of the calculations. We also acknowledge the hospitality of the IBM T.J.Watson Research Center Yorktown and the German Supercomputing Center HLRZ Juelich where parts of the work were accomplished. Most of the present calculations have been performed on the CRAY YMP at HLRZ Juelich as well as on the IBM RS/6000 "RIOS-Farm" Compute Clusters in Yorktown and Regensburg. This work was partially supported by "Bayrisches Hochschul-Verbundprojekt FORSUPRA".

9. References

/1/ M.Frick, I.Morgenstern, W.von der Linden, *Z.Phys.B - Cond.Matt.* 82 (1991) 339;
I.Morgenstern, M.Frick, W.von der Linden, *J.Phys.I. France* 2 (1992) 393 and references therein
/2/ D.M.Newns. P.C.Pattnaik, C.C.Tsuei, *Phys.Rev.B.*43 (1991) 3075;
P.C.Pattnaik, C.L.Kane, D.M.Newns, C.C.Tsuei, *Phys.Rev.B*45 (1992) 5714;
D.M.Newns, H.R.Krishnamurthy, P.C.Pattnaik, C.C.Tsuei, C.L.Kane,

      Phys. Rev. Lett. 69 (1992) 1264;

    R.S.Markiewicz, Preprint Boston 1992,

    D.M.Newns, C.C.Tsuei, P.C.Pattnaik, C.L.Kane,

    Comments Cond. Mat. Phys. 15 (1992) 273,

    D.M.Newns, H.R.Krishnamurthy, P.C.Pattnaik, C.C.Tsuei, C.C.Chi, C.L.Kane, Preprint 1993, and

    D.M.Newns, P.C.Pattnaik, Private Communications

/3/ H.de Raedt, *Comp.Phys.Rep.* 7 (1987) 3;

    S.E.Koonin et al., *J.Stat.Phys.* 44 (1986) 985;

    S.Sorella, S.Baroni, R.Car, M.Parinello, *Europhys.Lett.* 8 (1989) 663;

    W.von der Linden, I.Morgenstern, H.de Raedt, *Phys.Rev.* B (1990) 4669;

    W.von der Linden, *Physics Reports* 220(2& 3) (1992) 53

/4/ M.Suzuki, *Comm.Math.Phys.* 51 (1976) 183;

    M.Suzuki, *J.Stat.Phys.* 43 (1986) 883

/5/ J.E.Hirsch, *Phys.Rev.*B 28 (1983) 4059

/6/ I.Morgenstern, *Z.Phys.B - Cond.Matt.* 70 (1988) 291

/7/ Th.Husslein, Thesis, Univ.Regensburg 1993

/8/ P.C.Hohenberg, *Phys.Rev.* 158 (1967) 383

/9/ T.Schneider, M.Frick, M.Bormann, *Europhys.Lett.* 14 (1991) 101

/10/ C.N.Yang, *Rev.Mod.Phys.* 34 (1962) 694

/11/ A.Moreo, Univ. California, Preprint UCSBTH//91-33

/12/ M.Frick, I.Morgenstern, W.von der Linden, in *FERMION ALGORITHMS*, Ed. H.Herrmann, F.Karsch, World Scientific 1992

/13/ W.Kohn, D.Sherrington, *Rev.Mod.Phys.* 42 (1970) 1

/14/ I.Morgenstern, Th.Husslein, J.M.Singer, H.-G.Matuttis, *J.Phys. II France* 2 (1992) 1489;

    I.Morgenstern, Th.Husslein, J.M.Singer, H.-G.Matuttis, *J.Phys.I France* 3 (1993) 1043

/15/ J.M.Singer, Th.Husslein, D.M.Newns, I.Morgenstern, P.C.Pattnaik, in preparation, 1993

/16/ D.Dimos, P.Chaudhari, J.Mannhart, F.K.LeGoues, *Phys.Rev.Lett.* 61 (1988) 219;

    D.Dimos, P.Chaudhari, J.Mannhart, *Phys.Rev.B*41 (1990) 4083;

    J.Mannhart, Reprint, Toshiba Int'l. School of Superconductivity, 1991

/17/ J.M.Singer, Thesis, Univ. Regensburg 1992;

    J.M.Singer, I.Morgenstern, Th.Husslein, H.-G.Matuttis, Proc. Phys.Comp. 92, Prague

/18/ P.Chaudhari, Private Discussions

/19/ H.-G.Matuttis, Thesis, Univ.Regensburg 1992

/20/ I.Morgenstern, K.A.Mueller, J.G.Bednorz, *Z.Phys.B - Cond.Matt.* 69 (1987) 33

/21/ C.C.Tsuei, D.M.Newns, C.C.Chi, P.C.Pattnaik, *Phys. Rev. Lett.* 65 (1990) 2724

/22/ J.P.Franck et al., *Physica C* 185-189 (1991) 1379;
J.P.Franck et al., in: *"High Temperature Superconductivity"*,
edited by Ashkenazi et al., Plenum 1991;
J.P.Franck et al., *Phys. Rev.* B 44 (1991) 5318;
H.J.Bornemann, D.E.Morris, *Phys. Rev.* B 44 (1991) 5322

/23/ J.Mustre de Leon, I.Batistic, A.R.Bishop, S.D.Conradson, S.A.Trugman,
Phys. Rev. Lett. 68 (1992) 3236

/24/ G.Ruani and A.Bianconi, private communications and Proceedings of the International School of Crystallography, Erice 1993 (this book)

EVIDENCE FROM EXAFS FOR AN AXIAL OXYGEN CENTERED LATTICE INSTABILITY IN $YBa_2Cu_3O_{7-\delta}$?

J. RÖHLER*
Universität – GH Paderborn
Fachbereich Physik
33098 Paderborn
Fed. Rep. Germany

ABSTRACT. We address the problem of lattice anharmonicities in high-T_c superconductors. The polarized extended x-ray absorption- fine-structure (EXAFS) experiments at the Cu K-edge in $YBa_2Cu_3O_{7-\delta}$ showing evidence for a local two-site apex configuration are briefly reviewed. The analysis of beat structures in the photoelectron interference patterns is shown to be a useful tool for the analysis of structural disorder. Model calculations confirm the analysis of EXAFS data yielding two Cu1-O4 bond lengths different by $\simeq 0.1$ Å [Mustre et al., PRL 65, 1675 (1990); Stern et al., Physica C 209, 331 (1993)]. The data can be modelled with a bimodal anharmonicity of the apex oxygen along the c-axis, but a static model seems to be more appropriate. From the two-site apex-configuration, recently detected in the superstructure of the Ortho II phase, we conclude that similar local distortions have to occur also around isolated chain-oxygen-vacancies in the Ortho I phase. We find static two-site apex-configurations are consistent with the beat structure in the EXAFS. Thus a coherent picture for the data from superconducting and nonsuperconducting samples emerges.

1. Introduction

There is ample experimental evidence that lattice degrees of freedom are involved in the superconducting phase transition of the high-T_c's. The thermal expansion coefficient jumps over the superconducting to normal phase boundary, $\Delta\alpha$, along the a- and b-axis of detwinned $YBa_2Cu_3O_{7-\delta}$ crystals are about 0.6×10^{-6}/K, and exhibit opposite signs. Along the c-axis the thermal expansion coefficient jump is close to zero [1, 2].

Capacitance dilatometers are several orders of magnitude more sensitive than either x-ray or neutron diffraction. The data available from different laboratories compare well with each other and are even consistent with the available x-ray diffraction data [3, 4], taken the lower precison of the latter into account. The opposite sign jumps observed in $YBa_2Cu_3O_{7-\delta}$ imply a relatively large shear component [5] in this particular material. Since $YBa_2Cu_3O_{7-\delta}$ exhibits the largest orthorhombicity among

---

*and Universität zu Köln, Physikalisches Institut, Zülpicher Str. 77, 50937 Köln, FRG

E. Kaldis (ed.), Materials and Crystallographic Aspects of HTc-Superconductivity, 353–372.
© 1994 Kluwer Academic Publishers.

$Y_2Ba_4Cu_7O_{15}$ and $YBa_2Cu_4O_8$, non unexpected the thermal expansion coefficient jumps in the basal plane of the latter material exhibit same signs [6].

Thermal expansion jumps across T_c are not unknown in superconducting materials. The lattice of a superconductor adjusts itself below T_c thus minimizing the free enthalpy of the lattice and the superconducting electrons. The onset of the spontaneous strain results in second order jumps of the thermal expansion coefficient, which are proportional to $\rho(0)(k_BT_c)^2$. Here $\rho(0)$ denotes the density of states at the Fermi energy. A comparison between conventional and high-T_c materials shows [7], the lattice effects in the high-T_c's are not always qualitatively different or stronger than in conventional materials: in Al ($T_c = 1.12$ K) the thermal expansion coefficient jump across the superconducting phase transition was determined as $\simeq 1 \times 10^{-8}$/K, and in Pb as $\simeq 6 \times 10^{-8}$/K ($T_c = 7.2$) K [8]. No outstanding anharmonicity of the lattice has to be assumed for the explanation of the thermal expansion jumps in both, conventional and high-T_c materials; $\rho(0)(k_BT_c)^2$ is much bigger for a 90 K superconductor than for a 7 K one.

Admittedly, capacitive dilatometers are not a tool for structural analysis but probe lattice properties from macroscopically sized crystals, typically of mm^3. Single crystals of orthorhombic $YBa_2Cu_3O_{7-\delta}$ are usually twinned on $\{110\}$, thus reducing macroscopic stress. The stress occurs when the crystal is constrained in some way from outside during the formation of the orthorhombic phase. Thermal expansion measurements of such crystals average over the twin structure and therefore are expected to show a more isotropic thermal expansion behavior in the basal a-b plane than untwinned crystals. Indeed, same signs of $\Delta\alpha$ along both, a and b-axes, have been obtained from the thermal expansion of untwinned crystals [9].

The twin structure is only one type among many 'local structures', occuring in $YBa_2Cu_3O_{7-\delta}$. The twin boundaries are spaced by a few 10 nm and form a fairly regular superlattice. If one considers structural features on a mesoscopic scale are relevant to the mechanism of high-T_c superconductivity, this type of local structure consequently appears in the focus of interest [10].

Electron-lattice interactions and their possible relevance for high-T_c superconductivity, however, are mostly deduced (if at all) from microscopically small deviations of the lattice from the 'average structure'.

'Average structure' is usually defined as the crystallographic structure of 'real crystals' whose atomic positions deviate more or less from those defined by the spacegroup of the 'ideal crystal'. The average structure is thus an approximation (although an excellent one) to the real atomic structure of the solid. Usually the crystallographic structure describes deviations from the nominal sites by large thermal parameters. For instance, the large thermal ellipsoid of the apical oxygen (O4) in $YBa_2Cu_3O_{7-\delta}$ might point to a significant deviation of the O4-position from the nominal crystallographic site.

Outstanding lattice anharmonicities involving the apical oxygen have been reported from various experimental investigations using local structural probes: EXAFS (X-ray Absorption-Fine-Structure), neutron-radial-distribution function measurements, cf. the lecture of Egami [11]. The anharmonic lattice dynamics have been detected in the local structure but not in the average structure. Here 'local structure' means the atomic structure as determined by (mostly site selective) local techniques, which are not restricted by the symmetry conditions, underlying diffraction.

In this lecture we address the local structure in $YBa_2Cu_3O_{7-\delta}$ as determined from polarized Cu K x-ray absorption-fine-structure experiments by Mustre et al. [12, 13] and Stern et al. [14, 15]. The position of the apical oxygen O4 in $YBa_2Cu_3O_{7-\delta}$ has been found to deviate significantly from the crystallographically determined site. Both groups report the best fit to the bulk of their data can be obtained assuming two Cu1-O4 bond lengths differing by about 0.1-0.13 Å. As yet neutron and x-ray diffraction could not confirm the two-site position of the apical oxygen, or a related anomalous Debye-Waller factor of the apical oxygen.

While Stern et al. [15] find no dependence of the splitting on temperature and stoichimetry, Mustre et al. [12] report a significant temperature dependence around T_c. Their best fit to the EXAFS data outside the fluctuation region $(T_c - \delta T) < T < (T_c + \delta T)$ was obtained when the apical oxygen was equally distributed over two axial positions split by $\simeq 0.13$ Å. Inside the fluctuation region the two axial positions collapsed into a single one. Following the suggestion, that dynamic instabilities of a ferroelectric type might occur in the layered perovskite-like structures [16], the EXAFS have been interpreted dynamicly as arising from a bimodal axial anharmonicity of the apical oxygen.

We shall show that the double-site O4-configuration, observed by EXAFS in superconducting and nonsuperconducting $YBa_2Cu_3O_{7-\delta}$, has to arise from vacancy induced local static lattice distortions. We conclude, that there is no evidence for a bimodal lattice anharmonicity along the c-axis in $YBa_2Cu_3O_{7-\delta}$.

2. Aspects of the Local Structure in $YBa_2Cu_3O_{7-\delta}$

Most of the high-T_c cuprates exhibit a non-stoichiometric composition of oxygen, which controls the electronic and structural properties of the crystals. The order-disorder phenomena of the oxygen (O1) -vacancies in the chains of $YBa_2Cu_3O_{7-\delta}$, occurring through interchain and intrachain correlations have been studied very much in detail. For long-range ordering see e.g. [17] and for medium- and short-range ordering [18, 19]. 'Short' denotes the atomic scale of several Å up to nearest (nn) or next nearest neighbor distances, 'medium' mesoscopic lengths of several nm, 'long' macroscopic lengths of 1 μm or so.

In the following we refer to the average structures of oxygen deficient $YBa_2Cu_3O_{7-\delta}$ as determined by neutron powder diffraction for $0.07 < \delta < 0.91$ [20]. The structural properties of the samples exhibit the well known 'plateau' behavior of T_c versus δ. Deviations from these average structures are known to arise particularly from the break down of short-range order correlations. The '60 K plateau' appears to be completely absent in samples with the smallest degree of short-range vacancy ordering [19].

356

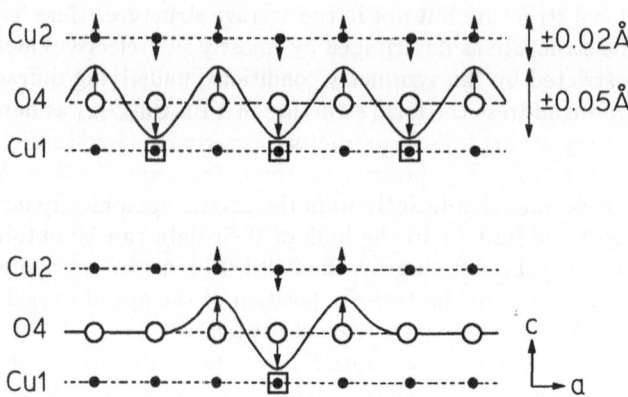

Fig. 1. *Top:* Schematic side view of the ordered Ortho II structure as determined from the crystallographic superstructure [23]. Squares indicate oxygen vacancies in the chains. *Bottom:* Expected local distortions around a single chain-oxygen-vacancy in Ortho I.

2.1 DISPLACEMENTS FROM THE AVERAGE IN THE ORTHO-II STRUCTURE

If deviations from the average structure exhibit translational invariance across lengths comparable to the coherent scattering length of x-rays or neutrons, $\xi \approx 100$ Å, diffraction techniques are proven to be an excellent tool to detect and to quantify them through the analysis of the resulting superstructure reflections. Recently the superstructure reflections, expected from long-range ordered Ortho II (OII) configurations in $YBa_2Cu_3O_{6.55}$, have been detected in a series of crystals from x-ray and neutron diffraction patterns [21, 22], *cf.* the lecture of Hohlwein [23]. Deviations from the average ordered Ortho II structure [20] were found to occur through displacements of Cu2 by ±0.02 Å, O4 by ±0.05 Å along the c-axis, of Ba by -0.04 Å and of Y by $+0.01$ Å along the a-axis [24]. These deviations from the average sites are of static nature and due to asymmetries of the Coulomb potential induced by the long-range ordered oxygen vacancies. From the linewidth of the $2a \times b$ superstructure reflections in several more oxygen deficient crystals with $\delta \leq 0.46$ a lower limit for the average size of the OII microdomains could be derived. It gets mesoscopically small as $5a \times b \times 1c$ in a $YBa_2Cu_3O_{6.7}$ single crystal ($T_c = 57(10)$ K).

Interestingly, the displacements from the average Ortho II structure given by the superstructure turn out to be in disagreement with the atomic positions, obtained from linear interpolation between the average structures of ortho I and tetragonal (T) crystals. From the Ortho II superstructure a double-site O4-configuration with Cu1-O4 bonds of 1.78 Å and 1.88 Å ($\Delta R = 0.1$ Å), and Cu2-O4 bonds of 2.37 Å and 2.41 Å ($\Delta R = 0.02$ Å) have been derived. Linear interpolation of bond lengths of average OI and T structures yields a double site O4 configuration with Cu1-O4 bonds of 1.80 Å and 1.85 Å ($\Delta R = 0.05$ Å), and Cu2-O4 bonds of 2.47 Å and 2.29 Å ($\Delta R = 0.18$

Å). We emphazise: the OII structure exhibits a double-site O4-configuration with strongly modulated Cu1-O4 bonds along the c-axis while the interpolation of average OI and T data gives a O4 double site configuration with extremely modulated Cu2-O4 bonds. Noteworthy also that Ba (located inside the layer between chains and planes) in OII is found displaced by -0.04 Å along the a-axis, and Y (located between the CuO_2 planes) displaced by $+0.01$ Å in the opposite direction along a. From the interpolated data no such displacements of Ba and Y are found.

2.2. LOCAL DISPLACEMENTS AROUND ISOLATED VACANCIES

The long-range ordered displacements of OII-type have to occur locally also in an optimum oxygenated Ortho I environment around isolated or small clusters of oxygen-vacancies in the chains, *i.e.* short-range deviations from the average OI-structure of comparable magnitude and with same signs. Fig. 1 *(top)* exhibits schematically the deviations of the apical O4 and of Cu2 along the c-axis from their average positions as determined from the superstructure. Squares indicate O1-vacancies along the b-axis. Fig.1 *(bottom)* displays the short-range ordered distortion along the c-axis expected from the presence of a single chain-oxygen-vacancy.

If a single vacancy may distort its nearest neighbor environment as proposed in Fig. 1 *(bottom)*, even weakly oxygen deficient Ortho I crystals, as for example optimum doped $YBa_2Cu_3O_{6.93}$, are expected to exhibit a heavily distorted local structure. Note that a single O1-vacancy displaces 12 apical O4-atoms and 12 Cu2-atoms, along the c-axis and several Ba- and Y-atoms along the a-axis. The optimum T_c in $YBa_2Cu_3O_{7-\delta}$ occurs for $\delta = 0.07$. Assuming the O1- vacancies are isolated from each other, 42% of the apical oxygen appear displaced from their average positions (not only 7% as expected from 'passive' vacancies). Certainly the detailed structure of the distorted nearest neighbor cluster around O1-vacancies requires a more elaborate analysis, which is beyond the scope of this lecture.

To shed light on the controversial double-site O4-configuration measured by EX-AFS in Ortho I crystals, we are going to discuss two simple vacancy configurations: a single isolated O1-vacancy, and two neighbored O1-vacancies ($YBa_2Cu_3O_6$) in a fully oxygenated environment. The latter configuration is shown in Fig. 2. The displacements of Ba and Y will not be considered.

A single isolated O1-vacancy creates a double-site O4-configuration with 4 apical O4 displaced towards Cu1 and 8 apical O4 displaced away from Cu1, giving a relative weight of 33:66. If we assume one cell of $YBa_2Cu_3O_6$ (with two vacant O1-sites), is diluted in a matrix of $YBa_2Cu_3O_7$ cells (see Fig. 2), we have 2 vacancies in neighbored chains: 8 apical O4 are displaced towards Cu1, 8 are displaced away from Cu1, yielding a relative weight of 50:50. The latter configuration will displace all apical O4 from their average positions for $\delta = 0.25$ (assuming an appropriate ordering of the double vacancies).

The case of $\delta = 0.17$ (at the borderline of the 90 K 'plateau' in the T_c versus δ diagram) is particularily important for the interpretation of the experimental EXAFS data. We expect up to 70% of the apical O4 are displaced from the average structure, but only up to 30% exhibit the positions of the average OI structure. As shown above, the relative weight of the two Cu1-O4 bonds may vary from 33:66 to 50:50, dependent on the detailed configuration of the vacancies creating the double-site O4-configurations. We shall show further below, that this configuration is still within the limit of detectibility of EXAFS.

In the next section we are going to discuss the potential of x-ray absorption spectroscopy for the exploration of deviations from the average structure.

3. EXAFS as a Probe of Local Distortions

We first give a brief introduction into the basic principles of the EXAFS as a structural tool. In particular we discuss the analysis of beats in the extended absorption-fine-structure yielding information on simple types of disorder. Then we analyze the beats arising in the linear configuration Cu1-O4-Cu2 on Cu K-excitation. The cases with single-site O4- and double-site O4-configurations will be considered.

3.1 THE PHOTOELECTRON INTERFERENCE PATTERN

The absorption x-ray-absorption fine-structure (XAFS) beyond the x-ray absorption thresholds of atoms in solids arises from the interference between freely propagating photoelectrons and photoelectrons backscattered from the atomic cluster to the excited central atom. X-rays are used for recording EXAFS spectra, but interfering electrons give the sinusoidal absorption pattern used for the structural analysis. Since electrons are strongly scattered by atoms, EXAFS is a very sensitive structural probe, even for the detection of low-Z scatterers as oxygen.

In particular EXAFS is a sensitive probe for local lattice distortions. The change of the momentum of the photoelectron after backscattering, $2k$, can be recorded up to 16 Å^{-1} or more. The high momentum-transfer limit of EXAFS is advantageous in discerning fine details of the local structure compared to that of diffraction with a momentum transfer limit of $q \approx 4.5$ Å^{-1}. For example, a resolution of about 0.1

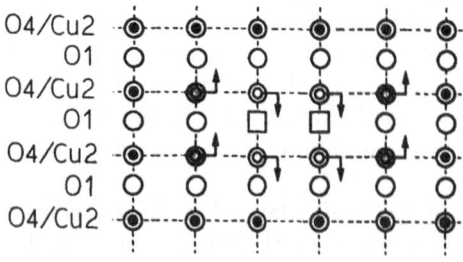

O4/Cu2
O1
O4/Cu2
O1
O4/Cu2
O1
O4/Cu2

Fig. 2. Schematic top view on a YBa$_2$Cu$_3$O$_{7-\delta}$ lattice (a-b plane) with two oxygen vacancies (squares) in the chain. Only Cu2- and O1-, O4-atoms are shown. Displacements of the apical O4 and the planar Cu2 expected along the c-axis are denoted by arrows. Blank double circles indicate Cu2 and O4 deviating from the average positions (full/blank double circles) *towards* Cu1, shaded double circles indicate Cu2 and O4 pushed *away* from Cu1.

Å in the radial distribution function achieved by EXAFS in the range of $4 \geq k \leq 14$ Å^{-1}, would require data for $q \approx 25\text{-}28$ Å^{-1} in the diffraction experiment.

The interference function $\chi(k)$ for a single backscattering configuration of $l = 1$ photoelectrons (K-absorption) can be expressed as

$$\chi(k) = -\mathrm{Im}\big[A(k, R_i)\exp[i2kR_i + i\phi(k, R_i)]\big], \tag{1}$$

where $A(k, R_i)$ is an envelope function given by

$$A(k, R_i) = \frac{NB(k, \pi)}{kR_i^2}. \tag{2}$$

k denotes the wavenumber of a free photoelectron given by $k = [(E - E_0)2m/\hbar^2]^{1/2}$ with E the photon energy, E_0 the ionization threshold. N is the number of scattering atoms at the distance R_i between the ith and the excited central atom. $B(k, \pi)$ is a complex amplitude factor associated with the backscattering process and including the inelastic losses suffered by the photoelectron. $\phi(k, R_i) = 2\delta_{l=1} + \arg[B(k, \pi)]$ denotes the complex net phaseshift of the photoelectron backscattered at the ith atom. Here $\delta_{l=1}$ is the central atom phase shift for the $l = 1$ partial wave and $\arg[B(k, \pi)]$ the backscattering phase.

The lattice vibrations are usually considered to contribute by an Debye-Waller–term $\exp(-2\sigma_i^2 k^2)$ to the interference function $\chi(k)$. The term σ_i^2 in the exponent is the second cumulant of the distribution of distances between the absorbing atom and the ith atom. It can be expressed to first order as

$$\sigma_i^2 = \langle[(\vec{u_i} - \vec{u_0}) \cdot \vec{r_i}]\rangle \tag{3}$$

where $\vec{u_0}$ are the equilibrium distance and $\vec{u_i}$ the instantaneous distance between the central atom and the ith atom. $\vec{r_i}$ is a unit vector at the origin pointing to the site i. The brackets denote a thermal average.

In the harmonic case cumulants beyond the second order are zero and the effects of lattice vibrations on the interference function are fully included by σ_i^2. For cases of weak to moderate anharmonicity a pertubative treatment must be applied and higher cumulants have to be included. The even cumulants contribute to the envelope function, $A(k, R_i)$, while the odd enter into the net phase of $\chi(k)$ [25]. This pertubative treatment of anharmonic lattice vibrations is not applicable to strong anharmonicities. At $k\sigma \simeq 1$ the cumulant expansion diverges and other non pertubative methods have to be applied.

A simple case of static disorder (or bimodal anharmonicity) is given by two closely neighbored shells of identical atoms with N_1 atoms at R_1, and with N_2 atoms at R_2 from the central atom. In this case one can extract the disorder directly from beats in the envelope function of $\chi(k)$.

3.2 EXTENDED X–RAY–ABSORPTION FINE–STRUCTURE BEATS

The superposition of two closely neighbored scattering shells produces beats in the envelope function of the extended x-ray-absorption fine-structure and a modulation in the scattering phases [26]. We consider an interference function for two shells, $\chi(k)$ of identical atoms at distances R_1 and R_2 from the excited central atom. The sum over the two shells $i = 1, 2$ can be written as

$$\chi(k) = -\text{Im}\left[\tilde{A}(k, R_2, R_2)\exp\left[i2k\tilde{R} + i\tilde{\phi}(k, R_1, R_2)\right]\right], \tag{4}$$

where $\tilde{R} = (R_1 + R_2)/2$ is the average distance of the shells from the absorbing atom and the relative distance is given by $\Delta R = R_2 - R_1$. The envelope function of the two shells takes the form

$$\tilde{A}(k, R_1, R_2) = A_1(k, R_1)[1 + C^2 + 2C\cos(2k\Delta R)]^{1/2}, \tag{5}$$

with $A_1(k, R_1)$ the envelope function of the first shell as defined in equ. 2, but multiplied with a Debye-Waller-term $\exp(2\sigma_i^2 k^2)$. C is the ratio of the envelope functions of the second and the first shell,

$$C = A_2/A_1 = (N_2/N_1)(R_1{}^2/R_2{}^2)exp[-2(\sigma_1{}^2 - \sigma_2{}^2)]. \tag{6}$$

The ratio of the envelope functions exhibit minima and maxima with a wavenumber determined by the relative distance ΔR. Assuming negligable differences in the mean-square displacements $\sigma_{1,2}$, beats occur at

$$k_b = (n + 1)\pi/2\Delta R, n = 0, 1, 2, 3..., \tag{7}$$

the minima at even n, the maxima at odd n.

The averaged phases $\tilde{\phi}(k)$ show up an arctan-shaped modulation with inflection points located at k_{min},

$$\tilde{\phi}(k, R_1, R_2) = \arctan\left[\frac{C-1}{C+1}\tan(\Delta Rk)\right]. \tag{8}$$

In 'simple' systems distinct minima in the envelope function and steps in the experimental phase shift may point directly to two closely neighbored neighbored distances. 'Simple' means: other possible reasons for the occurence of minima in the envelope function and modulations of the phase shift can be definitely ruled out, $e.g$ Ramsauer resonances from scattering by heavy atoms or multiple scattering contributions.

3.3 Cu-K EXAFS FROM THE LINEAR CONFIGURATION Cu1 $-$ O4 $-$ Cu2

The linear chain Cu1-O4-Cu2 along the c-axis of $YBa_2Cu_3O_{7-\delta}$ is such a 'simple' system. The scattering phase of low-Z oxygen decreases monotonously nearly as k^{-2} and exhibits no Ramsauer resonance in the relevant k-range. In R-space the single scattering Cu1-,Cu2-O4 contributions turn out well separated from the nearest neighbor Cu2-Cu2 contributions and from the strong multiple-scattering contributions appearing at $R \geq 3.5$ Å. Multiple-scattering contributions with up to 8 legs

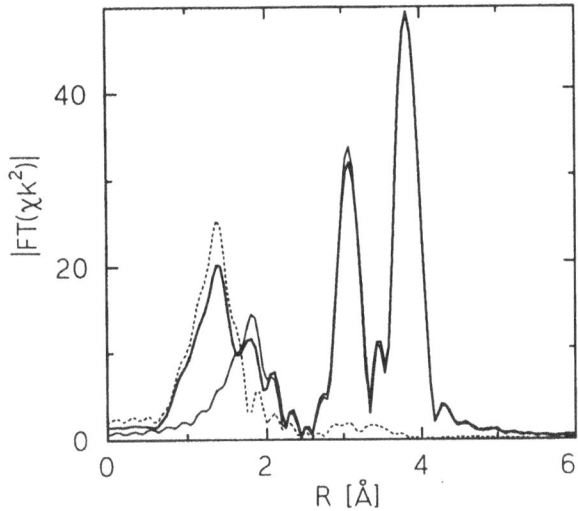

Fig. 3. Modulus of the Fourier transform of χk^2 from a full multiple scattering calculation of Cu K-excitation in a linear Cu1-O4-Cu2 chain with two O4-sites separated by 0.13 Å (see text). The radial distribution functions of the *nn* oxygens around 1.5 Å appear well separated from the *nn* Cu2 around 3.1 Å and from the multiple scattering starting at $R \geq 2.5$ Å. Thick drawn out line: the total signal from Cu1 and Cu2 (weighted 1:2). Drawn out line: partial contribution from excited Cu2. Thin dashed line: partial contribution from excited Cu1.

have been allowed, but only contributions with up to 4 legs turned out to be significant.

Fig. 3 shows the radial pair distribution functions (RDF) obtained from a full multiple scattering calculation of the Cu2 and Cu1 K-excitation in a linear chain

$$...Cu2 \rightleftharpoons O4 \rightleftharpoons Cu1 \rightleftharpoons O4 \rightleftharpoons Cu2 \rightleftharpoons Cu2...$$

This configuration corresponds to the Cu-O cluster sampled in an polarized Cu K-experiment with the electric field vector $\vec{E} \parallel c$. Here we neglect the scattering from the Ba- and Y-atoms adding their projected contributions to the polarized absorption. These contributions do not interfere with the single scattering Cu-O contributions, see also [27], but they pose a serious problem in a data analysis performed in R-space up to 3.2 Å [14].

The spectra are calculated for $k = 0\text{-}20$ Å^{-1} using the FEFF 5 multiple-scattering-code [28, 29] and transformed into R-space using a 10% broadened Gaussian window $k = 2\text{-}17$ Å^{-1}. In this example the thermal vibrations of all atoms are neglected. But the damping due to inelastic scattering is included. The detailed configuration of the cluster yielding the radial distribution functions in Fig. 3 is as follows: Two-site O4-configurations in an average Ortho I environment, $\delta = 0.07$, are assumed. The two O4-positions are separated by 0.13 Å, equally populated, and centered at average distances $R_{Cu1-O4} = 1.85$ Å. $R_{Cu2-Cu2} = 4.14$ Å is kept fixed. In Fig. 3 the thin drawn out line results from the average of short and long Cu1-O4 bonds, the dashed line from the average of short and long Cu2-O4 bonds. Interestingly, the Cu1 excitation produces only weak multiple-scattering contributions, while the Cu2 excitation generates strong multiple-scattering. As expected the spectra from infrared-active O4-Cu1-O4 configurations are nearly identical with the spectra from Raman-active

O4-Cu1-O4 configurations. Only weak differences appear in their multiple-scattering contributions.

In Fig. 4 we examine the EXAFS calculated from the average structures of the Ortho I, Ortho II and tetragonal phases. Only single scattering Cu1-,Cu2-O4 configurations are considered and one-site O4- configurations are assumed for the Ortho I and tetragonal phases. No thermal damping is included. Dashed lines show the spectra as calculated: the bold drawn out lines the same spectra, but after being Fourier transformed into R-space using a 10% broadened Gaussian window k=2−17 $Å^{-1}$. Backtransforming is performed using a rectangular window $R = 1$-2 Å. The latter narrow window has been also used in Ref. [27] for minimizing the Ba- and Y-contributions in the nn oxygen shell. As clearly visible in Fig. 4 (right), this narrow window intersects the RDF's of Cu2-O4 at $R = 2$ Å, but the beats turn out to survive in the filtered signals (left, drawn out lines).

In $YBa_2Cu_3O_{7-\delta}$ the apical oxygen are seen from two Cu-sites, Cu1 and Cu2. The average lengths of the Cu1-O4 and the Cu2-O4 bonds differ by 0.44 Å ($\delta = 0.07$) and produce beats for ($n = 1$) at 3.6 and for $n = 3$) at 10.7 $Å^{-1}$.

The Ortho I one-site O4-RDF in Fig.4 (right) exhibits two distinct peaks at $R \simeq$ 1.35 and 1.8 Å. In comparison the RDF of the two-site O4-configuration in Fig. 3 appears to be smeared out.

The ortho II spectrum is calculated from the average structure of $YBa_2Cu_3O_{6.55}$ [20] including displacements of O4 by ±0.05 Å and of Cu2 by ±0.02 Å [23]. Here we have two pairs of Cu1-O4 and Cu2-O4 distances as in the Ortho I two-site O4-configuration examined in Fig.3: one for the empty chains and one for the the filled chains. The two Cu1-O4 bonds differ by 0.1 Å, the two Cu2-O4 bonds by 0.04 Å. The beat arising from the latter is out of the experimentally accessible range, but the beat from the former is found around 14 $Å^{-1}$ (expected at 15.7 $Å^{-1}$). Obviously Fourier filtering distorts the signal at $k \geq 12$ $Å^{-1}$ (filtered and unfiltered data start to be out of phase). The broad minimum at 9 $Å^{-1}$ must be attributed to the average of four minima ($n = 2$): at 7.7, 8.3, 9.2, 10.0 $Å^{-1}$. We conclude that the atomic displacements from the average Ortho II structure ($\sigma^2 = 0$) appear as a particular type of disorder, which requires a more elaborate analysis.

The spectrum of a tetragonal one-site O4-configuration ($\delta = 0.91$) is shown on the bottom of Fig. 4. Beats from the relatively large difference of $\Delta R = 0.67$ Å between Cu2-O4 and Cu1-O4 appear clearly around 7, 11.8, 16.5 $Å^{-1}$. Here a window from $R = 1$-2 Å turns out to be a rather unsuited choice, wiping out the distinct minima in the envelope function. The split Cu2-O4 maximum is due to the strong k-dependence of the inelastic losses at high k's. The dashed RDF is obtained from a restricted k-range ≤ 12 $Å^{-1}$.

One recognizes, that the analysis of beats is a straightforward method to resolve the simple disorder, arising from two closely neighbored oxygen atoms with $\Delta R \geq 0.1$ Å, provided the bonds exhibit comparable thermal factors. Non unexpected the window functions, chosen for the Fourier filtering, affect the sharpness of the beats, but do not wipe them out. In order to isolate the nn oxygen, $R = 1$-2 Å turns out to be a reasonable choice.

The analysis of the disorder from four closely neighbored oxygens turns out to be beyond the potential of the beat method. Therefore the analysis of the disorder effects from two-site O4-configurations seems to require more elaborate methods.

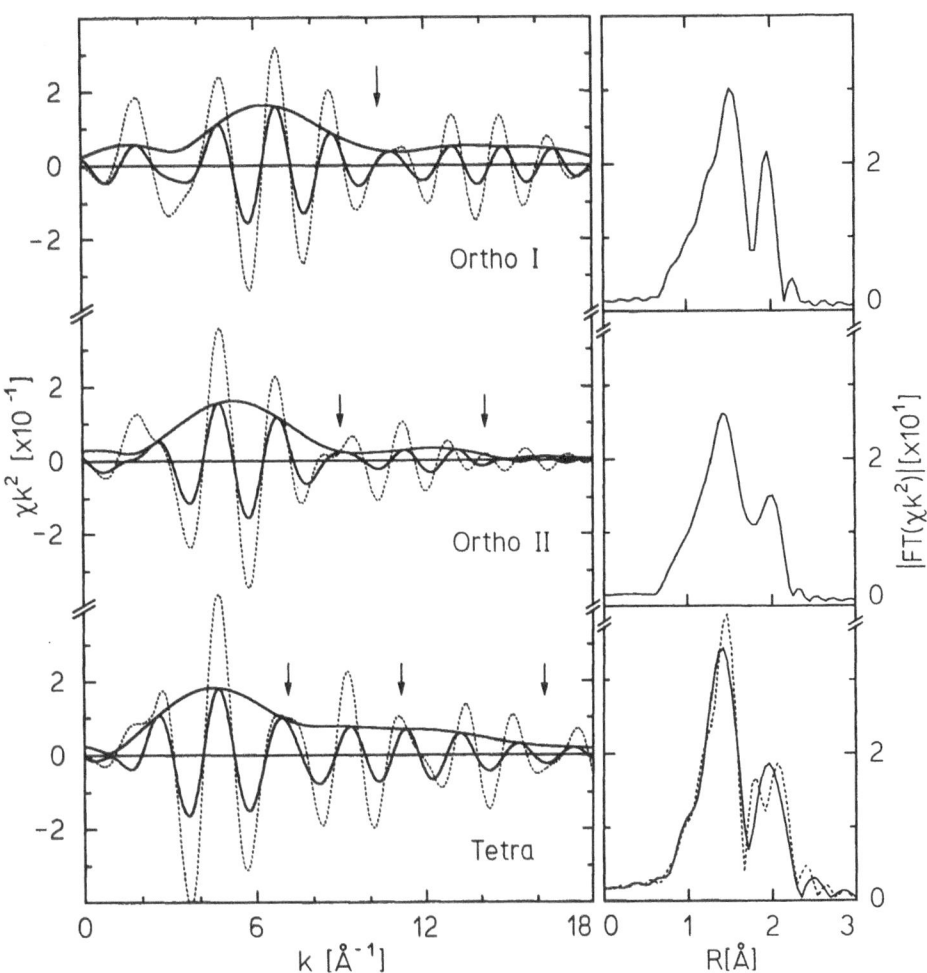

Fig. 4. *Left:* Calculated Cu K-EXAFS single scattering interference function $\chi(k)k^2$ (dashed lines) of Ortho I one-site O4 (top), Ortho II two-site O4 (middle) and tetragonal one-site O4 (bottom) $YBa_2Cu_3O_{7-\delta}$ as calculated from crystallographic data. Drawn out lines indicate the envelope function and the real part of the Fourier transform from the R-space using a rectangular window $R = 1\text{-}2$ Å. *Right:* Adjacent radial distribution functions (RDF). For the dashed line (right below) see text.

Fortunately largely different Debye-Waller factors for Cu1-O4 and Cu2-O4 bring back the potential of the beat analysis for the analysis of the available experimental data.

3.4 EFFECT OF THERMAL DAMPING TO THE BEATS

If the effective Debye-Waller factors of one pair turn out to be large compared to that of the other pair, the beats from the undamped pair may clearly appear at high k-numbers. Large Debye-Waller factors were found to damp the EXAFS of the Cu2-O4 pair in the two-site O4 configuration [30]. While the Cu1-O4 bonds in the two-site O4 configuration produce a beat at about 12 Å^{-1} (for $\delta R = 0.13$ Å), the Cu2-O4 EXAFS appear negligably small at this k-number. Due to their disorder the Cu2-O4 pairs do not interfere at high k's with the beat from the two-site Cu1-O4 configuration. Calculated spectra of a two-site O4-configuration with largely different σ^2 are shown in Fig. 5.

4. Comparison to Experimental Data

Due to the efforts of Stern *et al.* [15] very recently another set of EXAFS data from polarized Cu K-absorption in YBa$_2$Cu$_3$O$_{7-\delta}$ became available. The investigation includes a study on possible effects of sample- and absorber-preparation techniques on the data.

All scans were carried out in transmission geometry on magnetically textured powders embedded in epoxy or resin. The epoxy turned out to react with the grains, which came apparent in the EXAFS spectra. Chemical details of this reaction are not reported. The chemical reaction with the binder, neccesary for fixing the magnetically alligned grains, could be avoided by using resin.

The spectra were collected from three large grain samples (≈ 20 μm) and two fine grain samples ($\approx 5\mu$m). The large grain samples were made by standard ceramic grain techniques, with T_c's of 0 K, 56 K and 89 K. The starting material of the fine grain samples was made from sol-gel precipitated precursors and subsequently heat treated in oxygen. The T_c of this sample was found to be 92 K from magnetic susceptibility measurements. Another absorber was prepared from the same batch of the fine grain material but subjected to an additional annealing procedure in flowing oxygen. The T_c of this sample was 93 K.

The spectra of all samples with T_c's of 0, 56, 89, 92, 93 K exhibit a beat structure at 12-15 Å^{-1}, but *independent on temperature and on oxygen stoichiometry,* cf. the three spectra reproduced in Fig. 6. The observation of the beat structure confirms the existence of two-site Cu-O4 configurations separated of about 0.1 Å as reported previously. Fits confirmed the presence of rigid double Cu1-O4 pairs with $\sigma^2 = 0 \pm 1 \times 10^{-3}$ Å^2 and rather soft Cu2-O4 pairs as examined in Fig. 5. *The spectra of absorbers from the best oxygenated fine grains with the highest T_c embedded in chemically inert resin could be fitted with a one-site as well with a two-site Cu1-O4 configuration.* In disagreement with their previous results [14] Stern *et al.* [15] find

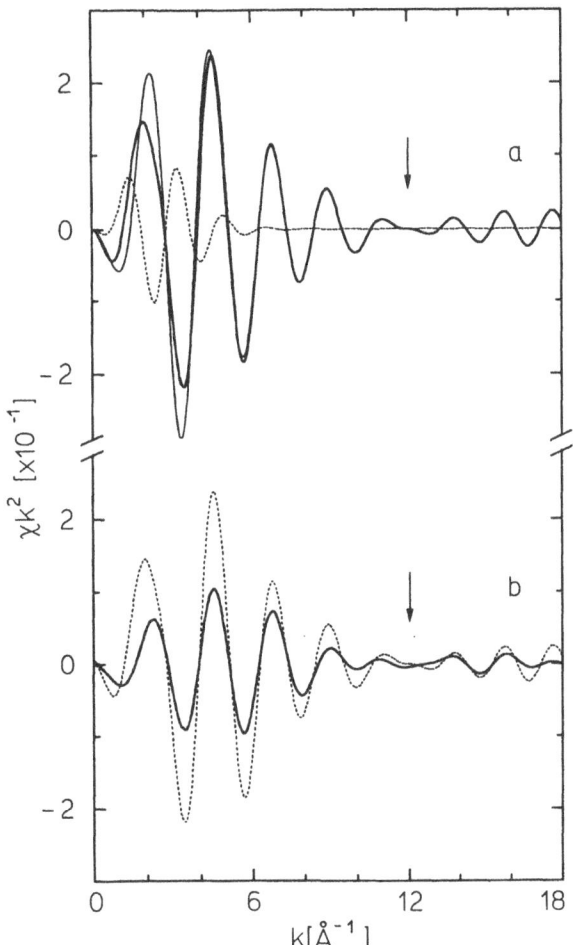

Fig. 5. *Top:* Calculated Cu K--EXAFS of a two-site O4-configuration (only single scattering). $\Delta R = 0.13$ Å, same structural parameters as used for the calculation of the radial distribution function in Fig. 3, but $\sigma^2 \neq 0$. The Cu2-O4 signals (dashed) are chosen strongly damped, $\sigma^2 = 0.043$ Å^2, while the Cu1-O4 contributions (thin drawn out line) are only weakly damped, $\sigma^2 = 0.001$ Å^2. The thick drawn out line is the weighted sum. The beat structure arising from the two Cu1-O4 signals is clearly observed at $k = 12$ Å^{-1} (arrow). Clearly the Cu2-O4 pairs do not interfere with the beat. *Bottom:* The same spectra, but Fourier transformed from $k = 2$-17 Å^{-1} using a 10% Gaussian window. The rectangular window of the backtransform into k-space was set to 1-2 Å. The beat structure is only slightly modified.

no anomaly around T_c. If at all there is any temperature anomaly in the splitting of the two O4-sites, it is found to be highly correlated with a large anomaly in the damping factor; fitting a collapsed two-site O4-configuration to the data around T_c results in a correlated increase of σ^2.

5. Assessment of a Static Origin of the Two-Site O4-Configuration

We summarize the current experimental situation as follows: There is strong evidence for a two-site O4-configuration in $YBa_2Cu_3O_{7-\delta}$, except for the 'best' optimum oxygenated sample.

The beats observed at $k = 12$-15 Å^{-1} cannot attributed to an artefact arising from filtering the nn oxygen shell with a relatively narrow window $R = 1\text{-}2$ Å, as stated in Ref. [31]. The interference between the Cu2-O4 and Cu2-O1 signals disappears in the relevant high-k-range due to relatively strong disorder in the two Cu2-O4 contributions. The origin of this strong disorder is not clear. We suggest to study the disorder of Cu2 in more detail.

The two-site O4-configuration is observed in both, superconducting and nonsuperconducting $YBa_2Cu_3O_{7-\delta}$, with a splitting independent on doping. No unambigous evidence is on the temperature dependence of the splitting around T_c. Altogether the EXAFS results point to the existence of two-site O4-configurations, created locally from chain-oxygen vacancies, not to a dynamic bistability of the apical oxygen.

5.1 THE DETECTABLE MINIMUM OF THE TWO-SITE O4-FRACTION

The two-site O4-configuration with $\Delta R = 0.1$ Å, reported from the superstructure reflections of long-range ordered Ortho II domains ($\delta = 0.45$), has to have appear as beat structure in the high-k-region, at least weakly around $k \simeq 15$ Å^{-1}, see Fig. 4. All samples with T_c's near the 60 K 'plateau' were found to exhibit a two-site O4-configuration.

The situation in the less oxygen deficient phases, particularily in highly doped OI crystals, needs a more detailed discussion. A large fraction of one-site O4-configurations obscures the beat structure from the two-site O4-configurations. In order to have an estimate for for the limits of detectibility, we have calculated a series of spectra, modelling materials with T_c's along the 90 K 'plateau'. We as-

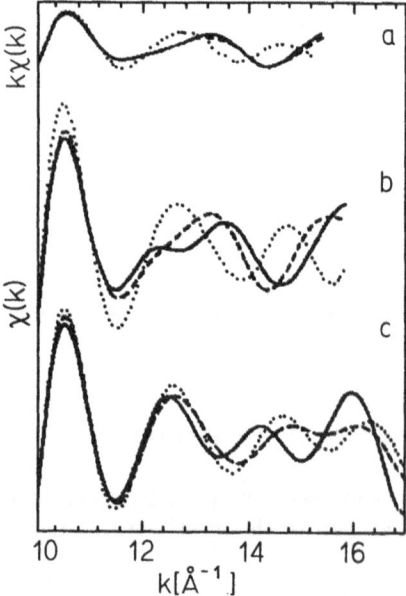

Fig. 6. Beats in the high k-range of the isolated first shell EXAFS of $YBa_2Cu_3O_{7-\delta}$ above the Cu K-edge for $\vec{E} \parallel c$, adapted from Stern et. al.[15]. (a) large grain sample with $T_c = 89$ K, (b) and (c) fine grain samples with $T_c = 92$ K and $T_c = 93$ K. Bold drawn out lines are experimental data, bold long dashed lines are fits using a two-site O4-configuration with $\Delta R = 0.1$ Å and $\sigma^2 \simeq \pm 0.001$Å^2. Dotted lines are fits of a one-site O4-configuration. (c) exhibits the spectrum of the 'best' absorber (see text). Note that the fits of both configurations are unsatisfactorily for $k > 13$ Å^{-1}.

sumed various mixing ratios between the one-site O4-configuration (describing a fully oxygenated matrix) and a two-site O4-configuration (describing a distorted cluster around isolated O1-vacancies). Some spectra from the model calculations are shown in Fig. 7. The structural parameters of the $\delta = 0.15$ average structure [20] and a two-site O4-separation of 0.13 Å were chosen. Fig. 7 shows spectra from mixing ratios (one-site O4) : (two-site O4) with 30:70, 40:60, 50:50. The beat from the two-site O4 -configuration appears clearly for the 30:70 mixing ratio. This spectrum may be compared to the experimental spectrum (b) in Fig. 6. Smaller fractions of the two-site O4-configurations turn out to be obscured by the one-site O4-signal. Our model calculations yield a lower limit of detectibility for the two-site O4-configuration created by vacancies in the Ortho I phase. As pointed out in section 2, a fraction of 70% two-site O4-configuration corresponds to an oxygen deficiency of $\delta \approx 0.17$. Thus we expect the two-site O4-configurations in optimum doped samples are below the limit of detectibility. This would explain the equivalent quality of one-site and two-site fits to the data of the 'best' sample in the measurements exhibited in Fig. 6.

5.2 DYNAMICS

The experimentally observed two-site O4-configuration has been modelled with a dynamic bistability within the O4-Cu1-O4 cluster [30]. Holes are assumed to tunnel inside the cluster O4-Cu1-O4 between two states and generate through the large polarizability of the apical O_2-ion a double-well potential. The characteristic life time of the two states comes out as about 10^{-13} s. EXAFS samples the local structure on a a time scale of $\ll 10^{-13}$ s and is believed to deliver 'snapshoots' pictures of the bimodal lattice dynamics. If one attributes a 'snapshoot' capability to EXAFS, one cannot discern from each other dynamicly generated configurations on one hand side, from staticly, $i.e.$ vacancy induced two-site O4-configurations, on the other hand side.

The dynamic interpretation of the two-site O4-configuration corresponds to an infrared mode and should be correlated with optical spectroscopy. The discrepany with optical results is controversially discussed [32, 33]. It is pointed out from Raman and infrared measurements, that there is no evidence for a double-well potential. Most importantly, the variations in the energy levels in the double well have singular character near T_c while the temperature dependence of the frequency, linewidth, and oscillator strength of the infrared-active vibration of the apical oxygen exhibits steplike discontinuities at T_c. A steplike discontinuity has been also detected in the temperature dependence of the Cu K-near-edge fine-structure around T_c [34], giving evidence for the shift of the chemical potential below T_c [35, 36].

One is tempted to compare the EXAFS pattern of the proposed dynamic 'two-level' system to the EXAFS patterns of other 'two-level' systems, for instance Rare Earth valence fluctuators. Most of the lanthanide ions can exhibit a mixed valent state in solids. These solids can be regarded as a dynamic alloy, composed by two different types of $4f$-ions with two adjacent valence states at crystallographically indistinguishable sites. In many cases the characteristic time scale of the valence fluctuation is $\approx 10^{-13}$ s. The timescale and the splitting of the nn distances from the largely different ionic radii of the $4f$-ions are quantitatively comparable to the parameters, given for the Cu1-O4-Cu1 double-well in the high-T_c's.

Numerous efforts have been undertaken to solve by EXAFS this so-called two-distance problem in Rare Earth valence fluctuators [37, 38, 39]. The result of these investigations can be summarized as following. Beat structures pointing to two different nn distances (where the lattice parameters from diffraction give only one average distance) are clearly observable in the so-called inhomogenous mixed valence systems, *i.e.* valence fluctuators in the static or 'slow' ($\approx 10^{-9}$s) limit. But they do not show up at all, or only weakly, in homogenously mixed valent systems.

Provided the ions can relax fully to their different nominal sizes, the EXAFS 'snapshoot' should capture the two-distance configuration. In all cases under investigation in which a fully relaxed configuration had been expected from independent measurements, the EXAFS 'snapshoots' yielded drastically reduced splittings of only 0.01 Å or less, instead the expected number of ≈ 0.1 Å. The reason for this effect is

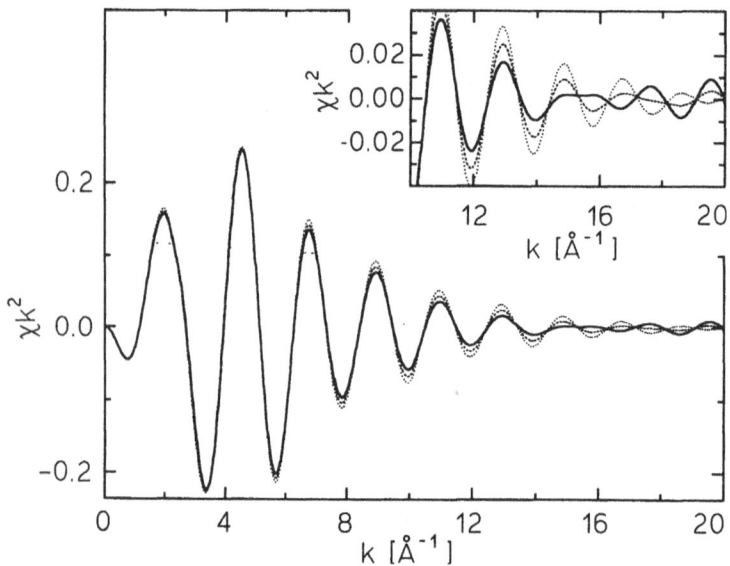

Fig. 7. Calculated EXAFS for a mixture of two local Ortho I configurations: one-site O4 and two-site O4: two-site ratios are 30:70 (drawn out line), 40:60 (dashed), and 50:50 (dotted). For the structural parameters see text. Clearly visible in the insert the known beat structure from the two-site O4-configuration, *cf.* Fig. 5. It survives in the spectrum of the 30:70 mixture, but weakens and shifts to higher k for 40:60 and seems to disappear at 20 Å^{-1} for 50:50. Note that the amplitude of the beat in the high-k region is only about 10^{-4} of the edge jump.

not yet clear; a possible explanation, although not fully satisfying, has been given in terms of hybridization of the instable wavefunctions with their nn-environment [40]. As a result from these studies we learnt, to handle EXAFS with special care as a tool for investigations of electronically driven lattice dynamics on a picosecond scale.

6. Concluding Remarks

While there is ample evidence that lattice degrees of freedom are involved in the superconducting phase transition of the high-T_c's, only poor evidence can be cited for strong anharmonicities in these materials. The two-site O4-configuration identified from polarized x-ray absorption-fine-structure measurements have been proposed to originate from a bimodal anharmonicity of the apical O4 along the c-axis. In this lecture we have tried to show, that the two-site O4 configurations in Ortho I materials may be related to the strong local distortions around single chain O1-vacancies. The basis of this interpretaion is in the deviations of O4 and Cu2 from the average structure as evidenced from the superstructure of long ordered Ortho II domains. We have argued, that isolated O1 vacancies in the Ortho I phase distort their local environment in a similar manner as has been experimentally proven for the Ortho II phase, but on a short-range order scale. Certainly this reasoning needs an independent experimental and theoretical verification. From straightforward electrostatic arguments, which were shown to explain satisfactorily the long-range ordered distortions in the Ortho II structure, we are fairly convinced, that single O1-vacancies do not behave as 'passive' impurities.

The interpretation of the observed two-site O4-configurations from the presence of a non-negligible amount of O1-vacancies in the superconducting materials yields a coherent picture explaining all available data, from both, superconducting and nonsuperconducting materials.

An open question seems to us the temperature behavior of the splitting. We feel, that a collapse of the two-site O4-configuration to a one-site O4-configuration around T_c is rather inprobable. But we would not rule out subtle effects on the two-site O4-configuration at the superconducting phase transition. Here we do not talk about possible anharmonicities or soft modes. Several experiments have evidenced the shift of the chemical potential in superconducting materials below T_c, which is expected from thermodynamic grounds, independent on the detailed pairing-mechanism [35]. In the layered high-T_c materials the shift of the chemical potential redistributes the charge between the doping block (chains) and the 'active' CuO_2-layer. Since the charge redistribution takes place at fixed O1 vacancy concentration, local structural distortions have to result, involving the Cu1-O4-Cu2 cluster (not to forget: Ba). However, we expect a step-like variation at T_c, not a singularity.

Due to the strong scattering power of atoms for electrons, the photoelectron interference patterns are extremly sensitive to the detection of local distortions. But missing restrictions through $e.g.$ symmetry conditions, require careful modelling of the possible structural configurations. A consistent picture can be achieved only on the basis of the structure provided by crystallographic studies, especially of superstructures. Thanks to the theoretical efforts undertaken during the last years, codes for the calculation of EXAFS data have become available, *e.g.* FEFF5, which yield theoretical spectra within the precision of the best experimental data. This overcomes the former difficult situation in the data analysis, arising from only approximate electron phase shifts and course approximations to the inelastic losses. A problem, which still needs more detailed theoretical consideration, seems to us the effects of non-standard lattice dynamics on the photoelectron interference patterns.

Acknowledgements

I thank the University-GH Paderborn for financial support and for the opportunity to use the CONVEX computer.

References

[1] C. Meingast, T. Wolf, H. Wühl, A. Erb and G. Müller-Vogt. Phys. Rev. Lett. **67** 1643 (1991).

[2] M. Kund and K. Andres. Physica C (1993). In press.

[3] P.M. Horn, D.T. Keane, G.A. Held, J.L. Jordan-Sweet, D.L. Kaiser and F. Holtzberg. Phys. Rev. Lett. **59**, 2772 (1987).

[4] H. You, U. Welp and Y. Fang. Phys. Rev. B **43**, 3660 (1991).

[5] S.J. Burns. Physica C **199**, 84 (1992).

[6] C. Meingast, J. Karpinski, E. Jilek and E. Kaldis. (1993). To be published.

[7] P.B. Allen. Superconductivity and Lattice Effects. In *Proc. of the Conference "Lattice Effects in High-T_c Superconductors", (Santa Fe, NM, U.S.A.), Jan. 13-15*, edited by Y. Bar-Yam, T. Egami, J. Mustre de Leon and A.R. Bishop p. 17, World Scientific (Singapore), 1992.

[8] H.R. Ott. J. Low Temp. Phys. **9**, 331 (1972).

[9] C. Meingast, B. Blank, H. Bürkle, B. Obst, T. Wolf, H. Wühl, V. Selvamanickam and K. Salama. Phys. Rev. **41**, 11299 (1990).

[10] D. Wohlleben. Superconductivity by Quantum Size Effect. In *Proc. of the NATO ASI (Bad Windsheim, Aug. 89)*. Kluwer, 1990.

[11] T. Egami. Local Structural Distortion: Implication to the mechanism of High Temperature Superconductivty. In *Materials and Crystallographic Aspects of High-T_c Superconductivity, Proc. 20th Course of the International School of Crystallography, (Erice, Italy)*, Edited by E. Kaldis, NATO Advanced Study Institute, Kluwer, 1993. This volume.

[12] J. Mustre de Leon, S.D. Conradson and I. Batistic. Phys. Rev. Lett. **65**, 1675 (1990).

[13] S.D. Conradson, I.D. Raistrick and A.R. Bishop. Science **248**, 1394 (1990).

[14] E.A. Stern, M. Qian, Y. Yacoby, S.M. Heald and H. Maeda. Unusual copper environment in $YBa_2Cu_3O_{7-\delta}$ superconductors as found by XAFS In *Proc. of the Conference "Lattice effects in High-T_c superconductors", (Santa Fe, NM, U.S.A.), Jan. 13-15,* Edited by Y. Bar-Yam, T. Egami, J. Mustre de Leon and A.R. Bishop, p. 51, World Scientific (Singapore), 1992.

[15] E.A. Stern, M. Qian, Y. Yacoby, S.M. Heald and H. Maeda. Physica C **209**, 331 (1993).

[16] A.R. Bishop, R.L. Martin, K.A. Müller and Z. Tesanovic. Z. Phys. B **76**, 17 (1989).

[17] D. de Fontaine, G. Ceder and M. Asta. Nature **343**, 5443 (1990).

[18] B.W. Veal, H. You, P. Paulikas, H. Shi, Y. Fang and J.W. Downey. Phys. Rev. B **42**, 4770 (1990).

[19] H. Claus, S. Yang, A.P. Paulikas, J.W. Downey and B.W. Veal. Physica C **171**, 205 (1990).

[20] J.D. Jorgensen, B.W. Veal, A.P. Paulikas, L.J. Nowicki, G.W. Crabtree, H. Claus and W.K. Kwok. Phys. Rev. B **41**, 1863 (1990).

[21] T. Zeiske, R. Sonntag, D. Hohlwein, N.H. Andersen and T. Wolf. Nature **353**, 542 (1991).

[22] T. Zeiske, D. Hohlwein, R. Sonntag, F. Kubanek and G. Collin. Z. Phys. B **86**, 11 (1992).

[23] D. Hohlwein. Superstructures in 123 compounds: x-ray and neutron diffraction. In *Materials and Crystallographic Aspects of High-T_c Superconductivity, Proc. 20th Course of the International School of Crystallography, Erice (Italy),* Edited by E. Kaldis, NATO Advanced Study Institute, Kluwer, 1993. This volume.

[24] T. Zeiske, D. Hohlwein, R. Sonntag, F. Kubanek and T. Wolf. Physica C **194**, 1 (1992).

[25] G. Bunker. Nucl. Instr. Meth. **207**, 437 (1983).

[26] G. Martens, P. Rabe, N. Schwentner and A. Werner. Phys. Rev. Lett. **39**, 1411 (1977).

[27] J. Mustre de Leon, S.D. Conradson, I. Batistic, A.R. Bishop, I.D. Raistrick, M.C. Aronson and F.H. Garzon. Phys. Rev. B, 2447 (1992).

[28] J.J. Rehr, R.C. Albers and S.I. Zabinsky. Phys. Rev. Lett. **69**, 3397 (1992).

372

[29] J. Mustre de Leon, J.J. Rehr, S.I. Zabinsky and R.C. Albers. Phys. Rev. **44**, 4146 (1991).

[30] J. Mustre de Leon, J. Batistic, A.R. Bishop and S.D. Conradson. S.A. Trugman. Phys. Rev. Lett. **68**, 3236 (1992).

[31] J. Röhler. Lattice distortions around T_c in $YBa_2Cu_3O_{7-\delta}$ studied by Cu K absorption. In *Proc. of the Conference "Lattice Effects in High T_c Superconductors", Santa Fe, NM (U.S.A), January 13-15*, Edited by Y. Bar-Yam, T. Egami, J. Mustre de Leon and A.R. Bishop, p. 77, World Scientific (Singapore), 1992.

[32] J. Mustre de Leon, I. Batistic, A.R. Bishop, S.D. Conradson and I. Raistrick. Phys. Rev. B **47**, 12322 (1993).

[33] C. Thomsen and M. Cardona. Phys. Rev. B **47**, 12320 (1993).

[34] J. Röhler, A. Larisch and R. Schäfer. Physica C **191**, 57 (1992).

[35] D.I. Khomskii and F.V. Kusmartsev. JETP Lett. **54**, 150 (1991).

[36] G. Rietveld, N.Y. Chen and D. van der Marel. Phys. Rev. Lett. **69**, 2578 (1992).

[37] H. Launois, M. Rawiso, E. Holland-Moritz, R. Pott and D. Wohlleben. Phys. Rev. Lett. **44**, 1271 (1980).

[38] G. Krill, J.P. Kappler, J. Röhler, M.F. Ravet, J.M. Léger and F. Gautier. New Information obtained from EXAFS Experiments on Intermediate Valent Systems. In *Valence Instabilties*, Proc. of the International Conference held in Zürich (Switzerland), p. 155, North-Holland (Amsterdam), Apr. 13-16 1982.

[39] E. Beaurepaire, D. Malterre, G. Krill, C. Godart, J.P. Kappler, B. Chevalier and J. Etourneau. Use of L_{III} XANES for the Determination of Atomic Relaxation in mixed-valent materials. In *Theoretical and Experimental Aspects of Valence Fluctuations and Heavy Fermions*, Edited by L.C. Gupta and S.K. Malik, Proc. of the Fifth International Conference on Valence Fluctuations (Bangalore, India), p. 667, Plenum (New York), Jan. 5-9 1987.

[40] W. Kohn, T.K. Lee and Y.R. Lin-Liu. Phys. Rev. B **25**, 3557 (1982).

SINGLE CRYSTAL GROWTH AND CHARACTERIZATION OF THALLIUM CUPRATE SUPERCONDUCTORS - A REVIEW

A.M. Hermann, M. Paranthaman, and H.M. Duan

Department of Physics, University of Colorado, Boulder CO 80309-0390

ABSTRACT

We review the single crystal growth and characterization of thallium cuprate superconductors. Reasonably large crystals (several mm size) have been grown by self-flux technique and KCl-flux methods. The anisotropic resistivities ρ_{ab} (along ab-plane) and ρ_c (along c-axis) were measured for all double-Tl-O compounds by Montgomery method. ρ_c is found to be 2 orders of magnitude greater than ρ_{ab} for all of these samples. Hence, these systems have high anisotropy. The paraconductivity data shows that Tl-2201 is a two-dimensional superconductor whereas Tl-2212 has a two-dimensional behavior at higher temperatures and some samples showed crossover from two-dimensional to three-dimensional at 6 degrees above T_c of 101 K. Hall data, annealing effects on different Tl systems, irradiation effects and pressure dependence of T_c are reported.

E. Kaldis (ed.), Materials and Crystallographic Aspects of HTc-Superconductivity, 373–398.

I. INTRODUCTION

The research on the single crystal growth and characterization of thallium cuprate superconductors has not yet been thoroughly carried out all over the world until now, even after four years of research on these systems. The reason could be due to the volatility of thallium during the crystal growth and hence the uncertainty in Tl content, lack of phase stability, and also inability to grow large crystals using standard flux methods. Many unusual normal state properties have also been reported in the literature. To understand the possible superconducting mechanisms in thallium systems, one needs to know the following characteristics: (i) how in-plane (ab-plane) resistivities are compared with out-of-plane (c-axis) resistivities (ii) whether the resistivity vs. temperature curve in c-axis is (metal-like) or semiconductor-like, (iii) interlayer interactions, (iv) fluctuation effects, (v) dimensionality, and (vi) anisotropy of these systems. Hence, there is a need to grow large single crystals and to carryout all the transport property measurements on those crystals. Therefore, we have attempted to review all the single crystal studies on thallium systems that are available to the present. We report here single crystal growth of thallium cuprate superconductors, anisotropic resistivity measurements, paraconductivity measurements, effects of annealing and corresponding magnetization, irradiation effects, and pressure dependence of T_c on different systems. In the text, we represent the double- Tl-O compounds $Tl_2Ba_2CuO_6$, $Tl_2Ba_2CaCu_2O_8$ and $Tl_2Ba_2Ca_2Cu_3O_{10}$ as Tl-2201, Tl-2212 and Tl-2223 respectively. We represent the mono-Tl-O compounds $TlBa_2CuO_5$, $TlBa_2CaCu_2O_7$ and $TlBa_2Ca_2Cu_3O_9$ as Tl-1201, Tl-1212 and Tl-1223 respectively.

II. SINGLE CRYSTAL GROWTH

Single crystals of Tl compounds have been grown primarily by the self-flux technique. Apart form the stoichiometric compositions, excess CaO and CuO were taken as the charge and they act like a flux to grow reasonably large crystals (of the order of several mm size). To compensate thallium loss during the crystal growth, excess Tl_2O_3 was also used. Gold crucibles,[1-6] platinum crucibles,[7-10] and alumina crucibles[11-14] were used as the containers for growth. Even though the melt is contaminated with Al_2O_3,[9,10] large crystals were obtained by using alumina crucibles.[11-14] The typical experimental set-up used for the crystal growth is shown in Fig. 1.[14] Either gold crucibles were sealed or alumina and platinum crucibles were covered with lids to protect Tl_2O_3 from direct evaporation. Sometimes, even the lids were sealed to the alumina crucibles by using ceramic adhesive (AREMCO, Model 569).[14] Typically 25 g of the sample was used as the charge.[14] Once the crucibles with charge are loaded in a vertical tube furnace, they are typically heated rapidly to 900-950 °C and held for 1-3 hours, then slowly cooled through the melt at the rate of 2-20 °C/hour to 750-780 °C, and finally cooled down to room temperature. Instead of using one-step self-flux technique, some researchers have used a two-step process where large crystals are often obtained.[14] In the two-step process, the same starting compositions were used and the bulk Tl samples were made by heating at 920-950 °C for 10 minutes and cooling fast, then grinding, pelletizing, and heating to the melting temperature and slowly cooling as in the above mentioned one-step process. Care must be taken to keep the furnaces inside the hood and handling by latex gloves is imperative. The speculated phase diagram for the double-Tl-O compounds is given in Fig. 2.[6] Table I summarizes the crystal growths of double- and mono-Tl-O compounds and their properties from different starting compositions (charge) in the self-flux technique. Oxygen was passed continuously through the tube during the crystal growth. Post-annealing of the crystals has been done in different atmospheres to increase or decrease the T_c. We will discuss this in detail in the following sections. Small (or no) crystals were obtained with the Tl concentration exceeding 4 and the Cu concentration exceeding 12 in the starting composition of 4:1:3:12 (Tl:Ba:Ca:Cu).[17] This suggests that there is a Tl-O and Cu-O composition window in which various phases can be grown. Ginley et al[7] appear to have produced predominantly of the Tl-2223 phase by using Ca:Ba ratio 3:1 in the charge (see also Table I). Hence by using Table I and Fig. 2, one can conveniently choose the starting compositions and temperatures to grow the desired phase. Also one needs precise temperature control to grow single crystals. Lee et al[21] have grown epitaxial Tl-2223 thin films and made bulk samples in reduced O_2 pressure ($\cong$0.03-0.15 atm of O_2 and 830-860 °C). The same treatment may be applied for growing single crystals too.

Recently, Manako et al[22,23] have grown single crystals of Tl-2201 by using a KCl flux method. The starting compositions of [Tl-2201]/[KCl] = 1-10 wt% were used as the charge in a gold crucible was heated at rate of 2-10 °C/hour. The thin plates were grown in the ab-plane with a typical size of 2x2x0.01 mm^3. Schneemeyer et al[24] have already demonstrated the crystal growth of Bi-2212 by using alkali chloride flux. Large crystals (3x3x1.5 mm^3) of

$YBa_2Cu_3O_7$ were grown by using a K_2CO_3-flux method.[25] Based on this information, one may attempt growing large crystals of Tl-compounds by either alkali chloride flux or K_2CO_3-flux methods.

III. TRANSPORT PROPERTY MEASUREMENTS

III.1. $Tl_2Ba_2CuO_6$ System

$Tl_2Ba_2CuO_6$ single crystals have been grown by the self-flux technique[13,14] and the KCl-flux methods.[22,23] The anisotropic resistivities, ρ_{ab} and ρ_c are plotted against temperature in Fig. 3 for Tl-2201 single crystals annealed in O_2 at 350 °C (metallic down to 4 K), in Ar at 300 °C (T_c = 10 K), and in Ar at 400 °C (T_c = 75 K).[22,23] The anisotropic resistivity ρ_c is much greater than ρ_{ab} by a factor of 2-3 orders of magnitude. These results are similar to those obtained by Duan et al.[13] The anisotropic resistivities were measured by the standard Montgomery method.[26] The temperature dependence of ρ_{ab} was fit to an expression of $\rho_{ab} = \rho_0 + AT^n$. The exponential factor "n" increased from 1.29 to 1.99 as T_c decreased. The room temperature in-plane resistivity is of the order of 4×10^{-4} Ω-cm. The normal state of ρ_c was metallic for all the samples. The anisotropy of resistivity ρ_c/ρ_{ab} is plotted against temperature in Fig. 4.[22] The anisotropy is about 600 at room temperature and it increases as the temperature decreases for all the samples. It is interesting to note that considerable resistivity anisotropy remains even in the normal metallic samples. This could be due to the large separation of CuO_2 sheets in these systems.

To find the dimensionality of the superconductors, the temperature dependence of paraconductivity can be fit to[11]

$$\Delta\sigma(T) = A\,(T/T_c - 1)^{-\alpha} \qquad (1)$$

where $\Delta\sigma$ is the excess conductivity, A is a constant, T_c is the critical temperature of the system, and α is equal to 1 and 1/2 for two-dimensional and three-dimensional superconductivity respectively. This can be rewritten as[11]

$$\log(-d(\Delta\sigma)/dT) = \log(\alpha A^{1/\alpha}/T_c) + (1+1/\alpha)\log(\Delta\sigma) \qquad (2)$$

From Eqn. 2, α, the dimensionality of the superconductors can be determined from the slope of $\log(-d(\Delta\sigma)/dT)$ vs. $\log(\Delta\sigma)$ plot. Fig. 5 demonstrates the plot of $\log(-d(\Delta\sigma)/dT)$ vs. $\log(\Delta\sigma)$ for Tl-2201 crystals.[13] The slope obtained was 2 indicating that Tl-2201 compounds are two-dimensional superconductors.

The temperature dependence of in-plane Hall coefficient for normal metallic T-2201 samples is shown in Fig. 6.[23] The Hall coefficient is about $+1\times10^{-3}$ cm^3/C and shows a broad maximum around 100 K. The ab-plane single crystal data are very similar to the data taken on ceramic samples. This suggests that the transport properties of ceramic samples mainly reflects the ab-plane nature.

This is confirmed by thermopower measurements. Recent measurements by our group on single crystals of Tl-2201 show significant anisotropy of thermopower. c-axis values are positive, and ab plane values are negative at room temperature becoming smaller in magnitude as T decreases to Tc. Using independently measured single crystal resistivity anisotropy ratios, excellent fits can be made to polycrystalline thermopower data (with no adjustable parameters) demonstrating definitively that the polycrystalline thermopower is dominated by ab-plane values. The thermopower/Hall effect sign discrepancy in the ab-plane is, however, not understood at present.

III.2. $Tl_2Ba_2CaCu_2O_8$ System

$Tl_2Ba_2CaCu_2O_8$ single crystals have been grown by the self-flux technique.[2,6,9,11,15] The anisotropic resistivities, ρ_{ab} and ρ_c are plotted against temperature in Fig. 7.[11] The out-of-plane resistivity, ρ_c is greater than in-plane resistivity, ρ_{ab} by about 2 orders of magnitude. This is similar to the results obtained for Tl-2201 single crystals.[13,22] Fig. 8 shows the plot of $\ln \Delta\sigma$ vs. $\ln(T/T_c-1)$. The paraconductivity is often discussed in such a plot according to Eqn. (1). Two-dimensionality is observed for all the samples in the high-temperature region. For some samples, the crossover from two-dimensional to three-dimensional occurs at few degrees above

T_c (for sample 1 in Fig. 8, the crossover occurs at 6 K above T_c of 101 K).[11] The same behavior is observed in the plot of $\log(-d(\Delta\sigma)/dT)$ vs. $\log(\Delta\sigma)$ in Fig. 9.[11] The crossover may be related to the formation of Josephson junctions between the Cu-O layers. The two-dimensional property in the 105-155 K region is consistent with results from Tl-2212 bulk samples and thin films.[27,28] No crossover was reported in these samples. The lack of crossover may also be due to the possible inhomogeneities of these polycrystalline sample.

The post-annealing effects of Tl-2212 single crystals are shown in Fig. 10.[19] The T_c (onset of Meissner signal) for the as-grown crystal is 104 K. The T_c decreased to 99 K after a vacuum anneal and increased to 106 K after annealing in O_2. This can be interpreted simply as an increase in the oxygen content and a corresponding increase in the number of holes. Similarly, when one removes oxygen by annealing in vacuum, T_c decreases because of the decrease in the hole concentration. But Morosin et al[19] observed for some of the other crystals, O_2 annealing decreases the T_c and vacuum annealing increases the T_c. This anomaly may be due to the possibility that the as-grown crystals may be already over-doped. Hence, when one anneals in O_2, T_c decreases. Morosin et al[19] also suggested that strain, Tl content, and Tl/Ca site disorder might also be important as the oxygen content in determining T_c.

The anisotropic thermoelectric power of Tl-2212 single crystals has been measured.[29] The room temperature value of S_{ab} (ab-plane) is about 14 μV/K and S_c (along c-axis) is around 30 μV/K. The onset superconductivity determined by thermopower measurements shows that the measured transition along c-axis is always lower than that in ab-plane.

The scattering of light in Tl-2212 single crystals has been studied experimentally and analytically.[10,30,31] Besides probing phonon effects, light scattering spectroscopy yields information about electronic excitations and their symmetry. The temperature dependence of the electronic scattering intensity gives evidence for a pronounced anisotropy of the superconducting gap ($2\Delta_{min} < 50$ cm^{-1}, $2\Delta_{max} \approx 300$ cm^{-1}).[31] The maximum value of the gap is estimated as $2\Delta_{max} \approx 300$ cm$^{-1} \approx 4 k_b T_c$.

Novel flux motion has also been observed in Tl-2212 crystals.[32] Krasnov et al[33] have extended the Bean critical state model for obtaining the bulk critical field, H_{c1} and the magnetically determined critical current density, J_c for Tl-2212 single crystals. This method is based on the original critical state model and accounts for the shape of the sample, the change of the demagnetization factor with flux penetration, the real distribution of the magnetic field, and the equilibrium flux density within the sample. They observed a small positive curvature of the H_{c1} (T) dependence at T < 50 K; this curvature was suggested to be a consequence of two-dimensional superconductivity. The critical current-density, J_c has also been measured for Tl-2212 crystals form I-V curves and its temperature dependence is plotted in Fig 11.[34] The J_c is observed to be 2.5×10^4 A/cm^2 at 77 K. Because of the insufficient data, we can not comment about the dimensionality.

III.3. $Tl_2Ba_2Ca_2Cu_3O_{10}$ System

$Tl_2Ba_2Ca_2Cu_3O_{10}$ single crystals can be grown using a self-flux technique with different starting compositions.[3,6,16,17,18] Ginley et al[7] have suggested that when one uses a Ca:Ba ratio 3:1 in the starting compositions, one obtains predominantly of Tl-2223 crystals. The temperature dependence of anisotropic resistivities, ρ_{ab} and ρ_c for Tl-2223 crystals are shown in Fig. 12.[17] ρ_c is about two orders of magnitude greater than ρ_{ab}. This is similar to the results observed on Tl-2201 and Tl-2212 single crystals. The anisotropy (ρ_c/ρ_{ab}) is quite high and is of the order of 50-60 at 300 K. Tigges et al[17] also observed from the resistivity data that the T_c and the width of the transition for Tl-2223 crystal plates grown by two different melt compositions were substantially different. A Tl-O and Cu-O rich flux yielded approximately stoichiometric crystals with sharp transitions beginning near 113 K, while a Tl-O rich flux produced crystals containing more Tl and less Ba with broad transitions starting near 103 K. These data demonstrate the extreme sensitivity of the superconductivity to cation site disorder in Tl systems.[17]

Cox et al[35] have refined the structure of Tl-2223 by neutron diffraction. The post-annealing effects on Tl-2223 single crystals are reported by Morosin et al.[18] They have suggested the possibility of the presence of small amounts Tl^{1+} in the Tl-O layer of as-grown crystals - from single crystal x-ray refinement data - which subsequently oxidizes during mild oxygen anneals to Tl^{3+}. The presence of mixed valency (Tl^{3+} and Tl^{1+} in double-Tl-O system has been reported already from band structure calculations[36] and from wet-chemical analysis data.[37-39] By using the wet-chemical procedures that are available in the literature,[37-41] one can conveniently measure the Tl content, oxygen content and hole concentrations for Tl single crystals.

The temperature dependence of the susceptibility with different applied magnetic fields for Tl-2223 crystals is shown in Fig. 13.[42] The measured diamagnetic-onset temperature is reduced from 118 to 113 K by the application of a magnetic field in the c-direction in the range 0.5 to 40 G. Also, there is a systematic increase in the width of the transition with increasing field. The strong suppression of the diamagnetic-onset temperature with smaller magnetic fields could be due to the intrinsic property of the bulk material. Such an effect in the polycrystalline samples is attributed to a weak coupling between grains.[43] Fig. 14 shows shielding, Meissner and flux trapping measurements on Tl-2223 single crystals in a field of 2 G along the c-direction.[42] Just below T_c (above 100 K), the fraction of Meissner to shielding increases to nearly 100%. Over the whole temperature range, the sum of the magnitudes of the Meissner and flux trapping signals equals to the magnitude of the shielding signal, similar to the observations in $YBa_2Cu_3O_{7-\delta}$.[42]

III.4. $TlBa_2Ca_2Cu_3O_9$ System

$TlBa_2Ca_2Cu_3O_9$ single crystals have been grown by the self-flux technique.[8,15,20] The annealing effects on Tl-1223 crystals are shown in Fig. 15.[20] The Meissner data shows the T_c's are 117 k after extended oxygen anneals and reach 121 K after nitrogen anneals. This suggests that there is only a small change in the hole concentration in the CuO_2 planes. Because of the recent interest in introducing pinning centers and achieving high J_c's in different Tl-1223 phases,[44-46] one should look forward to the growth of larger Tl-1223 single crystals by self-flux technique or alkali halide flux method and corresponding characterizations.

IV. IRRADIATION EFFECTS ON THALLIUM SYSTEMS

Venturini et al[15] have studied the magnetic relaxation (flux creep) for Tl-2212 and Tl-2223 single crystals by recording the diamagnetic shielding signal vs. time following a field change from 1 T to 50 mT applied normal to the crystal plate (along the c-axis). The temperature dependence of flux pinning potentials for as-grown Tl-2223 crystals, Tl-2223 after proton irradiation, and Tl-2212 after neutron irradiation are shown in Fig. 16.[15] Following the irradiation with high-energy protons and neutrons,[15] magnetic hysteresis at 50 K was observed above 1 T, and J_{cm} (magnetization critical current density) from the remanent moments increased by an order of magnitude to $5-8 \times 10^4$ A/cm^2.

V. PRESSURE DEPENDENCE OF T_c

The pressure dependence for a single crystal of $Tl_5Ba_5Ca_2Cu_6O_y$ (composition close to Tl-2212) with a T_c of 106 K is shown in Fig. 17.[47] The T_c seems to increase with the hydrostatic pressure with a slope of $dT_c/dP = 0.23(4)$ K/kbar in a gaseous He pressure system. Very recently, Berkley et al[48] have applied pressure on Tl-2223 single crystals and found a tremendous increase in T_c from 118 to 131.5 K under pressure. Also, Schirber et al[49] have obtained a relatively large positive value of dT_c/dP with the highest T_c of 93 K for Tl-2201 with $\delta \rightarrow 0$ in contrast with Tl-2201 with $\delta \sim 0.2$, where T_c is near 40 K and dT_c/dP is large and negative.

VI. CONCLUSIONS

Our review on the present status of the single crystal growth and characterization of thallium cuprate superconductors has led us to make the following conclusions:

(1) Reasonably large crystals (several mm size) can be grown by self-flux technique and KCl-flux methods.

(2) The anisotropic resistivity, ρ_c (along c-axis) is greater than ρ_{ab} (along ab-plane) by 2 orders of magnitude for all double-Tl-O compounds. Hence, these systems have high anisotropy.

(3) The paraconductivity data show that Tl-2201 is a two-dimensional superconductor. Tl-2212 was found to be two-dimensional too but some samples showed crossover from two-dimensional to three-dimensional at few degrees above T_c.

(4) The temperature dependence of Hall coefficient for normal metallic Tl-2201 crystals (along ab-plane) is found to be similar to that of ceramic bulk samples. This suggests that the transport property measurements on ceramic samples reflect the properties of the ab-plane.

(5) A Tl-O and Cu-O rich flux yielded stoichiometric Tl-2223 crystals with sharp transitions around 113 K by using the self-flux technique. Also a Ca:Ba ratio 3:1 in the starting composition was found to give predominantly Tl-2223 crystals.

(6) Post-annealing is necessary to alter the T_c of the as-grown crystals.

(7) The strong suppression of the diamagnetic-onset temperature with small (0.5 - 40 G) applied magnetic fields for Tl-2223 crystals could be due to the weak coupling between grains.

(8) The sum of the magnitudes of the Meissner and flux trapping signals equals to the magnitude of the shielding signal over the whole temperature range below T_c for Tl-2223 single crystals. This behavior is similar to that observed for $YBa_2Cu_3O_{7-\delta}$.

(9) Irradiation with high energy protons and neutrons on Tl-2223 crystals increases J_{cm} (magnetization critical current density) deduced from the remanent moments by an order of magnitude to $5-8 \times 10^4$ A/cm^2 at 50 K.

(10) The T_c seems to increase with the hydrostatic pressure with a slope of $dT_c/dP = 0.23(4)$ K/kbar in a gaseous He pressure system for a crystal of $Tl_5Ba_5Ca_2Cu_6O_y$ with a T_c of 106 K.

ACKNOWLEDGMENTS

We gratefully acknowledge the support of Office of Naval Research under ONR grant number N00014-90-J-1571.

REFERENCES

[1]C.C. Torardi, M.A. Subramanian, J.C. Calabrese, J. Gopalakrishnan, E.M. McCarron, K.J. Morrissey, T.R. Askew, R.B. Flippen, U. Chowdhry and A.W. Sleight, Phys. Rev. **B38**, 225 (1988).

[2]M.A. Subramanian, J.C. Calabrese, C.C. Torardi, J. Gopalakrishnan, T.R. Askew, R.B. Flippen, K.J. Morrissey, U. Chowdhry and A.W. Sleight, Nature **332**, 420 (1988).

[3]C.C. Torardi, M.A. Subramanian, J.C. Calabrese, J. Gopalakrishnan, K.J. Morrissey, T.R. Askew, R.B. Flippen, U. Chowdhry and A.W. Sleight, Science **240**, 631 (1988).

[4]A.W. Sleight, Science **242**, 1519 (1988).

[5]H. Takei, T. Kotani, T. Kaneko and K. Tada, Proceedings of the 1st International Symposium on Superconductivity, Nagoya, 1988, Ed. by K. Kitazawa and T. Ishiguro (Springer-Verlag, Tokyo, 1988), p. 229.

[6]T. Kotani, T. Kaneko, H. Takei, and K. Tada, Jpn. J. Appl. Phys. **28**, L1378 (1989).

[7]D.S. Ginley, B. Morosin, R.J. Baughman, E.L. Venturini, J.E. Schirber and J.F. Kwak, J. Crystal Growth **91**, 456 (1988).

[8]B. Morosin, D.S. Ginley, J.E. Schirber, and E.L. Venturini, Physica C **156**, 587 (1988).

[9]E.D. Bukowski and D.M. Ginsberg, J. Low Temp. Physics **77**, 285 (1989).

[10]S.E. Stupp and D.M. Ginsberg, in Physical properties of High Temperature Superconductors, Vol. III, Ed. by D.M. Ginsberg (World Scientific, Singapore, 1992) (in press) and references

therein.

[11]H.M. Duan, W. Kiehl, C. Dong, A.W. Cordes, M.J. Saeed, D.L. Viar and A.M. Hermann, Phys. Rev. **B43**, 12925 (1991).

[12]H.M. Duan, R.M. Yandrofski, T.S. Kaplan, B. Dlugosch, J.H. Wang and A.M. Hermann, Physica C **185-189**, 1283 (1991).

[13]H.M. Duan, R.M. Yandrofski, T.S. Kaplan, B. Dlugosch, J.H. Wang and A.M. Hermann, Chinese J. Phys. (1992) (in press).

[14]H.M. Duan, T.S. Kaplan, B. Dlugosch, A.M. Hermann, J. Swope, J. Drexler and P. Boni, preprint.

[15]E.L. Venturini, C.P. Tigges, R.J. Baughman, B. Morosin, J.C. Barbour, M.A. Mitchell and D.S. Ginley, J. Crystal Growth **109**, 441 (1991).

[16]T. Kajitani, K. Hiraga, S. Nakajima, M. Kikuchi, Y. Syono and C. Kabuto, Physica C **161**, 483 (1989).

[17]C.P. Tigges, E.L. Venturini, J.F. Kwak, B. Morosin, R.J. Baughman and D.S. Ginley, Appl. Phys. Lett. **57**, 517 (1990).

[18]B. Morosin, E.L. Venturini and D.S. Ginley, Physica C **175**, 241 (1991).

[19]B. Morosin, R.J. Baughman, D.S. Ginley, J.E. Schirber and E.L. Venturini, Physica C **161**, 115 (1990).

[20]B. Morosin, E.L. Venturini and D.S. Ginley, Physica C **183**, 90 (1991).

[21]W.Y. Lee, S.M. Garrison, M. Kawasaki, E.L. Venturini, B.T. Ahn, R. Beyers, J. Salem, R. Savoy and J. Vasquez, Appl. Phys. Lett. **60**, 772 (1992).

[22]T. Manako, Y. Shimakawa, Y. Kubo and H. Igarashi, Physica C **185-189**, 1327 (1991).

[23]T. Manako, Y. Shimakawa, Y. Kubo and H. Igarashi, Physica C **190**, 62 (1991).

[24]L.F. Schneemeyer, R.B. van Dover, S.H. Glarum, S.A. Sunshine, R.M. Fleming, B. Batlogg, T. Siegrist, J.H. Marshall, J.V. Waszczak and L.W. Rupp, Nature **332**, 422 (1988).

[25]P. Murugaraj, J. Maier and A. Rabenau, Solid State Commun. **71**, 167 (1989).
[26]H.C. Montgomery, J. Appl. Phys. **42**, 2971 (1971).

[27]N.P. Ong, Z.Z. Wang, S. Hagen, T.W. Jing and J. Hovarth, Physica C **153-155**, 1072 (1988).

[28]A. Poddar, P. Mandal, A.N. Das, B. Ghosh and P. Choudhury, Physica C **159**, 231 (1989).

[29]Lin Shu-yuan, Lu Li, Zhang Dian-lin, H.M. Duan and A.M. Hermann, Europhys. Lett. **12**, 641 (1990).

[30]V.B. Timofeev, A.A. Maksimov, O.V. Misochko and I.I. Tartakovskii, Physica C **162-164**, 1409 (1989) and references therein.

[31]A.A. Maksimov, I.I. Tartakovskii, V.B. Timofeev and L.A. Fal'kovskii, Sov. Phys. JETP **70**, 588 (1990).

[32]F. Zuo, M.B. Salamon, T. Dutta, K. Ghiron, H.M. Duan and A.M. Hermann, Physica C **176**, 541 (1991).

[33]V.M. Krasnov, V.A. Larkin and V.V. Ryazanov, Physica C **174**, 440 (1991).

[34]B. Dlugosch, H.M. Duan, T.S. Kaplan and A.M. Hermann, preprint.

[35]D.E. Cox, C.C. Torardi, M.A. Subramanian, J. Gopalakrishnan and A.W. Sleight, Phys. Rev.

B38, 6624 (1988).

[36]D. Jung, M. -H. Whangbo, N. Herron and C.C. Torardi, Physica C **160**, 381 (1989).

[37]M. Paranthaman, A. Manthiram and J.B. Goodenough, J. Solid State Chem. **87**, 479 (1990).

[38]A. Manthiram, M. Paranthaman and J.B. Goodenough, Physica C **171**, 135 (1990).

[39]M. Paranthaman, M. Foldeaki and A.M. Hermann, Physica C (in press).

[40]J. Gopalakrishnan, R. Vijayaraghavan, R. Nagarajan and C. Shivakumara, J. Solid State Chem. **93**, 272 (1991).

[41]A. Manthiram, M. Paranthaman and J.B. Goodenough, J. Solid State Chem. **96**, 464 (1992).

[42]J.Z. Liu, K.G. Vandervoort, H. Claus, G.W. Crabtree and D.J. Lam, Physica C **156**, 256 (1988).

[43]S.S.P. Parkin, V.Y. Lee, E.M. Engler, A.I. Nazzal, T.C. Huang, G. Gorman, R. Savoy and R. Beyers, Phys. Rev. Lett. **60**, 2539 (1988).

[44]J.A. DeLuca, M.F. Garbauskas, R.B. Bolon, J.G. McMullen, W.E. Balz and P.L. Karas, J. Mater. Res. **6**, 1415 (1991).

[45]T. Kamo, T. Doi , A. Soeta, T. Yuasa, N. Inoue, K. Aihara and S. Matsuda, Appl. Phys. Lett. **59**, 3186 (1991).

[46]R.S. Liu, D.N. Zheng, J.W. Loram, K.A. Mirza, A.M. Campbell and P.P. Edwards, Appl. Phys. Lett. **60** (1992).

[47]B. Morosin, D.S. Ginley, E.L. Venturini, P.F. Hlava, R.J. Baughman, J.F. Kwak and J.E. Schirber, Physica C **152**, 223 (1988).

[48]D.D. Berkley, E.F. Skelton, N.E. Moulton and D.H. Liebenberg, Paper presented at the APS March Meeting, Indianapolis (1992).

[49]J.E. Schirber, D.L. Overmyer, E.L. Venturini, D.S. Ginley and B. Morosin, Physica C **192** (in press).

Table I. Properties of various thallium cuprate single crystals grown by the self-flux technique.

Starting compositions Tl:Ba:Ca:Cu	Crystals obtained Tl:Ba:Ca:Cu	Lattice parameters		T_c (K)	Ref.
		a (Å)	c (Å)		
2202	2201	3.870	23.24	90	1
3202	2201	3.850	23.20	86	13, 14
2113	2212	3.855	29.32	112	2
2112	2212	--	--	114	9
2223	2212	3.850	29.30	110	6, 11, 15
2234	2223	3.850	35.90	125	3, 16
2223	2223	3.850	35.67	114	16
2266	2223	--	--	118	6
4136	2223	3.850	36.00	103	17
41310	2223	3.850	36.00	113	17
3148	2223	--	--	110	18
4148	2223	3.855	35.74	112	18
1112	1212	3.850	12.70	103	19
1234	1223	3.853	15.92	117	8
1115	1223	3.850	15.90	120	15
1133	1223	--	--	119	20
1135	1223	3.855	15.92	121	20

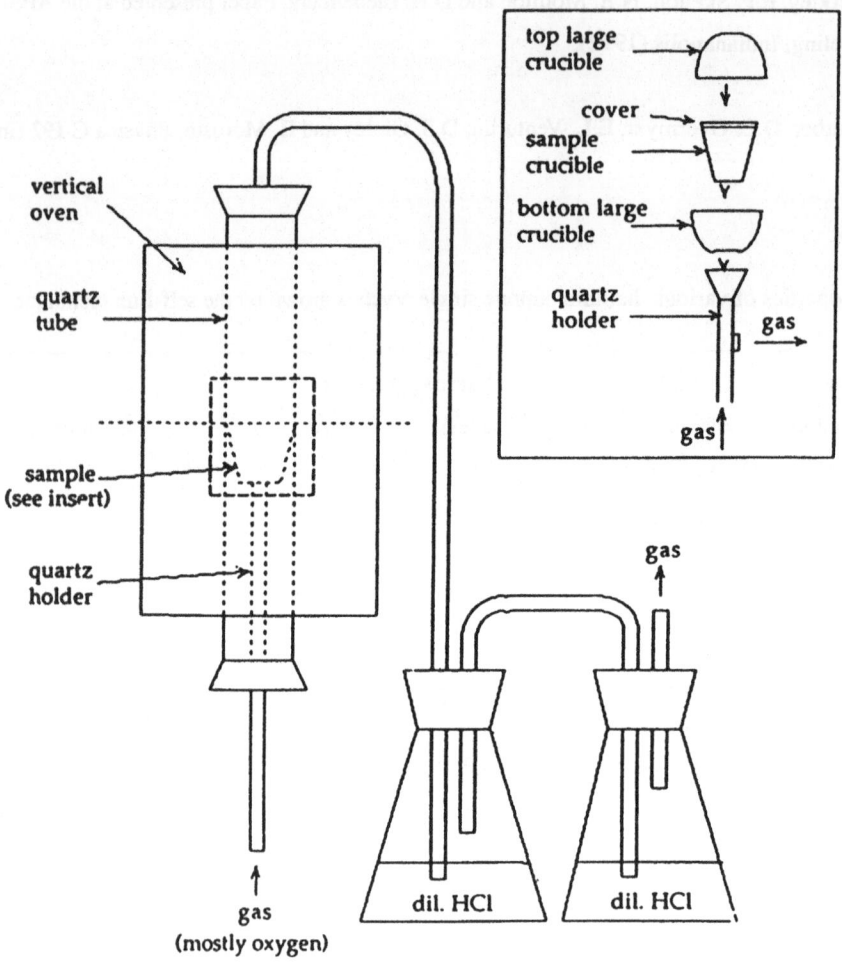

Figure 1 Typical experimental setup for crystal growth. The insert shows the enlarged picture of the crucible assembly inside the dashed rectangular portion. (Adapted from Ref. 14.)

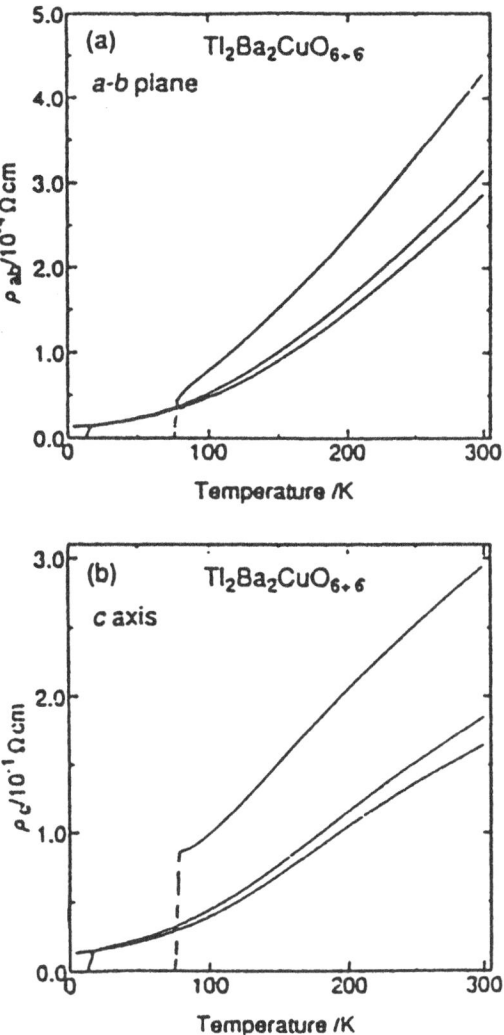

Figure 2   Temperature dependence of the anisotropic resistivities (a) ρ_{ab} and (b) ρ_c measured by the Montgomery method for Tl-2201 single crystals annealed at 350°C in O_2 (metallic), at 300°C in Ar ($T_c = 10$ K), and at 400°C in Ar ($T_c = 75$ K). (Adapted from Ref. 22.)

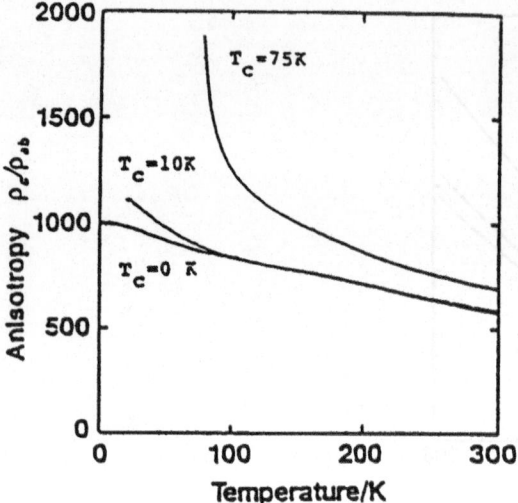

Figure 3 Anisotropy of resistivity plotted against temperature for Tl-2201 single crystals. The data points from Fig. 2 were replotted. (Adapted from Ref. 22.)

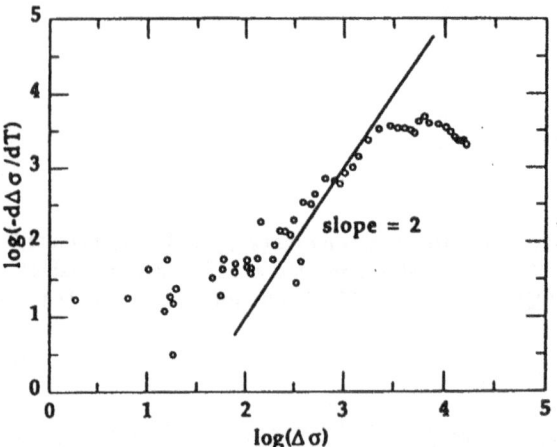

Figure 4 Plot of log $[-d(\Delta\sigma)/dT]$ versus log$\Delta\sigma$ for Tl-2201 single crystals. The slope of the line is 2, which corresponds to the two-dimensional superconductivity of the sample. (Adapted from Ref. 13.)

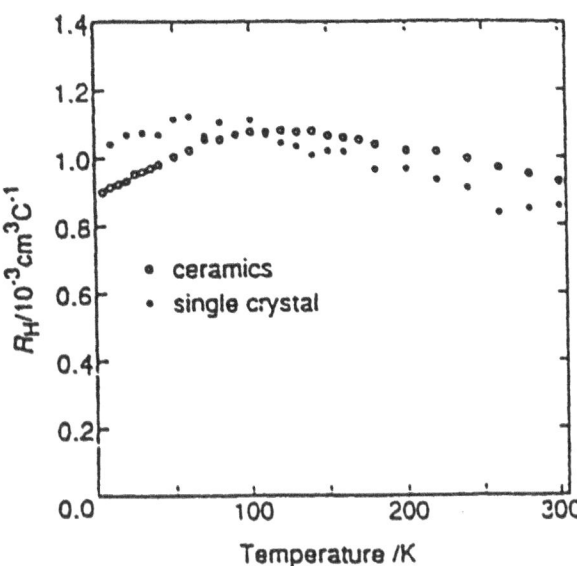

Figure 5 Temperature dependence of the Hall coefficients for normal metallic Tl-2201 samples. Single-crystal data in the *ab* plane (closed circles) and ceramic sample data (open circles) are compared. (Adapted from Ref. 23.)

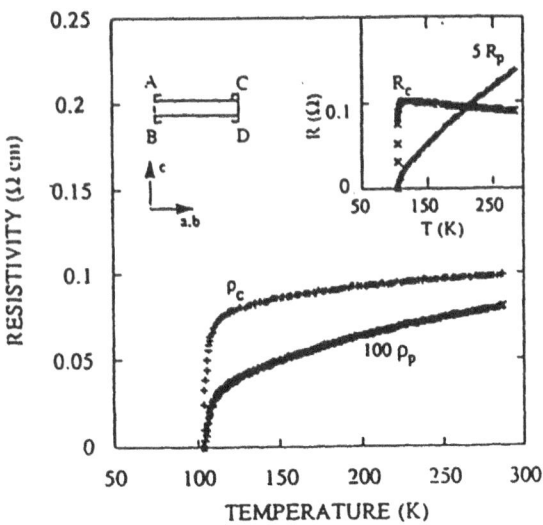

Figure 6 Temperature dependence of the resistivity for Tl-2212 single crystals. ρ_r is multiplied by 100. The insert shows the measured resistances by using the top-left configuration. (Adapted from Ref. 11.)

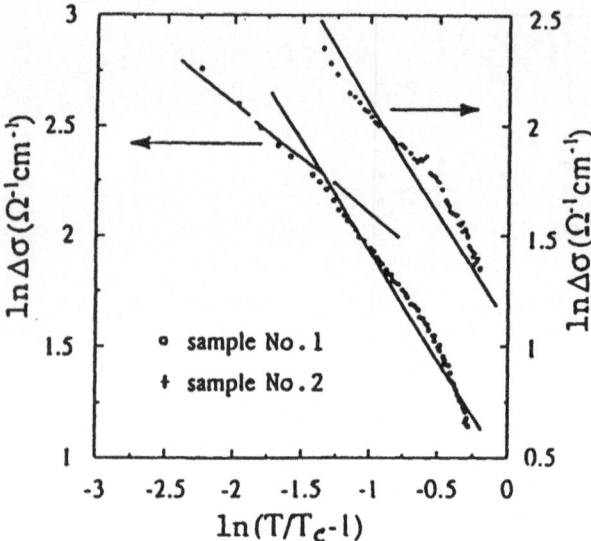

Figure 7 Plots of $\ln(\Delta\sigma)$ versus $\ln(T/T_c - 1)$ for Tl-2212 single crystals. The solid line and dashed line correspond to two- and three-dimensional superconductivity, respectively. In the high-temperature range, the superconductivity is two-dimensional. At low temperature, the superconductivity in some samples becomes three-dimensional. The T_c values used are the midpoints of the 10 to 90% resistance value. (Adapted from Ref. 11.)

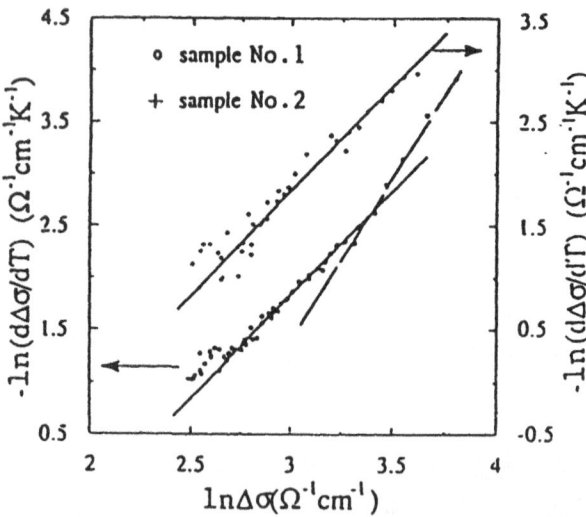

Figure 8 Plots of $\ln[d(\sim\Delta\sigma)/dT]$ versus $\ln\Delta\sigma$ for Tl-2212 single crystals. The solid and dashed lines correspond to two- and three-dimensional superconductivity, respectively. Superconductivity of sample 2 is two-dimensional. In the high-temperature range, the superconductivity of sample 1 is two-dimensional, and becomes three-dimensional in the low-temperature range. (Adapted from Ref. 11.)

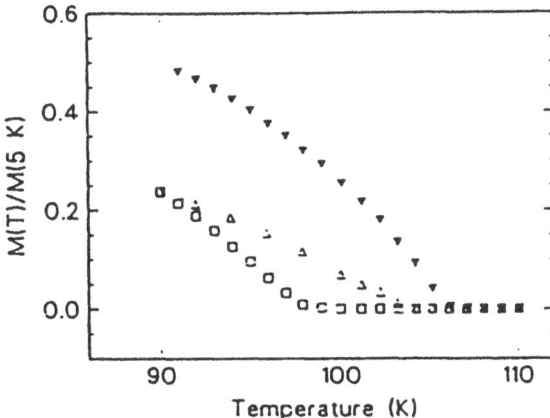

Figure 9   Meissner signal (normalized to its value at 5 K) versus temperature in 2.5 mT applied normal to a Tl-2212 single-crystal plate as-grown (open triangles), after a vacuum anneal (open squares), and then an oxygen anneal (solid triangles). (Adapted from Ref. 19.)

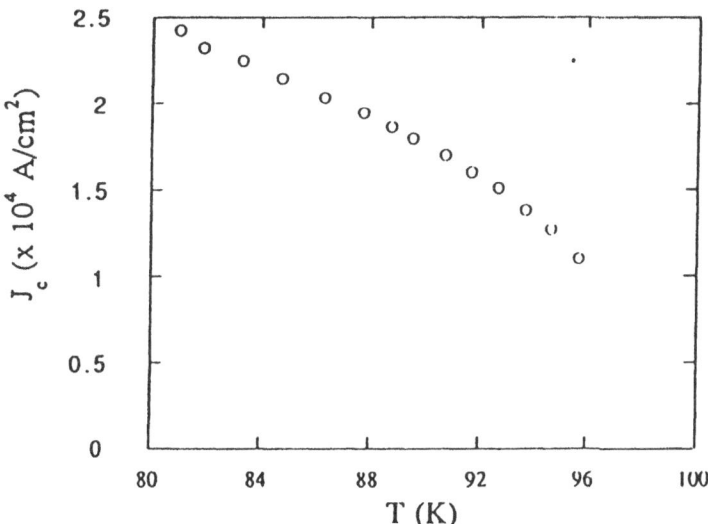

Figure 10   Temperature dependence of the transport critical current density J_c in the ab plane of a Tl-2212 single crystal. (Adapted from Ref. 34.)

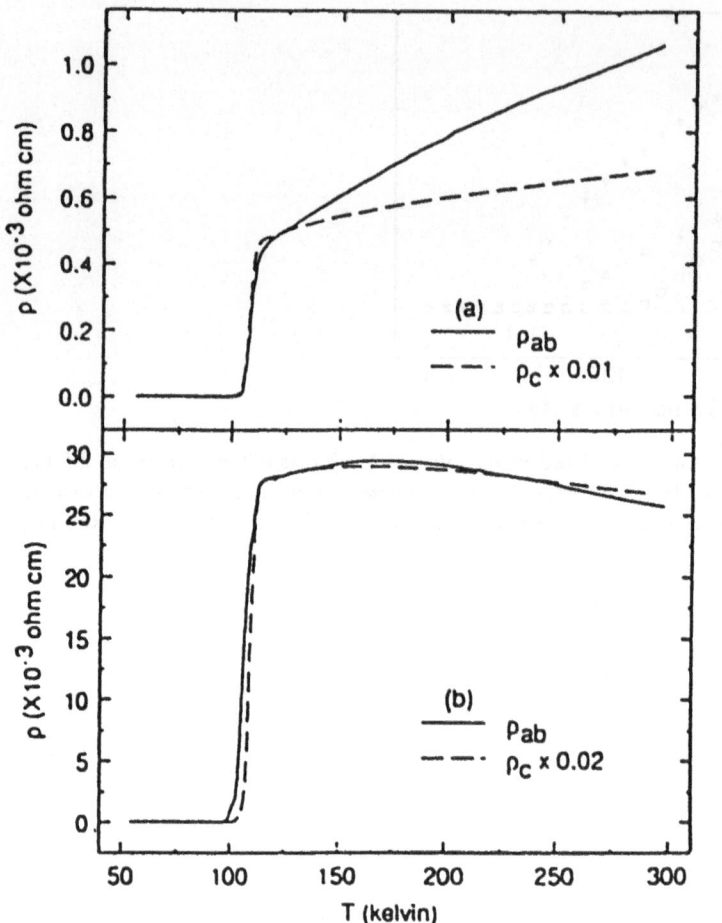

Figure 11 Temperature dependence of the anisotropic resistivities, ρ_{ab} and ρ_c for Tl-2223 single crystals. Crystal 1 (a) has a relatively sharp superconducting transition and high T_c value. For crystal 2 (b), the resistivity does not increase monotonically with temperature in the normal state and is about 25 times greater than for crystal 1. (Adapted from Ref. 17.)

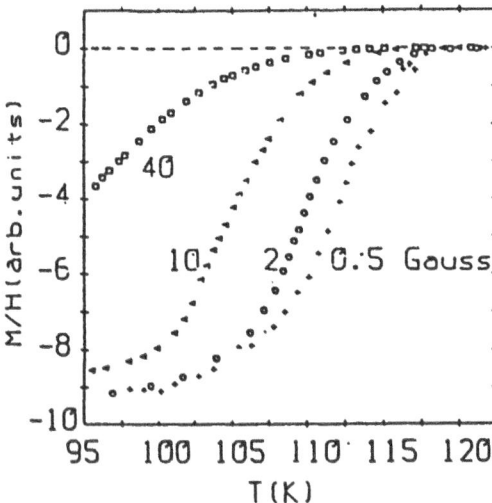

Figure 12   Shielding susceptibility M/H versus temperature for Tl-2223 single crystals. The numbers labeling the curves are the applied field in Gauss in the c direction. (Adapted from Ref. 42.)

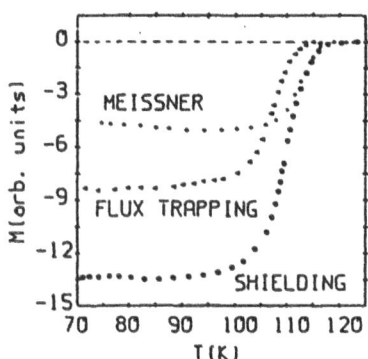

Figure 13   Flux trapping, Meissner, and shielding signals (in 2 G) versus temperature for Tl-2223 single crystals as in Fig. 12. One arbitrary unit corresponds to 0.031 G. For comparison, the positive flux trapping signal has been inverted. (Adapted from Ref. 42.)

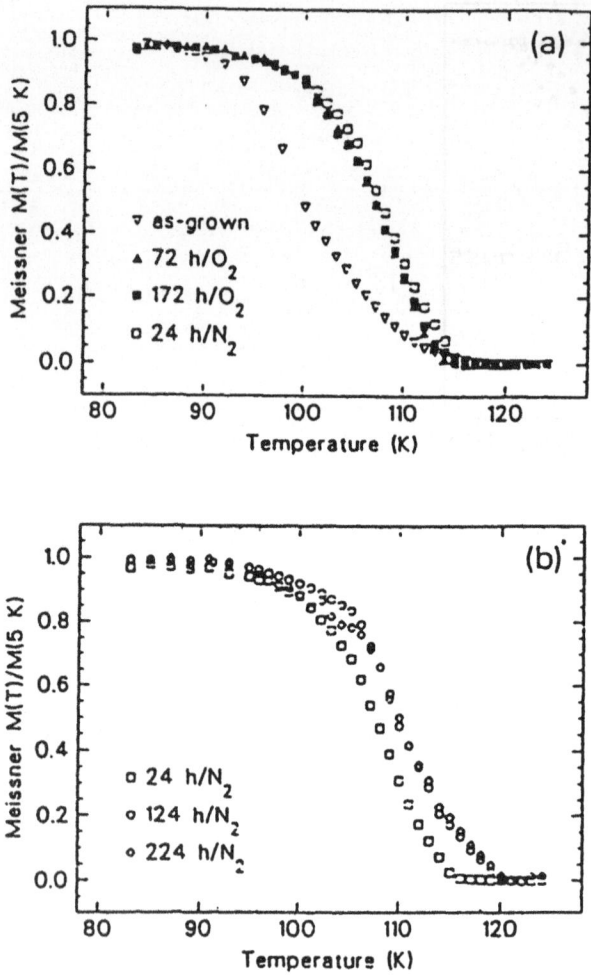

Figure 14 Comparison of the normalized Meissner signals for Tl-1223 single-crystal plates, (a) as-grown compared with initial oxygen (72 h), additional oxygen (172 h total) resulting in no change, and initial nitrogen (24 h) anneals, and (b) nitrogen for 24 h [same curve shown in (a)] and two additional 100-h nitrogen anneals. (Adapted from Ref. 20.)

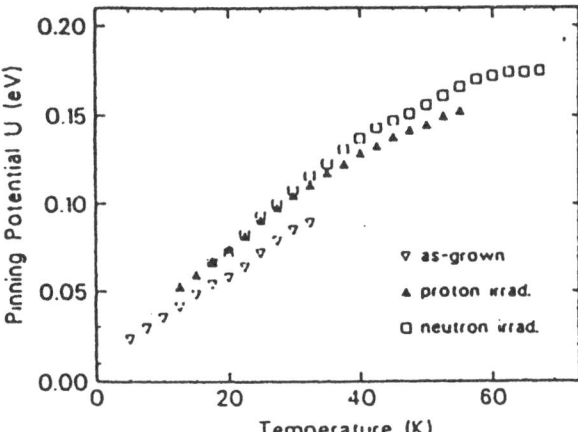

Figure 15 Comparison of flux pinning potential versus temperature determined from magnetization relaxation (flux creep) measurements in Tl-2223 single crystals as-grown (open triangles), Tl-2223 after proton irradiation (solid triangles), and Tl-2212 after neutron irradiation (open squares). (Adapted from Ref. 15.)

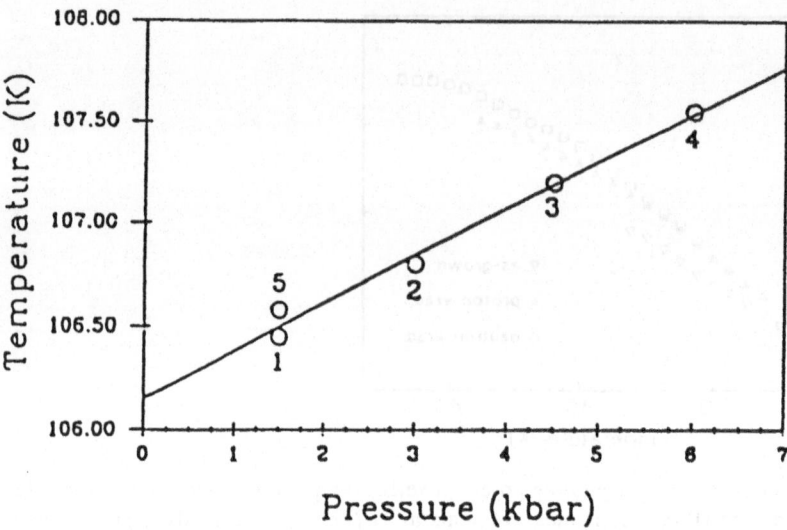

Figure 16 Pressure dependence of T_c value for Tl-5526 single crystals, showing the onset of superconductivity near 106 K. The points are numbered in the sequence they were taken. (Adapted from Ref. 47.)

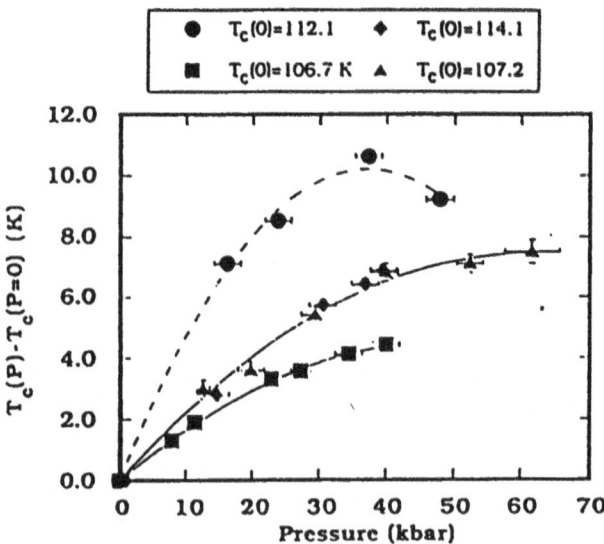

Figure 17 Pressure dependence of T_c for Tl-2212 single crystals. (Adapted from Ref. 49.)

Part III

**Flux Pinning,
Pinning Centers,
Applications**

FLUX PINNING IN HIGH-TEMPERATURE SUPERCONDUCTORS

P.H. KES

Kamerlingh Onnes Laboratory, Leiden University,
P.O. Box 9506, 2300 RA Leiden, The Netherlands

March 23, 1993

Abstract

Basic flux-pinning ingredients are reviewed and a selection of experimental results on high-temperature superconducting compounds is presented. Some recent results and properties which are central to the emerging understanding of the flux pinning mechanisms of these fascinating materials are emphasized.

1 Introduction

After the discovery of the high-temperature superconductors (HTS) [1] a great variety of superconducting applications at liquid nitrogen temperatures seemed within reach. However, as is now generally realized, a large critical temperature T_c and a strong upper critical field $H_{c2}(0)$ are only necessary conditions. In addition, a large critical current density J_c is required, e.g. for superconducting magnets a $J_c > 10^9$ Am^{-2} at fields B above 20 T is needed to be competitive with conventional superconductors like $(NbTi)_3Sn$. Although impressive progress has been made in the raising of J_c, it also became clear that some of the characteristic properties of the HTS, like the quasi two-dimensional nature related to the layered structure, or the short coherence length, lead to serious limitations of the high temperature applicability because they collaboratively increase the effectiveness of thermal fluctuations. This results in resistive behavior in a wide area of the B, T plane between the irreversibility line $B_i(T)$ and the mean-field transition line $B_{c2}(T)$. It is still under debate whether the irreversibility line marks a melting transition of the vortex lattice (VL) [2,3,4] or a transition from a liquid to a vortex glass [5,6] or that it is caused by thermally activated depinning [7,8,9]. In fact, these phenomena are intimately related [10,11,12,13]. From a practical point of view it is important to know if J_c can be improved or that the irreversibility line can be moved to higher T, by some treatment of the material. Such knowledge cannot be attained without a deep insight in the properties of a VL in an anisotropic superconductor, nor without fundamental investigations of flux pinning and flux creep in such materials. This paper is intended to give a brief overview of the progress made in the last few years.

E. Kaldis (ed.), Materials and Crystallographic Aspects of HTc-Superconductivity, 401–431.

The study of fundamental backgrounds requires both materials of high quality and systematics of the experiments. In view of the large anisotropy of the HTS it is also important to clearly define the configuration of field, current and crystal orientation. This paper is therefore restricted to investigations on single crystals and highly textured thin films. Work on $YBa_2Cu_3O_7$ (Y:123) and $Bi_2Sr_2CaCu_2O_8$ (Bi:2212) will be presented most as they are representative for a very and extremely anisotropic superconductor with, respectively, proximity and Josephson coupling between the supercon-ducting CuO2 double layers. In Section 2 some of the basic ingredients for pinning and creep will be dealt with as an introductory to the section about the experimental situation (Section 3) which is divided into many subsections in order to present results and discussions next to each other.

2 Basic ingredients

2.1 Properties of an anisotropic vortex lattice

The coupling between the superconducting CuO_2 double or triple planes of the high-T_c cuprates can be expressed in terms of an anisotropy parameter Γ being the ratio of the effective quasi-particle masses in the c and in the ab direction (we will ignore the anisotropy in the ab plane). The anisotropy is reflected in the anisotropy of upper and lower critical fields and the Ginzburg-Landau (GL) coherence lengths ξ_{ab}, ξ_c, and the penetration depths λ_{ab} and λ_c. In the latter case the subscript denotes the direction of the screening current. The following simple relations hold: $\Gamma = (\xi_{ab}/\xi_c)^2 = (\lambda_c/\lambda_{ab})^2$.

As first pointed out by Lawrence and Doniach [14] a quasi two-dimensional (2D) situation arises when $\xi_c(T)$ becomes much less than the interlayer spacing s. In the Bi and Tl:2212 compounds very large Γ values have been reported: ≥ 3000 and $\geq 10^5$ [15], respectively. Recently, it has been shown that the value for Bi:2212 probably depends on crystal quality and that it at least is 3×10^4 [16]. The crossover from 3D to 2D behavior then occurs a few mK below T_c. In Y:123 with $\Gamma \geq 29$ [17] the 3D regime is more extended, roughly 10 K [18]. Therefore, it seems reasonable to treat the VL properties of Y:123 in the framework of an anisotropic GL theory, whereas for the Bi and Tl compounds a quasi-2D approach is required [19]. In the first case the vortices are still considered as flux lines (tubes), buth with anisotropic lattice configurations and elastic constants.

With the field along the c axis a triangular Abrikosov lattice occurs with a lattice parameter $a_0 = 1.075(\Phi_0/B)^{\frac{1}{2}}$, denoted by the dashed lines in Fig. 1, the usual shear modulus c_{66}, and almost the same compression modulus c_{11}, but with a tilt modulus $c_{44}(k)$ which is reduced roughly by a factor Γ [4]. Here k denotes the wavevector of the deformation field. It should be noted that the reduction by Γ is fully obtained only in the limit of nonlocal elasticity [20], i.e. when $k^{-1} < \lambda/(1-b)^{\frac{1}{2}}$; for $k = 0$ there is no reduction. Both c_{66} and c_{44} are important parameters in describing the response to forces on the VL exerted by pinning centers [21,22,23]. Pinning by randomly distributed defects leads to a disordered VL which can be visualized most directly by decoration experiments [24]. The reduction of c_{44} causes a noticeable decrease of the longitudinal correlation length L_c of the VL [25].

With the field along the ab plane the VL adopts an isosceles struc- ture with

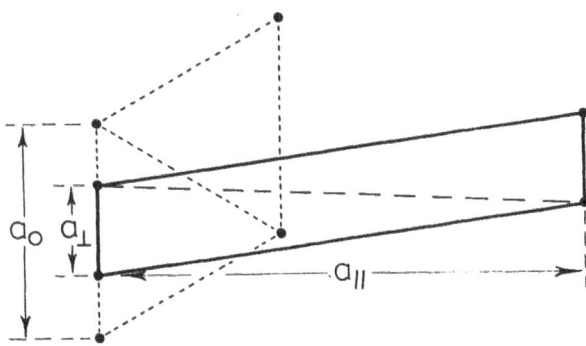

Figure 1: Vortex lattice unit cells in anisotropic superconductors with uniaxial symmetry. Dashed lines: the triangular VL for $H \parallel$ symmetry axis (c-axis); solid lines: the isosceles lattice for H along ab-planes and $\Gamma = 30$.

lattice parameters $\approx a_0 \Gamma^{\frac{1}{4}}$ and $a_0 \Gamma^{-\frac{1}{4}}$, Fig. 1 solid line, and very anisotropic c_{66} [26] and c_{44} [20,27] expressing the fact that deformations parallel to ab cost less energy. For arbitrary field orientations the screening currents mainly run in the ab planes leading to a torque whereby Γ can be determined [28]. New vortex configurations have been predicted for low fields and angles ϑ (between H and c axis) above ≈ 65 degrees [29] and have been evidenced by decoration experiments [30].

In the quasi-2D case the screening currents are confined to the CuO_2 layers leading to a segmentation of the vortices into 2D pancake vortices [31]. The coupling of the 2D vortices in adjacent layers determines the Josephson length $R_J \equiv \Gamma^{\frac{1}{2}} s$. Phenomena on a length scale smaller than R_J will appear as if the layers are totally decoupled, whereas for collective phenomena of larger size 3D behavior is expected [32]. At low temperatures a field $B_{2D} \approx \Phi_0 / \Gamma s^2$ marks the crossover from 3D (low) to 2D (high fields). When the field is within an angle of order Γ^{-1} parallel to the ab planes a lock-in transition to a Josephson VL takes place [33]. The properties of such a lattice have been first described by Bulaevski [19] and are presently intensively studied [34,35,36,37,38]. An important difference with an Abrikosov vortex is the absence of a normal core. The order parameter is uniform in the ab direction and varies periodically in the c direction with period s; it is large in the $CuO-2$ planes and almost zero between them. The screening currents between the planes are very weak so that the material is almost transparent for the field component along ab [39]. When the field makes an angle with the planes larger than Γ^{-1}, the relevant vortex properties are determined by the pancake vortices which can be treated as if a field $H \cos \vartheta$ is applied in the c direction [39,40,41]. A recent scaling theory explains this important feature [42].

Both for an anisotropic 3D and quasi-2D VL in the 3D regime the reduction of c_{44} leads to a dramatic increase of the squared-average displacement of the vortices $< u^2 >_T$ by thermal fluctuations [4]. When $< u^2 >_T \approx (a_0/10)^2$ the 3D VL melts by the development of dislocation loops [43]. The above condition determines a melting line in the $B - T$ plane [4,6].In the 2D regime thermal fluctuations give rise to the unbinding of dislocation pairs in the 2D VL according to the Kosterlitz-Thouless theory at a temperature $T_M = c_{66}a_0^2 s/4\pi k_B$. For Bi:2212 this amounts to 20-30 K when for $\lambda_{ab} \approx 140$ nm is assumed [32].

Because of the anisotropic penetration depth and GL coherence length layered superconductors have anisotropic critical fields H_{c1} and H_{c2} as well. The determination of H_{c1} is problematic because of surface and demagnetization effects and pinning, that of H_{c2} because of thermal fluc- tuations. The result is that at present reliable values are only available for Y:123. Measured with the field along c we have $-\mu_0 dH_{c2}/dT = 1.65$ T/K at T_c, yielding $\xi(0) = 1.7$ nm and $B_{c2}(0) = \mu_0 H_{c2}(0) = 110$ T [44,45]. As to the penetration depth direct measurements yield $\lambda_{ab}(0) = 140$ nm [46] from which $\mu_0 H_{c1}(0) \approx 40$ mT follows. In the ab plane $\mu_0 H_{c1}(0) \approx 7.5$ mT. Further, we have $\kappa = 69$ and $B_c(0) = 1$ T. For Bi:2212 the values for $\lambda_{ab}(0)$ range between 110 nm [16] and 185 nm [47]. In order to make estimates of pinning energies and critical currents it is necessary to decide on the value for ξ_{ab}. Rather arbitrary, we will henceforth assume $\xi_{ab}(0) = 2.3$ nm, $B_{c2}(0) = 44$ T, and $B_c(0) = 0.3$ T.

2.2 Defect-flux-line interaction

Crystal defects locally alter the material properties and consequently the superconducting parameters in their environment. These local changes may couple to the periodic variations of both the order parameter and the local field which are characteristic for the mixed state. In principle the interaction should follow from solving the Ginzburg-Landau equations with the proper boundary conditions imposed by the defects. Depending on the circumstances, i.e. flux density, size and character of the defect, etc., it is possible to classify the elementary interactions in terms of the predominant coupling mechanism, namely magnetic and core interactions, respectively.

2.2.1 Magnetic interaction

Examples of the magnetic interaction are the effect of surfaces parallel to the applied magnetic field (this might be the external surface as well as some large precipitate interface) and thickness variations of thin films for fields normal to the film. In the first case, at low flux densities, the effect of the order parameter may be ignored (London limit), sothat the interaction is determined by the boundary condition imposed on the supercurrents around the vortex core, i.e. the normal component at the surface should vanish. Theoretically, this is achieved by assuming an anti-vortex image at the other side of the boundary which results in an attractive surface-vortex interaction. This interaction is opposed by the repulsion due to the superconducting screening currents generated by the external field. The net effect is a potential barrier for flux entry and flux exit which is most effective just above the lower critical field. The barrier decreases with increasing field [48].

In the case of thin films with thickness variations, the vortices are trapped at the sites of smallest thickness where the line energy of the vortex is minimum. Evidently, the typical length scale related to the magnetic interaction is the penetration depth λ. In materials with a large Ginzburg-Landau parameter κ this kind of interaction is therefore small. It generally disappears with increasing magnetic field.

2.2.2 Core interaction

The coupling to the variation of the order parameter $|\Psi|^2$ (the density of Cooper pairs) is the origin of flux pinning for almost all defects, e.g. dislocations, point defects, voids, grain boun-daries, precipitates. Defects deviate from the surrounding material by a different density, elasticity, electron-phonon coupling, or electron mean free path. The first three properties give rise to a local change in T_c, whereas the latter predominantly leads to a variation in κ. Consequently, one may distinguish between δT_c- and $\delta\kappa$ -pinning.

The core pin mechanism follows from the Ginzburg-Landau free energy as derived from the microscopic theory, (see Ref. 48 pp. 171):

$$F = \int d^3r \left[A|\Delta|^2 + \frac{B}{2}|\Delta|^4 + C|\partial\Delta|^2 + \frac{h^2}{2\mu_0} \right] \tag{1}$$

with $A = N(0)(1-t)$, $t = T/T_c$, $B = 0.098N(0)/(kT_c)^2$, $C = 0.55\xi_0^2 N(0)\chi(\alpha)$, and $\partial = -i\Delta - (2e/\hbar)\vec{A}$. $\vec{A}$ is the vector potential related to the local field $\vec{h}$, $\Delta(\vec{r})$ the pair potential related to the order parameter via $|\Delta|^2 = \hbar^2|\Psi|^2/(4mC)$ with m the electron mass, $\alpha = 0.882\xi_0/l_{tr}$ the impurity parameter, i.e. the ratio of the BCS coherence length ξ_0 and the mean free path l_{tr}, $\chi(\alpha)$ the Gorkov impurity function, and $N(0)$ the density of states at the Fermi surface. The defects perturb the coefficients A,B, and C, as well as the vector and pair potentials. Because F is supposed to be minimized with respect to Δ and $\vec{A}$, the pin energy can in first order be obtained by only taking into account the variations of the coefficients [22,49]. However, this is only allowed if the relevant size of the defect is small compared to the distance over which Δ varies.

It is clear from Eq.(1) that the first two terms are involved in case of T_c-deviations, whereas l_{tr}-fluctuations only lead to perturbation of the third term. It is important to retain this distinction, because it gives rise to a different field and temperature dependence for the various elementary interactions which will help to determine the nature of the predominant pinning mechanism. In addition to temperature and field effects, also size and spatial distributions of the pinning centers are important. For extended defects the crystal anisotropy and the defect orientation with respect to the field direction play a role as well.

For electron scattering the theory has been worked out and reviewed by Thuneberg [49]. The concept is illustrated by deriving from Eq.(1) the elementary pinning potential

$$\Omega_p(r) = \mu_0 H_c^2 \xi^2 g(\alpha)\xi_0 \int d^3r' |\partial\Psi(\vec{r})|^2 \delta(l/l_{tr}(\vec{r}-\vec{r}')) \tag{2}$$

with $g(\alpha) = 0.882 d\ln(\chi)/d\alpha$ ($g(\alpha) = 0.85$ for $\alpha = 0$ and $g(\alpha) = l_{tr}/\xi_0$ for $\alpha \gg 1$), H_c the thermodynamic critical field, and δ_{tr}^{-1} the extra electron-scattering by the defect. A reduced order parameter has also been introduced: $|\psi|^2 = \hbar^2|\Psi|^2/(4m\xi^2\mu_0 H_c^2)$. The elementary pinning force f_p is obtained by computing the maximum value of the gradient of $\Omega_p(r)$. By the extra scattering of the electrons at the defect the Ginzburg-Landau coherence length ξ decreases locally, which means that this pin mechanism always results in an attractive interaction.

In case of a small void (size $D < (\xi_0^{-1} + l_{tr}^{-1})^{-1}$) the factor $\delta(l_{tr}^{-1})$ can be replaced by a delta-function times the scattering cross-section $\pi D^2/4$, thereby illustrating that the pin energy is proportional to the condensation energy multiplied by the defect volume enhanced by a factor δ_0/D or l_{tr}/D for the clean or dirty limit, respectively. For extended defects, such as grain boundaries or large, flat precipitates, parallel to the flux lines, the situation is more complicated, because the integral over the entire defect should be evaluated. In addition, the scattering probability of the defect should be determined from other experiments. A special case in which also the crystal anisotropy has to be considered, is that of the grain boundary in a bicrystal.

Regarding field and temperature dependence, two field regimes should be distinguished. In small fields ($B < 0.2B_{c2}$) the vortices are essentially isolated and the order parameter changes over a distance ξ. In large fields the vortices overlap which yields (near B_{c2})

$$|\Psi|^2 \approx 0.86(1 - b)\{1 - \frac{1}{3}[cos(x - \frac{y}{\sqrt{3}})k_0 + cos(\frac{2y}{\sqrt{3}})k_0 + cos(x + \frac{y}{\sqrt{3}})k_0]\} \quad (3)$$

Here $k_0 = 2\pi/a_0$, $a_0 = 1.075(\Phi_0/B)^{\frac{1}{2}}$ the VL parameter, and $b = B/B_{c2}$. For small fields the expression for f_p contains a factor $\xi^{-3}(\sim (1 - t)^{\frac{3}{2}})$, while for large fields a factor $(2\pi/a_0)^{-3}$ appears ($\sim B^{\frac{3}{2}}$). Since $a_0/\xi = 2.69b^{\frac{1}{2}}$, the difference is $12.7b^{\frac{3}{2}}$, which is of order 1 at b = 0.18. Accordingly, field and temperature dependences in both regimes are distinctly different, i.e. f_p for isolated vortices should be independent of B, but it is proportional to $b^{\frac{3}{2}}(1 - b)$ near $4B_{c2}$. Typically, in conventional superconductors for a void (D = 1 nm) and an isolated vortex $f_p \approx 0.3x10^{-14}N$, where $\mu_0 H_c = 0.1\ T$ and $\xi_0 = 10\ nm$ has been used.

Pinning by various kinds of dislocations is a typical example of δT_c pinning. This mechanism has extensively been studied in the past [22]. The formalism used differs from the one described above. Two effects are distinguished, both related to the normal character of the vortex core: a larger density (ΔV-effect) and larger elastic constants (ΔE-effect). These effects, though very small, give rise to a periodic stress and elasticity-field linked with the variation of $|\Psi|^2$. The pin potential arised via the coupling to the strain field around the defects and is linear in the strain for the ΔV-effect and quadratic for the ΔE-effect. Accordingly, these effects are referred to as first and second order interactions. Typical values of $\Delta V/V$ and $\Delta E/E$ are 10^{-7} and $3x10^{-5}$, respectively. Some typical values of f_p or f_{pl} (f_p per unit length) of an isolated vortex in conventional superconductors with $\mu_0 H_c = 0.1\ T$ and $\xi_0 = 10\ nm$ are: edge dislocation perpendicular to a flux line $f_p \approx 10^{-13}$N; edge or screw dislocation parallel to a flux line $f_{pl} \approx 5x10^{-10}$ and $10^{-5}N/m$, respectively; dislocation loop (D = 10 nm) perpendicular to the field direction $f_p \approx 3x10^{-14}N$. Moreover, $f_p \propto D^2$, see [50].

Precipitates may be considered as δT_c-defects as well. The pin potential can be estimated to be at most $\mu_0 H_c^2 V \xi_0/2D$ for dielectric inclusions of volume V. For conducting precipitates the proximity effect should be taken into account giving rise to a much lower value. No better theory exists at present. As to the field and temperature dependence for δT_c-pinning, f_p now contains a factor ξ^{-1} or $(2\pi/a_0)^{-1}$ in small and large fields, respectively. This kind of mechanism may lead to attractive or repulsive interactions depending on the defect characteristics.

2.2.3 Concluding remarks

From the above discussion it follows that a general formula for f_p cannot be given. For a specific material one should first determine the predominant defect structure and then try to estimate the pin interaction. Defect morphology, pin mechanism, characteristic length scales, and field orientation and regime are to be taken into account. One might say that reasonable estimates can now be made for the most important pinning centers, although the situation for precipitates is not yet firmly settled.

In the above, only single-effect interactions have been considered. Whether this simplification is allowed, depends on the concentration of pinning centers and the range of the pin interaction. The effectiveness of the pinning force will considerably decrease, when the pins strongly overlap, i.e. if the distance between the pins is much less than the interaction range. A nice example is an amorphous superconductor. Although the defect concentration is very large, the pinning is small, since on the relevant length scale ξ the material is homogeneous. Only density or stress modulations with wavelengths comparable to ξ will be effective.

2.3 Natural pin mechanisms in HTS

The elementary pin interaction of a single defect with a single vortex or with a VL has been computed for several potential pin mechanisms, e.g. for twin planes in YL:123 with orientations (110) and ($1\bar{1}0$) [25], see also [51], for oxygen vacancies [52], and for dislocation loops or stacking faults [50]. These are potentially the strongest pinning centers in bulks single crystals. In sintered bulk materials small precipitates of Y_2BaCuO_5 [53] or CuO [54] may be effective; their elementary interaction is only roughly known.

Recently, STM studies of sputtered Y:123 films revealed a large density of screw dislocations along the growth direction [55]. The pin interaction can be estimated by considering the core of the dislocation as a non-conducting cylinder of diameter s. This assumption is supported by the fact that the resistance and T_c of Y:123 are very sensitive to disorder. Further, it seems reasonable to take the pitch of the screw dislocation as the diameter of the disordered core region. For parallel vortices the interaction can be estimated from the cross-section for electron scattering making use of Thuneberg's formalism [49], see also [25]. One obtains per unit length in the single vortex limit

$$f_{pl} \approx (B_c^2/\mu_0)pD\xi_0/\xi \tag{4}$$

where ξ_0 is the BCS coherence length, p the scattering probability (p = 1 for a non-conducting defect) and D the diameter of the dislocation core, $D \approx s$.

An other pin mechanism one should consider is the surface roughness created by the screw dislocations. If one models the roughness by $\delta d \cdot sin(2\pi x/L_s)$ the pinning force per unit length for small field roughly is

$$f_{pl} \approx 2\pi\epsilon_0(\delta d/d)/L_s \tag{5}$$

with the line energy $\epsilon_0 = (\Phi_0^2/4\pi\mu_0\lambda^2)\ln\kappa$, L_s the mean distance between the dislocations, d the film thickness and $2\delta d$ the average thickness variation. Note that f_{pl} represents the maximum force. To obtain the bulk pinning force one still has to sum over the actual forces which depend on the mutual positions of the defects and the vortices taking into account the vortex deformations, see the next Section.

As mentioned above, the order parameter is modulated by the crystal structure being small in the area between the superconducting layers. Consequently, if the field is along the ab planes, the energetically favorable configuration is that with the vortex cores between the CuO_2 double layers. To move the cores across the layers needs a large driving force close to the ultimate value set by the depairing current. This kind of pinning caused by the crystal structure is denoted as intrinsic [34,35].

2.4 Summation models

The summation of the actual pinning forces of a realistic defect system has been a long standing problem which could be solved for a system of point pins with typical dimension smaller than ξ. It is assumed that each pin only gives rise to elastic strains and that the dislocation density of the VL is small enough to be ignorable. Finally, it should be remarked that many recent investigations have been carried out on HTS with artificial pinning centers produced by irradiation damage, especially by amorphous tracks caused by high-energy ions. Their effect will be discussed in Section 3. Even for weak pinning the positional order in the VL breaks down, given the system is sufficiently large. Correlated regions determined by the transverse and longitudinal correlation lengths R_c and L_c, with volume $V_c = R_c^2 L_c$, may be defined in which the collective effect of the pins gives rise to a net force $F_c = (n_p V_c < f^2 >)^{\frac{1}{2}} \equiv (WV_c)^{\frac{1}{2}}$. The average depinning force density follows from $F_p = F_c/V_c$ and leads to a critical current density $J_c = F_p/B$. This theory of collective pinning (CP) [56], has recently been reviewed [6,57,25]. Experiments have shown [58] that the CP theory applies to a 2D VL, but that for stronger pinning centers or for a soft VL, plastic deformations determine the disorder. In 3D the theoretical predictions are only valid in the amorphous limit defined by $R_c \approx a_0$. In layered superconductors and field normal to the layers the decoupling reduces the longitudinal correlation length L_c by a factor $\Gamma^{\frac{2}{3}}$ and J_c is enlarged by a factor $\Gamma^{\frac{1}{3}}$. Substituting experimental J_c values for Bi:2212 it can be shown that $L_c \approx s$, so that a 2D CP behavior can be expected.

For more extended defects in general only rough estimates can be made depending on the shape, orientation and nature of the defect and the vortex-defect coupling. A special example of such defect is a flat grain boundary or a twin plane. Decoration experiments [59] showed that the vortices are attracted to the twin planes while between the planes the VL is very disordered. Assuming the effect of the twin planes to be predominant a simple direct summation procedure seems appropriate, although limited to fields parallel to the planes and a geometry of mutually perpendicular twin planes [60]. Strong pins may disrupt the VL giving rise to a granular VL consisting of bundles separated by edge dislocations. This might be the case for the Y:123 films, see Eqs. (4) and (5). In such a situation the direct summation applies too.

For each of the pin mechanisms considered in the preceding section the situation at low fields is quite simple, because all vortices will be trapped by these strong pins.

J_c follows from

$$J_c = f_{pl}/\Phi_0 \tag{6}$$

In Section 3.1.4 we will further address this issue and compare with experimental results.

3 Pinning forces and critical currents of HTS

3.1 Pinning by natural defects

3.1.1 General features

To illustrate the state of the art an overview of $J_c(B)$ curves for wires and films of the Y and Bi HTS is given in Fig. 2, see [61] for references. For comparison data of conventional materials are shown as well. The performance of Y:123 films is very

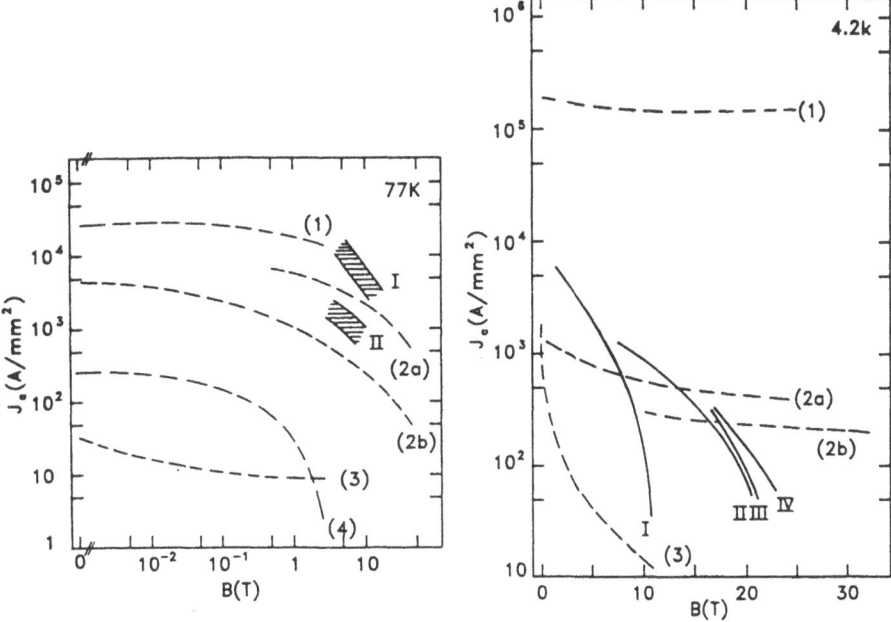

Figure 2: (a) Typical critical current densities as functions of field for various conventional and high-temperature superconductors at 4.2 K: (1) Y:123 film on MgO substrate, (2a) Bi:2212 Ag-sheated tape, (2b) Bi:2212 Ag-sheated wire, (3) Y:123 film on Ag tape, (I) NbTi, (II) Nb$_3$Sn, (III) V$_3$Ga, and (IV) (Nb,Ti)$_3$Sn. Source: Nikkei Superconductors, Japan, February 1990.

(b) Comparison of current densities of various HTS at 77 K and conventional materials at 4.2 K. Note the change to logarithmic scale of B. (I) NB$_3$Sn, (II) NbTi, (1) Sputtered Y:123 film, (2a) CVD deposited Y:123 film, $H_{\parallel}$ ab-planes, (2b) same as (2a), $H \parallel c$, (3) Y:123 wire, (4) Bi:2212 tape, $H \parallel ab$. Source: Nikkei Superconductors, Japan, December 1989.

promising [62], both at 4.2 K and 77 K close to the estimated optimal values [63]. Y:123 wire, however, struggles with the weak-link and grain-boundary problem [64], therefore Bi:2212 wires and tapes seem to offer the best perspectives, especially with respect to applications above 20 T [65]. That is, at low temperatures, T < 20 K, for the J_c collapses for B > 1 T at 77 K. It should be noted that the J_c's for the $H\|ab$ configurations can be more than an order of magnitude larger than with $H \perp ab$ [66]. It has been argued that this reduction is related to the large anisotropy [39,67]. A striking demonstration in support of this idea is given in Fig. 3 [67]. The drop in resistance at 5 T clearly shifts to lower T for larger Γ. A recent confirmation has been seen in a $a - Mo_3Ge/Ge$ multilayer model-system [68].

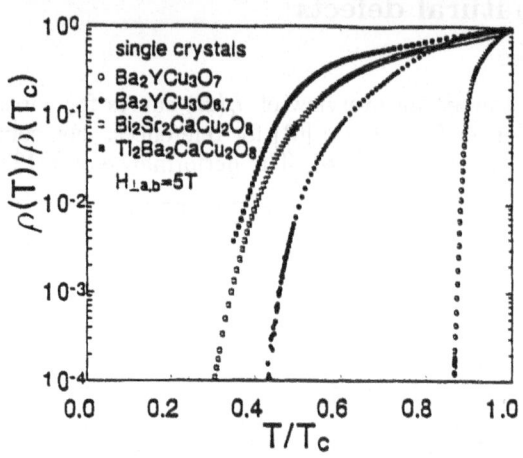

Figure 3: The resistive transitions in a magnetic field of 5 T for HTS single crystals of different anisotropies [67]. A clear correlation is seen between the drop of $\rho(T)$ and the values of γ.

From the very beginning it has been observed that the pinning force in the HTS in field goes to zero far below H_{c2} [69,8]. This penomenon has been connected to the existence of an irreversibility line. An interesting property of the $F_p(B)$ curves near this line is the scaling which resembles the well-known Kramer scaling law for flux shear [70], namely $F_p \propto (1 - b_0)^2$, with $b_0 = B/B_0$ in stead of B/B_{c2}. B_0 is the field at which $F_p = 0$ and can be interpreted as the dc irreversibility field. This scaling has been noticed in Y:123 [71], but recently also in Bi:2212 [72]. At low T the data display a linear decrease of F_p with field, see Fig. 4 for comparison. So it seems that the quadratic behavior is a high temperature feature. One may speculate about whether it reveals a shear behavior or is evidence for a melting transition.

3.1.2 Grain boundaries and twin planes

Grain boundaries (GB) in conventional superconductors are predominant pinning centers. In Y:123 they rather seem to act as weak links, which is not too surprising in view of the sensitivity to disorder and the short coherence length of YBCO. In a

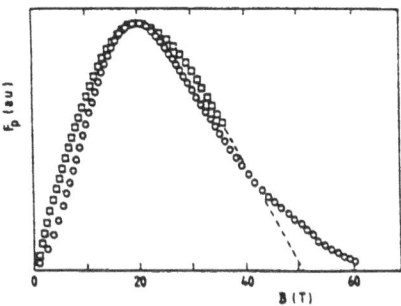

Figure 4: The magnetic-field dependence of the pinning force F_p at 20.4 K ($\square$) and 70 K ($\circ$) for $YBa_2Cu_3O_7$. The data for 70 K have been scaled in such a way that the maxima coincide with the data at 20.4 K [69]. The broader peak usually indicates linear decay [70].

series of elegant experiments on artificially grown, bicrystalline thin films the zero-field properties of GB have been systematically investigated [73]. The films were c-axis oriented with the GB also along this direction. The dependence of J_c at 4.2 K on the misalignment angle ϑ, as depicted in Fig. 5, shows a steep decrease between $5°$ and $15°$ leveling off to a value 50 times smaller than the J_c of $8 \times 10^{10} Am^{-2}$ at $\vartheta = 0$. High-resolution electron-microscope pictures of a $10°$ GB only revealed and small amount of disorder along the GB. The temperature dependence of J_c through the GB turned out to be consistent with SNS-type weak-link behavior. For large- angle GB J_c was inversely proportional to a three-half power of the boundary resistance [74].

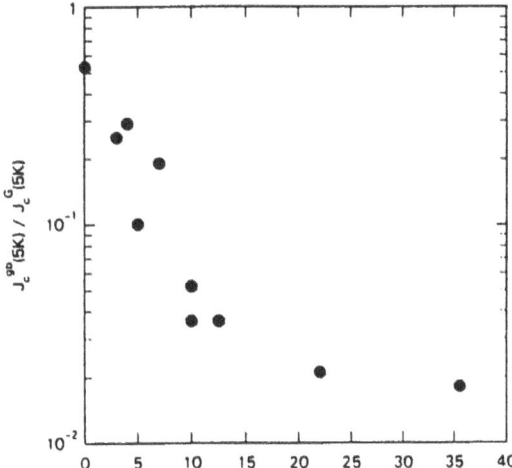

Figure 5: Plot of the ratio of the grain-boundary critical current density to the average value of the critical current density in the two grains at 4.2-5 K vs the misorientation angle in the basal plane [73].

The compositional changes at and near GB were studied by Babcock et al. on 91 % dense sintered YBCO samples [75]. They also looked at the 77 K field dependence of J_c and the pinning properties of high-angle GB in bulk bicrystals. For a 90° GB, i.e. mutual perpendicular c axis, and 22° GB with common c axis, clear evidence was found for flux-pinning controlled $J_c(b)$ which differed less than a factor of three with the single crystal J_c up to 7 T. No weak-link behavior was observed. For the 90° GB J_c at 5 T was as large as $10^7 Am^{-2}$.

A special kind of GB in Y:123 are the twin planes (TP) with (110) and ($1\bar{1}0$) orientations. Their pinning potential and force have been computed for $H\|c$ [25] using the electron scattering formalism [49], see Section 2.2.2. For a planar defect parallel to the yz plane and located at x', an exponential dependence of $\delta(l_{tr}^{-1})$ with the distance to the defect is assumed, see insert of Fig. 6.

$$\delta(\frac{1}{l_{tr}(x)}) = \frac{p}{2r_0} exp\left(-\frac{|x'-x|}{r_0}\right) \tag{7}$$

Here p denotes the scattering probability of the defect. The nonlocality range of the order parameter r_0 given by $r_0^{-1} = \xi_0^{-1} + l_{tr}^{-1}$, is a measure for the range of influence of the defect. In the limit of isolated vortices ($b < 0.2$), the expression suggested by Clem [76] is substituted for the gap function in the vicinity of the vortex core:

$$\Psi = \frac{\rho}{\sqrt{\rho^2 + \xi^2}} \; , \; \rho = \sqrt{x^2 + y^2} \tag{8}$$

For this gap function the generalized derivative can be calculated:

$$|\partial\Psi|^2 = \frac{\xi^4}{R^6} + \frac{1}{\xi^2}\frac{K_1^2(\frac{R}{\lambda})}{K_1^2(\frac{\xi}{\lambda})} \; , \; R = \sqrt{x^2 + y^2 + \xi^2} \tag{9}$$

where $K_1(x)$ is the first order modified Bessel function. The first and the second term result from the gradient and the vector-potential term of the generalized derivative, respectively. This yields for the pin potential per vortex (per unit length)

$$\Omega_{pl}(x') = g(\alpha)\mu_0 H_c^2 p\xi_0(\frac{\xi}{r_0})^2 I \tag{10}$$

$$I = \frac{r_0}{2}\int_{-\infty}^{\infty}\int_{-\infty}^{\infty} dxdy \left[\frac{\xi^4}{R^6} + \frac{1}{\xi^2}\frac{K_1^2(\frac{R}{y})}{K_1^2(\frac{\xi}{y})}\right] exp(-|x'-x|/r_0) \tag{11}$$

Eq.(11) is evaluated numerically. In this calculation, the experimental value of 50 T for $\mu_0 H_{c2}$ at 20.5 K is used [25]. This yields $\xi(0) = 2.44 \; nm$ and $\xi_0 = 2.76 \; nm$ (clean limit). For the penetration depth $\lambda(0) = 140 \; nm$ was used. Furthermore, we take: $r_0 = \xi_0, \lambda(t) = \lambda(0)(1 - t^4)^{-\frac{1}{2}}$ and $\xi(t) = 0.74\xi_0/\sqrt{h^*}(t)$ (clean limit) with t the reduced temperature $t \equiv T/T_c$ and $h^*(t) \equiv B_{c2}(ST_c)^{-1}$ [77]. S is given by $S \equiv -(\partial B_{c2}/\partial T)_{T_c}$. The maximum pinning force is determined by calculating the maximum in the derivative of Ω_{pl} with respect to x'. The results for several temperatures are plotted as black dots at $b = 0$ in Fig. 6. The contribution of the current density in $|\partial\Psi|^2$ appears to be responsible for about 70 % of the pinning force.

For the mixed state near $B_{c2}(b \to 1)$, we approximate the gap function by Eq.(3). Now the generalized derivative is replaced by the gradient term only, thus omitting the vector potential term. This has little effect on the computations for $b > 0.6$ [60], but for smaller b larger deviations may occur. This limits the validity of the calculation, hence the interpolation between the isolated vortex limit ($b = 0$, see above) and the large-b part of the curve only represents a trend. In addition, the effect of the gradient term is enlarged by a factor 1.28, caused by a perturbation in the magnetic field around the defect near H_{c2}, which in turn is due to the super currents around the vortex.

In principle, the VL can take two orientations with respect to the twin plane, namely with the closed-packed direction parallel and perpendicular to the twin plane, orientation I and II, respectively. For orientation II, we find no pin potential, in contradiction with the exact result by Thuneberg [49]. This is an artifact of our first order approximation for the vortex structure in the mixed state.

For orientation I (see Fig. 6) we obtain for the pin potential per unit length and per vortex width (note that for this orientation the x and y coordinate are interchanged with respect to Eq.(3)).

$$\Omega_{pl}(x") = 1.28(\frac{4\pi^2}{9})g(\alpha)\xi_0\mu_0 H_c^2 p(\frac{\xi}{r_0})^2(\frac{1-b}{\beta_A})\jmath(x") \qquad (12)$$

$$\jmath(x") = \int_{-\infty}^{\infty} dx_1 \int_{-\frac{1}{2}}^{\frac{1}{2}} dy_1 [\cos 2\pi\left(y_1 - \frac{x_1}{\sqrt{3}}\right) + \cos 2\pi\left(\frac{2x_1}{\sqrt{3}}\right) +$$

$$\cos 2\pi\left(y_1 + \frac{x_1}{\sqrt{3}}\right)]exp(-|x" - x_1|a_0/r_0) \qquad (13)$$

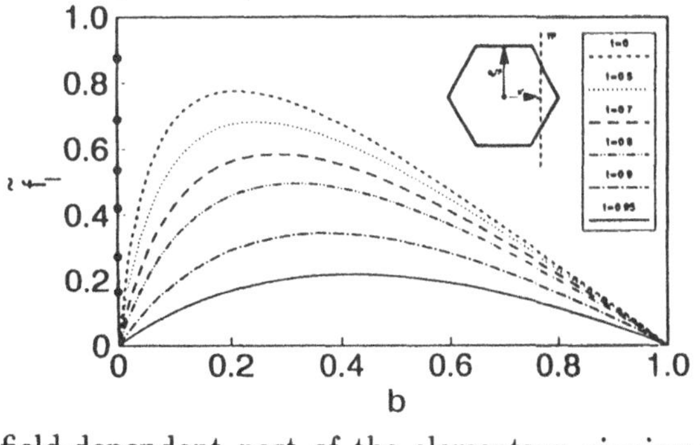

Figure 6: The field-dependent part of the elementary pinning force calculated for different reduced temperatures in the isolated vortex limit (filled circles) and in the mixed state ($b \to 1$) (lines). The curves and circles are calculated for the same reduced temperatures t, and a rough interpolation is necessary to extrapolate between them to obtain the full field dependence. The upper critical fields corresponding to the reduced temperatures are 55.2, 35.5, 23.0, 15.6, 7.74 and 3.87 T. Inset: Wigner-Seitz cell of the FLL with respect to a twin plane.

with $x_1 \equiv x/a_0$, $y_1 \equiv y/a_0$ and $x'' \equiv x'/a_0$. Via partial integration, the integral $\jmath$ can be solved analytically, resulting in

$$\jmath(x') = \frac{2a_0}{r_0} \frac{1}{(\frac{a_0}{r_0})^2 + \frac{16\pi^2}{3}} \cos\left(\frac{4\pi x'}{a_0\sqrt{3}}\right) \tag{14}$$

The maximum pinning force per unit length per flux is given by the gradient of Ω_{pl}

$$f_l(b) = 70.3 g(\alpha) \mu_0 H_c^2 p \xi_0 \left(\frac{\xi}{r_0}\right)^2 (1-b) \frac{1}{(\frac{a_0}{r_0})^2 + \frac{16\pi^2}{3}} \tag{15}$$

In Fig. 6 we plotted $\tilde{f}_l(b) = f_l(b)[g(\alpha)\mu_0 H_c^2 p \xi_0 (\xi/r_0)^2]^{-1}$, for the same values of t as in the isolated vortex limit.

Comparing our results with the results of Thuneberg, who calculates the elementary pinning force of a planar defect in the limit $b \to 1$ using a more fundamental method than our Eq.(7) to consider the scattering at the defect, it follows that our scattering probability p is about a factor 4 larger than Thuneberg's reflectivity $p^\perp$. For our choice of the order parameter and orientation of the FLL with respect to the planar defect, the diffusive scattering probability $(p^\parallel)$ plays no role.

As has been argued in Sec. 2.4, the macroscopic pinning force can be written, apart from a geometrical factor, as

$$F_p = \frac{f_l}{La_0} \tag{16}$$

where L is the average distance between TP. To calculate F_p for $YBa_2Cu_3O_{7-\delta}$, we substituted $\mu_0 H_c = 0.8(1-t^2)$ T, $L = 100$ nm, and $p = 0.4$. We thus obtain for $b > 0.2$

$$F_p = \frac{4.561 \times 10^{12}(1-t)(1+t)^2(1-b)\sqrt{bB_{c2}}}{(3 \cdot 14 \times 10^2/bB_{c2} + 52.6)} \tag{17}$$

In Fig. 7 the expected $F_p(b)$ is given for two temperatures, 20 and 75 K. For small fields the isolated vortex regime will give a small correc- tion. It yields a linear behavior, coinciding with the curve given at $b \approx 0.05$. For clarity this is not included in Fig. 7. Note here, that the twin plane mechanism does not contribute to pinning in the direction along the twin planes. Therefore, mutually perpendicular twin planes are necessary to give a net pinning force. To correct for the fact that F_p is in general not perpendicular to the twin planes, F_p was multiplied by a factor $\frac{1}{2}\sqrt{2}$. Such perpendicular twin structures are actually seen in $YBa_2Cu_3O_{7-k}$. The result depends linearly on the scattering parameter p for which 0.4 has been assumed to obtain reasonable agreement with experimental $F_p(B)$ curves. For GB this value of p is reasonable, but TP do not seem to effect the resistance, so that a much smaller value is more appropriate for TP, in agreement with more recent $F_p(B)$ data on TP-free single crystals [78] which did not show a distinct decrease of J_c. It should be mentioned that the effect of TP is only to be expected if H is within a small angle of the TP orientation [79]. This is indeed observed, both in single crystals and highly textured thin films [80,81,82] with the field along the ab planes and the driving

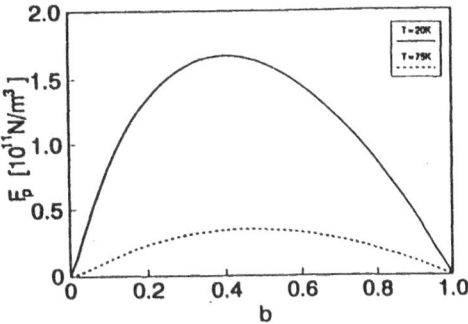

Figure 7: Twin plane contribution to the macroscopic pinning force vs. reduced field for $YBa_2Cu_3O_{7-\delta}$. Curves are calculated using Eq.(17) for T=20 K ($B_{c2} = 50$ T) and T=75 K (B_{c2}= 15 T). For other variables see text.

force parallel to the TP, see Fig. 8. For thin films this pin mechanism may give a relevant contribution to the overal J_c because of the well defined orientations. For YBCO wires it is hard to believe that TP pinning will ever play an important role, except if one considers the TP as an aggregation of other (point) defects, e.g. oxygen vacancies.

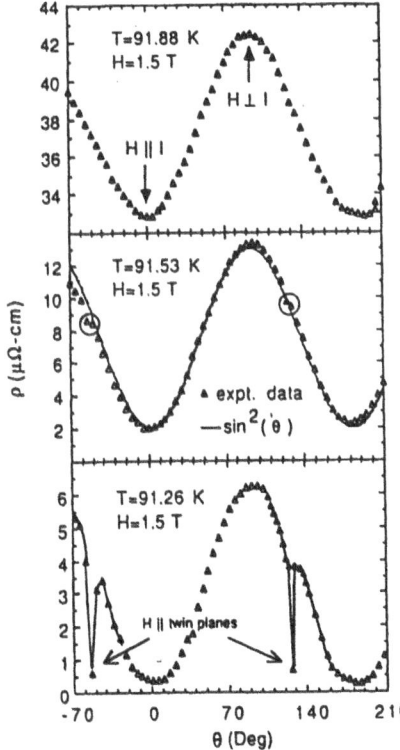

Figure 8: Angular dependence of the (flux flow) resistivity for three temperatures just below T_c in a Y:123 single crystal with one twin plane orientation. The applied field (1.5 T) is parallel to the ab planes and is rotated with respect to the current direction defining ϑ, top panel. Flux flow in the c direction should obey a $\sin^2(\vartheta)$ behavior, see middle panel. The circles indicate the onset of pinning due to twin planes when the field orientation coincides. At slightly lower T this pinning effect is very distinct, see lower panel from Kwok et al. [80]. The sudden disappearance with increasing temperature indicates a steep reduction of the shear modulus when the VL melts.

Recently, Gammel et al. [83] observed a peculiar vortex configuration, like a broken ladder, in decoration experiments in a field regime where the London approximation is valid. This configuration is caused by a saw-tooth twin plane structure with an angle $\beta \sim 5.1°$ between the twins. The pin potential derived in the London limit could satisfactorily explain these features.

3.1.3 Oxygen vacancies

Oxygen vacancies in the CuO_2 double layers pin most effectively because the pin interaction is caused by the local change of the GL coherence length and it is proportional to the maximum value of the order parameter. In the HTS the order parameter between the superconducting layers is either much smaller (Y:123) or almost zero and uniform (Bi:2212) [19]. Therefore pinning must be due to defects in the superconducting layers. The pinning force between a single vacancy of diameter D and a single vortex is computed in [52] again using Thuneberg's electron scattering formalism [49]. For $t = T/T_c < 0.6$ one gets

$$f_p \approx [10/(1+t)^4](B_c^2/\mu_0\xi)\sigma_{tr}\xi_0 \tag{18}$$

where $\sigma_{tr} = \pi D^2/4$. Taking for D the diameter of an O^{2-} ion, 0.29 nm, and $B_c = 0.3$ - 0.1 T, a maximum force at T = 0 is computed between 0.6 and $8 \times 10^{-13}N$. Note that this value does not depend on $\xi(0)(= 0.74\xi_0)$, but only on $B_c(0)$. Assuming that the distance between the defects is smaller than the range of the pinning force, i.e. $r_f \approx \xi$ in the single vortex limit, the collective effect should be estimated.

1. For the Bi compound in fields $B_{2D} < B < 0.1B_{c2}$ the 2D vortices are individually pinned and one finds for the total force F_v per 2D vortex

$$F_v \approx (0.9n_\square \pi\xi^2 f_p^2)^{\frac{1}{2}} \tag{19}$$

where $n_\square$ denotes the density of vacancies in the double layer. The pin energy follows from $U_v \approx F_v\xi$. Putting in numbers for the parameters [49] one obtains $U_v \approx 35 K$. F_p is computed by dividing F_v by the volume of the vortex segment, i.e. $a_0^2 s$ (s was the distance between the superconducting planes). This gives $J_c = F_p/B \approx 5 \times 10^{10} Am^{-2}$.

2. In case of Y:123 the 2D vortices are not decoupled and one should make an estimate for a vortex line segment of length L_c given by [25]

$$L_c \approx [\Phi_0^3 Br_f^2/(\Gamma^2\mu_0^2\lambda^4 W)]^{\frac{1}{3}} \tag{20}$$

with $W \equiv 0.5n_\square f_p^2\pi\xi^2/(a_0^2 s)$ the pinning strength, and Γ was the anisotropy parameter. Substituting $n_\square = 3.5 \times 10^{17} m^{-2}$ (this number was found for Bi:2212 [52] and means there is one vacancy per 80 oxygen atoms in each CuO_2 layer), $s = 1.2$ nm, $\xi = 1.5 nm$, $\lambda = 140nm$, $\Gamma = 29$ and $f_p = 8 \times 10^{-13}N$ one gets $W \approx 0.18B/[T]N^2 m^{-3}$ and $L_c \approx 6 nm$. This value is in good agreement with the length determined from a recent analysis of quantum creep at low temperatures [84]. Using $F_p = (W/L_c a_0^2)^{\frac{1}{2}}$ and $J_c = F_p/B$ one obtains $J_c \approx 1.2 \times 10^{11}\ Am^{-2}$ in reasonable agreement with experimental values at low temperatures. For the pin energy one finds $U_p = F_p L_c a_0^2 \xi \approx 140\ K$. This means that according to this model the creep barrier U at T = 0 is about 140 K. At higher temperatures the pinning will decrease due to thermal fluctuations yielding larger flux bundles and larger energy barriers. In addition, the ffect of the compression modulus should be taken into account which increases the size of the activated bundle with a factor $(c_{11}/c_{66})^{\frac{1}{2}}$ [10]. Therefore this result means that pinning by oxygen vacancies can account for the observed J_c's and U's in Y:123 single crystals.

The justification for taking the limit of single-vortex pinning follows from an estimate of the elastic shear energy $U_{el} \approx 3c_{66}(\xi/a_0)^2 a_0^2 L$ for a vortex line element L. This can be expressed as $U_{el} \approx 0.12(\epsilon_0 L/ln(\kappa))B/B_{c2}$. It turns out that for the HTS $\epsilon_0 = 10^3 ln(\kappa)\ K/nm$. Substituting $L = s$ and $L = L_c$ one gets $U_{el} = 180B/B_{c2}\ K$ and $U_{el} \approx 7.6 \times 10^2 B/B_{c2}\ K$, respectively. A comparison with the pin energies U_v and U_p shows that they both exceed U_{el} in the field regime we consider.

It should be noted that other non-conducting defects with $D < \xi$ act equivalently to vacancies. For instance, the effect of small defects created by electron irradiation damage has been successfully interpreted in terms of pinning by vacancies by Gianpintzakis et al. [85]. The pinning force of conducting defects, however, will be much smaller because of the proximity effect [86]. One may thus use the above considerations to estimate an optimal pinning force by taking $D \approx \xi(0)$ and $n_\square \approx 0.1\xi(0)^{-2}$. With this density of defects one may hope that T_c does not decrease due to proximity coupling but that an increase of J_c by an order of magnitude is still possible. More important is that U will increase by roughly the same amount which would reduce the creep process at 4.2 K considerably.

The (oxygen) vacancies in the CuO_2 layers are assumed to be much more effective pinning centers than the vacancies in the CuO planes. Fluctuation in the concentration of these latter defects are supposed to give rise to the so-called "fish-tail" effect in $J_c(B)$ curves. Weak superconducting particles become increasingly stronger pins when the field is increased [87]. Systematic studies of the oxygen concentration in YBCO on the critical current have been carried out by several groups [88]. Several effects play a role: T_c and therefore the condensation energy decrease, but in addition the coupling between the CuO_2 double layers decreases, i.e. the depleted material is more anisotropic and J_c drops.

3.1.4 Screw dislocations

In Section 2.3 two pin interactions due to screw dislocations in Y:123 films were proposed, namely the core interaction and the effect of thickness variations. The respective pinning forces per unit length were given by Eqs.(4) and (5). We now compare these forces with the possible effect of oxygen vacancies discussed above. Note, however, that $n_\square$ is not known from a direct measurement. Assuming that oxygen vacancies would give $J_c \approx 10^{11}\ Am^{-2}$ we estimate the line force on each vortex to be $2 \times 10^{-4}\ N/m$. This should be compared to the line force of the screw-dislocation core as determined from Eq.(4): $f_{pl} = 1.3 \times 10^{-3}\ N/m$, and the line force caused by the surface roughness, Eq.(5). Assuming $L_s = 300\ nm$, $d = 100\ nm$ and $\delta d = 5\ nm$ [55] the latter is estimated to be $f_{pl} = 6 \times 10^{-5}\ N/m$. The corresponding J_c values following from Eq.(6) are $J_c = 6 \times 10^{11}\ Am^{-2}$ and $3 \times 10^{10}\ Am^{-2}$, respectively. In addition, we should investigate the suggestion put forward by Hawley et al. [55] that grain boundaries related to the screw dislocations may cause the large J_c's in films. Using the expressions derived in Section 3.1.2 we estimate this contribution to be almost two orders of magnitude too small and therefore irrelevant.

It follows from the above analysis that the interaction of the screw dislocation cores, the effects of surface corrugation and oxygen vacancies are of roughly the same order of magnitude. At small fields the core mechanism is probably predominant since all vortices will be trapped by dislocation cores, while oxygen vacancies may take over at large fields. The combined effect provides a good explanation for the large J_c's reported in the literature. However, a definite conclusion cannot be given as long as crucial experimental evidence for the correlation between defect density and J_c is missing.

Till now only pin breaking has been considered. A quite different mechanism that might play a role here, is depinning by VL shear [70]. This means that weakly pinned regions of the VL start to flow along strongly pinned areas. When both the screw dislocations and the valleys between them act as such strong pins, this effect might occur. Upon increase of the driving force percolation paths of weak pinning regions would develop bridging the width of the film. The force to be exceeded is the flow stress. Suppose the percolation path-length is L_p, the width of this path is $(q/2)a_0$ (being q VL planes), W the width of the film, and $\tau = \alpha c_{66}$ the flow stress,

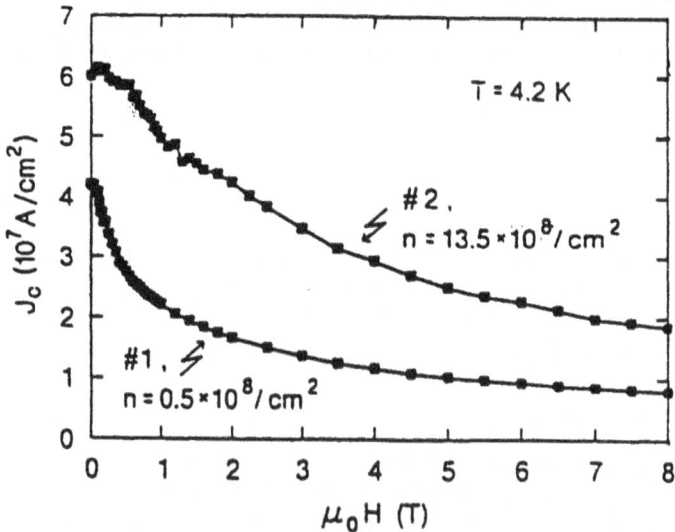

Figure 9: $J_c(H)$ at 4.2 K of two characteristic YBCO films with screw dislocation densities of $0.5 \times 10^8\ cm^{-2}$ and of $13.5 \times 10^8\ cm^{-2}$. The magnetic field was applied parallel to the c-axis, perpendicular to the transport current and to the sample surface [89].

the resulting critical shear force density $F_s = J_s B$ is according to Pruijmboom et al. [70]

$$F_s = (4\alpha L_p/qW)c_{66}/a_0 \tag{21}$$

In practice the prefactor will be of order unity yielding $J_s \approx c_{66}/(a_0 B) = (\Phi_0 B)^{\frac{1}{2}}$ / $(16\pi\mu_0\lambda^2)$. Thus, we estimate $J_s \approx 3.7 \times 10^{10}(B/[T])^{\frac{1}{2}} Am^{-1}$. Since this value is an order of magnitude smaller than the experimental value at low T and B being $J_c = 5 - 6 \times 10^{11} Am^{-2}$ [55], and because the $B^{\frac{1}{2}}$ dependence is not observed, the shear mechanism seems to be ruled out.

In Fig. 9 recent experimental results from Mannhart et al. [89] are shown. The $J_c(B)$ data of the sample with a screw- dislocation density $n = 13.5 \times 10^8 cm^{-2}$, corresponding to $L_s \approx 300 nm$, show a behavior that can be well understood in terms of the above mechanisms. Suppose that up to a field $B_{occ} = \Phi_0 n$ all vortices can be accomodated by the screw dislocations yielding $J_c(B < B_{occ}) = 6 \times 10^7 Acm^{-2}$. Above this field, the access vortices will be mainly pinned by a background mechanism supposedly provided by the oxygen vacancies leading to a high-field level of about $10^7 Acm^{-2}$. The observed gradual decrease of J_c will be related to the growing number of access vortices in combination with a constant voltage criterion for J_c. The field dependence of the flux-flow resistivity should give additional information by which this model can be tested.

The dislocation density would give $B_{occ} = 28 mT$, while from Fig. 9 one would rather estimate a value of 0.5 T. The density of screw dislocations is apparently not large enough and an equally strong pin mechanism is required to explain the

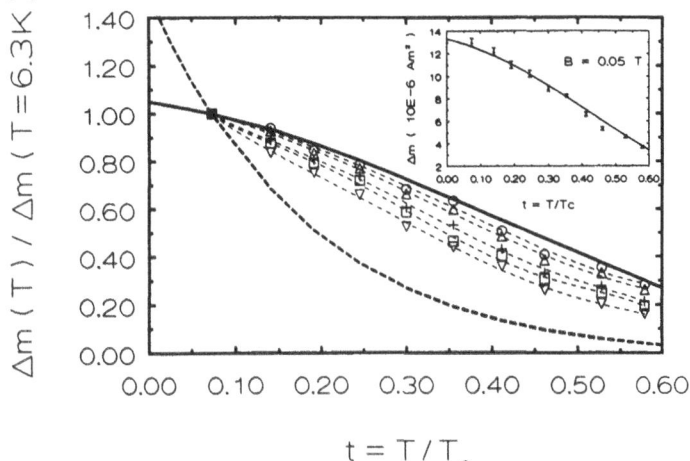

$$t = T / T_c$$

Figure 10: Low-field temperature dependence of normalized values of Δm compared to predictions for flux pinning by screw dislocations via $\delta\kappa$ interaction (solid line) and oxygen vacancies (broken line), for various magnetic fields: o, 0.05 T; Δ, 0.075 T; +, 0.15 T; $\square$, 0.2 T; ∇, 0.3 T. The light dashed lines are guides to the eye only. Inset: error estimate of $\Delta m(t)$ at 0.05 T together with the prediction for screw dislocation pinning via $\delta\kappa$ interaction (solid line) [91] .

420

discrepancy. It turns out that the STM picture also reveals the presence of a much larger density of edge dislocations parallel to the c-axis in agreement with other reports [90]. If these edge dislocations extend all the way down through the YBCO films the pinning force will be very similar to that of the dislocation cores.

In order to test the idea of two basic pin mechanisms, namely strong pinning from dislocation cores and a weaker background from oxygen vacancies or other point defects, one can investigate the temperature dependence of J_c at small and large fields, respectively. For small fields the dislocation cores are expected to dominate the critical current, whereas at large fields the background mechanism should take over. Such studies have been carried out by Douwes et al. [91]. The first analysis does not yet account for the effect of flux creep. More recent results justify this simplification. The temperature dependence for dislocation core-pinning is given by $J_c \propto (1+t)^2(1-t)^{\frac{5}{2}}$; that for vacancy pinning by $J_c \propto (1-t)^{\frac{5}{2}}/[(1+t)^3(1+t^2)^{\frac{1}{3}}]$.

Experimental results on Y:123 films are plotted on a reduced scale in Fig. 10 (small fields) and Fig. 11 (large fields), together with the expected temperature dependences for dislocation (solid line) and vacancy pinning (dashed line). It is seen in Fig. 10 that the 0.05 T data agree well with the solid line. For increasing field the data start to deviate from the small field behavior and gradually approach the background temperature dependence for fields above 6 T, as follows from Fig. 11. These results therefore seem to support the above assumptions.

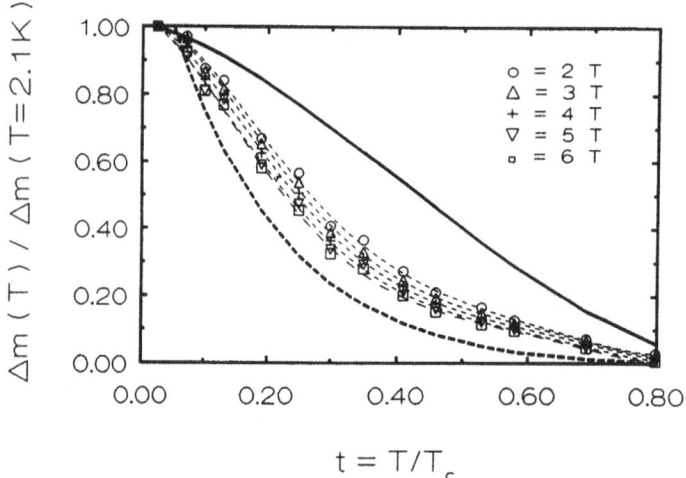

Figure 11: High-field temperature dependence of normalized values of Δm compared to predictions for flux pinning by screw dislocations (solid line) and oxygen vacancies (broken line), for various magnetic fields: o, 2 T; $\triangle$, 3 T; +, 4 T; ∇, 5 T; $\square$, 6 T. The light dashed lines are guides to the eye only [91].

3.1.5 Maximum critical currents

From the considerations in Section 3.1.4 one can easily derive the optimal value of J_c. Assuming that every vortex is pinned over its full length one gets $J_c < B_c^2 \xi_0 D/(\mu_0 \Phi_0 \xi) \approx 6 \times 10^{11}\ Am^{-2}$ at low temperature and $\approx 2 \times 10^{10}\ Am^{-2}$ at 77 K. The corresponding pin energy per unit length is $U_{pl} \leq B_c^2 D\xi_0/\mu_0 \approx 1.3 \times 10^{11}\ K/m$. For a 100 nm thick film this would yield a pin energy of $1.3 \times 10^4\ K$ at low T and $1.2 \times 10^3\ K$ at 77 K. In the latter case U/kT is small enough to give rise to thermally activated dissipation leading to a somewhat smaller J_c than just computed. Depending on the density of line pins the above values do not depend on B. As soon as all line pins are occupied, however, J_c begins to decrease to the value dictated by the shear mechanism or by a background pinning mechanism.

It is clear that for Bi:2212 the quasi 2D nature above B2D leads to much smaller pin energies, i.e. with d = 1.5 nm one obtains $U_p \leq 200\ K$. This would already give considerable creep effects at 10 K in view of the general observation that flux creep is seen for $U_p/kT \leq 25$.

3.2 Artificial pinning centers, ion tracks

Of crucial importance for applications is to know whether the depinning line can be shifted by increasing the defect density. Probably the answer is yes, but not much, except, of course, in cases where pinning was weak to begin with. The overall impression that emerges is that there exists a boarder line above which pinning disappears and that the position of this line depends mainly on the anisotropy.

To investigate this question one has to change the pin density in a controlled fashion, e.g. by substitutions or melt-growth processing [92,93], neutron [94,95] or ion irradiation [96,97,98], or radiation damage by fission products [99]. Ideally, the

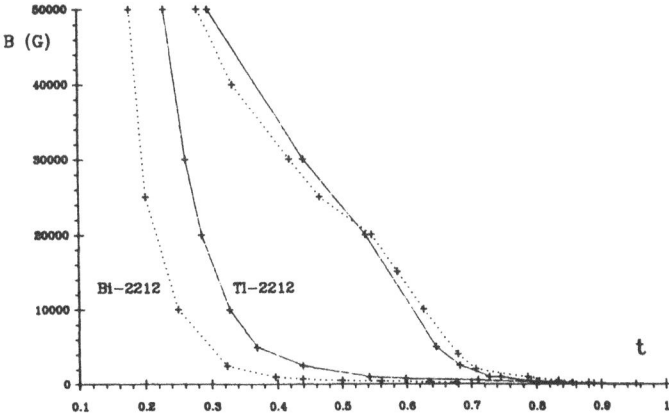

Figure 12: Irreversibility lines plotted versus reduced temperatures T/T_c for Bi-2212 (dotted line) and Tl-2212 (solid line), before and after 2×10^{11} Pb cm^{-2}. The lines are only guides for the eye (adopted from Hardy et al. [102]).

422

superconducting parameters do not change, otherwise one should scale the results; usually, in a first approximation, to reduced temperatures, but a true comparison becomes less reliable when the damage is too large. Because of the large effect of anisotropy and grain boundaries, these studies should be carried out on single crystals and highly textured or epitaxial films.

Much recent work has concentrated on the effect of radiation damage caused by high-energy ($\sim$ GeV) heavy-ion irradiation. This kind of irradiation causes cylindrical damage tracks (columnar defects) with typical diameter 5-10 nm consisting of amorphous material [100]. The length depends on the energy and can be as large as $100\mu m$ for 5 GeV Pb ions. Discontinuous tracks occur when the irradiation is carried out with 3.5 GeV xenon ions [101]. The continuous tracks are ideal pinning centers, because the amorphous core is non-conducting so that the pinning energy can be easily estimated. Gerhauser et al. [100] obtained for the interaction with an isolated vortex per unit length

$$U_{pl} = \frac{\mu_0}{2} H_c^2 \pi \xi^2 + \frac{\Phi_0^2}{4\pi\mu_0\lambda^2} ln\left(\frac{R}{\xi}\right) \tag{22}$$

where 2R is the diameter of the track. The resulting pinning force per unit length can be estimated from

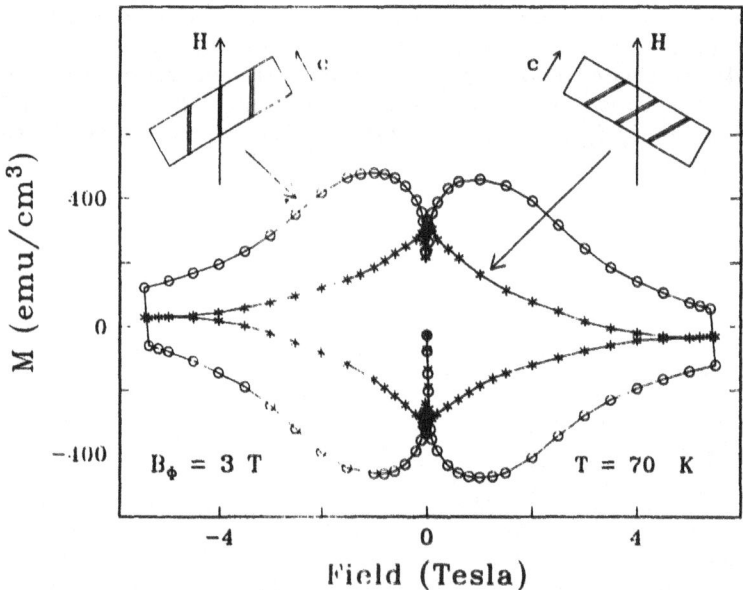

Figure 13: Hysteresis loops taken at 30 K for an YBCO crystal irradiated at 30° off the c axis. The hysteresis loops are shown with the applied field aligned ±30° with respect to the c axis. Inset: The relationship between the radiation and field directions (from Civale et al. [100]).

$$f_{pl} \approx U_{pl}/\lambda \qquad a_0 > \lambda \qquad\qquad (23)$$

$$f_{pl} \approx U_{pl}/a_0 \qquad a_0 < \lambda \qquad\qquad (24)$$

To get the pinning energy and pinning forces on a pancake vortex in Bi:2212 one has to multiply with $s = 1.5$ nm. In case of Y:123 films with the thickness of the film. Comparing with the estimates for screw dislocations (Section 3.1.4) we get $f_{pl} = 4 \times 10^{-4}$ N/m for $a_0 > \lambda$ and $f_{pl} \simeq 1 \times 10^{-3}$ N/m at 1 T which is almost equal to the effect of a screw dislocation, 1.3×10^{-3} N/m.

Obviously, the critical current is increased by the columnar defects when the field is oriented parallel to them. It drops when the flux density becomes larger than the "matching field" B_ϕ. This field is determined by $B_\phi = \phi_0 n_{cd}$ with n_{cd} the areal density of the defects which is linear proportional to the fluence. In concurrence with the increase of J_c, the irreversibility line is strongly shifted upwards, as can be seen in Fig. 12 for the Bi and Tl based compounds [102].

Civale et al. [100] observed that when in Y:123 the field is not parallel to the defects J_c is much less enhanced, see Fig. 13. However, for Bi:2212 there is no dependence on the field orientation at low temperatures, whereas above 60 K there is [103]. This different behavior is related to the different anisotropy of both HTS.

4   Concluding remarks

In the preceding sections an overview has been given of the state of the art regarding our knowledge of flux pinning and critical currents in the HTS. The comparison of model predictions and experimental results has been limited to experiments on single crystals and (nearly) epitaxial films. Because of space limitations thermal effects are only taken into account by means of the explicit temperature dependences of the superconducting parameters, i.e. $B_c(T)$, $\xi(T)$ and $\lambda(T)$. At high temperatures, however, the effect of thermal fluctuations on the mixed-state properties becomes predominant and leads to flux creep and VL melting. These phenomena will unfortunately reduce the effectiveness of the pinning. In case of large anisotropy this reduction is so large that so far applications seem to be restricted to temperatures below 20 K. On the other hand it turns out that the critical currents in the HTS at low temperatures are very large. This implicates that optimization by purposely adding pinning centers can still lead to improvements. In addition, the search for new HTS with possibly smaller anisotropy should be continued.

Acknowledgements

Stimulating conversations with many colleagues are gratefully acknowledged. In particular, I would like to thank Valerii Vinokur and Kees van der Beek for many helpful, clarifying and pleasant discussions.

References

[1] J.G. Bednorz and K.A. Müller, Z.Phys. **B64**.189 (1986).

[2] D.R. Nelson, Phys.Rev.Lett. **60**, 1973 (1988).

[3] E.H. Brandt, Phys.Rev.Lett. **63**, 1106 (1989).

[4] A. Houghton, R.A. Pelcovits, and A. Sudbø, Phys.Rev. **B40**, 6763 (1989).

[5] M.P.A. Fisher, Phys.Rev.Lett. **62**, 1415 (1989).

[6] D.S. Fisher, M.P.A. Fisher, and D. Huse, Phys.Rev. **B43**, 130 (1991).

[7] Y. Yeshurun and A.P. Malozemoff, Phys.Rev.Lett. **60**, 2202 (1988).

[8] A.P. Malozemoff, T.K. Worthington, Y. Yeshurun, F. Holtzberg, and P.H. Kes, Phys.Rev. **B38**, 7203 (1988).

[9] D. Dew-Hughes, Cryogenics **28**, 627 (1988).

[10] M.V. Feigelman, V.M. Vinokur, Phys.Rev. **B41**, 8986 (1990).

[11] M.V.Feigelman, V.B. Geshkenbein, A.I. Larkin, and V.M. Vinokur, Phys.Rev.Lett. **63**, 2303 (1989).

[12] V.M. Vinokur, M.V. Feigelman, V.B. Geshkenbein, and A.I. Larkin, Phys.Rev.Lett. **65**, 259 (1990).

[13] A.P. Malozemoff and M.P.A. Fisher, Phys.Rev. **B42**, 6784 (1990).

[14] W.E. Lawrence and S. Doniach, in Proceedings of the Twelfth International Conference on Low Temperature Physics, Kyoto, 1970, ed. E. Kanda (Kigaku, Tokyo, 1971) p.361.

[15] D.E. Farrell, R.G. Beck, M.F. Booth, C.J. Allen, E.D. Bukowski, and D.M. Ginsberg, Phys.Rev. **B42**, 6758 (1990).

[16] J.C. Martinez, S.H. Brongersma, A. Koshelev, B. Ivlev, P.H. Kes, R.P. Griessen, D.G. de Groot, Z. Tarnavski, and A.A. Menovsky, Phys.Rev.Lett. **69**, (1992).

[17] D.E. Farrell, C.M. Williams, S.A. Wolf, N.P. Bansal, and V.G. Kogan, Phys.Rev.Lett. **61**, 2805 (1988).

[18] D.E. Farrell, J.P. Rice, D.M. Ginsberg, and J. Liu, Phys.Rev.Lett. **64**, 1573 (1990).

[19] L.N. Bulaevski, Zh.Eksp., Teor.Fiz **64**, 2241 (1973) [Sov.Phys.JETP **37**, 1133 (1973)].

[20] A. Sudbø and E.H. Brandt, Phys.Rev. **43**, 10482 (1991).

[21] R. Labusch, Crystal Lattice Defects **1**, 1 (1969).

[22] A.M. Campbell and J.E. Evetts, Adv.Phys. **21**, 199 (1972).

[23] R. Schmucker and E.H. Brandt, Phys.Stat.Sol.(b) **79**, 479 (1977).

[24] H. Träuble and U. Essmann, J.App.Phys. **39**, 4052 (1968).

[25] P.H. Kes, J. van den Berg in: Studies of High Temperature Superconductors, Vol. 5, A. Narlikar, ed. (NOVA Science Publishers, New York, 1990) p.83.

[26] V.G. Kogan and L.J. Campbell, Phys.Rev.Lett. **62**, 1552 (1989); L.J. Campbell, M.N. Doria, and V.G. Kogan, Phys.Rev. **B38**, 2439 (1988).

[27] A. Sudbø and E.H. Brandt, Phys.Rev.Lett. **66**, 1781 (1991).

[28] V.G. Kogan, Phys.Rev. **B38**, 7049 (1988); Z. Hao and J.R. Clem, Phys.Rev. **B43**, 7622 (1991).

[29] A.I. Buzdin and A.Yu. Simonov, JETP Lett. **51**, 191 (1990); V.G. Kogan, N. Nakagawa, and S.L. Thiemann, Phys.Rev. **B42**, 2631 (1990); A.I. Buzdin, S.S. Krotov, and D.A. Kuptrov, Physica **C175**, 42 (1991).

[30] C.A. Bolle, P.L. Gammel, D.G. Grier, C.A. Murray, D.J. Bishop, D.B. Mitzi, A. Kapitulnik, Phys.Rev.Lett. **66**, 112 (1991); P.L. Gammel, D.J. Bishop, J.P. Rice, and D.M. Ginsberg, Phys.Rev.Lett. **68**, 3343 (1992).

[31] J.R. Clem, Phys.Rev. **B43**, 7837 (1991).

[32] V.M. Vinokur, P.H. Kes, and A.E. Koshelev, Physica **C168**, 29 (1990).

[33] D. Feinberg, C. Villard, Phys.Rev.Lett. **65**, 919 (1990).

[34] M. Tachiki and S. Takahashi, Sol.State Comm. **70**, 291 (1989).

[35] B.I. Ivlev and N.B. Kopnin, Phys.Rev. Lett. **64**, 1828 (1990); J.Low Temp.Phys. **80**, 161 (1990); G. Blatter, B.I. Ivlev, and J. Rhyner, Phys.Rev.Lett. **66**, 2395 (1991).

[36] D.I. Glazmann and A.E. Koshelev, Physica **C173**, 180 (1991); Sov.Phys. JETP **70**, 774 (1990); Phys.Rev. **B43**, 2835 (1991).

[37] S. Chakravarty, B.I. Ivlev, and Ya.N. Ovchinnikov, Phys.Rev.Lett. **64**, 3187 (1990); Phys.Rev. **B42**, 2143 (1990).

[38] S.N. Artemenko, I.G. Gorlova, and Yu.I. Latyshev, Phys.Lett. **A138**, 428 (1989); N.V. Zavaritskii et al., Pis'ma Zh.Eksp.Teor.Fiz **53**, 212 (1991).

[39] P.H. Kes, J. Aarts, V.M. Vinokur, and C.J. van der Beek, Phys.Rev.Lett. **64**, 1063 (1990).

[40] B.I. Ivlev and N.B. Kopnin, Europh.Lett. **15**, 349 (1991).

426

[41] J.R. Clem and M.W. Coffey, Phys.Rev. **B42**, 6209 (1990).

[42] G. Blatter, V.B. Geshkenbein and A.I. Larkin, Phys.Rev.Lett. **68**, 875 (1992).

[43] M.C. Marchetti and D.R. Nelson, Phys.Rev. **B41**, 1910 (1990); D.R. Nelson and P. Le Doussal, Phys.Rev. **B42**, 10, 113 (1990).

[44] U. Welp, W.K. Kwok, G.W. Crabtree, K.G. Vandervoort, and J.Z. Liu, Phys.Rev.Lett **62**. 1908 (1989); U. Welp et al., Phys.Rev.Lett. **67**, 3180 (1991).

[45] Z. Hao, J.R. Clem, M.W. McElfresh, L.Civale, A.P. Malozemoff, and F. Holtzberg, Phys.Rev. **B43**, 2844 (1991).

[46] S. Sriahar, D.-H. Wu, and W. Kennedy, Phys.Rev.Lett. **63**, 1873 (1989); Y.J. Uemura, L.P. Le, G.M. Luke, B.J. Sternlieb, W.D. Wu, J.H. Brewer, T.M. Riseman, C.L. Seaman, M.B. Maple, M. Ishikawa, D.G. Hinks, J.D. Jorgensen, G. Saito, and H. Yamochi, Phys.Rev.Lett. **66**, 2665 (1991); M. Weber et al., Supercond.Sci.Technol. **4**, 403 (1991).

[47] M. Weber et al., preprint.

[48] P.G. de Gennes, Superconductivity of Metals and Alloys, Benjamin, New York, 1966, pp. 76-80.

[49] E.V. Thuneberg, Cryogenics **29**, 236 (1989); J.Low Temp.Phys. **57**, 415 (1984).

[50] E.J. Kramer, Phil.Mag. **33**, 331 (1976); C.S. Pande, Appl.Phys.Lett. **28**, 462 (1976).

[51] , T. Matsushita et al., Jpn.J.Appl.Phys. **26**, L1524 (1987); M.M. Fang et al., Phys.Rev. **B37**, 2334 (1988).

[52] C.J. van der Beek and P.H. Kes, Phys.Rev. **B43**, 13032 (1991).

[53] M. Murakami, M. Morita, K. Doi, K. Miyamoto, Jpn.J.Appl.Phys. **28**, 1189 (1989).

[54] S. Jin, T.H. Tiefel, S. Kakahera, J.E. Graebner, H.M. O'Bryan, R.A. Fastnacht, and G.W. Kammlott, Appl.Phys.Lett. **56** 1287 (1990).

[55] M. Hawley, J.S. Raistrick, J.G. Beery, and R.J. Houlton, Science **251**, 1587 (1991); Ch. Gerber, D. Anselmetti, J.G. Bednorz, J. Mannhart and D.G. Schlom, Nature **350**, 279 (1991).

[56] A.I. Larkin, Yu.N. Ovchinnikov, J.Low Temp.Phys. **34**, 409 (1979).

[57] E.H. Brandt and U. Essmann, Phys.Stat.Sol.(b) **144**, 13 (1987).

[58] P.H. Kes and C.C. Tsuei, Phys.Rev.Lett. **47**, 1930 (1981); Phys.Rev. **B28**, 5126 (1983); R. Wördenweber, P.H. Kes, and C.C. Tsuei, Phys.Rev. **B33**, 3172 (1986); R. Wördenweber and P.H. Kes, Phys.Rev. **B34**, 494 (1986); Cryogenics **29**, 329 (1889).

[59] G.L. Dolan, G.V. Chandrashekhar, T.R. Dinger, C. Feild, and F. Holtzberg, Phys.Rev.Lett. **62**, 827 (1987); L.Ya. Vinnikov, L.A. Gurevich, I.V. Grigoryeva, A.E. Koshelev, and Yu.A. Osipyan, J.Less Comm.Met. **164& 165**, 1271 (1990).

[60] P.H. Kes, A. Pruijmboom, J. van den Berg, and J.A. Mydosh, Cryogenics **29**, 228 (1989).

[61] H. Krauth, K. Heine, and J. Tenbrink, in Proceedings ICMC'90 Topical Conf. on Materials Aspects of HTS, eds. H.C. Freyhardt, R. Flükiger and M. Peuckert, (DGM Informationsgesellschaft, Oberursel, 1991) **29**.

[62] J.W. Ekin, K. Salama, V. Selvamanickam, Nature **350**, 26 (1991).

[63] T. Matsushita, M. Iwakuma, Y. Sudo, B. Ni, T. Kisu, M. Funabi, M. Takeo, and K. Yamafuji, Jpn.J.Appl.Phys. **26**, L1524 (1987).

[64] S. Jin and J.E. Graebner, Materials Sc.Eng. **B**, preprint.

[65] K. Sato, T. Hikata, and Y. Iwasa, Appl.Phys.Lett. **57**, 1928 (1990).

[66] K. Watanabe, N. Kobayashi, H. Yamone, T. Hirai, and Y. Muto, ICMC'90, 965 (1991).

[67] T.T.M. Palstra, B. batlogg, R.B. van Dover, L.F. Schneemeyer, and J.V. Waszczak, Phys.Rev. **B41**, 6621 (1990); Phys.Rev. **B43**, 3756 (1991).

[68] W.R. White, A. Kapitulnik, and M.R, Beasley, Phys.Rev.Lett. **66**, 2826 (1991).

[69] P.H. Kes, Physica **C153-155**, 1121 (1988); J. van den Berg, C.J. van der Beek, P.H. Kes, J.A. Mydosh, A.A. Menovsky, and M.J.V. Menken, Physica **C153-155**, 1465 (1988); K. Kadowaki et al., Physica **B155**, 136 (1989).

[70] E.J. Kramer, J.Appl.Phys. **44**, 1360 (1976); D. Dew- Hughes, Phil.Mag. **B55**, 459 (1987); A. Pruijmboom, P.H. Kes, E. van der Drift, and S. Radelaar, Phys.Rev.Lett. **60**, 1430 (1988); Appl.Phys.Lett. **52**, 662 (1988); Cryogenics **29**, 232 (1989); C.J.G. Plummer and J.E. Evetts, IEEE Trans.Magn.Mag. **23**, 1179 (1987); W.H. Warnes and D.C. Larbalestier, IEEE Trans.Magn.Mag. **23**, 1183 (1987).

[71] J.N. Li, F.R. de Boer, L.W.Roeland, M.J.V. Menken, K. Kadowaki, A.A. Menovsky, and J.J.M. Franse, Physica **C169**, 81 (1990); V.M. Pan et al., Cryogenics **29**, 392 (1989).

[72] R. Rose, S.B. Ota, P.A.J. de Groot, and B. Jayaram, Physica **C170**, 51 (1990).

[73] D. Dimos, P. Chaudhari, J. Mannhart, and F.K. LeGoues, Phys.Rev.Lett. **61**, 219 (1988); J. Mannhart, P. Chaudhari, D. Dimos, C.C. Tsuei, and T.R. McGuire, Phys.Rev.Lett. **61**, 2476 (1988).

[74] R. Gross, P.Chaudhari, M. Kawasaki, and A. Gupta, Phys.Rev. **B42**, 10735 (1990).

[75] S.E. Babcock and D.C. Larbalestier, Appl.Phys.Lett. **55**, 393 (1989); S.E. Babcock, X.Y. Cai, D.L. Kaiser, and D.C. Larbalestier, Nature **347**, 167 (1990).

[76] J.R. Clem, J.Low Temp.Phys. **18**, 427 (1975).

[77] M.R. Werthamer, in "Superconductivity", ed. R.D. Parks, Marcel Dekker, New York, 1969, p.321.

[78] U. Welp, W.K. Kwok, G.W. Crabtree, and K.G. Vandervoort, Appl.Phys.Lett. **57**, 84 (1990); B.M. Lairson, S.K. Streiffer, and J.C. Bravman, Phys.Rev. **B42**, 10067 (1990).

[79] G. Blatter, J. Rhyner, and V.M. Vinokur, Phys.Rev. **B43**, 7826 (1991).

[80] W.K. Kwok, U. Welp, G.W. Crabtree, K.G. Vandervoort, R. Hulshcer, J.Z. Lui, Phys.Rev.Lett. **64**, 966 (1990).

[81] E.M. Gyorgy, R.B. van Dover, L.F. Schneemeyer, A.E. White, M.M. O'Bryan, R.J. Felder, J.V. Waszczak, and W.W. Rhodes, Appl.Phys.Lett. **56**, 283 (1989); J.Z. Liu, Y.X. Jia, R.N. Shelton, and M.J. Fluss, Phys.Rev.Lett. **66**, 1354 (1991); B. Janossy, R. Hergt, and L. Fruchter, Physica **C170**, 22 (1990).

[82] Y. Iye, S. Nakamura, T. Tamegai, T. Terashima, K. Yamamoto, and Y. Bando, Physica **C166**, 62 (1990); S. Fleshler, W.K. Kwok, U. Welp, V.M. Vinokur, M.K. Smith, J. Downey and G.W. Crabtree, Phys. Rev. **B**, (1993).

[83] P.L. Gammel, C.A. Duran, D.J. Bishop, V.G. Kogan, M. Ledvij, A.Yu Simonov, J.P. Rice and D.M. Ginsberg, Phys.Rev.Lett. **69**, 3808 (1992).

[84] R.P. Griessen, J.G. Lensink and H.G. Schnack, Physica **C185-189**, 337 (1991).

[85] J. Gianpintzakis et al., Phys.Rev. **B45**, 10667 (1992).

[86] E.J. Kramer and H.C. Freyhardt, J.Appl.Phys. **51**, 4930 (1980); T. Matsushita, J.Appl.Phys. **54**, 281 (1983); A.A. Golub et al., Fiz.Nizk.Temp. **10**, 258 (1984) [Sov.J.Low Temp.Phys. **10**, 133 (1984)].

[87] M. Däumling, J. Seuntjes, D.C. Larbalestier, Nature **346**, 332 (1990).

[88] M. Däumling, Physica **C184**, 13 (1991); R. Feenstra, D.K. Christen, C.E. Klabunde, and J.D. Budai, Phys.Rev. **B45**, 7555 (1992); J.G. Ossandon et al., Phys.Rev. **B46**, 3050 (1992); J.L. Vargas and D.C. Larbalestier, Appl.Phys.Lett. **60**, 174 (1992).

[89] J. Mannhart, D. Anselmetti, J.G. Bednorz, Ch. Gerber, K.A. Müller, and D.G. Schlom, Proceedings of the International Workshop on Critical Currents, Cambridge, July 1991, to be published in Superc.Sci.Technol.

[90] V. Pan et al., Proceedings of the International Workshop on Critical Currents, Cambridge, July 1991, to be published in Superc.Sci.Technol.; S.K. Streiffer, B.M. Lairson, C.B. Eom, B.M. Clemens, J.C. Bravman, and T.H. Geballe, Phys.Rev. **B43**, 13007 (1991).

[91] H. Douwes, P.H. Kes, C. Gerber and J. Mannhart, Cryogenics, May 1993.

[92] R. Wördenweber, K. Heinemann, G.V.S. Sastry, and H.C. Freyhardt, Su-perc.Sci.Technol. **2**, 207 (1989); S.H. Whang, Z.X. Li, D.X. Pang, M. Suenaga, D.O. Welch, S. Jin, and T.H. Tiefel, Physica **C168**, 185 (1990); Y. Xu and M. Suenaga, Phys.Rev. **B43**, 5516 (1991).

[93] D. Shi, M.S. Boley, U. Welp, J.G. Chen, and Y. Liao, Phys.Rev. **B40**, 5255 (1989); S. Jin. T.H. Tiefel, S. Nakahara, J.E. Graebner, H.M. O'Bryan, R.A. Fastnacht, and G.W. Kammlott, Appl.Phys.Lett. **56**, 1287 (1990); Z.J. Huang, Y.Y. Xue, J. Kulik, Y.Y. Sun and P.H. Hor, Physica **C174**, 253 (1991); M. Murakami, S. Gotoh, H. Fujimoto, K. Yamaguchi, N. Koshizuka, and S. Tanaka, Supercond.Sci.Technol. **4**, S43 (1991).

[94] H.W. Weber and G.W. Crabtree, to be published in Studies of High Temp. Superconductors, A.V. Narlikar ed. NOVA Science Publishers, New York, Vol. 9 (1991).

[95] R.B. van Dover, E.M. Gyorgy, L.F. Schneemeyer, J.W. Mitchell, K.V. Rao, R. Puzniak, and J.W. Waszczak, Nature **342**, 55 (1989); W. Schindler, B. Roas, G. Saemann-Ischenko, L. Schultz, and H. Gerstenberg, Physica **C169** 117 (1990); F.M. Sauerzopf, H.P. Wiesinger, W. Kritscha, H.W. Weber, G.W. Crabtree, and J.Z. Liu, Phys.Rev. **B43**, 3091 (1991).

[96] E.L. Venturini, J.C. Barbour, D.S. Ginley, R.J. Baughman, and B. Morosin, Appl.Phys.Lett. **56**, 2456 (1990); R.B. van Dover, E.M. Gyorgy, A.E. White, L.F. Schneemeyer, R.J. Felder, and J.V. Waszczak, Appl.Phys.Lett. **56**, 2681 (1990); L. Civale, A.D. Marwick, M.W. McElfresh, T.K. Worthington, A.P. Malozemoff, F.H. Holtzberg, J.R. Thompson, and M.A. Kirk, Phys.Rev.Lett. **65**, 1164 (1990).

[97] B. Roas, B. Hensel, S. Henke, S. Klaumnzer, B. Kabius, W. Watanabe, G. Saemann-Ischenko, L. Schultz, and K. Urban, Europhys.Lett. **11**, 669 (1990); F. Rullier-Albenque, A. Legris, S. Bouffard, E. Paumier, and P. Lejay, Physica **C175**, 111 (1991).

[98] H.-W. Neumüller, G. Ries, W. Schmidt, W. Gerhäuser, and S. Klaumünzer, Supercond.Sci.Technol. **4**, S370 (1990); H.-W.Neumüller, G. Ries, W. Schmidt, W. Gerhäuser, and S. Klaumünzer, Journ. of Les Comm.Met. **164& 165**, 1351 (1990); B. Roas, B. Hensel, G. Saemann-Ischenko, and L. Schultz, Appl.Phys.Lett. **5**, 1051 (1989).

[99] R.L. Fleischer, H.R. Hart, Jr., K.W. Lay, and F.E. Luborsky, Phys.Rev. **B40**, 2163 (1989); F.E. Luborsky, R.H. Arendt, R.L. Fleischer, H.R. Hart, Jr., K.W. Lay, J.E. Tkaczyk, and D. Orsini, J.Mater.Res. **6**, 28 (1991).

[100] L. Civale et al., Phys.Rev.Lett. **67**, 648 (1991); W. Gerhäuser et al., Phys.Rev.Lett. **68**, 879 (1992); V. Hardy, D. Groult, J. Provost and B. Raveau, Physica **C190**, 289 (1992).

[101] D. Bourgault et al. Phys.Rev. **B39**, 6549 (1989).

[102] V. Hardy et al., Physica **C191**, 85 (1992).

[103] L. Klein, E.R. Yacoley, Y. Yeshurun, M. Konczykowski, F. Holtzberg and K. Kishio, Physica C, to be published (1993).

Additional Note

P.H. Kes

Kamerlingh Onnes Laboratory, Leiden University

P.O. Box 9506, 2300 RA Leiden, The Netherlands

30 May 1993

Several participants raised the point that the occurrence of oxygen vacancies in the CuO_2 planes in a concentration of about 1 % seems unlikely. They suggested that vacancies in the apex oxygen layer would be better candidates for strong pinning centers. Let me again elucidate my point of view.

My arguments are based on the situation in Bi:2212. The estimate of about one percent vacancies follows when for $B_c(0)$ a value of 0.3 T is substituted. If the value for Y:123 would have been taken a much lower concentration of vacancies (about 2×10^{16} m^{-2}) is sufficient to account for a J_c of 4×10^{10} A/m^2 at $T = 0$. Such a low concentration is outside the range of any detection technique presently available as followed from the discussions at the session. Moreover, such low concentration may have been frozen in during the slow cool-down of disorder after the crystal-growth process. A small amount of disorder would lower the free energy. For a vacancy in the apex oxygen (SrO) layer to be an effective pinning center the magnitude of the order parameter in this layer should not be too much reduced. Unfortunately, it is not known how fast the order parameter decays outside the superconducting layers. If we speculate on a 10 times lower value, $n_\square$ should be at least 10^2 times larger, corresponding to a vacancy concentration in the SrO layer of some ten percents. The disorder would then be considerable, even on the scale of the Ginzburg-Landau coherence length $\xi(0)$. But it is well known that in materials with small length-scale disorder, i.e. amorphous alloys, the pinning is very weak because the effect of the disorder is averaged out over the scale of $\xi(0)$. Therefore, I maintain that some vacancy in the CuO_2 layers is the best candidate to explain the strong pinning observed in Bi:2212 single crystals. However, I do not insist on oxygen vacancies. From the pinning point of view the only requirement is that the defect should strongly scatter the quasi-particles in the superconducting layer. Any defect, e.g. substitutional elements, that has this property would be a good candidate as effective pinning center.

Additional Notes

P.H. Kes
Kamerlingh Onnes Laboratory, Leiden University
P.O. Box 9506, 2300 RA Leiden, The Netherlands

26 Aug 1992

FLUX PINNING OF HIGH TEMPERATURE SUPERCONDUCTORS AND THEIR APPLICATIONS

M. MURAKAMI
ISTEC, Superconductivity Research Laboratory
1-10-13, Shinonome, Koto-ku, Tokyo 135 Japan

ABSTRACT. In this paper, I first summarize how critical current density is determined in oxide superconductors, with an emphasis placed on flux pinning. Second, the defects which are considered to be effective in flux pinning in melt processed Y-Ba-Cu-O superconductors are reviewed, along with the methods to introduce effective pinning centers. Finally I introduce some applications of bulk oxide superconductors which exhibit a large flux pinning force.

1. Introduction

High temperature superconductors are highly attractive, since they can superconduct at liquid nitrogen temperature. A number of attempts have been made to utilize these superconductors for practical applications. However, it has been found that they have several problems to be solved before they can be used for any practical application. Those are followings. (1) Critical current density (J_c) is small. (2) Anisotropy is large. (3) Fracture toughness is very small. (4) Second phase or stacking faults are readily formed. (5) Superconducting properties are very sensitive to oxygen content.

Among these, low J_c values were once believed to be intrinsic to oxide superconductors[1], and therefore seemed to be the most serious problem for practical applications, which brought about very pessimistic prospects on the future of high temperature superconductors[2]. However, in contrast to the critical temperature, J_c is not intrinsic to a superconductor and is strongly dependent on its microstructure. Recent progress has shown that high J_c values are achievable even in high temperature superconductors, if proper microstructural control is performed[3-5]. Although, it is still difficult to fabricate long superconducting wires which exhibit large J_c values at 77K, Bi-Sr-Ca-Cu-O superconducting wires longer than 100m have already fabricated with reasonably high J_c below 30K, and many researchers started to try various applications of bulk Y-Ba-Cu-O with high J_c at 77K

In this review, I first summarize the factors which affect J_c values in oxide superconductors. Second, the defects which are believed to be effective in flux pinning and the methods to increase J_c values in melt processed Y-Ba-Cu-O superconductors are reviewed. Finally, applications of strong pinning bulk oxide superconductors are introduced.

E. Kaldis (ed.), Materials and Crystallographic Aspects of HTc-Superconductivity, 433–452.
© 1994 *Kluwer Academic Publishers.*

2. Factors to influence J_c values

2.1. Carrier density

It has been pointed out that low carrier density of oxide superconductors may inevitably lead to small J_c values. Carrier density (N) of oxide superconductors is two orders of magnitude lower than that of metals, being of order $10^{21} cm^{-3}$. However, it is now generally accepted that this density is large enough to ensure J_c values of 10^4-$10^6 A/cm^2$ which are required for practical applications.

2.2. Coherence length

It is known that the coherence length (ξ_0) of oxide superconductors is extremely short. According to the BCS theory[6], the coherence length is described by the following relation:

$$\xi_0 = 0.18 \; \hbar v_F/kT_c$$

where $2\pi\hbar$ is the Planck constant, v_F the Fermi velocity, k the Boltzman constant, and T_c the critical temperature. Therefore, high T_c leads to small ξ. Low carrier density (N) is also the source of small ξ_0, since v_F is proportional to $N^{1/3}$. Since ξ determines the size of Cooper pair, the defect of order ξ_0 can destroy superconductivity. It is believed that low critical current density of sintered oxide superconductors is attributed to small defects present at high angle grain boundaries, since the defect structure must be introduced as the misorientation angle between two grains increases.

There is another coherence length, which is known as Ginzburg-Landau coherence length (ξ_{GL}). This determines the size of fluxoid or quantized flux and is related to the upper critical field (H_{c2}) by the following equation.

$$\phi_0/2\pi\xi_{GL}^2 = \mu_0 H_{c2}$$

where ϕ_0 is the flux quantum and μ_0 is the permeability of free space. Although it is confusing, ξ_0 and ξ_{GL} are closely correlated by the following relation:

$$\xi_{GL} = 0.74 \; \xi_0 \; \{1-(T/T_c)\}^{-1/2}.$$

ξ_{GL} determines the distance over which the amplitude of electron-pair wave or the order parameter can change significantly, and thus it determines the size of fluxoid as mentioned above. It should also be noted that ξ_{GL} is temperature dependent, while ξ_0 is temperature independent. Considering the fact that H_{c2} with $H//c$ axis for Y-Ba-Cu-O is 10 - 20T at 77K, it is not concluded that ξ_{GL} is too small.

2.3. Depairing current[7]

The superconducting state is more stable than the normal conducting state at temperatures below T_c. Temperature dependence of the free energy of these two states is schematically illustrated in Fig. 1. The condensation energy, which is the free energy difference (f_s - f_n) between the superconducting and the normal states, is given by $(1/2)N(E_F)\Delta^2$, where $N(E_F)$ is the density of states at the Fermi energy and Δ is the energy gap.

More practically, the condensation energy is given by using the thermodynamical critical field (H_c). A superconductor repels magnetic field completely, which is known as the

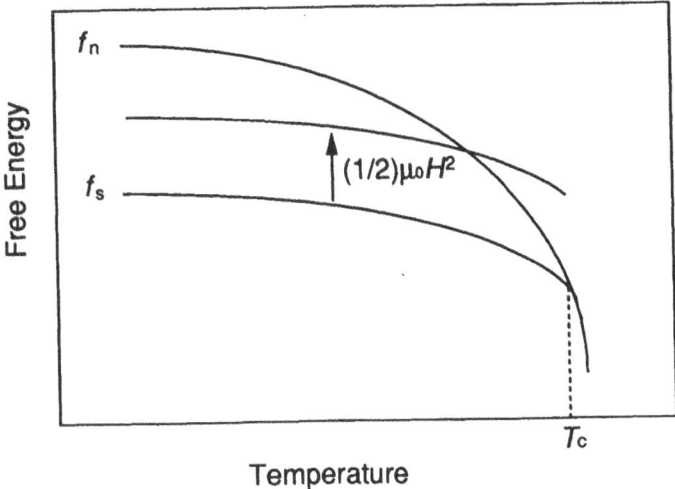

Fig. 1. Temperature dependence of the free energy for the superconducting state (f_s) and the normal conducting state (f_n). At temperature below T_c, f_s is smaller than f_n, and the difference gives the condensation energy. When magnetic field (H) is applied, magnetic energy of $(1/2)\mu_0 H^2$ is added to f_s.

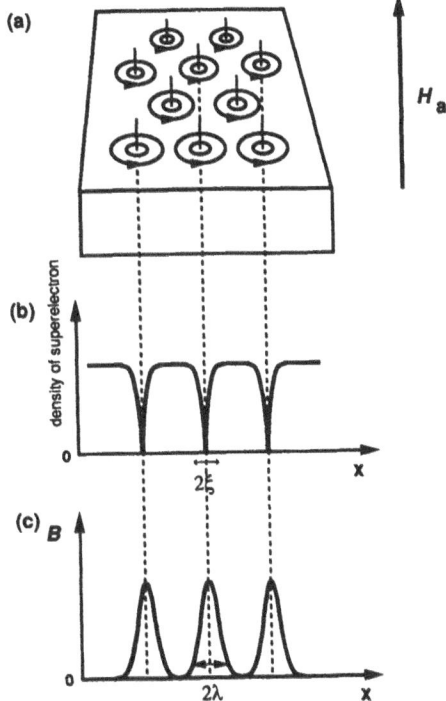

Fig. 2. Schematic illustration of the mixed state of type II superconductors.

Meissner effect. However, since the excess energy is needed to expel magnetic field, at a certain field (H_C), this energy becomes equal to the condensation energy, where the superconducting state is destroyed. Then the condensation energy per unit volume is given by

$$f_s - f_n = \int_0^{H_c} M dH = (1/2)\mu_0 H_c^2 = B_c^2/2\mu_0.$$

When the currents are passed through the superconductor, the kinetic energy $(1/2)n_s m v^2$ is added to f_s where n_s is the density of the superelectrons. When this kinetic energy becomes equal to the condensation energy, the superconducting state is no more stable. This gives us the ideal critical current or the depairing current.

$$(1/2)n_s m v^2 = (1/2)\mu_0 \lambda^2 J_c^2 = (1/2)\mu_0 H_c^2.$$

Then the depairing current is given by

$$J_c = H_c/\lambda$$

where λ is the penetration depth. This ideal J_c is observed only in very thin films or small sized wires and is much higher than practical J_c values. However, if the depairing current is smaller than the J_c values required for practical application, the superconducting materials are not suitable. At 77K, Y-Ba-Cu-O has H_c of 1.6×10^5 A/m ($B_c = 0.2$T) and λ of 10^{-7}m, which gives the depairing current density of 1.6×10^{12} A/m^2 ($= 1.6 \times 10^8$ A/cm^2), which is higher than the required J_c values, and thus indicating that the depairing current is large enough in this system.

2.4. Flux pinning[8]

In conventional superconductors, J_c values are mainly determined by flux pinning force. In a type II superconductor, where λ/ξ is larger than 0.71, the external field can penetrate into the superconductor in the form of quantized flux as schematically illustrated in Fig. 2. When electric currents are passed through a type II superconductor in the mixed state, Lorentz force acts on the fluxoid, by which the fluxoid will move. The voltage appears due to this flux motion along the direction of the current, leading to the appearance of resistance. In order to achieve zero resistance state, the flux motion must be prevented.

When a superconductor contains nonsuperconducting regions (inclusions, precipitates, and crystal imperfections) as schematically shown in Fig. 3, the fluxoid can lower its energy by interacting with such a region. As already mentioned, the superconducting state has lower free energy than the normal state at temperatures below T_c. Since the region where the fluxoid penetrates is driven normal, the free energy is increased by that corresponding to the condensation energy. (Though the total energy can be lowered by allowing field penetration). However, when the fluxoid threads nonsuperconducting region, there is no increment in the free energy. Conversely, when the fluxoid is moved from this region, the superconductivity must be destroyed per interaction volume, and this will lead to the increase in the free energy. This indicates that the fluxoid exhibits attractive interaction with nonsuperconducting region, which is known as flux pinning. Thus when flux pinning force is sufficiently large, the fluxoids remain stable and the zero resistance state is realized. J_c is then equivalent to the current density (J) at which the Lorentz force ($F_L = JB$) becomes equal to flux pinning force (F_p), and therefore is given by $J_c = F_p/B$.

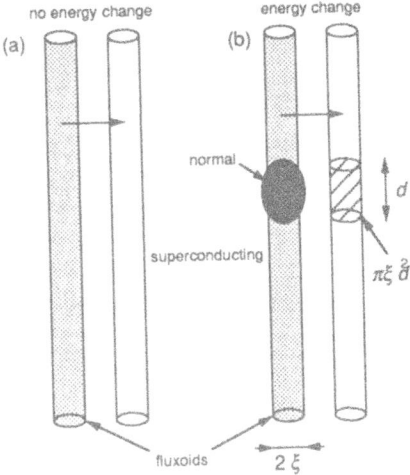

Fig. 3. Schematic illustration showing how the fluxoid is pinned by nonsuperconducting particle. There is no energy change so long as the fluxoid moves in the superconducting region. When the fluxoid sits on nonsuperconducting particle, the fluxoid can lower its energy per interaction volume which is $(1/2)\mu_0 H_c^2 \times \pi\xi^2 d$. In order to move the fluxoid, this amount of energy must be used, and therefore the fluxoid is pinned at nonsuperconducting region.

Fig. 4. Optical micrograph of microstructure of Y-Ba-Cu-O solidified from molten region. Note that large Y_2BaCuO_5 particles are trapped in $YBa_2Cu_3O_7$ matrix, which is evidence of the peritectic reaction.

Here F_p is dependent on kind, size and density of pinning centers. Suppose that the size of nonsuperconducting particle is d, the pinning energy U_p is then given by

$$U_p = \pi \xi_{GL}^2 d \, (B_c^2/2\mu_0)$$

where $\pi \xi_{GL}^2 d$ is the interaction volume and $B_c^2/2\mu_0$ is the condensation energy per unit volume. Since the force is the derivative of energy as a function of distance, the pinning force f_p is given by $f_p = dU_p/dx$. To first approximation, f_p is obtained by dividing U_p by the distance over which the order parameter changes.

$$f_p = U_p/\xi_{GL} = \pi \xi_{GL} d \, (B_c^2/2\mu_0)$$

This f_p is called the elementary pinning force. If ξ_{GL} is extremely small, f_p becomes small. But as mentioned before, ξ_{GL} is temperature dependent and ξ_{GL} at 77K is not small even compared to that of conventional superconductors (at 4.2K).

It should also be noted that the bulk pinning force (F_p) is obtained by adding up the contributions from various pinning centers.

$$F_p = \sum_i f_{pi}$$

This means that $J_c(=F_p/B)$ can be increased by introducing pinning centers even when f_p is small. Therefore, the problem is not intrinsic but an increase in J_c depends on microstructural control.

Although only nonsuperconducting regions were considered in this section, any inhomogeneities may act as pinning centers. But the pinning energy should be modified to $U_p=(\pi \xi_{GL}^2 d)\delta(B_c^2/2\mu_0)$ where $\delta(B_c^2/2\mu_0)$ is the difference in the condensation energy between the matrix and the pinning center. It is also important to note that nonsuperconducting regions are most effective in pinning.

2.5. Thermal energy

Oxide superconductors are attractive, since they can be used at high temperatures. However, the effects of thermal energy on critical current are significant.

2.5.1. *Flux creep*[9]. At nonzero temperatures, pinned fluxoids are not stable and can move from the pinning well with the help of thermal activation. This phenomenon is known as flux creep. Therefore, in a very strict sense, there is no perfect zero resistance state for type II superconductors in the mixed state. However, for most practical applications, flux creep is not a serious problem if the voltage due to flux creep is negligible.

It should also be noted that the critical current determined in an experiment depends on the voltage which can be measured. If this is E_c, then

$$J_c = J_{c0} \{1 - (kT/U)\ln(Bd\Omega/E_c)\}$$

where J_{c0} is the J_c without thermal activation, d the distance between pinning centers, Ω the attempt frequency of fluxoids. This shows that J_c is reduced as the temperature is raised.

Another serious effect of flux creep is that J_c decays logarithmically with time:

$$J_c(t) = J_{c0} \{1 - (kT/U)\ln t\}.$$

Hence at high temperatures, J_c is strongly affected by flux creep. However, we should notice that in both cases, J_c is governed by not only kT but also U, and if it is possible to enhance U, the detrimental effect of flux creep on J_c can be reduced, indicating that microstructural control which enables us to introduce effective pinning centers is again very important.

2.5.2. *Thermal fluctuation*[10]. The superconducting state is more stable than the normal state at temperatures below T_c. However, in a temperature region near T_c, where the condensation energy $(f_s - f_n)$ is small, the superconducting state may fluctuate due to thermal energy. The conditions for fluctuation to be observed is

$$kT > (4/3)\pi\xi_{GL}^3 \, (f_s - f_n).$$

So far, we have no quantitative data on this matter, however this fluctuation may affect J_c values at high temperatures.

2.6. Conclusions

Although, oxide superconductors are highly attractive due to high T_c, this attractive feature causes many problems for practical applications. The intrinsic problems are small ξ and large thermal agitation. Because of these, it has once been believed that J_c values of oxide superconductors are intrinsically small. However, it is possible to increase J_c values through a combination of grain alignment and the introduction of effective pinning centers. In the next chapter, I will review what kind of defects are considered to be pinning centers and how we can increase J_c values in melt processed Y-Ba-Cu-O.

3. Critical currents of melt processed Y-Ba-Cu-O superconductors

The problems related to low J_c values in oxide superconductors are weak links at high angle grain boundaries and insufficient flux pinning. These two problems have been solved by employing melt processes.

3.1. Melt Process[3-5]

When Y-Ba-Cu-O materials are solidified in temperature gradient, highly textured structure or the structure with all grains aligned unidirectionally can be obtained as shown in Fig. 4. The weak link problem was not observed in such samples, and J_c values exceeded 10^4A/cm^2 at 77K in zero field[3]. This result indicates that the weak links can be eliminated from bulk high temperature superconductors. However, J_c values dropped rather sharply in magnetic fields, suggesting that flux pinning is not sufficient.

3.2. Flux pinning enhancement

As presented in the former section, the introduction of pinning centers is necessary to improve J_c in magnetic fields. There are several candidates for pinning centers in melt processed Y-Ba-Cu-O. Those are twin planes, stacking faults, dislocations, cracks, oxygen deficient regions and nonsuperconducting inclusions.

3.2.1. *Twin planes*[11,12]. In the Y-Ba-Cu-O system, twins are formed along {110} planes to accommodate the strain associated with tetragonal to orthorhombic transformation. It is probable that the superconducting properties are altered at twin planes and thus they can act

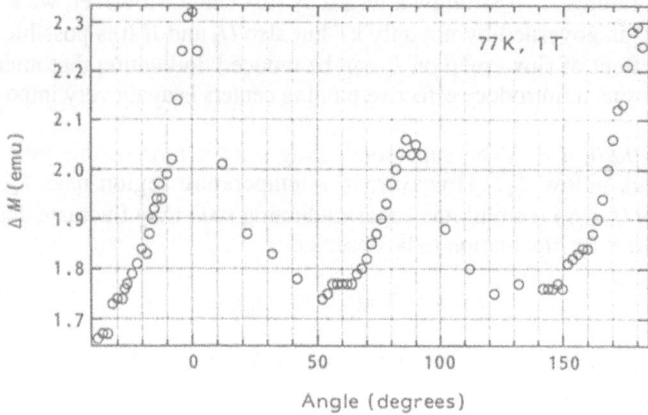

Fig. 5. Angular dependence of ΔM at 77K and 1T. The peak is observed when the field is parallel to twin planes.

Fig. 6. Transmission electron micrograph of melt processed Y-Ba-Cu-O observed from a direction perpendicular to the c axis. Note that a number of stacking faults are present, which have a structure as extra CuO plane inserted to the chain site.

as pinning centers. In the case of planar defects like twins, flux pinning occurs only when fluxoids are parallel to the defects. In fact, small but clear flux pinning enhancement was observed when the field is applied parallel to the c axis, as shown in Fig. 5. However, considering the fact that the number of twin planes cannot be significantly increased, since it is determined by the phase transformation strain, flux pinning enhancement by twin planes is limited. It should also be noted that the pinning energy of the twin planes will not be very large, since large supercurrents can flow across the planes.

3.2.2. Stacking faults.

Yamaguchi et al.[13] have found that melt processed Y-Ba-Cu-O contains a number of stacking faults as shown in Fig. 6. The stacking faults have a structure with an extra CuO plane inserted into the chain site of Y123 structure, therefore it is identical to that of Y124 structure. Since the stacking faults are planar defects, pinning occurs only when fluxoids are aligned parallel to the planes, namely the field is applied perpendicular to the c axis. Although qualitative analyses are difficult, I believe that the stacking faults can serve as pinning centers in this field direction. Very large J_c values exceeding 10^4 A/cm^2 at 77K and 20T may be partly attributed to the presence of stacking faults along with very large H_{c2}[14]. But their contribution to flux pinning for $H//c$ axis will be very small.

3.2.3. Dislocations.

It has been reported that melt processed Y-Ba-Cu-O contains a number of dislocations[15], which are thought to be generated during grain growth. However, recent observations have shown that the dislocation density is very small in melt grown Y-Ba-Cu-O[16]. Even when the dislocations are generated, most of them will be annealed out, since the samples are solidified at very slow cooling rate at high temperatures. It is also important to note that dislocation like defects are likely to be introduced during the process to prepare thin samples for TEM observation.

It is also known that flux pinning enhancement by dislocations is not significant in conventional superconductors[8]. Considering the fact that melt processed Y-Ba-Cu-O with low dislocation density exhibits high J_c values, dislocations do not seem to be important pinning centers. However, if it is possible to introduce a large number of dislocations, J_c can be improved by dislocations.

3.2.4 Cracks.

Cracks are also considered to be planar defects and certainly nonsuperconducting. Therefore they can serve as pinning centers when a current and the field is parallel to the cracks, although the current cannot flow across the cracks. In fact, Matsushita[17] has shown that a large J_c values are obtainable by the presence of cracks in thin films. However, in bulk melt grown Y-Ba-Cu-O, the amount of cracking is small. Considering the fact that the samples without cracks exhibit high J_c values, flux pinning enhancement by the cracks will not be important in bulk Y-Ba-Cu-O.

3.2.5. Oxygen deficient region.

Since ξ of oxide superconductors is very short, very small point defects such as oxygen deficiency have long been considered to be important pinning centers in Y-Ba-Cu-O superconductors. Dauemling et al.[18] have observed anomalous peaks in M-H loops in oxygen deficient Y-Ba-Cu-O single crystals and have attributed such peaks to flux pinning by oxygen deficient regions. Kupfer et al.[19] have also proposed that flux pinning enhancement due to neutron irradiation in bulk Y-Ba-Cu-O is caused by the production of oxygen defects.

However, in the case of small defects, the pinning energy becomes very small. Therefore, their number must be very large for significant pinning enhancement to be observed. It should also be noted that in the above two cases, weak links are inadvertently introduced, which results in low transport J_c values. Since it is very difficult to control

442

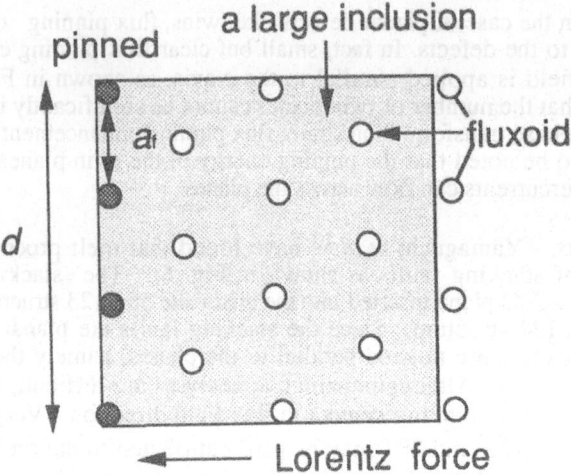

Fig. 7. Schematic illustration of the interaction of flux lines with a large nonsuperconducting inclusion. Note that the number of pinned fluxoids is d/a_f.

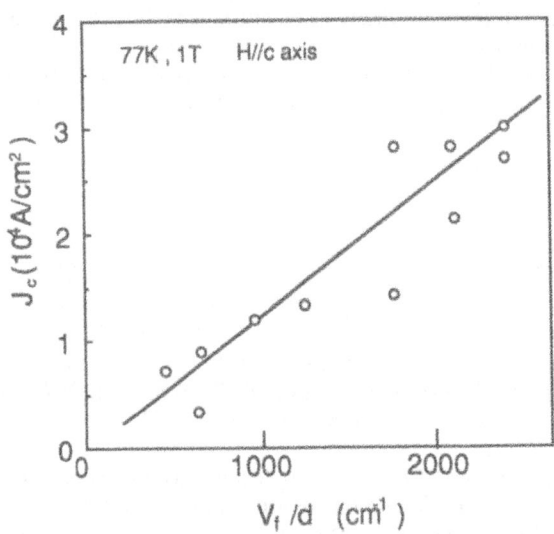

Fig. 8. J_c(77k,1T,$H//c$ axis) versus V_f/d (effective surface area of the 211/123 interface) for various melt processed Y-Ba-Cu-O superconductors.

oxygen deficiency, oxygen deficient regions are not practically good pinning centers, though they may serve as pinning centers. The fact that the sample, which is well-annealed in oxygen, can exhibit high J_c values also suggests that oxygen deficient regions are not necessarily important pinning centers in bulk Y-Ba-Cu-O.

3.2.6. *Nonsuperconducting inclusions.* In the Y-Ba-Cu-O system, when solidified from molten region, the superconducting $YBa_2Cu_3O_x$ phase is produced by the following peritectic reaction

$$Y_2BaCuO_5 + L \ (3BaCuO_2 + 2CuO) \ \rightarrow \ 2YBa_2Cu_3O_x$$

As a result of incomplete peritectic reaction, Y_2BaCuO_5 (211) phase is occasionally trapped in Y123 matrix, as already shown in Fig. 4. It is known that nonsuperconducing particles work as pinning centers in conventional superconductors. Therefore, they are also expected to be pinning centers in Y-Ba-Cu-O. Definitely their effectiveness strongly depends on their size and density.

Suppose that the size and density of 211 inclusions are d and N_p, respectively. Here, d should be much larger than ξ_{GL} since d cannot be reduced to the order of ξ_{GL} by metallurgical control. The pinning energy per interaction is given by

$$U_p = \pi\xi_{GL}{}^2 d \ (B_c{}^2/2\mu_0).$$

Therefore

$$f_p = U_p/\xi_{GL} = \pi\xi_{GL} d \ (B_c{}^2/2\mu_0).$$

In the case of large inclusions, only the interface is effective in flux pinning, since there is no energy difference for flux motion inside the inclusions so that the number of fluxoids which can be pinned by one inclusion is d/a_f as schematically illustrated in Fig. 7, where a_f is the flux line lattice (FLL) spacing and given by $a_f = 1.075(\phi_0/B)^{1/2}$. It is also known that FLL of oxide superconductors is very soft due to large κ ($=\lambda/\xi$), and therefore a direct summation seems to be reasonably good approximation to obtain bulk pinning force. Then

$$F_p = N_p f_p$$

From the relation $J_c = F_p/B$, we obtain

$$J_c = \pi B_c{}^2 \xi_{GL} N_p d^2/4\mu_0\phi_0{}^{1/2}B^{1/2}.$$

This relation indicates that J_c is proportional to $N_p d^2$ which is equal to V_f/d where V_f ($=N_p d^3$) is the volume fraction of the 211. Hence, J_c can be increased either by increasing V_f or by decreasing d of the 211.

An increase of V_f of 211 inclusions dispersed in the Y123 matrix can easily be performed by changing the starting composition toward the 211 rich region. However, when the volume exceeds 30%, agglomeration of the 211 is observed, leading to large d. It should also be noted that effective cross sectional area for supercurrent flow is reduced as the volume of 211 is increased.

The reduction of d is not easy to achieve, but several methods have been found to be successful. The key is the microstructural control at the point just above the peritectic temperature, where fine and uniform distribution of the 211 is desirable.

Starting with fine 211 powders and holding at low temperatures in the 211 + L region is effective in reducing the size of the 211. An addition of Pt[20], Ce[21], BaSnO3[22] has also been found to be effective in reducing the size of 211. Fig. 8 shows the relationship between J_c

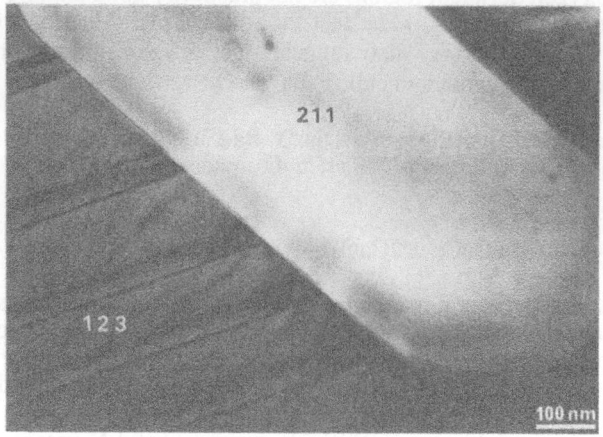

Fig. 9. Transmission electron micrograph of the 211/123 interface. Note that the interface is very sharp and without distortion.

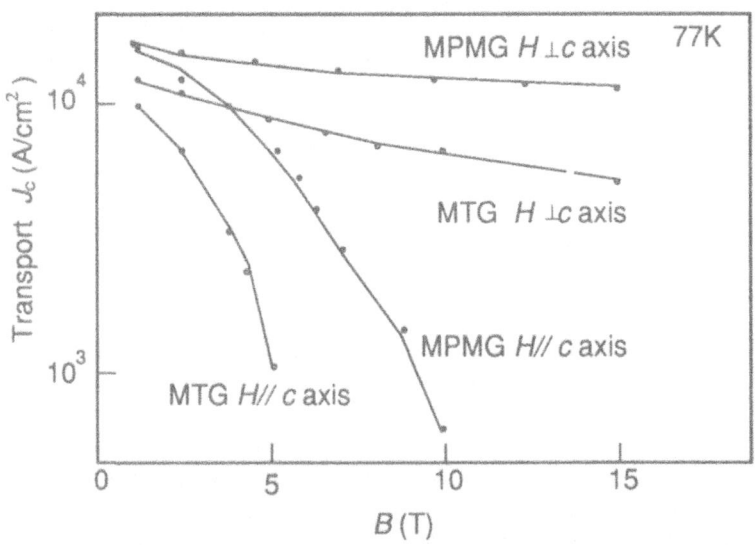

Fig. 10. Magnetic field dependence of J_c for two different melt processed Y-Ba-Cu-O samples. The size of 211 inclusions are finer in an MPMG processed sample than in a MTG processed sample.

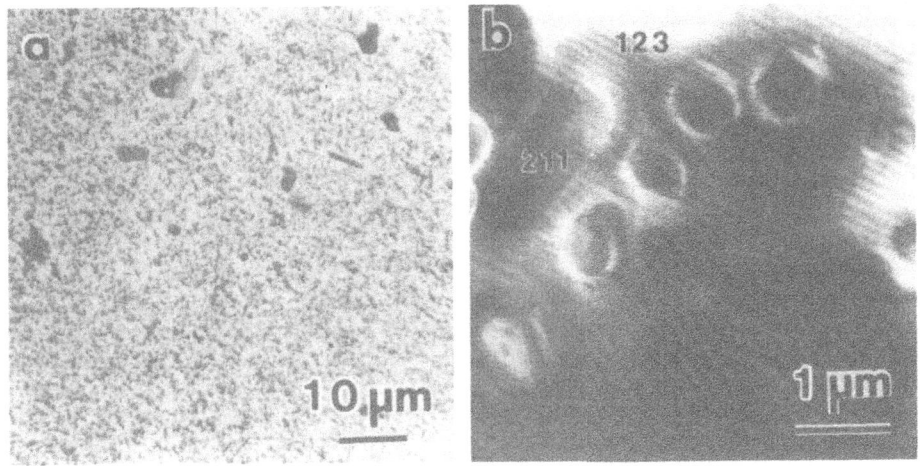

Fig. 11. Typical microstructure of MPMG processed Y-Ba-Cu-O sample: (a) Optical micrograph and (b) transmission electron micrograph.

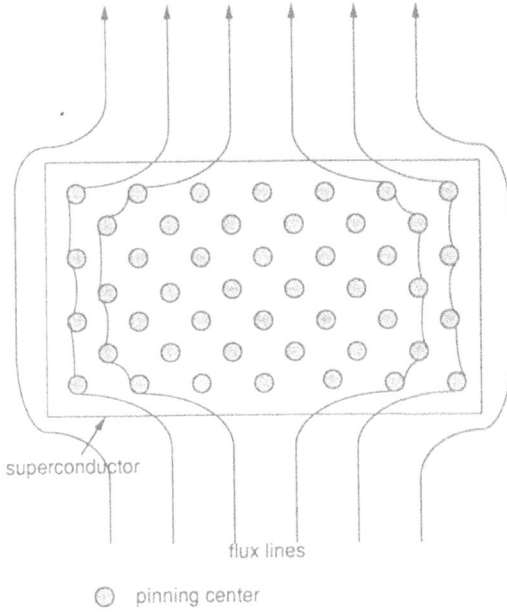

superconductor

flux lines

○ pinning center

Fig. 12. Schematic illustration showing how magnetic field is repelled by a pinned superconductor.

and V_f/d of the 211 for various melt processed Y-Ba-Cu-O. It is notable that J_c is proportional to V_f/d, which is in good agreement with the above estimates, supporting the fact that the 211/123 interface provides flux pinning. Although, the defect structures around 211 inclusions are reported to be responsible for flux pinning enhancement, the 211/123 interface is very sharp and without distortion as shown in Fig. 9. Here the sharpness of the interface is also important for pinning enhancement. As mentioned before, the pinning force is given by $f_p = dU_p/dx$, which indicates that f_p becomes small if the distance over which the pinning energy changes (dx) is large even when U_p is very large. The clean interface guarantees that the energy change is abrupt at the interface, and thus a large f_p is obtained.

3.3. Conclusions

In a type II superconductor, any inhomogeneity may act as pinning centers. Candidate defects for flux pinning centers in melt grown Y-Ba-Cu-O are twins, dislocations, stacking faults, oxygen deficient regions, cracks and 211 inclusions. Among these, the 211 seems to be most suitable pinning centers, since the size, volume and distribution can be controlled. Fig. 10 shows J_c-B properties (77K) for two melt processed Y-Ba-Cu-O samples. At present, the largest J_c values achieved are 30000 A/cm^2 at 77K and 1T for $H//c$ axis and 10000 A/cm^2 at 77K and 20T for $H//ab$ plane. Fig. 11 shows typical microstructure of the melt processed (MPMG processed) Y-Ba-Cu-O with high J_c values. It should be noted that the size of 211 is much finer and its distribution is more uniform than that (MTG processed Y-Ba-Cu-O) of Fig. 4.

4. Applications

Most applications of superconductors are based on zero resistance. The superconductors must be fabricated into the form of long wires or tapes for such applications. This is very difficult for brittle oxide superconductors. Powder metallurgical methods have been applied, and have made a great progress in Bi compounds, although J_c values are still lower than required level at 77K.

It is desirable if we can find some applications in which oxide superconductors can be used in the bulk form. In fact, many applications have been proposed using the response of bulk superconductors to magnetic field. The superconductors can repel magnetic field completely, which is known as the Meissner effect. A magnet can be levitated above the superconductor and vice versa, using the effect. However, the field penetrates into the superconductor when the applied field exceeds lower critical field (H_{c1}). Since H_{c1} is very small, the levitation force is very small, which will limit the application. On the other hand, strong pinning type II superconductors can repel magnetic field strongly. Fig. 12 shows schematic illustration showing how magnetic field is repelled by pinned type II superconductors. Fig. 13 presents results of direct observations of such effects in melt processed Y-Ba-Cu-O with fine 211 inclusions, where magnetic field penetration is observed using magneto-optical effect. At low fields, the superconductor can repel magnetic field and a concentration of the external field at the sample edge is observed. At fields above H_{c1}, the field enters the superconductor, but they cannot penetrate into the interior region due to flux pinning. It should also be noted that this observation clearly supports the fact that magnetic behavior of a good oxide superconductor can be explained by the critical state model.

Recently we have succeeded in levitating a person using melt processed Y-Ba-Cu-O superconductors containing fine Y_2BaCuO_5 inclusions as shown in Fig. 14. This demonstration has clearly shown that the levitation can be used for a variety of applications.

It is also important to note that strong pinning superconductors can attract large magnetic field due to flux pinning, which bring about a large attractive force. An heavy object therefore can be suspended below a magnet as shown in Fig. 15. The fact that we can

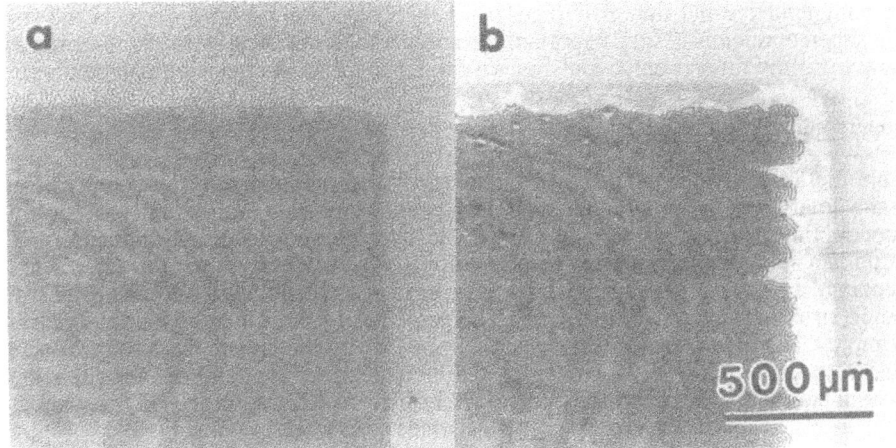

Fig. 13. Direct observation of field distribution within an MPMG processed Y-Ba-Cu-O using magneto-optical effect. The observations were performed at 10K with applied fields of (a) 50Oe and (b) 75Oe. Note that the field can penetrate into the surface region, which is in good agreement with a model in Fig. 12.

Fig. 14. Levitation of a person using a repulsive force of MPMG processed Y-Ba-Cu-O superconductor against 0.4T Fe-Nd-B magnets.

deduce both repulsive and attractive forces using the interaction between magnets and strong pinning superconductors is very important for applications. Especially in space, since there is no gravity, if the force is only repulsive, we cannot suspend an object without contact.

4.1. Superconducting bearing[23]

From the beginning of the discovery of oxide superconductors, a non-contact bearing using the levitation by superconductors has been proposed, and actually a prototype superconducting bearing has already been made by a number of research groups and its operation has been demonstrated at many exhibitions. However, since the force was not large enough, a practical bearing could not have been constructed until our recent success in producing strong pinning Y-Ba-Cu-O.   The principle of the superconducting bearing is very simple. The magnet rotor is suspended in midair by the superconducting stator, and thus the rotor can spin freely. However, we have realized that the construction of the practical superconducting bearing is not very simple. Since the rotor and stator must be cooled in a limited space, the attractive force becomes very small.

At temperatures above T_c, magnetic field can penetrate the superconductor and most of the field is trapped by the superconductor on cooling. This problem has been solved by using an attractive force. In contrast to the repulsive force, the attractive force is maximum when the superconductor is in contact with the magnet before cooling. The technique we used for supporting a rotor in a limited space is schematically explained in Fig. 16. Initially the rotor is pushed upward so as to be in contact with the stator. When the rotor is released after cooling, it is suspended by an attractive force from the upper stator and and is repulsed by the lower stator, and it is thus stably suspended in midair. Using this technique, the rotor which weighs 2.4kg could be safely rotated at 30000 rpm[24]. Fig. 17 shows an outlook of the rotor and stator of the superconducting bearing.

4.2. Flywheel

A flywheel is a very simple device to store energy by rotation. However, the stored energy dissipates in a relative short time due to the friction at the rotation axis. Therefore, if the levitation of superconductor is used, a highly efficient flywheel can be constructed. We have already constructed a flywheel in which an energy of.700J was successfully stored and discharged. Now, we are constructing 1m flywheel where we can store an energy of 100Wh.

4.3. Transport system

There is no friction for the motion of the superconductors on a magnet as long as they move along the direction where magnetic field is unchanged. Therefore, if we construct guide rails using uniform magnets, the superconductor can move without friction, which can be used for load transport. The test line has already been constructed to transport semiconductors in clean rooms as shown in Fig. 18.

4.4. Superconducting permanent magnet

Strong pinning superconductors can trap large magnetic fields, indicating that they become strong permanent magnet. We have found that small piece of melt processed Y-Ba-Cu-O with dimensions 1x1x3mm[3] exhibited BH_{max} of 80MGOe at 5K[25]. In the case of the superconductor, trapped field can be increased by increasing either J_c values or its size. It is also interesting to note that trapped field can be changed by changing magnetization proces. Even at 77K, a superconducting bulk magnet which can generate 1T has been fabricated using an MPMG processed Y-Ba-Cu-O sample 4cm in diameter and 2cm in thickness.

Fig. 15. Suspension of a 10kg globe below an MPMG processed Y-Ba-Cu-O superconductor.

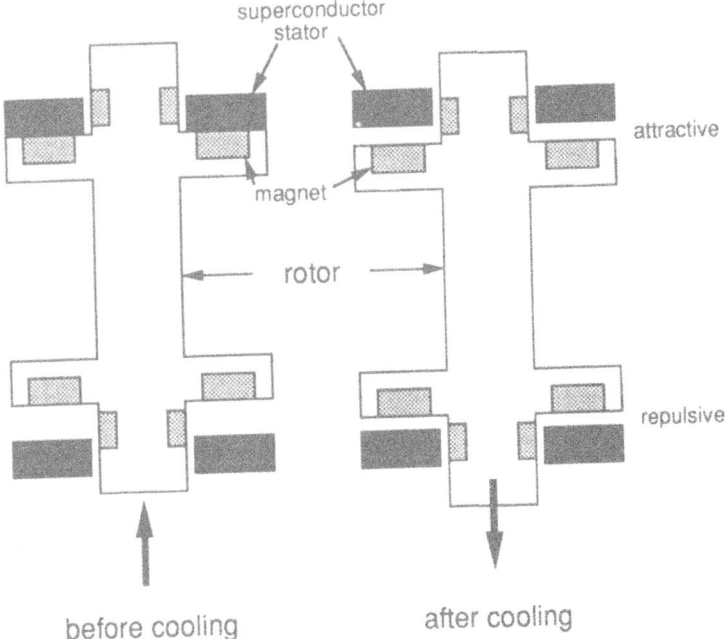

Fig. 16. The technique used in suspending a rotor in a limited space.

450

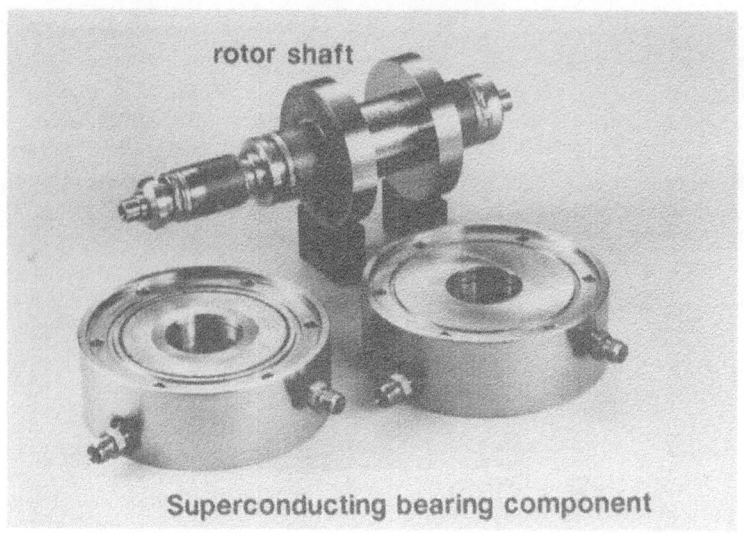

Fig. 17. Superconducting bearing rotor and stators. MPMG processed Y-Ba-Cu-O superconductors are capsulated in the stators.

Fig. 18. Transport system of semiconductors in a clean room constructed by Toshiba corporation (courtesy of H. Ogiwara).

4.5. Magnetic shield

In the case of a pinned superconductor, the penetration depth of the external field is described by $dx=dH/J_c$ (in MKSA units). Therefore, when J_c is high, a large magnetic field can be shielded by the superconductor, e. g., when J_c is 10^5A/cm^2, a 1mm superconductor wall can shield about 1000G. However, since the size of a superconductor without weak links is limited, process modification is needed to enlarge the seize of weak-link-free superconductor for a practical magnetic shielding device. Recently, joining of two superconducting plates has been tried. If it is possible to connect the superconductors, it is much easier to construct a shielding device.

4.5. Other applications

There are various applications besides those listed above. Those are physics experiment, magnetic clamp, decoration, attraction, modern art etc. In physics experiment, fluid dynamics can be analyzed under the ideal conditions, since an object can be lifted in midair without any contact. Levitation itself has been demonstrated in many fairs as an attraction. Now, the MPMG Y-Ba-Cu-O samples can be purchased from Japanese companies.

5. Conclusions

Flux pinning enhancement has the key to practical applications of oxide superconductors. Although very pessimistic prospects have been cast on J_c enhancement of oxide superconductors, it has been clarified that microstructural control enables us to enhance flux pinning, e. g., by dispersing fine nonsuperconducting 211 inclusions into Y123 matrix. Such bulk superconductors exhibit large repulsive and attractive forces by the interaction with magnets. Such forces can be used for a variety of applications, such as bearings, flywheels, transport systems and permanent magnet. These applications are very attractive since very brittle oxide superconductors can be used in the bulk form, where complicated processes are not necessary.

Acknowledgments
The author would like to thank S. Gotoh (now at Kawasaki Steel), H. Fujimoto (now at Railway Technical Institute), A. Kondoh (now at Kawasaki Heavy Industry), H. Takaichi, N. Sakai, Dr. D. Matthews, Dr. N. Koshizuka and Prof. S. Tanaka of Superconductivity Research Laboratory for their useful discussions.

References
[1] e. g., *Science* **244** (1989) 914.
[2] A. Khurana: *Physics Today* (1989) No. 3 p. 17.
[3] S. Jin, T. H. Tiefel, R. C. Sherwood, M. E. Davis, R. B. van Dover, G. W. Kammlott, R. A. Fastnacht and H. D. Keith: *Appl. Phys. Lett.* **52** (1988) 2074.
[4] M. Murakami, M. Morita, K. Doi and K. Miyamoto: *Jpn. J. Appl. Phys.* **28** (1989) 1189.
[5] K. Salama, V. Selvamanickam, L. Gao and K. Sun: *Appl. Phys. Lett.* **54** (1989) 2352.
[6] J. Bardeen, L. N. Cooper and J. R. Schrieffer: *Phys. Rev. B* **108** (1957) 1175.
[7] M. Tinkam: *Introduction to Superconductivity* (McGraw-Hill, New York, 1975).
[8] A. M. Campbell and J. E. Evetts: *Adv. Phys.* **21** (1972) 199.
[9] P. W. Anderson: *Phys. Rev. Lett.* **9** (1962) 309.

[10] D. R. Tilley and J. Tilley: *Superfluidity and Superconductivity* (Adma Hilger, Briston and Boston, 1986).

[11] P. H. Kes, A. Pruymboom, J. van den Berg, and J. A. Mydosh: *Cryogenics* **29** (1989) 228.

[12] W. K. Kwok, U. Welp, G. W. Crabtree, K. G. Vandervoort, R. Hulscher and J. Z. Liu: *Phys. Rev. Lett.* **64** (1990) 966.

[13] K. Yamaguchi, M. Murakami, H. Fujimoto, S. Gotoh, T. Oyama, Y. Shiohara, N. Koshizuka and S. Tanaka: *J. Mater. Res.* **6** (1991) 1404.

[14] M. Murakami: *Proc. Los Alamos Symposium on High Temperature Superconductors* (Addison-Wesley, 1992) p. 103.

[15] S. Nakahara, S. Jin, R. C. Sherwood and T. H. Tiefel: *Appl. Phys. Lett.* **54** (1989) 1926.

[16] K. Yamaguchi, M. Murakami, H. Fujimoto, S. Gotoh, N. Koshizuka and S. Tanaka: *Jpn. J. Appl. Phys.* **29** (1990) L1428.

[17] T. Matsushita: *Jpn. J. Appl. Phys.* **27** (188) L1712.

[18] M. Daeumliong, J. M. Seuntjens and D. C. Larbalestier: *Nature* **346** (1990) 332.

[19] H. Kupfer, C. Keller, R. Meier-Hirmer, K. Salama, V. Selvamanickam and G. P. Tartaglia: *IEEE Trans. Mag.* **27** (1991) 1369.

[20] N. Ogawa, I. Hirabayashi and S. Tanaka: *Physica C* **177** (1991) 101.

[21] N. Ogawa and H. Yoshida: *Advances in Superconductivity* 4 (1992) 455.

[22] P. McGinn, W. Chen, N Zhu, L. Tan, C. Varanasi and S. Sengupta: *Appl. Phys. Lett.* **59** (1991) 120.

[23] F. C. Moon, M. M. Yanoviak and R. Ware: *Appl. Phys. Lett.* **52** (1988) 1534.

[24] H. Fukuyama, K. Seki, T. Takizawa, S. Aihara, M. Murakami, H. Takaichi and S. Tanaka: *Advances in Superconductivity* 4 (1992) 1093.

[25] S. Gotoh, M. Murakami, N. Koshizuka and S. Tanaka: *Physica B* **165&166** (1990) 1379.

HIGH-T_c THIN FILMS.
GROWTH MODES – STRUCTURE – APPLICATIONS

J. MANNHART, J.G. BEDNORZ, A. CATANA*, Ch. GERBER and
D.G. SCHLOM**
IBM Research Division,
Zurich Research Laboratory,
8803 Rüschlikon, Switzerland

ABSTRACT. Using microscopic investigations, a basic understanding of the growth mode of high-T_c films has been obtained in the past couple of years. In this presentation a brief overview of the growth mechanism of $YBa_2Cu_3O_{7-\delta}$ films will be given, and the morphology of the films as well as their potential applications will be addressed.

1. Introduction

Thin solid films are the basis of countless electronic and optical devices and play a vital role in most fields of solid-state physics, including research in superconductivity.

Accordingly, soon after the discovery of high-T_c superconductivity many groups embarked on the fabrication of high-quality films of the new superconductors. However, epitaxial growth of high-T_c compounds was found to be a challenge, because the high-T_c materials have characteristic properties which require growth conditions and growth techniques different from those used to deposit films of classical superconductors like Nb or Al or for films of semiconductors like Si or GaAs. Compared to these materials, the high-T_c compounds have a far more complex unit cell, which involves at least four different types of atoms, and require growth temperatures of $600-900\,°C$ which is 0.7 to 0.85 times the melting temperature. Nevertheless, substantial progress has been made over the past few years in solving the problems associated with the epitaxial growth of the high-T_c compounds [1], and excellent high-T_c films are now fabricated on a routine basis, notably by laser-ablation, sputter deposition, molecular beam epitaxy (MBE) and chemical vapor deposition (CVD). After film deposition was mastered, an important question remained: How does the growth of high-T_c films proceed under such unusual conditions?

In the first part of this lecture this question will be discussed, based on microstructural investigations of high-T_c films with scanning tunneling microscopy (STM), atomic force microscopy (AFM), and transmission electron microscopy (TEM), as these techniques proved

---

* *Present address: EPFL, Prospective and Research, 1015 Lausanne, Switzerland*
** *Present address: Department of Materials Science and Engineering, The Pennsylvania State University, University Park, PA 16802-5005, USA*

E. Kaldis (ed.), Materials and Crystallographic Aspects of HTc-Superconductivity, 453–470.

to be sources of valuable information about film growth. To limit the scope of the discussion, we restrict ourselves to the growth properties of films for which epitaxial growth occurred at the time of deposition, rather than during a subsequent high-temperature annealing step.

The growth mechanism has a direct influence on a film's microstructure, which, in turn, controls the film's superconducting properties. For the high-T_c materials the superconducting characteristics are particularly sensitive to the microstructure, a consequence of their extremely short coherence lengths of only a few angstroms. Hence, due to its importance, the microstructure of the films will be discussed in some detail in the second chapter, with special attention given to growth-induced defects. In the third chapter we discuss the potential of epitaxial high-T_c films with respect to applications.

As the fields to be addressed are rather broad, the discussions will be centered on epitaxial $YBa_2Cu_3O_{7-\delta}$ films, which have been investigated in great detail and can serve to some extent as an example for the other high-T_c systems, too.

The discussions given are valid for films with a thickness below a critical value of, say, 0.5 μm, beyond which epitaxy becomes poor. We name these films "thin" to distinguish them from films that are much thicker than one micron, which are mostly polycrystalline and are at present intensively being studied as a substitute for bulk materials for current transport applications.

The purpose of this lecture is to give a brief introduction to the physics and materials science of high-T_c films; for a more extensive or a more complete understanding of the field, the reader is referred to the reference list.

2. Growth Modes

Most information about the growth mechanism of high-T_c films stems from investigations of the films' surfaces by AFM/STM and TEM, from cross-sectional TEM studies and from the observation of intensity oscillations of reflective high energy electron diffraction (RHEED) patterns in the early stages of film growth [2,3]. One of the prominent features in the AFM/STM images of the (001) surface of $YBa_2Cu_3O_{7-\delta}$ films is a rather high density (5×10^4 cm^{-1} in Fig. 1 and 5×10^5 cm^{-1} in Fig. 2) of steps one unit cell high. If the films are grown on well-oriented substrates, growth spirals emanating from screw dislocations with a dislocation core oriented parallel to the c-axis provide the surface steps (Fig. 1). If the substrates are oriented a few degrees away from (100), the steps run basically

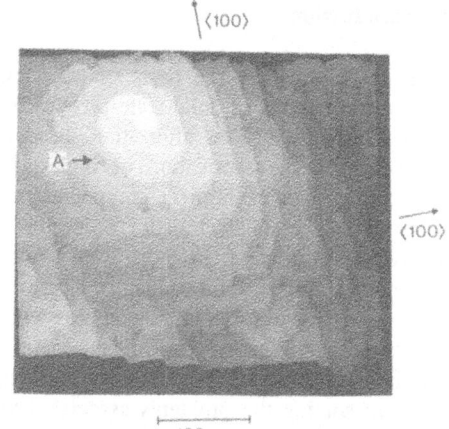

⟨100⟩

A →

→ ⟨100⟩

⊢————⊣ 100 nm

Figure 1: STM image of a $YBa_2Cu_3O_{7-\delta}$ film sputtered on a $SrTiO_3$ substrate at a substrate holder temperature of ≈755 °C. The image shows a growth spiral with a step height of one unit cell (≈11 Å). The small dark spots such as the one pointed to by arrow A are holes or insulating regions. Clumps of material with irregular shape, for example at arrow B, are present at the step edges (from [19]).

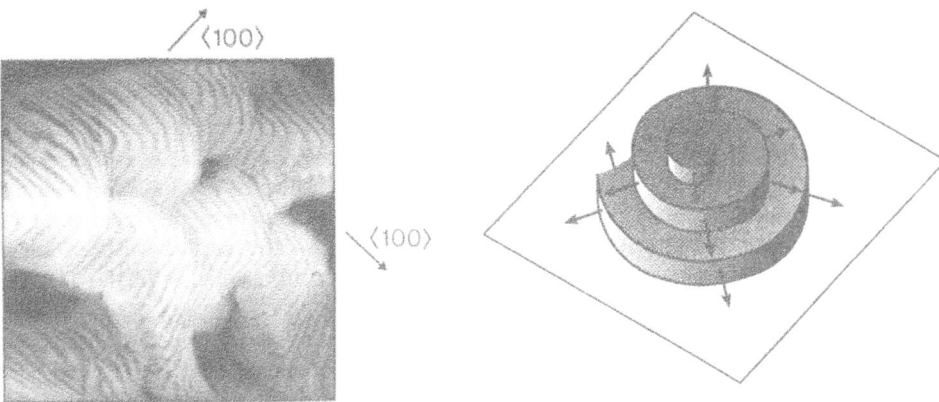

100 nm

Figure 3: Illustration of the screw dislocation-mediated growth mechanism.

Figure 2: STM image of a $YBa_2Cu_3O_{7-\delta}$ film sputtered on a vicinal $SrTiO_3$ substrate at a substrate holder temperature of $\approx$780 °C. The micrograph shows a surface structure dominated by growth steps originating from the substrate tilt (from [19]).

perpendicular to the substrate tilt (Fig. 2). These surface steps give clear evidence that the films grow by accommodating adatoms at growth steps, which are energetically favorable attachment sites.

If growth spirals are present, arriving species at the spiral ledges cause the spiral to rotate around the screw dislocation at its center. Thereby the spirals ascend one layer with each rotation, leading to self-perpetuating, spiral shaped surface steps, as illustrated in Fig. 3. This screw dislocation-mediated growth mechanism is well known from many different materials including NaCl [4], Ag [5], SiC [6] or organic materials such as paraffin [7] and β-methylnaphtalene [8] and was proposed by Frank as early as 1949 [9]. Of course, when this growth mechanism is dominant, a high density of screw dislocations $(\geq 10^7 - 10^8\,cm^{-2})$ is present in the films, which leads to the question of the screw dislocation nucleation process. This issue is currently unsettled, but AFM investigations of the early stages of film growth support mechanisms based on the coalescence of

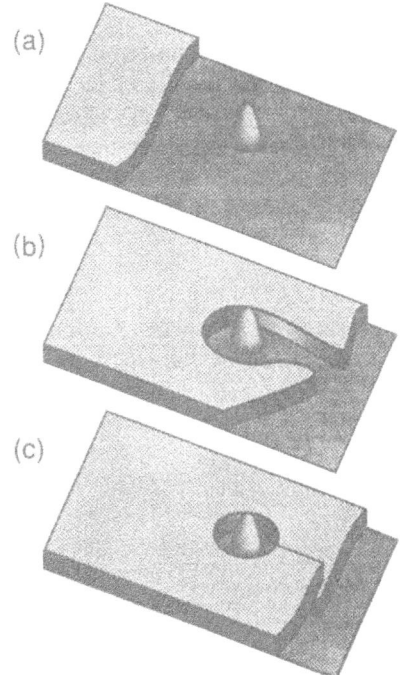

(a)

(b)

(c)

Figure 4: Illustration of screw dislocation nucleation by the incoherent joining of growth fronts.

growth fronts which are vertically offset and inclined with respect to each other. These growth fronts may consist either of two separate fronts or of two branches of a single growth front which became separated and offset during growth, for example by flowing around obstacles (see Fig. 4), [8, 10–12] such as defects or substrate imperfections. Such defects have been observed by TEM, for example in ultrathin films deposited on MgO [13], and AFM images of early stages of the growth of sputtered $YBa_2Cu_3O_{7-\delta}$ films onto $SrTiO_3$ substrates suggest that screw dislocations are indeed nucleated by these mechanisms [14]. Screw dislocations may also arise, at least to some extent, from the inheritance of substrate dislocations. But as screw dislocation densities of $10^8 - 10^{10}$ cm^{-2} have been found for $YBa_2Cu_3O_{7-\delta}$ films grown on virtually all available substrate materials ($SrTiO_3$ [15–25] MgO [16, 17, 19, 20, 24–30] $LaAlO_3$ [17, 31, 32], ZrO_2 (Y) [33], $NdGaO_3$ [25, 34], Mg_2TiO_4 [35]) which have dissimilar microstructural properties and low screw dislocation densities, it is unlikely that this nucleation mechanism is the dominant one.

When tilted a few degrees away from <100>, the microscopic surface structure of the substrate consists of a sequence of steps separating (100) terraces. These surface steps are energetically favorable sites for film nucleation and subsequent [19, 36, 37] growth. As growth proceeds by the incorporation of the arriving adatoms at the steps, the steps propagate laterally across the film surface (Fig. 5). The substrate tilt necessary to achieve growth by this step propagating growth mechanism is presumably that which causes the steps to be spaced sufficiently close together that the depositing species diffuse to and are accommodated by the existing ledges without reaching a supersaturation on the terrace for two-dimensional nucleation to occur [38]. Indeed, a crossover from screw dislocation-mediated growth to growth by step propagation has been observed experimentally on vicinal $SrTiO_3$ (100) surfaces, and the amount of tilt necessary was seen to decrease as the growth temperature was increased [19]. As the substrate steps guide the propagation of the growth fronts in a direction parallel to the substrate tilt and thereby reduce the chance for incoherent encounter between growth fronts, one may expect the screw dislocation density to decrease with increasing substrate misorientation. This was indeed observed as shown in Fig. 6, where the

Figure 5: Illustration of film growth by the propagation of growth steps caused by substrate tilt.

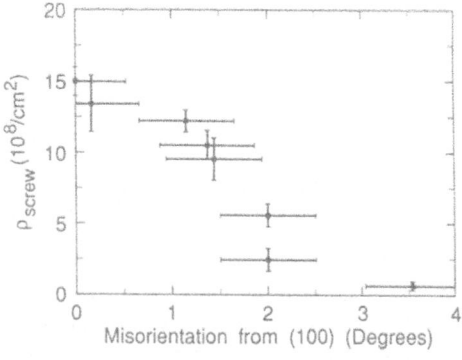

Figure 6: Screw dislocation density as a function of substrate misorientation angle for $YBa_2Cu_3O_{7-\delta}$ films sputter-deposited onto $SrTiO_3$ substrates. The films were grown at a substrate holder temperature of $\approx 750\,°C$ to a thickness of $1000 - 1500$ Å.

screw dislocation density is plotted as a function of misorientation of a $SrTiO_3$ substrate away from (100). The crossover from screw dislocation-mediated growth to growth by step propagation is illustrated by Fig. 7, which shows an AFM image of an $YBa_2Cu_3O_{7-\delta}$ film grown on a $SrTiO_3$ substrate misoriented $\approx 3°$ from (100). The surface of this film is clearly dominated by surface steps reflecting the substrate misorientation. However, as a secondary feature, screw dislocations are also observed on several terraces. But because of the finite substrate tilt, the screw dislocations could no longer give rise to well-developed growth spirals.

In the case where screw dislocations are lacking and growth occurs by step propagation, situations can arise in which the supersaturation on a terrace becomes high enough for two-dimensional nucleation to occur. For instance, the width of the top terrace of a vicinal film tends to increase steadily as growth proceeds. But once a terrace has attained a critical width, the supersaturation far from the terrace steps becomes large enough for two-dimensional nucleation to take place (Fig. 8). Thereby new ledges are created which serve as attachment sites and reduce the supersaturation again.

Frequently, the observation of RHEED intensity oscillations during film growth has been taken as evidence that the films

$\vdash\!\!-\!\!-\!\!-\!\!-\!\dashv$
250 nm

Figure 7: STM image of a $YBa_2Cu_3O_{7-\delta}$ film sputtered on a vicinal $SrTiO_3$ substrate at a substrate holder temperature of $\approx 750\,°C$ with a surface morphology characterized by growth steps originating from the substrate tilt and from screw dislocations.

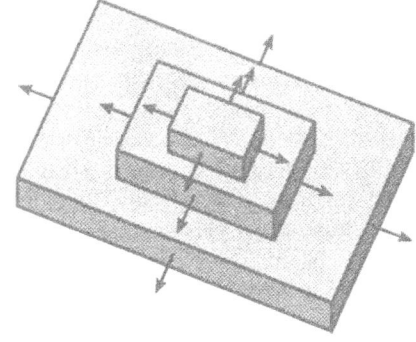

Figure 8: Illustration of the two-dimensional nucleation of growth fronts.

grow in a layer-by-layer mode, in the sense that the majority of one growth layer is completed before the next layer is nucleated. Such a growth mode would be highly interesting as it bears the intriguing possibility of growing truly flat films. Unfortunately, layer-by-layer growth does not appear to be reconcilable with the AFM/STM results described above. For film growth by ledge propagation, RHEED intensity oscillations are not expected, since the surface step density is essentially time-independent. Indeed, the observation of RHEED intensity oscillations has only been reported [39, 40] to occur during the initial stage of growth on bare substrates, indicative for the time necessary for a steady-state surface step density to become established.

Bauer [41] introduced three classifications for film growth: island growth (Volmer-Weber), layer-by-layer growth (Frank-van der Merwe) and initial layer growth followed by island growth (Stranski-Krastanov). These classifications are based only on the relative surface energies of the substrate and the overlayer, and on the strain energy in the overlayer. One of these growth modes should characterize a particular substrate/overlayer system in the absence of substrate surface defect sites. In the case of high-T_c films, however, as elaborated above, defect sites such as screw dislocations or step edges impact the crystal growth process greatly by supplying low-energy attachment sites for adatom incorporation. Since the observed STM/AFM or AFM images of the $YBa_2Cu_3O_{7-\delta}$ films do not indicate which of these classical growth modes would be operative in the *absence* of defects, we find it appropriate to use the term "ledge growth" [2, 13] rather than "island growth" or "layer-by-layer growth" to describe the growth mechanism leading to the observed microstructure.

In summary, the growth process of c-axis oriented $YBa_2Cu_3O_{7-\delta}$ films may be generally described as ledge growth, emphasizing the energetic preference of the depositing species to be incorporated at step edges which propagate laterally across the film surface. The ledges can be provided by two-dimensional nucleation, but more favorably – and more frequently – by screw dislocations and by substrate steps.

3. Microstructure

Compared to bulk single crystals, high-T_c films are grown at lower temperatures and in much shorter times. Therefore one expects high-T_c films to be rich in growth defects. Indeed, it has been found that all high-T_c films are abundant in defects of various types.

Many defects are already induced in the early stages of film growth by the substrates. For example, due to the finite lattice mismatch between substrate and film, the films are exposed to substantial elastic strain which, for films with a thickness exceeding a critical value of $100 - 200$ Å, is accommodated by the formation of misfit dislocations. By HREM [36, 42–45] interface dislocations have been observed directly, with densities as high as 8×10^5 cm^{-1}. In another study, for $YBa_2Cu_3O_{7-\delta}$ films grown on $LaAlO_3$, interface dislocations were found only near surface steps [46]. In addition, due to the high growth temperature, interdiffusion between the substrate and the film is often non-negligible. Interdiffusion leads to substitutional or interstitial point defects in the vicinity of the interface, which can destroy superconductivity if substrate materials like Si or GaAs are used.

Moreover, numerous types of defects are introduced during growth and thus are not substrate-related. Oxygen vacancies are point defects that are expected to be abundant in high-T_c films, with densities of the order of 10^{21} cm^{-3}, depending on the film's oxygen concentration. Unfortunately, due to the lack of instrumental resolution, point defects have not been directly observed yet.

Stacking faults are common defects in high-T_c films, too. They are usually caused by small deviations of the film's stoichiometry, and, for $YBa_2Cu_3O_{7-\delta}$, have been reported to consist frequently of extra Cu-O planes [44, 47]. Among the Bi- and Tl-based superconductors, stacking faults are particularly abundant. Frequently, the stacking faults form superconducting phases which are members of the same family of superconductors as the matrix material, i.e. for $Bi_2Sr_2CaCu_2O_{8+\delta}$, stacking faults consist of the $Bi_2Sr_2CuO_{6+\delta}$ and the $Bi_2Sr_2Ca_2Cu_3O_{10+\delta}$ phase.

Precipitates, like stacking faults, are frequently observed and are caused by deviations of the film stoichiometry. An entire zoo of precipitates has been observed both in the bulk and at the film surface, depending on growth conditions and stoichiometry. For instance, for

$YBa_2Cu_3O_{7-\delta}$ films, precipitates consisting of the following phases have been reported: CuO [44, 48, 49], Y_2O_3 [44, 50, 51], $YCuO_2$ [49, 52], $BaCu_2O_2$ [44], $Y_2Cu_2O_5$ [44], Y_2BaCuO_{5-x} [44, 48], where the precipitates consist either of single-phase material or are an agglomeration of phases. Precipitate morphology is a fascinating field of study, as intriguing growth phenomena result from the interaction between the precipitates and the film matrix as shall be demonstrated with the examples given in Figures 9-11, which are micrographs of precipitates in sputtered $YBa_2Cu_3O_{7-\delta}$ films.

Figure 9 shows a cross-sectional HREM micrograph of a Y_2O_3 surface precipitate where the [110] planes of the Y_2O_3 lie parallel to the (001) $YBa_2Cu_3O_{7-\delta}$ planes. The precipitate separates c-axis from a,b-axis oriented $YBa_2Cu_3O_{7-\delta}$ regions, a configuration which has frequently been observed [43, 49, 53] and is favored owing to good lattice matching between (001) Y_2O_3 and (001) $YBa_2Cu_3O_{7-\delta}$. The presence of a-axis oriented $YBa_2Cu_3O_{7-\delta}$ grains promotes the nucleation and enhances the growth rate of Y_2O_3 with this particular orientation.

Precipitates may also be incorporated epitaxially into the film matrix, an example of which is given in Fig. 10. Once more, this figure shows a Y_2O_3 precipitate with (001) Y_2O_3 parallel to (001) $YBa_2Cu_3O_{7-\delta}$, but this time within a c-axis oriented $YBa_2Cu_3O_{7-\delta}$ grain. With this orientation, epitaxial overgrowth by $YBa_2Cu_3O_{7-\delta}$ occurs. However, the lattice mismatch between these materials is 3% when parallel to the (001) $YBa_2Cu_3O_{7-\delta}$ planes, but 9% when perpendicular to it, as can be seen by looking at Fig. 10 at a grazing angle. As suggested by this micrograph, if overgrowth occurs in an imperfect manner, or if the lattice mismatch is significant, precipitates can be a source of additional lattice defects, such as stacking faults or edge and screw dislocations.

Conversely, precipitate formation is promoted by existing defects within the $YBa_2Cu_3O_{7-\delta}$ matrix, an example of which is given in Fig. 11. This figure shows an AFM image of the (001) surface of a $YBa_2Cu_3O_{7-\delta}$ film with various types of surface outgrowths (I: Y_2O_3, II: $Y_2O_3CuYO_2$, III: $YBa_2Cu_3O_{7-\delta}$ / Y_2O_3 / CuO). It is pointed out that one out-

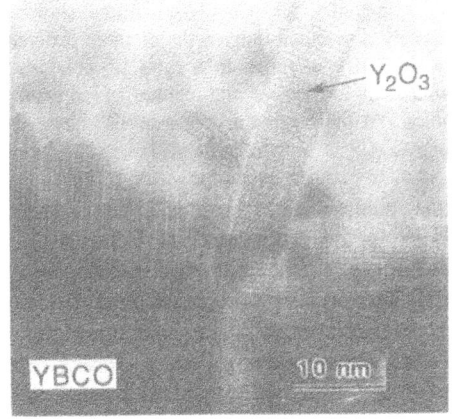

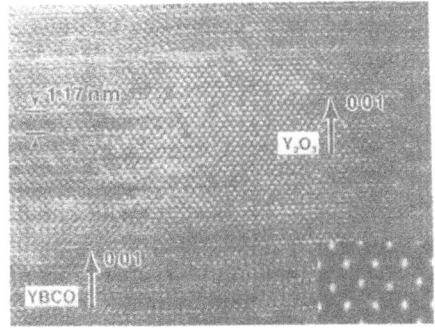

Figure 9: Cross-sectional TEM micrograph of an epitaxial Y_2O_3 outgrowth at an a,b-axis $YBa_2Cu_3O_{7-\delta}$ / c-axis $YBa_2Cu_3O_{7-\delta}$ domain boundary in a sputtered film (from [50]).

Figure 10: Cross-sectional TEM micrograph of a Y_2O_3 inclusion in a sputtered $YBa_2Cu_3O_{7-\delta}$ film. The (001) Y_2O_3 planes are parallel to the (001) $YBa_2Cu_3O_{7-\delta}$ planes (from [50]).

460

growth, which is labelled B in Fig. 11, is located at the center of a growth spiral. This interesting configuration can presumably be accounted for by high interfacial energies and internal stresses between the non-epitaxial precipitates and the $YBa_2Cu_3O_{7-\delta}$ matrix. This stress provides a driving force for the transport of the impurity species to energetically more favorable attachment sites, for example to locations where defects such as dislocations or grain boundaries intersect the film surface. Surface outgrowths induced by compressive stress are known to occur in thin films; these stresses are relieved by the transport of material along easy diffusion paths, such as dislocation cores or grain boundaries (compare also with Fig. 9) [49].

Thus we have shown that, during growth, precipitates frequently interact with other lattice defects. On the one hand, precipitate formation is fostered by existing defects, and on the other hand, precipitates promote the nucleation of additional lattice imperfections.

Grain boundaries are prominent defects that have already been mentioned several times. Low-angle grain boundaries, which for (001) oriented films consist in most cases of arrays of edge dislocations with cores parallel to [001] [54], are frequently caused by the coalescence of crystallites during the initial stages of growth on a substrate not perfectly lattice-matched, due

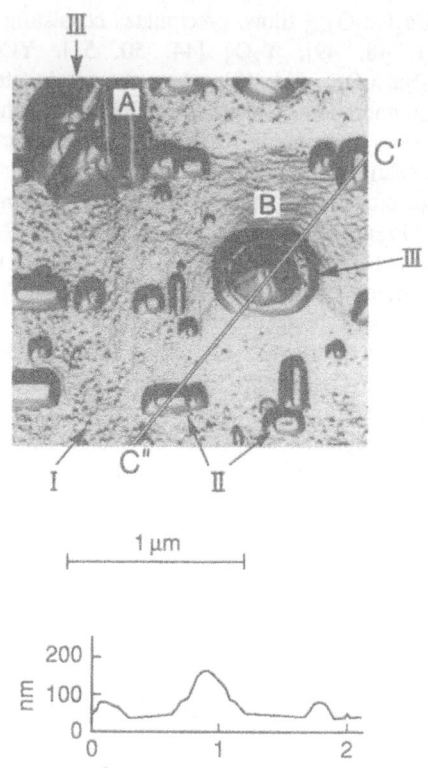

Figure 11: AFM image of the (001) surface of a sputtered $YBa_2Cu_3O_{7-\delta}$ film showing densely distributed outgrowths of three different types (I-III). A refers to a type III precipitate which is located at the center of a growth spiral. At the bottom a height profile along the line C'-C" is shown (from [49]).

to rotational or translational misalignment of the coalescing crystallites. For example, 10^{11} dislocations/cm^2 were observed to be generated in the coalescence of $YBa_2Cu_3O_{7-\delta}$ crystallites on MgO substrates [13]. Analogously, during screw dislocation-mediated growth, low-angle grain boundaries are formed if neighboring growth spirals are misaligned. Another source of low-angle grain boundaries is the substrate's mosaic structure, which reflects a network of subgrain boundaries inherited by the film during growth. As a consequence, higher critical current densities have been reported for films grown on substrates with smaller mosaic spread [55] because then the misorientation angle of the subgrain boundaries is lower and low-angle grain boundaries have higher critical current densities [56].

An example for substrate defects that cause high-angle grain boundaries is provided by substrate surface steps, as they can serve not only as nucleation sites for second phases [36]

Figure 13: STM image of a c-axis oriented, 1200 Å thick YBa$_2$Cu$_3$O$_{7-\delta}$ film grown by sputter deposition, showing left- and right-hand growth spirals emanating from screw dislocations.

Figure 12: Planar TEM image of an a-axis oriented YBa$_2$Cu$_3$O$_{7-\delta}$ film grown by sputter deposition. Small a-axis grains and various grain boundaries are shown (from [60]. © 1990 by the AAAS).

but also for 90° misoriented grains. For instance, c-axis oriented grains have been observed to nucleate at substrate steps within a,b-axis oriented films [36], and a,b-axis oriented grains within c-axis oriented films [57, 58]. The corresponding grain boundaries act as Josephson junctions, which is exploited in the step edge junction technology to fabricate Josephson junctions within epitaxial YBa$_2$Cu$_3$O$_{7-\delta}$ films by growing the films on substrates with surface steps [59].

Other well-known high-angle grain boundaries are the twin planes along [110] in (001)-oriented YBa$_2$Cu$_3$O$_{7-\delta}$ films and, for a-axis oriented YBa$_2$Cu$_3$O$_{7-\delta}$ films, the boundaries between grains with the c-axis parallel to the substrate [100] and [010], respectively (see Fig. 12) [60]. If not purposely avoided, for example by breaking the substrate symmetry between <100> and <010> with applied stress, twin planes occur with a density as high as 4×10^5 cm^{-1} [61]. As shown in Fig. 12, the density of the 90° grain boundaries in the a-axis oriented films can be even higher, as the grains have side lengths of some 10 nm.

Defects that appear at astonishingly high densities are dislocations. Edge dislocations with densities between 10^9 and 10^{11} cm^{-2} have been observed with TEM by a number of groups working with excellent films [13, 44]. Screw dislocation densities can be as high as 10^{10} cm^{-2} so that in some films they completely dominate the surface morphology, as illustrated by Fig. 13.

Dislocations are not only important for the growth process, they directly influence the superconducting properties of the films, too [62]. This is because the dislocations are line defects, which are excellent pinning sites for magnetic flux lines if these are oriented parallel to the dislocation core.

The morphology of the film surface is of prime importance for many electronic applications of high-T_c films. For rf devices, for example, the surface is required to have excellent superconducting properties; for other applications, like for planar tunnel junctions, a very flat surface is needed in addition. Both are tough requirements to meet with $YBa_2Cu_3O_{7-\delta}$ films.

Because the coherence length of the high-T_c compounds is so short, surface superconductivity is highly dependent on the quality of the top layers. Unfortunately, surface layers with excellent superconducting properties are difficult to obtain, because the film surface is chemically reactive and degrades quickly as soon as the films are removed from the deposition chamber. In addition, the surface relaxes by surface reconstruction processes, as recently shown for $GdBa_2Cu_3O_{7-\delta}$ films with HREM [63].

As demonstrated in the previous figures, surface outgrowths and growth steps govern the surface morphology. Surface outgrowths can be minimized by adjusting the film's composition to the exact stoichiometric values, or by limiting diffusion by lowering the growth temperature or by enhancing the growth rate, so that off-stoichiometric species are not able to agglomerate as surface precipitates. Relatively smooth films can be grown between the precipitates by adjusting growth rate and deposition temperature. Also it is observed that $YBa_2Cu_3O_{7-\delta}$ films grown on buffer layers of $PrBa_2Cu_3O_{7-\delta}$ have a reduced surface roughness. However, no way has yet been found to grow films that are free of growth steps over areas of several square millimeters, which is the size desired for applications or for studies which involve single unit cells of high-T_c materials.

Surface defects in c-axis oriented films, which appear as small dark spots such as the ones labelled "A" in Fig. 1, have been observed by many groups in STM investigations, but are not yet sufficiently understood. These spots measure about 5-10 nm in diameter and can appear at rather high densities of the order of $10^{10}\,cm^{-2}$ [15, 19]. They are most likely holes which are more than 2 nm deep, but may also consist of insulating regions. The growth fronts frequently envelop these spots, as though these spots impeded the crystal growth process. This observation implies that the spots are holes present during growth and are not due to tip-induced damage during STM imaging. Concerning the spots' origin, two explanations seem reasonable: First, the dots may be associated with c-axis oriented cores of edge dislocations. The dislocation cores could be hollow or filled with insulating material. Second, the spots may be caused by small (Y_2O_3) precipitates buried underneath. Disturbed overgrowth can lead to insulating regions at the film surface or even to holes. As the STM tip is not observed to touch the film surface during imaging of the spots, although it is moved ≈ 2 nm towards the film (which is more than its flying height), it is concluded that the spots are most likely caused by holes rather than by insulating regions.

In summary, high-T_c films have the tendency to be rich in defects. If the films are to be optimized with respect to applications, control of defect type, size, orientation and density is required. For some applications the defect density should simply be minimized: when the defects degrade the pertinent sample qualities for a particular application. For example, the surface microwave resistance of epitaxial films is increased by grain boundaries and precipitates. For other applications, such as planar tunnel junctions or devices which involve ultrathin films, the surface roughness is to be minimized. Thus, in this case, precipitates are deleterious and even surface steps have to be avoided.

On the other hand, defects such as dislocations, point defects or precipitates act as pinning sites for magnetic flux lines. Therefore, if a high critical current density is to be achieved, films are desired that contain a spectrum of defects optimized for vortex pinning.

4. Applications

When thinking of applications of high-T_c films, applications in electronic circuitry usually come to mind. But high-T_c films are also applied in various fields of research as model systems for bulk materials, as will be described in the first part of this chapter.

4.1 MODEL SYSTEMS

High-T_c films are grown under different conditions than for the growth of bulk materials. For films, growth occurs at higher supersaturations, so that they crystallize under conditions further from equilibrium. Thus it is possible to obtain phases in the films that cannot be achieved with bulk compounds. Further, finite lattice mismatch between the substrate and the film easily results in significant stress and strain; to achieve comparable stress in bulk materials requires the application of substantial pressure.

For example, bulk synthesis of the infinite layer compound $SrCuO_2$ requires the application of high pressures [64], whereas Sr_xCuO_2 films can be grown by alternate deposition of SrO_x and CuO_x layers in a low-pressure atmosphere of NO_2 [65]. Moreover, using multiple source deposition systems with adequate rate control, the film stoichiometry can be varied during growth to modulate the film composition at a scale of one unit cell or even below. This technique allows the growth of films with stacking sequences not attainable in bulk materials. For instance, by using shuttered molecular beam epitaxy, superconducting BiSrCaCuO films have been grown that are composed of 61 bilayers, each consisting of a half unit cell of $Bi_2Sr_2CuO_{6+\delta}$ followed by a half unit cell of $Bi_2Sr_2Ca_2Cu_3O_{10+\delta}$ [66].

Superlattices with longer modulation lengths, consisting of layers of a high-T_c compound (one or several unit cells thick) alternating with non-superconducting layers (e.g. $YBa_2Cu_3O_{7-\delta}$ alternating with

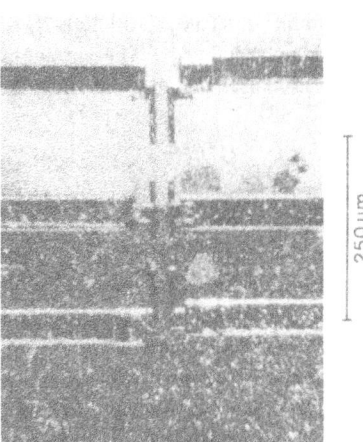

Figure 14: Photograph of an excimer-laser patterned $YBa_2Cu_3O_{7-\delta}$ film showing three superconducting bridges: two bridges within adjacent grains and one bridge straddling the grain boundary. By using an oblique-incidence illumination technique, the two adjacent grains are clearly distinguishable as dark and light regions (from [56]).

$PrBa_2Cu_3O_{7-\delta}$) [67] have been utilized to study the importance of c-axis coupling between CuO layers or to investigate whether a minimal thickness is required for high-T_c layers to be superconducting.

However, there are even more aspects of bulk materials that can be investigated by utilizing high-T_c films as model systems. For example, by using large grained $YBa_2Cu_3O_{7-\delta}$ films, it has been proved that the critical current densities in bulk high-T_c materials are limited by grain boundaries. This has been achieved by directly measuring the critical current of superconducting bridges straddling single grain boundaries as illustrated in Fig. 14 [68]. The value of the critical current across the grain boundary was always found to be lower than that within the adjacent grains (see Fig. 14). The grain boundary critical current can be easily depressed with applied magnetic fields, giving evidence for weak superconducting coupling across grain boundaries. Using a bicrystal technique, it is even possible to measure the grain boundary critical current as a function of a misorientation of the adjacent grains [56]. Clearly, such experiments are much more difficult to perform with bulk materials than with thin films; the latter may be utilized as convenient model systems to understand the critical current limitations of bulk materials [69]. Notably, the bicrystal experiments described have directly led to the development of bicrystalline high-T_c Squids [68, 70], with the possibility of selecting the critical current density of the Squids by adjusting the grain boundary misorientation.

4.2 ELECTRONIC APPLICATIONS

Electronic applications of high-T_c films can be divided into two groups: passive and active devices, respectively. The former usually consist of plain films, whereas the latter rely on heterostructures and/or weak links to be incorporated into the films.

4.2.1 Passive Devices

4.2.1.1 Interconnects. Epitaxial high-T_c films are of interest for interchip connects in digital applications and for general wiring purposes in (analog) rf devices. For these applications, superconductors offer numerous advantages over conventional conductors, such as signal transmission with low dispersion and low loss, high current densities (which permits a high wiring density) and negligible electromigration. Of course, these advantages have to outweigh the efforts required to integrate high-T_c films into the device fabrication process and the need for cryogenic operating temperatures. In general, however, superconducting interconnects for data transfer become more attractive with increasing frequency and with increasing line length. For example, for interconnection lengths of more than 1 cm, superconducting lines seem to have a clear edge over conventional ones for frequencies above 100 GHz [71]. But as clock frequencies of present computers are well below this value, there is no motivation at present or in the near future to incorporate high-T_c transmission lines into standard computers, which would otherwise be operated at room temperature.

For other high-frequency signal processing devices, high-T_c transmission lines offer advantages that are already clear. The potential of operating GaAs MESFETs or HEMTs at 77 K with high-T_c interconnects has been demonstrated recently by a number of groups [72, 73] by fabricating rf amplifiers and oscillators. Given present progress in the development of cryocoolers, the cooling of superconducting devices is not generally considered to be a major disadvantage, in particular as cooling provides additional benefits, such as the possibility to reduce noise or to enhance the performance of semiconducting devices.

4.2.1.2 Microwave Devices. Superconducting microwave devices benefit from the low surface resistance of superconducting materials, which allows the fabrication of devices with large quality factors (Q-values). For example, at 77 K and at 10 GHz the surface resistance of Cu is $R_S \approx 13$ mΩ, whereas for $Tl_2Ba_2CaCu_2O_8$ on $LaAlO_3$ a value of $R_S \approx 130$ μΩ has been reported [74]. We should point out in this context that the surface resistance of superconductors rises (theoretically) with the square of the operating frequency, compared to a square root increase for conventional conductors. They thus outperform superconductors beyond a certain crossover frequency. For example, at 77 K the surface resistance of $YBa_2Cu_3O_{7-\delta}$ exceeds that of Cu above ≈230 GHz. Low surface resistance not only reduces losses, it also offers the possibility of reducing the size and the weight of the respective microwave component drastically, which is an important issue for avionic and space applications. As the advantages of high-T_c microwave devices are obvious and their fabrication is comparatively simple, their development has already reached a relatively mature state. Devices under investigation, some of which are already on the market, encompass a broad spectrum including resonators, filters, antennas, delay lines, mixers, phase shifters and other signal processing units including convolvers, correlators and Fourier transformers. For an overview of the present status of this field, the reader is referred to [75].

4.2.1.3 Bolometers. High-T_c films can be used as efficient bolometers by operating them right at the superconducting transition temperature and by recording their resistance increase due to heating caused by the IR irradiation to be detected. These bolometers are of particular interest for the wavelength region > 20 μm for which no semiconducting detectors are available at present [76]. The responsiveness of high-T_c bolometers has been measured to be above 1000 V/W, and response times have been optimized to be as short as a few microseconds [77]. For high-quality films, noise equivalent powers of $(1 - 20) \times 10^{-12}$ W/Hz$^{-1/2}$ have been predicted [76], which compare favorably to the noise data of conventional 77 K pyroelectric detectors.

4.2.2 Active Devices

4.2.2.1 Two-Terminal Devices. All active, two-terminal high-T_c devices rely on some type of weak link, which is preferably a Josephson junction. There are numerous viable high-T_c thin film Josephson junctions available, a selection of which is sketched in Fig. 15.

As high-angle grain boundaries have been shown to act as Josephson junctions [56], one road to viable high-T_c junctions consists of adding grain boundaries to high-T_c films. The objective is to use single grain boundary junctions [68, 70], as each grain boundary is a source of noise. There are several ways to achieve this goal, one is to fabricate bicrystalline films by deposition on bicrystalline substrates [56], another is to use seed layers to change the film orientation on part of the substrate [78]. Grain boundaries can also be induced by taking advantage of 90° misoriented grains which nucleate at surface steps, as illustrated in Fig. 15 [59].

In another family of high-T_c junctions, the weak link, is caused by a thin layer of non-high-T_c material separating two superconducting electrodes (preferably in the *a,b*-direc-

466

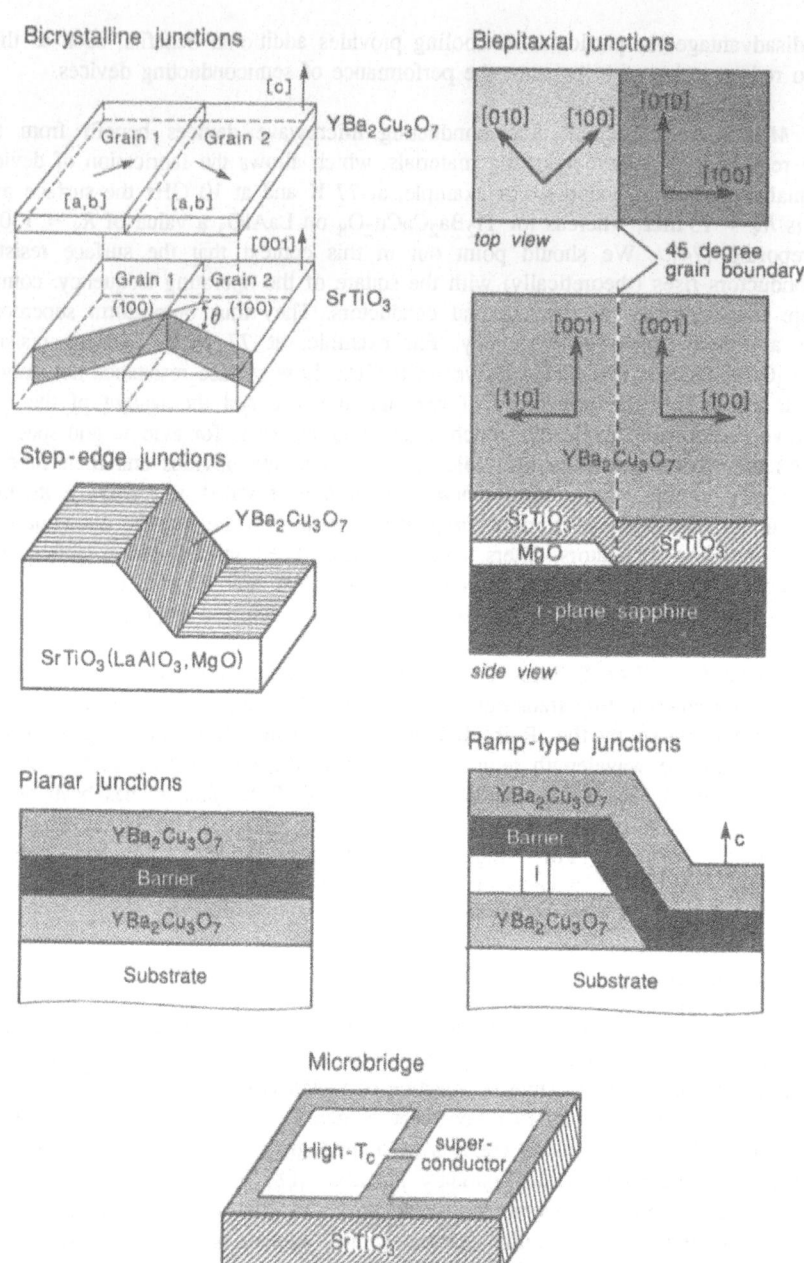

Figure 15: Sketches of various types of high-T_c Josephson junctions.

tion) like in the ramp-type junctions [79, 80] or in the more conventional planar junctions [81].

Finally, microbridges, which are narrow constrictions structured into the films, can be used as Josephson junctions.

These junctions are the basis of applications within digital electronics, but also of Squid-based sensors. Although the high-T_c Squids can be improved further, especially with respect to $1/f$ noise and the reproducible fabrication of low-noise Squids, the noise data of available high-T_c thin film Squids are quite appealing. For dc-Squids based on step-edge junctions, energy sensitivities of 2×10^{-29} J/Hz at 1 Hz have been achieved [82], which compare favorably to the resolution of commercial rf Squids operated at 4.2 K. If the Squids are to be used as magnetometers, the magnetic field sensitivity is the figure of merit, because it also reflects the quality of the flux transformer and the magnetic coupling of the transformer to the Squid loop. Although the integration of low-noise flux transformers into the Squids is still in its infancy, magnetic field sensitivities of $170\,fT\ Hz^{-1/2}$ have been reported at 77 K for frequencies exceeding 1 Hz [83], which is sufficient to monitor the human heartbeat magnetically.

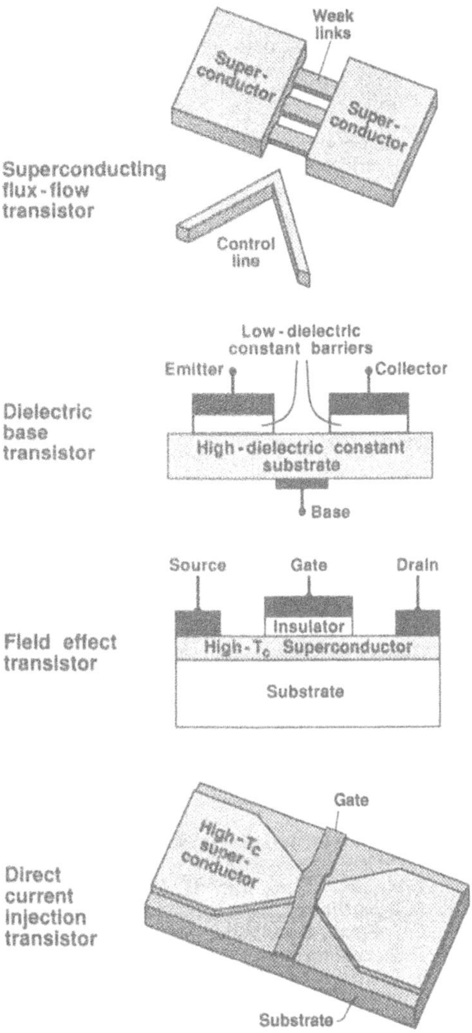

4.2.2.2 Three-Terminal Devices. Various concepts for true high-T_c transistors are being investigated at present (see Fig. 16). In these devices a supercurrent flowing in the drain-source channels is controlled either by applying magnetic fields (superconducting flux flow transistor) [84], by

Figure 16: Sketches of various sample structures used to investigate high-T_c three-terminal devices.

electric fields (superconducting field effect transistor [85], or by injection of quasiparticles (direct current injection transistor [86]). In another approach, normal currents flowing between $YBa_2Cu_3O_{7-\delta}$ contacts on the surface of a $SrTiO_3$ substrate are controlled by means of an electric field (dielectric base transistor) [87].

Compared to the high-T_c Josephson junctions, these three-terminal devices are in an early stage of development, and at present it is far from clear whether these approaches will lead to viable transistors. But much exciting progress has been made within a relatively short

468

time in this dynamic field, as demonstrated in Fig. 17 for a field effect sample. The curve shows the resistance vs. temperature characteristic of a weakly linked $YBa_2Cu_3O_{7-\delta}$ channel; the parameter is an applied electric field. As shown, by applying the electric field, the critical temperature of the $YBa_2Cu_3O_{7-\delta}$ film can be suppressed by more than 10 K [88] which is quite an interesting number for device applications.

Acknowledgment. Many thanks to the American Association for the Advancement of Science for permission to reprint Figure 12.

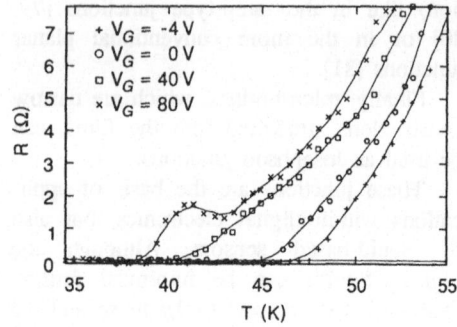

Figure 17: *R(T)* curve of a weakly linked $YBa_2Cu_3O_{7-\delta}$ film of an electric field effect structure as shown in Fig. 16 with four different gate voltages V_G applied. The $YBa_2Cu_3O_{7-\delta}$ film was ≈80 Å thick.

5. REFERENCES

1. P. Chaudhari, R.H. Koch, R.B. Laibowitz, T.R. McGuire, and R.J. Gambino, Phys. Rev. Lett. **58**, 2684 (1987).
2. D.G. Schlom, D. Anselmetti, J.G. Bednorz, Ch. Gerber, and J. Mannhart, "Defect-Mediated Growth of $YBa_2Cu_3O_{7-\delta}$ Films," Proc. MRS Fall Meeting, Boston, MA, (1992) in press.
3. D.G. Schlom, "Epitaxial Growth of High Temperature Superconductors from the Gas Phase," in Int'l Workshop on Superconductivity, Co-sponsored by ISTEC and MRS: Controlled Growth of Single- and Polycrystals of High-Temperature Superconductors (ISTEC/MRS, Honolulu, 1992) p. 34.
4. B. Mutaftschiev, "Crystal Growth and Dislocations" in *Dislocations in Solids*, F.R.N. Nabarro, ed. (North Holland, Amsterdam, 1980).
5. T. Suzuki, J. Cryst. Growth **20**, 202 (1973).
6. W. Dekeyser and S. Amelinckx, "Les Dislocations et la Croisance des Cristaux," Masson, Paris (1955).
7. C.M. Heck, Phys. Rev. **51**, 690 (1937).
8. M.I. Kozlovskii, Sov. Phys. Crystallogr. **3**, 236 (1958).
9. F.C. Frank, Disc. Farad. Soc. **5**, 48 (1949).
10. G.G. Lemmlein and E.D. Dukova, Sov. Phys. Crystallogr. **1**, 269 (1956).
11. M.I. Kozlovskii, Sov. Phys. Crystallogr. **3**, 206 (1958).
12. A. Baronnet, J. Cryst. Growth **19**, 193 (1973).
13. S.K. Streiffer, B.M. Lairson, C.B. Eom, B.M. Clemens, J.C. Bravman and T.H. Geballe, Phys. Rev. **B43**, 13007 (1991).
14. J. Burger, PhD thesis, University of Erlangen (1992), (in German).
15. Ch. Gerber, D. Anselmetti, J.G. Bednorz, J. Mannhart, and D.G. Schlom, Nature **350**, 279 (1991).
16. M. Hawley, I.D. Raistrick, J.G. Beery, R.J. Houlton, Science **251**, 1587 (1991).
17. I.D. Raistrick, M. Hawley, J.G. Beery, F.H. Garzon, and R.J. Houlton, Appl. Phys. Lett. **59**, 3177 (1991).
18. I. Maggio-Aprile, A.D. Kent, Ph. Niedermann, Ch. Renner, L. Antognazza, L. Mieville, O. Brunner, J.M. Triscone and O. Fischer, Ultramicroscopy **42-44**, 728 (1992).
19. D.G. Schlom, D. Anselmetti, J.G. Bednorz, R.F. Broom, A. Catana, T. Frey, Ch. Gerber, H.-J. Güntherodt, H.P. Lang, and J. Mannhart, Z. Phys. B **86**, 163 (1992).
20. J. Burger, P. Bauer, M. Veith, and G. Saemannn-Ischenko, Ultramicroscopy **42-44**, 721 (1992).
21. N. Chandrasekhar, V. Agrawal, V.S. Achutharaman, and A.M. Goldman, Appl. Phys. Lett. **60**, 2424 (1992).
22. M. Kawasaki, J.P. Gong, M. Nantoh, T. Hasegawa, K. Kitazawa, M. Kumagai, K. Hirai, K. Horiguchi, M. Yoshimoto, and H. Koinuma, to be published in Jpn. J. Appl. Phys.

23. L. Luo, M.E. Hawley, C.J. Maggiore, R.C. Dye, R.E. Muenchhausen, L. Chen, B. Schmidt, and A.E. Kaloyeros, Appl. Phys. Lett. **62**, 99 (1993).

24. X.-Y. Zeng, D.H. Lowndes, S. Zhu, J.D. Budai, and R.J. Warmack, Phys. Rev. B **45**, 7584 (1992).

25. H.S. Wang, D. Eissler, W. Dietsche, A. Fischer, and K. Ploog, J. Cryst. Growth **126**, 565 (1993).

26. J. Moreland, P. Rice, S.E. Russek, B. Jeanneret, A. Roshko, R.H. Ono, and D.A. Rudman, Appl. Phys. Lett. **59**, 3039 (1991).

27. H.P. Lang, T. Frey, H.-J. Güntherodt, Europhys. Lett. **15**, 667 (1991).

28. H.U. Krebs, Ch. Krauns, X. Yang, and U. Geyer, Appl. Phys. Lett. **59**, 2180 (1991).

29. F. Baudenbacher, K. Hirata, P. Berberich, H. Kinder, W. Assmann, and H.P. Lang, Physica C **185-189**, 2177 (1991).

30. J.R. Sheats and P. Merchant, Appl. Phys. Lett. **62**, 99 (1993).

31. H.P. Lang, H. Haefke, G. Leemann, and H.-J. Güntherodt, Physica C **194**, 81 (1992).

32. M. McElfresh, T.G. Miller, D.M. Schaefer, R. Reifenberger, R.E. Muenchhausen, M. Hawley, S.R. Foltyn, and X.D. Wu, J. Appl. Phys. **71**, 5099 (1992).

33. H. Olin, G. Brorsson, P. Davidsson, Z.G. Ivanov, P.A. Nilsson, and T. Claesson, Ultramicroscopy **42-44**, 734 (1992).

34. M.E. Hawley, I.D. Raistrick, R.J. Houlton, F.H. Garzon, and M. Piza, Ultramicroscopy, **42-44**, 705 (1992).

35. H. Haefke, H.P. Lang, R. Sum, H.-J. Güntherodt, L. Berthold, and D. Hesse, Appl. Phys. Lett. **61**, 2359 (1992).

36. R. Ramesh, A. Inam, D.M. Hwang, T.S. Ravi, T.D. Sands, X.X. Xi, X.D. Wu, Q. Li, T. Venkatesan, and R. Kilaas, J. Mater. Res. **6**, 2264 (1991).

37. M.G. Norton and C.B. Carter, J. Cryst. Growth **110**, 641 (1991).

38. T. Nishinaga, T. Shitara, K. Mochizuki, and K.I. Cho, J. Cryst. Growth **99**, 482 (1990).

39. T. Frey (private communications) for the growth of $YBa_2Cu_3O_{7-\delta}$ by PLD.

40. N. Chandrasekhar (private communications) for the growth of $YBa_2Cu_3O_{7-\delta}$ by MBE.

41. E. Bauer, Z. Kristallogr. **110**, 372 (1958).

42. M.G. Norton and C.B. Carter, "Growth of $YBa_2Cu_3O_{7-\delta}$ Thin Films – Nucleation, Heteroepitaxy and Interfaces," submitted to Scanning Microscopy.

43. R. Ramesh, A. Inam, D.M. Hwang, T.D. Sands, C.C. Chang, and D.L. Hart, Appl. Phys. Lett. **58**, 1557 (1991).

44. O. Eibl and B. Roas, J. Mater. Res. **5**, 2620 (1990).

45. S.N. Basu, A.H. Carim, and T.E. Mitchell, J. Mater. Res. **6**, 1823 (1991).

46. J.G. Wen, C. Traeholt, and H.W. Zandbergen, Physica C **205**, 354 (1993).

47. T.E. Mitchell, "Microstructure of Epitaxial $YBa_2Cu_3O_{7-\delta}$ Films," in *Ceramic Superconductors*, 2nd Int'l Conf. Ceramic Science and Technol., Orlando, FL, November 12-15, 1990, (American Ceramic Society).

48. M.J. Casanove, A. Alimoussa, C. Roucau, C. Escribe-Filippini, P.L. Reydet, and P. Marcus, Physica C **175**, 285 (1991).

49. A. Catana, J.G. Bednorz, Ch. Gerber, J. Mannhart, and D.G. Schlom, "Surface Outgrowths on Sputtered $YBa_2Cu_3O_{7-\delta}$ Films: A Combined Atomic Force Microscopy and Transmission Electron Microscopy Study," submitted to Appl. Phys. Lett.

50. A. Catana, R.F. Broom, J.G. Bednorz, J. Mannhart, and D.G. Schlom, Appl. Phys. Lett. **60**, 1016 (1992).

51. T.I. Selinder, U. Helmersson, Z. Han, J.E. Sundgren, H. Sjöström, and L.R. Wallenberg, Physica C **202**, 69 (1992).

52. A.F. Marshall, V. Matijasevic, P. Roesenthal, K. Shinohara, R.H. Hammond, and M.R. Beasley, Appl. Phys. Lett. **57**, 1158 (1990).

53. A. Catana, D.G. Schlom, J. Mannhart and J.G. Bednorz, Appl. Phys. Lett. **61**, 720 (1992).

54. Y. Gao, K.L. Merkle, G. Bai, H.L.M. Chang, and D.J. Lam, Physica C **174**, 1 (1991).

55. W. Schauer, X.X. Xi, V. Windte, O. Meyer, G. Linker, Q. Li, and J. Geerk, Cryogenics **30**, 586 (1990).

56. D. Dimos, P. Chaudhari, J. Mannhart, and F.K. LeGoues, Phys. Rev. Lett. **61**, 219 (1988).

57. T.E. Mitchell, S.N. Basu, M. Nastasi, and T. Roy, MRS Symp. Proc. **183**, 357 (1990).

58. H. Takahashi, Y. Aoki, T. Usui, R. Fromknecht, T. Morishita, and S. Tanaka, Physica C **175**, 381 (1991).

59. K.P. Daly, W.D. Dozier, J.F. Burch, S.B. Coons, R. Hu, C.E. Platt, and R.W. Simon, Appl. Phys. Lett. **58**, 543 (1991).

470

60. C.B. Eom, A.F. Marshall, S.S. Laderman, R.D. Jacowitz, and T.H. Geballe, Science, **249**, 1549 (1990).
61. D.H. Kim, D.J. Miller, J.C. Smith, R.A. Holoboff, J.H. Kang, and J. Talvacchio, Phys. Rev. **B 44**, 7607 (1991).
62. J. Mannhart, D. Anselmetti, J.G. Bednorz, A. Catana, Ch. Gerber, K.A. Müller, and D.G. Schlom, Z. Phys. B **86**, 177 (1992).
63. H.W. Zandbergen, "Reconstruction and Relaxation of the Surface of (001) Cleaved $GdBa_2Cu_3O_{7-\delta}$," in Proc. 1992 TCSUH Workshop on HTS Materials, Bulk Processing and Bulk Applications, Houston, Texas, Feb. 27-18, 1992 (World Scientific, Singapore).
64. M. Takano, Y. Takeda, H. Okada, M. Miyamoto, and T. Kusada, Physica C **159**, 375 (1989).
65. X. Li, M. Kanai, T. Kawai, and S. Kawai, Jpn. J. Appl. Phys. **31**, 217 (1992).
66. D.G. Schlom, A.F. Marshall, J.S. Harris, I. Bozovic, and J.N. Eckstein, "Growth of Metastable Phase and Superlattice Structures of Bi-Sr-Ca-Cu-O Compounds by an Atomic Layering MBE Technique," Proc. 3rd Int'l Symp. on Superconductivity, Sendai, Japan, Nov. 6-9, 1990.
67. J.M. Triscone, M.G. Karkut, L. Antognazza, O. Brunner, and O. Fischer, Phys. Rev. Lett. **63**, 1016 (1989); U. Poppe, P. Prieto, J. Schubert, H. Soltner, K. Urban, and Ch. Buchal, Sol. State Comm. **71**, 569 (1989).
68. P. Chaudhari, J. Mannhart, D. Dimos, C.C. Tsuei, J. Chi, M.M. Oprysko, and M. Scheuermann, Phys. Rev. Lett. **60**, 1653 (1988).
69. J. Mannhart and C.C. Tsuei, Z. Phys. B **77**, 53 (1989).
70. C.C. Tsuei, J. Mannhart, and D. Dimos, "Limitations on Critical Currents in High-Temperature Superconductors," Proc. Topical Conf. on High-T_c Superconducting Thin Films, Devices and Applications, Atlanta, GA, 1988, G. Mraraitondo, R. Joint, and M. Onellion, eds. (American Institute of Physics, New York, 1989) pp. 194-207.
71. S.K. Tewksbury, L.A. Hornak, and M. Hatamian, "High-T_c Superconductors for Digital System Interconnections," preprint.
72. J.W. Smuk, M.G. Stubbs and J.S. Wight, "Hybrid Semiconductive/High Temperature Superconductive Ku-Band Oscillator and Amplifier MICs," Proc. IEEE MTT Symposium, Albuquerque, NM, June 1992, in press.
73. N.J. Rohrer, M.A. Richard, G.J. Valco, and K.B. Bhasin, "A 10 GHz YBCO/GaAs Hybrid Oscillator Proximity Coupled to a Circular Microstrip Patch Antenna," IEEE Transactions on Applied Superconductivity, March 1993.
74. W.L. Holstein, L.A. Parisi, C. Wilker, and R.B. Flippen, Appl. Phys. Lett. **60**, 2014 (1992).
75. M. Nisenoff, J.C. Ritter, G. Price, and S.A. Wolf, "High Temperature Superconductivity Space Experiment: HTSSE I – Components and HTSSE II – Subsystems and Advanced Devices," Proc. ASC '92.
76. P.L. Richards, J. Clarke, R. Leoni, Ph. Lerch, S. Verghese, M.R. Beasley, T.H. Geballe, R.H. Hammond, P. Rosenthal, and S.R. Spielman, Appl. Phys. Lett. **54**, 283 (1989).
77. K. Li, J.E. Johnson, and B.W. Aker J. Appl. Phys. **73**, 1531 (1993).
78. K. Char, M.S. Colclough, S.M. Garrison, N. Newman, and G. Zaharchuck, Appl. Phys. Lett. **59**, 733 (1991).
79. M.S. Dilorio, S. Yoshizumi, M. Maung, K.Y. Yang, J. Zhang, and N.Q. Fan, Nature **354**, 513 (1991).
80. Yu.M. Boguslavskij, J. Gao, A.J.H.M. Rijnders, D. Terpstra, G.J. Gerritsma, and H. Rogalla, Physica C **194**, 268 (1992).
81. G.F. Virshup, M.E. Klausmeier-Brown, I. Bosovic, and J.N. Eckstein, Appl. Phys. Lett. **60**, 2288 (1992).
82. G. Friedl, M. Vildic, B. Roas, D. Uhl, F. Bömmel, M. Römheld, B. Hillenbrand, B. Stritzker, and G. Daalmans, Appl. Phys. Lett. **60**, 3048 (1992).
83. A.I. Braginski, Cryogenics **32**, ICEC Supplement 562 (1992); for this device, a superconducting washer was used instead of a flux transformer for flux focusing.
84. J.S. Martens, G.K.G. Hohenwarter, J.B. Beyer, and J.E. Nordman, J. Appl. Phys. **65**, 4057 (1989).
85. J. Mannhart, J.G. Bednorz, K.A. Müller, and D.G. Schlom, Z. Phys. B **83**, 307 (1991).
86. H. Higashino, K. Setsune, and K. Wasa, "Three Terminal Devices of High Temperature Superconductors," preprint.
87. H. Tamura, A. Yoshida, and S. Hasua, Appl. Phys. Lett. **59**, 298 (1991).
88. J. Mannhart, J. Ströbel, J.G. Bednorz, and Ch. Gerber, Appl. Phys. Lett. **62**, 630 (1993).

ENGINEERED GRAIN BOUNDARY JUNCTIONS - CHARACTERISTICS, STRUCTURE, APPLICATIONS

J. ALARCO, YU. BOIKOV, G. BRORSSON, T. CLAESON,
G. DAALMANS, J. EDSTAM, Z. IVANOV, V.K. KAPLUNENKO,
P-Å. NILSSON, E. OLSSON, H.K. OLSSON, J. RAMOS,
E. STEPANTSOV, A. TZALENCHUK, D. WINKLER, Y-M. ZHANG
Physics Department, Chalmers Univ. Technology and Göteborg University,
41296 Göteborg, Sweden

ABSTRACT. The Josephson junctions using high T_c superconductors are presently, to a large extent, based upon grain boundary junctions. We discuss examples of such from our own experience: bi-crystal, bi-epitaxial, and step type junctions. Their electrical transport properties and structure are reviewed. The junction technology is sufficiently advanced to allow the development of electronic devices.

1. Introduction

Josephson junctions are crucial in active electronics devices using superconductors. The performance is determined by the parameters of the junctions. One important figure of merit is often considered to be the product of the critical current and the normal state resistance, $I_c R_n$. A tunnel junction has often to be shunted by a small resistor in order to obtain a non-hysteretic behavior of the current-voltage, I-V, characteristic. This means that the presence of a shunting quasi-particle current in a superconducting weak link is no great hindrance if a sufficiently high $I_c R_n$ product can be realized. So not only superconductor-insulator-superconductor tunnel junctions but also weak links connecting superconductors can be considered. This is a definite advantage as no appropriate conventional tunnel junctions in the new high T_c superconductors have been developed yet. Several different types of weak links have been realized. They are generally of the superconductor/normal metal/superconductor (SNS) type characterized by an RSJ (resistively shunted junction) I-V curve. Although none of them is ideal, each has its merits and some are promising for applications.

Several requirements have to be fulfilled. Besides having a high $I_c R_n$, the junction must have controlled parameters. I_c, the shunting resistance and capacitance must be chosen to give stable operation, i.e. a correct β_c-value ($\beta_c = 2\pi I_c R^2 C/\Phi_0$, $\Phi_0 = h/2e$ is the flux quantum). Operating at 77 K, the critical current has to be sufficiently large not to be affected by thermal fluctuations. This, in turn, means that the inductance of the structure containing the junction must be small, i.e. a small physical size to fulfill $I_c L \leq \Phi_0$. The noise level should be low for many applications. The SQUID (Superconducting QUantum Interference Device) generally demands a low 1/f noise. In other applications, like digital ones, the reproducibility is at premium; many junctions have to be made within close tolerances. High frequency operation demands high cut-off frequencies.

E. Kaldis (ed.), Materials and Crystallographic Aspects of HTc-Superconductivity, 471–490.

We will limit our considerations to thin film devices. The first weak links in high T_c films were microbridges, i.e. narrow constrictions connecting wider banks. They would contain a number of grain boundaries between typically sub-micron size grains in epitaxially grown films. The microbridge can be weakened further by thinning it or by depositing a narrow strip of a normal material like Al or Si which is diffused into a neighboring grain boundary during a mild anneal. A drawback is usually a lack of reproducibility. Other weak links may be formed across a step etched into the substrate. The superconductors on either side of the step can be connected either by the same superconductor that forms a continuos, but weakened, connection, or by a normal metal bridge. Both deep and shallow steps have been used. Another type, an edge junctions is formed by (i) etching out a strip in the high T_c superconductor, that has been covered by an insulator, (ii) depositing or forming, in some way, a thin insulating or semiconducting layer on the slanted edge, and (iii) depositing and patterning a new layer of a high T_c superconductor. In a bi-crystal junction, a well defined grain boundary in a superconducting microbridge has been reproduced by epitaxial deposition on a substrate that has been cut from a crystal sintered together from two differently oriented single crystals. The junction properties are determined by the misorientation angle that can be chosen within a range of 0-45°. A 45° misorientation between superconducting films on either side of the grain boundary can be obtained in a bi-epitaxial junction. By using a seed layer, it is possible to make the superconducting film to orient differently as compared to the substrate. A weak link is formed at the grain boundary between differently oriented parts of the film.

This paper will discuss a few examples of weak links that have been investigated by our group: bi-crystal, bi-epitaxial, and step junctions, as well as their applications in dc-SQUIDs, high frequency oscillators and digital circuits.

2. Artificial Grain Boundary Junctions Based on Bi-Crystals

The IBM group[1] pioneered studies of bi-crystal junctions. These are based upon the controlled grain boundary reproduced in a superconducting film that is epitaxially grown on a bi-crystal which has been formed by sintering together two cut and polished single crystals under a misorientation angle θ. The first experiments with YBaCuO on SrTiO3 substrates gave a critical current density j_c that decreased exponentially with θ up to about 30 degrees where it flattened out. Later experiments on yttria stabilized zirconia[2] (YSZ), on which we will concentrate, as well[3] as on SrTiO3, have shown that j_c continues to decrease up to the largest angles, $\theta=45°$. An enhancement of j_c with ozone treatment indicates an oxygen depletion in the boundary region. $I_c R_n$ varies roughly like $\sqrt{j_c}$ and is typically of the order of up to 8-10 mV at 4.2 K for a low angle boundary junction. Bi-crystal junctions have also been fabricated using several other high temperature superconductors[4].

An advantage of a bi-crystal junction is that the current transport is along the copper oxide planes, in the ab-direction where the critical current density is high and the superconducting coherence length is relatively large (as compared to transport in the c direction). Furthermore, there is a good, intimate contact between the main superconducting banks and the weak region as the junction is formed in situ during the deposition process without a processing of the interlayer that might give damage. A disadvantage is that the location of the weak link is fixed by the substrate and it will be difficult to use bi-crystal junctions in truly integrated circuits.

2.1. ELECTRICAL TRANSPORT IN BI-CRYSTAL JUNCTIONS

The resistive transition to the superconducting state of a high angle boundary is generally rounded, presumably due to thermally activated phase slips[5]. The current-voltage (I-V) curve of a junction (that is narrow compared to the Josephson penetration depth) displays an RSJ (Resistively Shunted Junction) behavior[6] for θ larger than 20 degrees (10° has been quoted[3] for SrTiO3). For smaller misorientation angles, the I-V is characteristic of flux flow with increasing dynamic resistance at large bias. At large θ, or at a temperature close to T_c, where the critical current, I_c, is small, the RSJ type curve is thermally rounded close to I_c.

An anomalous temperature dependence has been noted for $\theta=40$ and 45° junctions. The I_c increases as the temperature decreases but goes through a maximum at about 6 K and has decreased about 30% at 4 K. The behavior can be fitted with a theoretical expression for a dirty semiconductor barrier material[7]. These junctions, however, sometimes show a strong magnetic field dependence that may complicate the interpretation.

The most notable feature of bi-crystal grain boundary junctions is the reproducible, exponential dependence of the critical current density upon misorientation angle. An example of j_c vs. θ at T=77 K is given in Fig. 1. Similar dependencies have been found for YBCO and Tl cuprate junctions on SrTiO3, MgO and Si bi-crystals[3,8,9]. The characteristic $I_c R_n$ product depends upon j_c, $I_c R_n \sim j_c^a$, where $a \approx 0.5$.

The magnetic field dependence of the critical current shows overall a Fraunhofer diffraction pattern. This has been interpreted as if the current flow through the junction were homogeneous. $\theta=40$ and 45° junctions display a distorted Fraunhofer pattern which indicates that the current is carried through a limited number of channels[2,7,10]. Recent measurements[11] show a significant residual critical current in a magnetic field of several Tesla. This has been interpreted as the grain boundary being comprised of superconducting microbridges on a very fine scale separated by normal regions[11].

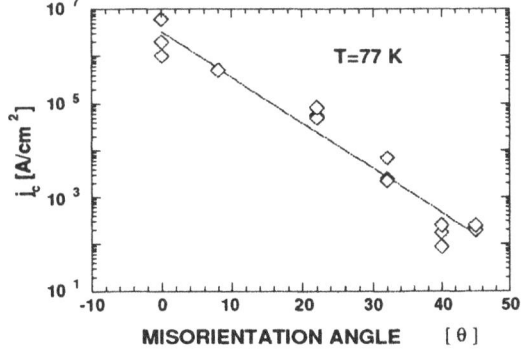

Fig.1.   Critical current density vs. misorientation angle for weeak links in YBCO on YSZ bi-crystals.   T=77 K. Similar logarithmic relations have been obtained for bi-crystals of SrTiO3, ranging from $5 \cdot 10^6$ A/cm^2 ($\theta=0$) to $3 \cdot 10^3$ A/cm^2 ($\theta=45^\circ$) at 4.2 K, Ref.3, and of MgO, 10^6 A/cm^2 ($\theta=0$) to $2 \cdot 10^3$ A/cm^2 ($\theta=30^\circ$), T=77 K, Ref.8.

2.2. NOISE IN BI-CRYSTAL JUNCTIONS

The low frequency 1/f noise is more of a hindrance to the applications of high T_C junctions than the white noise. As no drastic drop in the 1/f noise has been noted at low temperature, thermally activated flux jumps in the adjoining superconducting electrodes can be disregarded (except close to T_C). Instead we have to consider noise caused by fluctuations in the junction parameters. An oscillating reverse biasing mode of a SQUID gives a decrease of the 1/f noise pointing upon fluctuations in the bias current[12,13]. These are determining both the fluctuations in I_c at bias currents close to I_c and fluctuations in R_n for bias currents well above I_c.

Treatments in ozone has given a lowered noise level[1,3]. This points upon a non-uniform, non-stoichiometric distribution of oxygen in the junction and a connection to fluctuating traps. There have been claims of telegraph noise in small junctions, like the situation in small low T_C junctions, but other studies see no effect in the noise properties as the junction size is decreased[1]. The latter would indicate a separation of scattering centers less than 270 nm.

2.3 MICROSTRUCTURE OF BI-CRYSTAL JUNCTIONS

A scanning electron microscope (SEM) study of the grain boundary in YBCO on a YSZ bi-crystal showed that the boundary was not seen when the misorientation angle θ was less than 32°. The surface of the YBCO films was often covered by particles with dimensions of the order of 200 nm (which is typical for a laser ablation process). For $\theta >$ 32°, the particles tended to line up along the boundary. It was estimated that the particles covered less than 10% of the boundary length. (Fig.2, see the inset).

Transmission electron microscopy (200 kV, JEOL 2000FX TEM/STEM with a Link AN10000 EDX detector) was used to study the structure of the boundary region in more detail. TEM of cross sections indicated a thin interface layer of $BaZrO_3$ between the film

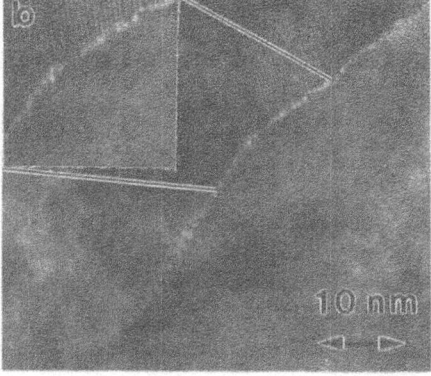

Fig.2 (a) TEM picture of a 0-45° bi-crystal grain boundary; YBCO on YSZ. The insert shows an SEM picture of a patterned 4 μm wide strip across the boundary.(b) Lattice image of the boundary region obtained with the electron beam travelling along the [001] direction.

475

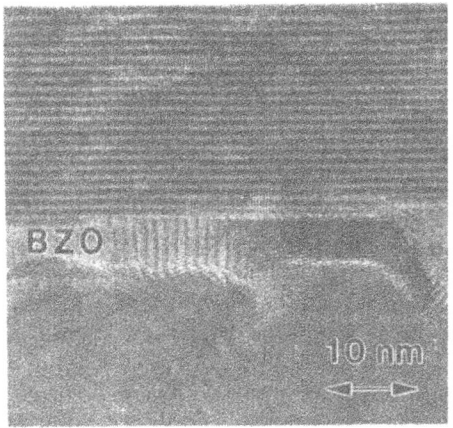
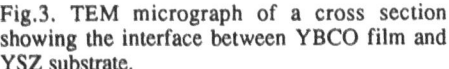

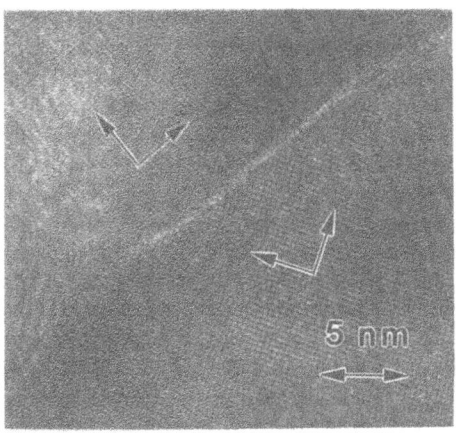

Fig.3. TEM micrograph of a cross section showing the interface between YBCO film and YSZ substrate.

Fig.4. TEM photo of a boundary region (in YBCO) of a 0-32⁰ bi-crystal (YSZ).

and the substrate (Fig.3). A detailed study[14] revealed that this layer is a consequence of Ba diffusing into the YSZ substrate and reacting with Zr. The $BaZrO_3$ thickness depended on the substrate temperature, T_S, during the growth and was 5-6 nm in a region far from the boundary in case of optimal deposition parameters. There were indications of enhanced Ba diffusion along the boundary resulting in a thicker $BaZrO_3$ layer.

TEM confirmed that YBCO films grew epitaxially on both halves of the bi-crystal. Several structural inhomogeneities were observed at the grain boundary. The 200-300 nm large particles, that were observed at the surface in the SEM, were mostly a-axis oriented and EDX showed that they were often surrounded by a thin copper-rich amorphous layer. Twin walls that are clearly seen in TEM pictures (Fig.2a) are parallel to (110) planes Their separation was about 20 nm.

The grain boundary was clean and no inter-granular secondary phases were observed. Note, however, from Fig.2b, that the grain boundary, in that case for a 45o misorientation, is not straight. The substrate grain boundary itself is relatively straight, but the YBCO film grain boundary is faceted. Two types of fine structure are often seen: one at the level of a few atom plane separations, another at a scale of up to about 15 nm. In the latter case, there are regions, or micro-facets, of the film parallel to the {100} plane of one of the adjacent grains but matching to the underlying {110} plane. Conversely, there are regions matching to the other side extending onto the opposite side of the substrate grain boundary. These micro-facets typically extend about ±5 nm from the underlying grain boundary. In some cases there was a bending in the lattice confined to the vicinity of the boundary within a distance of a few basal-plane lattice spacings from the boundary. This gives rise to the finest zig-zag structure on the grain boundary as successively larger bending occurs from plane to plane to match lattices until, after a few plane separations, there is a discontinuity of a plane distance perpendicular to the boundary and the process starts over.

If the interactions responsible for the epitaxial growth of the film on either side of the bi-crystal substrate constrained the film to adopt the orientation of the substrate, the film boundary should inherit the one of the substrate. Orientation relationships between YBCO and YSZ were [001]YBCO//[001]YSZ and [110]YBCO//[100]YSZ. An island mechanism governs the growth of YBCO on YSZ surfaces[15]. One may first assume that bulging is promoted by a 45° misorientation. In those regions where one of the epitaxial grains has crossed the substrate grain boundary, the orientation relationship becomes [100]YBCO//[100]YSZ. This relationship occurs for grains of polycrystalline films[16] and can be dominant for films deposited at relatively high substrate temperature[17]. Thus it is not too surprising that a growing island will cross the substrate grain boundary some distance into a region where its orientation is almost as favorable in the case of a 45° misorientation. Though the relationships between the orientations of YBCO grains and the YSZ substrate can be explained in terms of near coincidence site lattices of their respective oxygen sub-lattices[18], the situation is further complicated by the presence of the $BaZrO_3$ intermediate layer. Hence it is interesting to note that a similar wiggling of the boundary line is present also in a 0-32° YBCO grain boundary on a YSZ bi-crystal, see Fig.4. There, a simple relationship between the oxygen sub-lattices does not exist and one may have expected a straight boundary following the one of the substrate. But this was not the case. The faceting is thus a more general property of bi-crystal boundaries in this system.

2.4. IMPURITIES IN THE BARRIER REGION

There is a desire of controlling the junction parameters. The misorientation angle is one way. Another way may be to introduce impurities in the grain boundary in a controlled way. (A controlled study of the latter effect may also be of value in assessing how processing, that may give additional impurities, affects grain boundary resistances in bulk material.) Early studies[19] of "poisoning" a grain boundary by diffusing foreign material like Al or Si from a well defined narrow strip (at a relatively low temperature) into the closest grain boundary gave reasonable values of parameters like $I_c R_n$ but poor reproducibility. Networks of Al have been used in an attempt to form 2D arrays of junctions and to study microwave radiation emitted from such a network.

We have prepared bi-crystal substrates with Pt or Fe layers enclosed in the interface during the sintering process. When YBCO then is deposited on the hot substrate, impurity atoms diffuse preferably into the grain boundary region. For Pt, there was no noticeable effect on the critical current density or resistance as compared to a similarly prepared bi-crystal grain boundary junction with no impurities. There was a pronounced effect, however, with Fe in the substrate grain boundary. The j_c decreased and the resistance increased substantially. There was no big difference between junctions of different θ, and results discussed here are valid for θ=22°. A mapping of the magnetic field periodicity in the Fraunhofer type response of the critical current gave an effective area which was much smaller than the geometrical one. It appeared as if the current were transported through one narrow filament (or several) with a width of the order of tens of nm. After thermocycles to room temperature, the Josephson current was suppressed (the effective area decreased until the zero voltage current completely disappeared) and the resistance had increased further. A pronounced Coulomb blockade[20] occurred as the junction resistance became of the order of the quantum resistance value, $R_Q = h/e^2 \approx 6.5$ kΩ, see Fig.5. The Coulomb blockade was smeared by temperature but was well seen at nitrogen temperature. The behavior indicates that there is a filament (or a few filaments) breaking up leaving a small enclosed particle through which electrons tunnel, one by one, in two-step processes. Fe is known to attract oxygen into its local environment in YBCO and to strongly suppress superconductivity.

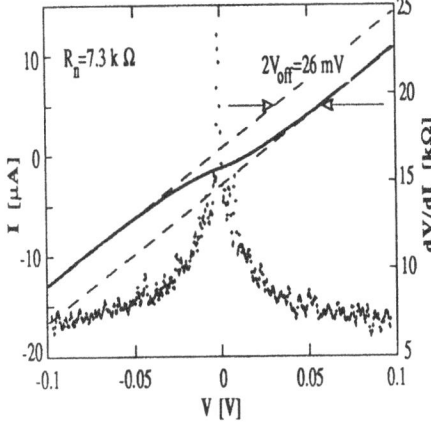

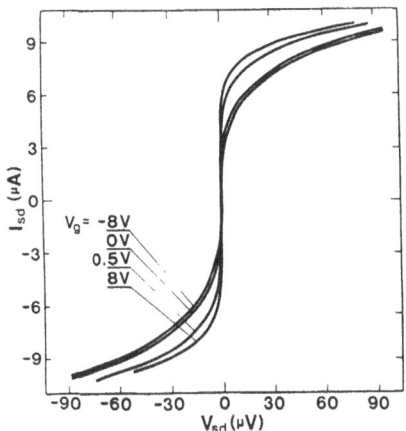

Fig.5. Current-voltage curve of an YBCO bi-crystal grain boundary junction with Fe introduced by diffusion from the substrate grain boundary during the deposition. Note the Coulomb blockade and the off-set voltage at large bias. T=77 K.

Fig.6. Gate voltage dependence of the current-voltage curve for a 0-45° bi-crystal YBCO junction with a gate on top of the grain boundary. T=4 K. Note the shift induced by, in particular, positive gate voltages.

The experiments did not give any reliable junctions but the results indicate that it is possible to influence the grain boundary junctions. Further experiments are carried out where more complicated structures, like multilayers, have been deposited on the bi-crystals before the sintering.

2.5 GATES ON 45° BI-CRYSTAL JUNCTIONS

We have already noted the peculiar temperature dependence that has been seen in high misorientation angle, 40 and 45 degree, junctions. The pronounced lowering of the critical current below a "freezing out" temperature may be a token of semiconducting behavior. If this were the case, it might be possible to control the carrier density, and the superconductivity, of the barrier region by applying a voltage to a gate that has been placed close to, but insulated from the barrier region and the electrodes. This was, indeed, the case[7] as exemplified in Fig.6. There it is seen that the I-V curve of a 45° bi-crystal YBCO junction is changed by the gate voltage. The critical current could be increased up to 50% by applying a positive voltage, while a negative gate voltage had a small effect. Most of the change occurs for gate voltages up to 0.5 V, but there is still some change up to 8 V in this particular case.

The effect was interpreted in the context of a changed charge carrier density in a semiconducting type barrier region[7]. However, it is not possible to completely discard the possibility of an influence by a multi-channel, fine filamentary weak link, see the discussion above. If YBCO is made thin enough, such that the electric field can penetrate into a substantial part of the film and cause a depletion or accumulation of charge carriers in the main part of the cross section of the film, superconductivity can be affected by a gate voltage[21-23]. In such a case, however, T_c generally decreases with a positive gate voltage.

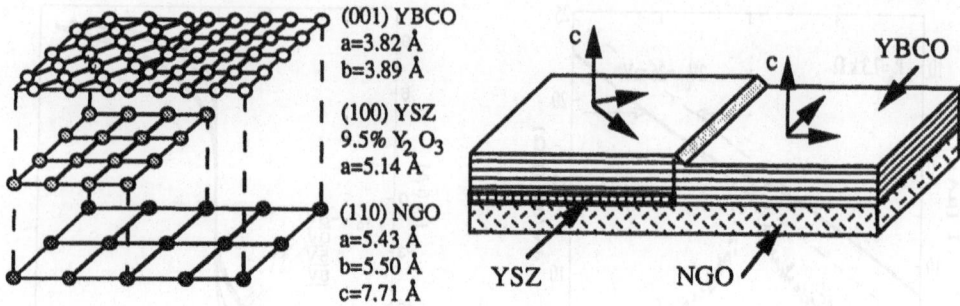

Fig.7. Lattice matches of NdGaO$_3$ substrate, Y-ZrO$_2$ seed (template) layer, and YBa$_2$Cu$_3$O$_{7-\delta}$ superconductor.

3. Bi-epitaxial Junctions Based on YBa$_2$Cu$_3$O$_{7-\delta}$ on NdGaO$_3$

Seed layers that are able to rotate the plane of epitaxial growth of a superconducting film (or an intermediate buffer layer) relatively the substrate can be used to form artificial grain boundary junctions. First experiments[24] used r-plane sapphire ([1$\bar{1}$20], in plane orientation) as the substrate, MgO[100] as a template (or seed layer) and SrTiO$_3$ ([110] on sapphire and [100] on the MgO part) as the buffer layer as well as SrTiO$_3$[100] as substrate layer, MgO[100] as template and CeO$_2$ ([100] on MgO[100], [110] on SrTiO$_3$[100]) as the buffer layer on which YBCO was deposited with two different orientations. The misorientation angle between the two parts is 45°, which results in a low j_c and a small I_cR_n product (typically 100-1000 A/cm^2 at 77 K and up to 1 mV at 4.2 K at the best). An 18° in-plane rotation has been reported[25] for YBCO deposited on an Ar ion etch structured film of CeO$_2$ on a MgO substrate but no correspondingly large values of j_c have yet been reported.

We will limit the discussion to a new bi-epitaxial combination using YSZ as a template on NdGaO$_3$. The advantage with NdGaO$_3$[110] is that it has a very small lattice mismatch both to YBCO[001] (only 0.3%) and to 9.5%-YSZ[100] (also 0.3%). YBCO grows epitaxially on YSZ[100] but with a 45° rotation. The build-up of the layers is illustrated in Fig.7. X-ray Φ-scans have shown that there is, indeed, a 45° rotation between the two parts of the superconducting film. Fig.8 shows diffraction peaks for the NGO substrate, YSZ template, and YBCO film.

Transport measurements were performed on a 4 μm wide bridge crossing the boundary between YSZ covered and non-covered NdGaO$_3$. (The YSZ was Ar ion etched to define the boundary.) A typical I-V curve is shown in Fig.9. It follows an RSJ dependence. Microwave radiation gave rise to Shapiro steps. Critical current densities up to 200 A/cm^2 were registered at 77 K. Rather poor values of the I_cR_n product were registered and one may suspect that there is a diffusion of Ga, Ba, or Zr along the YBCO grain boundary, widening the barrier.

In order to decrease the diffusion of material along the grain boundary from the substrate during the deposition process, we introduced a 150 Å thick buffer layer of SrTiO$_3$ on top of a previously deposited thin YBCO layer (on top of YSZ and NGO) before a second layer of YBCO (2000 Å) was laser deposited. This gave rise to a considerably larger R_n, a

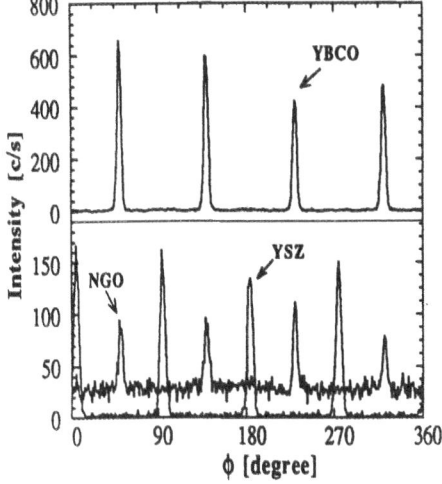

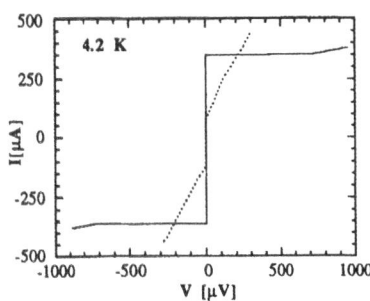

Fig.9. The non-hysteretic I-V curve (dashed line) is taken for a bi-epitaxial junction of YBCO on NdGaO3 with YSZ as a template. It is of RSJ type. Higher resistance and capacitance and a hysteretic I-V curve resulted when the first junction structure was covered by a SrTiO3 (150 Å) buffer layer and then by a thick YBCO (2000 Å) film. T=4.2 K.

Fig.8. X-ray Φ scans of NdGaO3{333}, Y-ZrO2{117} and YBa2Cu3O7-δ{113} reflexes.

higher capacitance, and a hysteretic I-V curve. The $I_C R_n$ increased to about 2 mV at 4.2 K. Increasing the substrate temperature during deposition of the top YBCO layer, the capacitance and the resistance of the junction decreased and a non-hysteretic I-V curve was recovered. These results indicate that it is possible to influence the junction parameters by changes in the growth conditions.

The NdGaO3 based bi-epitaxial junctions displayed a distorted Fraunhofer magnetic field dependence (beyond the second lobe) and gave a SQUID response up to 82 K with a modulation voltage of 1 µV at 77 K.

4. Step Junctions; YBa2Cu3O7-δ on Deep and Shallow Steps in MgO

Another type of grain boundary junction is based on a step etched into the substrate giving rise to grain boundaries between epitaxially grown films on either side of the step (and on the step itself). The step can be
-deep or shallow;
-steep or mildly slanting;
-covered by the same high T_C superconductor or by a lower T_C one, a normal metal or a semiconducting film.

There are a number of advantages with step junctions. They can more or less be placed at will on the substrate such that the latter is optimally used for dense packing. J_C can be controlled within several orders of magnitude by varying the junction parameters (like the step height, the angle of the step, the deposition parameters). Values up to 10^3-10^4 A/cm^2 at 77 K can readily be obtained. The $I_C R_n$ product, in turn, may vary between 0.01 and 1 mV at 77 K. The white noise level of step junctions (as measured, for example, by the SQUID energy sensitivity) seems to be somewhat lower than the one of other types of junctions (but one has to be cautious in such a statement as the quality of the superconducting films has to be comparable and the procedure of measurement similar).

However, there are also disadvantages. The reproducibility and the stability are somewhat poor. The spread of junction parameters can be up to one-to-two orders of magnitude over a (10 mm x 10 mm) chip and it may be even an order of magnitude larger between junctions on different chips.

Controllable weak links form on relatively steep ($\theta > 45^\circ$) and deep steps (at least of the same order as the film thickness) in materials like $SrTiO_3$[26], $LaAlO_3$[27], MgO[28] and Al_2O_3[29]. The YBCO film has grown c-axis oriented on the substrate on either side of the step while the c-axis is along the substrate in the step region, see Fig.10a. Hence two 90° grain boundary junctions form around the step region[30]. Such a 90° c-axis grain boundary can have a high j_c, of the order of the one of the grain itself[31]. RSJ type I-V curves with $I_c R_n$ scaling like $j_c^{1/2}$ are seen in $SrTiO_3$ step junctions.

It is somewhat questionable if a similar $\sqrt{j_c}$ dependence is present for deep step junctions based on MgO[100]. The microstructure of such a junction[28] is different from those on $SrTiO_3$ or $LaAlO_3$ substrates. There are at least two grain boundaries between portions of the YBCO film where the ab-planes in this case follow the surface of the substrate also in the edge region. The orientations are sketched in Fig.10b and the actual microstructure for an MgO step junction is given as a TEM cross section in Fig.11. In that particular junction, the step edge is slightly rounded. The YBCO planes have a couple of different orientations giving at least three grain boundaries of the planar type more characteristic of a bi-crystal junction. One of these grain boundaries may dominate the electrical transport. The magnetic field dependence of the critical current deviates from a Fraunhofer diffraction pattern. This indicates that the current distribution is not uniform in the barrier. The noise level of an MgO step junction can be as low as the one of a $SrTiO_3$ step junction, i.e. among the lowest obtained for high T_c junctions.

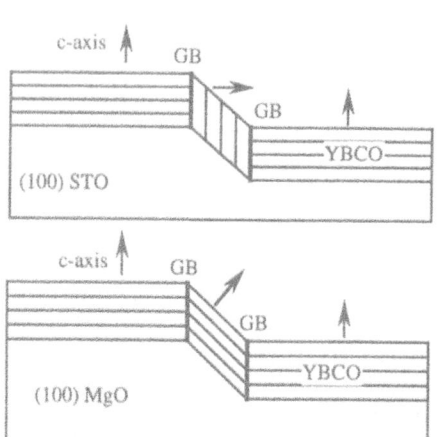

Fig.10. The c-axis orientations of an YBCO film deposited on top of a deep step etched in $SrTiO_3$ and MgO, resp.

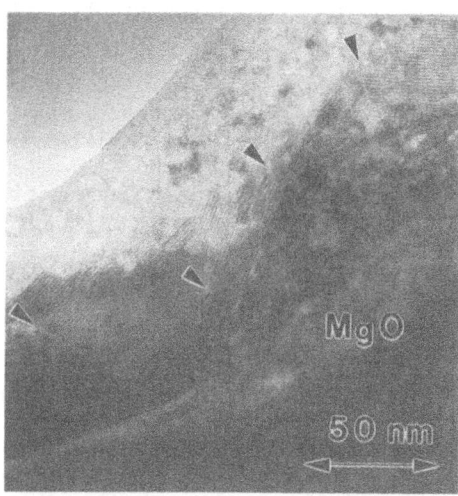

Fig.11. A TEM micrograph of YBCO film on a deep MgO step showing the presence of multiple grain boundaries (arrowed) along the step.

Our first experiments with shallow steps in MgO (where the step height, h=1-10 nm and the step angle $\theta \leq 30°$) gave a beautiful logarithmic decrease in jc (and I_cR_n) with step height, h. However, extended measurements gave a much larger spread in jc and I_cR_n when plotted versus h. Still, a plot of I_cR_n versus jc for a large number of fabricated junctions shows a relationship $I_cR_n \sim j_c^{0.6}$ as shown in Fig.12. The study had been motivated by a report of a 45° rotation of the in plane YBCO axes with respect to those of an MgO substrate which had been ion milled by Ar+ ions[32,33]. We found, indeed, that it was possible to form a 45° rotation in the area that had been shadowed by a resist mask during the ion milling that was performed under an angle of 60° to the substrate normal, see Fig.13. The reason of such a rotation is not known, but we speculate that the morphology of the surface of the substrate may be changed by a re-deposition in the shadow area or/and by diffracted ions coming there. The misoriented area was detected by TEM work. Two 45° planar grain boundaries are formed, one of which may dominate the electrical transport. The critical current density can be much higher than expected from bi-crystal or bi-epitaxial junctions. The magnetic field dependence of I_c indicates an uneven current distribution in the junction that possibly was formed by a number of superconducting filaments. The shallow MgO step junctions seem to be noisier than corresponding deep step junctions as inferred from SQUID measurements.

5. Discussion of Grain Boundary Junctions

Grain boundary junctions have been the favored Josephson elements up till now. The current transport is in the ab plane and it is possible to obtain a practically high current density. The elements are reasonably well protected and survive processing steps, but still there is an interaction with the environment, exemplified, for example, by the influence of ozone treatment.

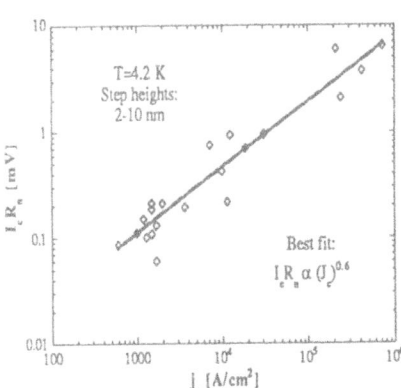

Fig.12. The I_cR_n product vs critical current density for YBCO junctions based upon shallow steps (1-10 nm) etched in MgO by a 60° to normal Ar ion beam. T=4 K.

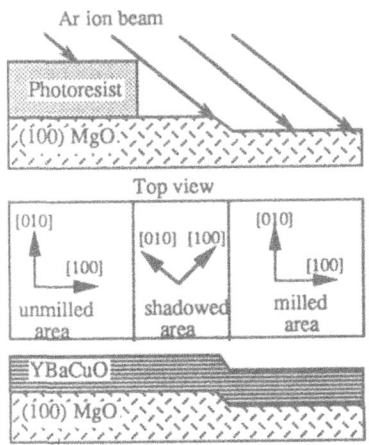

Fig.13. The build-up of a junction based upon a shallow MgO step etched by Ar ions. A 45 degree misoriented region may form in the shadowed area.

Reproducibility is an important property, particularly for integrated circuits where design margins in junction parameters are narrow. Bi-crystal junctions show a spread in critical currents of about 10% for not too remotely spaced elements on a chip, about 30% between chips. Bi-epitaxial junctions are also reasonably reproducible. Step junctions presently show a large scatter in critical currents, typically one order of magnitude within a chip and another order for junctions on different chips. A remarkable reproducibility has been reported by Martens and co-workers[34] for another type of junction, a nanobridge, that we have not discussed hitherto. About 60 junctions in a digital shift register gave a spread of only 8% despite the fact that the cross sections of these bridges were merely 20 nm x 20-25 nm and lengths 20 nm - the spread in sizes should be much larger than 8%. The functioning must be of another origin than a usual (Dayem) microbridge; possibly like an oxygen depleted region[34].

Sufficiently high j_C can be obtained in low angle bi-crystal junctions and step junctions. The former can give values close to the one of the film itself if the misorientation angle is made small enough. 45° bi-epitaxial junctions have difficulties reaching high j_C values. Low angle misorientations are needed also for these.

The characteristic $I_C R_n$ value generally scales with critical current density, roughly like $I_C R_n \sim j_C^{0.5-0.6}$. Sufficiently high values can be obtained for practical applications.

Better (lower) noise values have been obtained with grain boundary junctions than with other high-T_C junctions. SQUIDs based on bi-crystal junctions (particularly ozone treated ones) and step junctions have given very low noise properties. The deep step junctions seem to be more favorable than the shallow ones. The 1/f noise is the limiting factor. Traps in the grain boundary region, probably associated with oxygen deficiency, seem to be the cause of current fluctuations.

The type of symmetry change at the grain boundary does not seem to determine the junction properties. Differently oriented bi-crystals: symmetric or asymmetric combinations of crystals, different twists or tilts or combinations of these, different substrate materials, bi-crystal or bi-epitaxial junctions, they all seem to give similar junction parameters depending mainly upon angle. One is led to the conclusion that it is the grain boundary itself that determines the properties of the weak link. Important boundary properties are dislocations (the density of which is given by θ), charged dislocations, impurities, inclusions, voids, a-axis oriented particles, oxygen variation, traps, resonant states, etc.

Judging from the magnetic field variation of I_C, one finds that the current distribution within the boundary junction generally is non-uniform. Recent measurements[11] have shown that even up to fields of 4-5 T, there remains a residual zero voltage supercurrent and that the current oscillates with magnetic field around an average value. The period of oscillation is much smaller than expected from the junction dimension. Ozone treatment rather increased than decreased the residual current. It was suggested that the current transport in bi-crystal junctions occurs in the form of a periodic array of microbridges separated by normal regions. It may be difficult to apply such a filamentary model for a narrow, high angle grain boundary junction where the critical current has been depressed, say, four orders of magnitude. Tunneling via resonant states, via inclusions etc, which would give a spatial inhomogeneity, has also been suggested[3].

Besides the effect and degree of non-uniformity, it is not clear if the grain boundary junctions are SNS or SIS devices (or SNINS, or SS´NS´S, or SS´IS´S, etc, where S=superconductor, S´=superconductor with weakened order, N=normal metal, and I=insulator). Both models can account for some of the observed characteristics, like $I_C R_n$ scaling with j_C and arguments have been given pro and con (for a review, see Ref.3). The high resistivity of the grain boundary and the high 1/f noise level may favor tunneling through a highly doped insulator with localized states. Further discussion follows in 6.2.

6. Applications of Artificial Grain Boundary Junctions

A number of devices have been designed and tested with the decent quality Josephson junctions that now are available. Here, we give a few examples colored by our own experience.

6.1 DC-SQUIDs

The superconducting quantum interference device is the best developed application. It has been on the market during a long time in its low T_c version and also in a high T_c version (Mr. SQUID, Conductus) since a couple of years. A dc-SQUID based on thallium cuprate microbridges gave good noise properties several years ago[35] (flux noise power spectrum, $S_\Phi = \leq 10^{-9}\Phi_0^2$/Hz, minimum energy resolution per Hz, $\varepsilon = 2 \times 10^{-29}$ J/Hz at 10 Hz and 77 K). More recently, low noise dc-SQUIDs have been developed using bi-crystal junctions and step junctions, for example:

• IBM, ref.36, a small junction, ozone treated SQUID loop, bi-crystal, $\theta = 24°$ SrTiO$_3$, voltage modulation 250 μV/Φ_0 at 77 K, $S_\Phi = 2 \times 10^{-12}\Phi_0^2$/Hz, $\varepsilon = 3 \times 10^{-31}$ J/Hz at 71 kHz and 77 K, 1/f noise setting in at all lower frequencies,

• Chalmers, ref.2,13, bi-crystal $\theta = 32°$ YSZ, voltage modulation 6.5 μV/Φ_0 at 77 K, $S_\Phi = 1.5 \times 10^{-9}\Phi_0^2$/Hz, $\varepsilon = 4.5 \times 10^{-29}$ J/Hz at 10 Hz and 85 K,

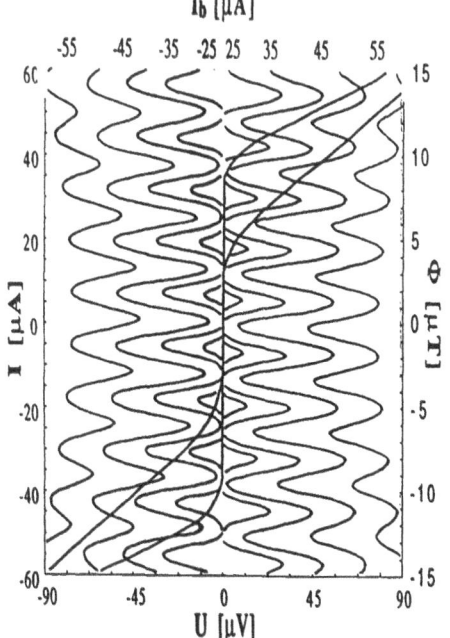

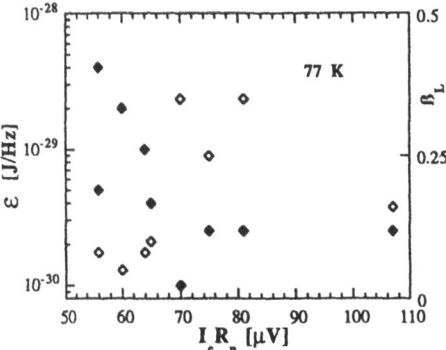

Fig.14. Response of a dc-SQUID based on YBCO step junctions in MgO. The two extremes of the I-V curve as a function of magnetic field are given. The periodic voltage-flux responses for different bias currents are also given. T=77 K.

Fig.15. The minimum energy sensitivity (filled) and β_L as functions of I_cR_n for several YBCO on MgO dc-SQUIDs fabricated on two substrates at different occasions. Values for T=77 K.

• Siemens, ref.37, deep step junction, SrTiO3, voltage modulation 15 µV ptp, transfer function dV/dF of 50 µV/Φ_0 at 77 K, SΦ= $10^{-10}\Phi_0^2$/Hz, ε= x10^{-29} J/Hz at 10 Hz and 77 K (about same performance at 1 Hz).

• The Jülich group has obtained similar performance for step junctions and has also reported[38] good performance for an rf-SQUID based upon a deep SrTiO3 step junction, L=125 pH, voltage modulation 40µV/Φ_0 at 77K, SΦ=$10^{-8}\Phi_0^2$/Hz, ε=1.4x10^{-28} J/Hz at 0.3 Hz and 77 K.

Recent results for a dc-SQUID based on 6 µm wide, 350 nm deep step junctions in MgO (L$_{SQUID}$=20 pH) will be published separately[39]; voltage modulation 65 µV/Φ_0 and dV/dΦ=240 µV/Φ_0 at 77 K, SΦ(white noise)=1.6·$10^{-11}\Phi_0^2$/Hz, ε$_w$=1.6·10^{-30} J/Hz, ε$_{1Hz}$=6·10^{-29} J/Hz at 77 K An example of the response of the Siemens-Chalmers SQUID is given in Fig.14. To optimize the sensitivity of a SQUID, junction stability parameters like β$_C$=2π$I_C$$R^2$C/$\Phi_0$ and β$_L$=I_CL/Φ_0 have to be kept within limited ranges. Hence, the sensitivity does not scale inversely with the $I_C R_n$ product for all values as indicated in Fig.15, but seems to have a minimum for certain design parameters.

The white noise level of a laboratory high T$_C$ SQUID is competetive with the one of a low T$_C$ commercial SQUID, but the 1/f knee has still to be pushed to lower frequencies (note the knee at 0.3 Hz for the ac-SQUID). Computer simulations indicate that the performance of coupled high T$_C$ SQUIDs should be of the order of the low T$_C$ ones if the SQUID loop and the coupling transformers are designed together taking into account restraints by high T$_C$ superconductors[40].

6.2 HIGH FREQUENCY EFFECTS

A number of high performance high frequency devices are based upon tunnel junctions using low T$_C$ superconductors. Receivers based on SIS (quasiparticle) mixers have lowest noise figures in the mm- and sub-mm wavelength bands and are used routinely in radio astronomy observatories[41]. The voltage standard of most national laboratories is based upon the ac Josephson effect. High frequency, small linewidth, large bandwidth, flux flow oscillators, that may be integrated into a sub-mm wave receiver, have recently been developed[42]. Basic Josephson effects have been demonstrated using grain boundary junctions in high T$_C$ superconducting films, but the junction quality is not yet good enough for demanding high frequency detectors. However, the Josephson oscillations that are generated in a high T$_C$ grain boundary junction can be utilized to study the high frequency properties of such superconductors, like surface resistances and penetration depths.

An example of the performance of a bi-crystal grain boundary junction coupled to a high frequency microstrip resonator is shown in Fig.16. From the singularities in the I-V curves connected with resonances in the 80-220 GHz range, it was possible to extract characteristic superconducting data at these high frequencies[43].

Internal geometric resonances, so called Fiske steps, have been detected[44] in the I-V curve of a 20 µm long bi-crystal (0-32° tilt on YSZ) junction at voltages corresponding to resonant frequencies at 280, 490, and 640 GHz. Examples of the magnetic field dependence of the step height is given in Fig.17. The existence of well defined geometrical resonances indicates the presence of a not too lossy dielectric layer in the region between the superconductiong electrodes that form the resonator. From the measurements, we extract a value of λ=1400 Å for the penetration depth, t/ϵ_r=4 Å for the ratio of the thickness of the dielectric layer divided by the relative dielectric constant of the layer, and R_S=10 μΩ for the surface resistance at 1 GHz. The values of λ and R_S are in good accord with other types of measurements.

An increased critical current density may give rise to a flux flow mode in long Josephson junctions where a magnetic field has been applied perpendicular to the current flow. Fluxons are affected by a Lorentz force and are driven towards one end of the junction. They are not damped if the losses in the junction and the superconducting electrodes are acceptably low. Characteristic are velocity matching steps in the I-V curve that occur when the phase velocity of the propagating fluxons equals the velocity of electromagnetic waves on the transmission line formed by the tunnel junction. Flux flow current steps have recently been observed[45] in 10, 20, 30, and 40 μm long bi-crystal YBCO junctions (0-32°, j_c=20 kA/cm², non-uniform current distribution as judged from the magnetic Fraunhofer pattern). The voltage corresponding to the maximum current of the step varied proportionally to the applied magnetic field. From the slope it is possible to extract the average effective width, t/ϵ_r≈2.9 Å, in the frequency range of 200 to 1000 GHz.

The high frequency effects support a model of a dielectric layer in the grain boundary, possibly with filaments of superconducting regions (for the high current density junctions).

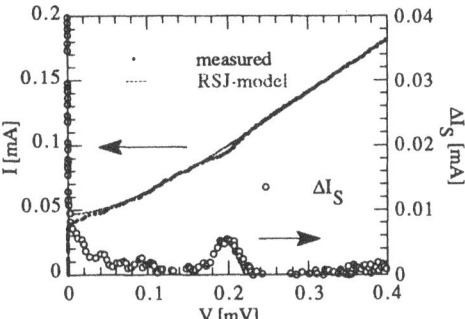

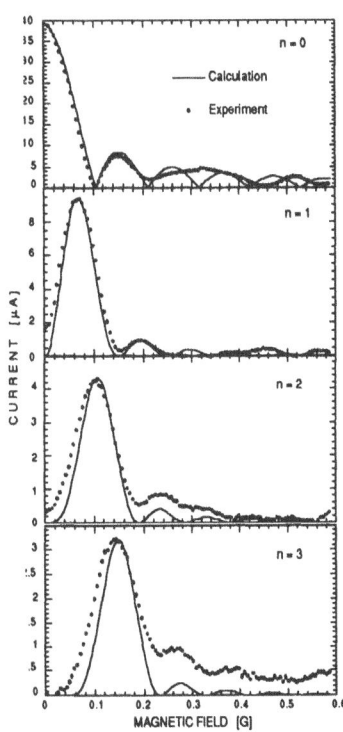

Fig.16. An RSJ mode fit to the measured I-V curve gives the junction I-V curve without external circuitry connected. The difference current, ΔI_S, contains information on the frequency response (f=2eV/h) of the circuitry connected to the junction (mainly a micro-strip resonator). YBCO junction on YSZ bi-crystal. T=4 K. Resonator length = 0.25 mm.

Fig.17. Magnetic field dependence of the critical current (n=0) and of the first three Fiske modes (n=1-3) for a 20 μm long YBCO on YSZ bi-crystal junction. The solid line gives the best theoretical fit to the experimental data (dots). T=15 K.

6.3 GRAIN BOUNDARY JUNCTION TRANSISTOR

Components based on the controlled flux flow in superconducting films are of interest both for analog and digital applications. Vortices can be induced into a long grain boundary junction by applying a control current to a loop on top of, but isolated from the junction. Experiments with a 0-32° YBCO bi-crystal junction with an Au control loop showed that it, indeed, was possible to modulate the maximum current by varying the gate current[46].

A 30 μm long junction gave a larger response than one that was 20 μm long. The gain, which is the ratio of the junction current to the gate current, increased with the junction length and with the critical current density, roughly like $\sqrt{j_c}$, when j_c was increased by lowering the temperature. Gains larger than unity was registered for T<70 K for the 30 μm long junction[46].

6.4 DIGITAL APPLICATIONS, RSFQ

It may be possible to realize digital electronics circuits using high T_c junctions. These could, in principle, be used in future supercomputers, but more specialized utilizations, like in very fast analog-to-digital converters coupled to superconducting sensors or fast switches for telecommunication, and limited to a large scale integrated level seem to be more likely today. We have already mentioned the recent demonstration of fluxon transfer in a large array shift register by Conductus[34]. Rapid Single Flux Quantum (RSFQ) digital circuits are based upon the transfer of single flux quanta through loops containing Josephson junctions and are considered to have many advantages as compared to competing digital elements[47]. They are operated in a non-latching mode allowing very high frequency operation. The parameter margin, which is crucial for Josephson logics, may be decent for RSFQ circuits. There are two main advantages of using high T_c superconductors: a working temperature of 77 K and an increase of the high frequency limit (coupled to I_cR_n) for circuits operating at 4 K. It is not obvious that high T_c material can be used for SFQ circuits. Flux pinning in films and a considerable spread in junction parameters may impede the application.

We have designed and tested a simple high T_c RSFQ circuit based on YBCO junctions to test the concept. It consists (see Fig.18) of two dc-to-SFQ converters (J1-J2 and J3-J4), two Josephson transmission lines (J5-J8 and J9-J12), and a memory cell (J15-J16 and L25-L26) combined with an SFQ-to-dc output circuit (J17 and J18). The signal and the clock (two triangle waveforms with frequency ω and 2ω, resp.) are applied to the inputs of the signal dc-to-SFQ and the clock dc-to-SFQ converters. The signal and the clock currents are converted into fluxons that are transmitted through the signal and clock Josephson transmission lines to the memory cell / SFQ-to-dc converter (J15-J18). The signal pulse triggers J15 that gives a dc voltage at the output of the SFQ / dc converter. That state is then reset by next clock pulse. Simulation results are shown in Fig.18.

The circuit was realized by using YBCO junctions on a 0-32° asymmetric YSZ bi-crystal substrate. The chip contains 18 junctions aligned along the grain boundary, see Fig.19. There are 6 film layers, including the YBCO circuit, two SrTiO3 amorphous insulating layers, a superconducting Pb/Au/Cr ground plane, and Au/Cr wiring. The performance, at 4.2 K, is illustrated in Fig.20, which shows that the circuit functions as expected.

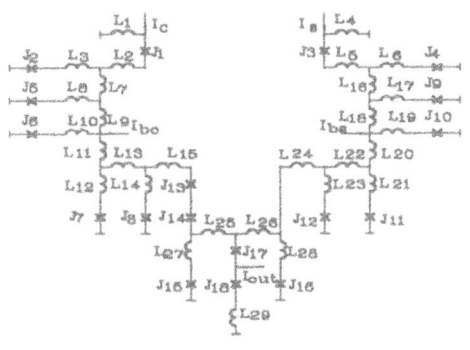

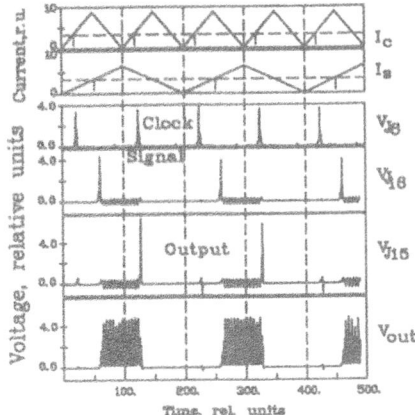

Fig.18. Circuit diagram of test RSFQ circuit based on high T_c junctions.

Fig.19. Simulations of the test RSFQ circuit shown in Fig.18. Parameters are chosen from experimentally achievable values for bi-crystal junctions.

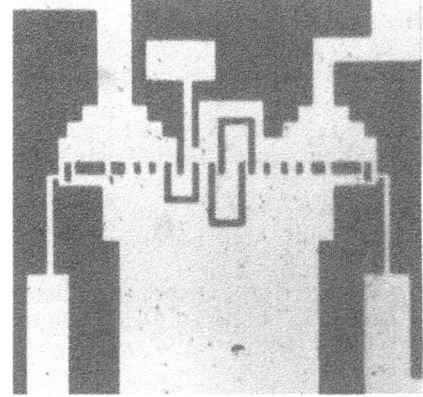

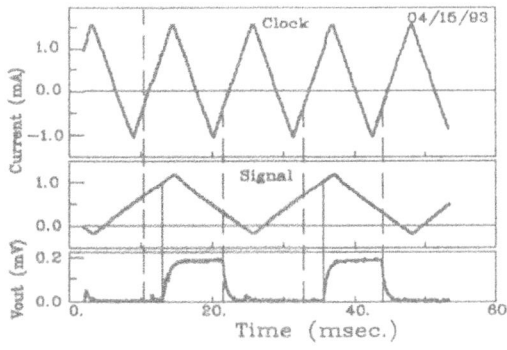

Fig.20. Picture of a fabricated RSFQ test circuit according to Fig.18. The junctions follow the grain boundary line in the YSZ bi-crystal.

Fig.21. Results of tests with the RSFQ circuit of Fig.18. Clock, signal, and output signals are shown. The experiment was performed at 4.2 K.

Acknowledgements

Communication of data by J.S. Martens is acknowledged. Support has been given by several sources, in particular the Swedish Natural Sciences Research Council and the Board of Technical Development (partly through the Materials Consortia), the Technical Sciences Research Council, the Royal Academy of Sciences Cooperation with institutes within the former Soviet Union, and the Swedish Institute.

References

1) Chaudhari, P., "Critical currents, grain boundaries and SQUIDs in the high temperature cuprate superconductors", Physica C, 185-189, 292 (1991). This review gives references to original works.

2) Ivanov, Z.G., Nilsson, P.Å., Winkler, D., Alarco, J.A., Claeson, T., Stepantsov, E.A., and Tzalenchuk, A.Ya., "Weak links and dc SQUIDs on artificial nonsymmetric grain boundaries in YBa$_2$Cu$_3$O$_{7-d}$", Appl. Phys. Lett., 59, 3030 (1991).

3) Gross, R., "Grain boundary Josephson junctions in the high temperature superconductors" in *Interfaces in Superconducting Systems* (Shinde, S.L, and Rudman, D., eds.) Springer Verlag, New York, 1992. The review gives several references.

4) Tomita, N., Takahashi, Y., and Ishida, Y.,"Preparation of bi", Jpn. J. Appl. Phys.29, 130 (1990); Kawasaki, M., Sarnelli, E., Chaudhari, P., Gupta, A., Kussmaul, A., Lacey, J., and Lee, W.,"Weak link behavior of grain boundaries in Nd-, Bi, and Tl-based cuprate superconductors", Appl.Phys.Lett. 62, 417 (1993); Cardona, A.H., Suzuki, H., Yamashita, T., Young, K.H., and Bourne, L.C., "Transport characteristics of Tl$_2$Ba$_2$CaCu$_2$O$_8$ bicrystal grain boundary junctions at 77 K", Appl.Phys.Lett. 62, 411 (1993).

5) Ambegaokar, V. and Halperin, B., "Voltage due to thermal noise in the dc Josephson effect", Phys.Rev.Lett. 22, 1364 (1969); Gross, R., Chaudhari, P., Dimos, D., Gupta, A., and Koren, G., "Thermally activated phase slippage in high-T$_C$ grain boundary Josephson junctions", Phys.Rev.Lett. 64, 228 (1990).

6) Stewart, W.C.,"Current-voltage characteristics of Josephson junctions", Appl.Phys. Lett. 12, 277 (1968); McCumber, D.E.,"Effect of ac impedance on dc voltage-current characteristics of superconductor weak-link junctions", J.Appl.Phys. 39, 3113 (1968).

7) Ivanov, Z.G., Stepantsov, E.A., Tzalenchuk, A.Ya., Shekhter, R.I., and Claeson, T., "Field effect transistor based on a bi-crystal grain boundary Josephson junction", IEEE Trans. Appl. Supercond.3, 2925 (1993).

8) Lu, H.B., Huang, T.W., Wang, J.J., Liu, J., Tu, S.L., Yang, S.J., and Hsu, S.E., "Artificial grain boundaries of YBa$_2$Cu$_3$O$_{7-x}$ on MgO bicrystals", IEEE Trans. Appl. Supercond.3, 2325 (1993).

9) Chen, J., Yamashita, T., Suzuki, H., Myoren, H., Nakajima, K., and Osaka, Y., "YBCO artificial grain boundary junctions on Si", IEEE Trans. Appl. Supercond.3, 2333 (1993).

10) Ivanov, Z.G., Alarco, J.A., Claeson, T., Z.G., Nilsson, P.Å., Olsson, E., Olsson, H.K., Stepantsov, E.A., Tzalenchuk, A.Ya., and Winkler, D., "Artificial grain boundary Josephson junctions - properties and applications", Proc. Beijing Int. Conf. High-T$_C$ Supercond. (1992).

11) Sarnelli, E., Chaudhari, P., and Lacey, J., "Residual critical current in high Tc bicrystal grain boundary junctions", Appl.Phys.Lett. 62, 777 (1993); Däumling, M., Sarnelli, E., Chaudhari, P., Gupta, A., and Lacey, J., "Critical current of a high T$_C$ Josephson junction grain boundary junction in high magnetic field", Appl.Phys.Lett. 61, 1335 (1992).

12) Koch, R.H., Eidelloth, W., Oh, B., Robertazzi, R.P., Andrek, S.A., and Gallaher, W.J., "Identifying the source of 1/f noise in SQUIDs made from high-T$_C$ grain-boundary Josephson junctions", Appl.Phys.Lett. 60, 507 (1992).

13) Olsson, H.K., Koch, R.H., Nilsson, P.Å., Ivanov, Z., Stepantsov, E.A., and Tzalenchuk, A.Ya., "Low 1/f noise in YBa$_2$Cu$_3$O$_{7-\delta}$ dc-SQUID on (Y)ZrO$_2$ bicrystal substrates", Appl.Phys.Lett. 61, 861 (1992); Olsson, H.K., Koch, R.H., Nilsson, P.Å.,

and Stepantsov, E.A., "Dc-SQUIDs with low noise and large β_L-values on (Y)ZrO$_2$ bicrystal substrate", IEEE Trans. Appl. Supercond.$\underline{3}$, 2426 (1993).

14) Alarco, J.A., Olsson, E., Ivanov, Z.G., Nilsson, P.Å., Winkler, D., Stepantsov, E.A., and Tzalenchuk, A.Ya., "Microstructure of an YBa$_2$Cu$_3$O$_{7-\delta}$ thin film artificial grain boundary weak link grown on a 0-45º [001] tilt Y-ZrO$_2$ bicrystal", Ultramicroscopy $\underline{51}$, 239 (1993).

15) Alarco, J.A., Brorsson, G., Olin, H., and Olsson, E., "Early stages of growth of YBa$_2$Cu$_3$O$_{7-\delta}$ high T_c superconducting films on (001) Y-ZrO$_2$ substrates", to be publ.

16) Tietz, L.A., Carter, C.B., Lathrop, D.K., Russek, S.E., Buhrman, R.A., and Michael, J.R., "Crystallography of YBa$_2$Cu$_3$O$_{6+x}$ thin film-substrate interfaces", J.Mater.Res. $\underline{4}$, 1072 (1989).

17) Garrison, S.M., Newman, N., Cole, B.F., Char, K., and Barton, R.W., "Observation of two in-plane epitaxial states in YBa$_2$Cu$_3$O$_{7-\delta}$ films on yttria-stabilized ZrO$_2$", Appl.Phys.Lett. $\underline{58}$, 2168 (1991).

18) Singh, R.K., Narayan, J., and Singh, A.K., "*In situ* fabrication of epitaxial YBa$_2$Cu$_3$O$_7$ films on lattice-mismatched (100) YS-ZrO$_2$ substrates by the pulsed laser evaporation method", Appl.Phys.Lett $\underline{67}$, 3452 (1990).

19) Ivanov, Z.G., Brorsson, G., and Claeson, T., "Fabrication and properties of HTS diffusion type weak links", IEEE Trans.Magn., $\underline{MAG-27}$, 3324 (1991).

20) See, e.g., *Single Charge Tunneling - Coulomb Blockade Phenomena in Nano-structures* (Grabert, H. and Devoret, M.H., eds.) NATO ASI Series B: Physics Vol. 294, Plenum, N.Y. (1992).

21) Mannhart, J., Bednorz, J.G., Müller, K.A., and Schlom, D.G., "Electric field effect on superconducting YBa$_2$Cu$_3$O$_{7-\delta}$ films", Z.Phys.B $\underline{83}$, 307 (1991).

22) Xi, X.X., Li, Q., Doughty, C., Kwon, C., Bhattacharya, S., Findikogly, A.T., and Venkatesan, T., "Electric field effect in high T_c superconducting ultrathin YBa$_2$Cu$_3$O$_{7-x}$ films", Appl.Phys.Lett. $\underline{59}$, 3470 (1991).

23) Brorsson, G., Boikov, Yu., Ivanov, Z.G., and Claeson, T., "Field effect devices based on metal-insulator-YBa$_2$Cu$_3$O$_{7-x}$ films", IEEE Trans. Appl. Supercond.$\underline{3}$, 2922 (1993).

24) Char, K., Colclough, M.S., Garrison, S.M., Newman, N, and Zaharchuk, G., "Bi-epitaxial grain boundary junctions in YBa$_2$Cu$_3$O$_7$", Appl. Phys. Lett., $\underline{59}$, 733 (1991); Char, K., Colclough, M.S., Lee, L.P., and Zaharchuk, G., "Extension of the bi-epitaxial Josephson junction process to various substrates", Appl. Phys. Lett., $\underline{59}$, 2177 (1991).

25) IJsselstein, R.P.J., Hilgenkamp, J.W.M, Eisenber, M., Terpstra, D., Flokstra, and Rogalla, H., "Multilayer studies and applications in template bi-epitaxial dc SQUIDs", IEEE Trans. Appl. Supercond.$\underline{3}$, 2321 (1993).

26) Siegel, M., Herrmann, K., Copetti, C., Jia, C.L., Kabius, B., Schubert, J., Zander, W., and Braginski, A.I., "Investigation of YBCO step-edge Josephson junctions", IEEE Trans. Appl. Supercond.$\underline{3}$, 2369 (1993).; Vildic, M., Friedl, G., Uhl, D., Daalmans, G., Köhler, H., Meyer, H., Bömmel, F., and Saemann-Ishenko, G., "Transport porperties of YBaCuO step edge junctions", IEEE Trans. Appl. Supercond.$\underline{3}$, 2357 (1993)..

27) Jia, C.L., Kabius, B., Urban, K., Herrmann, K., Schubert, J., Zander, W., and Braginski, A.I., "The microstructure of epitaxial YBa$_2$Cu$_3$O$_7$ films on steep steps in LaAlO$_3$ substrates", Physica C $\underline{196}$, 211 (1992).

28) Tanaka, S., Kado, H., Matsuura, T., and Itozaki, H., "Step-edge junction of YBCO thin films on MgO substrates", IEEE Trans. Appl. Supercond.$\underline{3}$, 2365 (1993).

29) Yuan, C.W., Berezin, A.B., and de Lozanne, A.L., Step-edge YBCO dc SQUIDs on sapphire substrates", Appl.Phys.Lett. 60, 2552 (1991).

30) Jia, C.L., Kabius, B., Urban, K., Herrman, K., Cui, G.J., Schubert, J., Zander, W., Braginski, A.I., and Heiden, C., "Microstructure of epitaxial $YBa_2Co_3O_7$ films on step-edge $SrTiO_3$ substrates", Physica C, 175, 545 (1991).

31) Eom, C.B., Marshall, A.F., Suzuki, Y., Geballe, T.H., Boyer, B., Pease, R.F.W., van Dover, R.B., and Phillips, J.M., "Growth mechanisms and properties of 90° grain boundaries in $YBa_2Cu_3O_7$ thin films", Phys.Rev. B 46, 11902 (1992).

32) Moeckly, B.H., Russek, S.E., Lathorp, D.K., Buhrman, R.A., Li Jian, and Mayer J.W., "Growth of $YBa_2Co_3O_7$ thin films on MgO: The effect of substrate preparation", Appl. Phys. Lett., 57, 1687 (1990).

33) Chew, N.G., Goodyear, S.W., Humphreys, R.G., Satchell, J.S., Edwards, J.A., and Keene, M.N., "Orientation control of $YBa_2Cu_3O_7$ thin films on MgO for epitaxial junctions", Appl. Phys. Lett., 60, 1516 (1992).

34) Martens, J.S., Char, K., Johansson, M.E., Whiteley, S.R., Wendt, J.R., Hietala, V.M., Plut, T.A., Ashby, C.I.H., Hou, S.Y., and Phillips, J.M., "A high temperature superconducting shift register operating over 100 GHz", subm. Appl.Phys.Lett. (1993).

35) Koch, R.H., Gallagher, W.J., Bumble, B., Lee, W.Y.,"Low noise thin film TlBaCaCuO dc SQUIDs operated at 77 K", Appl.Phys.Lett 54, 951 (1989).

36) Kawasaki, M., Chaudhari, P., Newman, T.H., and Gupta, A., "Submicron $YBa_2Cu_3O_{7-\delta}$ grain boundary junction dc SQUIDs", Appl.Phys.Lett. 58, 2555 (1991).

37) Friedl, G., Vildic, M., Roas, B., Uhl, D., Bömmel, F., Römheld, B., Hillenbrand, B., Stritzker, B., and Daalmans, G., "Low 1/f noise single-layer $YBa_2Cu_3O_x$ DC-SQUID at 77 K", Appl.Phys.Lett.60,3048 (1992).

38) Zhang Yi, Mück, H.-M., Herrmann, K., Schubert, J., Zander, W., Braginski, A.I., and Heiden, C., "Low-noise $YBa_2Cu_3O_7$ rf SQUID magnetometer", Appl. Phys. Lett., 60, 645 (1992).

39) Ramos, J. Seitz, M., Dallmans, G.M., Uhl, D., Ivanov, Z., and Claeson, T., to be publ.

40) Zarembinski S. and Claeson, T., "Design of multi-loop input circuits for high T_C superconducting quantum interference magnetometers", J. Appl. Phys., 72, 1918 (1992)

41) For a review, see, e.g., Winkler, D., Ivanov, Z., and Claeson, T., "Superconducting detectors for mm and sub-mm waves", in *Superconducting Technology - 10 Case Studies* (ed. Fossheim, K.), World Sci., Singapore (1991), p.51.

42) Zhang, Y.M., Winkler, D., and Claeson, T., "Linewidth measurements of flux-flow Josephson oscillators using a CAD designed integrated sub-mm wave receiver", Proc. 4th Int. Symp. Space THz Techn., 485 (1993); "Linewidth measurements of Josephson flux-flow oscillators in the band 280-330 GHz, Appl.Phys.Lett.62, 3195(1993).

43) Edstam, J., Nilsson, P.-Å., Stepantsov, E.A., and Olsson, H.K., "100 GHz oscillations on monolithic high T_C chips", Appl.Phys.Lett.62, 896 (1993); "On-chip diagnostics of high-T_C superconductors at mm-wave frequencies", to be publ.

44) Winkler, D., Zhang, Y.M., Nilsson, P.Å., Stepantsov, E., and Claeson, T., "Self-induced resonances in YBCO bicrystal grain boundary Josephson junctions", subm LT-20.

45) Zhang, Y.M., Winkler, D., Nilsson, P.-Å., Claeson, T., and Stepantsov, E., "Velocity-matching steps in YBCO grain boundary long Josephson junctions", to be publ.

46) Zhang, Y.M. et al., to be publ.

47) For a recent review, see Likharev, K.K., "Rapid single-flux-quantum logic" in *The New Superconducting Electronics*, Kluver, Dordrecht (1993), in print.

Part IV

Organic Superconductors

HIGH-ENERGY SPECTROSCOPIC STUDIES OF FULLERENE AND CUPRATE SUPERCONDUCTORS

J. FINK, P. ADELMANN, M. ALEXANDER, K.-P. BOHNEN,
M.S. GOLDEN, M. KNUPFER, M. MERKEL, N. NÜCKER,
E. PELLEGRIN, H. ROMBERG, M. ROTH, AND E. SOHMEN
Kernforschungszentrum Karlsruhe
Institut für Nukleare Festkörperphysik
Postfach 3640
W-7500 Karlsruhe
Federal Republic of Germany

ABSTRACT. In order to try to understand the mechanism of superconductivity in doped fullerenes and cuprates we have performed high-energy spectroscopic studies of the electronic structure of both these superconductors in the normal state and of related compounds. In particular, we report measurements by photoemission, electron energy-loss and x-ray absorption spectroscopy.

1.   Introduction

1.1.   ELECTRONIC STRUCTURE AND THE MECHANISMS OF SUPER-CONDUCTIVITY

The microscopic understanding of the mechanism of superconductivity in any system requires detailed knowledge of the electronic structure in the normal state. Within the conventional BCS theory of superconductivity, the superconducting transition temperature T_c is given by

$$kT_c = 1.14\, \hbar\omega_c\, e^{-\frac{1}{\lambda}} \qquad (1)$$

where λ is the electron-phonon coupling strength which is given by the product $\lambda = N(E_F)V$ of the single spin density of states at the Fermi surface, $N(E_F)$, with the pairing potential, V. ω_c is the cut-off frequency of the phonon spectrum, which is of the order of the Debye frequency. Thus the density of states at the Fermi level is an important parameter in electron-phonon based theories of superconductivity. In the previous high-T_c superconductors such as the A15 compounds, there is a particularly high and stongly varying density of states at E_F, leading to a strong electron-phonon coupling. So not only the value $N(E_F)$, but also the detailed shape of the density of states in the energy range of the Debye frequency is important for the understanding of the high transition temperatures in these compounds. In some superconductors, such as

E. Kaldis (ed.), Materials and Crystallographic Aspects of HTc-Superconductivity, 493–520.
© 1994 *Kluwer Academic Publishers.*

vanadium, $N(E_F)$ is so high that the Stoner condition for magnetism is almost satisfied. Then, T_c is supressed by spin fluctuations. Heavy fermion superconductors have a very high density of states as derived by specific heat measurements but they have low critical temperatures, T_c. Their low-energy electronic structure is determined by a Kondo-like weak hybridization of f electrons with s, p, and d conduction electrons.

In the doped fullerenes, the width of the conduction band is comparable to the high vibrational energies of the C_{60} molecule. A crucial point to the validity of the BCS theory or more refined theories, such as the Eliashberg theory for strong coupling superconductors, is the applicability of Migdal's theorem, which implies a slowly varying electronic density of states around E_F on the scale of $\hbar\omega_c$. In the doped fullerenes, there is lively debate whether or not the Migdal theorem can be applied. Presently, it is under discussion whether coupling of the conduction electrons is to molecular vibrations, low-energy phonons or whether electron-correlation based couplings are operative.

For the cuprate supercoductors a low density of states at E_F has been determined. This may suggest that the pairing interaction in the cuprates is probably not mediated by a coupling of the electrons of a Fermi liquid to phonons. At present there is a strong discussion whether these compounds are Fermi liquid systems at all, or rather whether they behave as marginal Fermi liquids or Luttinger liquids. There are numerous theories on the mechanism of superconductivity in the cuprate compounds. Some theories attribute the condensation into the superconducting state to excitonic (charge transfer, acoustic plasmons, d-d excitions) or magnetic degrees of freedom, others start from Hubbard like Hamiltonians. One reason for the development of such a variety of theories is that even the electronic structure in the normal state of these compounds is still rather unclear.

1.2. HIGH-ENERGY SPECTROSCOPIES

High-energy spectroscopies are particular well suited to obtain information on the electronic structure of superconductors in the normal state. Recently, important results have also been derived for the superconducting state. In Fig. 1 we have illustrated the principle of these spectroscopies. Ultraviolet photoemission spectroscopy (UPS) and angle resolved UPS (ARUPS) give information on the occupied states and their dispersion in k-space by analyzing photoelectrons produced by the absorption of photons from gas discharge lamps or from a synchrotron radiation source. Since the energy of the emitted electrons is fairly low, this method is extremely surface sensitive. In order to obtain results representing bulk properties, careful preparation of surfaces is needed. The best energy resolution which can presently be achieved is about 20 meV. Therefore, detailed information of the electronic structure close to the Fermi level can be derived. X-ray induced photoelectron spectroscopy can be used to determine the density of occupied valence states as well as the binding energy of core levels relative to the Fermi energy. Inverse photoemission (IPES) and Bremsstrahlen Isochromat Spectroscopy (BIS) probe the unoccupied states by detecting photons which are emitted after bombarding the sample with

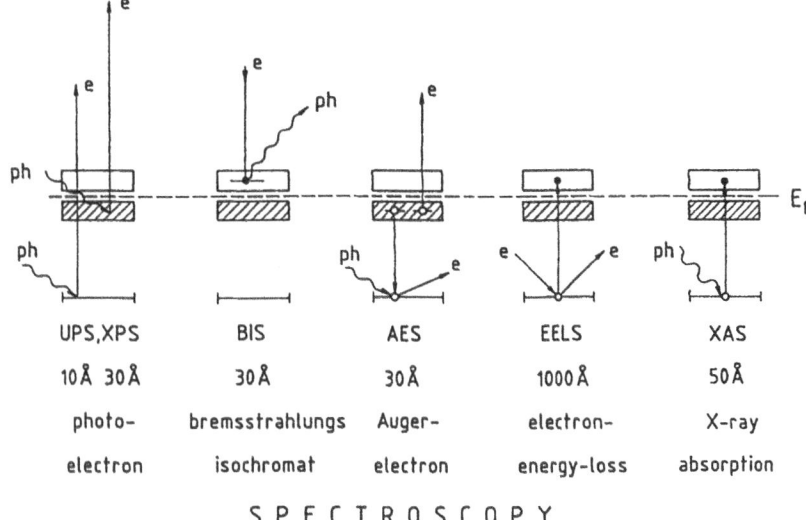

Fig. 1: Illustration of the principle of various high-energy spectroscopies. In addition, approximate probing depths are given.

monochromatic electrons. Using Auger electron spectroscopy (AES), a two-hole state in the valence band can be probed and when comparing with a one-hole state as measured by photoemission, information on the Coulomb repulsion, U, of two holes in the valence band can be derived. This is an important parameter in highly correlated systems such as the cuprates and the fullerenes.

In electron energy-loss spectroscopy (EELS) high-energy electrons ($30\,\mathrm{keV} < E_{kin} < 300\,\mathrm{keV}$) are transmitted through samples having a thickness of about 1000 Å. By recording the energy-loss of the inelastically scattered electrons, excitation of electrons from core levels and from valence bands into unoccupied states can be studied, yielding information on the unoccupied and the joint density of states, respectively. Starting from core levels of different atoms in the sample, EELS (and XAS) offers the unique possibility to obtain site-selective information on the unoccupied states at the different atoms. Furthermore, when performing angle-dependent inelastic electron scattering experiments on single crystals, the orbital symmetry of unoccupied states can be determined. We emphasize here that EELS is not a surface sensitive method.

Similar core-electron excitations into unoccupied states can be performed by x-ray absorption spectroscopy. When measuring the absorption of photons by recording Auger and secondary electrons emitted after the absorption process, the technique is rather surface sensitive (~ 50 Å). On the other hand, when recording the fluorescence radiation produced on core-hole decay, a probing depth of the order of 1 μm is expected. In this case, essentially bulk properties are measured.

2.   Fullerenes

2.1.   ELECTRONIC STRUCTURE OF UNDOPED FULLERENES

The fullerene molecules were first discovered by Kroto et al. [1] during laser evaporation of graphite. The prototype of the fullerenes, the soccerball-like carbon cluster, C_{60}, has the symmetry of a truncated icosahedron. In a first approximation, the four valence electrons of carbon are hybridized in a sp^2 configuration. Three strong σ bonds within the surface of the molecule essentially form the cage. In addition, there are weaker π bonds from p_z orbitals which are perpendicular to the surface of the molecules. A calculation of the π-electron

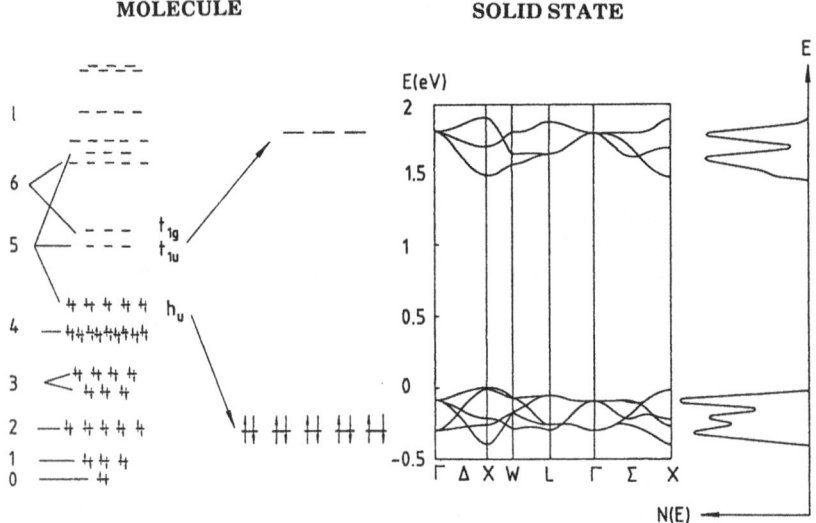

Fig. 2:   Electronic structure of the π-electrons in C_{60}. Extreme left panel: molecular orbital levels of the C_{60} molecule. Left panel: highest occupied molecular orbital and lowest unoccupied molecular orbital of the C_{60} molecule. Right panel: group of the five highest π valence bands and group of the three lowest $π^*$ conduction bands in solid C_{60} (from band structure calculations [3]). Extreme right panel: schematic density of states of the highest valence bands and the lowest conduction bands.

level scheme is shown in the extreme left panel of Fig. 2. The molecular orbital levels of the σ electrons, which are further away from the gap are not shown. Since the molecule is almost spherical, the orbitals can be classified by the angular momentum, l, as in a hydrogen atom. But due to deviations of the truncated icosahedron from the spherical shape, there is a splitting of the molecular levels for l ≥ 3. The highest occupied molecular level (HOMO) is five-fold degenerate and has h_u symmetry. The lowest unoccupied molecular orbital (LUMO) and the second lowest unoccupied molecular orbital (LUMO + 1) have t_{1u} and t_{1g} symmetry, respectively, and are both threefold degenerate. After the synthesis of macroscopic amounts of fullerenes by Krätschmer et al. [2], it has been possible to investigate solid C_{60}. The molecules condense to form an fcc lattice. In the solid the molecular levels are slightly broadened due to the weak overlap of π orbitals between neighbouring molecules. The band structure of the highest group of five valence bands and the lowest group of three conduction bands as derived from calculations in the local density appoximation (LDA) [3] are also shown in Fig. 2. The width of the bands is close to 0.5 eV and the gap has been estimated to be about 2 eV. The extreme right panel of Fig. 2 shows schematically the density of states for the highest group of five valence bands and the lowest group of three conduction bands. The question is whether this picture for the electronic structure of solid C_{60} is applicable or whether correlation effects which are not exactly taken into account in LDA band structure calculations or strong coupling to phonons, invalidate this model.

In Fig. 3 (a) we show high-resolution valence band photoemission profiles [4] of undoped C_{60}. In Fig. 3 (b) and (c) we compare these experimental results with calculations of the density of states in the local density approximation (LDA) using the linear-muffin-tin-orbitals (LMTO) method beyond the atomic spheres approximation [5] and using the pseudopotential method with a mixed basis set [6], respectively. At lower binding energies, E_B, the bands have mainly π character while for $E_B > 5$ eV the sharp features can be probably assigned to σ bands. The peak with the lowest binding energy is due to the HOMO-derived highest group of five valence bands. There is a qualitative agreement between the calculated and the measured density of states. However, the calculated density of states show well pronounced structures which could not be resolved in the experiment although the energy resolution in the experiment is an order of magnitude smaller than the separation of the calculated van Hove singularities. Probably these singularities are washed out in the experiment due to phonon excitations produced during the photoemission process. The spectra do not change as a function of temperature, although there are structural phase transitions at 250 and 90 K. We emphasize that in particular in the energy range $E_B < 4$ eV there are considerable differences between the various calculated density of states.

Information on the unoccupied density of states is obtained from C1s absorption edges measured by EELS [7]. Those spectra are shown in Fig. 4 for solid C_{60} and C_{70}. Well resolved transitions from the C1s core level into unoccupied π* levels are observed between 284 and 290 eV. By a comparison with similar data on graphite, the transitions above 292 eV can be assigned to tran-

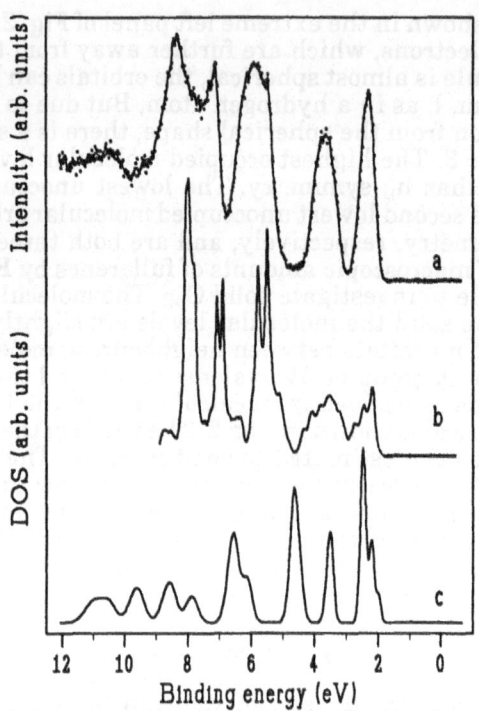

Fig.3: (a) Ultraviolet photoemission spectra of solid C_{60} taken with an energy resolution of 25 meV. (b) LDA band structure calculations of the density of states (DOS) using the linear-muffin-fin-orbitals method beyond the atomic spheres approximation. (c) LDA band structure calculations, using the pseudopotential method with a mixed basis set.

sitions into unoccupied σ^* bands. For C_{60}, peaks at 284.4 and at 285.8 eV correspond to the LUMO derived lowest group of conduction bands and the (LUMO + 1)-derived bands, respectively. The C_{70} molecule is no longer quasi spherical and therefore has a lower symmetry. This leads to a partial lifting of the degeneracy of the molecular orbitals in C_{70} and to a different π electronic structure when compared with C_{60}. Therefore, for C_{70} the energy positions of the π^* bands and the widths of these bands are different from those of C_{60} (see Fig. 4).

The results for the electronic structure shown in this review and many similar results by other groups support the picture for the electronic structure derived from LDA band structure calculations. However, recent Auger experiments by Lof et al. [8] yield a high Coulomb repulsion, $U = 1.6$ eV, of two electrons (or holes) on a C_{60} molecule. These results are supported by the large gap measured by UPS and IPES. Since U is larger than the width of the bands ($W = 0.5$ eV) these solids should be highly correlated systems. This should be of particular importance in systems with a partially filled band, e.g., the doped fullerenes.

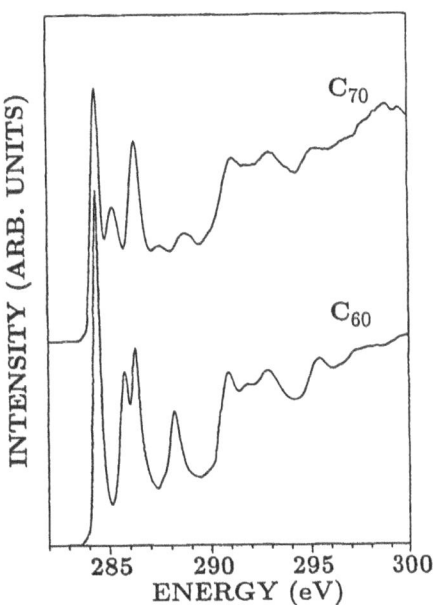

Fig. 4: C1s absorption edges, measured by electron energy-loss spectroscopy for solid C_{60} and C_{70}.

2.2. ELECTRONIC STRUCTURE OF ALKALI-METAL DOPED FULLERENES

In the salt-like compounds A_3C_{60} (A = K, Rb, and mixtures of alkali metals) superconductivity has been discovered [9] with transition temperatures of up to $T_c = 33$ K. In these compounds, a charge-transfer of the three alkali metal electrons to the C_{60} molecule is expected leading to a half-filled LUMO-derived conduction band. In the compounds A_6C_{60}, the lowest group of conduction bands should be completely filled, in agreement with the observation of insulating behaviour for these compounds. For the A_3C_{60} phases several photoemission studies have observed a sharp Fermi edge at the maximum of the occupied LUMO-derived bands. In constrast, Takahashi et al. [10] claim that K_3C_{60} exhibits a pseudogap of about 0.5 eV at the Fermi level, due to either electron correlations or polaron formation. This claim raises again the question about the role correlation plays in the electronic properties of fullerenes, and has been cited by Lof et al. [8] as support for their contention that stoichiometric K_3C_{60} is a Mott-Hubbard insulator with a gap of 0.7 eV.

In order to clarify these differences we have performed a series of UPS measurements accompanied by a detailed determination of the surface concentration by XPS [4]. In Fig. 5 we show photoemission energy distribution curves of

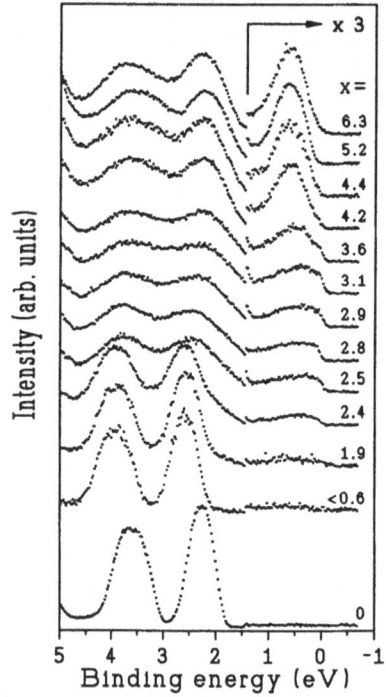

Fig. 5:   Room-temperature photoemission spectra of K_xC_{60} as a function of K doping level, x. The measurements were performed with a photoenergy of 21.2 eV and an energy resolution of 25 meV. The occupied part of the LUMO-derived bands is expanded vertically by a factor of 3.

K_xC_{60} for various x measured at room temperature. On doping with K, besides the occupied π bands, one observes for x = 3 the half-filled LUMO-derived conduction band with a clear Fermi edge. For x = 6, this band is completely filled and the Fermi edge has disappeared. In the intermediate doping ranges there is probably a mixture of two phases, which is evidence for the existence of only three phases in the K_xC_{60} system under growth conditions, used here.

In Fig. 6 we show the density of states at E_F as a function of x. Within the experimental uncertainty, the maximum in the density of the states is found in the half-doped, K_3C_{60}. These findings indicate that there is no pseudogap at the Fermi level for K_3C_{60}. Although the existence of a Mott-Hubbard insulator with a gap of 0.7 eV for exactly x = 3 cannot ruled out, the data shown in Fig. 6 suggest that this eventuality is unlikely.

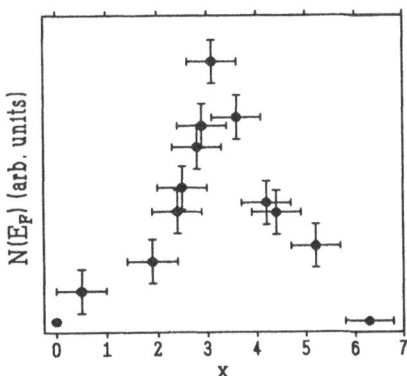

Fig. 6:   The density of states at E_F as a function of x for K_xC_{60} determined from the height of the Fermi edges observed in the spectra of Fig. 5.

Evaluating the absolute value of the density of states at E_F for K_3C_{60}, one obtains $N(E_F) \sim 3$ states eV spin C_{60}. This value is about a factor of 3 smaller than that evaluated by LDA band structure calculations. Moreover, the width of the half-filled conduction bands ($W_{1/2} = 1.3$ eV) indicates a total width of 2.6 eV and represents a real puzzle, since LDA band structure calculation predict a total width of 0.5 eV. In a recent temperature dependent UPS study of the conduction bands of K_3C_{60} and Rb_3C_{60} it was realized that the spectral profiles exibit at 10 K fine structure (see Fig. 7) not observed at higher temperatures [11]. Also shown in Fig. 7 are calculations of the photoemission spectra taking into account the excitation of a charge carrier plasmon at 0.5 eV and the A_g and H_g vibrational modes of the C_{60} molecule during the photoemission process. For a certain electron-phonon coupling constant, $\lambda/N(E_F) = 0.091$, good agreement between calculated and experimental photoemission spectra could be achieved. Using the values $N(E_F) = 8.5$ and 10 from LDA band struc-

502

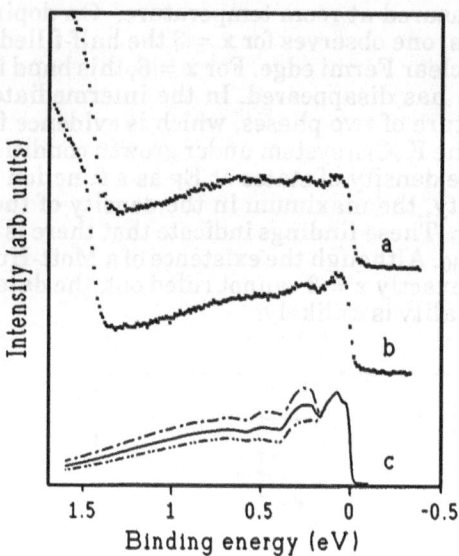

Fig. 7: Conduction band photoemission spectra of (a) K_3C_{60} and (b) Rb_3C_{60} recorded at 10 K. Also shown in (c) are calculated photoemission spectra with electron phonon coupling constants, $\lambda/N(E_F)$, of 0.069 (dotted-dashed line), 0.091 (solid line) and 0.114 (dashed line).

ture calculations, $\lambda = 0.77$ and 0.91 for K_3C_{60} and Rb_3C_{60} is obtained, respectively, which are of the correct order of magnitude to explain the superconducting transition temperatures within a BCS mechanism (see Eq. 1). *Ab initio* calculations of the electron phonon coupling strength yield similar values.

This analysis of the spectra solves the problem of the unusual width of the conduction bands detected in photoemission studies. This large width is caused by phonon and plasmon satellites. Moreover, this study indicates, as do many other experimental results, that superconductivity in the A_3C_{60} compounds is probably related to the coupling of the conduction electrons to the vibrational modes of the C_{60} molecule. Finally, these results cannot exclude the existence of some spectral weight in the satellite region due to correlation effects. However, the analysis of the data clearly shows, that most of the satellite spectral weight is due to a coupling to phonon and plasmons. The spectra also show that about 20 % of the total spectral weight of the conduction band is in coherent states at the Fermi level with a width of about 0.2 eV. Therefore, the system cannot be completely condensed in a polaronic or strongly correlated state in which no coherent states would be expected.

The unoccupied states of doped fullerenes have also been studied by recording C1s absorption spectra using EELS. In Fig. 8 we show those spectra for the system Rb_xC_{60} [12, 13]. For x = 0, four peaks are observed, corresponding to transitions from the C1s level into groups of unoccupied π^* bands. For x = 6, only three peaks are observed indicating the complete filling of the LUMO-derived group of π^* bands. When comparing the x = 6 with the x = 0 spectrum, there is a shift of about 1 eV to lower energy due to a reduction of the 1s core-level binding energy caused be a screening of the nuclear potential by the six additional conduction electrons. For x = 3, the intensity of the first peak (the 1s → LUMO transition) is about one half of that for x = 0. This shows again that for Rb_xC_{60} the conduction band is half filled.

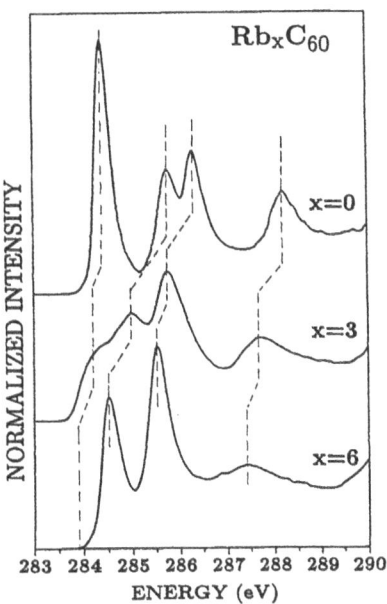

Fig. 8: C1s absorption edges of Rb_xC_{60} for x = 0, 3, and 6. The spectra were recorded by electron energy-loss spectroscopy with an energy resolution of 0.12 eV.

In Fig. 9 we show low-energy loss functions (Im(-1/ε), ε being the dielectric function) for Rb_xC_{60} (x = 0, 3, and 6). In the undoped case, a gap of about 2 eV is seen, followed by several maxima due to π plasmons caused by π-π^* transitions. For x = 3, new transitions appear in the gap. The peak at 0.5 eV has been assigned to a charge carrier plasmon related to the three conduction electrons in Rb_3C_{60}. The second peak at 1.2 eV is due to a transition from the LUMO-derived bands to the (LUMO + 1)-derived bands. This transition is enhanced for x = 6 since there are now more electrons in the initial state, i.e. the LUMO-derived bands. But no charge carrier plasmon is observed for x = 6, indicating again the complete filling of the LUMO-derived bands at this doping level.

The loss data for the superconducting Rb_3C_{60} clearly show a Drude-like behaviour of the low-energy excitations in this system. However, we cannot exclude the existence of additional low-energy excitations for E < 100 meV, which were recently observed by optical spectroscopy [14].

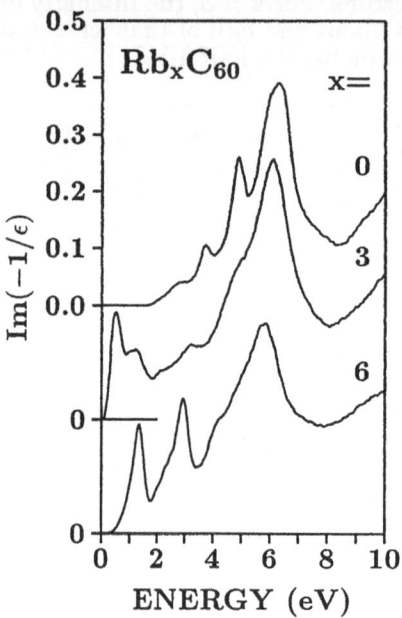

Fig. 9: Low-energy loss function of Rb_xC_{60} for x = 0, 3, and 6 as measured by electron energy-loss spectroscopy.

3.   Cuprates

3.1.   INTRODUCTION TO THE ELECTRONIC STRUCTURE OF CUPRATES

The cuprate high-T_c superconductors have one structural element in common, namely the two-dimensional CuO_2 planes. They are located in between block layers from which p- and n-type doping can be achieved. The undoped cuprates, e.g., La_2CuO_4 are antiferromagnetic insulators with a gap of about 2 eV.

505

Upon doping, long-range antiferromagnetism is lost and superconductivity appears only to disappear itself at higher dopant concentrations (x ~ 0.25). LDA band structure calculations [15] predict undoped La₂CuO₄ to be a paramagnetic metal. The electronic structure near the Fermi level should be determined by an antibonding Cu-O band formed from in-plane Cu3d$_{x^2-y^2}$ and O2p$_{x, y}$ orbitals, which form σ bonds with Cu. The Fermi level should be in the center of this antibonding σ* band leading to a half-filled band. This band is illustrated in Fig. 10 (a). Neglecting the impurity potential of the dopants (e.g., for La₂CuO₄, divalent Sr replaces trivalent La) the Fermi level should be shifted

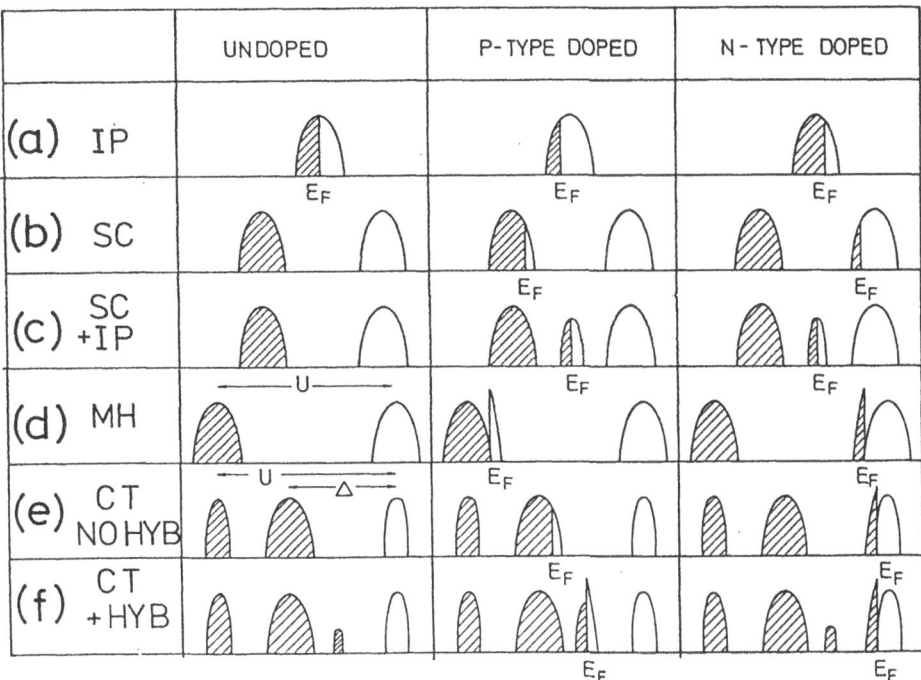

Fig. 10: Models, being discussed for the electronic structure of unodped, p-type doped, and n-type doped CuO₂ planes. (a) antibonding Cu3d-O2p σ* band as derived in an independent particle (IP) model or from LDA band structure calculations. (b) SC: normal semiconductor model with small impurity potential of the dopant atoms. (c) SC + IP: semiconductor model with strong impurity potential pinning the Fermi level in the center of the gap. (d) MH: Mott-Hubbard model (e) CT: charge transfer insulator without hybridization. (f) CT + HYB: charge-transfer insulator with hybridization.

to lower (higher) energies on p(n)-type doping. In Fig. 10 (b) we show the usual semiconductor model. As long as the impurity potential is small, the Fermi level would move into the valence (conduction) band on p(n)-type doping. However, from band structure calculations, there is no indication for the formation of a normal semiconducting gap in the CuO_2 planes. In Fig. 10 (c) we show a semiconductor model in which the acceptor or donor states formed on p- or n-type doping, respectively, are pinned by a strong impurity potential.

A way to explain the insulating behaviour of undoped CuO_2 planes would be to assume a Mott-Hubbard picture in which, due to strong on-site correlation effects on the Cu-sites, a lower (LHB) and an upper Hubbard band (UHB) are formed. In an ionic picture, on each Cu site there would be one hole in the Cu3d shell. When creating a second hole on the Cu site by photoemission or p-type doping, in addition to the binding energy of the Cu3d electrons, we have to spend the repulsive Coulomb interaction energy, $U = 8 - 10$ eV, between two holes on a Cu site. This is the reason for the formation of the LHB. It should be noted that for the strongly correlated models the states shown in Fig. 10 are no longer a density of states in the sense of an independent particle model. Rather they are spectral weights relevant for experiments in which electrons are removed (shaded area) from the sample by, e.g., photoemission spectroscopy or electrons are added (non shaded area) to the sample by, e.g., inverse photoemission or by core excitations into unoccupied states (EELS, XAS), when neglecting the interaction with the core hole. Upon n-type doping, part of the Cu sites would have a completely filled 3d shell. To remove an electron from these sites, the Coulomb energy U is no longer needed. The spectral weight of these sites must have the same energy as the UHB. Therefore, this spectral weight of the n-type doped Cu sites must appear at the bottom of the UHB (see Fig. 10 (d)). Note that the spectral weight of the UHB loses one state while the spectral weight of the occupied states at the bottom of the UHB gains two states. This is different from the semiconductor model (Fig. 10 (b)). Similarly, on p-type doping, unoccupied spectral weight appears at the top of the LHB. Therefore, in this model, the chemical potential moves across the insulating gap as one goes from electron to hole doping.

Since the gap observed in cuprates is only of the order 2 eV, the simple Mott-Hubbard model is not applicable to these compounds since U is between 8 and 10 eV. The smaller gap can be explained by the charge transfer model [16] in which a less correlated O2p band is located between the LHB and the UHB. The gap is now related to the charge transfer energy, Δ. Without hybridization between Cu and O orbitals, the valence band has O2p character and the UHB has Cu3d character. Upon p-type doping the system would form holes in the O2p band as in the semiconductor model (Fig. 5 (b)). Upon n-type doping, the system would react as in the Mott-Hubbard model.

When switching on the hybridization between O2p and Cu3d states, Cu3d states are admixed to the O2p band and O2p states are admixed to the UHB. According to exact diagonalization cluster calculations [17], with increasing hybridization the charge transfer model transforms increasingly to an effective single-band Mott-Hubbard picture with an effective U of the order of Δ, as has already been proposed by Anderson [18] and in the t-J model [19].

For p-type doping it is important to take into accont the magnetic exchange interaction between holes on O sites and holes on Cu sites, leading to singlet and triplet states. According to cluster calculations, the singlet states are split off from the O2p band. Upon p-doping the Fermi level moves into the singlet band.

Up to now, we have assumed that the low-energy properties of cuprates can be described only by three in-plane orbitals, namely the $Cu3d_{x^2-y^2}$ and the $O2p_{x,y}$ orbitals. At present there is lively debate as to whether other orbitals such as $Cu3d_{3z^2-r^2}$, $O2p_z$ orbitals from apical O sites above and below Cu sites, or in-plane $O2p_{x,y,z}$ (π-bonded to Cu3d) are important for transport properties and the mechanism for high-T_c superconducitvity in cuprates [20 - 24].

3.2. THE CHARACTER OF THE CHARGE CARRIERS IN HIGH-T_c SUPERCONDUCTORS

The first high-T_c superconductor $La_{2-x}Ba(Sr_x)CuO_{4+\delta}$ which was discovered by Bednorz and Müller [25] may be regarded as a paradigm for these materials. Only one CuO_2 plane is situated between two La(Sr)O planes. In Fig. 11 we show O1s absorption edges of $La_{2-x}Sr_xCuO_{4+\delta}$ recorded by EELS [26]. According to Fermi's golden rule and to dipole selection rules they measure the unoccupied density of states at the O sites which have p character. For the undoped sample ($x = 0$, $\delta = 0$), there is a "pre-edge" (labelled with C) at 530.2 eV. It is assigned to transitions into O2p states which, by hybridization, are admixed to the UHB, which has predominantly Cu3d character. In the p-type doped samples ($x > 0$, $\delta > 0$), a second pre-edge (labeled with V) appears at 528.7 eV. Its intensity is roughly proportional to the dopant concentration. The peak can be assigned to holes on O sites in the O2p band or, to be more specific, in the singlet band. This indicates that the holes formed upon doping have a large fraction of O2p character.

When regarding only the pre-edge V, the spectra could be explained in the charge transfer model without hybridization (see Fig. 10 (e)). However, without hybridization the peak C should not exist and, in contrast to the Mott-Hubbard model, in a non-correlated system such as a normal semiconductor, the intensity should not decrease with doping. The decrease is therefore a key signature of the experiment indicating that, due to hybridization, the system has transformed to a highly correlated one which probably can be described in an effective single-band Hubbard model.

The threshold of the pre-edge V is about 1.4 eV lower than the threshold of the pre-edge C, a value which is close to the gap measured by optical spectroscopy ($E_g \sim 1.7$ eV). This indicates that the Fermi level for p-type doped cuprates is close to the top of the valence band. This agrees with the charge transfer models or with the Mott-Hubbard model but is at variance with some of the photoemission data [27] which derive the chemical potential to be in the center of the gap for p- and n-type doping (analogous to the model shown in Fig. 10 (c)).

508

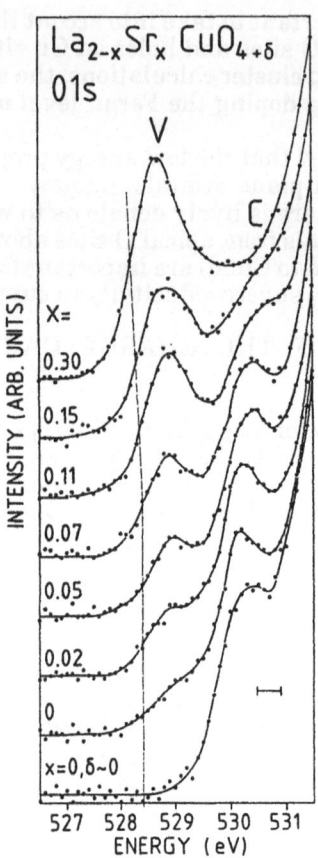

Fig. 11: O1s absorption edges of $La_{2-x}Sr_xCuO_{4+\delta}$ as measured by electron energy-loss spectroscopy.

When regarding the spectra in Fig. 11 it should be noted that there is a continuous change of the spectral weight as a function of dopant concentration. There is no discontinuity between $x = 0.05$ and 0.07 at the insulator-metal transition near $x = 0.06$. Since holes on O sites are already seen for $x \leq 0.05$ in the insulating range, these holes must be localized, e.g., by the impurity potential of the disordered Sr atoms or by magnetic interactions. The UHB can be clearly seen in the superconducting range ($x = 0.11$ and 0.15) again indicating that correlation effects are important for high-T_c superconductors. Even in the overdoped range ($x = 0.3$), the UHB has still 30-50 % of its spectral weight for $x = 0$.

Similar investigations have been performed [28] on polycrystalline samples of the n-type doped system $Nd_{2-x}Ce_xCuO_{4-\delta}$. In Fig. 12 we show the pre-edge structures of EELS measurements of O1s absorption edges on this system.

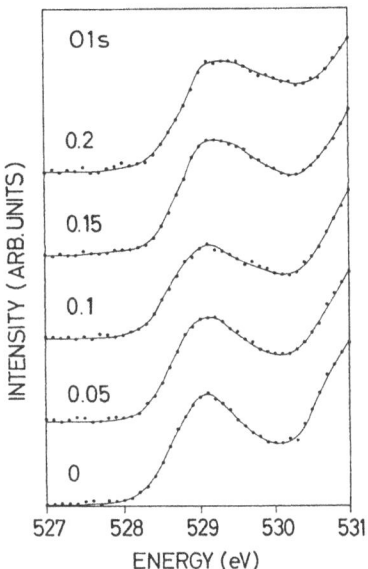

Fig. 12: O1s pre-edge absorption structures of $Nd_{2-x}Ce_xCuO_4$ for various dopant concentrations, x. The measurements were performed by electron energy-loss spectroscopy.

From comparison with the $La_2CuO_{4+\delta}$ data, for undoped $Nd_2CuO_{4-\delta}$ the peak at 529 eV can be assigned to transitions from the O1s level into O2p states admixed to the UHB which has predominantly Cu3d character. The existence of this peak again clearly shows that the charge transfer model without hybridization (see Fig. 10 (e)) is not adequate to describe the electronic structure of the n-type doped cuprates. On n-type doping, there is almost no change of the pre-edge feature. This is consistent with the Mott-Hubbard model (see Fig. 10 (d)) and with the charge transfer models with hybridization (see Fig. 10 (f)) which predicts the occupied spectral weight to appear at the bottom of the UHB having predominantly Cu3d character. Thus, the additional electron does not go onto the O sites. With increasing x, there is a small increase of spectral weight at about 1.5 eV above threshold, filling the valley between the pre-edge and the main absorption edge. According to band structure calculations [29] on Ce doped Nd_2CuO_4, this spectral weight can be assigned to Ce5d states hybridized with O2p states. On the other hand, the existence of Ce4f states at this energy hybridized with O2p states cannot be excluded. The assignment of the additional spectral weight at 1.5 eV above threshold is supported by the fact that in Th doped Nd_2CuO_4 the filling of the valley is not observed.

It is important to note that for Ce doped compounds no spectral weight appears at energies below the threshold of the undoped system. This indicates that there are no partially filled mid-gap bands (see Fig. 10 (c)) which should cause a further pre-edge at lower energies. Even if this band would have predominantly Cu3d character, the strong hybridization between Cu3d and O2p states would cause some admixture of O2p states to this band and therefore a further pre-edge at lower energies should appear. Therefore, the data shown in Fig. 12 clearly support the model of a n-type doped charge transfer insulator with hybridization (see Fig. 10 (f)). Furthermore, in agreement with that model, the data show that the Fermi level in the n-type doped compounds is at the bottom of the UHB. This is again at variance with photoemission data which suggested that the Fermi level is in the centre of the gap [27].

According to the charge transfer models, shown in Figs. 10 (e) and (f), in the n-type doped compounds the additional electron should go onto Cu sites, i.e. the number of unoccupied Cu3d states should be reduced. This can be tested by

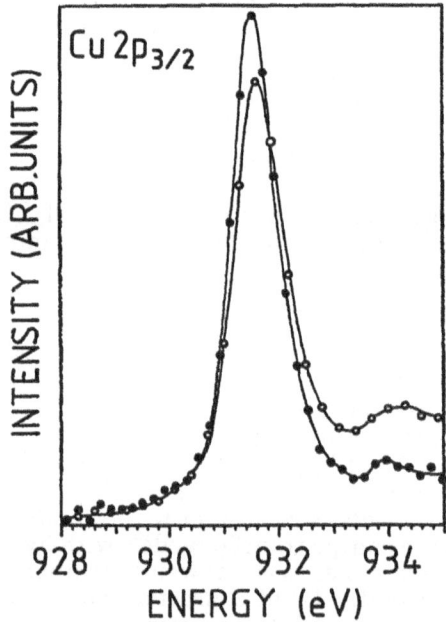

Fig. 13: Cu2p$_{3/2}$ absorption edges of Nd_2CuO_4 (closed circles) and $Nd_{1.85}Th_{0.15}CuO_{4-\delta}$ (open circles). The spectra, recorded by electron energy-loss spectrscopy, are normalized to the Nd3d$_{5/2}$ absorption line.

recording Cu2p absorption edges. Due to a strong interaction with the core hole these edges do not provide detailed information on the density of unoccupied 3d states but give just an excitonic line, the intensity of which should be proportional to the number of unoccupied Cu3d states. In Fig. 13 we show $Cu2p_{3/2}$ absorption edges of $Nd_{2-x}Th_xCuO_{4-\delta}$ for $x = 0$ and $x = 0.15$. There is a 14 % reduction of the absorption line on going from the undoped Nd_2CuO_4 to the n-type doped superconductor $Nd_{1.85}Th_{0.15}CuO_{4-\delta}$. This clearly shows that the additional electrons formed upon doping occupy Cu3d states and that the charge carriers in this superconductor have predominantly Cu3d character.

3.3. THE SYMMETRY OF THE STATES CLOSE TO THE FERMI LEVEL

As already mentioned in Sect. 1.2, angle dependent EELS and polarization dependent XAS measurements of core excitations in single crystals provide important information on the symmetry of unoccupied states. In Fig. 14 we show angle dependent measurements of O1s absorption edges on a superconducting single crystal of $Bi_2Sr_2CaCu_2O_8$ [30]. In the two spectra shown in Fig. 14, scattering angles were chosen in such a way that the momentum transfer, q, is parallel and perpendicular to the CuO_2 (a, b) planes. For q ∥ a, b and q ∥ c, $O2p_{x,y}$ and $O2p_z$ states are probed, respectively. A pre-edge at 528 eV is only observed for q ∥ a, b but not for q ∥ c. Therefore, the data clearly show that the

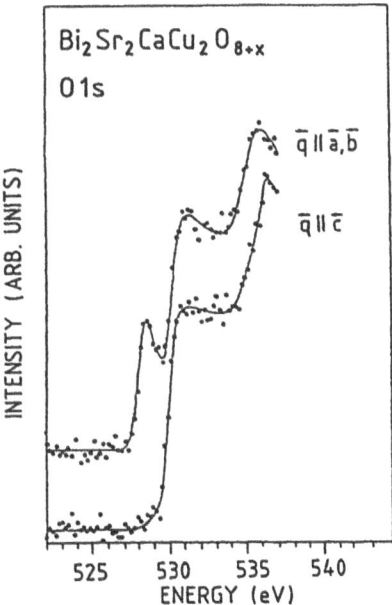

Fig. 14: Angle dependent O1s absorption edges of $Bi_2Sr_2CaCu_2O_8$ measured by electron energy-loss spectroscopy.

hole states on O sites have only $O2p_{x, y}$ and not $O2p_z$ symmetry, i.e. holes are only formed in in-plane $O2p_{x, y}$ orbitals. These results rule out all models for high-T_c superconductivity based on out-of-plane п-holes in CuO_2 planes and $2p_z$ holes on the apical O atoms above and below the Cu sites.

Similar measurements have been performed on the $Cu2p$ absorption edges of $Bi_2Sr_2CaCu_2O_8$ (see Fig. 15) [30]. They probe the symmetry of unoccupied

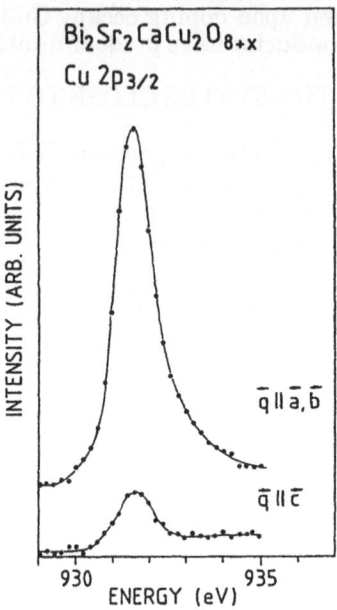

Fig. 15: Angle dependent $Cu2p_{3/2}$ absorption edges of $Bi_2Sr_2CaCu_2O_8$ measured by electron energy-loss spectroscopy.

$Cu3d$ states, i.e. the $Cu3d$ states in the UHB. For $q \parallel a, b$ mainly $3d_{x^2-y^2}$ states and $3d_{3z^2-r^2}$ states are reached, while for $q \parallel c$ only $3d_{3z^2-r^2}$ can be detected. As expected, the spectra reveal that the unoccupied Cu states in the UHB have predominantly $Cu3d_{x^2-y^2}$ symmetry. From the $q \parallel c$ spectrum, a 10 % admixture of $Cu3d_{3z^2-r^2}$ states to the UHB is derived.

Recently, we have performed polarization and concentration dependent XAS measurements on the $La_{2-x}Sr_xCuO_{4+\delta}$ system [31, 32]. In Fig. 16 we show the pre-edges of $La_{2-x}Sr_xCuO_{4+\delta}$ for the polarization vector of the synchrotron radiation, E, parallel and perpendicular to the CuO_2 planes.

For $E \perp c$ and $E \parallel c$, unoccupied $O2p_{x, y}$ and $O2p_z$ states are probed, respectively. As for the data on polycrystalline samples (shown in Fig. 13), for $E \perp c$ two pre-edges are observed. That one at higher energy which is well pronounced for $x = 0$ and decreases with increasing x and shifts to higher energies

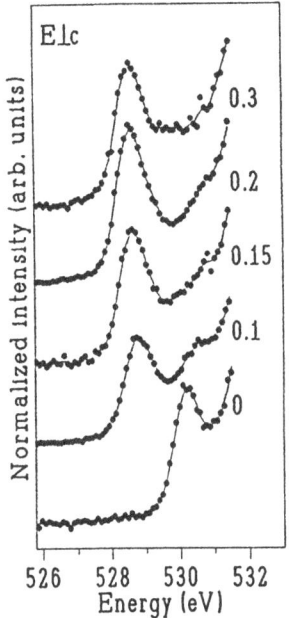 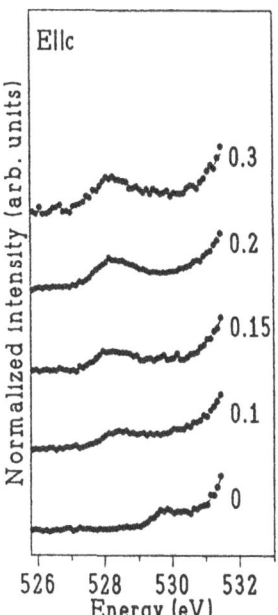

Fig. 16: Polarization and concentration dependent O1s x-ray absorption spectra in the pre-edge region of $La_{2-x}Sr_xCuO_{4+\delta}$ for $E \perp c$ and $E \parallel c$. The spectra are labelled with the concentration, x.

is again assigned to transitions into $O2p_{x, y}$ states admixed to the UHB. The pre-edge at lower energy is due to hole states on O sites with $O2p_{x, y}$ symmetry, formed on p-type doping. In contrast to the $Bi_2Sr_2CaCu_2O_8$ system, hole states with $O2p_z$ symmetry are observed in the $E \parallel c$ spectra. Moreover, even the UHB has some admixture of $O2p_z$ states. Since there is an energy shift between the pre-edges for $E \perp c$ and $E \parallel c$, indicating different binding energies of the O1s level on different O sites relative to the Fermi level, the unoccupied $O2p_{x, y}$ states have been assigned to in-plane O sites while the $O2p_z$ states are assigned to apical O sites. The intensities of the pre-edge for $E \perp c$ and $E \parallel c$ as a function of dopant concentration, x, are displayed in Fig. 17 (a). In addition, the fraction, R, of unoccupied $O2p_z$ states with respect of the total number of unoccupied O2p states in the valence (or singlet band) is shown in Fig. 17 (b) as a function of x, together with analogous data by Chen et al. [32]. For x < 0.2, the number of hole states in the valence (singlet band) with both $O2p_{x, y}$ and $O2p_z$ symmetry increases roughly in proportion to the dopant concentration. For x = 0.3, a saturation is observed, probably because of an O deficiency of the single crystals. The number of unoccupied O states ad-

514

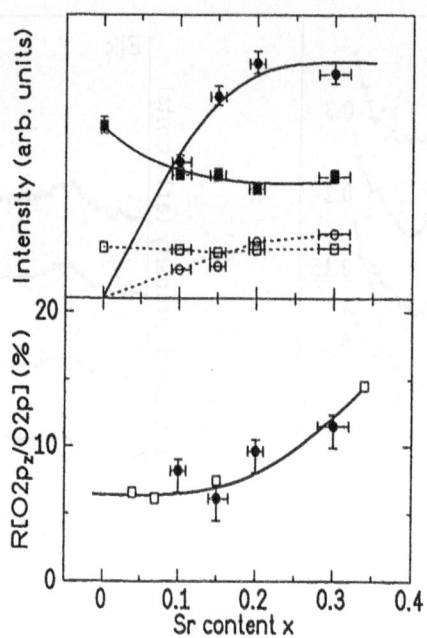

Fig. 17: (a) Intensity of hole states on O sites in $La_{2-x}Sr_xCuO_{4+\delta}$ as a function of Sr content x. Closed (open) circles: valence (singlet) band, $O2p_{x, y}$ ($O2p_z$) symmetry. Closed (open) squares: upper Hubbard band, $O2p_{x, y}$ ($O2p_z$) symmetry. (b) Fraction R of unoccupied $O2p_z$ states with respect to the total number of unoccupied $O2p$ states in the valence (singlet) band of $La_{2-x}Sr_xCuO_{4-\delta}$. Closed circles: data from Ref. 31. Open squares: data from Ref. 32.

mixed to the UHB both with $O2p_{x, y}$ and $O2p_z$ symmetries decreases slightly with increasing x.

For $x < 0.2$, about 7 % of the total number of holes in the valence (singlet) band have $O2p_z$ character. The fact that the number of $O2p_z$ states is considerably lower in $Bi_2Sr_2CaCu_2O_8$ ($T_c = 80$ K) compared to $La_{1.85}Sr_{0.15}CuO_{4+\delta}$ ($T_c = 35$ K) can be taken as an evidence that holes on apical O sites are detrimental to superconductivity. This would support theories which claim that holes on apical O sites will destabilize the Zhang-Rice singlet states and therefore suppress superconductivity [23]. Although the data shown in Fig. 17 do not provide a clear correlation between the number of $O2p_z$ holes and T_c, the fact that the number of these holes increases at high dopant concentration ($x \sim 0.3$) may support such theories.

In Fig. 18 we show polarization dependent $Cu2p_{3/2}$ absorption edges of insulating $La_2CuO_{4+\delta}$ and superconducting $La_{1.85}Sr_{0.15}CuO_{4+\delta}$ for $E \perp c$ and

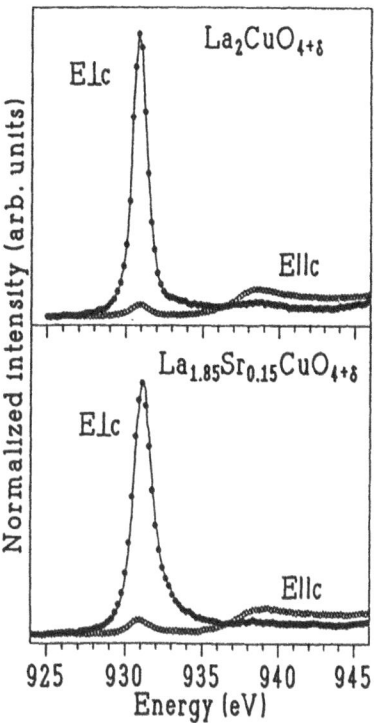

Fig. 18: Polarization-dependent $Cu2p_{3/2}$ x-ray absorption spectra of La cuprate single crystals: (a) insulating $La_2CuO_{4+\delta}$; (b) superconducting $La_{1.85}Sr_{0.15}CuO_{4+\delta}$. Closed circles: $E \perp c$; open diamonds: $E \parallel c$.

$E \parallel c$. For $E \perp c$ at both Sr concentrations, there is a strong excitonic line due to mainly unoccupied $Cu3d_{x^2-y^2}$ states and some $Cu3d_{3z^2-r^2}$ states. For $E \parallel c$ a much smaller excitonic line due to unoccupied $Cu3d_{3z^2-r^2}$ states is observed. The number of holes close to E_F on Cu sites with $3d_{x^2-y^2}$ and $3d_{3z^2-r^2}$ symmetry, as derived from the intensity of the excitonic lines, is shown as a function of x in Fig. 19 (a). In addition, the fraction, R, of $3d_{3z^2-r^2}$ hole states with respect to the total number of Cu3d hole states is shown in Fig. 19 (b), together with analogous data from Chen et al. [32]. Within error bars, the intensity of unoccupied $Cu3d_{x^2-y^2}$ states in the excitonic line is independent of the dopant concentration. The Cu states of the singlet band and of the UHB contribute to the spectral weight of the excitonic line. In the charge transfer model with hybridization there is a a transfer of Cu3d spectal weight to the singlet band, but

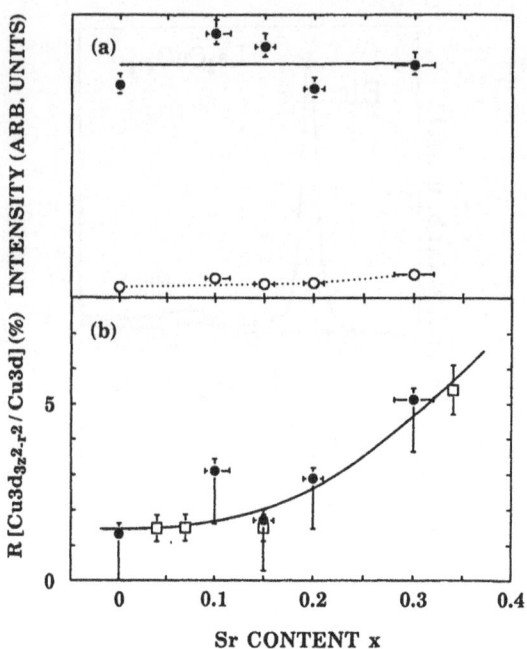

Fig. 19: Intensity due to hole states on Cu sites contributing to the excitonic line in La$_{2-x}$Sr$_x$CuO$_{4+\delta}$ for various dopant concentrations, x: (a) closed circles: Cu3d$_{x^2-y^2}$ states; open circles: Cu3d$_{3z^2-r^2}$ states. (b) Fraction, R, of unoccupied Cu3d$_{3z^2-r^2}$ states with respect to the total number of unoccupied Cu3d states (near the Fermi level). Closed circles: results from Ref. 31; open squares: results from Ref. 32.

mainly to the unoccupied part of this band [17]. Therefore, the observed independence of the total amount of unoccupied Cu3d$_{x^2-y^2}$ states from the dopant concentration is compatible with the model shown in Fig. 10 (f).

The fraction, R, of unoccupied Cu3d$_{3z^2-r^2}$ states with respect to the total number of unoccupied Cu3d states for x $\leq$ 0.15 is only about 1.5 % - a figure which is even considerably lower than that in the Bi$_2$Sr$_2$CaCu$_2$O$_8$ system. These results rule out theories for high-T$_c$ superconductivity which are based on a large fraction of Cu3d$_{3z^2-r^2}$ holes. The large fraction, R, for x = 0.3 supports the simple picture that due to the increasing number of holes on O sites in the CuO$_2$ plane, the crystal field at the Cu sites changes in such a way as to lower Cu3d$_{x^2-y^2}$ states and raise the Cu3d$_{3z^2-r^2}$ states. The increase of both Cu3d$_{3z^2-r^2}$ states and O2p$_z$ states at the Fermi level at high dopant concentrations indicates a stronger coupling between the CuO$_2$ planes which is consistent with the observation that La$_{2-x}$Sr$_x$CuO$_{4+\delta}$ transforms from a quasi two-dimensional metal at x ~ 0.15 to a more three-dimensional metal for x = 0.3.

Similar studies [31] on single crystalline n-type doped materials also show unoccupied $O2p_z$ states (3 - 10 %) in the pre-edge region (UHB), which is difficult to understand since there are no apical O sites in these compounds. Part of this spectral weight increases with x and had been assigned to impurity states, formed by a hybridization of $O2p_z$ states with $Ce5d$ and $4f$ states (see the discussion on polycrystalline samples). For $Cu3d_{3z^2-r^2}$ states R values similar to those in the $La_{2-x}Sr_xCuO_{4+\delta}$ system have been detected.

Recently, we have performed polarization-dependent XAS measurements on detwinned single crystals of $YBa_2Cu_3O_x$ (x = 6.10 - 6.95) [33, 34]. In Fig. 21 we show polarization dependent O1s absorption spectra in the pre-edge region for E ∥ a, E ∥ b and E ∥ c, for O concentrations ranging from the insulating compound x = 6.10 to the superconductor with x = 6.95. In the high-T_c supercon-

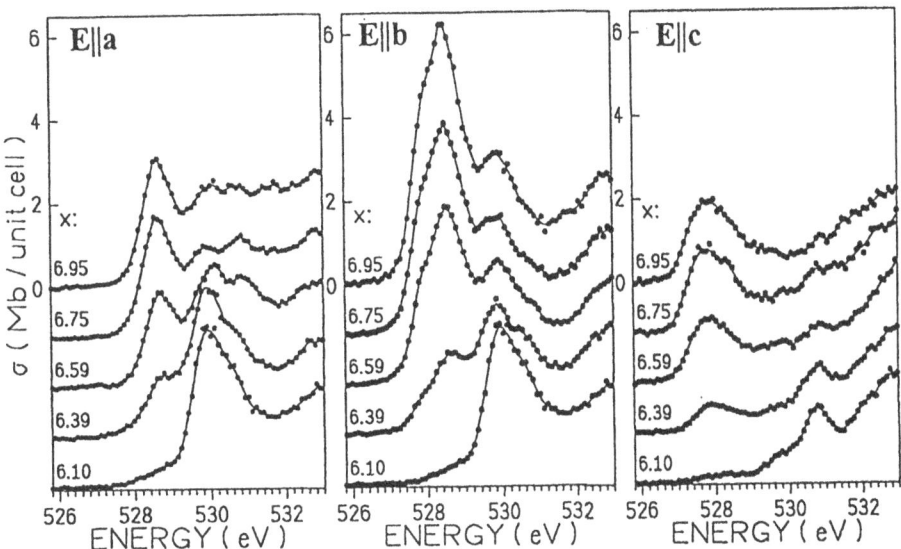

Fig. 20: Polarization and concentration-dependent O1s x-ray absorption spectra in the preedge region of $YBa_2Cu_3O_x$ for E ∥ a, E ∥ b and E ∥ c. The spectra are labelled by the concentration x.

ductor (x = 6.95) there are quasi one-dimensional CuO chains along the b-axis which act as acceptors removing electrons from the two-dimensional CuO_2 planes. According to band structure calculations, O2p and Cu3d states from both the CuO_2 planes and the CuO chains are expected at the Fermi level.

Assuming only σ bonding between these states, for E ‖ a, only O2p$_x$ states from the O(2) sites in the plane are reached on core level excitation. For x = 6.10 a well pronounced peak at 530 eV is assigned to O2p$_x$ states admixed to the UHB. This peak decreases with increasing dopant concentration as in La$_{2-x}$Sr$_x$CuO$_{4+\delta}$. The peak at 528.5 eV is assigned to holes in the valence (singlet) band. For E ‖ b we expect contributions from the O(3) sites in the planes and from the O(1) site in the chain. While the spectral weight of the UHB at 531 eV has almost the same intensity as for E ‖ a, the intensity due to hole states in the valence (singlet) band is considerably increased for E ‖ b. The difference between the spectral weights for E ‖ b and E ‖ a, is attributed to hole states on a the O(1) sites in the chain. Finally, for E ‖ c contributions from O2p$_z$ states from the apical O(4) sites, strongly bonded to the Cu atoms in the chains, are expected. The low-energy pre-edge is again due to holes in the valence (singlet) band. The number of holes on the O(4) sites increases with increasing dopant concentration. The peak at 530.8 eV is probably not related to the UHB, but is due to the CuO$_2$ dumb-bell along the c-axis, existing at x = 6 and being replaced at higher dopant concentrations by the CuO$_3$ entities of the chain.

A detailed analysis of the hole concentration on the four different O sites and Cu2p absoption edges as a function of dopant concentration will be published elsewhere [33].

4. Summary

High-energy spectroscopy has provided detailed information on the electronic structure of fullerene and cuprate superconductors which is the prerequisite for the understanding of the mechansim for superconductivity in these compounds.

In the fullerenes it is at present not clear whether correlation effects are important for the compounds A$_3$C$_{60}$ or whether electron phonon coupling is dominant. The analysis of the recent low-temperature photoemission data indicates that possibly the coupling to phonons is stronger than the coupling to the other electrons of the molecule. There are many further experiments which indicate that superconductivity occurs by a coupling of the conduction electrons to vibrations in the C$_{60}$ molecule.

In the cuprates, there is clear evidence for strong correlation effects due to the Coulomb repulsion between two holes on the Cu sites. The situation is even more complicated due to the strong hybridization between Cu3d and O2p orbitals. Although there is now a much better understanding of the electronic structure in the normal state, the mechanism for high-T$_c$ superconductivity is still rather unclear. Some experimental results have enabled the exclusion of certain models.

5. Acknowledgements

M.S. G. is grateful for the award of a SERC / NATO Advanced Postdoctoral Fellowship.

References

1. H.W. Kroto, J.R. Heath, S.C. O'Brien, R.F. Curl, R.E. Smalley, Nature 318, 162 (1985)
2. W. Krätschmer, L.D. Lamb, K. Fostiropoulos, D.R. Huffman, Nature 347, 354 (1990)
3. S. Saito and A. Oshima, Phys. Rev. Lett. 66, 2637 (1991)
4. M. Merkel, M. Knupfer, M.S. Golden, J. Fink, R. Seemann, and R.L. Johnson, Phys. Rev. B 47, (1993), in print
5. S. Satpathy, V.P. Antropov, O.K. Andersen, O. Jepsen, O. Gunnarsson, and A.I. Lichtenstein, Phys. Rev. B 46, 1773 (1992); (private communication)
6. K.-P. Bohnen, unpublished
7. E. Sohmen, J. Fink, and W. Krätschmer, Z. Phys. B 86, 87 (1992)
8. R.W. Lof, M.A. von Veenendaal, B. Koopmans, H.T. Jonkman, and G.A. Sawatzky, Phys. Rev. Lett. 68, 3924 (1992)
9. A.F. Hebard, M.J. Rosseinsky, R.C. Haddon, D.W. Murphy, S.H. Glarum, T.T.M. Palstra, A.P. Ramirez, and A.R. Kortan, Nature 350, 600 (1991)
10. T. Takahashi, S. Suzuki, T. Morikawa, H. Katayama-Yoshida, S. Hasegawa, H. Inokuchi, K. Seki, K. Kikuchi, S. Suzuki, K. Ikemoto, and Y. Achiba, Phys. Rev. Lett. 68, 1232 (1992)
11. M. Knupfer, M. Merkel, M.S. Golden, J. Fink, O. Gunnarsson, and V.A. Antropov, Phys. Rev. B., in print
12. E. Sohmen, J. Fink, and W. Krätschmer, Europhys. Lett. 17, 51 (1992)
13. E. Sohmen and J. Fink, Phys. Rev. B 47, (1993), in print
14. L. Degiorgi, G. Grüner, P. Wachter, S.-M. Huang, J. Wiley, R.L. Whetten, R.B. Kaner, K. Holczer, and F. Diederich, Phys. Rev. B 46, 11250 (1992)
15. L.F. Mattheiss, Phys. Rev. Lett. 58, 1028 (1987)
16. J. Zaanen, G.A. Sawatzky, and J.W. Allen, Phys. Rev. Lett. 55, 418 (1985)
17. H. Eskes, M.B.J. Meinders, and G.A. Sawatzky, Phys. Rev. Lett. 67, 1035 (1991); M.B.J. Meinders, H. Eskes, and G.A. Sawatzky, preprint
18. P.W. Anderson, Science 235, 1196 (1987)
19. F.C. Zhang and T.M. Rice, Phys. Rev. B 37, 3759 (1988)
20. K.A. Müller, Z. Phys. B 80. 193 (1990)
21. W. Weber, Z. Phys. B 70, 323 (1988)
22. H. Kamimura and M. Eto, J. Phys. Soc. Jpn. 59, 3053 (1990)
23. Y. Ohta, T. Tohyama, and S. Maekawa, Phys. Rev. B 43, 2968 (1991)
24. L.F. Feiner, M. Grilli, and C. Di Castro, Phys. Rev. B 45, 10647 (1992)

520

25. J.G. Bednorz and K.A. Müller, Z. Phys. B **64**, 189 (1986)
26. H. Romberg, M. Alexander, N. Nücker, P. Adelmann, and J. Fink, Phys. Rev. B **42**, 8768 (1990)
27. J.W. Allen, C.G. Olson, M.B. Maple, J.-S. Kang, L.Z. Liu, J.-H. Park, R.O. Anderson, W.P. Ellis, J.T. Markert, Y. Dalichaouch, and R. Liu, Phys. Rev. Lett. **64**, 595 (1990)
28. M. Alexander, H. Romberg, N. Nücker, P. Adelmann, J. Fink, J.T. Markert, M.B. Maple, S. Uchida, H. Takagi, Y. Tokura, A.C.W.P. James, and D.W. Murphy, Phys. Rev. B **43**, 333 (1991)
29. G.Y. Guo, Z. Szotek, and W.M. Temmermann, Physica C **162**, 1351 (1989)
30. N. Nücker, H. Romberg, X.X. Xi, J. Fink, B. Gegenheimer, and Z.X. Zhao, Phys. Rev. B **39**, 6619 (1989)
31. E. Pellegrin, N. Nücker, J. Fink, S.L. Molodtsov, A. Gutiérrez, E. Navas, O. Strebel, Z. Hu, M. Domke, G. Kaindl, S. Uchida, Y. Nakamura, J. Markl, M. Klauda, G. Saemann-Ischenko, A. Krol, J.L. Peng, Z.Y. Li, and R.L. Greene, Phys. Rev. B **47**, 3354 (1993)
32. see also: C.T. Chen, L.H. Tjeng, J. Kwo, H.L. Kao, P. Rudolf, F. Sette, and R.M. Fleming, Phys. Rev. Lett. **68**, 2543 (1992)
33. N. Nücker et al., unpublished
34. see also: A. Krol, Z.H. Ming, Y.H. Kao, N. Nücker, G. Roth, J. Fink, G.C. Smith, K.T. Park, J. Yu, A.J. Freeman, A. Erb, G. Müller-Vogt, J. Karpinski, E. Kaldis, and K. Schönmann, Phys. Rev. B **45**, 2581 (1992)

ELECTRON MICROSCOPY AND THE STRUCTURAL STUDIES OF SUPER-CONDUCTING MATERIALS AND FULLERITES.

G. VAN TENDELOO, S. AMELINCKX
EMAT, University of Antwerp (RUCA)
Groenenborgerlaan 171
B 2020 Antwerpen
Belgium

ABSTRACT. The possibilities of electron microscopy in the study of high temperature ceramic superconductors and fullerites or fullerite related materials are discussed. For the superconductors, particular attention is devoted to the structural changes introduced by different substitutions in the CuO chain plane. The defect structure of the pure C_{60} and C_{70} fullerites is studied by high resolution electron microscopy, while the phase transitions at low temperature are studied by electron diffraction. Different electron microscopy techniques also allow a detailed study of the texture of carbon tubes.

1. Introduction

Many of the physical, chemical or mechanical properties of a material are governed by defects present in the material. It is therefore clear that apart from the crystal structure also the knowledge of the *local structure* is very important. Since the presence of these defects only implies very limited and very local structural changes on an atomic scale, high resolution electron microscopy has proven to be an excellent technique in the study of these defects. In many cases it is the only technique by which the nature as well as the spatial distribution of the defects can be determined.

One of the main advantages of transmission electron microscopy (TEM) is to combine different imaging modes with electron diffraction as well as with different analytical techniques such as X-ray energy dispersive spectroscopy (EDS) in order to characterize defects structurally and chemically.

In the high resolution mode (HREM) the crystal is mostly oriented such that the electron beam is incident along a low order zone axis, so as to produce a symmetrical diffraction pattern. A number of diffracted beams together with the incident beam are then allowed to interfere and contribute to the final image. The number of contributing reflections is determined by the choice of the objective aperture, positioned in the back focal plane of the objective lens. In the present commercial HREM instruments, where information up to the 0.10 nm level can be transferred to the image, a large number of reflections is generally allowed to contribute to the image. In some situations however, when one is interested in particular structural features, it can be useful to restrict intentionally the number of reflections. This is the case for instance when the ordering on an underlying sublattice is important, rather than the information about the lattice itself[1]. By chosing the aperture in such a way that it includes the ordering reflections, but excludes the lattice reflections, one intentionally limits the resolution, but emphasizes the ordering phenomena.

E. Kaldis (ed.), Materials and Crystallographic Aspects of HTc-Superconductivity, 521–538.
© 1994 *Kluwer Academic Publishers.*

Most of the published HREM images of known structures show a striking correspondence with the projected crystal potential. This however is only the case when the images are recorded at a particular defocus value of the objective lens: the so called Scherzer defocus. For this defocus value one can easily show that for very thin samples there is a one to one correspondence between the projected crystal potential and the HREM image (see e.g. [2,3]). This one to one correspondence holds up to the Scherzer resolution limit (which e.g. for a Jeol 4000 EX microscope is approximately 0.17 nm). However the instrumental resolution is better, and with the present generation of electron microscopes equiped with a field emission gun, the instrumental resolution of some instruments (e.g. the Philips CM 30 FEG Ultra Twin) extends beyond 0.1 nm. The problem of retrieving *correct* structural details in this obscure zone between the Scherzer resolution limit and the instrumental resolution limit is not an easy problem. Recently several techniques have been proposed; the most promising one seems to be the paraboloid method (PAM) proposed by Van Dyck [4,5], eventually in combination with a new variant of the Kirkland method [6] as proposed by Coene [7]. This method requires a large series of images taken at slightly different defocus values, but then produces a final projected structure image with more detail than any individual image. An example is shown in fig 1. for the $YBa_2Cu_4O_8$ superconductor imaged along the [100] orientation. The oxygen positions indicated by arrows are detected, although their projected separation is only 0.14 nm and the Scherzer resolution limit of the microscope used was "only" 0.20 nm. This technique is highly promising to study structures and crystal defects on the 0.1 nm scale. It should be noted however that a dedicated instrument with a field emission gun, a CCD camera and a computer with the necessary software are required to make this information accessible.

Since the discovery in 1986 of ceramic superconductors with critical temperatures exceeding these of the "classical" intermetallic superconductors [8], a large number of new superconducting materials has been discovered and most of them have been studied by electron microscopy. The idea of this paper is not to review them all, but to illustrate with a

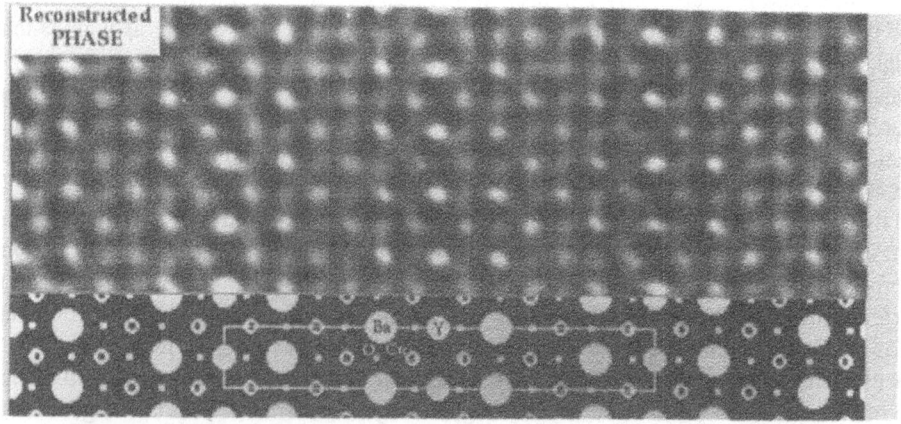

Fig.1.(a) Processed HREM image along the [100] orientation of $YBa_2Cu_4O_8$; the result of applying the paraboloid method, .(b) Projected crystal potential of $YBa_2Cu_4O_8$ along [100].

few examples the power (and the shortcommings) of electron microscopy. For more complete information on new superconducting materials and studies by HREM of such materials, we can refer e.g. to the Physica C journal which is exclusively devoted to superconducting materials and which almost in each issue contains structural or micro-structural studies of (new) superconducting materials.

Recently high resolution electron microscopy and electron diffraction have played an essential role in elucidating structures and textures of fullerites [9, 10] and of related objects such as whiskers[11], tubes[12,13], onions[14], ... Because these techniques require only minute crystal fragments, as compared to other diffraction techniques, they were among the first to be used in the early structural studies. Moreover, the relationship between direct and reciprocal space of the same crystal fragment can easily be established, which is especially helpful for the interpretation of the observed textures and defects. Cooling and heating the material inside the electron microscope allows to study the structural aspects of phase transitions. However due to the fact that in an electron microscope the specimen temperature is difficult to measure with a reasonable precision, most of these studies are only qualitative.

2. Structural studies of YBCO based superconductors

2.1 VACANCY SUBSTITUTION FOR OXYGEN IN 1-2-3 BASED MATERIALS

In the fully oxidated state $ABa_2Cu_3O_{7-\delta}$ (with $\delta=0$), the basic orthorhombic structure (Ortho I) is formed for the Y compound as well as for all substituted compounds (A = Yb, Er, Sm, Nd,...). CuO chains are formed along the **b**-axis by the ordering of oxygen atoms on sites between every two Cu atoms of the square Cu lattice. In the Ortho II phase, which has an ideal oxygen content of 6.5, the oxygen atoms of every other copper-oxygen chain are missing; the unit cell doubles along the **a**-axis and therefore this superstructure is sometimes also called the $2a_0$ structure. The appearence of this superstructure is often related to the presence of a plateau at about 60K in the Tc versus oxygen content curve of well annealed samples[15,16]. In the case of quenched samples no plateau is observed and the critical temperature decreases monotonically with increasing oxygen deficiency.[17] In such material, no or very ill defined Ortho II superstructures are detected.[16].

Recently, Buchgeister et al. replaced in a systematic way the Y ion by different rare earth ions and established a close relationship between the ion size of the substituted rare earth and the different structural and electronic properties of the material [18]. Apart from a weak dependence of orthorhombicity on the ionic radius of the rare earth ion, a variation was observed of the width of the 60 K plateau. A pronounced plateau appeared for the smaller ions (Yb, Er, Y) while for the larger ions (Nd,La) no plateau was observed at all (fig.2a).

Electron diffraction of these materials was carried out at 100 kV [19]. For the smaller ions, very clear and well defined extra Ortho II or $2a_0$ ordering reflections were detected in the [001] zone pattern (fig.3a). Best defined ordering occurred (for all compounds studied) at oxygen contents around $O_{6.6}$. Dark field imaging in these sharp reflections reveals the

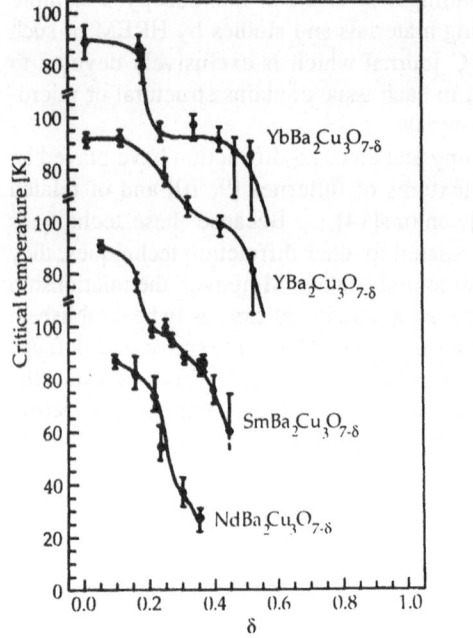

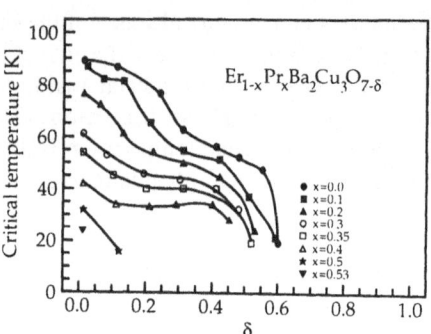

Fig. 2. (a) T_c versus oxygen concentration δ for $ABa_2Cu_3O_{7-\delta}$ with A = Yb, Y, Sm and Nd (b) Same graph for $Er_{1-x}Pr_xBa_2Cu_3O_{7-\delta}$.

$2a_0$ ordered domains as bright specks (fig.4); the domain shape is fairly isotropic and the size exceeds several tens of nm. Usually the extra diffraction spots are elongated along a*, particularly for oxygen concentrations corresponding to the edges of the 60K plateau. The corresponding $2a_0$ domains are now lenticular in shape and their size depends on the oxygen concentration; it decreases with increasing difference of the latter from that corresponding with the middle of the plateau. For the larger ions (e.g. Nd) the corresponding [001] pattern only showed a weak diffuse scattering along the [100]* and [010]* directions (fig.3b). Attempts to image the domains corresponding to such weak diffuse intensity failed in all cases.

In these measurements a clear correlation between the presence of the Ortho II structure and the presence of a 60 K plateau is established. This correspondence is further supported by experiments performed on Pr-doped samples. With increasing Pr-content the critical temperature of the higher Tc plateau (93 K) is suppressed relative to the lower Tc plateau to a level where the Tc's of both plateaus meet (fig.2b).[18,19] The Tc of the lower plateau is only slightly affected by the doping and also the Ortho II superstructure is not affected.

In these measurements a clear correlation between the presence of the Ortho II structure and the presence of a 60 K plateau is established. This correspondence is further supported by experiments performed on Pr-doped samples. With increasing Pr-content the critical temperature of the higher Tc plateau (93 K) is suppressed relative to the lower Tc plateau to

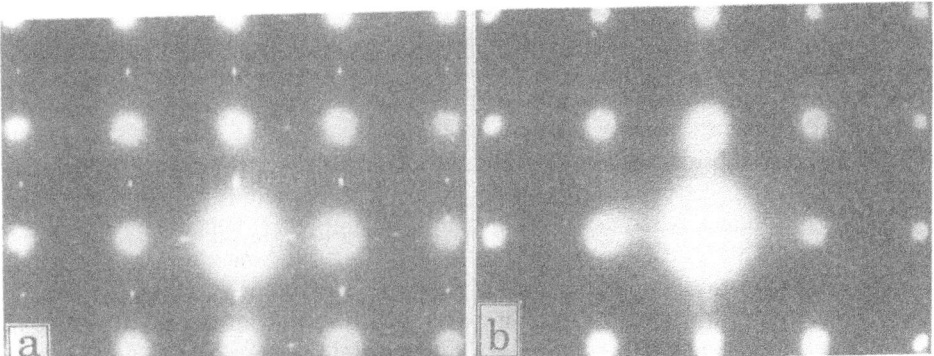

Fig. 3 [001] diffraction patterns from different $ABa_2Cu_3O_{7-\delta}$ materials. (a) Sharp superlattice reflections are present in well ordered Ortho II material for A = Yb or Y. (b) Only very ill defined [100]* streaking is observed when A = Nd.

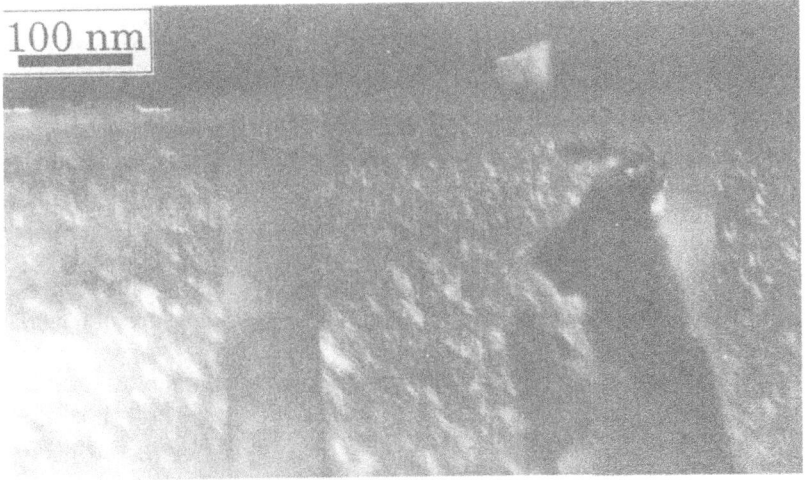

Fig. 4 Dark field image showing Ortho II ordered domains as bright specles.

a level where the Tc's of both plateaus meet (fig.2b).[18,19] The Tc of the lower plateau is only slightly affected by the doping and also the Ortho II superstructure is not affected.

We conclude that an unambiguous correlation exists between the superconducting 60 K phase and the structural Ortho II phase.

2.2. SO_4 SUBSTITUTION FOR Cu IN 1-2-3 BASED MATERIALS

The basic 1-2-3 $YBa_2Cu_3O_7$ structure contains an orthorhombic Cu(1)O chain layer, with CuO chains along the basic cell **b**- direction. Efforts to stabilize the Ba-free compound have led to a variety of cation substitutions of the Cu(1) chain site. Ga and Co substitutions in compounds such as $YSr_2GaCu_2O_7$ and $YSr_2CoCu_2O_7$ have led to a new type of perovskite like block with GaO- or CoO-chains in which the metal-atoms are tetrahedrally coordinated by oxygen and which run diagonally with respect to the perovskite cell [20-25].

Also recently substitutions of carbon for $Cu(1)$ have been reported in the $YSr_2Cu_3O_7$ compound [26]; in this case the CO_3 triangles align in chains along the perovskite b-axis. It is therefore not surprising that scientists have tried the substitution of $Cu(1)$ by sulphate groups [27,28], phosphate groups or nitride groups [29]. We will restrict our discussion here to the substitution by tetrahedral SO_4-goups in compounds with composition $[Y_{1-y}Sr_y]Sr_2[Cu_{3-x}(SO_4)_x]O_{7-\delta}$, where $y = 0.16$ and $x = 0.22$. Neutron data reveal an orthorhombic tri-perovskite structure with lattice parameters very close to those for $YBa_2Cu_3O_7$, but with a somewhat shorter c-axis [27]. The sulphur atoms are located on the $Cu(1)$ positions and the SO_4-groups have a tetrahedral oxygen configuration, with the S atoms slightly excentric.

The electron diffraction patterns from this material confirm this average structure but strong satellites are associated with every basic reflection. The [010] zone diffracton pattern of fig. 5a shows the presence of bright first order satellites and weaker second order satellites. The positions of first and second order satellites of a Bragg spot at $(h, 0, l)$ are $(h \pm q_s, 0, l \pm 1/2)$ and $(h \pm 2q_s, 0, l \pm 1)$ with $q_s \approx 0.3$. The length of the satellite-q-vector is approximately $1 nm^{-1}$. The [001]-zone diffraction pattern of figure 5b shows that satellite spots are present in the $(010)^*$ planes only. The two second order satellites at positions $h \pm 2q_s$ appear in between every two basic spots, along the a^*-direction. The structure can therefore be considered as composition modulated with a wavevector in the (010) plane, inclined over an angle α ($\alpha \approx 32°$)with respect to the [100]-direction and with a wavelength $\lambda \approx 1 nm$.

The angle α is not constant; a range of angles α has been observed between 27° and 32°, as well as a range of satellite separations, in different grains of the same batch of material, suggesting the incommensurability of the superstructure.

Although the observed modulation reduces the symmetry to orthorhombic, most [001]

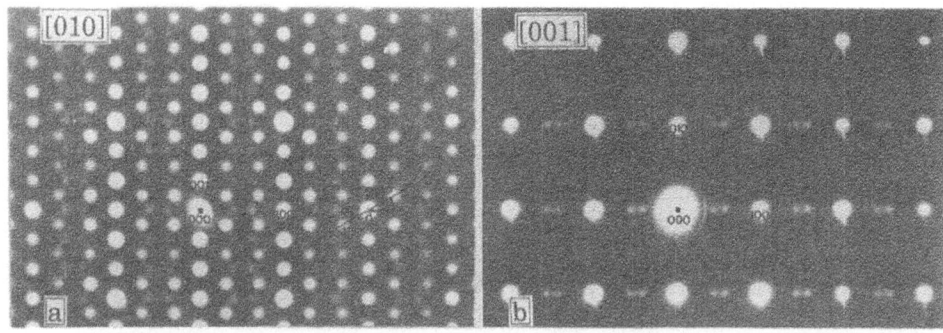

Fig.5.: Diffraction patterns of $[Y_{0.84}Sr_{0.16}]Sr_2Cu_{2.78}(SO_4)_{0.22}O_{6.12}$.
(a) [010] pattern showing satellites at every basic spot; numbers indicate spot order, the angle indicated is $\alpha=29°$;
(b) [001] pattern with only weaker second order spots; indexing refers to the basic unit cell.

observed diffraction patterns show tetragonal symmetry, with superstructure spots along **a***** as well as along **b***. This is due to the presence of two orthorhombic variants, related by the usual (110) twin law common in $YBa_2Cu_3O_{7-\delta}$. The small orthorhombicity $(b_0-a_0)/a_0=0.005$ [27, 28] (versus 0.016 in $YBa_2Cu_3O_{7-\delta}$) results in a fragmentation of the material in small domains of the two orthorhombic variants, exhibiting a texture known as "tweed"[28].

High resolution images along the [010]-zone (fig.6), show the presence of a modulation corresponding to the satellites in the corresponding diffraction pattern of figure 5a. The image shows, on the average, the same dot configuration as in undoped $YBa_2Cu_3O_{7-\delta}$ material with the Cu(1)-O-layer imaged as the brighter dot row. However, one aspect, different from the undoped material is rather striking : every three or four repeat distances along the **a**-direction, brighter squares of dots are formed. The observed contrast is similar to that observed in CO_3-containing samples [26] and, according to image simulations, is to be associated with the presence of SO_4-groups in this layer. The bright squares indicate **b**-oriented columns with a high SO_4-content. Along the **c**-direction the stacking of these columns or "chains"is usually staggered in such a fashion that their positions in every second Cu(1)-S-O-plane coincide vertically. Within a single Cu(1)-S-O-plane the alternation is incommensurate but the intensity maxima change roughly as $-4a_0-3a_0-3a_0-$. Looking along the directions indicated by heavy lines in figure 6b, planes containing a high density of SO_4-clusters are revealed by a modulation in dot brightness. These planes can be considered as the maxima of a planar concentration wave. The normal to this

Fig. 6: High resolution image along the [010]-zone showing S-rich columns as brighter squares of four dots centered on the Cu(1)-S-O-layers. (a) Low magnification image, clearly showing the intensity maxima. (b) High magnification, showing atomic resolution. The maxima of the modulating wave are stressed by two sets of heavy lines with normals inclined over α and $-\alpha$ with respect to the [100] direction. The brighter dot configurations are indicated in part of the image by small crosses. The separation between these crosses along the **a** direction is $3a_0$ or $4a_0$. The stacking mode along **c** is such that, in a projection along the **c**-direction, bright crosses are generally superimposed vertically every second Cu(1)-S-O-layer.

concentration wave makes an angle α of about $30°$ with the [100]-direction and the wavelength is about 1nm. The equivalent modulation, making an angle of $-\alpha$ with the [100] direction, is present in figure 6a and indicated by a second set of lines in figure 6b.

These observations can be explained by a simple model of SO_4-chain formation and ordering. The [001] high resolution image suggests the presence of chains oriented along the b-direction, the nature of which must differ from that of CuO chains by their S-content. The stacking-rule of the SO_4-rich chains can be derived from the observation of the concentration waves in fig.6.

$1°$ The SO_4-chains substitute the sites of the Cu(1)-sublattice that are closest to the maxima of the concentration waves.

$2°$ The integer l-coordinates of the second order satellites as well as direct observations in high resolution images suggest that the SO_4-chain-stacking coincides vertically in every second plane along the [001] direction.

Based on the observations in real space as well as on the analysis in reciprocal space we can propose an atomistic model for the modulation in $[Y_{1-y}Sr_y]Sr_2[Cu_{3-x}(SO_4)_x]O_{7-\delta}$. Figure 7 represents the Cu(1) sublattice only; dashed lines are the maxima of the concentration waves with angles $\alpha = \pm 30°$ and $\lambda \approx 1$nm. SO_4-chains (circles) are positioned at the Cu(1) column sites (dots) that are closest to the wave maxima. Due to the incommensurability of the basic lattice with the modulation, the ordering along the a-direction is essentially aperiodic, which is in agreement with the absence of superstructure spots in the diffraction patterns along this direction.

A physical explanation for the observed SO_4-chain arrangement is not straightforward but one can speculate. The chains occupied by SO_4^{2-} ions cause cylindrically symmetrical stress fields as a consequence of the difference in size of the sulphur-ions and the copper-ions that they replace. Parallel chains thus interact by mutually repulsive elastic forces which depend on the separation r as $1/r$. The equilibrium configuration of such a set of repelling parallel chains, when confined to a finite area, or with a specified concentration, consists of a triangular array. With the restriction that the SO_4-chains can only be located at positions in the Cu(1)O-chain layers, the observed distribution of SO_4-chains is the one that yields the largest average separations between different chains, taking into account the concentration of SO_4 chains.

The restriction on the allowed sites for the SO_4 chains, also restricts the possible wavevectors of the concentration waves. The c*-component of the wavevector q must be a simple fraction of c_0^*. Since a staggered arrangement of repelling chains is energetically more favourable than a vertically aligned one, it can be understood that the c*-component is $1/2\ c_0^*$ and not c_0^*. On the other hand, the a* component of q is SO_4 concentration dependent, and leads to a variable q-vector.

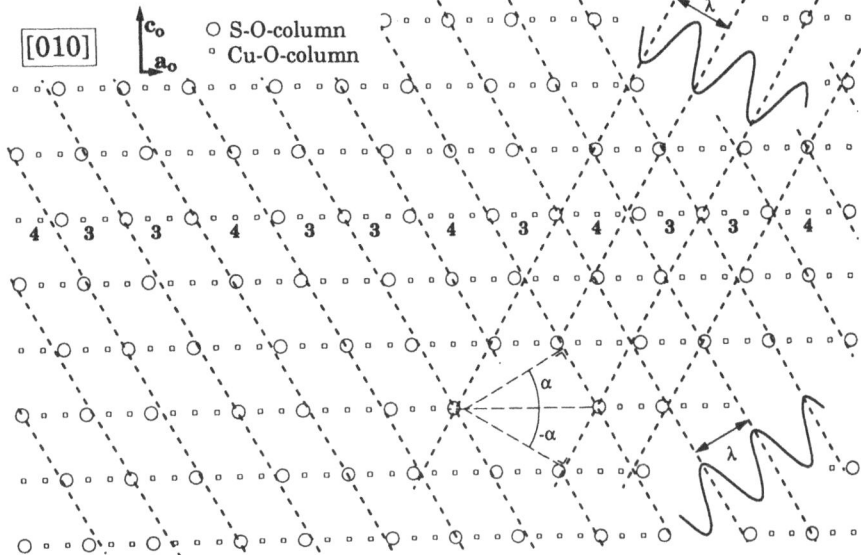

Fig.7: Schematic representation of the incommensurate modulation of the Cu(1)-column-sublattice. Squares represent the Cu(1)-column sublattice, circles represent SO_4-chains and are located on Cu(1)-column-sites closest to the maxima of a SO_4-occupancy wave with wavevector inclined over α with respect to [100] and with wavelength λ. Both symmetry related modulations inclined $+\alpha$ and $-\alpha$ are represented.

3. Structural studies of C_{60} and C_{70} fullerites

3.1. C_{60}

3.1.1 The room temperature structure. The room temperature crystal structure of the C_{60} fullerite has initially been the subject of some discussion until it was realized that the sample purity is of overall importance. Crystals grown from an organic liquid such as toluene are predominantly faulted hexagonally close packed [30, 9]. However the observation in the microscope vacuum under electron irradiation causes a transformation of such crystals into a faulted cubic close packed structure. It is assumed that the solvent molecules occupy the channels along c, formed by the unoccupied interstitial positions in the HCP structure. Under vacuum these solvent molecules are removed and a shear transformation converts the faulted hexagonal into a faulted cubic structure. Sublimation grown crystals of sufficient purity are always cubic at room temperature. The isotropically rotating quasi-spherical molecules occupy the node points of a face centered cubic lattice with $a_0 = 1.41$ nm [31, 32, 9]. The molecules can be modelled as spherical shells of diffracting material with an average radius $r_0 = 0.353$ nm. Due to the fact that fortuitously

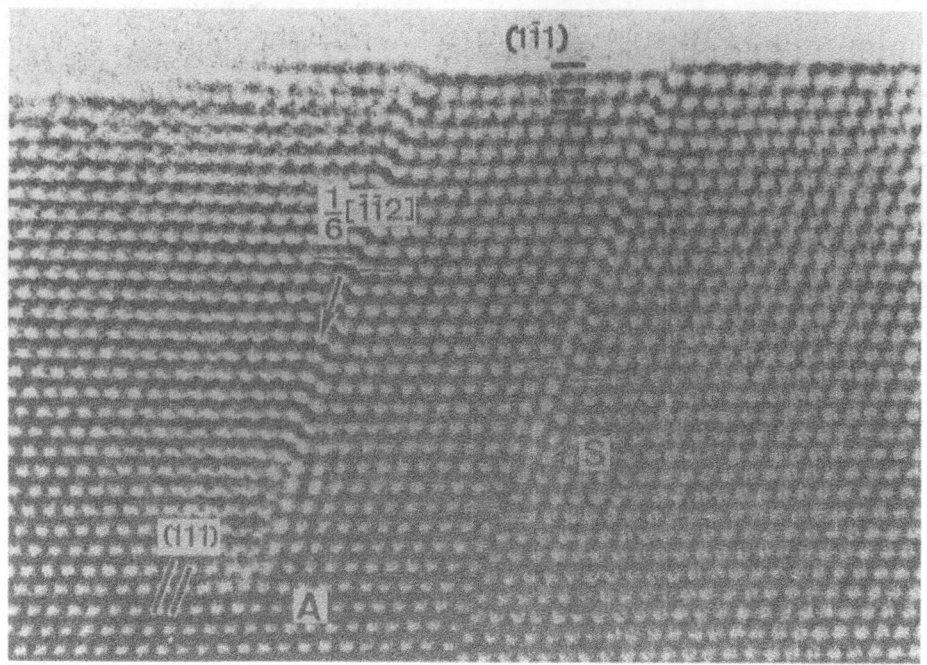

Fig. 8: Defect configuration of C_{60}. In S an intrinsic stacking fault changes from one (111) layer to the next, giving rise to a dipole of stair rod dislocations. In A a Frank partial dislocation and its associated intrinsic stacking fault are present.

the lattice parameter is four times the radius of the spherical shell, the h00 reflections have vanishingly small intensities for all values of h. This can be understood intuitively, as shown in [33, 34], by noting that although the structure is periodic the average density of diffracting material is very nearly constant in planes perpendicular to the cube directions. In terms of reciprocal space it can be shown that if $a_0 = 4r_0$ the zero's of the molecular Fourier transform, for reasonable models of the spherical shell, occur at positions in reciprocal space corresponding with the h00 nodes of the FCC lattice [34]. Experimentally the kinematical intensities of the h00 reflections are found to be negligibly small, but under conditions of dynamical diffraction this is no longer strictly true. With increasing thickness the intensity of the 200 reflection becomes non negligible.

The structure was visualized using the high resolution imaging mode. The columns of molecules produce ring shaped images of which the characteristics sensitively depend on the specimen thickness because at small thickness the 220 lattice fringes are predominantly excited, whereas at larger thickness also the h00 type of fringes contribute significantly to the image because of dynamical effects [34, 10].

3.1.1. *Lattice defects in the face centered cubic structure.* The defects in the room temperature structure can conveniently be studied by means of high resolution images, especially since single molecular columns can easily be resolved along the [110] zone, i.e. along the close packed rows of molecules.

The most abundant defects are coherent twin interfaces and intrinsic stacking faults on (111) planes. Occasionally also extrinsic stacking faults may occur. All these defects can easily be identified in high resolution images since the stacking of the close packed layers is directly imaged along this zone. Many of the complex defects, occurring in low stacking fault energy metallic alloys have also been described in C_{60} crystals.

Intrinsic stacking faults, jogging from one (111) plane to the adjacent one, producing a dipole of stair rod dislocations have been revealed.

Lomer - Cottrell barriers, consisting of V-shaped arrangements of two stacking faults on intersecting (111) planes, forming a stair-rod dislocation along their intersection line, have been observed as well. Stacking faults were often found to terminate when meeting a twin interface.

Different types of dislocations were observed. The image of fig.8 shows for instance a Frank partial dislocation with Burgersvector 1/3[111] and its associated intrinsic stacking fault.

Most of these defects have been studied earlier in low stacking fault face centered cubic alloys, either by diffraction contrast or under high resolution imaging conditions, but their

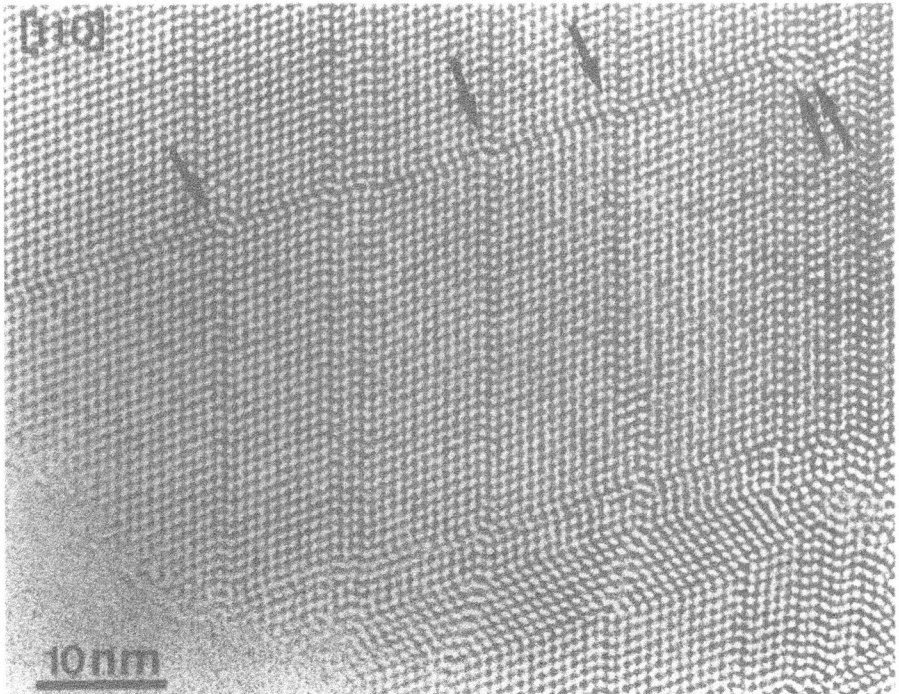

Fig. 9: High resolution image of C_{60} along [110] showing several intersections of two intrinsic stacking faults along different (111) planes (see arrows). Note the rectangular configuration of bright dots at every intersection.

geometry is more clearly resolved in C60 because of the large size of the molecules as compared to that of atoms. The geometrical features of the defects in such alloys, in C60 and in bubble rafts are amazingly similar because the bonding of metallic atoms, molecules and soap bubbles has some common characteristics: it is non-directional and short range.

A line defect, which was predicted, but so far not observed in a high resolution image of any material is shown in fig.9 . It consists of a row of fractional vacancies, formed along the intersection line of two intrinsic faults. In metallic alloys the stacking faults usually do not intersect, but one terminates on meeting the other. In C60 the stacking fault energy is very small and moreover the formation energy of vacancies is sufficiently small to allow the formation of such line defects. For a different combination of displacement vectors of the stacking faults one could geometrically also envisage the formation of a line of fractional interstitials; however the formation energy is presumably prohibitive.

The room temperature diffraction patterns often exhibit additional spots at fractional reciprocal lattice positions. They can be shown to be due to the intersection of Ewald's sphere with long <111>* streaks resulting from the presence of a high density of coherent twin interfaces, stacking faults, or both on (111) planes. Phase transitions clearly have to be studied on crystal fragments which do not exhibit such additional spots at room temperature.

3.1.3. *Phase Transitions in C60* "In situ" electron diffraction has confirmed the orientational order-disorder transition at around 255 K, from the room temperature face centerd cubic structure into a primitive cubic structure, with approximately the same lattice parameter.The diffraction conditions allow to establish unambiguously the space group as

$Pa\overline{3}$ [33, 36]. The reduction of the fourfold symmetry axis to a two fold one on transformation gives rise to two orientation variants, related by a 90° rotation about a cube axis. The loss of the symmetry translations of the type $1/2<111>$FCC leads to four translation variants of the primitive structure for each orientation variant. The phase transformation thus causes a severe domain fragmentation of the primitive structure (SC), even when starting with a perfect FCC single crystal. The orientationally ordered structure, proposed in [33] is consistent with all these observations.

Fig.10 shows the [001] diffraction pattern of the primitive structure; in (a) and (b) a single orientation variant is present in the selected area, whereas in (c) both variants are present.

The domain fragmentation was studied using dark field diffraction constrast imaging in a superlattice reflection in [10]. It confirmed the presence of the two types of orientation domains and showed that domain walls often coincide with twin interfaces.

At a somewhat lower temperature certain crystal fragments which were initially free of additional spots, were found to exhibit not only the spots due to the primitive lattice, but moreover also spots characteristic of a face centered cubic structure with a lattice parameter $2a_0$. This pattern was attributed to a structure in which the rows of molecules along the cube directions of the simple cubic structure adopt alternatingly two different orientations. Whereas the observation of the additional spots is unambiguously established, the interpretation is still controversial, because the additional spots are not always observed in

X-ray or neutron diffraction experiments. On the other hand low temperature e.p.r. experiments confirm the doubling of the lattice parameters [37]. Different reasons for the abscence of additional spots in some diffraction experiments can be imagined; the most probable one being the small size of coherent domains in this phase.

Recent theoretical work [38] indicates the existence at room temperature of minima of the interaction energy for two different orientations of the C_{60} molecules. Starting from the standard orientation of the molecule with respect to the crystal axis these two orientations are obtained by rotations of 38° and 98° about the [111] axis i.e. the two orientations differ by 60°, which is a quasi-symmetry operation. At room temperture one of these energy minima is an absolute minimum (98°) whilst the other one (38°) is only a shallower relative minimum. It is possible that at low temperature this second minimum becomes deeper, thus leading to two equally occupied sublattices, associated with each of the <111> type rotation axis. It is clear that the occurrence of such two sublattices is consistent with the proposed $2a_0$ structure. Low temperature neutron diffraction evidence also suggested the occurrence of two different molecular unequally occupied orientations on the same sublattice of the primitive structure, but no ordering was found [38].

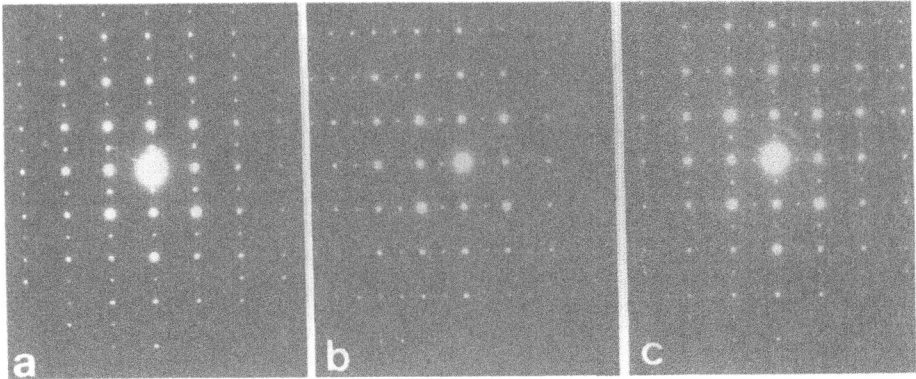

Fig. 10: Diffraction patterns taken across a 90° domain boundary
(a) to the left of the boundary
(b) to the right of the boundary
(c) across the boundary.

3.2. C_{70}

Vapour grown crystals of very pure C_{70} (> 99.9%) exhibit, even within the same batch, two different morphologies: cubic and hexagonal. At room temperature the corresponding predominant crystal structures are face centered cubic and hexagonal close packed. However both types of structures are always faulted; respectively on $(111)_{FCC}$ or $(0001)_{HCP}$ planes. The face centered cubic structure is moreover heavily twinned. The hcp phase exhibits orientationally ordered superstructures of the hcp stacking at low temperatures.

534

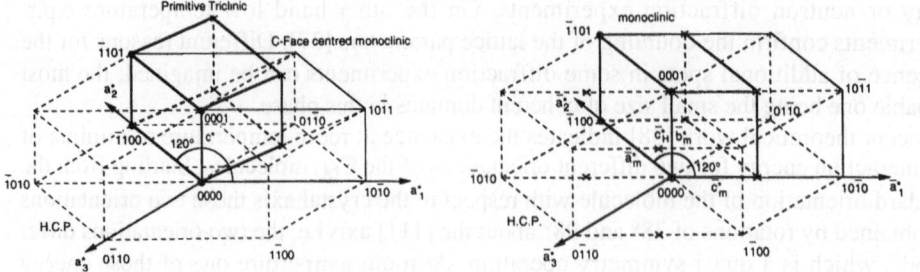

Fig.11: a) Reciprocal lattice of the monoclinic I phase as obtained from electron diffraction data.
b) Reciprocal lattice of the monoclinic II phase as obtained from X-ray data and electron diffraction measurements.

The reciprocal lattices of the two low temperature superstructures, as deduced from electron diffraction patterns at liquid nitrogen temperature, are represented in fig.11 a and b. In both structures one of the a-parameters of the hcp basic lattice is doubled, breaking the hexagonal symmetry. Moreover in one of them the c-parameter is doubled as well. The two structures can be referred to monoclinic unit cells with angles close to 120°; one unit cell is primitive (fig.12a) the second one is C-centered (fig.12b). The corresponding structures are closely related; they contain the same type of layers which are orientationally ordered variants of close packed layers of molecules with their long axis parallel to the c-direction. The relative azimuthal orientation of the molecules within one layer is determined by the same packing principle as in C_{60}. Along the "equator" where the intermolecular separation is smallest, electron rich regions along the 6 - 6 or 6 - 5 edges of the cage should face electron poor regions in the centre of the hexagons of the neighbouring molecule.

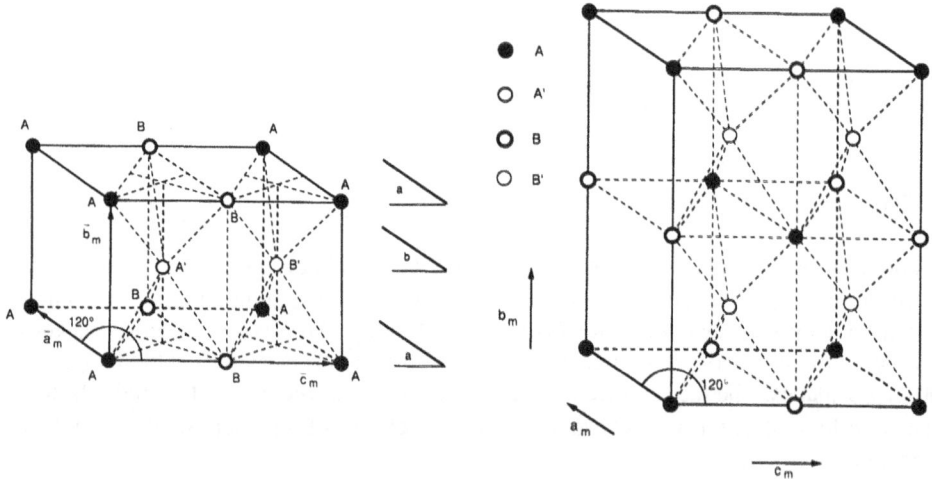

Fig. 12: 3D arrangement of the stacking sequences of the mon. I and mon. II phases.

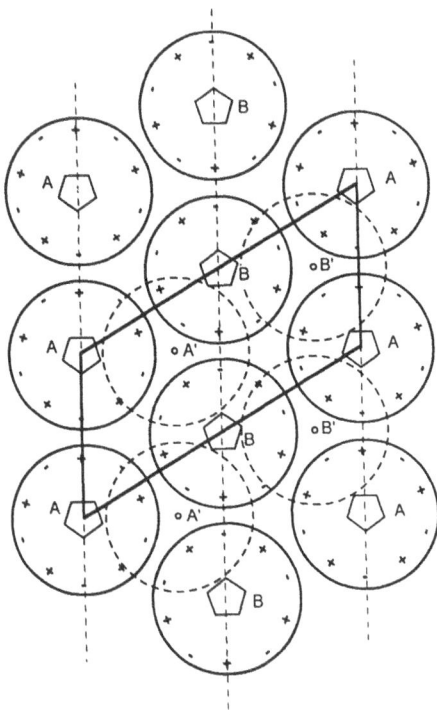

Fig. 13 : Basal plane arrangement of C70 molecules in the low temperature monoclinic phases.

In fig.13 electron rich regions are indicated by + signs and electron poor regions by - sign. Applying the mentionned stacking principle the resulting packing of molecules, with their long axis parallel to c, leads to the layer structure represented in fig.13. The molecules in neighouring rows thus adopt two different azimuths which differ by 180°, this leads to doubling of one of the edges of the unit mesh in the layer planes. The AB spatial arrangement of such layers can give rise to various stackings with different c-parameters. The structures suggested by the two types of diffraction patterns are represented in fig.12. The stackings only differ by the orientation of the molecules in successive layers; they could be called orientational polytypes. One structure, the most abundant one , has a primitive monoclinic lattice, whereas the other structure is based on a C-face centered monoclinic lattice; both lattices have a monoclinic angle of approximately $\beta = 120°$. Both structures can be formed in six orientation variants out of one hcp crystal, simulating hexagonal symmetry in polydomain crystals.

On heating the low temperature orientationally ordered phase gradually disorders by successive excitations of different rotational degrees of freedom [39-41].

In the first transition (276K) the superstructure spots disappear; the structure becomes hcp with a c/a = 1.82-3. This ratio is close to the ideal ratio for a hcp stacking of spheres, corrected for the aspect ratio of the elongated C_{70} molecules. In this phase the molecules have cylindrical symmetry since they are nearly freely rotating about their long axis; they are packed with their long axis parallel to c.

Around 337 K the hcp structure acquires a smaller c/a ratio, which is still somewhat larger than, but close to the c/a ratio for spheres (c/a≈1.63). This indicates that the molecules now also rotate about axes other than the long axis, leading to an effective, time averaged shape for the molecules which appraoches a spherical shell.

It is assumed that slightly above this temperature the stacking tends to change into an ABC stacking with a rhombohedral lattice. This requires a shear transformation on the $(0001)_H = (1\bar{1}1)_R$ planes. The rhombohedral deformation suggests the persistance of a

536

Table I: Schematic representation of the different phases occurring in C_{70}.

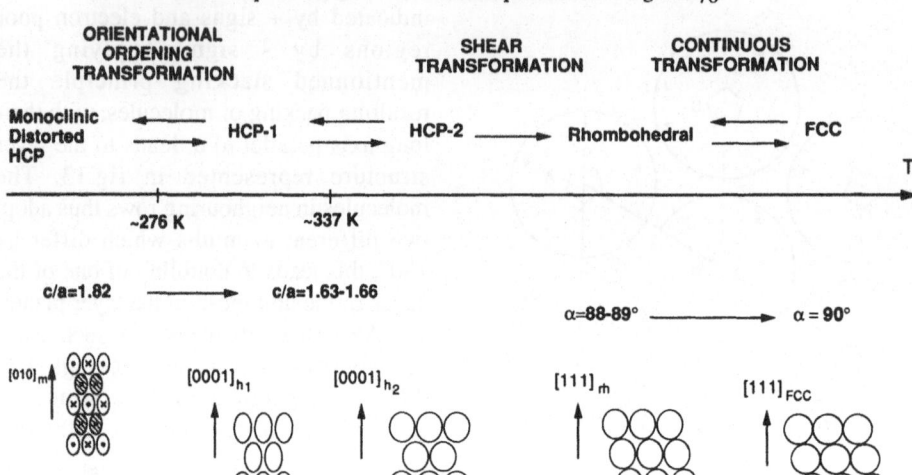

slightly preferential orientation of the molecules with their long axis along the long diagonal of the elongated rhombohedron.

At still higher temperature the structure becomes FCC. The transformation rhombohedral - FCC is reversible; it can be driven in one or the other sense by changing the electron beam current. The shear transformation on the other hand does not seem to be reversible; cooling an ABC stacked crystal does not produce an AB stacking; This may be related to the fact that the most stable rotation axes of the elongated C_{70} molecule are those perpendicular to the long axis. The formation of the hcp stacking starting from an ABC stacking would require the realignment of the long rotation axis, which seems a difficult cooperative process.

Acknowledgements

The authors would like to acknowledge the following scientists for their collaboration: T. Krekels, S. Muto, C. Van Heurck, J. Van Landuyt of the University of Antwerp (RUCA), O. Milat of the University of Zagreb, D. Wagener, M. Buchgeister, S.M. Hosseini, P. Herzog of the University of Bonn, T.N.G. Babu, P.R. Slater and C. Greaves of the University of Birmingham, M.A. Verheijen, P.H.M. van Loosdrecht, G. Meijer of the University of Nijmegen.

The work has been performed in the framework of an IUAP-48 contract and with financial help of the National Impulse Programme on High T_c Superconductivity (SU/03/17).

References

[1] G. Van Tendeloo, J. of Microscopy **119** (1980) 125.
[2] J.H. Spence "Experimental High Resolution Electron Microscopy" Clarendon Press Oxford (1981)
[3] "High Resolution Electron Microscopy and Associated Techniques" Eds. P.R. Buseck, J.M. Cowley, L. Eyring, Oxford University Press, New York, Oxford, 1988
[4] D. Van Dyck, M. Op de Beeck, Proceedings of the XIIth International Congress for Electron Microscopy, (1990) 26.
[5] M. Op de Beeck, D. van Dyck, W. Coene, Ultramicroscopy (1993) submitted
[6] E. Kirkland, Ultramicroscopy **15** (1984) 151.
[7] W. Coene, A. Janssen, M. Op de Beeck, D. Van Dyck, Phys. Rev. Lett. **69** (1992) 3743.
[8] G. Bednorz, K.A. Muller, Z. Physik, **64** (1986) 189
[9] G. Van Tendeloo, M. Op de Beeck, S. Amelinckx, J. Bohr, W. Krätschmer, Europhys. Lett. **15** (1991), 295
[10] G. Van Tendeloo, C. Van Heurck, J. Van Landuyt, S. Amelinckx, M.A. Verheijen, P.H.M. van Loosdrecht, G. Meijer, J. Phys. Chem. **96** (1992) 7424
[11] S. Amelinckx, W. Luyten, T. Krekels, G. Van Tendeloo, J. Van Landuyt, J. Crystal Growth **121** (1992) 543
[12] S. Iijima, Nature 354 (1991) 56
[13] P.M. Ajayan, S. Iijima, Nature, **361** (1993) 333
[14] D. Ugarte, Nature **359** (1992) 707
[15] R. Beyers , B. T. Ahn, G. Gorman, V. Y. Lee, S. S. P. Parkin, M. L. Ramirez, K. P. Roche, J. E. Vazquez, T. M. Gür and R. A. Huggings, Nature **340** (1989) 619.
[16] J. Reyes-Gasga, T. Krekels, G. Van Tendeloo, J. Van Landuyt, S. Amelinckx, W. H. M. Bruggink, H. Verweij, Physica C 159 (1989) 831.
[17] C. N. R. Rao, R. Nagarajan, A. K. Ganguli, G. N. Subbana and S. V. Bhat, Phys. Rev. B **42** (1990) 6765.
[18] M. Buchgeister, W. Hiller, S. M. Hosseini, K. Kopitzki and D. Wagener, Proceedings of the International Conference on Transport Properties of Superconductors, April 29 - May 4 1990, Rio de Janeiro, Brazil, ed. R. Nicolsky, (World Scientific Publishing, Singapore, 1990) 511.
[19] T. Krekels, G. Van Tendeloo, S. Amelinckx, D. Wagener, M. Buchgeister, S.M. Hosseini, P. Herzog, Physica C **196** (1992) 363
[20] G. Roth, P. Adelmann, G. Heger, R. Knitter, Th. Wolf, J. de Physique **1** (1991) 721.
[21] J.T. Vaughey, J.P.Thiel, E.F. Hasty, D.A. Groenke, C.L. Stern, K.L. Poeppelmeier, B. Dabrowski, P. Radaelli, A.W. Mitchel, D.G. Hinks, Chem. Mater, **3** (1991) 935 .
[22] O. Milat, T. Krekels, G. Van Tendeloo, S. Amelinckx, J. de Physique (1993) in the press

538

[23] Q. Huang, R.J. Cava, A. Santoro, J.J. Krajewski, W.F. Peck, Physica C **193** (1992) 196 .

[24] R.J. Cava, H. Zandbergen, J.J. Krajewski, W.F. Peck Jr., B. Hessen, R.B. Van Dover, S.-W. Cheong, Physica C **198** (1992) 27.

[25] T. Krekels, O. Milat, G. Van Tendeloo, S. Amelinckx, T. G. N. Babu, A. J. Wright, C. Greaves, J. Sol. State Chem. (1993) in the press.

[26] Y. Miyazaki, H. Yamane, N. Ohnishi, T. Kajitani, K. Hiraga, Y. Morii, S. Funahashi, T. Hirai, Physica C **198** (1992) 7.

[27] P. R. Slater, C. Greaves, M. Slaski, C. M. Muirhead, Physca C (1993) in the press.

[28] T. Krekels, O. Milat, G. Van Tendeloo, J. Van Landuyt, S. Amelinckx, P.R. Slater, C. Greaves, Physica C (1993) in the press.

[29] B. Raveau et al., this volume.

[30] W. Krätschmer, K. Fostiropoulos, D.R. Huffman, Chem. Phys. Lett. **170** (1990) 167.

[31] R. M. Fleming, T. Siegrist, P. M. Marsh, B. Hessen, A. R. Kortan, D. W. Murphy, R. C. Haddon, R. Tycko, G. Dabbagh, A. M. Mujsce, M. L. Kaplan and S. M. Zahurac, Mat. Res. Soc. Symp. Proc. **206**, (1991), 691

[32] D. Huffman, Physics Today, november (1991), 22.

[33] P. A. Heiney , J.E. Fisher, A.R. Mc Ghie, W.J. Romanòw, A.M. Denenstein, J.P. Mc Cauley, Jr., A.B. Smith, III, D.E. Cox, Phys. Rev. Lett. **66** (1991) 2911.

[34] S. Amelinckx, C. Van Heurck, D. Van Dyck, G. Van Tendeloo, Phys. Stat. Sol. (a) **131** (1992) 589.

[35] S. Muto, G. Van Tendeloo, S. Amelinckx, Phil. Mag. (1993) in the press.

[36] G. Van Tendeloo, S. Amelinckx, M.A. Verheijen, P.H.M. van Loosdrecht, G. Meijer, Phys. Rev. Lett. **69** (1992) 1065

[37] E.J.J. Groenen, O.G. Poluektov, M. Matsushita, J. Smidt, J.H. Van der Waals, G. Meijer, Chem Phys. Lett. **197**, 314 (1992)

[38] D. Lamoen, K. H. Michel, Phys. Rev.B. (1993) in the press

[39] G.B.M. Vaughan, P.A. Heiney, J.E. Fischer, D.E. Luzzi, D.A. Ricketts-Foot, A.R. McGhie, Y.W. Hui, A.L. Smith, D.E. Cox, W.J. Romanow, B.H. Allen, N. Coustel, J.P. McCauley, Jr., A.B. Smith,III, Science, **254**, (1991) 1350

[40] G. Van Tendeloo, S. Amelinckx, J.L. De Boer, S. van Smaalen, M.A. Verheijen, G. Mijer, Europhys. Lett. **21** (1993) 329

[41] M. A. Verheijen, H. Meekes, G. Meijer, P. Bennema, J. L. De Boer, S. van Smaalen, G. Van Tendeloo, S. Amelinckx, S. Muto, J. Van Landuyt, Chem. Phys. **166** (1992) 287

STRUCTURE-PROPERTY RELATIONSHIPS IN RADICAL-CATION (ELECTRON-DONOR MOLECULE) AND ANION-BASED (INCLUDING FULLERIDES) ORGANIC SUPERCONDUCTORS AND THEIR USE IN THE DESIGN OF NEW MATERIALS

JACK M. WILLIAMS, K. DOUGLAS CARLSON, ARAVINDA M. KINI,
H. HAU WANG, URS GEISER, JOHN A. SCHLUETER and
ARTHUR J. SCHULTZ
Argonne National Laboratories
Chemistry and Materials Science Divisions
9700 S. Cass Avenue
Argonne, IL 60439

JAMES E. SCHIRBER, EUGENE L. VENTURINI and
DONALD L. OVERMYER
Sandia Laboratories
Albuquerque, NM 87185

MYUNG-HWAN WHANGBO
Department of Chemistry
North Carolina State University
Raleigh, NC 27695

Abstract

The presently known structure-property relationships that have been developed for organic superconductors based on the ET molecule (β-phases and κ-phases), and the C_{60}-anion-based fullerides, and their use in the structural design of new superconducting materials are discussed.

Introduction

Whereas most organic substances are insulators, there is a class of conducting organic materials known as "synthetic metals" or "organic metals." Amongst the thousands of existing organic substances a very few, but steadily rising number, are superconducting at ambient pressure or under modest applied pressures.[1] The "molecular components" from which organic superconductors are derived are given in Figure 1, and a listing of these superconductors other than those derived from C_{60} is given in Table 1.

E. Kaldis (ed.), Materials and Crystallographic Aspects of HTc-Superconductivity, 539–551.
© *1994 Kluwer Academic Publishers.*
The U.S. Government right to vetain a non-exclusive,
royalty free licence in and to any copyright is acknowledged.

Fig.1. Molecular components of organic superconductors.

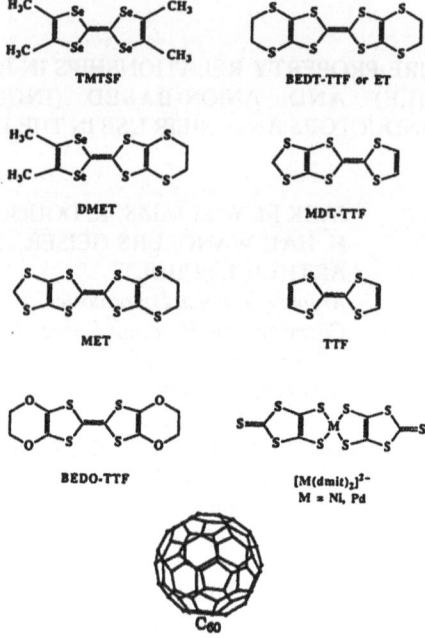

Table 1. Radical-cation based organic superconductors and their critical temperatures.

Compound	T_c† (K)	Compound	T_c† (K)
TMTSF compounds			
$(TMTSF)_2PF_6$	0.9 (12 kbar)	κ-$(ET)_4Hg_{3-\delta}Cl_8$	1.8 (12 kbar);
$(TMTSF)_2AsF_6$	1.1 (12 kbar)		5.3 (29 kbar)
$(TMTSF)_2SbF_6$	0.4 (11 kbar)	κ-$(ET)_4Hg_{2.89}Br_8$	4.3;
$(TMTSF)_2TaF_6$	1.4 (12 kbar)		6.7 (3.5 kbar)
$(TMTSF)_2ReO_4$	1.3 (9.5 kbar)	$(ET)_2Hg_{1.41}Br_4$	2.0
$(TMTSF)_2FSO_3$	2.1 (6.5 kbar)	α-$(ET)_2[(NH_4)Hg(SCN)_4]$	1.15
$(TMTSF)_2ClO_4$	1.4	$(ET)_3Cl_2 \cdot 2H_2O$	2.0 (16 kbar)
ET compounds		κ-$(ET)_2Cu(NCS)_2$	10.4
$(ET)_2ReO_4$	2.0 (4.5 kbar)	κ-$(ET)_2Ag(CN)_2 \cdot H_2O$	5.0
β-$(ET)_2I_3$‡	1.4	κ-$(ET)_2Cu[N(CN)_2]Br$	11.6
β^*-$(ET)_2I_3$§	8.0 (0.5 kbar)	κ-$(ET)_2Cu[N(CN)_2]Cl$	12.8 (0.3 kbar)
γ-$(ET)_3(I_3)_{2.5}$	2.5	*DMET compounds*	
θ-$(ET)_2I_3$	3.6	$(DMET)_2Au(CN)_2$	0.8 (5 kbar)
κ-$(ET)_2I_3$	3.6	$(DMET)_2I_3$	0.47
α_t-$(ET)_2I_3$	7–8	$(DMET)_2IBr_2$	0.59
(α/β)-$(ET)_2I_3$	2.5–6.9	$(DMET)_2AuCl_2$	0.83
β-$(ET)_{1.96}(MET)_{0.04}I_3$	4.6	$(DMET)_2AuI_2$	0.55 (5 kbar)
β-$(ET)_2IBr_2$	2.8	$(DMET)_2AuBr_2$	1.0 (1.5 kbar)
β-$(ET)_2AuI_2$	4.98	κ-$(DMET)_2AuBr_2$	1.9
		Other compounds	
		κ-$(MDT\text{-}TTF)_2AuI_2$	4.5
		β_m-$(BEDO\text{-}TTF)_3Cu_2(SCN)_3$	1.06
		$(TTF)[Ni(dmit)_2]_2$	1.6 (7 kbar)
		$Me_4N[Ni(dmit)_2]_2$	5.0 (7 kbar)
		α'-$(TTF)[Pd(dmit)_2]_2$	6.42 (20.7 kbar)
		α-$(TTF)[Pd(dmit)_2]_2$	1.7 (21.75 kbar)

†At ambient pressure in those entries where no pressure is indicted in parentheses.

‡Also referred to as β_L-$(ET)_2I_3$ §Also referred to as β_H-$(ET)_2I_3$.

The first organic superconductor, quasi-1D (TMTST)$_2$PF$_6$ (T$_c$ = 0.9 K) required considerable pressure (p = 9 kbar) to suppress a metal-insulator antiferromagnetic ordering transition (spin density wave, SDW) that occurred at ~ 16 K.[2] Competition between magnetic (SDW) and superconducting ground states are often observed in organic superconductors and SDW driven transitions occur when the on-site electron-electron Coulomb repulsion exceeds the stabilization gained from delocalization. In such cases electrons are localized to reduce the Coulomb repulsion and the ground state is that of an antiferromagnetically-coupled insulator (Mott-Hubbard insulator).[3] The original discovery of organic superconductors[2] based on TMTSF (Bechgaard salts) resulted in rapid developments in the field and, to date, organic superconductors have been derived from electron-*donor* molecule (non-C$_{60}$ based) and anion-based [M(dmit)$^{n-}$ and C$_{60}^{3-}$] species (Fig. 1). Only the ET and C$_{60}$-based systems have yielded a sufficiently large number of superconductors to allow the development of useful structure-property correlations (vide infra). The structures of these materials and these correlations, where they exist, will be the focus of this paper in so much that they can provide predictive power in the design of new superconducting organic systems.

β- and κ-phase ET-based Organic Superconductors

β-PHASE SYSTEMS

The β-(ET)$_2$X materials, X = triatomic and monovalent anion, are ambient-pressure superconductors with T$_c$'s as follows, viz., 2.8 K (X = IBr$_2^-$), 4.9 K (X = AuI$_2^-$) and 1.5 K (and 8 K at p >0.5 kbar) for X = I$_3^-$. An important feature of these systems is that they are *isostructural*, with the *linear-triatomic* anions occupying a cavity created by the ET-molecule ethylene-group hydrogen atoms formed from the "corrugated-sheet" layer-like packing of the electron-donor molecules, as shown in the stereodiagram in Fig. 2. The β-phase salts are actually *layered superconductors*, with the repeat unit being one layer of the conducting electron-donor molecules (in a general crystallographic position in the triclinic space group P $\overline{1}$, see Fig. 3) and one layer formed by the anions (located on inversion symmetry centers). In these systems the organic molecules are oriented parallel to each other and the molecular stacking axis occurs along the unit cell diagonal direction. The short S···S intermolecular contacts less than 3.6Å, the van der Waals S···S radius sum (see Fig. 3), are formed between molecules on neighboring stacks and the resulting honeycomb network they create in β-(ET)$_2$X systems is the electron conduction pathway in the donor-molecule layers (the anions are insulators). The effect of varying the linear-anion length is to change the unit cell volume in the same fashion, while also altering the network system of *intra-* and *inter*molecular S···S contacts. A representative set of linear anions that yield β or β-like structures is given in Table 2, in which anion length, unit cell volume, and T$_c$ (if appropriate) are presented.

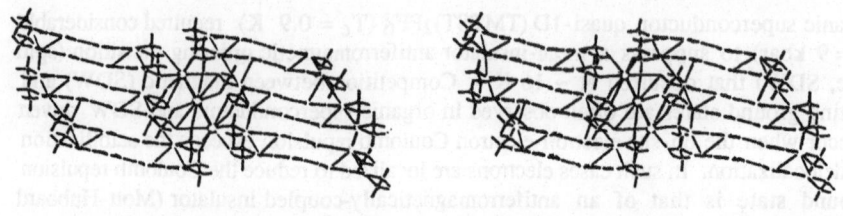

Fig. 2. Persepective view of a single (X-X-X)⁻ anion surrounded by ET molecule –CH₂ groups, in the H-pocket made up of 12 ET molecules in the β-(ET)₂X structure.

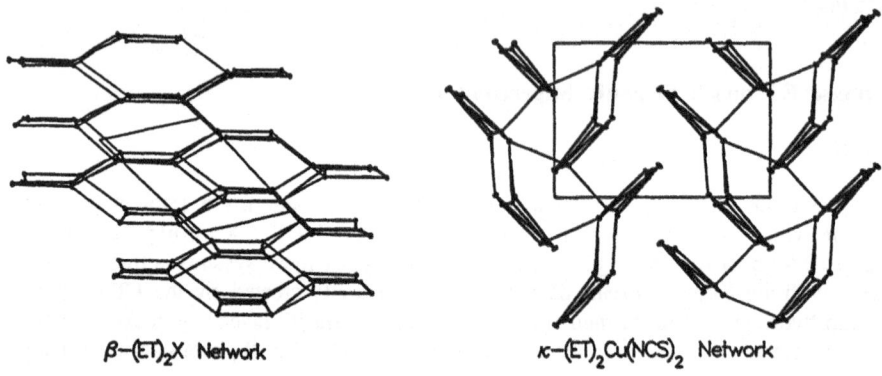

β–(ET)₂X Network   κ–(ET)₂Cu(NCS)₂ Network

Fig. 3.   ET molecule networks in β– and κ–type organic superconductors. Thin lines indicate short intermolecular S···S contacts, and the unit cell boundaries are also indicated by faint lines.

Table 2. β-(ET)$_2$X Salts (Anions and Properties)

X		Anion Length (Å)	V_e (Å^3 at 298 K)	T_c (K)
I_3^-	(I)(I)(I)	10.1	855.9	1.5
AuI_2^-	(I)(Au)(I)	9.4	845.2	5
I_2Br^-	(I)(I)(Br)	9.7	842.3	none
IBr_2^-	(Br)(I)(Br)	9.4	828.7	2.80
$ClIBr^-$	(Cl)(I)(Br)	9.0	821.3	?
ICl_2^-	(Cl)(I)(Cl)	8.7	814.3	$T_{MI} = 22.1$
$AuCl_2^-$	(Cl)(Au)(Cl)	8.1	800.7	$T_{MI} \approx 32$

It is noteworthy that for the shortest anions, i.e., ICl_2^- and $AuCl_2^-$ the salts undergo metal-insulator transitions below 40 K and the structures are only β-like due to the distortions of the regular β-lattice arising from Cl$\cdots$H and anion-ET molecule interactions. It is important to further point out that anion-donor molecule contacts in the ET-anion systems are through weak $-CH_2\cdots X$ (X = halide) hydrogen bonding interactions, and it is known the T_c's of the superconducting β-salts increase as these interactions increase in "softness"[1] of these contacts as longer anions are used due to an increase in electron-phonon coupling constants (λ) in agreement with BCS theory. Softness increases as the length of the anion is increased due to a weakening of hydrogen bonding $-CH_2\cdots X$ interactions as the atomic number and electronegativity of the halogen (X) increases, and also from a similar lengthening of the S$\cdots$S contacts between ET molecules that parallels the increase in anion length. While the *centrosymmetric* anions (I_3^-, AuI_2^-, and IBr_2^-) provide ambient pressure superconductors, the *un*symmetric I_2Br^- and $ClIBr^-$ anions produce nonsuperconducting products due to the disorder they induce in the lattice by random occupation of the crystallographic center of symmetry, which increases electron scattering and inhibits Cooper-pair formation. Again, β-(ET)$_2$I$_3$ is novel because a slight pressure of only 0.5 kbar[4] is sufficient to convert it from a modulated structure (below ~175 K, T_c = 1.5 K) to a fully ordered structure [β^*-(ET)$_2$I$_3$, T_c = 8 K][5] which has the highest T_c found in the β-phase organic superconductors. The three isostructural and centrosymmetric linear-anion salts (I_3^-, AuI_2^- and IBr_2^-) form the basis of a structure-property correlation that suggests T_c's as high as 40 K might result from the use of complex anions such as (NCS-M-SCN)$^-$, (NCSe-M-SeCN)$^-$, M = metal, etc., if they could be synthesized, were linear and could be incorporated into the β-(ET)$_2$X structure (see Fig. 4).

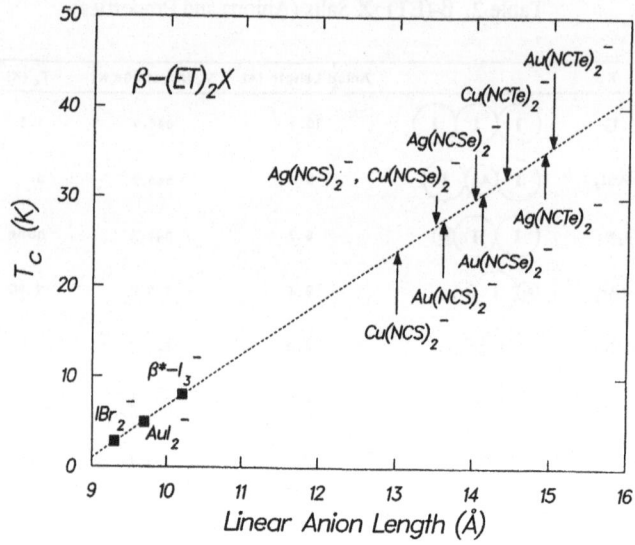

Fig. 4. Correlation of linear anion length in β-(ET)$_2$X salts with superconducting critical temperature T$_c$ and possible new anionic derivatives with higher T$_c$'s.

Finally, in the β-phase superconductors, the dependence of T$_c$ on X$^-$ and pressure[6] (p) is also related to changes in the lattice softness[1] as a function of X$^-$ and p, and a plot of T$_c$ vs. p is given in Figure 5.

Fig. 5. Pressure dependence of the T$_c$ of β-(ET)$_2$X (X$^-$ = I$_3^-$, AuI$_2^-$, IBr$_2^-$).

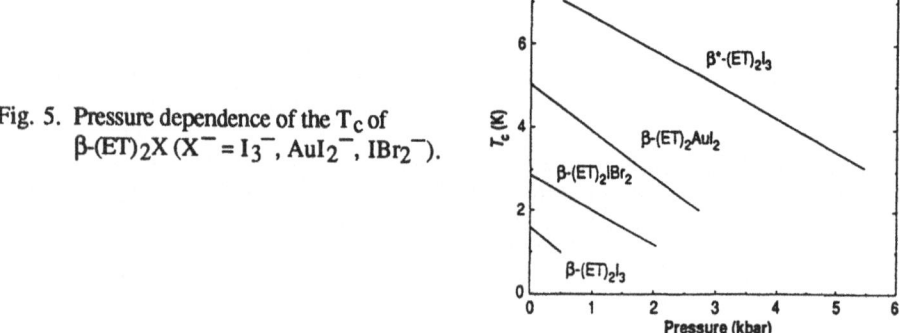

κ-PHASE SYSTEMS

Another important type of molecular packing motif occurs in the "κ-phase" ET-based materials, which contain orthogonally oriented ET-molecule dimers (see Fig. 3) and form layered superconductors possessing alternating layers of radical-cations and anions (the anions may be polymeric, see Fig. 6).

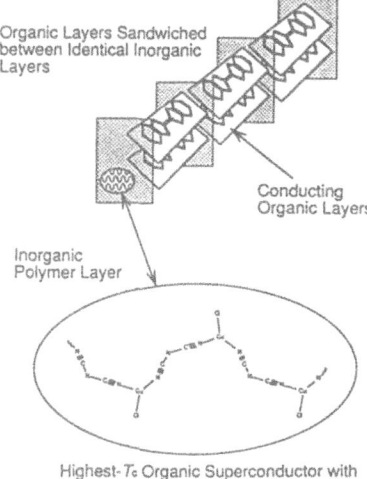

Fig. 6. Layered structure of κ-phase organic superconductors.

Organic Layers Sandwiched between Identical Inorganic Layers

Conducting Organic Layers

Inorganic Polymer Layer

Highest-T_c Organic Superconductor with T_c = 12.8 K, κ-$(ET)_2Cu[N(CN)_2]Cl$

The resulting superconductors have T_c's as high as 12.8 K (the maximum presently observed for radical-cation systems). Attempts to prepare β-phase salts with the X = "$Cu(NCS)_2^-$" (see Fig. 4) anion led to the first superconducting ET salt with a $T_c > 10$ K, but with a κ-phase, rather than β-packing motif. Furthermore, the "$Cu(SCN)_2^-$" anion is *not* linear, but is trigonal and *polymeric*.[7] In fact, the highest T_c's observed in the ET-based superconductors are found in κ-phase systems containing "self-assembling" polymeric anions derived from complexes of Cu^+ (vide infra).

The κ-$(ET)_2Cu(NCS)_2$ salt was unusual due to the presence of a polymeric anion derived from a trigonally coordinated Cu^+ ion. While attempting to prepare substituted salts of Cu^+ and the dicynamide $[N(CN)_2^-]$ anion in place of $(SCN)^-$, three novel κ-phase derivatives, again containing polymeric anions (see Fig. 7),[8,9] were discovered, viz., κ-$(ET)_2Cu[N(CN)_2]X$, X = Cl^-, Br^- and I^-. The X = Br^- and Cl^- salts exhibit the highest T_c's discovered in radical-cation based organic superconductors[1] whereas, surprisingly, the I^- derivative is a semiconductor. More importantly, even although these salts are *isostructural* they differ considerably in their electrical properties, i.e., the Br-salt[8] is an ambient pressure superconductor with $T_c = 11.6$ K, the Cl-salt[9] becomes superconducting ($T_c = 12.8$ K, the highest T_c found in any ET-based superconductor) under the smallest pressure (0.3 kbar) required to suppress a SDW transition for any organic superconductor. Furthermore, magnetization measurements at ambient pressure of the Cl^- salt reveal an antiferromagnetic transition near 45 K and, *for the first time in this class of materials*, a transition near 22 K to a state exhibiting weak ferromagnetic hysteresis effects.[10]

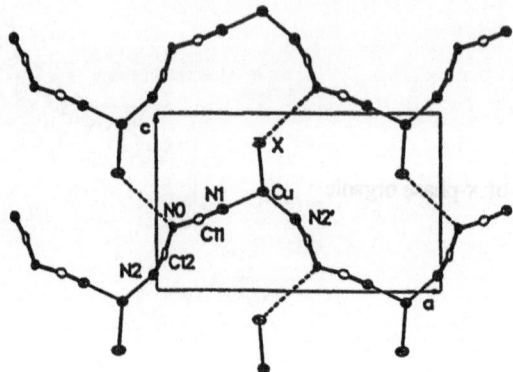

Fig. 7. Cu[N(CN)$_2$]X$^-$ anions κ–(ET)$_2$Cu[N(CN)$_2$]X, X = Cl$^-$, Br$^-$, I$^-$.

Calculations revealed that the band structures of the three Cu[N(CN)$_2$]X$^-$ salts are virtually the same and do not account for the differences in the observed electrical properties.[1] The principal structural difference between the superconducting Cu[N(CN)$_2$]X, X = Cl$^-$, Br$^-$ salts, and the semiconducting X = I$^-$ derivative, is that whereas the former are crystallographically *ordered* at low temperatures, the X = I$^-$ salt (which is not superconducting at p < 5 kbar) is not due to the presence of ET-molecule –CH$_2$-group disorder even at 16 K.[1] An analysis based on the "lattice softness" concept,[1] which arises from short intermolecular C-H···anion and C-H···H contacts in both the β- and κ-phase systems and it's relationship to T$_c$, has also been applied to the Cu[N(CN)$_2$]X, X = Cl$^-$, Br$^-$, I$^-$ salts with limited success.[11] Finally, future anionic substitution reactions on trigonal Cu$^+$ ions may continue to provide new κ-phase derivatives with unusual electrical properties.

Attempts to develop structure-property relationships for the superconducting κ-phase derivatives of ET have proven difficult, yet a first attempt at an empirical correlation is shown in Figure 8 by use of the data listed in Table 3.

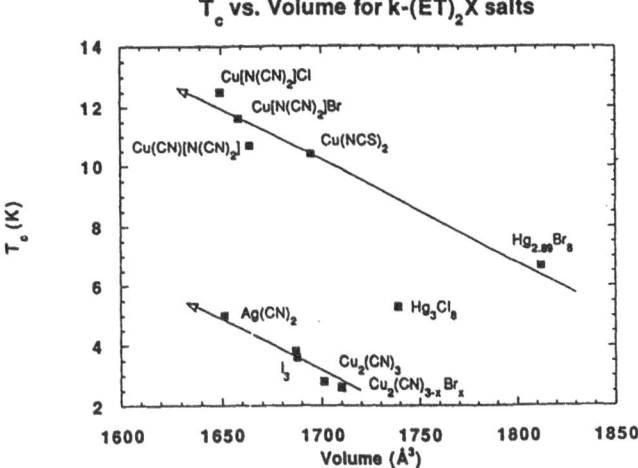

Fig 8. T_C vs. V_C for κ-phase ET salts.

Table 3. The unit cell constants and T_C's for κ-(ET)$_2$X salts.

Compound	a (Å)	b (Å)	c (Å)	T_c (K)	Volume (Å^3)
CuN(CN)$_2$I*	12.960	8.680	30.340	-	1706.0
Cu(NCS)$_2$	13.143	8.456	16.256	10.4	1694.8
CuN(CN)$_2$Br*	12.942	8.539	30.016	11.6	1658.5
CuN(CN)$_2$Cl *	12.977	8.480	29.979	12.5	1649.5
Ag(CN)$_2$(H$_2$O)	12.593	8.642	16.080	5.00	1651.2
I$_3$	12.832	8.466	16.387	3.60	1687.6
Hg$_3$Br$_8$*	11.219	8.706	37.105	6.70	1812.0
Hg$_3$Cl$_8$*	11.062	8.754	35.920	5.30	1739.0
Cu$_2$(CN)$_3$	13.400	8.586	16.117	2.80	1701.2
Cu$_2$(CN)$_{3-d}$Br$_d$	13.400	8.590	16.250	2.60	1710.0
Cu[N(CN)$_2$]CN	12.887	8.647	15.987	10.7	1664.0
Cu[N(CN)$_2$]CN	13.324	8.543	16.093	3.80	1687.0

*The volume listed = $\frac{real\ volume}{2}$ because they contain double layer structures.

In the κ-phases there appear to be separate correlations for different anion configurations based on either structurally 1-D anion (zig-zag) chains, such as found in Cu[N(CN)$_2$]X$^-$ salts (top line, Fig. 8), or structurally 2-D anion layers (2-D networks) in salts such as that observed in κ-(ET)$_2$I$_3$ (bottom line, Fig. 8). However, in surprising contrast to the β-(ET)$_2$X series, T_C *increases*, rather than decreases, with *decreasing* unit cell volume. The basis of these empirical relationships is under intense study currently and a better understanding of their origin is required before rational design criteria can be developed for κ-phase materials. Finally, based on years of experience gained by various investigators world-wide, the radical-cation (ET, etc.) superconductors are air-stable over long periods of time (years). However, they are brittle materials and not readily amenable to mechanical bending and forming.

C_{60}-based Organic Superconductors

In distinct contrast to the nearly planar radical-cations derived from the ET molecule and similar electron-donor molecules given in Fig. 1, are the remarkable new superconductors derived from the essentially *spherical* and all carbon-containing electron-accepting buckminsterfullerene molecule, C_{60}. Whereas the ET-based systems are usually 2D and of low symmetry (triclinic), the M_3C_{60}, M = alkali metal, organic superconductors are derived from the packing of spherical species (M^+ and C_{60}^{3-}) to form cubic (isotropic) structures and, therefore are truly 3D conductors. In addition, the T_c's of the fullerides are substantially higher (~33 K in Rb_2CsC_{60}) than those presently known for the ET-based superconductors (present maximum 12.8 K). Table 4 contains a list of relevant unit cell data and T_c's for the FCC (face centered cubic) alkali metal-based superconductors derived from C_{60}^{3-}. Based on existing data, it appears that no superconductors have been derived from purely Li or Na alone whereas pure and alloyed materials can be prepared by use of K, Rb or Cs.

Table 4. Unit cell and T_c's for FCC M_xC_{60} salts (from D. W. Murphy, et al.).[12]

	Lattice parameter(s) (Å)	T_c(K) (%)†
Na_2C_{60}‡	14.189 (1)	—
Na_3C_{60}	14.183 (3)	—
Na_6C_{60}	14.380 (8)	—
Na_2KC_{60}	14.120 (4)	—
Na_2RbC_{60}	14.091 (6)	—
$Na_2Rb_{0.5}Cs_{0.5}C_{60}$	14.148 (3)	8.0 (3)
Na_2CsC_{60} No. 1§	14.132 (2)	10.5 (8)
Na_2CsC_{60} No. 2§	14.176 (9)	14.0 (9)
K_3C_{60}	14.253 (3)	19.3 (30)
K_2RbC_{60}	14.299 (2)	21.8 (32)
Rb_2KC_{60} No. 1§	14.336 (1)	24.4 (34)
Rb_2KC_{60} No. 2§	14.364 (5)	26.4 (32)
Rb_3C_{60}	14.436 (2)	29.4 (35)
Rb_2CsC_{60}	14.493 (2)	31.3 (48)

†T_cs and shielding fractions (%) were measured by d.c. magnetization.
‡Sample is simple cubic.
§Samples labeled No. 1 and No. 2 have the same nominal composition.

The structures of the M_3C_{60} salts are described as intercalation compounds of the FCC structure derived from C_{60} itself, or hypothetical BCC (body centered cubic) or BCT (body centered tetragonal) structures (vide infra).[12] For the FCC structures the octahedral and tetrahedral interstices are located on special positions in the space group with the octahedral site being *larger* than any alkali metal ion whereas the tetrahedral site most closely resembles Na^+ in size. The C_{60} structure and that of different $M_x^+C_{60}^{n-}$ materials, including the M_3C_{60} superconductors, are shown in Figure 9.

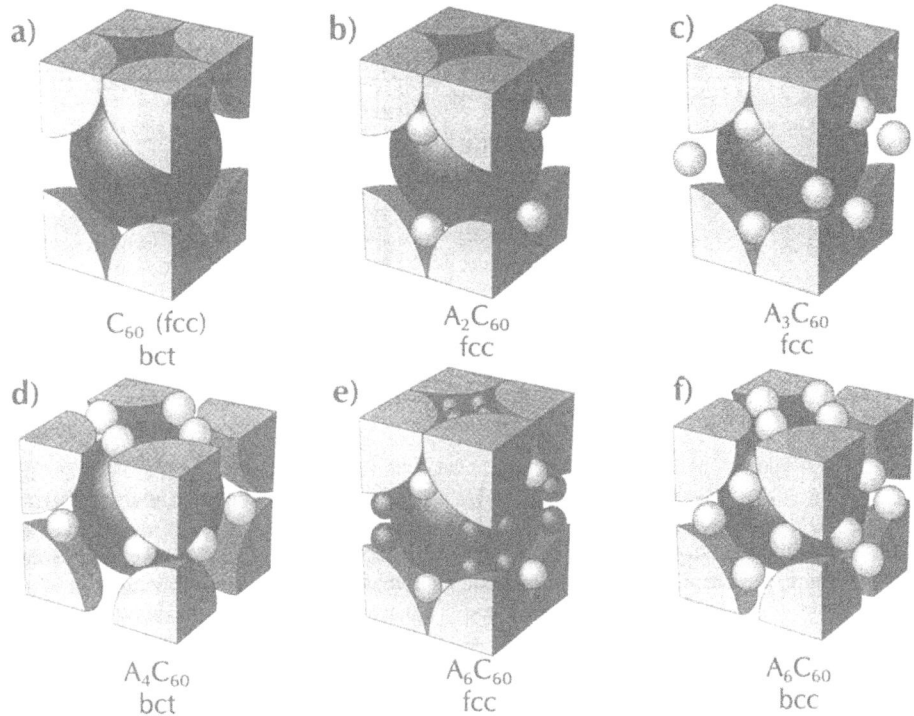

Fig. 9. Schematic structures of C_{60} and A_xC_{60} with C_{60}'s as large spheres and A as the smaller spheres. (a) FCC C_{60} drawn in an equivalent BCT representation. (b) The structure of Na_2C_{60} with Na ions in tetrahedral interstices. (c) A_3C_{60} with A ions in both tetrahedral and octahedral interstices. (d) The A_4C_{60} structure exhibited by K, Rb and Cs. (e) The FCC A_6C_{60} structure (A = Na, Ca) with the darker Na atom sites 50% occupied. (f) The BCC A_6C_{60} structure of K, Rb and Cs. (From D. W. Murphy, et al.).[12]

From a detailed analysis of the salts of C_{60}^{n-}, the structures they adopt agree with those one would predict on the basis of ion size (M^+ and C_{60}^{n-}) and from electrostatic calculations.[12] Since the M_3C_{60} salts are isostructural, the T_c's scale with cation volume and unit cell size (see Fig. 10).

550

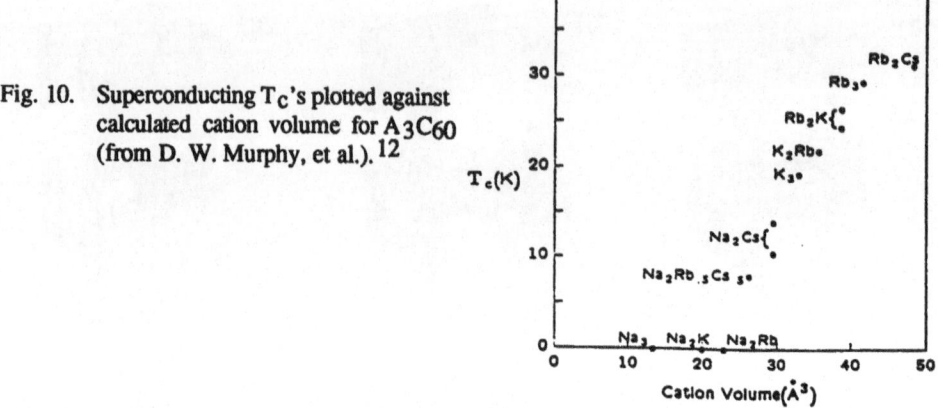

Fig. 10. Superconducting T_C's plotted against calculated cation volume for A_3C_{60} (from D. W. Murphy, et al.).[12]

One of the surprising structural contrasts between the ET and M_3C_{60} salts is that structural disorder in the former systems either reduces T_c or destroys superconductivity altogether (e.g., β-(ET)$_2$I$_2$Br),[1] whereas the octahedral and tetrahedral sites of the alkali metal ions, and the C_{60}^{3-} spheres themselves, are highly disordered in the M_3C_{60} superconductors. However, one structural feature common to both ET and C_{60}-based systems is that up to a certain point the T_c's increase steadily as either the ET-molecules, or the C_{60}^{3-} spheres, are increased in separation, i.e., as molecular overlap is reduced and, for example, as the interball C_{60}–C_{60} interactions soften. This argues for a BCS mechanism for the electron-pairing interaction in which the electronic density of states at the Fermi surface, which scales with T_c, increases as molecular overlap is decreased in these organic systems. However, the mechanism of superconductivity in organic superconductors remains an intensely studied mystery at this time!

Finally, one last important note is that the T_c's of the superconducting organic systems are rising at a rapid rate suggesting that they may rival those of the ceramic copper-oxide systems within this decade (see Fig. 11).

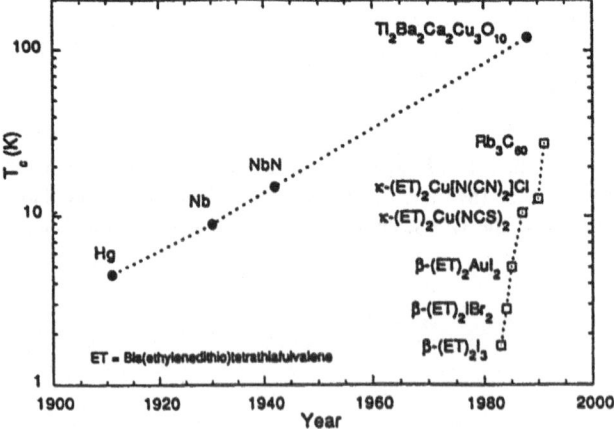

Fig. 11. T_c vs year.

Acknowledgements

Work at Argonne National Laboratory, Sandia National Laboratories, and North Carolina State University is supported by the Office of Basic Energy Sciences, Division of Materials Sciences, U. S. Department of Energy, under contracts W-31-109-ENG-38 and DE04-76DP00789 and grant DE-FG05-86ER45259, respectively.

References

1. J. M. Williams, A. J. Schultz, U. Geiser, K. D. Carlson, A. M. Kini, H. H. Wang, W.-K. Kwok, M.-H. Whangbo, J. E. Schirber, *Science*, **252**, 1501 (1991).
2. D. Jérome, A. Mazaud, M. Ribault, K. Bechgaard, *J. Phys. (Paris) Lett.* **41**, L95 (1980).
3. (a) N. F. Mott, *Proc. Phys. Soc. London Sect. A* **62**, 416 (1949)
   (b) J. Hubbard, *Proc. R. Soc. London Ser. A* **281**, 401 (1964).
4. (a) V. N. Laukhin, E. E. Kostyuchenko, Y. V. Sushko, I. F. Shchegolev, E. B. Yagubskii, *JETP Lett.* **41**, 68 (1985).
   (b) K. Murata, M. Tokumoto, H. Anzai, H. Bando, G. Saito, K. Kajimura, T. Ishiguro, *J. Phys. Soc. Jpn.* **54**, 1236 (1985).
5. (a) A. J. Schultz, M. A. Beno, H. H. Wang, J. M. Williams, *Phys. Rev. B* **33**, 7823 (1986).
   (b) A. J. Schultz, H. H. Wang, J. M. Williams, A. Filhol, *J. Am. Chem. Soc.* **108**, 7853 (1986).
6. (a) J. E. Schirber, L. J. Azevedo, J. F. Kwak, E. L. Venturini, P. C. W. Leung, M. A. Beno, H. H. Wang, J. M. Williams, *Phys. Rev. B* **33**, 1987 (1986).
   (b) J. E. Schirber, L. J. Azevedo, J. E. Kwak, E. L. Venturini, M. A. Beno, H. H. Wang, J. M. Williams, *Solid State Commun.* **59**, 525 (1986).
7. H. Urayama, H. Yamochi, G. Saito, K. Nozawa, T. Sugano, M. Kinoshita, S. Sato, K. Oshima, A. Kawamoto, J. Tanaka, *Chem. Lett.* 55 (1988).
8. A. M. Kini, U. Geiser, H. H. Wang, K. D. Carlson, J. M. Williams, W. K. Kwok, K. G. Vandervoort, J. E. Thompson, D. L. Stupka, D. Jung, M.-H. Whangbo, *Inorg. Chem.* **29**, 2555 (1990).
9. J. M. Williams, A. M. Kini, H. H. Wang, K. D. Carlson, U. Geiser, L. K. Montgomery, G. J. Pyrka, D. M. Watkins, J. M. Kommers, S. J. Boryschuk, A. V. Strieby Crouch, W. K. Kwok, J. E. Schirber, D. L. Overmyer, D. Jung, M.-H. Whangbo, *Inorg. Chem.* **29**, 3272 (1990).
10. U. Welp, S. Fleshler, W. K. Kwok, G. W. Crabtree, K. D. Carlson, H. H. Wang, U. Geiser, J. M. Williams, V. M. Hitsman, *Phys. Rev. Lett.* **69**, 840 (1992).
11. U. Geiser, A. J. Schultz, H. H. Wang, D. M. Watkins, D. L. Stupka, .J. M. Williams, J. E. Schirber, D. L. Overmyer, D. Jung, J. J. Novoa, M.-H. Whangbo, *Physica C* **174**, 475 (1991).
12. D. W. Murphy, M. J. Rosseinsky, R. M. Fleming, R. Tycko, A. P. Ramirez, R. C. Haddon, T. Siegrist, G. Dabbagh, J. C. Tully, R. E. Walstedt, *J. Phys. Chem. Solids* **53**, 1321 (1992).

Acknowledgments

Work at Argonne National Laboratory, Sandia National Laboratories, and North Carolina State University is supported by the Office of Basic Energy Sciences, Division of Materials Sciences, U. S. Department of Energy, under contracts W-31-109-ENG-38 and DE-AC04-76DP00789 and grant DE-FG05-88ER45375, respectively.

References

1. E. M. Williams, A. I. Schindler, U. Gonser, K. D. Gibson, A. M. Kini, R. H. Wang, W.-K. Kwok, M.-H. Whangbo, J. B. sohner, *Science* **252**, 1501 (1991).
2. D. Jerome, A. Mazaud, M. Ribault, K. Bechgaard, *J. Phys. (Paris) Lett.* **41**, L95 (1980).
3. D. W. du Mott, *Proc. Opt. Soc. London Ser. A* **82**, 43 (1963).

Part V

Phase Diagrams of HTSC

PHASE DIAGRAM, SYNTHESIS AND CRYSTAL GROWTH OF YBaCuO PHASES AT HIGH OXYGEN PRESSURE Po₂<3000 bar.

J.Karpinski, K.Conder, Ch.Krüger, H.Schwer, I.Mangelschots, E.Jilek, E.Kaldis

Laboratorium für Festkörperphysik ETH 8093–Zürich.

ABSTRACT. An overview is presented of the phase diagram, crystal growth and the structural and physical characterization studies of the double chain compounds $YBa_2Cu_4O_8$ (124) and $Y_2Ba_4Cu_7O_{14+x}$ (247).
The stability regions of the $YBa_2Cu_3O_{7-x}$ (123), 247 and 124 phases ($Po_2 \leq 3000$ bar) depend on the metal ratio. Different compositional sections of the P–T–X phase diagram have been investigated. The thermodynamics of the oxygen exchange reaction of the orthorhombic 123 with gaseous oxygen has been studied in the pressure range 10^{-3}–100 bar. Highly perfect ceramic 247 material with Tc of 93 K has been obtained. X–ray single crystal structure analysis of 124 and 247 shows, that structural changes and incorporation of impurities occur only in the "123" unit, whereas the "124" unit remains stable. In the case of the 247 compound not only oxygen nonstoichiometry, but also misorientation of oxygen, copper nonstoichiometry together with the incorporation of impurities, like C and Al, strongly affect Tc. Furthermore, the phase diagram for the crystal growth process at high Po₂ is also presented. The 124 single crystals were grown in Al_2O_3 (Tc=72 K) and in Y_2O_3 and ZrO_2 crucibles (Tc=80 K). Doping with Ca gives rise to an increase of Tc (Tc=84 K). Melting of bulk 124 samples at $Po_2 = 1000$–2700 bar results in textured ingots consisting of several mm large 124 grains with $Jc=5x10^4 A/cm^2$ at 5K.

1.Introduction

Since the discovery of $YBa_2Cu_3O_{7-x}$ the YBaCuO system has been studied intensively in a large oxygen pressure range ($10^{-4} \leq Po_2 \leq 3x10^3$ bar).
In the course of these investigations two other superconducting phases belonging to this system, $YBa_2Cu_4O_8$ and $Y_2Ba_4Co_7O_{14+x}$ have been synthesized both as ceramics and single crystals [1,2,3,4]. The 124 structure differs from that of 123 in

E. Kaldis (ed.), Materials and Crystallographic Aspects of HTc-Superconductivity, 555–583.

556

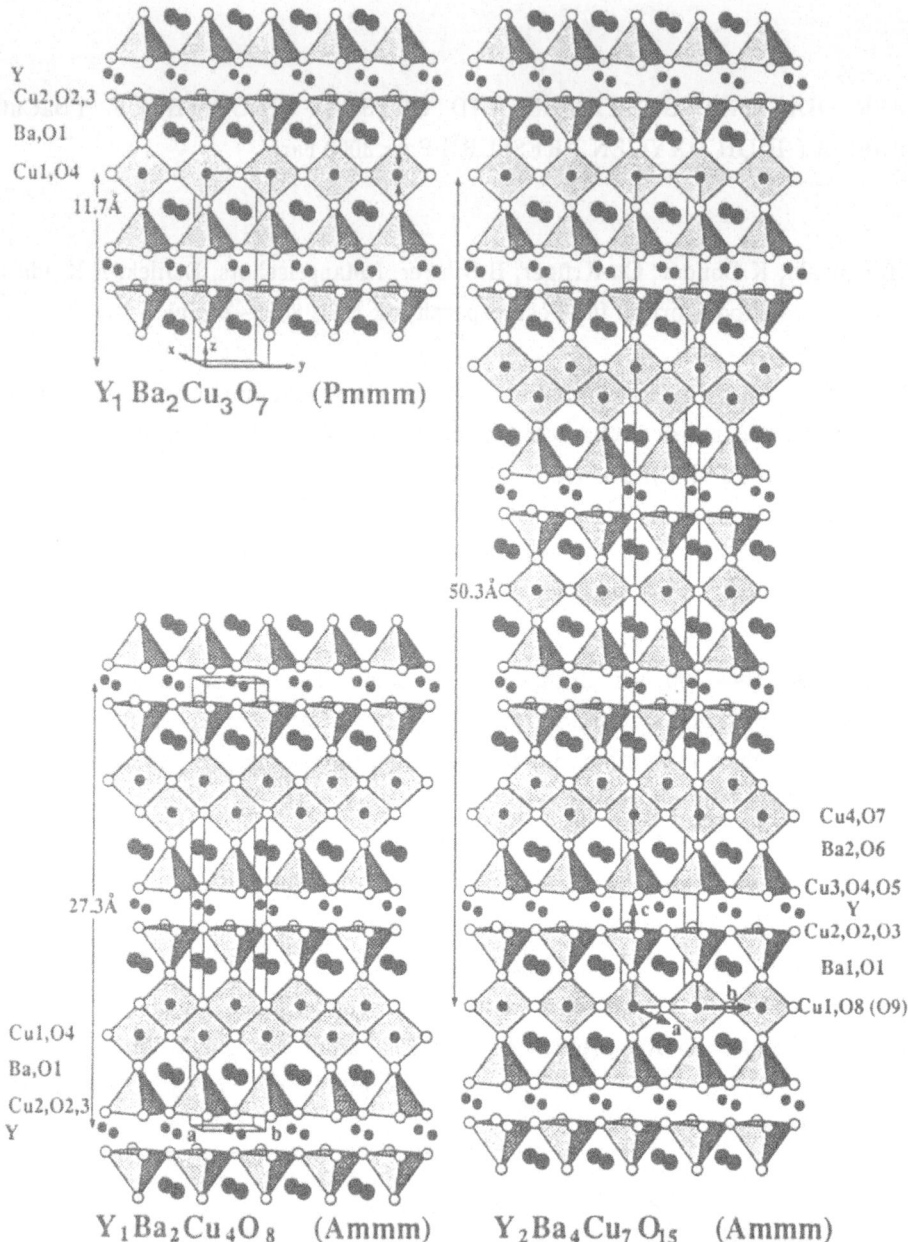

Y
Cu2,O2,3
Ba,O1
Cu1,O4
11.7Å

x z y

$Y_1 Ba_2 Cu_3 O_7$ (Pmmm)

50.3Å

Cu4,O7
Ba2,O6
Cu3,O4,O5
Y
Cu2,O2,O3
Ba1,O1
Cu1,O8 (O9)

27.3Å

Cu1,O4
Ba,O1
Cu2,O2,3
Y

$Y_1 Ba_2 Cu_4 O_8$ (Ammm) $Y_2 Ba_4 Cu_7 O_{15}$ (Ammm)

Fig.1. Three phases existing up to now in the $Y_2 Ba_4 Cu_{6+n} O_{14+n}$ family: a)123, b)247, c)124.

having a double CuO chain with a higher oxygen coordination by copper. Each oxygen atom in the 124 double chain has three neighboring Cu atoms whereas there are only 2 neighboring Cu atoms in 123. Therefore, the 124 structure is more stable, giving use to a constant oxygen stoichiometry which is independent on pressure and temperature. On the other hand, the 123 compound exists in a large oxygen nonstoichiometry range. A structural phase transition for $YBa_2CuO_{6.6}$ at about 700°C for $Po_2=1$ bar is observed. The superconducting, orthorhombic $YBa_2Cu_3O_{7-x}$ phase $(0 < x < 0.4)$ is metastable at temperatures below 700°C at $Po_2=1$ bar and the decomposition kinetic is very slow.

Recently, a modification of the tetragonal 123 structure with splitted equilibrium positions for the apical oxygen in Cu–deficient $YBa_2Cu_{2.78}O_{7-x}$ crystals has been observed [5,6]. Some of those crystals, synthesized at high oxygen pressure display a diamagnetic transition at 120K.

The main disadvantage of the 124 compound is the Tc of 80 K, lower than Tc of 93 K of the 123 phase. Ca–doping, however, increases Tc up to 90 K [7]. The reason of this increase is still not clear. Structural investigations by neutron powder diffraction [8,9] and x–ray refinement performed on small Ca– doped 124 single crystal [10] show evidence for the substitution of the trivalent Y by the divalent Ca which possibly leads to an increase of the hole carrier concentration. However, electron energy–loss experiments made by Knupfer et al [11] suggest no increase of the hole concentration upon doping and that the increase of Tc is rather related to a charge transfer between the CuO chains and the CuO_2 plains. Also NQR/NMR experiments in Ca doped 124 did not find any evidence for an increase of the carrier concentration in the CuO_2 planes, since the NQR frequency, the magnetic shift and the spin–lattice relaxation rate at the Cu2 site remained unchanged upon Ca–doping [12]. Except of Ca, all other up to now investigated substitutes in 124 decrease Tc. One of such elements is Al, which can be dissolved in the 124 crystals during growth from the alumina crucibles, decreases Tc to 72–73 K. The use of ZrO_2 and Y_2O_3 crucibles increases Tc up to 80 K [13] and doping with Ca enhances Tc up to 84 K [14].

Due to the difficulties in synthesis at the normal pressure the third compound of the YBaCuO family: $Y_2Ba_4Cu_7O_{14+x}$ (247) has been less investigated despite of many interesting properties. The 247 structure might be considered as a bulk superlattice containing blocks of both 123 and 124. Its unit cell is much larger (c=50.59 A) than that of 124 (c=27.26 A) or 123 (c=11.68 A). Due to the existence of single chains the oxygen content can vary between 14.3 and 15, changing Tc from 20 up to 95 K [5,15,16,17]. As the 247 structure is closely related to that of 123, it is interesting to compare the Tc dependence on oxygen nonstoichiometry in 247 and 123. The two–plateau Tc behaviour of 123 has been usually correlated with Ortho I and Ortho II phases corresponding to the a_0 and $2a_0$ superstructures, respectively. In the stoichiometric 123 all available oxygen sites in the chains are filled leading to the Ortho I phase which is believed to have Tc=93 K. The Ortho II phase with every second chain empty and ideal stoichiometry with x=0.5 is the Tc=60 K phase. Transmission Electron Microscopy investigations of 247 [18] show the existance of both the Ortho II ($2a_0$) and the Ortho III ($3a_0$) superstructures in the single CuO chains. The ideal stoichiometries are x=0.5 and 0.67 for the Ortho II and Ortho III, respectively. At first sight a similar dependence of Tc on oxygen stoichiometry in 247 as in 123 might be expected. However, the existing measurements show a linear behaviour [5,15,16,19]. The correlation between the superstructure, oxygen stoichiometry and Tc needs to be elaborated on carefully equilibrated 247 samples with various oxygen contents by using neutron and X–ray diffraction, exact volumetric analyses of oxygen contents, TEM and susceptibility measurements. Such experiments are foreseen in the near future. X–ray investigations on 247 single crystals performed recently [17,20], show that not only oxygen but also copper nonstoichiometry strongly influences on Tc. Vacancies or substitutions by Al or CO_3^{2-} ions at the Cu1 positions might lead to the decrease of Tc down to 4 K in case of a 75% occupation of Cu1 positions. Lower Cu1 occupation and CO_3^{2-} substitution leads to the misorientation of the O8 oxygen in single chain and the occupation of the O9 interstitial positions.

2. Phase diagrams

For last years the phase diagram of the YBaCuO system has been of interest to many laboratories, as its knowledge is of great importance for any preparation procedure [17,21–25]. The YBaCuO is a quaternary system, which makes its investigations rather complicated. In order to make phase–relationships more clear, we will also discuss sections of the ternary Y_2BaCuO (211), $Ba_2Cu_3O_5$ (023), CuO (001) system along the 123 – CuO tie line. Figure 2 shows the ternary system for Po_2=1 bar at 800°C. At this temperature all three superconducting compounds, 123, 124 and 247, are stable as single phases being in equilibrium with the O_2 gas phase.

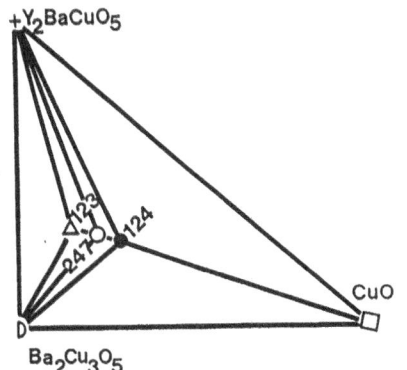

Fig.2. Part of the ternary Y_2O_3–BaO–CuO system at T=800°C and Po_2=1bar.

However, in the presence of CuO at these P,T–conditions, 123 and 247 react with CuO forming 124 which indicates the importance of the composition of the sample for the phase stability. Therefore, we have investigated the P,T phase diagrams for six various compositions lying on the tie line 123–CuO. Based on these P,T–diagrams, we were able to construct the T–x sections of ternary system for various pressures ($1 \leq Po_2 \leq 3000$ bar). Another important section of the ternary phase diagram includes 123, 124 or 247 and the ternary eutectic. This pseudobinary diagram displays the conditions for the crystal growth.

2.1. P–T PHASE DIAGRAMS OF the 123–CuO SYSTEM.

2.1.1. *Experimental*. High pressure experiments have been performed in a two chamber autoclave [26], which allows pressures up to 3000 bar at temperatures up to 1500°C. The sample weighting up to 20g has been placed in the internal ceramic chamber, which is the alumina crucible. The oxygen atmosphere of the sample is separated from the furnace argon atmosphere. The total oxygen volume in the hot zone is small — only a few cm³, improving the safety of the setup.

The starting material for the phase diagram studies was a mixture of 123 and CuO, corresponding to the compositions: 123, $123^1/_4$,$123^1/_3$,$123^1/_2$,$123^2/_3$ and 124.The experimental procedure consisted of annealing for 50–100 h at various Po_2 and T, followed by fast cooling or quenching. At $Po_2 \leq 1$ bar, the thermodynamic stability range requires lower temperatures resulting in slow kinetics of the synthesis reaction (123 with CuO producing 124 or 247). Therefore, in order to determine the stability conditions of 124 and 247, the decomposition reaction has been studied rather than synthesis.

560

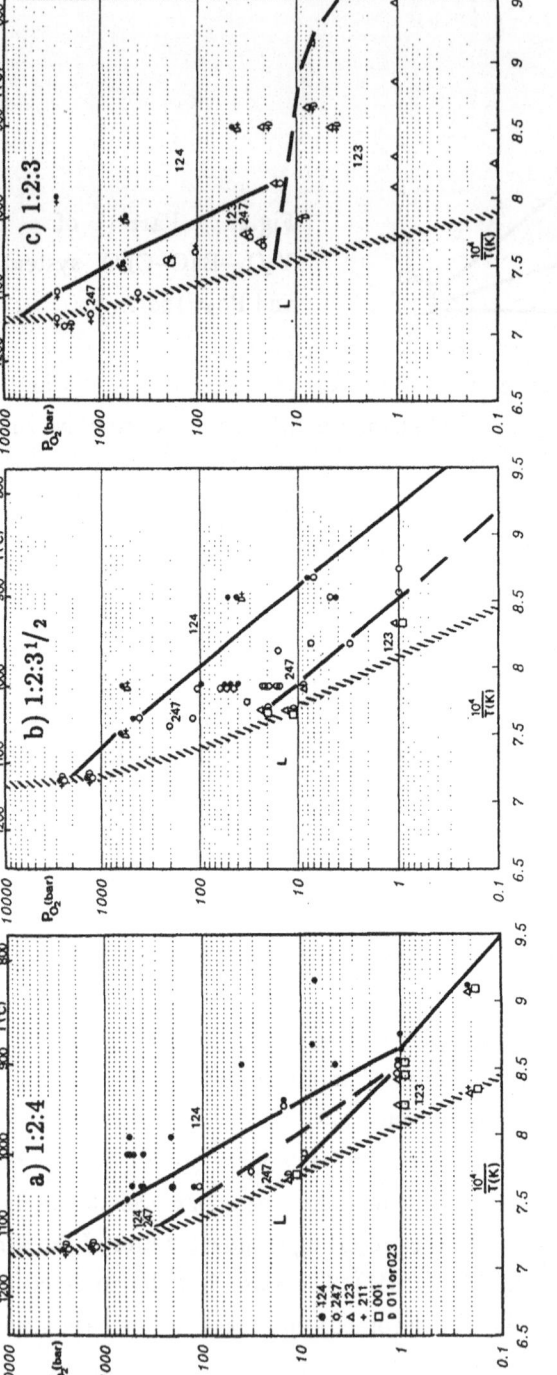

Fig.3. $LogP_{O_2}-1/T$ cross–section of the P_{O_2}–T–x phase diagram for the composition Y:Ba:Cu; a) 1:2:4, b) 1:2:3½ and c) 1:2:3.

2.1.2. *Results.* After annealing at various pressures the samples have been investigated by means of x–ray, magnetic susceptibility and occasionally by TEM. The results of phase determination by x–ray are shown in Figure 3. The various observed phases are marked with graphic symbols. The stability fields of 123, 247 and 124 are quite different and strongly dependent on the composition of the samples [17,27].

Composition 124:

The stability field of the 124 phase in the phase diagram (Fig.3 a)) in the high temperature range is limited by the following decomposition reactions:

(1) $2YBa_2Cu_4O_8 = Y_2Ba_4Cu_7O_{14+x} + CuO + (1–x)/2\ O_2$.

(2) $YBa_2Cu_4O_8 = YBa_2Cu_3O_{7-x} + CuO + x/2\ O_2$.

Since the Gibbs free energies of these reactions are not equal, these two lines cross around the point $Po_2=1$ bar, T=890°C indicating following reaction:

(3) $YBa_2Cu_4O_8 + YBa_2Cu_3O_{7-x} = Y_2Ba_4Cu_7O_{15-x}$.

Our experimental results show, that 124 decomposes to 247 + CuO and 123 + CuO at $Po_2 \geq 1$ bar and $Po_2 \leq 1$ bar, respectively. This possibility was also obtained by Voronin and Degterov [25], but the proposed temperature of such intersection was lower, namely 800°C. We have observed an coexistence of the 124 and 247 phases in several 124 samples annealed at high oxygen pressure and temperature above the decomposition of 124. This is quite unexpected, as the remaining products should be 247 + CuO. However, by cooling 247 together with a fine dispersed CuO, 124 is again formed. The decomposition line of 124 intersects the peritectic decomposition line at $Po_2 \cong 3000$ bar, indicating that at this pressure it is possible to obtain 124 bulk by peritectic reaction from the melt. We have not observed a low temperature limitation of the stability range of 124. Therefore, we concluded that 124 is the only superconducting phase in YBaCuO system, which is thermodynamically stable at room temperature. Several attempts to form new phases with higher CuO content such as 249 or 125 at high pressure up to 3000 bar were not successful.

Composition 247:

The stability field of 247 in the phase diagram (Fig.3 b)) is limited at the high temperature side by the peritectic decomposition (4) at higher pressures and the decomposition to 123 (5) at lower pressures:

(4) $Y_2Ba_4Cu_7O_{14+x} \rightarrow Y_2BaCuO_5 + L$

(5) $Y_2Ba_4Cu_7O_{14+x} \rightarrow 2YBa_2Cu_3O_{7-y} + [(x–1)/2+y]O_2$

The low temperature boundary is governed by the reaction:

(6) $3Y_2Ba_4Cu_7O_{14+x} = 2Y_2BaCuO_5 + 2YBa_2Cu_4O_8 + 3Ba_2Cu_3O_5 + 2CuO + (3x-1)O_2$

We were able to observe reaction (6) at high pressures for temperatures $\geq$ 900°C, but not at $Po_2=1$ bar due to kinetic reasons.

The annealings of 247 at $Po_2 = 1$bar have been performed at various temperatures 500°C $\leq$ T $\leq$ 920°C for t = 100 h. The decomposition of 247 to 123 + CuO has been observed above 900°C. Similar as in the case of the 124 phase diagram the existence of both 124 and 247 phases are detected in several samples annealed at high pressure, which is not supposed to take place. This result can be easily understood based on the T–x sections of the ternary diagram. Figure 5 b) shows the T–x section along the 123–CuO tie line at the pressure $Po_2=20$ bar, which was later used for the synthesis of the 247 samples. At $Po_2=20$ bar the stability conditions of the 247 phase are limited to the temperature range 900<T<1020°C, 123 to 900<T<1070°C, and 124 up to 950°C. By heating a mixture of 123+CuO from room up to high temperature through the stability range of 124+211+023 phases, a part of the sample is transformed in 124. Due to the fact, that both the 124 and 247 phases are stable in a certain temperature range, 124 remains in the samples until the decomposition temperature of 124 is reached. For $Po_2=20$ bar this temperature range is 900–950°C. Then, after the decomposition of 124 phase, 247 remains present. After the synthesis of 247 at high temperature, cooling must take place across the stability range of 124. Therefore, quenching is necessary in order to avoid intergrowth with 124.

Composition 123:

It is rather difficult to determine the stability range of 123 due to its metastability. In Figure 3 c), the results of annealing of the 123 samples are displayed for various P,T conditions. In spite of the long annealing times the decomposition reaction did often not reach equilibrium so that 123 as well as products of the decomposition remained in the sample. The 123 compound decomposes to 124 or 247 at increased pressure, following the reaction:

$24YBa_2Cu_3O_{7-x} = 9Y_2BaCuO_5 + 6YBa_2Cu_4O_8 + 13Ba_2Cu_3O_5 + BaO + (30-24x)O_2$

However, it is difficult to define exactly the phase boundaries between the stability regions of the 124, 247 and 123 phases of the average 123 composition of the sample.

Therefore, some of the phase boundaries in this phase diagram are tentative. Many authors indicated the instability of 123 also at normal pressure at temperature $\leq$ 700°C [28,29]. The extremely slow decomposition rate strongly hinders further investigation of the 123 decomposition at these conditions.

2.2. T–x SECTIONS OF THE TERNARY SYSTEM.

We will present phase equilibria using T–x sections of the ternary system along the 123–CuO tie line for various pressures. The composition parameter, x corresponds to the ratio of CuO to all metals in at.%. Up to now, only the T–x diagram at P_{O_2} = 1 bar, have been published [24,30]. Figure 4 shows a part of the ternary system with vertical T axes at P_{O_2} = 1 bar. In the Figure 5 five sections of such a three–dimensional system are shown for various pressures. One can observe the variations of the phase stability ranges with P_{O_2}. All three superconducting phases are stable at P_{O_2}=1 bar for a limited temperature range. The stability field of 123 decreases with P_{O_2} and completely disappears at P_{O_2} = 100 bar. Both 124 and 247 phases exist up to 3000 bar. The stability field of 247 is limited from both the high

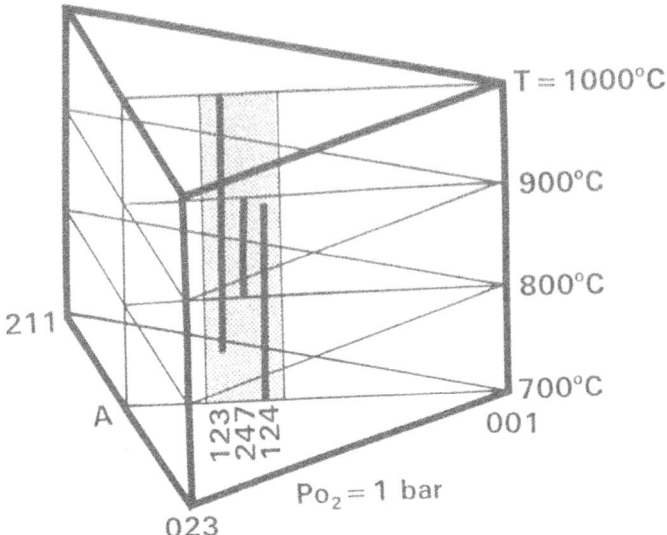

Fig.4. T–x section of ternary Y_2BaCuO_5–$Ba_2Cu_3O_5$–CuO system at P_{O_2}=1 bar. Parts of such sections for various P_{O_2} are presented in Fig.5.

564

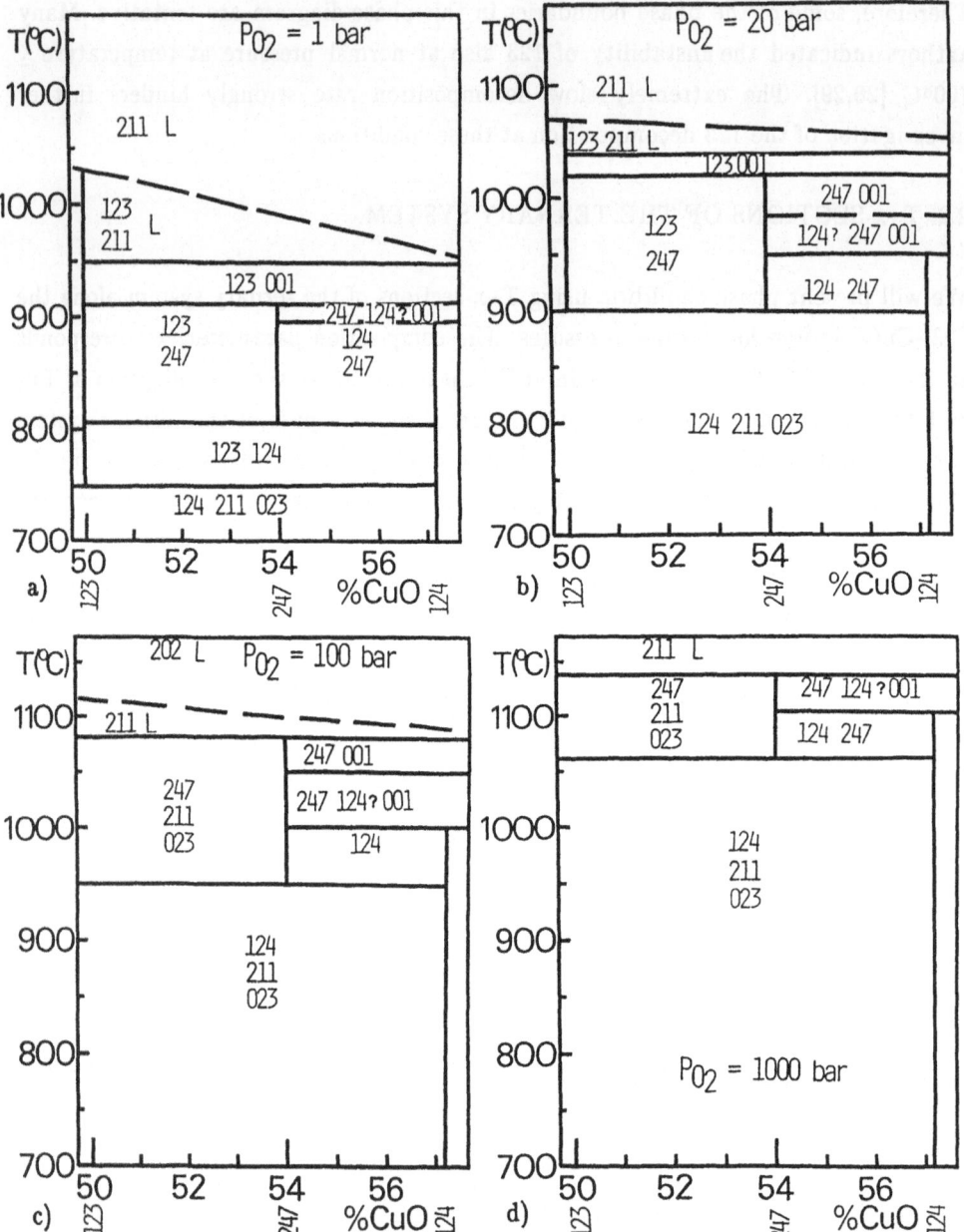

Fig.5. T–x sections of ternary system for various pressures. x – ratio of CuO to all metals in atomic %.

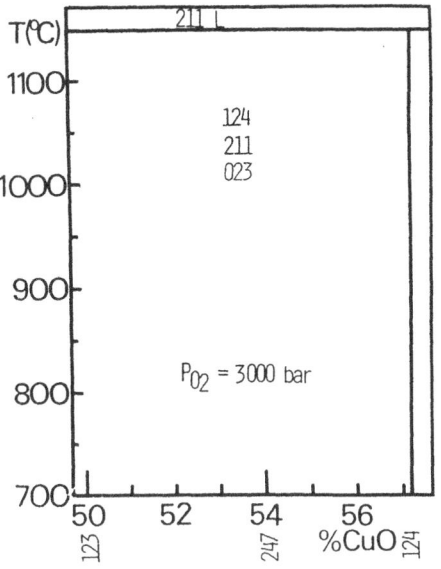

Fig.5 e) T–x section for P_{O_2}=3000 bar.

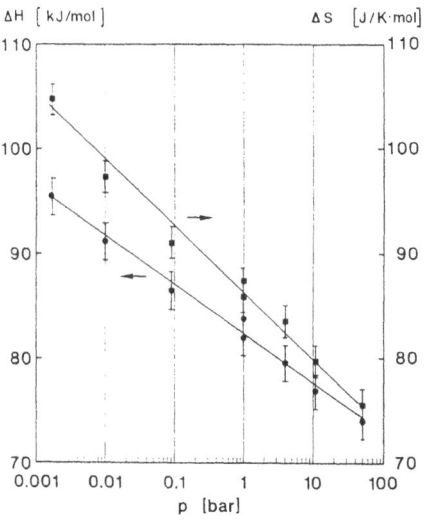

Fig.7. Dependence of ΔH and ΔS for the reaction of oxygen exchange in 123.

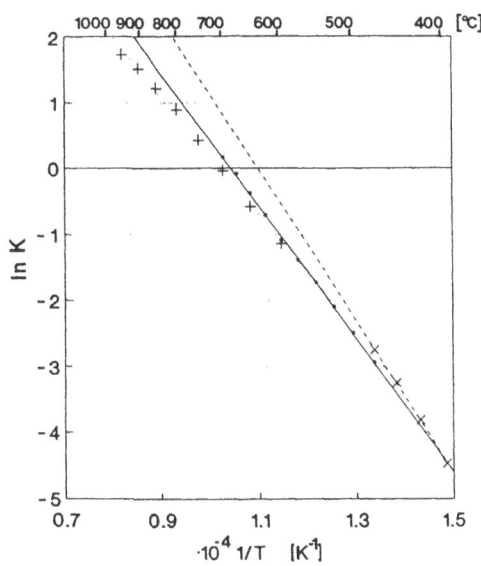

Fig.6. Relationship of lnK vs.1/T for the oxygen pressures 50 bar (+), 1 bar (·) and 0.0017 bar (x).

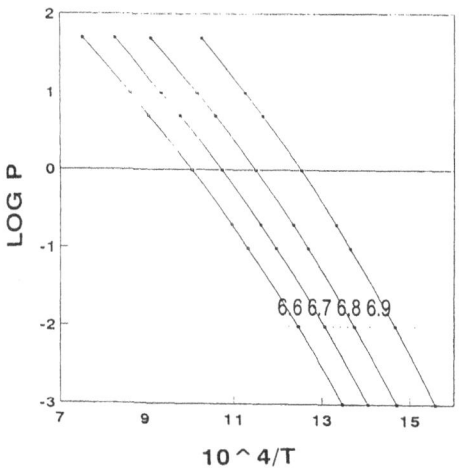

Fig.8 P–T van't Hoff diagram for 123 with x=const lines. Variations of ΔH and ΔS with pressure results in nonlinearity of the isocompositional lines of 123.

and low temperature side. Its decomposition temperature is always higher than that of 124. It is clearly shown that in the $Po_2 = 3000$ bar diagram (Fig.5 e)), 124 remains as the only stable phase. Therefore, it is possible to obtain 124 bulk by peritectic reaction from melt at this pressure.

2.3. T–x PHASE DIAGRAM OF THE 123 COMPOUND.

The nonstoichiometry of oxygen in the 123 compound has been widely investigated at pressures $Po_2 \leq 1$ bar. Exploiting our high oxygen pressure facility, we extended this study to the pressure range of $10^{-3} - 100$ bar [17,31,32]. The pressures below 1 bar were achieved by mixing oxygen with argon. The stoichiometry was calculated from the weight– loss at various temperatures in thermogravimetric experiments. In the pressure range $1 \leq Po_2 \leq 100$ bar the composition of 123 has been measured in the high pressure autoclave. To measure the changes in the oxygen content the pressure in the crucible was kept constant during heating, and the excess O_2 gas was collected in the reservoir with calibrated volume. During cooling, the decrease of the pressure was compensated by the oxygen from the reservoir. The system was calibrated prior to experiments in a blank run using an alumina rod of the same volume, so that the pressure changes resulting from the thermal contraction could be subtracted. The oxygen exchange with the gas phase can be described by the reaction of the oxygen vacancy formation:

$$O_O^x + Cu_{Cu}^x \rightleftharpoons V_O^{\cdot} + Cu_{Cu}^{\lambda} + 1/2 O_2$$

The concentrations of the reactants are stoichiometry dependent:

$$[O_O^x] = (1-x)/2, \quad [V_O^{\cdot}] = (1+x)/2$$

$$[Cu_{Cu}^x] = (1-x)/1, \quad [Cu_{Cu}^{\lambda}] = x/1$$

so that the mass action low is given by:

$$K = \frac{x(1+x)}{(1-x)^2} \cdot p_{O_2}^{1/2}$$

The logarithm of this equilibrium constant versus reciprocal temperature displays different slopes for different pressures (Fig.6). Therefore, as clearly shown in the Fig.7, the entropy and enthalpy of the oxygen vacancy formation are pressure dependent. Consequently, the isocompositional lines for 123 in the coordinate system log $Po_2 = f(1/T)$ (Fig.8) are not linear but have a curvature. Based on these results, we have calculated the T–x equilibrium curves in the range $6.4 < x < 7$ for temperatures between 300 and 1200 °C (Fig.9). These results are in good agreement with the experimental data, in the stability range of the orthorhombic phase.

It is necessary to point out that all these experiments were made in the metastability range of 123. Experiments under pressures above 50 bar and at higher temperatures show the phase transformation to 124, 247, 211 (Y_2BaCuO_5) and 023 ($Ba_2Cu_3O_5$). Depending on the equilibrium time, this reaction can start above 20 bars.

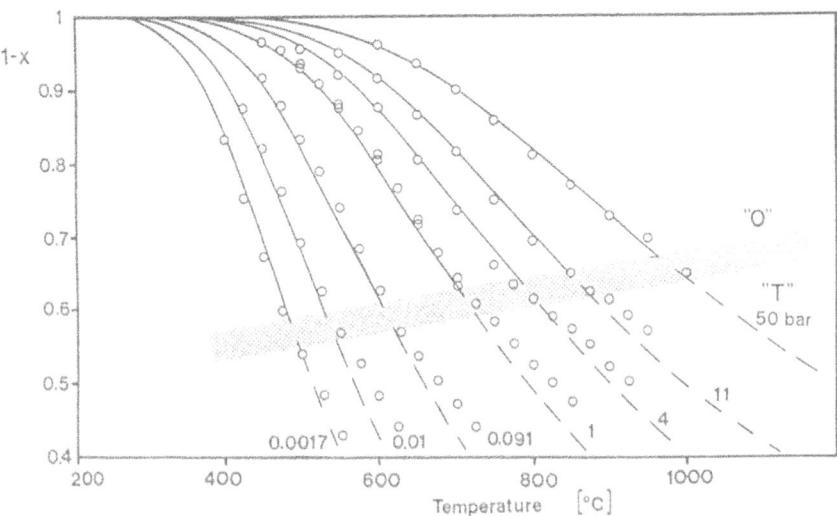

Fig.9. The oxygen nonstoichiometry in orthorhombic 123 as a function of temperature, with Po_2 as parameter. The transition from the orthorhombic to the tetragonal structure is marked.

3. Solidification of the YBaCuO phases from the stoichiometric melt or solution.

All three compounds decompose peritectically during melting at oxygen pressure up to 3000 bar. Therefore two alternative ways are possible for obtaining solidified YBaCuO samples.

(a) Solidification of stoichiometric melt by peritectic reaction.

(b) Crystal growth from solution above eutectic melting point.

3.1. MELTING OF 123 AND 124 AT HIGH PRESSURE.

During the past years many papers concerning the solidification of 123 from melt have been published [33,34,35]. Using this technique textured samples with large critical current densities (Jc $\leq$ 10^5 A/cm^3) have been obtained. It has been commonly assumed that 123 crystals grow by peritectic reaction from 211 crystals which act as heterogeneous nucleation sites in a temperature interval between T_{perit}–T_{eut}. Recently, several authors investigated the formation mechanism of textured 123 [36,37,38] and concluded, that not a peritectic reaction is responsible for the crystal growth of 123, but a direct growth from the melt. We have investigated the influence of oxygen pressure on the solidification of YBaCuO melt in the pressure range 1000 $\leq$ Po$_2$ $\leq$ 3000 bar.

3.1.1. *Experimental.*
Melting of 123 samples:
Some 10–20g. pellets of 123 have been heated above the peritectic decomposition temperature and cooled at oxygen pressure 100 $\leq$ Po$_2$ $\leq$ 2500 bar. The experimental parameters were:
(1) heating 5°C/min → 1150°C, (2) dwelling 1–2 h 1150°C, (3) cooling 5°C/min → room temperature.
Melting of 124 samples:
The 124 pellets have been heated at oxygen pressure 1000 $\leq$ Po$_2$ $\leq$ 2700 bar with the following procedure:
(1) heating 2–5°C/min → 1170°C, (2) dwelling 1–2 h 1170°C, (3) cooling 1–5°C/h → 1050–1100°C, (4) cooling 2–5°C/min → room temp.
The samples have been characterized by means of x–ray, SEM, EDAX, magnetic susceptibility, Jc – measurements and TEM.

3.1.2. *Results.*
During heating of all samples above the peritectic melting, an increase of Po$_2$ in the crucible has been observed, due to the release of oxygen by the peritectic decomposition. Because of the high viscosity of the melt all samples contain some voids after the release of oxygen. The concentration of these voids is lower in the samples, solidified at Po$_2$ $\geq$ 2000 bar (Fig.10 a)), than at Po$_2$= 1200 bar (Fig.10 b)). The samples molten at lower pressure or fast cooled are porous. The molten samples of 123 consist after solidification of 247, 211 and barium cuprate

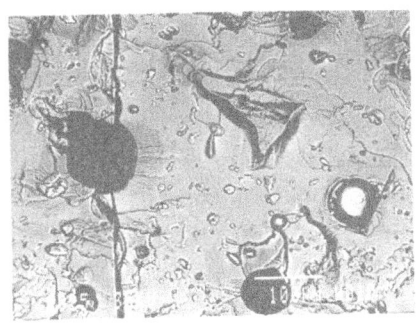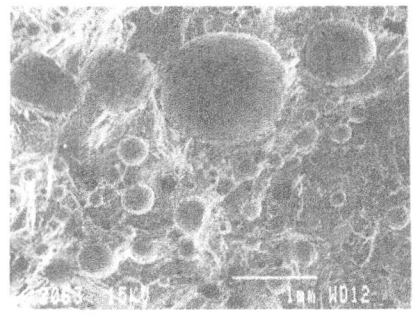

Fig. 10. Cleaved surface of the 124 samples molten at a) P_{O_2}=2500 bar, b) P_{O_2}=1300 bar.

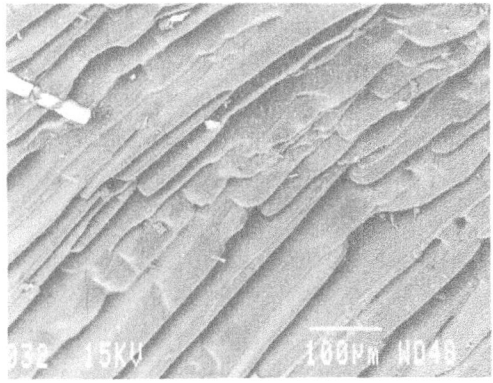

Fig.11. As grown 124 ingot with plate–oriented grains.

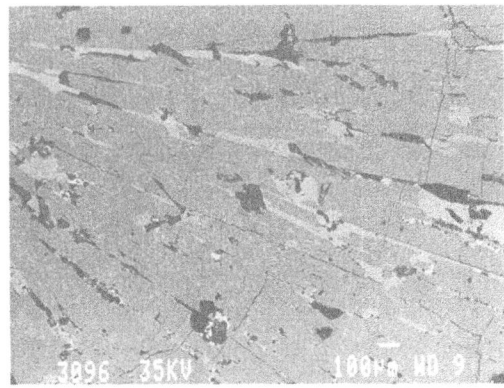

Fig.12. SEM picture of polished, 124 samples molten at P_{O_2}=2500 bar.

(023). The 124 samples solidified at $Po_2 \geq 2000$ bar contain as a major phase 124 with some amount of 247, 023, 001 and 211. The samples slowly cooled with a rate of 1–3°C/h consist of several mm large grains, most of them oriented in the ab plane. Figure 11 shows the surface of as grown 124 ingot with plate–like oriented grains. On the Figure 12 one can see an SEM picture of a polished, 124 sample molten at $Po_2 = 2500$ bar. Lighter areas are macroprecipitations of barium cuprate and darker area CuO. Small bright particles are Y_2BaCuO_5 precipitations. The matrix consists of 124 with some amounts of 247. The molten sample has Tc=70 K, i.e. 10 K lower than 124 ceramic, probably due to reaction of the melt with the alumina crucible. Magnetic measurements of Jc at 4K show 4.3×10^4 A/cm², weakly dependent on the field in the range 100–3000 G. This is one order of magnitude lower than Jc measured on the 124 single crystals [39]. Figure 13 shows Jc as a function of T. The strong increase of Jc at lower temperatures, points clearly towards a macroscopic current flow which is not weak link governed. On the other hand, the almost simultaneous onset of diamagnetism and irreversibility, as well as the stability of Jc in magnetic field indicates that the flux pinning is also good [17].

3.2. CRYSTAL GROWTH OF 124 AND 247 FROM SOLUTION.

The availability of single crystals of 124 and 247 is important, due to the anisotropic character of these materials. Due to the lack of twinning in the a–/b– directions, the physical properties in both directions can be separately studied. The 124 crystals are stable up to 890°C and do not experience a phase transition during cooling, as in case of 123.

3.2.1. *Experimental.* The growth of 124 and 247 single crystals requires a high oxygen pressure atmosphere. However, 124 melts incongruently. Therefore, in order to decrease the melting temperature and avoid peritectic decomposition use of flux is necessary. In our experiments the following composition was used: $123:BaCuO_2:CuO$ in the ratio (1) 1:1:3.5 or (2) 1:4:8. The oxygen pressure has been varied for different experiments from 60 to 2800 bar. Three kinds of crucibles have been used for the crystal growth: Al_2O_3, ZrO_2 and Y_2O_3. The crystal growth runs typically consist of: (a) heating up to 1060–1120°C with rate 5°C/min, (b) dwelling for 1 to 48 h at maximum temperature, (c) cooling 1–3°C/h down to 1050°C, (d) cooling 1°C/min down to room temperature [4,17,27]. Due to the high reactivity between ZrO_2 and the YBaCuO, the dwell time has been reduced to 0.5 h.

3.2.2. Results.

The crystals grow on the walls of the crucible as well as on the partially molten pellets. A proper adjustment of the dwell time is very important for the quality of the obtaining crystals. Too long time leads to the disappearance of the crystals by dissolving or covering by the flux. Only crystals growing out of the flux are accessible. The composition, number and size of the crystals depend on the various experimental parameters, (P_{O2}, T_{max}, time, cooling rate and composition of the flux). Tc of 124 crystals varies as a function of doping during crystal growth. The 124 crystals grown in Al_2O_3 crucible have Tc = 72–73 K. Due to the comparable ionic radii of Al and Cu^{+2} (R(Al) = 0.5 A, $R(Cu^{2+})$ = 0.69 A) Al prefers substituting the Cu1 positions, causing a decrease of Tc in the 124 single crystals to 72 K at the doping level of 0.4 at.% Figure 14 shows the corresponding d.c. susceptibility curve with a transition width of 1.5 K. Note, that d.c. measurements always show a larger transition width than a.c., due to the volume effect of magnetization, hence an a.c. signal corresponds more to the surface effect.

Unfortunately, the alternative crucible material, ZrO_2, reacts very strongly with the YBaCuO, creating $BaZrO_3$. The mixture of ZrO_2 and 123 completely reacts at 950°C within 6 min. forming $BaZrO_3$ [40]. Nevertheless, a very fast crystal growth procedure with ZrO_2 crucibles allowed to grow 124 single crystals with Tc ≈80 K. The best results have been recently obtained with an Y_2O_3 crucible [14] which allowed an increase of Tc up to 84 K after doping with Ca, or 80 K in undoped crystals (Fig.15 and 16, respectively). Since the Zr and Y ions are larger (0.8 and 0.93 A respectively) than Cu, no substitution takes place.

The structure of several crystals has been refined by single crystal diffractometry; the refinements converged at small residual index R < 0.03. The structural parameter and bondlengths are in good agreement with those given by Bordet et al. [41]: all atomic positions have a full occupation and no impurities and misorientations could be detected. The Cu–O double chains stabilize the 124 crystal structure in a very effective way and do not permit any structural change.

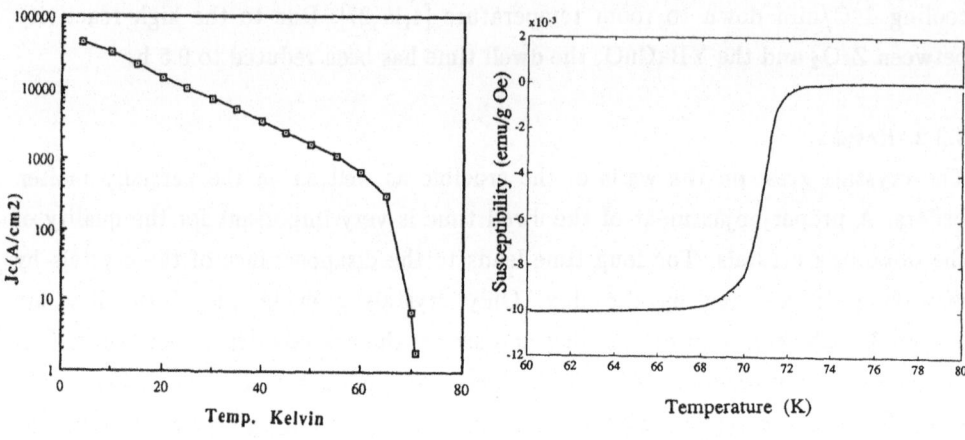

Fig.13. Jc of 124 sample, molten at Po₂=2500 bar.

Fig.14. D.C.Susceptibility curve of 124 single crystal grown in Al₂O₃ crucible.

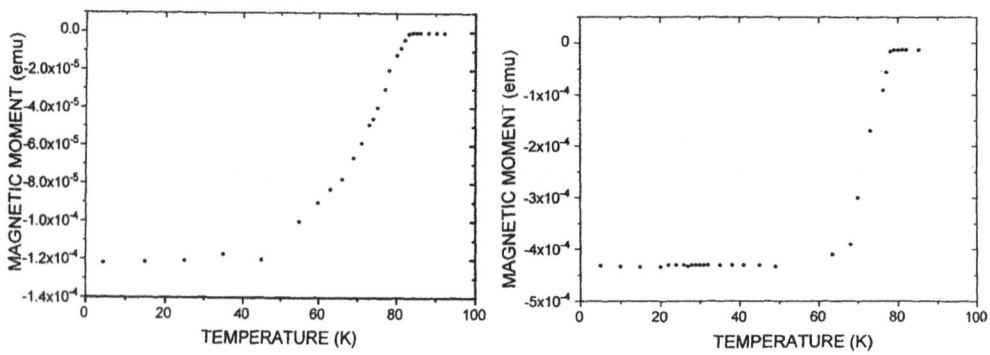

Fig.15. D.C.Susceptibility curve of Ca doped 124 single crystal grown in Y₂O₃ crucible.

Fig.16. D.C.Susceptibility curve of 124 single crystal grown in Y₂O₃ crucible.

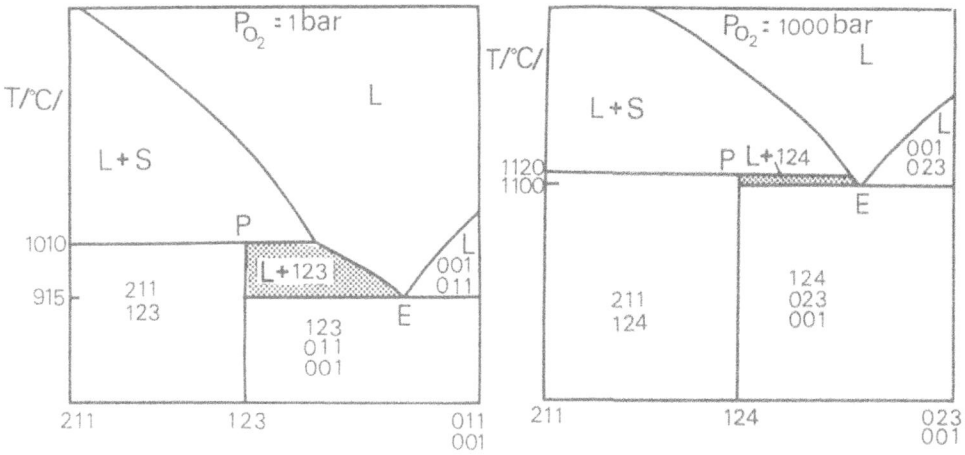

Fig.17. Simplified T–x phase diagram for the crystal growth for Po₂ = 1 bar and Po₂ = 1000 bar. One can see the increase of eutectic and peritectic melting temperatures with pressure.

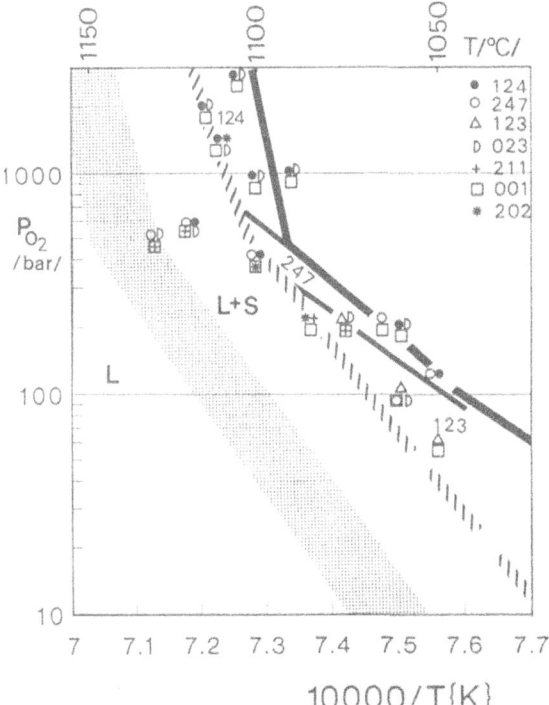

Fig.18. P–T phase diagram for total flux composition $YBa_3Cu_{7.5}O_x$.

3.3. P–T–X PHASE DIAGRAM AND CRYSTAL GROWTH.

In order to determine the conditions for the crystal growth one has to consider the P–T–x phase diagram corresponding to the flux composition. Figure 17 shows T–x section of the ternary system for two pressures (a) $Po_2=1$ bar, (b) $Po_2=1000$ bar. The increase of the eutectic and peritectic melting temperatures with pressure are evident from this diagram. A lack of data for various flux compositions does not allow to give more details on these diagrams as a function of pressure, however further experiments are planned in the future. Figure 18 shows P–T phase diagram for the total flux composition $YBa_3Cu_{7.5}O_x$ (composition (1) in chapter 3.2.1). Note the P–T conditions for the growth of 124, 247 and 123 crystals. Since the 247 crystal growth field is very narrow, it is rather difficult to grow 247 crystals without the intergrowth of 123 or 124. In some flux samples used for 247 crystal growth, partial transition to the 124 phase occured, due to the slow cooling through the 124 stability region. The eutectic melting temperature of the flux increases rapidly with pressure from $\cong 900$°C at $Po_2=1$ bar to $\cong 1100$°C at $Po_2=1000$ bar. Simultaneously, the peritectic melting temperature of the YBaCuO system increases from 1010°C up to 1130°C. At $Po_2=1$ bar, the difference between peritectic decomposition and eutectic melting is $\cong 110$°C, allowing the growth of 123 in this wide temperature range. At $Po_2=1000$ bar this difference is only 30°C. This narrow temperature range makes crystal growth very difficult. However, growth rate at high pressure and

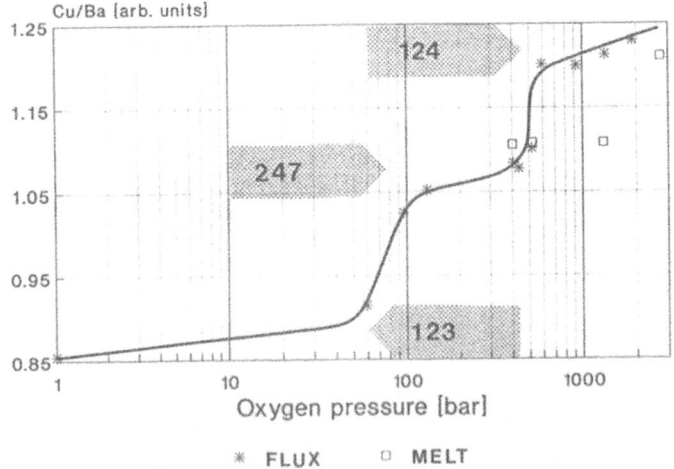

Fig.19. Cu/Ba ratio measured in the crystals grown at various Po_2.

temperature is much faster than at normal pressure. Crystals of several mm (1–5 mm) size were grown in the time of 12 to 24 hours. Fig.19 shows, that the increase of the pressure leads to a step wise increase of the copper content in the crystals. Each step corresponds to another phase of the $Y_2Ba_4Cu_{6+n}O_{14+n}$ family. In the vicinity of the phase boundaries some crystals contain intergrowths of two phases (124 and 247 or 247 and 123). Therefore the steps in the figure are not orthogonal, but they show certain curvature.

4. Synthesis of 247 ceramic samples.

This compound has been synthesized from a mixture of 123+CuO obtained from (1) $BaCO_3$, Y_2O_3 and CuO or (2) Ba,Y and Cu metals oxidized in a closed system without contact with air [17,42]. According to our phase diagram the P,T conditions for synthesis were: Po_2=20 bar, T=1000°C. Under normal pressure the synthesis temperature can not exceed 870°C, which makes the reaction very sluggish. The application of the increased oxygen pressure allows higher temperature and therefore leads to higher reaction rates. Cooling was done in the high pressure autoclave by dropping the sample out of the hot zone.

Scanning Electron Microscopy and X–ray investigations show pure 247 phase. Despite of the high oxygen content (x=14.89) samples prepared from $BaCO_3$ (1) have a low Tc of 64K (with relatively sharp transition), almost 30K below the expected value. X–ray investigations performed on single crystals, taken from this batch show a misorientation of the single chains in the structure, indicated by a partial occupation of the O8 and O9 positions in the chain. Simultaneously the Cu1 single chain positions contain several defects. This is supported by NQR measurements, which show strong disorder in the single chains. Annealing at Po_2=500 bar at 500 and 300°C for 100 h increases the Tc only by 4K and the oxygen content up to x=14.95. Suprisingly, annealing at Po_2=1 bar and 900°C for 60 h, with cooling rate 1°C/min, leads to an increase of Tc onset up to 93K in spite of the decrease of the oxygen content down to x=14.80. The transition, however is very broad. The reason of the increase of Tc is a decrease of the CO_2 content in the sample due to high temperature annealing. This has been proved with mass spectrometry [42]. Samples prepared also under high pressure but using starting material synthesized by the oxidation of the metals (Y, Ba and Cu) or from Ba, CuO and Y_2O_3 without carbonate, have a sharp Tc onset at 93K (Fig.20).

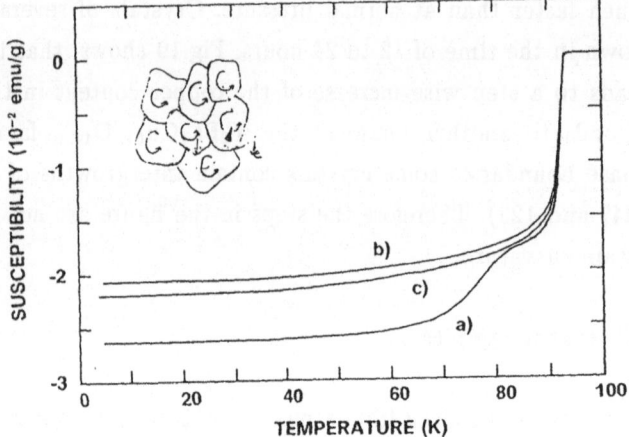

Fig.20. Susceptibility of 247 powder sample a) sintered, b) pulverized, c) additionally annealed at 300°C.

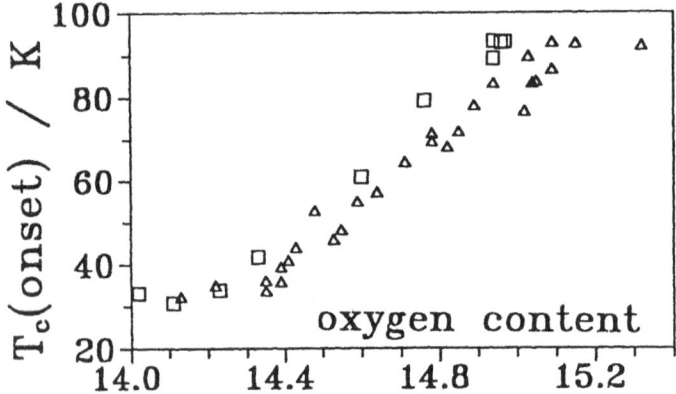

Fig.21. Tc dependence on oxygen content in 247 ceramic samples. □ this work and Δ [16] data.

The oxygen content of such samples is x=14.94. Single crystal x–ray diffraction shows almost ideal structure with full occupation of all sites, without defects and no disorder in the single chains. Annealing at 500 and 300°C for 100h, at P_{O_2}=1 bar reduces the transition width of the susceptibility curve and slightly increased the oxygen content up to x=14.97. It is worth noting, that these samples achieved Tc ≥ 90 K directly after quenching in autoclave and do not require additional high pressure annealing to increase Tc. The NQR measurements showed very narrow frequency lines of Cu which indicate high perfection of the samples [42]. Nevertheless in most sintered samples prepared in that way a step in χ(T) at

T=80K is still present (Fig.20 a)). To our surprise this step disappeared after pulverizing the sample, or when the field was increased up to 100 Oe (Fig.20 b)). This indicates that the intergrain shielding current of the sintered material is suppresed by pulverising the sample or by higher field.

The 247 ceramic samples have been annealed at 800°C in closed ampoules together with Cu spans in order to decrease the oxygen content. As a results a series of 247 samples with various oxygen contents between 14 and 15 have been obtained. Figure 21 shows Tc dependence on x. Data of J.–Y.Genoud et al [16] are also shows for comparison. The deviation between our and Genoud et al [16] data at higher oxygen content could be caused by different methods of oxygen content determination. We have measured oxygen content using volumetric analyses, Genoud et al [16] have used thermobalance.

Comparison of the results obtained with and without carbonate suggest, that carbonate remained in the lattice (probably in the single chains of 247), causing disorder and lower Tc. Even annealing at 900°C does not remove these impurities. In order to prove this, we have performed thermogravimetric measurements combined with mass spectrometry, which detected larger CO_2 signals in the samples prepared from $BaCO_3$ [17,42].

5. The structural studies of $Y_2Ba_4Cu_7O_{14+x}$ single crystals.

Our structural studies of 247 single crystals show remarkable structural changes with increasing transition temperatures T_c [20]. The changes we are discussing here, are not due to different oxygen contents, but mainly to the amount of defects in the Cu1 sites (single chains) of the crystal structures. In particular, the following changes can be observed with increasing T_c:

1. Shortening of the Cu2–O1 apical bondlength in the single chain units, caused by the mutual displacements of Cu2 towards the CuO–chain, and O1 towards the CuO_2–plane (Fig.23).

2. Shift of the Ba1 atom in the single chain 123–unit in direction to the chain, and smoothing of the corresponding Ba1–O1 plane.

3. Reduction of the defect concentration at the single chain Cu1–position (Fig.24).

Crystals with $T_c > 90$ K have the ideal 247 structure: all site occupancies are close to 100 %, and the O9 position remains empty; i.e. all CuO single chains are aligned parallel to the b–axis.

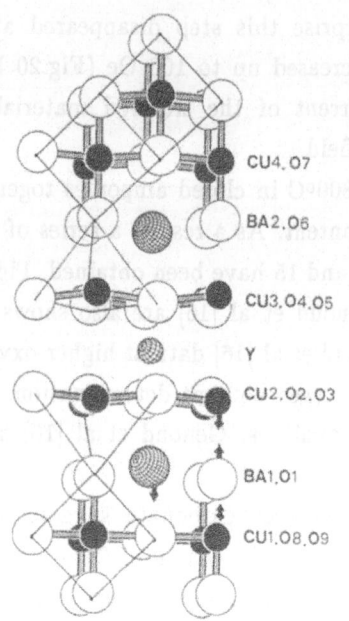

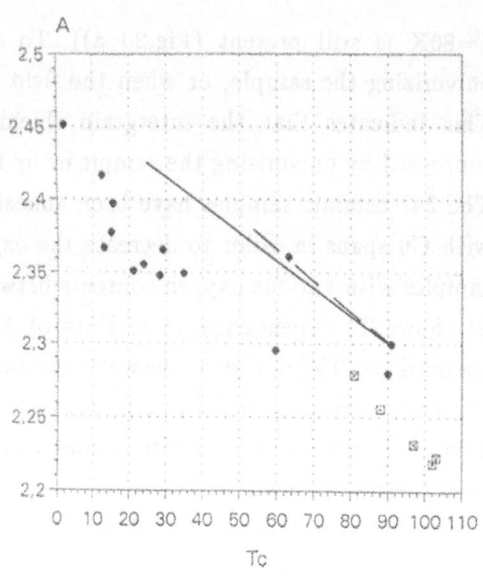

Fig.22. Segment of the 247 crystal structure, with the "124" unit at the upper part, and the "123" unit at the lower one. Arrows indicate the direction of the atomic displacements and the change of the bondlengths occurring with increasing Tc [20].

Fig.23. The apical bond length Cu2–O1 as a function of Tc in: 123 — —[43], 123 — [44],124 ⊠ [45], and 247 ● [17,20].

Fig.24. Occupancy of Cu1 in 247 single crystals as a function of Tc [20].

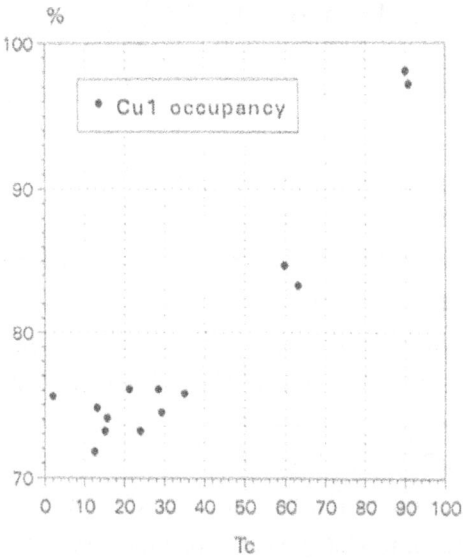

These crystals have sharp diamagnetic transitions. Crystals with lower T_C show a smaller occupancy of the Cu1 position. This deficiency is explained by the introduction of defects at the Cu1 site which destroy the local order and lead to a misorientation of the CuO single chains, indicated by partial occupations of both the O8 and O9 oxygen sites (Fig.25). The Cu atom might be substituted by aluminium or by a carbonate group. A third case conceivable is the removal of copper without substitution, i.e. the introduction of vacancies. Most likely, in the 247 crystals, there exists a mixture of these three kinds of defects. It is not possible to determine by X–ray diffraction the amount of each of these defects separately. If the sum of defects exceeds about 25 % of all Cu1 atoms, the superconductivity vanishes. As a consequence for the preparation of $Y_2Ba_4Cu_7O_{14+x}$, these results require the avoidance of carbonate and air during synthesis, and the substitution of Al_2O_3 crucible material in order to obtain defect–poor high–T_C superconductors. It is remarkable that all structural changes, which affect T_C only occur in the single chain 123–unit, whereas the double chain 124–unit is not susceptible to defects and distortions. This may be explained by the fact that the 124–unit is the stable element in the 247 structure, whereas the 123 unit is responsible for the wide range of varying T_C from 0 – 93 K.

The transition temperature of 124 ceramics is increased by about 10 K by doping with calcium. We performed x–ray structure refinements of both doped and undoped 124 single crystals [10]. The Ca–substituted sites are determined by releasing the occupation factors of the Y– and Ba–positions. A lower heavy atom occupancy corresponds to an incorporation of Ca at this site. It is clearly shown that most of the calcium ($>$ 80 %) substitutes yttrium (Fig.26). A structural consequence of the substitution of the small Y (r = 1.02 Å) by the bigger Ca (r = 1.12 Å) is the widening of the neighbouring O2,O3–planes in c–direction. At the same time the Ba atom moves towards the CuO_2–plane, which is expressed by the decreasing z/c fractinonal coordinate of Ba with increasing Ca–content.

The Ca–substitution in $Y_2Ba_4Cu_7O_{14+x}$ has no effect on T_C. In 247 there are three possible sites for the incorporation of Ca, namely Y, Ba1 (123–unit), and Ba2 (124–unit). Ca substitutes at Ba1 and Y in the ratio 3 : 2, whereas at Ba2 in the 124 double–chain unit is not affected (Fig.26). The O2,O3–oxygen layer between Y and Ba1 is shifted towards Ba1, because Ba1 is bigger and Y is smaller than the substituting Ca–atom. With increasing Ca–doping the single chain 123–block, containing the calcium, is compressed; again, the 124–unit remains unchanged.

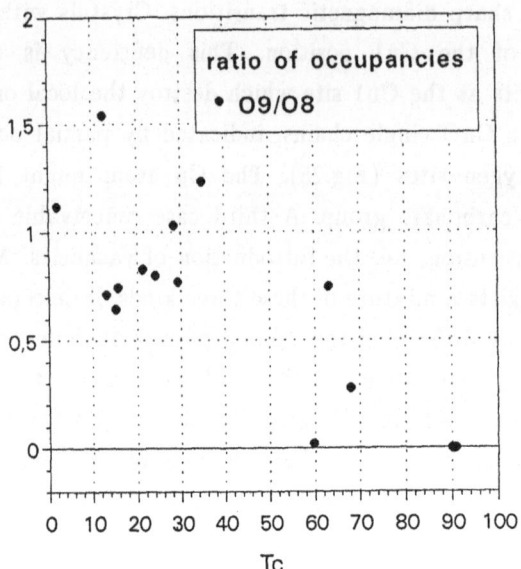

Fig.25. Misorientation of CuO single chains, expressed by the ratios of O9/O8 occupanties.

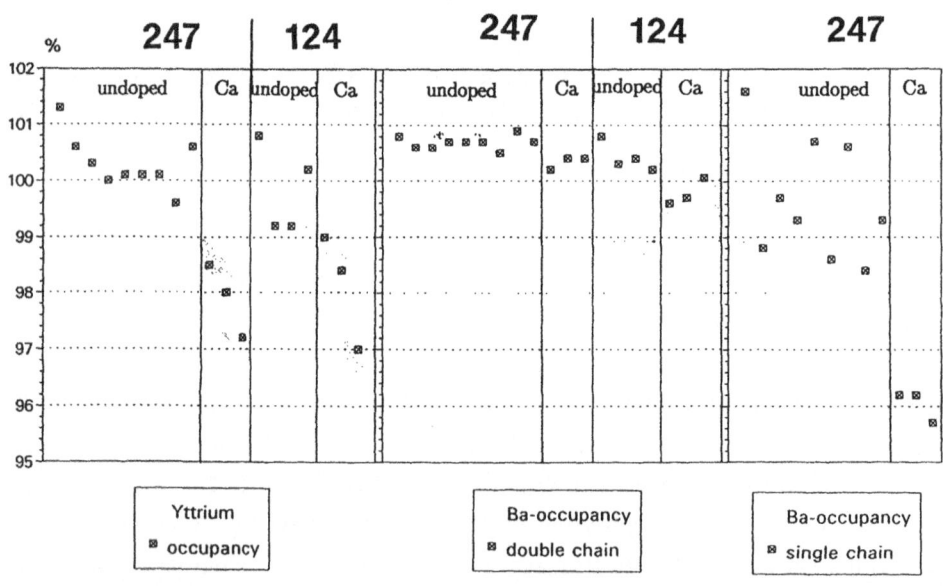

Fig.26. Occupancies of the Ba– and Y–positions in Ca–doped and undoped 124 and 247 single crystals; occupancies lower than 100 % (grey areas) correspond to Ca–substitution at that site [10].

6. Conclusions.

We have investigated P–T–x phase diagram of Y–Ba–Cu–O system at $P_{O_2} \leq 3000$ bar and determined the P–T synthesis conditions of various superconducting phases. After optimization of the synthesis all phases in the YBaCuO family reach similar maximum values of Tc $\simeq$ 90–95 K.

On the example of 247 compound one can see how important are the following elements of the preparation path:

(a) Proper P,T conditions of synthesis, because of very limited stability range.

(b) Thermal conditions of annealing or the way of cooling which influence the oxygen ordering (slow cooling prefered), but also intergrowth of other phases (like 124) by cooling the sample through the instability fields (fast cooling prefered by going through stability fields of other phases).

(c) Using a proper precursor material in order to avoid CO_3^{2-} ions incorporation in the structure which lead to disorder and low Tc.

(d) Using a proper crucible material which has low reactivity towards the melt for the crystal growth from the liquid phase, because of substitutions lowering Tc.

References.

1. J.Karpinski, J.Beeli, E.Kaldis, A.Wisard, E.Jilek, Physica C 153 (1988) 830.

2. J.Karpinski, E.Kaldis, S.Rusiecki, B.Bucher, E.Jilek, Nature 336(1988)660.

3. P.Bordet, C.Chaillout, J.Chenavas, J.L.Hodeau, M.Marezio, J.Karpinski, E.Kaldis, Nature 334(1988)596.

4. J.Karpinski, S.Rusiecki, E.Kaldis, E.Jilek, J.Less–Common Metals, 164&165 (1990) 3–19.

5. E.Kaldis, J.Karpinski, S.Rusiecki, B.Bucher, K.Conder, E.Jilek, Physica C 185–189 (1991) 190–197.

6. P.Bordet, C.Chaillout, T.Fournier, M.Marezio, E.Kaldis, J.Karpinski, E.Jilek, Physical Review B vol.47, 6 (1993) 3465.

7. T.Miyatake, S.Gotoh, N.Koshizuka, S.Tanaka, Nature vol.341 (1989) 41.

8. T.Sakurei, T.Wada, N.Suzuki, S.Koriyama, T.Miyatake, H.Yamauchi, N.Koshizuka, S.Tanaka, Phys.Rev. B 42 (1990) 8030.

9. P.Fischer, E.Kaldis, J.Karpinski, S.Rusiecki, E.Jilek, V.Trunov, A.W.Hewat, Physica C 205 (1993) 259–265.

10. H.Schwer, J.Karpinski, E.Kaldis, to be published.

11. M.Knupfer, N.Nücker, M.Alexander, H.Romberg, P.Adelmann, J.Fink, J.Karpinski, E.Kaldis, S.Rusiecki, E.Jilek, Physica C 182 (1991) 62–66.

12. I.Mangelschots, M.Mali, J.Ross, D.Brinkmann, S.Rusiecki, J.Karpinski, E.Kaldis, Physica C 194 (1992) 277–286.

13. B.Dabrowski, K.Zhang, J.J.Pluth, J.L.Wagner, D.G.Hinks, Physica C 202 (1992) 271–276.

14. J.Karpinski, I.Mangelschots, C.Krüger, E.Kaldis, to be published.

15. J.L.Tallon, D.M.Poke, R.G.Buckley, M.R.Presland, F.J.Blunt, Phys.Rev.B 41(1990)7220.

16. J.–Y.Genoud, T.Graf, A.Junod, J.Muller, Physica C 192(1992) 137.

17. J.Karpinski, H.Schwer, K.Conder, E.Jilek, E.Kaldis, C.Rossel, H.P.Lang, T.Ba mann, Applied Superconductivity, Vol.1,Nos 3–6 (1993) 333–349.

18. T.Krekels, G.Van Tendeloo, S.Amelinckx, J.Karpinski, E.Kaldis, S.Rusiecki, Appl.Phys.Lett. 59(23) 1991 3048.

19. J.Karpinski, S.Rusiecki, B.Bucher, E.Kaldis, E.Jilek, Physica C 161 (1989) 618–625.

20. H.Schwer, E.Kaldis, J.Karpinski, C.Rossel, Physica C 211 (1993) 165.

21. J.Karpinski, S.Rusiecki, E.Kaldis, B.Bucher, E.Jilek, Physica C 160 (1989) 449.

22. M.R.Chandrachood, D.E.Morris, A.P.B.Sinha, Physica C 171 (1990)

23. T.Graff, J.L.Jorda, J.Muller, J.Less–Comm.Met.146 (1989) 49.

24. T.B.Lindemer, F.A.Washburn, C.S.McDougall, R.Feenstra, O.B.Cavin, Physica C 178 (1991) 93.

25. G.F.Voronin, S.A.Degterov, Physica C 176 (1991) 387.

26. J.Karpinski, E.Kaldis, J.Cryst. Growth, 79 (1986) 477–483.

27. J.Karpinski, E.Kaldis, K.Conder, S.Rusiecki, E.Jilek, MRS Symp.Proc. Vol.251, (1992)291.

28. T.Wada, N.Suzuki, A.Ichinose, Y.Yaegashi, H.Yamauchi, S.Tanaka, Jap.Journal Appl.Phys. 29 (1990) 915.

29. H.Murakami, T.Suga, T.Noda, Y.Shiohara, S.Tanaka, Jap.Journal Appl.Phys. 29 (1990) 2720.

30. R.S.Roth, J.P.Cline, J.J.Ritter, Proceedings of The International Workshop on Superconductivity Co–Sponsored by ISTEC and MRS 1992 Honolulu.

31. J.Karpinski, E.Kaldis, S.Rusiecki, J.Less–Comm.Met, 150 (1989) 207–210.

32. K.Conder, J.Karpinski, E.Kaldis, S.Rusiecki, E.Jilek, Physica C 196(1992)164.

33. S.Jin, T.H.Tiefel, R.C.Sherwood, R.B.van Dover, M.Davis, G.Kammlott,

R.Fastnacht, Phys.Rev.B, vol.37 (1988) 7850.

34. M.Murakami, M.Morita, K.Doi, K.Miyamoto, J.Appl.Phys. 7(1989) 1189.

35. K.Salama, V.Selvamanickam, L.Gao, K.Sun, Appl.Phys.Lett 54 (1989) 2352.

36. C.Bateman, L.Zhang, H.Chan, M.Harmer, J.Am.Ceram.Soc. 75 (1992) 1281.

37. M.Murakami, in Processing and Properties of High—Tc Superconductors vol.1 edited by S.Jin, World Sci.Publ.

38. S.Karabashev, Th.Wolf, to be published in Materials Letters.

39 J.Martinez,J.Prejean,J.Karpinski,E.Kaldis,P.Bordet, Physica B 169(1991)669.

40. S.W.Filipczuk, Physica C 173 (1991) 1—8.

41. P.Bordet, J.Hodeau, R.Argoud, J.Muller, M.Marezio, J.Martinez, J.Prejean, J.Karpinski, E.Kaldis, S.Rusiecki, B.Bucher, Physica C 162—164 (1989) 524.

42. J.Karpinski, K.Conder, H.Schwer, Ch.Krüger, M.Maciejewski, C.Rossel, M.Mali, D.Brinkmann, E.Kaldis, to be published

C.Rossel, J.Karpinski, E.Kaldis, Proceedings of Low Temperature Conference Eugene 93.

43. R.J.Cava, A.Hewat, E.Hewat, B.Batlog, M.Marezio, K.Rabe, J.Krajewski, W.Peck, L.Rupp, Physica C 165 (1990) 419.

44. J.Jorgensen, B.Veal, A.Paulikas, L.Nowicki, G.Crabtree, H.Claus, W.Kwok, Phys.Rev.B 41 (1990) 1863.

45. R.J.Nelmes, J.Loveday, E.Kaldis, J.Karpinski, Physica C172 (1990) 311.

THERMODYNAMIC STABILITY OF SUPERCONDUCTORS IN THE Y-Ba-Cu-O SYSTEM

G.F.VORONIN
Department of Chemistry,
Moscow State University
119899 Moscow, Russia

ABSTRACT. The problem of thermodynamic stability of high-temperature superconductors in the Y-Ba-Cu-O system is discussed here. It has been found that these materials are thermodynamically unstable under application conditions. The effect of carbon dioxide and zirconium dioxide on their stability have been established for a wide range of external conditions as well. The methods of theoretical and experimental thermodynamic investigations and available data are briefly reviewed.

1. Introduction

This report attempts to review what has been achieved so far in the field of chemical thermodynamics of high-temperature superconductors (HTSC). The most important and attractive problem here is the thermodynamic stability of these substances. It can only be solved by means of thermodynamic investigations, including experimental studies and calculations of equilibria, that require lots of thermodynamic data. At present such data are only available for solid phases in the Y-Ba-Cu-O system (YBCO). Three superconductors from the family of phases $Y_2Ba_4Cu_{6+n}O_{14+n}$ with $n = 0, 1, 2$ have been obtained in bulk in this system, namely $YBa_2Cu_3O_{6+z}$ (123), $Y_2Ba_4Cu_7O_{14+w}$ (247) and $YBa_2Cu_4O_8$ (124) (e.g., see review by Kaldis and Karpinski [1]), and mainly their stabilities will be discussed further. However, it is necessary to bear in mind, that those depend not only on the properties of the HTSC themselves, but on the equilibrium behavior of all the other phases in this system and its surroundings, which can interact with the HTSC under consideration.

When the HTSC are studied in chemical thermodynamics, the special physical properties of a superconductor, such as its zero electric resistance at low temperatures, are insignificant. The methods of thermodynamics are applied to superconductors, as they are applied to any other substance and material, for the prediction of their stability under various external conditions: in contact with the ambient air, with the crucibles and substrates, and when subjected to heating, pressure, mechanical loads or external force fields. The ultimate aim is to select the optimal conditions for synthesizing the material and for using the final product to minimize the expenditure of time and resources.

The principal advantages of the thermodynamic approach to solving these problems are seen especially at low temperatures or high pressures when equilibria can hardly be studied by a direct experiment while being essential for understanding many special features of the HTSC. Here thermodynamic calculations of the phase equilibria are feasible and valuable, because chemical thermodynamics is provided with powerful methods of estimate and extrapolation of thermodynamic functions to the range of variables where experimental data are lacking. Although, there exist some serious limitations in putting thermodynamic considerations into practice because of applying the model of equilibrium state at only high relaxation rate of system (see below section 7).

All solutions of the above problems are called briefly in the chemical thermodynamics as equilibria calculations. Thermodynamics has an effective method to represent the results of these determinations and to show the stability conditions of substances by means of phase diagrams. They allow to see the conditions of all stable and metastable equilibria in the system at once, as well as the results of chemical and phase transformations, if any. These diagrams will be the main objects of our discussion below.

E. Kaldis (ed.), Materials and Crystallographic Aspects of HTc-Superconductivity, 585–602.

2. Thermodynamic Equilibrium and Stability

It is worthwhile to begin the considerations of the HTSC stability by giving some remarks to the notion of stability itself, because it varies mostly depending on the field of its application. This term is often used when copper valency in the HTSC, their superconducting states or for instance crystal lattice vibrations are discussed. Contrary to this we deal with stability of thermodynamic equilibrium state in chemical systems. Therefore we take a particular interest in the stability of each phase relative to the other phases and, what is the same, relative to all possible chemical reactions and phase transitions in the system under consideration. Such "chemical" stability or its loss is significant for the HTSC production because superconducting properties depend critically on how the material is processed.

There are stable and metastable thermodynamic equilibria, systems and phases. They cannot be unstable because of always existing fluctuations of physical quantities. Therefore a loss of thermodynamic stability actually means the beginning of processes of relaxation in a system. Naturally, a system can exist in such a labile position for a long time if relaxation is very slow, but this time is limited. The stable equilibrium is resistant to any fluctuations and the metastable one only to fluctuations with limited magnitudes. Unfortunately, it is impossible to use these rigorous notions to distinguish between stable and metastable states in practice and their interpretation seems desirable.

When using the terms "thermodynamic equilibrium" and "thermodynamic stability", one must always imply the equilibrium and stability with respect to the definite processes, i.e., foresee the final state of a system when losing its stability. Discussion about equilibrium and stability relative to all conceivable processes or absolute equilibrium and stability is senseless for no other reason than most atoms of chemical elements are unstable with respect to nuclear reactions. Thus, discussing stability of thermodynamic system, we have to know all processes, which are possible in a system of interest in principal. These are any chemical and phase transformations between known substances if they are consistent with the component composition of a system.

However, some of these possible processes do not occur at the given conditions indeed for a variety of *nonthermodynamic* reasons, and we can be provided with this information. For example, some phases cannot be formed at the given time of the experiment because of slow diffusion, chemical reaction or crystal nucleation. The equilibrium state of a system is usually called metastable if it exists only with these known, thermodynamically possible, but really not occurring processes. Accordingly, the equilibrium phase is metastable if it exists only at the certain forbidden processes. The equilibrium without these restrictions as well as thermodynamic system and all its phases are stable. For example in fig.1 the calculated stability conditions of the $YBa_2Cu_3O_{6+z}$ - CuO system are shown.

Fig. 1 (a) shows the equilibria of all possible phases in this quasi-binary system having known thermodynamic properties. The equilibria with the phases 249 and 125 are hypothetical, because, as far as we know, these phases are not synthesized and investigated yet. Their properties were evaluated from the available properties of the 123, 247 and 124 phases, as described below. However, a long annealing of both the 123 and 124 phases or their heterogeneous mixture at various temperatures and pressures of oxygen near atmospheric did not result in the formation of the 247 phase, although this phase coexists with 123 or 124 under other different conditions with the certain efforts taken to form it. This was clearly shown by Zhang and Osamura [2] and by Grenoud, Graf et al.[3]. Fig.1 (b) shows the metastable equilibria of the 123 and 124 phases, or the metastable reaction of the 124 phase decomposition into 123 and CuO, when formations of the other phases are impossible. The metastable state forms here easier than the stable one. In one's turn some of the phases in figs.1 (a) may also be metastable out of this quasi-binary section, i.e., in other phases presence (see below section 6.1).

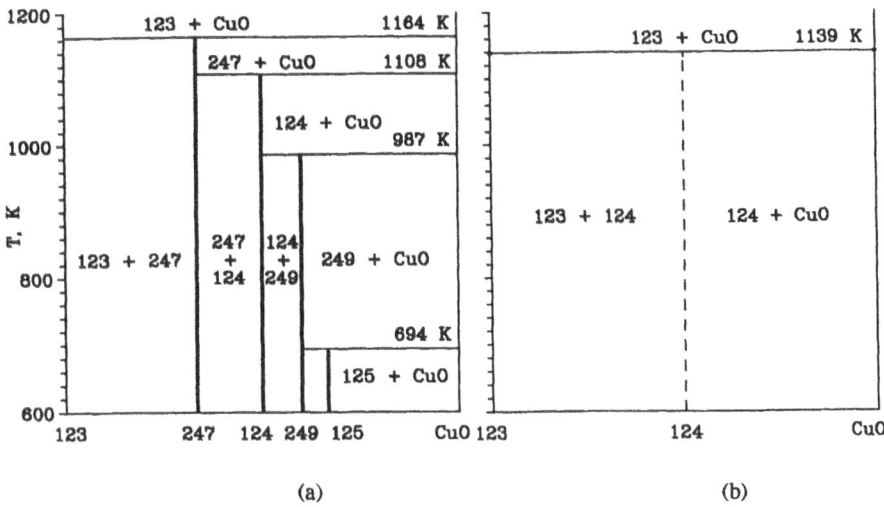

Fig. 1 Phase diagram of the system YBa$_2$Cu$_3$O$_{6+z}$ -CuO at 1 atm of oxygen pressure. 123, 247, 124, 249 and 125 denote YBa$_2$Cu$_3$O$_{6+z}$, Y$_2$Ba$_4$Cu$_7$O$_{14+w}$, YBa$_2$Cu$_4$O$_8$, Y$_2$Ba$_4$Cu$_9$O$_{17}$ and YBa$_2$Cu$_5$O$_9$ respectively. The diagram (a) is calculated without any limitations on the listed phases formation, (b) with the forbidden formation of the 247, 249 and 125 phases.

3. Methods of Calculations

3.1. EXISTING APPROACHES

There are different approaches to obtaining the temperature, pressure and composition of a multi-component system at equilibria in order to study the appropriate phase diagrams. The first way is a purely experimental one, this is for instance a direct measurement of phase composition at equilibrium when temperature and pressure are known. Several experiments with different temperatures or pressures allow a phase diagram of a system to be obtained. Both the stable and some metastable phase boundaries are usually observed, but there are simple thermodynamic rules for distinguishing between these two kinds of equilibria on the phase diagrams. Direct experimental study of stability fields is often laborious and time-consuming, and the most severe problem is an attainment of a true equilibrium.

The thermodynamic approach consists in investigating thermodynamic properties the phases in a specified system followed by calculating phase diagrams. This way cannot be regarded as a theoretical prediction of stability regions, because the thermodynamic properties are obtained mostly by measurements including the experimental determination of the phase diagrams. The calculations of phase equilibria have no need for any model assumptions by itself and cannot introduce an error into the results. Therefore, if the thermodynamic properties of the phases are determined well, the thermodynamic calculations define the conditions for thermodynamic stability of a superconductor properly. It is notable that different kinds of phase diagrams also their sections and projections may be calculated with the same set of thermodynamic data.

3.2. THERMODYNAMIC FUNDAMENTALS

The problem of equilibria and phase diagrams calculation may be divided distinctly into two parts. These are (a) a calculation of the equilibrium phase composition of the whole system under given external conditions (temperature, total pressure and total number of moles of components in the system) and a set of phases, which can exist in this system in principal (under conditions of every kind), and (b) a calculation of chemical composition of every equilibrium phase at the given phase composition and external conditions. The solution of the first part (a) gives a set of phases and numbers of their moles in the equilibrium heterogeneous mixture . The second part (b) of this problem is important obviously only if nonstoichiometric phases or solutions are present because the chemical composition of individual compounds does not depend upon the composition of mixture.

The most general method of the phase diagrams calculation is based directly on the fundamental property of the Gibbs energy function of a closed system G to have a minimum in equilibrium at the fixed temperature T and pressure P. This minimum exists in a manifold of internal variables. There are two kinds of variables in accord with the mentioned two parts of the whole problem. For the YBCO these are a set of the numbers of moles of phases $n^{(r)}$ and a set of the nonstoichiometric indexes of oxygen in its solutions $\delta^{(r)}$, where r refers to each phase in a system. The last set of the variables is not in use if a system consists only of individual stoichiometric compounds. In fact, the problem is to express G as a function of the $n^{(r)}$, $\delta^{(r)}$ and to seek those values of $n^{(r)} \geq 0$, $\delta^{(r)} \geq 0$ that make G a minimum subject to the constraints of mass conservation in a closed system:

$$n_i = \sum_r \alpha_i^{(r)} n^{(r)}, \ (i=1,...,k). \tag{1}$$

Here n_i represents the total number of moles of the component i and k the number of components in a system, $\alpha_i^{(r)}$ is a stoichiometric index of the component i in the phase r. For instance if phase r is $YBa_2Cu_3O_{6+z}$ (phase 123), then $\alpha_Y^{123}=1$, $\alpha_{Ba}^{123}=2$, $\alpha_{Cu}^{123}=3$, $\alpha_O^{123}=6+z$, $\delta^{123}=z$ and $k=4$. For any heterogeneous system

$$G(n^{(r)},\delta^{(r)}) = \sum_r n^{(r)} \Delta G_f^{(r)}(\delta^{(r)}) + constant, \tag{2}$$

where $\Delta G_f^{(r)}$ is a molar Gibbs energy of phase r formation. These functions have to be known for all phases beforehand.

3.3. PROBLEMS OF COMPUTATIONS

Any computer program of nonlinear constrained minimization can be used. If an equilibrium is stable, then a solution to this problem is unique. But at the loss of stability and in the case of some metastable equilibria several local minima in the G as a function of the composition can exist. It causes some practical difficulties in the calculations by means of most general programs. Fortunately it is easy to predict in advance when these complications occur. According to thermodynamics the heterogeneous system will be stable if and only if all phases in this system are individually stable. Thus the local minima in the total Gibbs energy can arise only for a system having solutions, which are unstable at some values of variables, and each of these phases must have $\Delta G_f(\delta)$ function also with several local minima. In the book by Smith and Missen [4] one can find some details and computer programs of that kind of computations. At the fixed partial pressure of oxygen in the YBCO the calculations are simplified owing to the existence in this system of only oxygen solutions and consequently fixation of all δ variables (see below section 3.5).

3.4. PHASE DIAGRAMS OF CLOSED SYSTEMS

Thus, if a set of the Gibbs energies of phases as functions of T, P and composition is available, it enables the whole system phase composition as well as chemical compositions (nonstoichiometry) of each phase at equilibrium to be determined. Simultaneously the chemical potentials of all components and equilibrium partial pressures of oxygen are calculated. Several such calculations with different temperatures and the component compositions n_i permit the whole phase diagram to be obtained. For example the calculated isothermal section of the temperature-composition phase diagram of the ternary BaO-Cu$_2$O-O$_2$ system is shown in fig.2. There are two independent composition variables on such a diagram. The partial pressure of oxygen remains constant within the boundaries of each heterogeneous field and changes abruptly at the transition between these two fields. This type of diagrams is useful for analyzing equilibria and processes in closed systems. For example, the sequence of equilibrium phase assemblages can easily be derived

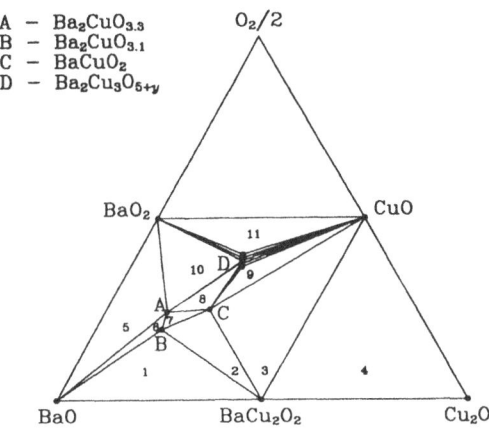

A − Ba$_2$CuO$_{3.3}$
B − Ba$_2$CuO$_{3.1}$
C − BaCuO$_2$
D − Ba$_2$Cu$_3$O$_{5+y}$

Fig 2. Calculated phase equilibria in the BaO-Cu$_2$O-O$_2$ system at 1100K. Equilibrium pressures of oxygen in three-phase fields 1-11 are $1.46 \cdot 10^{-7}$, $4.09 \cdot 10^{-5}$, $7.21 \cdot 10^{-3}$, $2.32 \cdot 10^{-3}$, 1.17, 0.301, 0.301, 40.6, 0.350, 127 and $1.90 \cdot 10^5$ atm, respectively [5].

from this diagram if coulonometric titration at fixed temperature is used to alter the oxygen content in the volume initially filled with specified quantities of BaO and CuO.

3.5. PHASE DIAGRAMS OF OPEN SYSTEMS

The YBCO system may be often considered as an open system with respect to oxygen because of its almost free exchange with environment. Then chemical potential of oxygen and its content in the system depend on the oxygen partial pressure in the ambient gas. Here the general criterion for thermodynamic equilibrium and stability is a minimum of $(G - n_o \mu_o)$ difference subject to the fixed cation composition in (1), n_o and μ_o are the number of moles and chemical potential of oxygen in the whole system. Because of the fixed μ_o both the goal function $(G - n_o \mu_o)$ and the restrictions (1) are linear with respect to unknown $n^{(r)}$ (all $\delta^{(r)}$ are calculated here for each solution phase individually and before the minimization). The calculations get then simpler and it is possible to use the linear programming.

Fig.3 shows the example of the phase diagram for the same system as in fig.2 but open with respect to oxygen. This type of diagrams is useful to keep track of phase relations in the system, which is heated under the fixed partial pressure of oxygen for instance. One- and two-phase regions showed by vertical lines and areas respectively. Horizontal lines represent univariant three-phase equilibria. They can be calculated still easier, because here the equilibrium set of phases is already known and we deal only with the mentioned second part of the whole problem.

For example, one can see on fig.2 that in field 8 the phases BaCuO$_2$, Ba$_2$CuO$_{3.3}$ and Ba$_2$Cu$_3$O$_{5+y}$ coexist. At arbitrary temperature $\Delta G_f(\text{BaCuO}_2) = \Delta\mu_{Ba} + \Delta\mu_{Cu} + 2\Delta\mu_O$, $\Delta G_f(\text{Ba}_2\text{CuO}_{3.3}) = 2\Delta\mu_{Ba} +$

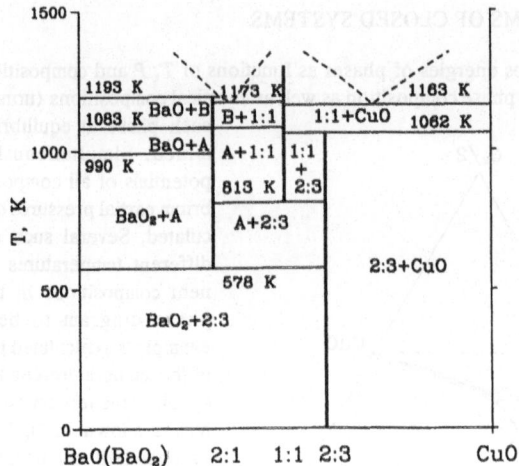

Fig. 3. Phase diagram of the system BaO(BaO$_2$)-CuO at 0.21 atm of oxygen pressure. A and B denote the Ba$_2$CuO$_{3.3}$ and Ba$_2$CuO$_{3.1}$ phases, respectively [5].

$\Delta\mu_{Cu}+3.3\Delta\mu_O$ and $\Delta G_f((Ba_2Cu_3O_{5+y})=2\Delta\mu_{Ba}+3\Delta\mu_{Cu}+(5+y)\Delta\mu_O$, where $\Delta\mu_i$ is the standard equilibrium chemical potential of the component i. These equations yield to

$$4\Delta G_f(BaCuO_2)+(0.3+y)(RT/2)\ln P(O_2) = \Delta G_f(Ba_2Cu_3O_{5+y})+\Delta G_f(Ba_2CuO_{3.3}) \qquad (3)$$

for the corresponding decomposition reaction

$$4BaCuO_2+ (0.3+y)O_2/2 = Ba_2Cu_3O_{5+y} + Ba_2CuO_{3.3} . \qquad (4)$$

If one of the variables T or $P(O_2)$ is fixed, a solution of (3) allows the other variable to be obtained. For instance, if $P(O_2)=0.21$ atm then $T=813$ K as in fig.3, and if $T=1100$ K then $P(O_2)=4.09\cdot10^{-5}$ as in fig.2.

3.6. STABILITY DIAGRAMS

Another important type of phase diagrams is shown in fig.4. This type is exemplified by the so-called stability (or predominance) diagram. Stability fields of phases in fig.4 are superimposed over each other on this plot. Each line represents conditions for the corresponding two-phase or three-phase equilibrium, and the phases stable on each side of the line are showed in fig.4. For example, the line denoted 2:1+2:3 from above and 1:1 from below represents the reaction (4), the 2:1 phase is stable at temperature and pressures between the bold solid line, the 1:1 phase is stable between the thin solid lines, 2:3 is stable above the short dashed line and 1:2 below the chained-dot line.

4. Experimental Methods

A minimum set of data that is sufficient for the calculation of equilibria and stability is formed by the values of the Gibbs energies of the main phases in the system as functions of temperature, pressure, and chemical composition. There are many good reviews of experimental thermodynamic methods, which can be used to obtain these data (e.g., see review by Komarek [6]). Therefore here is only a

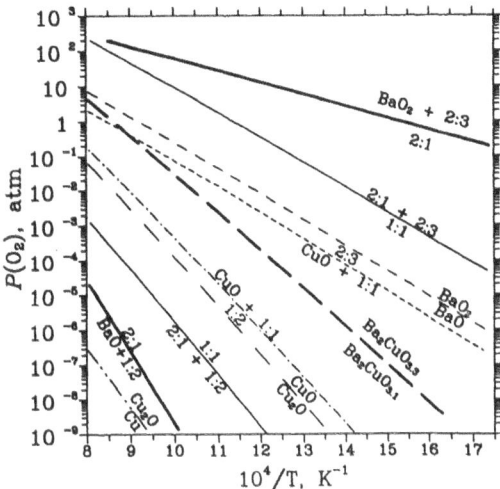

Fig. 4. Stability diagram of the Ba-Cu-O system. The curves correspond to univariant equilibria, which limit the stability ranges of the phases in this system. The notations 2:1, 1:1 and 2:3 indicate the Ba:Cu ratio of the mixed oxides [5].

synopsis of those peculiar for the study of the HTSC. As a rule these systems have to be classed as open with respect to ambient oxygen. They "breathe" oxygen and can exist as equilibrium specimens only at quite large values of oxygen chemical potential or partial pressure. For this reason, many of traditional experimental methods are not applicable for thermodynamic study of these substances. For example, this is the case with the methods based on vaporization or effusion of species into vacuum, such as high-temperature mass-spectrometry. A special experimental technique has to be used while working with the HTSC specimens at high temperature and high oxygen pressure owing to its extreme chemical activity, such as two chamber autoclave described by Karpinski and Kaldis [7]. Calorimetry faces the similar problems since quenched and obviously nonequilibrium specimens are involved or it is necessary to provide for conservation of oxygen content in the specimen. As a rule, the results obtained between these extreme opportunities cannot be strictly interpreted. This is shown in fig.5: the measured specific heats of the 123 phase are scattered between two curves describing equilibria in the open and closed system.

Most of the experimental thermodynamic studies of the HTSC materials concerned the sintered ceramic specimens of the 123 phase in the Y-Ba-Cu-O system. The gas phase equilibria were most often used for these purposes. The specimen was annealed in controlled atmosphere, quenched and its phase and chemical composition determined. X-ray diffraction was generally used for phase identification and chemical analyses - for determination of oxygen content. The partial thermodynamic properties of oxygen in the 123 solid solution and the equilibrium conditions of invariant reactions in a heterogeneous mixture of compounds were determined in this way. An electrochemical oxygen pump was often used to create the appropriate gas atmosphere. To change the composition of a sample precisely, the coulometric titration and thermogravimetry were used with success in several works.

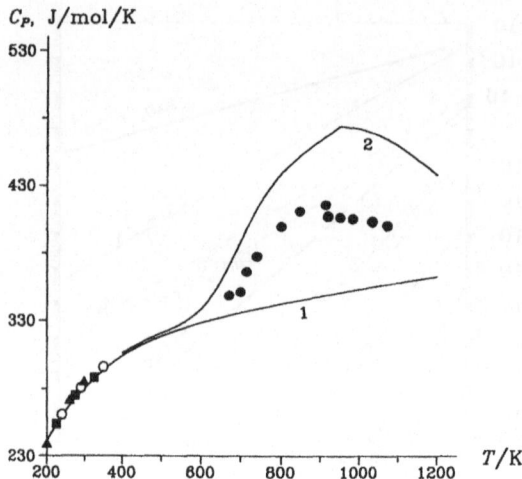

Fig. 5. Specific heat of $YBa_2Cu_3O_{6+z}$ as a function of temperature at the fixed composition, $z = 0.9$, (line 1) and at 1 atm of equilibrium oxygen pressure, (line 2). The points represent the data from refs. [8-10]. The lines are calculated (see below section 5).

The most complete and useful integral thermodynamic properties of the YBCO, the Gibbs energy, enthalpy and entropy of individual compounds and solutions, were measured in a few studies by an electromotive force method over the range of about 950 to 1150 K. For example, the following electrochemical cells were assembled for this purpose in a series of papers by Pasin, Skolis et al. [11, 12]:

(Pt) O_2 | $BaCuO_2$,CuO,BaF_2 | | BaF_2 | | CaO,CaF_2 | O_2 (Pt)   (I)

(Pt) O_2 | Y_2O_3,YOF | | CaF_2 | | $Y_2Cu_2O_5$,CuO,YOF | O_2 (Pt)   (II)

(Pt) O_2 | Y_2BaCuO_5,$Y_2Cu_2O_5$,CuO,BaF_2 | | BaF_2 | | CaO,CaF_2 | O_2 (Pt)   (III)

(Pt) O_2 | Y_2BaCuO_5,$Y_2Cu_2O_5$,Y_2O_3,BaF_2 | | BaF_2 | | CaO,CaF_2 | O_2 (Pt)   (IV)

(Pt) O_2 | BaO,BaF_2 | | BaF_2 | | CaO,CaF_2 | O_2 (Pt)   (V)

(Pt) O_2 | $BaCu_2O_2$,Cu_2O,BaF_2 | | BaF_2 | | CaO,CaF_2 | O_2 (Pt)   (VI)

(Pt) O_2 | $YBa_2Cu_3O_{6+z}$,Y_2BaCuO_5,$BaCuO_2$,BaF_2 | | BaF_2
     | | $YBa_2Cu_3O_{6+z}$,Y_2BaCuO_5,CuO,BaF_2 | O_2 (Pt)   (VII)

(Pt) O_2 | $BaCuO_2$,CuO,BaF_2 | | BaF_2 | | $YBa_2Cu_3O_{6+z}$,Y_2BaCuO_5,CuO,BaF_2 | O_2 (Pt)   (VIII)

(Pt) O_2 | $YBa_2Cu_3O_{6+z}$,Y_2BaCuO_5,CuO,BaF_2 | | BaF_2 | | CaO,CaF_2 | O_2 (Pt)   (IX)

The electromotive force was measured on each cell at several tens of temperatures. The measurements on the cells (I), (VI), and (VIII) were made at various oxygen pressures ranging from 0.001 to 1 atm. In the other cases, the oxygen pressure was equal to 1 atm. This study makes it possible to obtain thermodynamic functions of almost all the major phases in the vicinity of 123 via several independent ways combining data for different galvanic cells. For example, the Gibbs energy changes a reaction

$(1/2) Y_2O_3 + 2BaO + 3CuO + [(z-0.5)/2] O_2 = YBa_2Cu_3O_{6+z}$ (5)

is the following:

$$\Delta G_f^{ox}(YBa_2Cu_3O_{6+z}) = 3F[E(I) - E(V) - 0.4E(VII)] + \omega =$$
$$3F[E(I) - E(V) - E(VIII)] + \omega = 3F[E(IX) - E(V)] + \omega ,$$

where $\omega = \Delta G_f^{ox}(Y_2BaCuO_5)/2$, $E(J)$ is the electromotive force of the cell (J), F is the Faraday constant. Here and below the indexes (f) and (ox) denote the formation of a compound from the above oxides and oxygen.

A thermodynamic method for the determination of oxygen nonstoichiometry in the HTSC by Skolis, Kovba et al. [13] is of particular interest. It allows one to find the oxygen content in nonstoichiometric compounds or solutions during the measurements of their thermodynamic properties. This method is based on the dependence of the electromotive force of an electrochemical cell on the oxygen pressure over the cell when oxygen takes part in a net electrochemical reaction. A formula for the calculation of nonstoichiometric index (z) is

$$z = \pm(\partial \Delta G_r / \partial \ln P(O_2)) / RT ,$$

where ΔG_r is the Gibbs energy of the net cell reaction. Figure 6 shows the application of this method for the $Sr_{14}Cu_{24}O_{38+z}$ compound at 1150 K based on two different cells. The first cell is $(Pt)O_2$

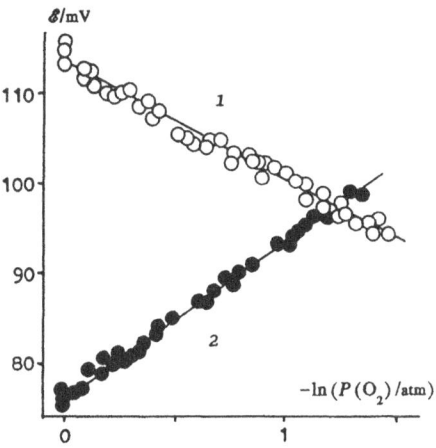

$|SrO,SrF_2|$ $|SrF_2|$ $|Sr_{14}Cu_{24}O_{38+z},CuO,SrF_2|O_2$ (Pt), and its net reaction is $14SrO+24CuO+(z/2)O_2=Sr_{14}Cu_{24}O_{38+z}$; the second one is $(Pt)O_2|SrO,SrF_2|$ $|SrF_2|$ $|SrCuO_2,SrF_2$, $Sr_{14}Cu_{24}O_{38+z}|O_2(Pt)$ with the net reaction $Sr_{14}Cu_{24}O_{38+z}+10SrO=24SrCuO_2+(z/2)O_2$. As seen in Fig. 6, at oxygen pressure from 0.1 to 1 atm, the composition of the phase under consideration remains constant, so that it has a formula $Sr_{14}Cu_{24}O_{41}$.

A lot of erroneous results that have not withstood rechecking have been also obtained. This takes place in particular owing to errors in the real phase composition determination. Frequently, poor reproducibility results from incomplete conversion of the initial substances during the synthesis or the pronounced influence of the surrounding gaseous medium on the composition and properties of the specimen and its interactions with water vapor, carbon dioxide, and the materials of crucibles and substrates.

Fig. 6. Oxygen content in $Sr_{14}Cu_{24}O_{41}$ at 1150 K obtained by electromotive force measurements [13]. For cell 1, $z = 28F(\partial E/\partial P(O_2))/RT = 3.03 \pm 0.15$, and for cell 2, $z = -20F(\partial E/\partial P(O_2))/RT = 3.06 \pm 0.08$.

5. Thermodynamic Data

The success of each thermodynamic calculation entirely depends on the completeness and quality of the used thermodynamic data. Unfortunately, the thermodynamic properties of the HTSC have been little studied. It is true even for the most investigated the YBCO system. Moreover, the above complications in the experiments with these subjects do not permit, as a rule, the initial experimental data to be used for thermodynamic calculations. For example, in fig.7 the scattering of enthalpies of the 123 phase formation is seen [14]. The errors of different experimental data are not correlated with each other, although the values of the thermodynamic properties them selves are united by means of thermodynamic relations. This allows to diminish the errors of data and get a set of so-called self - consistent data. This is the most important part of thermodynamic calculations of equilibria. In order to obtain the thermodynamic data that are fit into the equilibria calculations usually the following work is to be done:

-the collection of information about compositions, crystal structures, chemical bonds, thermodynamic properties, phase diagrams and chemical reactions equilibria for substances in systems
with the HTSC,

-critical evaluation of the completeness, reliability and accuracy of available thermodynamic data
and definition of the lacking ones,

-extraction of the necessary data from phase diagrams, their assessment with the help of ab initio,
semi-empirical and other methods,

-analytical representation of dependencies of the thermodynamic properties of solutions under
consideration on the composition, temperature, pressure etc.,

-computer-assisted "optimization" analysis to obtain a set of data with the best fitting to the
experimental results and fulfillment of thermodynamic relations. In this stage each system for which
some thermodynamic data, phase diagrams and other thermodynamically coupled information
are available is investigated iteratively and the resultant computed thermodynamic functions, phase
di agram etc. are traced as a function of the parametric input. The "optimal" parameters are chosen to
give the best possible description of the whole observed information.

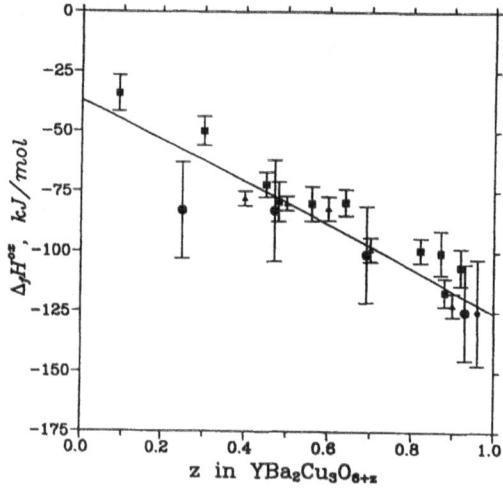

Fig. 7. Enthalpies of $YBa_2Cu_3O_{6+z}$ formation according to reaction (5) versus composition z at 298.15 K.
The points represent the experimental calorimetric data ●-[15], ▲-[16], • -[17], ■-[18].
The data [15] and [17] were increased to 54.5 kJ/mol compared with the initial values (see [19]). The
inaccuracy boundaries from the original works are shown. The curve is calculated (see below).

Such a self-consistent set of thermodynamic functions was derived in the YBCO from available
experimental data and phenomenological models of phases in this system in our earlier work [19-21].
But this is not the work to be done once. It has to be continual in order to be useful. Since then, this
set of thermodynamic functions has been improved with the experimental data obtained during the
last two years taken into account. It includes the data for the individual compounds $BaCuO_2$,

$BaCu_2O_2$, BaO_2, Y_2BaCuO_5, $YCuO_2$, $Y_2Cu_2O_5$, Y_2BaO_4, $YBa_2Cu_4O_8$, $YBa_2Cu_5O_9$, $Y_2Ba_4Cu_9O_{17}$ and for the oxygen solutions $YBa_2Cu_3O_{6+z}$ ($0 \le z \le 1$), $Y_2Ba_4Cu_7O_{14+w}$ ($0 \le w \le 1$), $Ba_2Cu_3O_{5+y}$ ($0 \le y \le 1$), $YBa_4Cu_3O_{8.5+q}$ ($0 \le q \le 0.5$) [5, 14, 22]. These data are useful for the YBCO equilibria and stability calculations at temperatures 298-1250 K and pressure up to several thousands atm.

For example, the parameters of the model for the 123 phase were determined with the help of more than 600 experimental data points including different thermodynamic and structural measurements. The thermodynamic functions of formation of the 123 phase according to reaction (5) at temperatures 298-1250 K are the following:

$$\Delta G_f^{ox}(YBa_2Cu_3O_{6+z}) = \Delta H_f^{ox} - T\Delta S_f^{ox}, \tag{6}$$

$$\Delta H_f^{ox}/R = A_1 + A_2 z + z(1-z)a_1^h + (c^2 - x^2)b_1^h, \tag{7}$$

$$\Delta S_f^{ox}/R = B_1 + B_2 z + z(1-z)(a_1^s + a_2^s(1-z)) + (c^2 - x^2)b + S_d/R, \tag{8}$$

$$S_d/R = -(c+x)\ln(c+x) - (c-x)\ln(c-x) - (1-c+x)\ln(1-c+x) - (1-c-x)\ln(1-c-x) - z\ln z - (1-z)\ln(1-z), \tag{9}$$

where x is the order parameter that depends on the oxygen ordering in the 123 phase ($x=0$ for the tetragonal phase and $x=c$ for the completely ordered orthorhombic phase), $c = z/2$; $R = 8.314$ J/mol/K; $A_1 = -4477.6$ K; $B_1 = -4.117$; $A_2 = -10669.8$ K; $B_2 = -10.4058$; $a_1^h = 886.731$ K; $a_1^s = 0.4871$; $a_2^s = -1.4204$; $b_1^h = 1158.53$ K, $b_1^s = -3.33514$. The equilibrium values of x, z and $P(O_2)$ are interrelated by equations

$$T\ln((c+x)(1-c+x)/((c-x)(1-c-x))) = 2(b_1^h - Tb_1^s)x, \tag{10}$$

$$\ln P(O_2) = \ln((c+x)(c-x)z^2/((1-c-x)(1-c+x)(1-z)^2)) + 2[A_2 - TB_2 + (a_1^h - Ta_1^h)(1-2z) - Ta_2^s(1-z)(1-3z) + c(b_1^h - Tb_1^s)]/T. \tag{11}$$

This model permits other thermodynamic properties of the 123 phase to be calculated. So, it shows, that depending on the conditions of the calorimetric experiment, three different specific heats can be obtained. These are the following. The specific heat of completely frozen state, C_x, with the fixed z and x in (7), equal the specific heat of the heterogeneous mixture of initial oxides and oxygen in the left part of equation (5). C_x can be measured only at low temperatures, since oxygen is quite mobile in the basal plane of the 123 structure. The specific heat C_z with the fixed z (only intrinsic equilibrium state) can be measured at high temperatures only in the closed ampoules. In accord with (7)

$$C_z/R = C_x/R - 2xb_1^h(\partial x/\partial T)_z, \tag{12}$$

while by total equilibrium of 123 including atmosphere at a given $P(O_2)$,

$$C_p/R = C_z/R + [A_2 + (1-2z)a_1^h + zb_1^h/2](\partial z/\partial T)_{P(O_2)}. \tag{13}$$

In fig.5 the difference between C_z and C_p shows. Fig.7 represents how the formula (7) describes the experimental data on the enthalpies of the 123 solution. By means of (6)-(11) model some nonthermodynamic equilibrium properties of the 123 phase can be obtained, such as oxygen site occupancies in the basal $-CuO_z-$ plane of its crystal structure [19].

Properties of the Ba_2CuO_{3+v} ($0 \le v \le 0.5$) solid solutions with phase transition at about 1083 K were described by including two individual compounds, $Ba_2CuO_{3.1}$ and $Ba_2CuO_{3.3}$ [5].

Some lacking entropies of phases in the HTSC systems at 298 K were evaluated by adding the following contributions of oxides: Y_2O_3 (107.8), BaO (57.5), CuO (44.6), Cu_2O (125.9), SrO (51.8), La_2O_3 (134.0) in J/mol/K. For example, such an estimate gives $S_{298.15}^O = 107.8/2 + 2(57.5) + 2(44.6) + 125.9/2 = 321.0$ for $YBa_2Cu_3O_7$ compound in good accordance with experimental values. In this way the entropies of the end-member compounds of the $Ba_2Cu_3O_{5+y}$ solution as well as of the $Ba_2CuO_{3.1}$ and $Ba_2 CuO_{3.3}$ compounds were found.

The properties of the new $Y_2Ba_4Cu_9O_{17}$ and $YBa_2Cu_5O_9$ phases were obtained by means of extrapolation of the known Gibbs energies of the stoichiometric 123, 247 and 124 phases to the 249 and 125 phases composition at certain temperatures. These phases belong to the same homologous series of phases and their structural models can be reduced at 298 and 1000 K to formulas

$$\Delta G_f^{qx}/R = -17297.1 - 85.2m + 4110.0/m, \quad \Delta G_f^{qx}/R = -17434.0 + 1439.2m + 7136.4/m \qquad (14)$$

respectively, where ΔG_f^{qx} is the Gibbs energy of formation of the phase having m CuO layers located in the perovskite basal plane from Y_2O_3, BaO, CuO and oxygen. For the 123, 247, 124, 249 and 125 phases $m = 1$, 1.5, 2, 2.5 and 3 respectively. The numerical coefficients were obtained in these formulas following ΔG_f^{qx} of three earliest members of this series.

6. Stability of Superconductors

Phase equilibria in the YBCO were calculated by minimizing the Gibbs energies and evaluating equilibrium conditions for univariant reactions between the phases, as described above. The mentioned set of the self-consistent thermodynamic functions of the phases that can occur in equilibrium with the superconducting compounds was used. Detailed comparisons of the calculated equilibria with the primary experimental data will be given in refs. [5, 14, 22, 24].

6.1. THE SYSTEM Y-BA-CU-O

Stability fields of the 123 phase are shown in fig. 8-10. All the lines in these figures satisfy the equilibrium between solid phases and oxygen in the gas. Long-dashed isobars in fig. 8 represent the temperature dependence of the equilibrium composition of the solid solution, short-dashed lines represent metastable equilibria of the orthorhombic and tetragonal phases as well as a metastable miscibility gap of solid solution. Bold lines in fig. 8 show the equilibrium conditions for several decomposition reactions of 123, which outline its thermodynamic stability range. These reactions are:

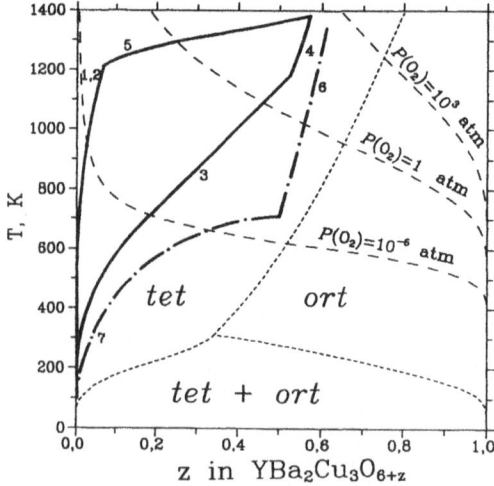

Fig. 8. Stability fields of the 123 phase. This phase is thermodynamically stable between the bold lines 1-5. The lines 1-7 represent decomposition reactions according to equations (15-21).

$$9YBa_2Cu_3O_{6+z} = 4Y_2BaCuO_5+YBa_4Cu_3O_{8.5+q} +10BaCu_2O_2+[(9z+5.5-q)/2]O_2, \tag{15}$$

$$2YBa_2Cu_3O_{6+z} = Y_2BaCuO_5+BaCuO_2+2BaCu_2O_2+[(2z+1)/2]O_2, \tag{16}$$

$$6YBa_2Cu_3O_{6+z} = Y_2BaCuO_5+3BaCuO_2+2Y_2Ba_4Cu_7O_{14+w}+[(6z-2w-3)/2]O_2, \tag{17}$$

$$6YBa_2Cu_3O_{6+z} = 2Y_2BaCuO_5+3Ba_2Cu_3O_{5+y}+Y_2Ba_4Cu_7O_{14+w}+[(6z-3-3y-w)/2]O_2, \tag{18}$$

$$YBa_2Cu_3O_{6+z} \longrightarrow Y_2BaCuO_5+ \text{Liquid} + O_2, \tag{19}$$

$$4YBa_2Cu_3O_{6+z} = 2Y_2BaCuO_5+3Ba_2Cu_3O_{5+y}+CuO+[(4z-2-3y)/2]O_2, \tag{20}$$

$$2YBa_2Cu_3O_{6+z} = Y_2BaCuO_5+3BaCuO_2+2CuO+[(2z-1)/2]O_2. \tag{21}$$

All the boundaries were calculated from the thermodynamic properties of the phases except the line (19) of incongruent melting of the 123 phase taken from the work of Lindemer, Washburn et al.[23]. Phase diagram in Fig.9 shows the same stability region of the 123 phase in other variables. The 123 phase is thermodynamically unstable at all temperatures and compositions outside the mentioned field of stability. In particular, it is true for the superconducting orthorombic structure of 123. By the way, synthesis of this superconductor usually starts at the conditions inside its stability field. Then cooling of the sample and relative quick transition into unstable area of the phase diagram follow. It is notable that decomposition of the 123 phase can occur if and only if this phase exists together with other participants of decomposition reactions, such as 247, 211 and others. When the nucleation of some of them is kinetically suppressed, the 123 phase may exist outside of its stability field as a metastable one. For the comparison the right boundary of 123 stability field shown in the figs.8 and 9, when the 247 phase does not form during the time of experiment (lines 6 and 7). This case may be realized indeed under certain conditions of the 123 phase synthesis, for instance at the short specimens annealing. A similar possibility was under consideration concerning fig. 1(b). The area between the new boundary and the old one is in this case the area of the metastable equilibria of the 123 phase with oxygen. We can use both stable and metastable boundary depending on the time of the process of interest.

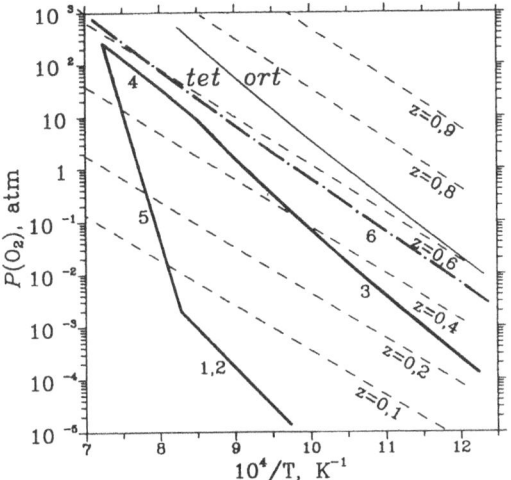

Fig. 9. Stability field of the 123 phase in the YBCO. This phase is stable between the bold lines, which represent the conditions for equlibrium of the same reactions, as in fig.8. Metastable equilibrium of the ortorombic and teragonal phases is shown by the thin line. The dashed lines indicate the oxygen stoichiometry, and the chainded-dot one shows metastable decomposition of 123.

We have to deal with stable equilibrium if this time is long enough for the formation of the 247 phase from 123. Otherwise we have to use metastable boundary.

At low temperatures the same limitations take place for the formations of the other phases in the right side of (15)-(21) and 123 exists in the whole field of phase diagram. It is exactly its state, which we obtain almost always. Sometimes they obtain several ordered phases at low temperatures in the area of miscibility gap in fig.8. These phases as well as miscibility gap belong evidently to the intermediate metastable state of the system at its step by step transformation into the phases written above.

Phase boundaries corresponding to different decomposition reactions are often close to one another. Owing to the experimental uncertainties in the values of the thermodynamic properties of the phases used for the calculations, it is sometimes hard to separate the boundaries into stable and metastable, although, as a rule, the location and dimensions of the stability field are practically independent of this selection. So, the lines corresponding to reactions 1 and 2 in figs. 8, 9 are very closely placed and cross each other twice. The distance in temperature between them is only a few degrees and less than the error of the calculations, so we cannot reliably establish which of them is stable or metastable.

Evidently, the superconducting orthorhombic phase can be obtained only if the nucleation and/or growth of phases with cation stoichiometry other than 1:2:3 is kinetically suppressed. This result was obtained by us about three years ago with another set of thermodynamic data of the YBCO. Those data were based on the limited experimental study of the 123 phase and theoretical models of thermodynamic functions of other superconductors, because there were not any measurements of their thermodynamic properties at those days. The present results are in quite good agreement with the previous ones in spite of different boundary reactions. The stability of other superconductors in this system will be discussed just below.

6.2. THE SYSTEM Y-BA-CU-C-O

The conclusion about the metastability of the 123 superconductor has to be true, if we take into account the other substances from the ambient atmosphere or materials as well, since interactions of 123 with those may only diminish the stability field of the 123 solid solution. The most possible impurity in this solution is carbon from carbon dioxide. Figs. 10 and 11 show the effect of carbon dioxide, CO_2 as an additional component of the nominal YBCO on the thermodynamic stability of the 123 phase

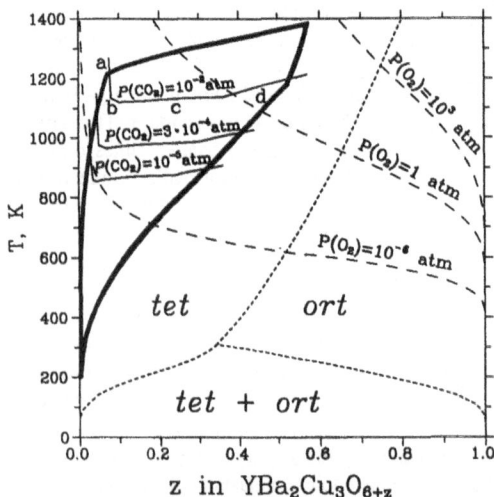

[24]. When carbon is added to the YBCO, some stability boundaries or their portions become metastable, and several new phase boundaries appear.

Equilibria in the Y-Ba-Cu-O-CO_2 system were calculated without taking into account any oxycarbonates formation, except $BaCO_3$. Nevertheless, we believe that the consideration of the other oxycarbonates does not affect the stability fields of the superconductors in essence because of the supposed relative low thermodynamic stabilities of these substances as compared to $BaCO_3$. Although they can play a significant role at low temperatures and metastable states of the system.

The boundary corresponding to a decomposition reaction is a surface in the spaces of three independent variables: temperature, partial pressure of oxygen and partial pressure of CO_2. Therefore, only some sections of this surface at the fixed CO_2 partial pressure can be depicted in the two-dimensional plot. The

Fig. 10. Stability of the 123 phase in the O_2+CO_2 atmosphere. Bold solid lines correspond to the similar lines and reactions in fig.8, i.e., at $P(CO_2)=0$. Thin solid lines represent additional stability boundaries at the fixed CO_2 partial pressure. The 123 phase is stable between the bold lines and above the thin solid lines.

additional reactions (a),(b),(c) and (d) in figs. 10, 11 are correspondingly:

$$4YBa_2Cu_3O_{6+z} +CO_2= 2Y_2BaCuO_5+5BaCu_2O_2+BaCO_3+[(4z+3)/2]O_2, \qquad (22)$$
$$4YBa_2Cu_3O_{6+z} +6CO_2= 2Y_2BaCuO_5+5Cu_2O+6BaCO_3+[(4z+3)/2]O_2, \qquad (23)$$

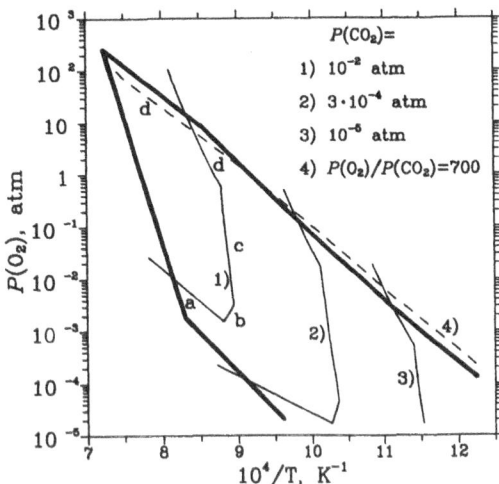

Fig. 11. Stability field of the 123 phase in the O_2+CO_2 atmosphere. This phase is thermodynamically stable between the bold lines, below the dashed line and to the left from the thin solid lines. The reactions of 123 with CO_2 are the same as in fig.10.

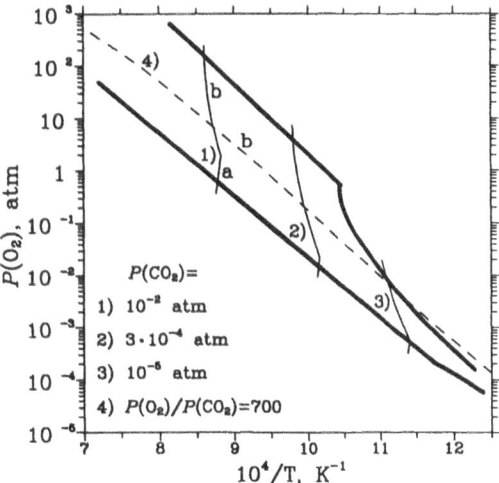

Fig. 12. Stability field of the 247 phase. This phase is thermodynamically stable between the bold lines, below the dashed line and to the left from the thin solid lines. The reactions (a)-(f) are (26-31) ones correspondingly.

$$2YBa_2Cu_3O_{6+z} +3CO_2 = Y_2BaCuO_5+5CuO+3BaCO_3+[(2z-1)/2]O_2, \tag{24}$$

$$12YBa_2Cu_3O_{6+z}+3CO_2 = Y_2BaCuO_5+5Y_2Ba_4Cu_7O_{14+w}+3BaCO_3+[(12z-5w-6)/2]O_2. \tag{25}$$

The CO_2 content in the ambient air is about 300 ppm or $P(CO_2)=3\cdot10^{-4}$ atm. Represented by the dashed line in fig. 11 is another section for the constant percentage of CO_2 in oxygen which equals that of the ambient air.

Fig. 12 represents the results of the stability fields calculation of the 247 phase. In the atmosphere without CO_2 this phase is thermodynamically stable between bold lines. The reactions of its decomposition are:

$$2Y_2Ba_4Cu_7O_{14+w} = YBa_2Cu_3O_{6+z} +CuO+[(1+w-2z)/2]O_2, \tag{26}$$

$$Y_2Ba_4Cu_7O_{14+w} = YBa_2Cu_3O_{6+z} +YBa_2Cu_4O_8+[(w-z)/2]O_2, \tag{27}$$

$$4Y_2Ba_4Cu_7O_{14+w} = 6YBa_2Cu_4O_8+Y_2BaCuO_5+3BaCuO_2+[(4w-3)/2]O_2, \tag{28}$$

$$5Y_2Ba_4Cu_7O_{14+w} = 6YBa_2Cu_4O_8+2Y_2BaCuO_5+3Ba_2Cu_3O_{5+y}+[(5z-3-3y)/2]O_2. \tag{29}$$

In the presence of CO_2 the boundary reactions add:

$$Y_2Ba_4Cu_7O_{14+w} +3CO_2 = Y_2BaCuO_5 +6CuO+3BaCO_3 +(w/2)O_2, \tag{30}$$

$$7Y_2Ba_4Cu_7O_{14+w} +3CO_2 = 12YBa_2Cu_4O_8 +Y_2BaCuO_5 +3BaCO_3 +[(7w-6)/2]O_2. \tag{31}$$

So, the 247 superconductor has a behavior similar to the 123 one, they both are thermodynamically metastable under the application conditions.

Fig. 13 shows stability of the 124 phase with and without CO_2 in the equilibrium gas mixture. Here are two decomposition reactions without participation of carbon

$$2YBa_2Cu_4O_8 = Y_2Ba_4Cu_7O_{14+w} +CuO+[(1-w)/2]O_2, \tag{32}$$

$$YBa_2Cu_4O_8 = YBa_2Cu_3O_{6+z} +CuO+[(1-z)/2]O_2, \tag{33}$$

and two boundary reactions with CO_2 :

$$2YBa_2Cu_4O_8 +3CO_2 = Y_2BaCuO_5 +7CuO+3BaCO_3 +(1/2)O_2, \tag{34}$$

$$2YBa_2Cu_4O_8+4CO_2 = Y_2Cu_2O_5 +6CuO+4BaCO_3 +(1/2)O_2. \tag{35}$$

The 124 phase may be the only superconductor in this system that is thermodynamically stable at low T and usual P. However, it is not clear the existence of other superconductors in this system yet, and we have only preliminary data for the thermodynamic properties of high members of homological series of compounds including the known 123, 247 and 124 phases. The stability range of 124 may somewhat decrease above the bold line in fig. 13 because of decomposition into the 249, 125 and other phases, as it showed in fig.1 (a).

Evidently, CO_2 modifies the stability field of the phases substantially even at its very low pressure. The influence of CO_2 on the stability of the 247 and 124 phases is even stronger than for 123. It limits the stability fields at low temperatures. As shows fig.13, at the fixed percentage of CO_2 in O_2, 124 is stable only in the narrow band between the bold and the dashed line. The present results are in good qualitative agreement with the available experimental data (see ref.[24]).

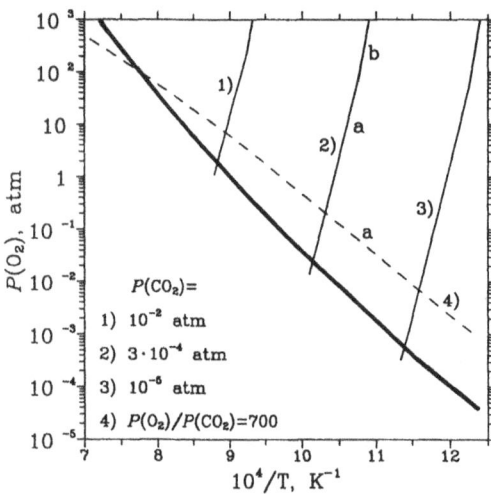

Fig. 13. Stability field of the 124 phase. This phase is stable above the bold lines, below the dashed line and to the left from the thin lines. The reactions corresponding to the bold lines and (a), (b) lines (32, 33) and (34, 35) respectively.

6.3. THE SYSTEM Y-BA-CU-ZR-O

In practice they often use superconductors in the presence of zirconium dioxide, ZrO_2. This material is used as a substrate or buffering layer at the superconducting thin films production, as a solid electrolyte in the thermodynamic YBCO studies, for enhancement of the 123 superconductor mechanical strength etc. The equilibria calculations in the above system show that all superconducting also the other Ba-containing phases have to react with ZrO_2 forming $BaZrO_3$ at any temperatures up to 1300 K. This result is in accord with lots of experimental ones.

7. Concluding remarks

What the loss of thermodynamic stability of superconductors means for the practice ? At high temperature this conclusion is very useful for choosing the synthesis conditions of proper materials because the easy setting equilibria. At low temperature, as we know at present, materials on the basis of the YBCO can be used under proper conditions several mounts or, say, years without problem, but in principle this time is limited. Thermodynamics cannot give any guarantee that metastable material will exist. It means that the time of this material existence depends on the kinetical reasons, and, for example, superconductivity of transcontinental cable, worked from this material, can suddenly disappear because of mentioned chemical reactions. Although such degradation of thermodynamically stable material is impossible.

This was one of the questions about the sense and the area of use of the thermodynamic conclusions. Sometimes they overstate their potentialities forgetting limitations of the equilibrium models. The stability fields and phase equilibria of the superconductors cannot exactly specify the conditions needed to make superior materials. Effective properties of a particular material depend not only on the phase and chemical compositions but on the overall microstructure, defects, etc. as well. But often the fact that some phases exist for a very long time outside their stability range without any detectable changes may drive one to the conclusion that the thermodynamic stability is of little practical importance and only kinetics play a significant role. This is the other extremity. We have already seen that analyzing the thermodynamic stability, we can take into account some qualitative kinetic information. It is more probable to obtain the reproducible properties of a material when the synthesis approaches some thermodynamic limit and is not controlled by kinetics only. In the latter case, the material is very sensitive to small fluctuations of the parameters of the process. The stability fields can enable researchers working in all application areas to avoid unnecessary empirical searches for suitable materials by eliminating the most thermodynamically unfavorable choices.

The available values of the thermodynamic properties used for the above calculations are known only with the limited precision. How does this affect on the reliability of the final results? The answer can vary in every single case, but it may say in general, that the calculated phase diagrams are stable enough with respect to the variations of some thermodynamic data, if only they belong to the mentioned self-consistent sets of data. A low sensitivity of the stability conditions of 123 to this inaccuracy in the thermodynamic data can be seen by comparison of fig.8 or 9 with the similar results of our previous work [20], as is said above. That work was done without any experimental information about the thermodynamic functions of the 247 and 124 phases. Further, the equilibrium dissociation of 247 into 123 and CuO at $P(O_2)=1$atm was found occurring at 1136 K and dissociation of 124 into 247 and CuO at 1099 K. As can be seen in fig. 1 (a), these temperatures are 1164 and 1108 K respectively at present. These differences are comparable with the errors of the ordinary experimental phase stability measurements.

602

8. References

1. E.Kaldis and J.Karpinski, Eur.J.Solid State Inorg.Chem.,27 (1990) 143.
2. W.Zhang and K.Osamura, Physica C, 190 (1992) 396.
3. J.-Y.Grenoud, T.Graf, G.Triscone, A.Junod and J.Muller, Physica C, 192 (1992) 137.
4. W.R.Smith and R.W.Missen, Chemical Reaction Equilibrium Analysis.John Wiley and Sons, New York, 1982.
5. G.F.Voronin and S.A.Degterov, J.Solid State Chemistry, in press.
6. K.L.Komarek, Pure and Appl.Chemistry, 64 (1992) 93.
7. J.Karpinski and E.Kaldis, J.Cryst.Growth, 79 (1986) 477.
8. A.Junod, D.Eckert, T. Graf, E.Kaldis, J.Karpinski, S.Rusiecki, D.Sanches, G.Tiscone and J.Miller, Physica C, 168 (1990) 47.
9. R.Shaviv, E.F.Westrum, R.J.C.Born, M.Sayer, X.Yu and R.D.Weir, J.Chem.Phys., 92 (1990) 6794.
10. G.A.Sharpataja, Z.P.Ozerova, I.A.Konovalova, V.B.Lazarev and I.S.Shapligin, Neorganicheskie materiali, 27 (1991) 1674.
11. S.F.Pashin and Yu.Ya. Skolis, Zhur.Fiz.Khim., 65 (1991) 256.
12. S.F.Pashin, E.V.Antipiv, L.M.Kovba and Yu.Ya.Skolis, Superconductivity: Phys.Chem. Eng., 2 (1989) 102.
13. Yu.Ya.Skolis, M.L.Kovba and L.A.Chramzova, Zhur.Fiz.Khim., 65 (1991) 1070.
14. G.F.Voronin and S.A.Degterov, Zhur.Fiz.Khim., in press.
15. L.S.Morss, D.C.Sonnenberger and R.J.Thorn, Inorg.Chem., 27 (1988) 2106
16. N.I Matskevich, V.A.Titov, T.L.Popova and V.P.Shaburova, Abstr. Int. Symp. on Calorimetry and Chem. Thermod., Moscow, June 23-28,1991, p.44.
17. F.H.Garson, I.D.Raistrick, D.S.Ginley and J.W.Halloran, J.Mater.Res., 6 (1991) 885.
18. Z.Zhou and A.Navrotsky, Mater.Res., in press.
19. G.F.Voronin and S.A.Degterov, Physica C, 176 (1991) 387.
20. S.A.Degterov and G.F.Voronin, Physica C, 178 (1991) 213.
21. G.F.Voronin, S.A.Degterov and Yu.Ya.Skolis, Proc. 3rd German-Soviet Bilateral Seminar on High-Temperature Superconductivity, October 8-12, 1990, Karlsruhe, Germany, p.562.
22. S.A.Degterov and G.F.Voronin, Zhur.Fiz.Khim., in press.
23. T.B.Lindemer, F.A.Washburn, C.S.McDougall, R.Feenstra and O.B.Cavin, Physica C, 178 (1991) 93.
24. S.A.Degterov and G.F.Voronin, Physica C, 208 (1993) 403.

The manufacturer's authorised representative in the EU is Springer
Nature Customer Service Centre GmbH, Europaplatz 3, 69115 Heidelberg,
Germany. If you have any concerns regarding our products, please
contact ProductSafety@springernature.com

Printed and bound by CPI Group (UK) Ltd, Croydon, CR0 4YY
23/04/2026
02095628-0008